VOLUME FOUR HUNDRED AND FIFTY-ONE

METHODS IN ENZYMOLOGY

Autophagy: Lower Eukaryotes and Non-Mammalian Systems, Part A

METHODS IN ENZYMOLOGY

Editors-in-Chief

JOHN N. ABELSON AND MELVIN I. SIMON

Division of Biology
California Institute of Technology
Pasadena, California

Founding Editors

SIDNEY P. COLOWICK AND NATHAN O. KAPLAN

VOLUME FOUR HUNDRED AND FIFTY-ONE

METHODS IN ENZYMOLOGY

Autophagy: Lower Eukaryotes and Non-Mammalian Systems, Part A

EDITED BY

DANIEL J. KLIONSKY
Life Sciences Institute
University of Michigan
Ann Arbor, Michigan, USA

[redacted]

ELSEVIER

AMSTERDAM • BOSTON • HEIDELBERG • LONDON
NEW YORK • OXFORD • PARIS • SAN DIEGO
SAN FRANCISCO • SINGAPORE • SYDNEY • TOKYO
Academic Press is an imprint of Elsevier

Academic Press is an imprint of Elsevier
525 B Street, Suite 1900, San Diego, CA 92101-4495, USA
30 Corporate Drive, Suite 400, Burlington, MA01803, USA
32 Jamestown Road, London NW17BY, UK

For information on all Elsevier Academic Press publications
visit our Web site at elsevierdirect.com

ISBN-13: 978-0-12-374548-4

PRINTED IN THE UNITED STATES OF AMERICA
08 09 10 11 9 8 7 6 5 4 3 2 1

CONTENTS

Contributors

Hagai Abeliovich
Department of Biochemistry and Food Science, Hebrew University of Jerusalem, Rehovot, Israel

David S. Askew
Department of Pathology and Laboratory Medicine, University of Cincinnati College of Medicine, Cincinnati, USA

Misuzu Baba
Department of Chemical and Biological Sciences, Faculty of Science, Japan Women's University, Tokyo, Japan

Attila Baksa
Cell Physiology Laboratory, Department of Anatomy, Cell and Developmental Biology, Eötvös Loránd University, Budapest, Hungary

János Barna
Department of Genetics, Eötvös Loránd University, Budapest, Hungary

Torsten Barth
Interfaculty Institute of Biochemistry, University of Tübingen, Tübingen, Germany

Bjoern Bassarak
Interfaculty Institute of Biochemistry, University of Tübingen, Tübingen, Germany

Wanda Buzgariu
Department of Zoology and Animal Biology, Faculty of Sciences, University of Geneva, Geneva, Switzerland

Margarita Cabrera
Department of Biology, Biochemistry Section, University of Osnabrück, Osnabrück, Germany

Nadine Camougrand
Institut de Biochimie et Génétique Cellulaires, CNRS, Université de Bordeaux2, Bordeaux, France

Eduardo Cebollero
University Medical Center Utrecht, Department of Cell Biology, Utrecht, The Netherlands

Heesun Cheong
Life Sciences Institute and Departments of Molecular, Cellular and Developmental Biology and Biological Chemistry, University of Michigan, Ann Arbor, Michigan, USA

Simona Chera
Department of Zoology and Animal Biology, Faculty of Sciences, University of Geneva, Geneva, Switzerland

Corinne Clavé
Department of Molecular Biology and Biochemistry, Rutgers, The State University of New Jersey, Piscataway, New Jersey, USA

Robert C. Cumming
Department of Biology, University of Western Ontario, London, Ontario, Canada

Yi Zhen Deng
Fungal Patho-Biology Group, Temasek Life Sciences Laboratory and Department of Biological Sciences, National University of Singapore, Singapore

Viola Denninger
Interfaculty Institute of Biochemistry, University of Tübingen, Tübingen, Germany

Rodney J. Devenish
Department of Biochemistry and Molecular Biology, and ARC Centre of Excellence in Structural and Functional Microbial Genomics, Monash University, Clayton Campus, Melbourne, Victoria, Australia

Michael Duszenko
Interfaculty Institute of Biochemistry, University of Tübingen, Tübingen, Germany

Judit Fehér
Cell Physiology Laboratory, Department of Anatomy, Cell and Developmental Biology, Eötvös Loránd University, Budapest, Hungary

Kim D. Finley
BioScience Center, San Diego State University, San Diego, California, and Cellular Neurobiology Laboratory, Salk Institute for Biological Studies, La Jolla, California, USA

Kozo Fujisaki
Laboratory of Emerging Infectious Diseases, Department of Frontier Veterinary Medicine, Kagoshima University, Kagoshima, Japan

Brigitte Galliot
Department of Zoology and Animal Biology, Faculty of Sciences, University of Geneva, Geneva, Switzerland

Jack Gibbons
Division of Biological Sciences, University of Illinois at Chicago, Chicago, Illinois, USA

Corinne Giusti
Centre d'Immunologie de Marseille-Luminy, Marseille, France

Pierre Golstein
Centre d'Immunologie de Marseille-Luminy, Marseille, France

Cristina González-Estévez
Department of Developmental Genetics and Gene Control, Institute of Genetics, Queen's Medical Centre, University of Nottingham, United Kingdom

Ramón González
Instituto de Ciencias de la Vid y del Vino (CSIC-UR-GR) C. Madre de Dios, 51, 26006 Logroño, Spain

David H. Hall
Center for *C. elegans* Anatomy, Albert Einstein College of Medicine, Bronx, New York, USA

Malene Hansen
Burnham Institute for Medical Research, Program of Development and Aging, La Jolla, California, USA

Guowu Hu
Section of Infectious Diseases, Department of Medicine, University of Illinois at Chicago, Chicago, Illinois, USA

Yuko Inoue
Life Sciences Institute, University of Michigan, Ann Arbor, Michigan, USA

Dikla Journo
Department of Biochemistry and Food Science, Hebrew University of Jerusalem, Rehovot, Israel

Jan A. K. W. Kiel
Molecular Cell Biology, University of Groningen, The Netherlands

Bruno Kubata Kilunga
NEPAD/Biosciences Eastern and Central Africa, ILRI Campus, Nairobi, Kenya

Ingrid Kiššová
Comenius University, Faculty of Natural Sciences, Department of Biochemistry, Bratislava, Slovak Republic

Ida J. van der Klei
Kluyver Centre for Genomics of Industrial Fermentation, Delft, The Netherlands, and Molecular Cell Biology, University of Groningen, The Netherlands

Daniel J. Klionsky
Life Sciences Institute and Departments of Molecular, Cellular and Developmental Biology and Biological Chemistry, University of Michigan, Ann Arbor, Michigan, USA

Angela Kohler
Alfred Wegener Institute for Polar and Marine Research, Department of Ecotoxicology and Ecophysiology, Bremerhaven, Germany

Rudolf Koopmann
Interfaculty Institute of Biochemistry, University of Tübingen, Tübingen, Germany

Artemis Kosta
Centre d'Immunologie de Marseille-Luminy, Marseille, France

János Kovács
Cell Physiology Laboratory, Department of Anatomy, Cell and Developmental Biology, Eötvös Loránd University, Budapest, Hungary

Attila L. Kovács
Cell Physiology Laboratory, Department of Anatomy, Cell and Developmental Biology, Eötvös Loránd University, Budapest, Hungary

Fu-Cheng Lin
State Key Laboratory for Rice Biology, Institute of Biotechnology, Zhejiang University, Huajiachi Campus, Hangzhou, Zhejiang, China

Tong-Bao Liu
State Key Laboratory for Rice Biology, Institute of Biotechnology, Zhejiang University, Huajiachi Campus, Hangzhou, Zhejiang, China

Xiao-Hong Liu
State Key Laboratory for Rice Biology, Institute of Biotechnology, Zhejiang University, Huajiachi Campus, Hangzhou, Zhejiang, China

David Lowe
Plymouth Marine Laboratory, Prospect Place, The Hoe, Plymouth, United Kingdom

Marie-Françoise Luciani
Centre d'Immunologie de Marseille-Luminy, Marseille, France

Davide Malagoli
Department of Animal Biology, University of Modena and Reggio Emilia, Modena, Italy

Tomohide Matsuo
Department of Infectious Diseases, Kyorin University School of Medicine, Tokyo, Japan

Ken Matsuoka
Faculty of Agriculture, Kyushu University, Fukuoka, Japan

Andreas Mayer
Département de Biochimie, Chemin des Boveresses 155, CH-1066 Epalinges, Switzerland

Alicia Meléndez
Queens College-CUNY, Department of Biology, Flushing, New York, USA

Dalibor Mijaljica
Department of Biochemistry and Molecular Biology, and ARC Centre of Excellence in Structural and Functional Microbial Genomics, Monash University, Clayton Campus, Melbourne, Victoria, Australia

Michael N. Moore
Plymouth Marine Laboratory, Prospect Place, The Hoe, Plymouth, United Kingdom

Yuji Moriyasu
Department of Regulatory Biology, Faculty of Science, Saitama University, Saitama, Japan

Khalid Muhammad
Interfaculty Institute of Biochemistry, University of Tübingen, Tübingen, Germany

Kumiko Nakada-Tsukui
Department of Parasitology, National Institute of Infectious Diseases, Tokyo, Japan, and Department of Parasitology, Gunma University Graduate School of Medicine, Maebashi, Japan

Naweed I. Naqvi
Fungal Patho-Biology Group, Temasek Life Sciences Laboratory and Department of Biological Sciences, National University of Singapore, Singapore

Taras Y. Nazarko
Section of Molecular Biology, University of California, San Diego, La Jolla USA, and Institute of Cell Biology, NAS of Ukraine, Lviv, Ukraine

Thomas P. Neufeld
Department of Genetics, Cell Biology and Development, University of Minnesota, Minneapolis, Minnesota, USA

Ioannis P. Nezis
Faculty of Biology, Department of Cell Biology and Biophysics, University of Athens, Athens, Greece, and Centre for Cancer Biomedicine, University of Oslo and Institute for Cancer Research, Department of Biochemistry, The Norwegian Radium Hospital, Montebello, N-0310, Oslo, Norway

Takeshi Noda
Department of Cellular Regulation, Research Institute for Microbial Diseases, Osaka University, Osaka, Japan

Tomoyoshi Nozaki
Department of Parasitology, National Institute of Infectious Diseases, Tokyo, Japan, and Department of Parasitology, Gunma University Graduate School of Medicine, Maebashi, Japan

Masahide Oku
CREST, Japan Science and Technology Agency, Japan, and Division of Applied Life Sciences, Graduate School of Agriculture, Kyoto University, Kyoto, Japan

Zsolt Pálfia
Cell Physiology Laboratory, Department of Anatomy, Cell and Developmental Biology, Eötvös Loránd University, Budapest, Hungary

Glen E. Palmer
Department of Oral and Craniofacial Biology, LSUHSC School of Dentistry, New Orleans, Louisiana, USA

Issidora Papassideri
Faculty of Biology, Department of Cell Biology and Biophysics, University of Athens, Athens, Greece

Gabriella Pásti
Cell Physiology Laboratory, Department of Anatomy, Cell and Developmental Biology, Eötvös Loránd University, Budapest, Hungary

Karina Picazarri
Department of Parasitology, Gunma University Graduate School of Medicine, Maebashi, Japan

Bérangère Pinan-Lucarré
Laboratoire de Génétique Moléculaire des Champignons, Institut de Biochimie et de Génétique Cellulaires, Université de Bordeaux 2 et CNRS, Bordeaux, France

Mark Prescott
Department of Biochemistry and Molecular Biology, and ARC Centre of Excellence in Structural and Functional Microbial Genomics, Monash University, Clayton Campus, Melbourne, Victoria, Australia

Tanja Prick
Georg-August-University, Center of Biochemistry and Molecular Cell Biology, Goettingen, Germany

Marilou Ramos-Pamplona
Fungal Patho-Biology Group, Temasek Life Sciences Laboratory and Department of Biological Sciences, National University of Singapore, Singapore

M. Teresa Rejas
Servicio de Microscopía Electrónica, Centro de Biología Molecular Severo Ochoa (CSIC-UAM)

Fulvio Reggiori
Department of Cell Biology and Institute of Biomembranes, University Medical Centre Utrecht, Utrecht, The Netherlands

Daryl L. Richie
Department of Pathology and Laboratory Medicine, University of Cincinnati College of Medicine, Cincinnati, USA

Yasuyoshi Sakai
CREST, Japan Science and Technology Agency, Japan, and Division of Applied Life Sciences, Graduate School of Agriculture, Kyoto University, Kyoto, Japan

Benedicte Salin
Institut de Biochimie et Génétique Cellulaires, CNRS, Université de Bordeaux2, Bordeaux, France

Dan Sato
Center for Integrated Medical Research, School of Medicine, Keio University, Tokyo, Japan, and Institute for Advanced Biosciences, Keio University, Yamagata, Japan, and Department of Parasitology, Gunma University Graduate School of Medicine, Maebashi, Japan

Anne Simonsen
Center for Cancer Biomedicine, University of Oslo and Department of Biochemistry, Norwegian Radium Hospital, Oslo, Norway

Oleh V. Stasyk
Institute of Cell Biology, NAS of Ukraine, Lviv, Ukraine

Andriy A. Sibirny
Department of Biotechnology and Microbiology, Rzeszów University, Rzeszów, Poland, and Institute of Cell Biology, NAS of Ukraine, Lviv, Ukraine

Caroline Schönfeld
Interfaculty Institute of Biochemistry, University of Tübingen, Tübingen, Germany

Peter Schu
Georg-August-University Göttingen, Zentrum für Biochemie und Molekulare Zellbiologie, Biochemie II, Göttingen, Germany

Takahiro Shintani
Laboratory of Bioindustrial Genomics, Graduate School of Agricultural Science, Tohoku University, Sendai, Japan

Timea Sigmond
Laboratory of Developmental Genetics, Department of Genetics, Eötvös Loránd University, Budapest, Hungary

Krisztina Takács-Vellai
Laboratory of Developmental Genetics, Department of Genetics, Eötvös Loránd University, Budapest, Hungary

Gianluca Tettamanti
Department of Structural and Functional Biology, University of Insubria, Varese, Italy

Michael Thumm
Georg-August-University, Center of Biochemistry and Molecular Cell Biology, Goettingen, Germany

Márton L. Tóth
Department of Genetics, Eötvös Loránd University, Budapest, Hungary

Emilie Tresse
Centre d'Immunologie de Marseille-Luminy, Marseille, France

Kristina Turcic
Department of Biochemistry and Molecular Biology, and ARC Centre of Excellence in Structural and Functional Microbial Genomics, Monash University, Clayton Campus, Melbourne, Victoria, Australia

Rika Umemiya-Shirafuji
Laboratory of Emerging Infectious Diseases, Department of Frontier Veterinary Medicine, Kagoshima University, Kagoshima, Japan

Christian Ungermann
Department of Biology, Biochemistry Section, University of Osnabrück, Osnabrück, Germany

Tibor Vellai
Laboratory of Developmental Genetics, Department of Genetics, Eötvös Loránd University, Budapest, Hungary

Aldo Viarengo
Department of Environmental and Life Science, Universita di Piemonte Orientale (Amadeo Avogaddro), Alessandria, Italy

Peter R. Williamson
Jesse Brown VA Medical Center, Chicago, Illinois, USA, and Section of Infectious Diseases, Department of Medicine, University of Illinois at Chicago, Chicago, Illinois, USA

Gal Winter
Department of Biochemistry and Food Science, Hebrew University of Jerusalem, Rehovot, Israel

Tim van Zutphen
Molecular Cell Biology, University of Groningen, The Netherlands

PREFACE

The later the molecular details of a particular subcellular pathway are elucidated relative to known pathways, the greater is the likelihood of an explosive advance in knowledge as the network of connections is made to these other established processes. Such is certainly the case with autophagy. The molecular details of autophagy are little more than a decade old, and the field has been expanding almost logarithmically, fueled in part by increasing links with other areas of study. For example, ten years ago, or, in some cases, just five years ago, nothing was known about the role of autophagy in cellular immunity, microbial pathogenesis, neurodegeneration or cancer, let alone its importance in differentiation in protists and kinetoplastida, regeneration in hydra and planarians, resistance to toxic chemicals in aquatic animals, starvation in ticks, egg development in flies, mitophagy in reticulocytes, ventricular remodeling, liver disease, and so on.

Accompanying the expansion of autophagy research is a need to learn the available methodologies for monitoring autophagy. Of course, that is the rationale for these volumes of *Methods in Enzymology*. Because the molecular analysis of autophagy is relatively new, I expect additional techniques to be developed and modifications made to the existing ones. Nonetheless, the protocols and information presented in these volumes represent an important and very useful starting point for researchers new to this area, for those currently studying autophagy who wish to expand their repertoire of techniques, as well as for those who simply desire to compare their current methods with those of others working in this field. For information on the applicability of these and other methods to study autophagy, I refer readers to the "Guidelines for the use and interpretation of assays for monitoring autophagy in higher eukaryotes," published in the journal *Autophagy* in 2008.

These volumes have been arranged according to the model system/organism being used rather than by technique. Therefore, the first volume starts with baker's yeast and concludes with insects, whereas the second volume is concerned with autophagy in mammals and ends with procedures that are clinically relevant. One outcome of this approach is that certain techniques or variations on those techniques appear in multiple chapters. I think this is useful because there are specific aspects of some methods that must be followed to make them applicable to a particular organism. For example, electron microscopy was the first method used to analyze autophagy, and it remains one of the most versatile in that it can be applied to essentially any organism. A researcher interested in using electron microscopy to study autophagy will find variations of this technique described in several chapters in both part A and part B. Similarly, other methods, including assays for total protein turnover, the use of fluorescence microscopy, and the analysis of selective types of autophagy, are covered in more than one chapter. Therefore, I encourage readers to peruse the entirety of these volumes to uncover relevant information.

These volumes of *Methods in Enzymology* represent the first time that many of these protocols, in particular in multiple organisms, have been brought together in one place. This will facilitate comparisons among protocols and model systems, and it may stimulate further modifications and advances in the technologies used to study autophagy. I thank all of the authors who have contributed to these volumes, and I apologize to anyone who was unintentionally left out. As the field continues to grow and new techniques are added, I am certain that there will be a need for a follow-up volume in the not-too-distant future.

Daniel J. Klionsky

Methods in Enzymology

Volume I. Preparation and Assay of Enzymes
Edited by Sidney P. Colowick and Nathan O. Kaplan

Volume II. Preparation and Assay of Enzymes
Edited by Sidney P. Colowick and Nathan O. Kaplan

Volume III. Preparation and Assay of Substrates
Edited by Sidney P. Colowick and Nathan O. Kaplan

Volume IV. Special Techniques for the Enzymologist
Edited by Sidney P. Colowick and Nathan O. Kaplan

Volume V. Preparation and Assay of Enzymes
Edited by Sidney P. Colowick and Nathan O. Kaplan

Volume VI. Preparation and Assay of Enzymes *(Continued)*
Preparation and Assay of Substrates
Special Techniques
Edited by Sidney P. Colowick and Nathan O. Kaplan

Volume VII. Cumulative Subject Index
Edited by Sidney P. Colowick and Nathan O. Kaplan

Volume VIII. Complex Carbohydrates
Edited by Elizabeth F. Neufeld and Victor Ginsburg

Volume IX. Carbohydrate Metabolism
Edited by Willis A. Wood

Volume X. Oxidation and Phosphorylation
Edited by Ronald W. Estabrook and Maynard E. Pullman

Volume XI. Enzyme Structure
Edited by C. H. W. Hirs

Volume XII. Nucleic Acids (Parts A and B)
Edited by Lawrence Grossman and Kivie Moldave

Volume XIII. Citric Acid Cycle
Edited by J. M. Lowenstein

Volume XIV. Lipids
Edited by J. M. Lowenstein

Volume XV. Steroids and Terpenoids
Edited by Raymond B. Clayton

Volume XVI. Fast Reactions
Edited by Kenneth Kustin

Volume XVII. Metabolism of Amino Acids and Amines (Parts A and B)
Edited by Herbert Tabor and Celia White Tabor

Volume XVIII. Vitamins and Coenzymes (Parts A, B, and C)
Edited by Donald B. McCormick and Lemuel D. Wright

Volume XIX. Proteolytic Enzymes
Edited by Gertrude E. Perlmann and Laszlo Lorand

Volume XX. Nucleic Acids and Protein Synthesis (Part C)
Edited by Kivie Moldave and Lawrence Grossman

Volume XXI. Nucleic Acids (Part D)
Edited by Lawrence Grossman and Kivie Moldave

Volume XXII. Enzyme Purification and Related Techniques
Edited by William B. Jakoby

Volume XXIII. Photosynthesis (Part A)
Edited by Anthony San Pietro

Volume XXIV. Photosynthesis and Nitrogen Fixation (Part B)
Edited by Anthony San Pietro

Volume XXV. Enzyme Structure (Part B)
Edited by C. H. W. Hirs and Serge N. Timasheff

Volume XXVI. Enzyme Structure (Part C)
Edited by C. H. W. Hirs and Serge N. Timasheff

Volume XXVII. Enzyme Structure (Part D)
Edited by C. H. W. Hirs and Serge N. Timasheff

Volume XXVIII. Complex Carbohydrates (Part B)
Edited by Victor Ginsburg

Volume XXIX. Nucleic Acids and Protein Synthesis (Part E)
Edited by Lawrence Grossman and Kivie Moldave

Volume XXX. Nucleic Acids and Protein Synthesis (Part F)
Edited by Kivie Moldave and Lawrence Grossman

Volume XXXI. Biomembranes (Part A)
Edited by Sidney Fleischer and Lester Packer

Volume XXXII. Biomembranes (Part B)
Edited by Sidney Fleischer and Lester Packer

Volume XXXIII. Cumulative Subject Index Volumes I-XXX
Edited by Martha G. Dennis and Edward A. Dennis

Volume XXXIV. Affinity Techniques (Enzyme Purification: Part B)
Edited by William B. Jakoby and Meir Wilchek

Volume XXXV. Lipids (Part B)
Edited by John M. Lowenstein

Volume XXXVI. Hormone Action (Part A: Steroid Hormones)
Edited by Bert W. O'Malley and Joel G. Hardman

Volume XXXVII. Hormone Action (Part B: Peptide Hormones)
Edited by Bert W. O'Malley and Joel G. Hardman

Volume XXXVIII. Hormone Action (Part C: Cyclic Nucleotides)
Edited by Joel G. Hardman and Bert W. O'Malley

Volume XXXIX. Hormone Action (Part D: Isolated Cells, Tissues, and Organ Systems)
Edited by Joel G. Hardman and Bert W. O'Malley

Volume XL. Hormone Action (Part E: Nuclear Structure and Function)
Edited by Bert W. O'Malley and Joel G. Hardman

Volume XLI. Carbohydrate Metabolism (Part B)
Edited by W. A. Wood

Volume XLII. Carbohydrate Metabolism (Part C)
Edited by W. A. Wood

Volume XLIII. Antibiotics
Edited by John H. Hash

Volume XLIV. Immobilized Enzymes
Edited by Klaus Mosbach

Volume XLV. Proteolytic Enzymes (Part B)
Edited by Laszlo Lorand

Volume XLVI. Affinity Labeling
Edited by William B. Jakoby and Meir Wilchek

Volume XLVII. Enzyme Structure (Part E)
Edited by C. H. W. Hirs and Serge N. Timasheff

Volume XLVIII. Enzyme Structure (Part F)
Edited by C. H. W. Hirs and Serge N. Timasheff

Volume XLIX. Enzyme Structure (Part G)
Edited by C. H. W. Hirs and Serge N. Timasheff

Volume L. Complex Carbohydrates (Part C)
Edited by Victor Ginsburg

Volume LI. Purine and Pyrimidine Nucleotide Metabolism
Edited by Patricia A. Hoffee and Mary Ellen Jones

Volume LII. Biomembranes (Part C: Biological Oxidations)
Edited by Sidney Fleischer and Lester Packer

Volume LIII. Biomembranes (Part D: Biological Oxidations)
Edited by Sidney Fleischer and Lester Packer

Volume LIV. Biomembranes (Part E: Biological Oxidations)
Edited by Sidney Fleischer and Lester Packer

Volume LV. Biomembranes (Part F: Bioenergetics)
Edited by Sidney Fleischer and Lester Packer

Volume LVI. Biomembranes (Part G: Bioenergetics)
Edited by Sidney Fleischer and Lester Packer

Volume LVII. Bioluminescence and Chemiluminescence
Edited by Marlene A. DeLuca

Volume LVIII. Cell Culture
Edited by William B. Jakoby and Ira Pastan

Volume LIX. Nucleic Acids and Protein Synthesis (Part G)
Edited by Kivie Moldave and Lawrence Grossman

Volume LX. Nucleic Acids and Protein Synthesis (Part H)
Edited by Kivie Moldave and Lawrence Grossman

Volume 61. Enzyme Structure (Part H)
Edited by C. H. W. Hirs and Serge N. Timasheff

Volume 62. Vitamins and Coenzymes (Part D)
Edited by Donald B. McCormick and Lemuel D. Wright

Volume 63. Enzyme Kinetics and Mechanism (Part A: Initial Rate and Inhibitor Methods)
Edited by Daniel L. Purich

Volume 64. Enzyme Kinetics and Mechanism (Part B: Isotopic Probes and Complex Enzyme Systems)
Edited by Daniel L. Purich

Volume 65. Nucleic Acids (Part I)
Edited by Lawrence Grossman and Kivie Moldave

Volume 66. Vitamins and Coenzymes (Part E)
Edited by Donald B. McCormick and Lemuel D. Wright

Volume 67. Vitamins and Coenzymes (Part F)
Edited by Donald B. McCormick and Lemuel D. Wright

Volume 68. Recombinant DNA
Edited by Ray Wu

Volume 69. Photosynthesis and Nitrogen Fixation (Part C)
Edited by Anthony San Pietro

Volume 70. Immunochemical Techniques (Part A)
Edited by Helen Van Vunakis and John J. Langone

VOLUME 71. Lipids (Part C)
Edited by JOHN M. LOWENSTEIN

VOLUME 72. Lipids (Part D)
Edited by JOHN M. LOWENSTEIN

VOLUME 73. Immunochemical Techniques (Part B)
Edited by JOHN J. LANGONE AND HELEN VAN VUNAKIS

VOLUME 74. Immunochemical Techniques (Part C)
Edited by JOHN J. LANGONE AND HELEN VAN VUNAKIS

VOLUME 75. Cumulative Subject Index Volumes XXXI, XXXII, XXXIV–LX
Edited by EDWARD A. DENNIS AND MARTHA G. DENNIS

VOLUME 76. Hemoglobins
Edited by ERALDO ANTONINI, LUIGI ROSSI-BERNARDI, AND EMILIA CHIANCONE

VOLUME 77. Detoxication and Drug Metabolism
Edited by WILLIAM B. JAKOBY

VOLUME 78. Interferons (Part A)
Edited by SIDNEY PESTKA

VOLUME 79. Interferons (Part B)
Edited by SIDNEY PESTKA

VOLUME 80. Proteolytic Enzymes (Part C)
Edited by LASZLO LORAND

VOLUME 81. Biomembranes (Part H: Visual Pigments and Purple Membranes, I)
Edited by LESTER PACKER

VOLUME 82. Structural and Contractile Proteins (Part A: Extracellular Matrix)
Edited by LEON W. CUNNINGHAM AND DIXIE W. FREDERIKSEN

VOLUME 83. Complex Carbohydrates (Part D)
Edited by VICTOR GINSBURG

VOLUME 84. Immunochemical Techniques (Part D: Selected Immunoassays)
Edited by JOHN J. LANGONE AND HELEN VAN VUNAKIS

VOLUME 85. Structural and Contractile Proteins (Part B: The Contractile Apparatus and the Cytoskeleton)
Edited by DIXIE W. FREDERIKSEN AND LEON W. CUNNINGHAM

VOLUME 86. Prostaglandins and Arachidonate Metabolites
Edited by WILLIAM E. M. LANDS AND WILLIAM L. SMITH

VOLUME 87. Enzyme Kinetics and Mechanism (Part C: Intermediates, Stereo-chemistry, and Rate Studies)
Edited by DANIEL L. PURICH

VOLUME 88. Biomembranes (Part I: Visual Pigments and Purple Membranes, II)
Edited by LESTER PACKER

Volume 89. Carbohydrate Metabolism (Part D)
Edited by Willis A. Wood

Volume 90. Carbohydrate Metabolism (Part E)
Edited by Willis A. Wood

Volume 91. Enzyme Structure (Part I)
Edited by C. H. W. Hirs and Serge N. Timasheff

Volume 92. Immunochemical Techniques (Part E: Monoclonal Antibodies and General Immunoassay Methods)
Edited by John J. Langone and Helen Van Vunakis

Volume 93. Immunochemical Techniques (Part F: Conventional Antibodies, Fc Receptors, and Cytotoxicity)
Edited by John J. Langone and Helen Van Vunakis

Volume 94. Polyamines
Edited by Herbert Tabor and Celia White Tabor

Volume 95. Cumulative Subject Index Volumes 61–74, 76–80
Edited by Edward A. Dennis and Martha G. Dennis

Volume 96. Biomembranes [Part J: Membrane Biogenesis: Assembly and Targeting (General Methods; Eukaryotes)]
Edited by Sidney Fleischer and Becca Fleischer

Volume 97. Biomembranes [Part K: Membrane Biogenesis: Assembly and Targeting (Prokaryotes, Mitochondria, and Chloroplasts)]
Edited by Sidney Fleischer and Becca Fleischer

Volume 98. Biomembranes (Part L: Membrane Biogenesis: Processing and Recycling)
Edited by Sidney Fleischer and Becca Fleischer

Volume 99. Hormone Action (Part F: Protein Kinases)
Edited by Jackie D. Corbin and Joel G. Hardman

Volume 100. Recombinant DNA (Part B)
Edited by Ray Wu, Lawrence Grossman, and Kivie Moldave

Volume 101. Recombinant DNA (Part C)
Edited by Ray Wu, Lawrence Grossman, and Kivie Moldave

Volume 102. Hormone Action (Part G: Calmodulin and Calcium-Binding Proteins)
Edited by Anthony R. Means and Bert W. O'Malley

Volume 103. Hormone Action (Part H: Neuroendocrine Peptides)
Edited by P. Michael Conn

Volume 104. Enzyme Purification and Related Techniques (Part C)
Edited by William B. Jakoby

VOLUME 105. Oxygen Radicals in Biological Systems
Edited by LESTER PACKER

VOLUME 106. Posttranslational Modifications (Part A)
Edited by FINN WOLD AND KIVIE MOLDAVE

VOLUME 107. Posttranslational Modifications (Part B)
Edited by FINN WOLD AND KIVIE MOLDAVE

VOLUME 108. Immunochemical Techniques (Part G: Separation and Characterization of Lymphoid Cells)
Edited by GIOVANNI DI SABATO, JOHN J. LANGONE, AND HELEN VAN VUNAKIS

VOLUME 109. Hormone Action (Part I: Peptide Hormones)
Edited by LUTZ BIRNBAUMER AND BERT W. O'MALLEY

VOLUME 110. Steroids and Isoprenoids (Part A)
Edited by JOHN H. LAW AND HANS C. RILLING

VOLUME 111. Steroids and Isoprenoids (Part B)
Edited by JOHN H. LAW AND HANS C. RILLING

VOLUME 112. Drug and Enzyme Targeting (Part A)
Edited by KENNETH J. WIDDER AND RALPH GREEN

VOLUME 113. Glutamate, Glutamine, Glutathione, and Related Compounds
Edited by ALTON MEISTER

VOLUME 114. Diffraction Methods for Biological Macromolecules (Part A)
Edited by HAROLD W. WYCKOFF, C. H. W. HIRS, AND SERGE N. TIMASHEFF

VOLUME 115. Diffraction Methods for Biological Macromolecules (Part B)
Edited by HAROLD W. WYCKOFF, C. H. W. HIRS, AND SERGE N. TIMASHEFF

VOLUME 116. Immunochemical Techniques
(Part H: Effectors and Mediators of Lymphoid Cell Functions)
Edited by GIOVANNI DI SABATO, JOHN J. LANGONE, AND HELEN VAN VUNAKIS

VOLUME 117. Enzyme Structure (Part J)
Edited by C. H. W. HIRS AND SERGE N. TIMASHEFF

VOLUME 118. Plant Molecular Biology
Edited by ARTHUR WEISSBACH AND HERBERT WEISSBACH

VOLUME 119. Interferons (Part C)
Edited by SIDNEY PESTKA

VOLUME 120. Cumulative Subject Index Volumes 81–94, 96–101

VOLUME 121. Immunochemical Techniques (Part I: Hybridoma Technology and Monoclonal Antibodies)
Edited by JOHN J. LANGONE AND HELEN VAN VUNAKIS

VOLUME 122. Vitamins and Coenzymes (Part G)
Edited by FRANK CHYTIL AND DONALD B. MCCORMICK

VOLUME 123. Vitamins and Coenzymes (Part H)
Edited by FRANK CHYTIL AND DONALD B. MCCORMICK

VOLUME 124. Hormone Action (Part J: Neuroendocrine Peptides)
Edited by P. MICHAEL CONN

VOLUME 125. Biomembranes (Part M: Transport in Bacteria, Mitochondria, and Chloroplasts: General Approaches and Transport Systems)
Edited by SIDNEY FLEISCHER AND BECCA FLEISCHER

VOLUME 126. Biomembranes (Part N: Transport in Bacteria, Mitochondria, and Chloroplasts: Protonmotive Force)
Edited by SIDNEY FLEISCHER AND BECCA FLEISCHER

VOLUME 127. Biomembranes (Part O: Protons and Water: Structure and Translocation)
Edited by LESTER PACKER

VOLUME 128. Plasma Lipoproteins (Part A: Preparation, Structure, and Molecular Biology)
Edited by JERE P. SEGREST AND JOHN J. ALBERS

VOLUME 129. Plasma Lipoproteins (Part B: Characterization, Cell Biology, and Metabolism)
Edited by JOHN J. ALBERS AND JERE P. SEGREST

VOLUME 130. Enzyme Structure (Part K)
Edited by C. H. W. HIRS AND SERGE N. TIMASHEFF

VOLUME 131. Enzyme Structure (Part L)
Edited by C. H. W. HIRS AND SERGE N. TIMASHEFF

VOLUME 132. Immunochemical Techniques (Part J: Phagocytosis and Cell-Mediated Cytotoxicity)
Edited by GIOVANNI DI SABATO AND JOHANNES EVERSE

VOLUME 133. Bioluminescence and Chemiluminescence (Part B)
Edited by MARLENE DELUCA AND WILLIAM D. MCELROY

VOLUME 134. Structural and Contractile Proteins (Part C: The Contractile Apparatus and the Cytoskeleton)
Edited by RICHARD B. VALLEE

VOLUME 135. Immobilized Enzymes and Cells (Part B)
Edited by KLAUS MOSBACH

VOLUME 136. Immobilized Enzymes and Cells (Part C)
Edited by KLAUS MOSBACH

VOLUME 137. Immobilized Enzymes and Cells (Part D)
Edited by KLAUS MOSBACH

VOLUME 138. Complex Carbohydrates (Part E)
Edited by VICTOR GINSBURG

Volume 139. Cellular Regulators (Part A: Calcium- and Calmodulin-Binding Proteins)
Edited by Anthony R. Means and P. Michael Conn

Volume 140. Cumulative Subject Index Volumes 102–119, 121–134

Volume 141. Cellular Regulators (Part B: Calcium and Lipids)
Edited by P. Michael Conn and Anthony R. Means

Volume 142. Metabolism of Aromatic Amino Acids and Amines
Edited by Seymour Kaufman

Volume 143. Sulfur and Sulfur Amino Acids
Edited by William B. Jakoby and Owen Griffith

Volume 144. Structural and Contractile Proteins (Part D: Extracellular Matrix)
Edited by Leon W. Cunningham

Volume 145. Structural and Contractile Proteins (Part E: Extracellular Matrix)
Edited by Leon W. Cunningham

Volume 146. Peptide Growth Factors (Part A)
Edited by David Barnes and David A. Sirbasku

Volume 147. Peptide Growth Factors (Part B)
Edited by David Barnes and David A. Sirbasku

Volume 148. Plant Cell Membranes
Edited by Lester Packer and Roland Douce

Volume 149. Drug and Enzyme Targeting (Part B)
Edited by Ralph Green and Kenneth J. Widder

Volume 150. Immunochemical Techniques (Part K: *In Vitro* Models of B and T Cell Functions and Lymphoid Cell Receptors)
Edited by Giovanni Di Sabato

Volume 151. Molecular Genetics of Mammalian Cells
Edited by Michael M. Gottesman

Volume 152. Guide to Molecular Cloning Techniques
Edited by Shelby L. Berger and Alan R. Kimmel

Volume 153. Recombinant DNA (Part D)
Edited by Ray Wu and Lawrence Grossman

Volume 154. Recombinant DNA (Part E)
Edited by Ray Wu and Lawrence Grossman

Volume 155. Recombinant DNA (Part F)
Edited by Ray Wu

Volume 156. Biomembranes (Part P: ATP-Driven Pumps and Related Transport: The Na, K-Pump)
Edited by Sidney Fleischer and Becca Fleischer

Volume 157. Biomembranes (Part Q: ATP-Driven Pumps and Related Transport: Calcium, Proton, and Potassium Pumps)
Edited by Sidney Fleischer and Becca Fleischer

Volume 158. Metalloproteins (Part A)
Edited by James F. Riordan and Bert L. Vallee

Volume 159. Initiation and Termination of Cyclic Nucleotide Action
Edited by Jackie D. Corbin and Roger A. Johnson

Volume 160. Biomass (Part A: Cellulose and Hemicellulose)
Edited by Willis A. Wood and Scott T. Kellogg

Volume 161. Biomass (Part B: Lignin, Pectin, and Chitin)
Edited by Willis A. Wood and Scott T. Kellogg

Volume 162. Immunochemical Techniques (Part L: Chemotaxis and Inflammation)
Edited by Giovanni Di Sabato

Volume 163. Immunochemical Techniques (Part M: Chemotaxis and Inflammation)
Edited by Giovanni Di Sabato

Volume 164. Ribosomes
Edited by Harry F. Noller, Jr., and Kivie Moldave

Volume 165. Microbial Toxins: Tools for Enzymology
Edited by Sidney Harshman

Volume 166. Branched-Chain Amino Acids
Edited by Robert Harris and John R. Sokatch

Volume 167. Cyanobacteria
Edited by Lester Packer and Alexander N. Glazer

Volume 168. Hormone Action (Part K: Neuroendocrine Peptides)
Edited by P. Michael Conn

Volume 169. Platelets: Receptors, Adhesion, Secretion (Part A)
Edited by Jacek Hawiger

Volume 170. Nucleosomes
Edited by Paul M. Wassarman and Roger D. Kornberg

Volume 171. Biomembranes (Part R: Transport Theory: Cells and Model Membranes)
Edited by Sidney Fleischer and Becca Fleischer

Volume 172. Biomembranes (Part S: Transport: Membrane Isolation and Characterization)
Edited by Sidney Fleischer and Becca Fleischer

Volume 190. Retinoids (Part B: Cell Differentiation and Clinical Applications)
Edited by Lester Packer

Volume 191. Biomembranes (Part V: Cellular and Subcellular Transport: Epithelial Cells)
Edited by Sidney Fleischer and Becca Fleischer

Volume 192. Biomembranes (Part W: Cellular and Subcellular Transport: Epithelial Cells)
Edited by Sidney Fleischer and Becca Fleischer

Volume 193. Mass Spectrometry
Edited by James A. McCloskey

Volume 194. Guide to Yeast Genetics and Molecular Biology
Edited by Christine Guthrie and Gerald R. Fink

Volume 195. Adenylyl Cyclase, G Proteins, and Guanylyl Cyclase
Edited by Roger A. Johnson and Jackie D. Corbin

Volume 196. Molecular Motors and the Cytoskeleton
Edited by Richard B. Vallee

Volume 197. Phospholipases
Edited by Edward A. Dennis

Volume 198. Peptide Growth Factors (Part C)
Edited by David Barnes, J. P. Mather, and Gordon H. Sato

Volume 199. Cumulative Subject Index Volumes 168–174, 176–194

Volume 200. Protein Phosphorylation (Part A: Protein Kinases: Assays, Purification, Antibodies, Functional Analysis, Cloning, and Expression)
Edited by Tony Hunter and Bartholomew M. Sefton

Volume 201. Protein Phosphorylation (Part B: Analysis of Protein Phosphorylation, Protein Kinase Inhibitors, and Protein Phosphatases)
Edited by Tony Hunter and Bartholomew M. Sefton

Volume 202. Molecular Design and Modeling: Concepts and Applications (Part A: Proteins, Peptides, and Enzymes)
Edited by John J. Langone

Volume 203. Molecular Design and Modeling: Concepts and Applications (Part B: Antibodies and Antigens, Nucleic Acids, Polysaccharides, and Drugs)
Edited by John J. Langone

Volume 204. Bacterial Genetic Systems
Edited by Jeffrey H. Miller

Volume 205. Metallobiochemistry (Part B: Metallothionein and Related Molecules)
Edited by James F. Riordan and Bert L. Vallee

Volume 223. Proteolytic Enzymes in Coagulation, Fibrinolysis, and Complement Activation (Part B: Complement Activation, Fibrinolysis, and Nonmammalian Blood Coagulation Factors)
Edited by Laszlo Lorand and Kenneth G. Mann

Volume 224. Molecular Evolution: Producing the Biochemical Data
Edited by Elizabeth Anne Zimmer, Thomas J. White, Rebecca L. Cann, and Allan C. Wilson

Volume 225. Guide to Techniques in Mouse Development
Edited by Paul M. Wassarman and Melvin L. DePamphilis

Volume 226. Metallobiochemistry (Part C: Spectroscopic and Physical Methods for Probing Metal Ion Environments in Metalloenzymes and Metalloproteins)
Edited by James F. Riordan and Bert L. Vallee

Volume 227. Metallobiochemistry (Part D: Physical and Spectroscopic Methods for Probing Metal Ion Environments in Metalloproteins)
Edited by James F. Riordan and Bert L. Vallee

Volume 228. Aqueous Two-Phase Systems
Edited by Harry Walter and Göte Johansson

Volume 229. Cumulative Subject Index Volumes 195–198, 200–227

Volume 230. Guide to Techniques in Glycobiology
Edited by William J. Lennarz and Gerald W. Hart

Volume 231. Hemoglobins (Part B: Biochemical and Analytical Methods)
Edited by Johannes Everse, Kim D. Vandegriff, and Robert M. Winslow

Volume 232. Hemoglobins (Part C: Biophysical Methods)
Edited by Johannes Everse, Kim D. Vandegriff, and Robert M. Winslow

Volume 233. Oxygen Radicals in Biological Systems (Part C)
Edited by Lester Packer

Volume 234. Oxygen Radicals in Biological Systems (Part D)
Edited by Lester Packer

Volume 235. Bacterial Pathogenesis (Part A: Identification and Regulation of Virulence Factors)
Edited by Virginia L. Clark and Patrik M. Bavoil

Volume 236. Bacterial Pathogenesis (Part B: Integration of Pathogenic Bacteria with Host Cells)
Edited by Virginia L. Clark and Patrik M. Bavoil

Volume 237. Heterotrimeric G Proteins
Edited by Ravi Iyengar

Volume 238. Heterotrimeric G-Protein Effectors
Edited by Ravi Iyengar

VOLUME 257. Small GTPases and Their Regulators (Part C: Proteins Involved in Transport)
Edited by W. E. BALCH, CHANNING J. DER, AND ALAN HALL

VOLUME 258. Redox-Active Amino Acids in Biology
Edited by JUDITH P. KLINMAN

VOLUME 259. Energetics of Biological Macromolecules
Edited by MICHAEL L. JOHNSON AND GARY K. ACKERS

VOLUME 260. Mitochondrial Biogenesis and Genetics (Part A)
Edited by GIUSEPPE M. ATTARDI AND ANNE CHOMYN

VOLUME 261. Nuclear Magnetic Resonance and Nucleic Acids
Edited by THOMAS L. JAMES

VOLUME 262. DNA Replication
Edited by JUDITH L. CAMPBELL

VOLUME 263. Plasma Lipoproteins (Part C: Quantitation)
Edited by WILLIAM A. BRADLEY, SANDRA H. GIANTURCO, AND JERE P. SEGREST

VOLUME 264. Mitochondrial Biogenesis and Genetics (Part B)
Edited by GIUSEPPE M. ATTARDI AND ANNE CHOMYN

VOLUME 265. Cumulative Subject Index Volumes 228, 230–262

VOLUME 266. Computer Methods for Macromolecular Sequence Analysis
Edited by RUSSELL F. DOOLITTLE

VOLUME 267. Combinatorial Chemistry
Edited by JOHN N. ABELSON

VOLUME 268. Nitric Oxide (Part A: Sources and Detection of NO; NO Synthase)
Edited by LESTER PACKER

VOLUME 269. Nitric Oxide (Part B: Physiological and Pathological Processes)
Edited by LESTER PACKER

VOLUME 270. High Resolution Separation and Analysis of Biological Macromolecules (Part A: Fundamentals)
Edited by BARRY L. KARGER AND WILLIAM S. HANCOCK

VOLUME 271. High Resolution Separation and Analysis of Biological Macromolecules (Part B: Applications)
Edited by BARRY L. KARGER AND WILLIAM S. HANCOCK

VOLUME 272. Cytochrome P450 (Part B)
Edited by ERIC F. JOHNSON AND MICHAEL R. WATERMAN

VOLUME 273. RNA Polymerase and Associated Factors (Part A)
Edited by SANKAR ADHYA

VOLUME 274. RNA Polymerase and Associated Factors (Part B)
Edited by SANKAR ADHYA

VOLUME 275. Viral Polymerases and Related Proteins
Edited by LAWRENCE C. KUO, DAVID B. OLSEN, AND STEVEN S. CARROLL

VOLUME 276. Macromolecular Crystallography (Part A)
Edited by CHARLES W. CARTER, JR., AND ROBERT M. SWEET

VOLUME 277. Macromolecular Crystallography (Part B)
Edited by CHARLES W. CARTER, JR., AND ROBERT M. SWEET

VOLUME 278. Fluorescence Spectroscopy
Edited by LUDWIG BRAND AND MICHAEL L. JOHNSON

VOLUME 279. Vitamins and Coenzymes (Part I)
Edited by DONALD B. MCCORMICK, JOHN W. SUTTIE, AND CONRAD WAGNER

VOLUME 280. Vitamins and Coenzymes (Part J)
Edited by DONALD B. MCCORMICK, JOHN W. SUTTIE, AND CONRAD WAGNER

VOLUME 281. Vitamins and Coenzymes (Part K)
Edited by DONALD B. MCCORMICK, JOHN W. SUTTIE, AND CONRAD WAGNER

VOLUME 282. Vitamins and Coenzymes (Part L)
Edited by DONALD B. MCCORMICK, JOHN W. SUTTIE, AND CONRAD WAGNER

VOLUME 283. Cell Cycle Control
Edited by WILLIAM G. DUNPHY

VOLUME 284. Lipases (Part A: Biotechnology)
Edited by BYRON RUBIN AND EDWARD A. DENNIS

VOLUME 285. Cumulative Subject Index Volumes 263, 264, 266–284, 286–289

VOLUME 286. Lipases (Part B: Enzyme Characterization and Utilization)
Edited by BYRON RUBIN AND EDWARD A. DENNIS

VOLUME 287. Chemokines
Edited by RICHARD HORUK

VOLUME 288. Chemokine Receptors
Edited by RICHARD HORUK

VOLUME 289. Solid Phase Peptide Synthesis
Edited by GREGG B. FIELDS

VOLUME 290. Molecular Chaperones
Edited by GEORGE H. LORIMER AND THOMAS BALDWIN

VOLUME 291. Caged Compounds
Edited by GERARD MARRIOTT

VOLUME 292. ABC Transporters: Biochemical, Cellular, and Molecular Aspects
Edited by SURESH V. AMBUDKAR AND MICHAEL M. GOTTESMAN

VOLUME 293. Ion Channels (Part B)
Edited by P. MICHAEL CONN

Volume 312. Sphingolipid Metabolism and Cell Signaling (Part B)
Edited by Alfred H. Merrill, Jr., and Yusuf A. Hannun

Volume 313. Antisense Technology (Part A: General Methods, Methods of Delivery, and RNA Studies)
Edited by M. Ian Phillips

Volume 314. Antisense Technology (Part B: Applications)
Edited by M. Ian Phillips

Volume 315. Vertebrate Phototransduction and the Visual Cycle (Part A)
Edited by Krzysztof Palczewski

Volume 316. Vertebrate Phototransduction and the Visual Cycle (Part B)
Edited by Krzysztof Palczewski

Volume 317. RNA–Ligand Interactions (Part A: Structural Biology Methods)
Edited by Daniel W. Celander and John N. Abelson

Volume 318. RNA–Ligand Interactions (Part B: Molecular Biology Methods)
Edited by Daniel W. Celander and John N. Abelson

Volume 319. Singlet Oxygen, UV-A, and Ozone
Edited by Lester Packer and Helmut Sies

Volume 320. Cumulative Subject Index Volumes 290–319

Volume 321. Numerical Computer Methods (Part C)
Edited by Michael L. Johnson and Ludwig Brand

Volume 322. Apoptosis
Edited by John C. Reed

Volume 323. Energetics of Biological Macromolecules (Part C)
Edited by Michael L. Johnson and Gary K. Ackers

Volume 324. Branched-Chain Amino Acids (Part B)
Edited by Robert A. Harris and John R. Sokatch

Volume 325. Regulators and Effectors of Small GTPases (Part D: Rho Family)
Edited by W. E. Balch, Channing J. Der, and Alan Hall

Volume 326. Applications of Chimeric Genes and Hybrid Proteins (Part A: Gene Expression and Protein Purification)
Edited by Jeremy Thorner, Scott D. Emr, and John N. Abelson

Volume 327. Applications of Chimeric Genes and Hybrid Proteins (Part B: Cell Biology and Physiology)
Edited by Jeremy Thorner, Scott D. Emr, and John N. Abelson

Volume 328. Applications of Chimeric Genes and Hybrid Proteins (Part C: Protein–Protein Interactions and Genomics)
Edited by Jeremy Thorner, Scott D. Emr, and John N. Abelson

Volume 329. Regulators and Effectors of Small GTPases (Part E: GTPases Involved in Vesicular Traffic)
Edited by W. E. Balch, Channing J. Der, and Alan Hall

Volume 330. Hyperthermophilic Enzymes (Part A)
Edited by Michael W. W. Adams and Robert M. Kelly

Volume 331. Hyperthermophilic Enzymes (Part B)
Edited by Michael W. W. Adams and Robert M. Kelly

Volume 332. Regulators and Effectors of Small GTPases (Part F: Ras Family I)
Edited by W. E. Balch, Channing J. Der, and Alan Hall

Volume 333. Regulators and Effectors of Small GTPases (Part G: Ras Family II)
Edited by W. E. Balch, Channing J. Der, and Alan Hall

Volume 334. Hyperthermophilic Enzymes (Part C)
Edited by Michael W. W. Adams and Robert M. Kelly

Volume 335. Flavonoids and Other Polyphenols
Edited by Lester Packer

Volume 336. Microbial Growth in Biofilms (Part A: Developmental and Molecular Biological Aspects)
Edited by Ron J. Doyle

Volume 337. Microbial Growth in Biofilms (Part B: Special Environments and Physicochemical Aspects)
Edited by Ron J. Doyle

Volume 338. Nuclear Magnetic Resonance of Biological Macromolecules (Part A)
Edited by Thomas L. James, Volker Dötsch, and Uli Schmitz

Volume 339. Nuclear Magnetic Resonance of Biological Macromolecules (Part B)
Edited by Thomas L. James, Volker Dötsch, and Uli Schmitz

Volume 340. Drug–Nucleic Acid Interactions
Edited by Jonathan B. Chaires and Michael J. Waring

Volume 341. Ribonucleases (Part A)
Edited by Allen W. Nicholson

Volume 342. Ribonucleases (Part B)
Edited by Allen W. Nicholson

Volume 343. G Protein Pathways (Part A: Receptors)
Edited by Ravi Iyengar and John D. Hildebrandt

Volume 344. G Protein Pathways (Part B: G Proteins and Their Regulators)
Edited by Ravi Iyengar and John D. Hildebrandt

Volume 345. G Protein Pathways (Part C: Effector Mechanisms)
Edited by Ravi Iyengar and John D. Hildebrandt

VOLUME 363. Recognition of Carbohydrates in Biological Systems (Part B)
Edited by YUAN C. LEE AND REIKO T. LEE

VOLUME 364. Nuclear Receptors
Edited by DAVID W. RUSSELL AND DAVID J. MANGELSDORF

VOLUME 365. Differentiation of Embryonic Stem Cells
Edited by PAUL M. WASSAUMAN AND GORDON M. KELLER

VOLUME 366. Protein Phosphatases
Edited by SUSANNE KLUMPP AND JOSEF KRIEGLSTEIN

VOLUME 367. Liposomes (Part A)
Edited by NEJAT DÜZGÜNEŞ

VOLUME 368. Macromolecular Crystallography (Part C)
Edited by CHARLES W. CARTER, JR., AND ROBERT M. SWEET

VOLUME 369. Combinational Chemistry (Part B)
Edited by GUILLERMO A. MORALES AND BARRY A. BUNIN

VOLUME 370. RNA Polymerases and Associated Factors (Part C)
Edited by SANKAR L. ADHYA AND SUSAN GARGES

VOLUME 371. RNA Polymerases and Associated Factors (Part D)
Edited by SANKAR L. ADHYA AND SUSAN GARGES

VOLUME 372. Liposomes (Part B)
Edited by NEJAT DÜZGÜNEŞ

VOLUME 373. Liposomes (Part C)
Edited by NEJAT DÜZGÜNEŞ

VOLUME 374. Macromolecular Crystallography (Part D)
Edited by CHARLES W. CARTER, JR., AND ROBERT W. SWEET

VOLUME 375. Chromatin and Chromatin Remodeling Enzymes (Part A)
Edited by C. DAVID ALLIS AND CARL WU

VOLUME 376. Chromatin and Chromatin Remodeling Enzymes (Part B)
Edited by C. DAVID ALLIS AND CARL WU

VOLUME 377. Chromatin and Chromatin Remodeling Enzymes (Part C)
Edited by C. DAVID ALLIS AND CARL WU

VOLUME 378. Quinones and Quinone Enzymes (Part A)
Edited by HELMUT SIES AND LESTER PACKER

VOLUME 379. Energetics of Biological Macromolecules (Part D)
Edited by JO M. HOLT, MICHAEL L. JOHNSON, AND GARY K. ACKERS

VOLUME 380. Energetics of Biological Macromolecules (Part E)
Edited by JO M. HOLT, MICHAEL L. JOHNSON, AND GARY K. ACKERS

VOLUME 381. Oxygen Sensing
Edited by CHANDAN K. SEN AND GREGG L. SEMENZA

Volume 401. Glutathione Transferases and Gamma Glutamyl Transpeptidases
Edited by Helmut Sies and Lester Packer

Volume 402. Biological Mass Spectrometry
Edited by A. L. Burlingame

Volume 403. GTPases Regulating Membrane Targeting and Fusion
Edited by William E. Balch, Channing J. Der, and Alan Hall

Volume 404. GTPases Regulating Membrane Dynamics
Edited by William E. Balch, Channing J. Der, and Alan Hall

Volume 405. Mass Spectrometry: Modified Proteins and Glycoconjugates
Edited by A. L. Burlingame

Volume 406. Regulators and Effectors of Small GTPases: Rho Family
Edited by William E. Balch, Channing J. Der, and Alan Hall

Volume 407. Regulators and Effectors of Small GTPases: Ras Family
Edited by William E. Balch, Channing J. Der, and Alan Hall

Volume 408. DNA Repair (Part A)
Edited by Judith L. Campbell and Paul Modrich

Volume 409. DNA Repair (Part B)
Edited by Judith L. Campbell and Paul Modrich

Volume 410. DNA Microarrays (Part A: Array Platforms and Web-Bench Protocols)
Edited by Alan Kimmel and Brian Oliver

Volume 411. DNA Microarrays (Part B: Databases and Statistics)
Edited by Alan Kimmel and Brian Oliver

Volume 412. Amyloid, Prions, and Other Protein Aggregates (Part B)
Edited by Indu Kheterpal and Ronald Wetzel

Volume 413. Amyloid, Prions, and Other Protein Aggregates (Part C)
Edited by Indu Kheterpal and Ronald Wetzel

Volume 414. Measuring Biological Responses with Automated Microscopy
Edited by James Inglese

Volume 415. Glycobiology
Edited by Minoru Fukuda

Volume 416. Glycomics
Edited by Minoru Fukuda

Volume 417. Functional Glycomics
Edited by Minoru Fukuda

Volume 418. Embryonic Stem Cells
Edited by Irina Klimanskaya and Robert Lanza

Volume 419. Adult Stem Cells
Edited by Irina Klimanskaya and Robert Lanza

Volume 420. Stem Cell Tools and Other Experimental Protocols
Edited by Irina Klimanskaya and Robert Lanza

Volume 421. Advanced Bacterial Genetics: Use of Transposons and Phage for Genomic Engineering
Edited by Kelly T. Hughes

Volume 422. Two-Component Signaling Systems, Part A
Edited by Melvin I. Simon, Brian R. Crane, and Alexandrine Crane

Volume 423. Two-Component Signaling Systems, Part B
Edited by Melvin I. Simon, Brian R. Crane, and Alexandrine Crane

Volume 424. RNA Editing
Edited by Jonatha M. Gott

Volume 425. RNA Modification
Edited by Jonatha M. Gott

Volume 426. Integrins
Edited by David Cheresh

Volume 427. MicroRNA Methods
Edited by John J. Rossi

Volume 428. Osmosensing and Osmosignaling
Edited by Helmut Sies and Dieter Haussinger

Volume 429. Translation Initiation: Extract Systems and Molecular Genetics
Edited by Jon Lorsch

Volume 430. Translation Initiation: Reconstituted Systems and Biophysical Methods
Edited by Jon Lorsch

Volume 431. Translation Initiation: Cell Biology, High-Throughput and Chemical-Based Approaches
Edited by Jon Lorsch

Volume 432. Lipidomics and Bioactive Lipids: Mass-Spectrometry–Based Lipid Analysis
Edited by H. Alex Brown

Volume 433. Lipidomics and Bioactive Lipids: Specialized Analytical Methods and Lipids in Disease
Edited by H. Alex Brown

Volume 434. Lipidomics and Bioactive Lipids: Lipids and Cell Signaling
Edited by H. Alex Brown

Volume 435. Oxygen Biology and Hypoxia
Edited by Helmut Sies and Bernhard Brüne

Volume 436. Globins and Other Nitric Oxide-Reactive Protiens (Part A)
Edited by Robert K. Poole

Volume 437. Globins and Other Nitric Oxide-Reactive Protiens (Part B)
Edited by Robert K. Poole

Volume 438. Small GTPases in Disease (Part A)
Edited by William E. Balch, Channing J. Der, and Alan Hall

Volume 439. Small GTPases in Disease (Part B)
Edited by William E. Balch, Channing J. Der, and Alan Hall

Volume 440. Nitric Oxide, Part F Oxidative and Nitrosative Stress in Redox Regulation of Cell Signaling
Edited by Enrique Cadenas and Lester Packer

Volume 441. Nitric Oxide, Part G Oxidative and Nitrosative Stress in Redox Regulation of Cell Signaling
Edited by Enrique Cadenas and Lester Packer

Volume 442. Programmed Cell Death, General Principles for Studying Cell Death (Part A)
Edited by Roya Khosravi-Far, Zahra Zakeri, Richard A. Lockshin, and Mauro Piacentini

Volume 443. Angiogenesis: *In Vitro* Systems
Edited by David A. Cheresh

Volume 444. Angiogenesis: *In Vivo* Systems (Part A)
Edited by David A. Cheresh

Volume 445. Angiogenesis: *In Vivo* Systems (Part B)
Edited by David A. Cheresh

Volume 446. Programmed Cell Death, The Biology and Therapeutic Implications of Cell Death (Part B)
Edited by Roya Khosravi-Far, Zahra Zakeri, Richard A. Lockshin, and Mauro Piacentini

Volume 447. RNA Turnover in Prokaryotes, Archae and Organelles
Edited by Lynne E. Maquat and Cecilia M. Arraiano

Volume 448. RNA Turnover in Eukaryotes: Nucleases, Pathways and Anaylsis of mRNA Decay
Edited by Lynne E. Maquat and Megerditch Kiledjian

Volume 449. RNA Turnover in Eukaryotes: Analysis of Specialized and Quality Control RNA Decay Pathways
Edited by Lynne E. Maquat and Megerditch Kiledjian

Volume 450. Fluorescence Spectroscopy
Edited by Ludwing Brand and Michael Johnson

Volume 451. Autophagy: Lower Eukaryotes and Non-mammalian Systems (Part A)
Edited by Daniel J. Klionsky

CHAPTER ONE

Biochemical Methods to Monitor Autophagy-Related Processes in Yeast

Heesun Cheong* *and* Daniel J. Klionsky*

Contents

Abstract

An increasing number of reports have elucidated the importance of macroautophagy in cell physiology and pathology. Macroautophagy occurs at a basal level and participates in the turnover of cytoplasmic constituents including long-lived proteins to maintain cellular homeostasis, but it also serves as an adaptive response to protect cells from various intra- or extracellular stresses. In addition, macroautophagy plays a role in development and aging and acts to protect against cancer, microbial invasion, and neurodegeneration.

The machinery involved in carrying out this process, the autophagy-related (Atg) proteins were identified and characterized in various fungal systems, in particular because of the powerful tools available for genetic manipulation and the relative abundance of good biochemical assays in these model organisms.

* Life Sciences Institute and Departments of Molecular, Cellular and Developmental Biology and Biological Chemistry, University of Michigan, Ann Arbor, Michigan, USA

Methods in Enzymology, Volume 451

ISSN 0076-6879, DOI: 10.1016/S0076-6879(08)03201-1

The analysis of these Atg proteins has allowed us to begin to understand the molecular mechanism of this process. Furthermore, many of the autophagy genes are functionally conserved in higher eukaryotes, including mammals, allowing the findings in fungi to be applied to other systems.

Here, we discuss three biochemical methods to measure autophagy-related activities and to examine individual steps of the corresponding process. These methods rely on the detection of different modification states of certain marker proteins. Processing of the precursor form of the resident vacuolar hydrolase aminopeptidase I (Ape1) is applicable to fungi, whereas cleavage of the GFP-Atg8 and Pex14-GFP chimeras can be used in a wide array of systems.

1. Introduction

Macroautophagy (hereafter autophagy) is one of the major cellular catabolic processes and is evolutionarily conserved in eukaryotic cells from yeast to human (Klionsky, 2005; Reggiori and Klionsky, 2002). This process occurs constitutively at basal levels but is dramatically stimulated by various extra- and intracellular stresses, such as nutrient or growth factor limitation, oxidative stress, and accumulation of damaged organelles or misfolded proteins (Mizushima and Klionsky, 2007).

During autophagy, cytoplasmic constituents including entire organelles are sequestered within double-membrane vesicles, termed *autophagosomes*. The completed autophagosomes are then transported to degradative compartments (e.g., vacuoles or lysosomes). The inner vesicles, named *autophagic bodies*, are released into the vacuole lumen (or exposed to the lumen of the lysosome, which is too small to encompass the vesicle) and broken down through hydrolysis (Klionsky, 2005; Xie and Klionsky, 2007). After the cargo has been degraded in the lysosome/vacuole, the breakdown products are exported to the cytosol for reuse (Yang *et al.*, 2006).

Autophagy is generally considered a nonspecific degradative process. Various studies have reported that autophagy and related processes (all of which are grouped under the term *autophagy-related*) have multiple functions in cell physiology. In particular, autophagy-related processes play critical roles in cellular remodeling through degradation, thus affecting normal developmental stages (Levine and Klionsky, 2004). Autophagy is are also involved in various pathophysiological conditions such as cancer, pathogen invasion, and certain types of neurodegeneration (Mizushima and Klionsky, 2007; Mizushima *et al.*, 2008; Rubinsztein *et al.*, 2005).

Most of the Atg proteins that are involved in different types of autophagy have been characterized by molecular genetic studies using fungal systems (Harding *et al.*, 1995; Thumm *et al.*, 1994; Titorenko *et al.*, 1995; Tsukada and Ohsumi, 1993). Up to now, 31 Atg proteins have been identified in various fungi. The functions of these proteins are involved in distinct steps of autophagy-related pathways including induction, vesicle nucleation and expansion, retrieval of Atg proteins from the forming vesicle, breakdown of the autophagic body and efflux of the degradation products. However, the exact functions of these components are still being actively investigated.

A subset of proteins including the Atg1 kinase form complexes that play roles in various steps of the autophagy-related pathways. The Atg1 complexes are particular candidates for regulating autophagy induction, which may be mediated directly or indirectly by Tor kinase (Kamada *et al.*, 2000). Most Atg proteins in yeast reside transiently at the phagophore assembly site (PAS), which is considered the potential nucleating site for sequestering vesicle formation, and a subgroup of Atg1 complexes has a critical role in initiating the PAS recruitment of other Atg proteins that function during nonspecific (starvation-induced) autophagy (Cheong *et al.*, 2008). On the other hand, Atg11, which is an Atg1-interacting protein, is important for PAS assembly to generate the sequestering vesicles that are used for the biosynthetic autophagy-related cytoplasm to vacuole targeting (Cvt) pathway that operates under vegetative conditions (Shintani and Klionsky, 2004).

In addition to the Atg1 complexes, most of the Atg proteins can be separated into three functional subgroups. (1) Atg9 is a transmembrane protein that may be involved in delivering membrane to the expanding sequestering vesicle. Atg9 appears to cycle between the PAS and non-PAS pools; the latter may represent the source of the membrane(s) that contribute to autophagosome and Cvt vesicle formation. Atg1, Atg13, Atg2, and Atg18 are required for retrograde movement of Atg9 away from the PAS and back to the peripheral sites; Atg23 and Atg27 (and Atg11, which acts at multiple steps) are involved in anterograde movement (He *et al.*, 2006; Reggiori *et al.*, 2004; Reggiori *et al.*, 2005). (2) The sole yeast phosphatidylinositol 3-kinase (PtdIns3K), Vps34, and its modulating proteins have multiple functions for vesicular transport (Kihara *et al.*, 2001). The core components include Vps34, Vps15, and Atg6/Vps30, which play a role in vesicle formation both for autophagy and the vacuolar protein sorting (Vps) pathway (Kihara *et al.*, 2001; Stack *et al.*, 1993). In addition to these components, two different proteins specify the function of each PtdIns3K complex within their respective pathways; Atg14 directs PtdIns3K to function in autophagy-related processess and Vps38 directs the complex into the Vps pathway. (3) Two ubiquitin-like (Ubl) proteins, Atg12 and Atg8, are involved in autophagy-related mechanisms (George *et al.*, 2000; Huang *et al.*, 2000; Ichimura *et al.*, 2000; Kirisako *et al.*, 2000; Mizushima *et al.*, 1998). These proteins are conjugated to Atg5 and the lipid phosphatidylethanolamine (PE), respectively. Atg8 is initially synthesized with a C-terminal arginine residue that is proteolytically removed through the action of Atg4. The E1-like activating enzyme Atg7 activates both Atg12 and Atg8, then transfers them to E2-like conjugating enzymes, Atg10 and Atg3, which attach them to their respective targets. The Atg12–Atg5 conjugate binds to Atg16 to form a tetrameric structure of Atg12–Atg5-Atg16 (Kuma *et al.*, 2002). Recent data indicate that the Atg12–Atg5-Atg16 complex acts as an E3-like enzyme for conjugation of Atg8 to PE, although conjugation can occur in the absence of these proteins. The Atg12–Atg5 conjugation appears to be irreversible, whereas Atg8–PE is subsequently deconjugated through a second Atg4-dependent cleavage (Kirisako *et al.*, 2000) and the liberated Atg8 is presumably reused for another round of conjugation.

Next, additional proteins such as SNAREs and Rab GTPases play a role in the fusion of autophagosomes and Cvt vesicles with the vacuole, but these proteins are common to all other vacuolar fusion events and thus are not considered specific Atg proteins (Gutierrez *et al.*, 2004; Klionsky, 2005). The membrane of the autophagic/Cvt bodies within the vacuolar lumen is lysed in an Atg15-dependent manner, allowing access to the cargo; Atg15 is a putative yeast lipase (Epple *et al.*, 2001; Teter *et al.*, 2001). Most autophagic cargo molecules are broken down by vacuolar hydrolases. Subsequently, Atg22 and other permeases allow efflux of the resulting macromolecules into the cytosol for recycling (Yang *et al.*, 2006).

As indicated previously, autophagy is generally considered a nonspecific process for the bulk degradation of cytoplasmic constituents, but various types of specific autophagy-related processes have been reported, including the Cvt pathway. Another well-characterized example is the degradation of peroxisomes in methylotrophic yeasts; shifting cells from conditions where peroxisomes are needed to one in which they are superfluous results in their specific degradation to adapt to the changing metabolic demands of the cell (Monastyrska and Klionsky, 2006). In addition, damaged organelles such as mitochondria might be removed by selective autophagy (Kanki and Klionsky, 2008; Kiššova *et al.*, 2007; Zhang *et al.*, 2007). See the chapters by Camougrand *et al.*, Oku and Sakai, and van Zutphen *et al.* in this volume for detailed information on specific organelle degradation.

The cytoplasm-to-vacuole-targeting (Cvt) pathway is a unique example of selective autophagy in fungi, because it is the only known biosynthetic autophagy-like pathway. As cargo proteins, the oligomeric form of the vacuolar hydrolase α-mannosidase (Ams1) and precursor aminopeptidase I (prApe1) are sequestered by double-membrane vesicles and transported from the cytosol to the vacuole through this alternative (i.e., not through a portion of the secretory pathway as occurs with most resident vacuolar hydrolases) targeting route that extensively overlaps with bulk autophagy (Hutchins and Klionsky, 2001; Klionsky *et al.*, 1992).

Although these various autophagy-related processes mostly share the same Atg protein machinery, there are clear morphological and mechanistic differences between specific and nonspecific pathways. For example, the Cvt pathway occurs constitutively during growth in nutrient-rich conditions. The main cargo, prApe1 and Ams1, are sequestered within double-membrane vesicles; Cvt vesicles are approximately 150 nm in diameter and appear to exclude bulk cytosol (Baba *et al.*, 1997; Hutchins and Klionsky, 2001). In contrast, nonspecific autophagy is usually induced by starvation but also occurs at a basal level under vegetative conditions. Autophagosomes are relatively larger than Cvt vesicles, approximately 300–900 nm in diameter and include bulk cytosol and even entire organelles (Baba *et al.*, 1994).

Accordingly, each pathway shows unique features, which suggests that certain types of Atg proteins may have critical roles to differentiate one pathway from another. For example, the adapter protein Atg11 determines the cargo selectivity for various types of specific autophagy-related pathways (Shintani *et al.*, 2002), whereas an autophagy-specific protein, Atg17, has a role in determining the magnitude of the nonspecific autophagic response, which might allow the cell to initiate the uptake of bulk cytoplasm when shifting to autophagy-inducing conditions (Cheong *et al.*, 2005).

To understand the nature of different types of autophagy-related pathways at the molecular level, a significant number of assays have been developed in various eukaryotic systems, including yeast (Klionsky *et al.*, 2007). Autophagosomes were first identified by electron microscopy (EM) based on the unique double-membrane morphology of the vesicle in mammalian cells. In yeast, the identification of the Cvt pathway that occurred concomitant with the analysis of autophagosomes by EM provided an extremely useful marker protein, precursor aminopeptidase I, for analysis of selective autophagy (Harding *et al.*, 1996; Klionsky *et al.*, 1992; Takeshige *et al.*, 1992). Examination of this protein has provided a means to generate quantitative molecular and biochemical information that provides insight into the specific functions of Atg proteins at individual steps of autophagy-related processes.

In this chapter we describe several conventional assays in baker's yeast to monitor certain types of selective and nonselective autophagy that rely on the detection of different marker proteins.

2. Assays to Monitor the Cvt Pathway

2.1. Precursor Ape1 processing

2.1.2. Background

Precursor Ape1 is a specific cargo of Cvt vesicles or autophagosomes, depending on the nutrient conditions. In contrast to most vacuolar proteins, prApe1 does not have a signal sequence to allow for translocation into the endoplasmic reticulum (ER) (Klionsky *et al.*, 1992). After synthesis in the cytosol, prApe1 rapidly oligomerizes to become a dodecamer and further assembles into a higher-order oligomeric structure, which is called the Ape1 complex (Oda *et al.*, 1996) (Fig. 1.1). This complex is selectively recognized and packaged through the action of certain Atg proteins. Atg19 functions as a receptor and binds the prApe1 propeptide to form the Cvt complex (Scott *et al.*, 2001). Ams1 also binds Atg19 (at a site distinct from the prApe1-binding site), but it is present at much lower levels than prApe1 and does not undergo a specific cleavage on vacuolar delivery, and accordingly is not a convenient marker (Hutchins and Klionsky, 2001). Atg11 connects the Cvt complex with the PAS by interacting with Atg19. In the absence of Atg11, the Cvt complex forms but does not localize at the PAS, suggesting Atg11 acts as an adaptor protein that links the cargo with the vesicle-forming machinery (Shintani *et al.*, 2002). In addition to Atg11, Atg8 also interacts with Atg19, which may allow the recognition of the Cvt complex by the sequestering membrane. After delivery to the vacuole, the N-terminal propeptide of prApe1 is cleaved off by vacuolar proteinase B. Thus, the vacuolar delivery of prApe1 can be followed through this maturation process, showing a molecular mass shift from the approximately 61-kDa precursor form to the 50-kDa mature form when resolved by SDS-PAGE and Western blotting. Precursor Ape1 processing can be used for monitoring vacuolar transport of a Cvt vesicle and even an autophagosome during starvation (although bulk autophagy is nonspecific, the Cvt complex is taken up specifically because, as noted above, of the participation of Atg19 and Atg11). To determine the kinetics of prApe1 processing, a pulse/chase analysis of radiolabeled prApe1 can also be utilized.

In contrast to the prApe1 processing that occurs in wild-type cells, certain mutants that are defective in the Cvt pathway and autophagy show prApe1 accumulation in vegetative and/or starvation conditions. In mutants that are defective in the Cvt pathway but not completely blocked in autophagy, prApe1 accumulates only under vegetative conditions, whereas most of the prApe1 becomes mature when the cells are shifted to starvation conditions (Kim *et al.*, 2001b; Scott *et al.*, 2000). For example, the *atg1Δ* strain accumulates only the precursor form of Ape1 in either vegetative or starvation conditions, which suggests this mutant is completely defective for both the Cvt and autophagy pathways (Fig. 1.2). In contrast, the *vac8Δ* mutant is blocked in the Cvt pathway but shows complete reversal of prApe1 accumulation under starvation conditions, as a result of the induction of autophagy. It is important to note that processing of prApe1 under starvation conditions is not by itself an indication that nonspecific autophagic activity is normal, because uptake occurs through a selective mechanism involving Atg19 and Atg11. For example, the abnormally small autophagosomes that are generated in the absence of Atg17 still allow efficient packaging of prApe1 (Cheong *et al.*, 2005).

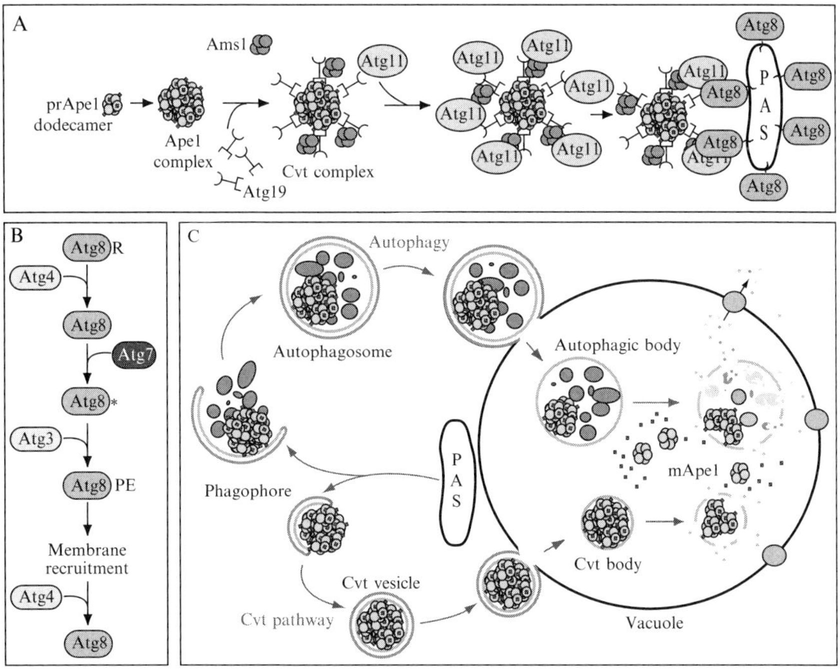

Figure 1.1 **Schematic overview of the Cvt pathway and macroautophagy.** (A) The steps of cargo packaging in the Cvt pathway. Precursor Ape1 forms a dodecamer in the cytosol, and multiple dodecamers assemble into an Ape1 complex. The receptor Atg19 binds the prApe1 propeptide to form the Cvt complex; Ams1 binds Atg19 at a separate site. Atg11 acts as an adaptor to link the Cvt complex with the phagophore assembly site (PAS) containing Atg8–PE. Atg19 subsequently binds Atg8–PE. The Atg8–PE proteins initially line both sides of the PAS, the phagophore, and the completed autophagosome or Cvt vesicle. The Atg8–PE located on the exterior (cytosolic) surface of the vesicles is subsequently removed by Atg4 and reused. (B) Processing of the ubiquitin-like protein Atg8. The Atg8 protein is synthesized with a C-terminal arginine residue that is removed by the Atg4 cysteine protease. Atg8 is next activated by a ubiquitin activating E1 homolog, Atg7, which then transfers it to a ubiquitin-conjugating E2 analog, Atg3, which covalently attaches it to PE, causing it to become membrane associated. Atg4 can remove Atg8 from PE through a second cleavage reaction. (C) The overall morphology of autophagy and the Cvt pathway is similar. Both processes use double-membrane vesicles to sequester cargo. Autophagy is generally nonspecific and sequesters bulk cytoplasm, whereas the Cvt pathway is specific and Cvt vesicles exclude bulk cytoplasm; prApe1 can be taken up as a specific cargo by autophagy because it uses the Atg19 and Atg11 proteins. The PAS gives rise to the phagophore, the initial sequestering compartment. The phagophore expands, presumably by vesicular fusion, to generate the completed Cvt vesicle or autophagosome. These vesicles then fuse with the vacuole and release the Cvt or autophagic body into the lumen. After breakdown of the Cvt or autophagic body, prApe1 is processed by removal of the N-terminal propeptide, whereas nonspecific cargos are degraded and the resulting macromolecules are released into the cytosol through permeases in the vacuole membrane.

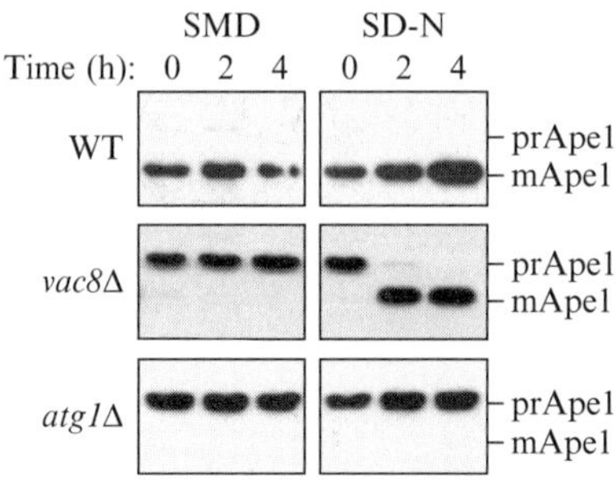

Figure 1.2 The prApe1 processing assay can be used to monitor vesicle transport by the Cvt and autophagy pathways. Wild-type, *vac8Δ* and *atg1Δ* strains were grown in SMD and shifted to SD-N for the indicated times to induce autophagy. The wild-type strain is normal for processing of prApe1 in both vegetative and starvation conditions, whereas the *atg1Δ* mutant shows a complete block in delivery of prApe1 in both vegetative and starvation conditions. The *vac8Δ* cells are defective for processing of prApe1 by the Cvt pathway but show normal transport of prApe1 when autophagy is induced. This figure was modified from data previously published in Cheong *et al.* (2005) and is reproduced by permission of the American Society for Cell Biology (copyright 2005).

2.1.1. Methods

1. *S. cerevisiae* strains (25–50 ml) are grown in 250-ml or 500-ml DeLong culture flasks at 30 °C to early mid-log phase at $O.D_{600} = 0.5 - 0.8$ in YPD (1% yeast extract, 2% peptone, and 2% glucose) or synthetic minimal medium (SMD; 0.67% yeast nitrogen base, 2% glucose) containing the appropriate auxotrophic nucleosides and amino acids and lacking those that are necessary to maintain plasmid selection, if any.

   Note: It is important to keep the cells in log phase to analyze the Cvt pathway. If the cells enter starvation conditions, they will begin to induce autophagy.

   For starvation to monitor prApe1 processing by nonspecific autophagy, cultured cells from step 1 are collected by centrifugation at 5000×*g* for 3 min and are washed once with water. Then cells are cultured in SD-N medium (0.17% yeast nitrogen base without ammonium sulfate or amino acids, containing 2% glucose). Cells are typically maintained in SD-N for 0 (first time point) to 4 h, depending on the experiment.

   Note: As an alternative to starvation, it is possible to treat the cells with rapamycin (200 ng/ml) to induce autophagy. The primary advantage is that rapamycin treatment can be done on cells in rich medium such as YPD, however, the magnitude of the response may be somewhat reduced compared with SD-N.

2. For protein extraction, aliquots corresponding to 1.0 to 2.0 $O.D._{600}$ units (1 ml of cells at an $O.D._{600} = 1.0$ is equivalent to 1 unit of cells) of the cultured cells are removed from the flask at the desired time points

and placed in plastic 1.7-ml microcentrifuge tubes. Trichloroacetic acid (TCA) is added to 10% final concentration, and the samples are incubated for 20 min on ice or at −20 °C. The samples can be kept at −20 °C indefinitely, until sample preparation is continued.

3. The precipitated proteins are pelleted by centrifugation at 15,000×*g* for 3 min. After washing the pellet twice with 1 ml of ice-cold acetone, the pellet is air-dried. The samples may again be stored indefinitely at this point, or in the presence of acetone.

   Note: For the acetone wash, after addition of the acetone the pellet is resuspended by vortex or water-bath sonication; the latter allows better disruption of the pellet. It may facilitate recovery if the sample is then kept at −20 °C for at least 30 min prior to centrifugation.

4. When ready to proceed, the dry cell pellet is resuspended in MURB buffer, (50 m*M* Na_2HPO_4, 25 m*M* 2-[N-morpholino]ethanesulfonic acid (MES), pH 7.0, 1% SDS, 3 *M* urea, 0.5% 2-mercaptoethanol, 1 m*M* NaN_3, and 0.05% bromophenol blue) and disrupted by vortex with an equal volume of acid-washed glass beads for 5 min. The samples are incubated at 70 °C for 10 min, and unlysed cells are removed by centrifugation at 10,000×*g* for 1 min.

   Note: If the sample turns yellow after addition of the MURB buffer it indicates the presence of residual TCA from the acid precipitation step. The sample can be neutralized by adding approximately 1–2 μl of 1 N NaOH or 1*M* Tris-HCL, pH 9.4, until the sample becomes blue.

   Note: It is important to keep the glass beads below the level of the liquid in the microcentrifuge tube to avoid excessive bubbling, which reduces the efficiency of lysis. In addition, it is difficult to transfer the glass beads from the storage container to the sample tube. We find it convenient to make a small scoop by almost cutting off the bottom portion of a microcentrifuge tube, leaving a thin section of the wall all the way to the top of the tube to serve as a ladle.

   Note: It may be helpful to add 1 m*M* of phenylmethylsulfonyl fluoride (PMSF; Sigma) to the samples along with MURB and during all subsequent steps to eliminate nonspecific cleavage of the prApe1 propeptide, which is relatively sensitive to degradation. For this reason, it is important to include a *pep4Δ* strain as a control; prApe1 should remain in the precursor form and the presence of mature Ape1 indicates background cleavage that may have occurred during sample preparation.

5. Protein extracts equivalent to $OD_{600} = 0.2$ U of yeast cells are loaded on an 8% gel and resolved by SDS-PAGE.
6. A standard Western blotting procedure is performed using semi-dry-transfer equipment (Bio-Rad, Hercules, CA). After blotting, PVDF membranes are probed with antiserum against Ape1 (1:5,000 dilution)

by incubation for 2 h at room temperature or overnight at 4 °C. After washing the membranes 3 times with TTBS (1% Tween 20, 50 mM Trizma Base, pH 7.6, 0.9% NaCl), a primary incubation is performed with HRP-conjugated goat anti-rabbit IgG (1:10,000 dilution) for 1 h. The signal for Ape1 is detected using the enhanced chemiluminescent (ECL) kit (Pierce, Rockford, IL). Antiserum against Ape1 has been described previously (Klionsky *et al.*, 1992).

2.2. Protease protection assay for vesicle completion

2.2.1. Introduction

The phagophore is the initial sequestering compartment, which expands to form the double-membrane autophagosome or Cvt vesicle. The protease-protection assay is used to determine whether a certain Atg protein is involved in the process of cargo packaging or vesicle completion. Precursor Ape1 can be analyzed as a marker to monitor vesicle completion because the propeptide is sensitive to protease treatment, and it is a cargo both of the Cvt vesicle and autophagosome. When prApe1 is already enclosed within completed vesicles, IT is protected from degradation resulting from treatment of lysed spheroplasts with exogenous protease. In contrast, if the vesicle formation is not completed, the propeptide of prApe1 can be cleaved by protease. The mature form of Ape1 is relatively resistant to degradation by proteases because it is a resident vacuolar hydrolase and normally exists in a protease-rich environment. Thus, in *atg* mutants that are defective in vesicle formation, prApe1 shows a molecular mass shift to the size of the mature form following protease treatment, a size change that is detected easily by Western blotting. Finally, the accessibility of prApe1 in vesicles is compared between control and experimental strains or is compared to precursor Pho8 as an internal control (Fig. 1.3).

The protease protection assay requires a preparation of spheroplasts that can be gently lysed; it is necessary to rupture the plasma membrane to allow access to the exogenous protease while retaining the integrity of the Cvt vesicles and autophagosomes. Two problems inherent in this approach are insufficient or excessive lysis. If the spheroplasts are not sufficiently lysed, prApe1 will not be accessible to exogenous protease even if the vesicle sequestration step is not complete. On the other hand, if there is excessive lysis, completed vesicles may be ruptured, resulting in a false signal of accessibility. Therefore, two important control strains are necessary for this assay. The first control utilizes a mutant in which prApe1 is protease protected; in this mutant, prApe1 is within the enclosed vesicles so cleavage of prApe1 by protease should not occur. The other control is a protease-sensitive mutant, which shows that prApe1 is cleaved by the protease treatment. Many *atg* mutants can be used for protease-sensitive controls,

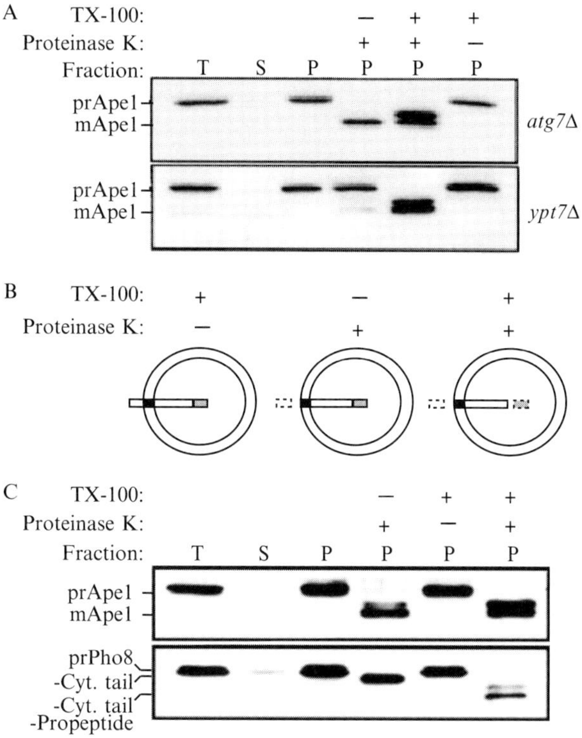

Figure 1.3 **Protease-protection assay for prApe1 to examine the completion of Cvt vesicles and autophagosomes.** Cells were grown in SMD and enzymatically converted to spheroplasts by removal of the cell wall. Spheroplasts were osmotically lysed, and a fraction of the lysate was taken for the total (T) fraction. Precursor Ape1 was separated from the soluble protein fraction (S) by centrifugation at $5000 \times g$ for 5 min. The pellet fraction (P) was resuspended and subjected to proteinase K treatment in the absence (−) or presence (+) of 0.2% Triton X-100. Samples were examined by western blot analysis with antiserum against Ape1. (A) Protease protection assay using a fusion-defective mutant for an external control. When the pellet fraction from the *atg7*Δ strain is subjected to proteinase K treatment, prApe1 is digested to the size of the mature form in the absence of detergent, indicating that the sequestering vesicle is not complete, even though prApe1 is present in the pellet fraction. In contrast, prApe1 in the *ypt7*Δ strain is protected from exogenous protease treatment in the absence of detergent and is accessible only when detergent is added. This figure was modified from data previously published in Kim *et al.* (1999) and is reproduced by permission of the American Society for Cell Biology (copyright 1999). (B) Schematic diagram showing the state of precursor Pho8 in the vacuole during the protease protection assay. The N-terminal transmembrane domain is shown as a black box, and the C-terminal propeptide is in gray; the white boxes depict the N-terminal cytosolic tail and the lumenal portion of the enzyme. The cytosolic tail is accessible to exogenous protease if there is efficient lysis of the spheroplasts, whereas the lumenal propeptide is protected from digestion if the vacuole integrity is retained. (C) Protease protection assay using Pho8 as an internal control. *atg21*Δ *pep4*Δ cells were converted to spheroplasts and osmotically lysed as in (*A*). The pellet fraction from the *atg21*Δ *pep4*Δ cells was subjected to proteinase K treatment in the

whereas certain mutants defective in vesicle fusion with the vacuole can be used for the protease-protected control. For example, Ypt7 is a member of the Rab family, a guanosine triphosphatase (GTPase) that is required for vesicle fusion with the vacuole, and thus the *ypt7Δ* strain accumulates prApe1 in a membrane-enclosed state (Kim *et al.*, 1999). In contrast, in the *atg7Δ* strain, prApe1 is accessible to exogenous protease treatment (Fig. 1.3A).

The main disadvantage of using control strains is that they are external to the experimental strain. That is, there is no way to be certain that these strains behave the same as the experimental strain or that differences in the experimental conditions do not account for observed differences between the strains. Therefore, a preferred approach to control for lysis conditions is to use precursor Pho8 (prPho8) as an internal control. The precursor form of Pho8 is delivered to the vacuole through a portion of the secretory pathway. It contains an internal uncleaved signal sequence at the N terminus that allows translocation into the ER, and that orients Pho8 as a type II integral membrane protein; there is an N-terminal cytosolic tail and a C-terminal propeptide within the vacuole lumen (Fig. 1.3B). The latter normally undergoes a Pep4-dependent proteolytic cleavage to activate the zymogen. When strains carry the *pep4Δ* mutation, the C-terminal propeptide of Pho8 is not cleaved unless the vacuole is lysed and exogenous protease is added. Therefore, following the cleavage state of the propeptide provides information on the integrity of the vacuole. This likely reflects the state of the sequestering vesicles, because the vacuole is considered one of the most fragile organelles during osmotic lysis. In contrast, the N-terminal cytosolic tail is accessible to exogenous protease digestion following osmotic lysis of the plasma membrane without detergent treatment. Therefore, cleavage of the cytosolic tail of prPho8 in the absence of detergent can be used to monitor the efficiency of spheroplast lysis (Fig. 1.3C).

2.2.2. Methods

1. *S. cerevisiae* strains are grown in YPD or SMD medium containing the appropriate auxotrophic nucleosides and amino acids and lacking those

absence and presence of detergent as indicated, and prApe1 and prPho8 were examined by Western blot. The prApe1 was accessible to proteinase K in the absence of detergent. The prPho8 showed removal of the cytosolic tail in the absence of detergent, indicating efficient spheroplast lysis; the propeptide that faces the vacuole lumen was degraded only on the addition of detergent, indicating that the integrity of the vacuole (and presumably the integrity of other subcellular compartments including Cvt vesicles and autophagosomes) was retained. This figure was modified from data previously published in Stromhaug *et al.* (2004) and is reproduced by permission of the American Society for Cell Biology (copyright 2004).

that are necessary to maintain plasmid selection, if any. Cells (25–50 ml) are grown at 30 °C to mid-log phase ($O.D._{600} = 0.8–1.0$).

Note: To monitor vesicle completion during nonspecific autophagy, the cells may be shifted to SD-N medium for at least one hour to induce starvation conditions. Incubation in starvation conditions, however, may increase the strength of the cell wall, making it difficult to convert the cells to spheroplasts. Therefore, rapamycin (200 ng/ml concentration) may be used to induce autophagy as an alternative approach.

2. To examine the protease sensitivity of prApe1 in the mutant of interest, *ypt7*Δ and *atg7*Δ strains are prepared as protease-resistant and protease-sensitive controls. Alternatively, the *PEP4* gene is deleted in the strain of interest to allow monitoring of prPho8.
3. 20–30 $O.D._{600}$ units of cells are harvested by centrifugation at 5,000×*g* for 3 min and resuspended in 2.5 ml of 0.1 *M* Tris-SO_4, pH 9.4, 10 m*M* DTT. Shake at 30 °C for 5 min. This step reduces disulfide bonds in the cell wall and improves the efficiency of the subsequent enzymatic digestion.
4. The cells are harvested by centrifugation and resuspended in 2–3 ml spheroplasting medium (1% yeast extract, 2% peptone, 0.5% glucose, 1.2 *M* sorbitol, 20 m*M* Tris-HCl, pH 7.5) in a 15-ml conical tube. A 100-μl aliquot of the cells is removed, diluted into 1.0 ml of water, and the optical density determined (time zero). A solution of yeast lytic enzyme is prepared fresh (1 mg/ml in spheroplasting medium) and added to the resuspended cells at a final concentration of approximately 20 μg/ml; the appropriate concentration of lytic enzyme is strain-dependent and must be determined empirically. Conversion of spheroplasts is carried out by incubation at 30 °C for 30 min with mild agitation. The efficiency of spheroplasting is monitored starting at approximately 20 min of incubation by removing a 100-μl aliquot as above and diluting into 900 μl water; the optical density should drop to approximately 10% of the initial value.

Note: There are various preparations of yeast lytic enzymes commercially available. The main differences are purity (these enzyme preparations can contain both glycosidases and proteases, the latter potentially being problematic) and price. It is probably not necessary to use the highest-purity preparation, but this will need to be determined empirically along with the required amount of enzyme.

5. Spheroplasts are collected by centrifugation at 1,500×*g* for 5 min, and lysed by resuspension in 1 ml of osmotic lysis buffer (20 mM Pipes/KOH, pH 6.8, 0.2 *M* sorbitol, 0.5 m*M* PMSF, and protease inhibitor cocktail (Roche Molecular Biochemicals, Indianapolis, IN) at a spheroplast density of 20$O.D._{600}$ units/ml.
6. After a preclearing step by centrifugation at 500×*g* for 5 min at 4 °C, the resulting lysate (20 $O.D._{600}$ units) is separated into S5 supernatant and P5 pellet fractions by centrifugation at 5,000×*g* for 5 min at 4 °C.

7. The P5 pellet fraction is resuspended in osmotic lysis buffer in the presence or absence of proteinase K (50 μg/ml; Roche Molecular Biochemicals, Indianapolis, IN) and 0.2% Triton X-100. The resuspended P5 pellet fractions are incubated on ice for 30 min.
8. TCA is added to the spheroplasts to 10% final concentration, and incubated on ice for 20 min. The precipitated proteins are pelleted by centrifugation at 15,000$\times g$ for 3 min at 4 °C, acetone washed and air-dried. The dried pellets are resuspended by water-bath sonication in SDS sample buffer and resolved by SDS-PAGE. Western blot analysis is performed with antiserum against Ape1 (and Pho8 if using this protein as an internal control) as primary antibody. Antibody against Pho8 is from Molecular Probes (Eugene, OR). To resolve the different forms of prPho8 following protease treatment, large (16 × 16 cm; R. Shadel, San Francisco, CA) SDS-PAGE gel analysis is used instead of mini gels (8 × 7.3 cm; Bio-Rad, Hercules, CA).

3. Assays to Monitor Pexophagy

3.1. Pex14-GFP processing assay

3.1.1. Background

In addition to the Cvt pathway, another type of selective autophagy is the degradation of excess peroxisomes, termed *pexophagy*, which is mostly studied in various fungi (Dunn *et al.*, 2005). When fungi grow on carbon sources that require peroxisome function, such as oleic acid (*Saccharomyces cerevisiae* is not methylotrophic) or methanol (for *Pichia pastoris* or *Hansenula polymorpha*), the peroxisomes are proliferated. Then, if a preferred carbon source such as glucose is added to the medium, the excess peroxisomes are degraded selectively by an autophagy-related mechanism. In *S. cerevisiae*, it has not been determined whether pexophagy occurs by a micro- or macropexophagic process, or both. One caveat is that synthetic medium containing glucose (SD) still allows normal cell growth, making it difficult to measure peroxisome degradation accurately. Therefore SD-N medium can be used to induce peroxisome degradation, as the cells typically double only once after shifting to these conditions. There is some controversy as to whether the degradation of peroxisomes in SD-N occurs by pexophagy or nonspecific autophagy; however, the kinetics of degradation suggest that it still occurs by a specific process (Hutchins *et al.*, 1999).

Methods have been developed to monitor pexophagy by Western blot and fluorescence microscopy. First, degradation of Fox3 (3-ketoacyl-CoA thiolase), a marker of the peroxisome matrix, can be used to follow the degradation of peroxisomes (Hutchins *et al.*, 1999). The main disadvantage of this type of approach is that the analysis detects the loss of the Fox3 signal. In general, it is preferable to detect the (positive) appearance of a signal rather than the (negative) disappearance.

A second way to visualize the degradation process is by observing a GFP-tagged peroxisomal protein. The large number of fluorescent puncta representing induced peroxisomes shows a rapid decrease upon shifting to nitrogen-starvation conditions (Guan *et al.*, 2001). Although this is a reliable method for visualizing peroxisome degradation by microscopy, it is problematic for quantifying the degradation level. In addition, peroxisomes in *S. cerevisiae* are much smaller than those in methylotrophic fungi, making analysis by microscopy less convenient.

The third approach relies on the detection of enzymatically cleaved Pex14-GFP. Pex14 is a peroxisomal membrane protein which is a component of the docking site for the peroxisomal targeting signal receptor of matrix proteins (Monastyrska and Klionsky, 2006). When Pex14 is C-terminally fused with GFP, the GFP moiety is exposed on the cytosolic face of the peroxisome. Once the peroxisomes containing Pex14-GFP are delivered to the vacuole, the GFP moiety is cleaved off by vacuolar hydrolases, which allows pexophagy to be monitored through the appearance of free GFP. Under pexophagy-inducing conditions (e.g., SD-N medium), the 27-kDa free GFP band is detected during a time course, whereas *atg* mutants which are blocked for pexophagy do not show the appearance of free GFP (Fig. 1.4). We use strains that have GFP integrated at the *PEX14* locus because this allows expression under the endogenous *PEX14* promoter.

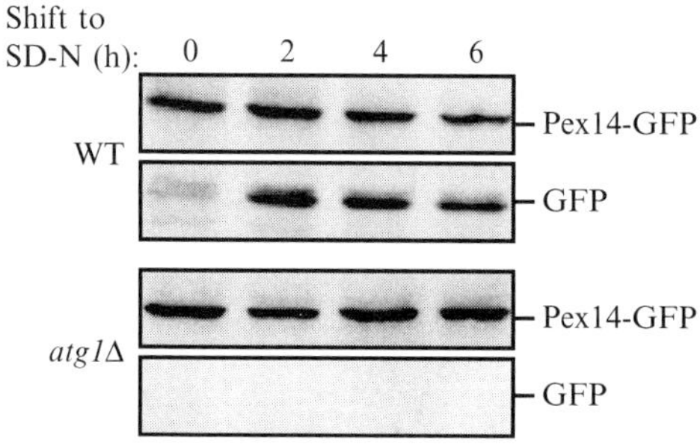

Figure 1.4 The Pex14-GFP processing assay to monitor pexophagy. The Pex14-GFP and Pex14-GFP *atg1*Δ strains were grown in oleic acid–containing medium to induce peroxisome proliferation and shifted to starvation medium, SD-N. Protein extracts were prepared from cells at each indicated time point, resolved by SDS-PAGE, and probed with monoclonal anti-GFP antibody. The positions of full-length Pex14-GFP and free GFP are indicated. The *atg1*Δ mutant harboring Pex14-GFP is completely defective in the processing of Pex14-GFP by pexophagy. This figure was modified from data previously published in Yen *et al.* (2007) and is reproduced by permission of the American Society for Cell Biology (copyright 2007).

3.1.2. Methods

1. *S. cerevisiae* strains (25–50 ml) are grown at 30 °C to early mid-log phase, $O.D_{600} = 0.8$, in YPD or SMD medium containing the appropriate auxotrophic nucleosides and amino acids and lacking those that are necessary to maintain plasmid selection, if any.

   Note: If using chromosomally integrated Pex14-GFP, it is not necessary to maintain selection for the fusion construct.

2. The cells are then shifted for 12 h to SGd medium (0.67% yeast nitrogen base, 3% glycerol, 0.1% glucose) at $O.D._{600} = 0.3–0.5$. Glycerol is a suboptimal carbon source for yeast and this incubation step results in better peroxisome proliferation when cells are shifted to oleic acid.
3. 10x YP is added to cultured cells at a final concentration of 1% yeast extract and 2% peptone and the cells are further cultured for 4 h.
4. The cells are collected by centrifugation at $5{,}000 \times g$ for 3 min, washed in sterile water and resuspended to $O.D._{600} = 0.8$ in YTO medium (0.67% yeast nitrogen base without amino acids, 0.1% Tween-40, 0.1% oleic acid), then further cultured for 19 h. During incubation in YTO medium containing oleic acid as the primary carbon source, peroxisomes proliferate but growth will be extremely slow.
5. To induce peroxisome degradation, cells are harvested from YTO medium by centrifugation, washed with sterile water and shifted to SD-N medium (0.17% yeast nitrogen base without ammonium sulfate or amino acids containing 2% glucose).
6. Approximately 1.0 $O.D._{600}$ unit of cells is removed from the culture at different time points (e.g., 0, 3, 6, and 12 h).
7. Protein extraction and SDS-PAGE are carried out as described previously.
8. After standard Western blotting, the PVDF membranes are probed with anti-GFP monoclonal antibody (Covance, Princeton, NJ) by incubation overnight at 4 °C. After washing the membranes, the secondary incubation is performed with HRP-conjugated rabbit anti-mouse IgG (1:10,000 dilution) for 1 h. The signal for GFP is detected using the ECL kit (Pierce, Rockford, IL).

4. Assays to Monitor Nonspecific Autophagy

4.1. GFP-Atg8 processing assay

4.1.1. Introduction

Atg8 is a ubiquitin-like protein that is conjugated to phosphatidylethanolamine (PE; see Fig. 1.1). After protein synthesis, a pool of Atg8 associates with the phagophore/autophagosome membrane as a result of the lipid modification. Atg8 is the only Atg protein, aside from Atg19, which remains associated with the completed autophagosome or Cvt vesicle (Huang *et al.*, 2000; Ichimura *et al.*, 2000; Kirisako *et al.*, 1999; Kirisako *et al.*, 2000). Atg8–PE is proteolytically removed from the outer surface of the sequestering vesicle through a second cleavage reaction mediated by Atg4, but the chimera located on the inside of the vesicle remains trapped. Accordingly, N-terminal GFP-tagged Atg8 can be used as a functional marker for monitoring delivery of vesicles to the vacuole by fluorescence microscopy (Kim *et al.*, 2001a; also see the chapter in this volume by Shintani and Reggiori). Note that the C terminus of Atg8 is proteolytically processed by Atg4 (see Fig. 1.1), so that the C-terminal fusion cannot be used as an autophagosome marker, but it can be used for studying the lipid modification of Atg8, particularly the initial processing step (Kirisako *et al.*, 2000; Kim *et al.*, 2001a).

In a wild-type strain, GFP-Atg8 is observed within the vacuole (although in nutrient-rich conditions it is difficult to detect a signal because of the relatively low level of GFP-Atg8), but in most autophagy mutant strains, a vacuolar signal of GFP-Atg8 is not detected, which suggests that the transport of this protein into the vacuole occurs by the Cvt pathway and autophagy. A GFP-Atg8 signal is also detected inside Cvt or autophagic bodies that accumulate within the vacuole in *pep4Δ* or *atg15Δ* strains that are defective in intravacuolar vesicle breakdown (see Fig. 4.2 in the chapter by Shintani and Reggiori). Therefore, GFP-Atg8 can be used as a marker of autophagosomes or Cvt vesicles (or the resulting autophagic/Cvt bodies) to monitor their localization (e.g., in the cytosol in mutants defective in fusion with the vacuole). Upon lysis of Cvt or autophagic bodies containing GFP-Atg8, the GFP moiety is proteolytically removed from Atg8 in the vacuole lumen. The released GFP moiety remains relatively stable from vacuolar hydrolysis, whereas Atg8 is rapidly degraded. Accordingly, the appearance of free GFP can be a useful marker to monitor vesicle delivery and lysis of the Cvt/autophagic body using a biochemical analysis (Cheong *et al.*, 2005; Shintani and Klionsky, 2004).

In practice, the level of Atg8 expressed from the endogenous promoter is quite low during vegetative growth, which may make it difficult to detect GFP-Atg8 or free GFP under these conditions. Therefore, this assay is best suited to monitor nonspecific autophagy unless GFP-Atg8 is overexpressed, which does not appear to interfere with the Cvt pathway. In general, the strains of interest are transformed with a plasmid encoding GFP-Atg8 and shifted from growth in rich medium to starvation conditions to induce autophagy. Finally, the appearance of free GFP on Western blots represents lysis of the membrane of the autophagic body and breakdown of the cargo. In wild-type cells, the amount of free GFP increases during the course of starvation, whereas the *atg1Δ* negative control accumulates only the full-length GFP-Atg8 (Fig. 1.5). Usually, the GFP moiety in the yeast vacuole is

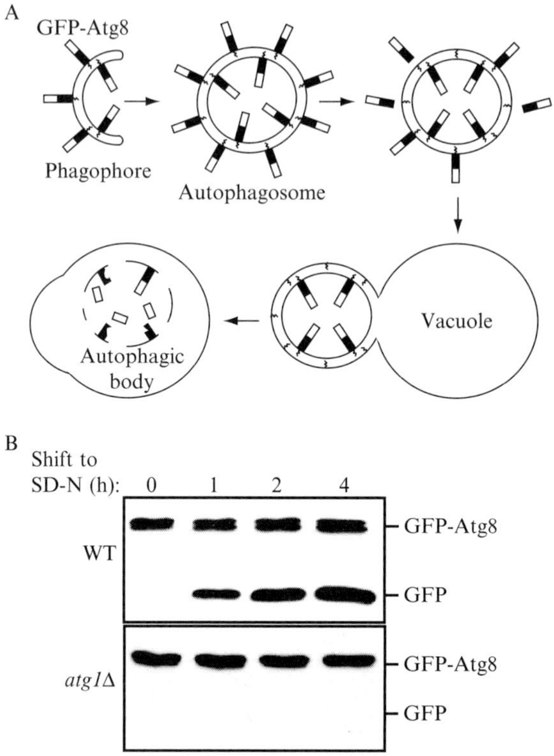

Figure 1.5 The GFP-Atg8 processing assay to monitor autophagosome delivery and autophagic body lysis. (A) Schematic diagram of the GFP-Atg8 processing assay. Initially, GFP-Atg8 lines both sides of the phagophore and the completed autophagosome. The GFP-Atg8 on the surface of the autophagosome is proteolytically removed from PE by a second Atg4 cleavage (see Fig. 1.1); GFP-Atg8 inside the autophagosome is delivered to the vacuole following fusion with the limiting membrane. Lysis of the autophagic body allows access of GFP-Atg8 to vacuolar hydrolases. Atg8 is rapidly degraded, whereas free GFP is relatively stable and accumulates in the lumen. (B) Wild-type and *atg1*Δ strains expressing plasmid-based GFP-Atg8 under the control of the endogenous *ATG8* promoter were grown in SMD medium containing the appropriate auxotrophic nucleosides and amino acids and lacking those that are necessary to maintain plasmid selection, and then shifted to SD-N medium. At the indicated times, aliquots were removed and protein extracts prepared and resolved by SDS-PAGE. Full-length GFP-Atg8 and free GFP were detected by Western blot using anti-GFP antibodies. The band running just below full-length GFP-Atg8 is a cross-reacting contaminant. The positions of full-length GFP-Atg8 and free GFP are indicated. The *atg1*Δ mutant containing GFP-Atg8 does not show generation of free GFP under starvation conditions. This figure was modified from data previously published in Cheong *et al.* (2005) and is reproduced by permission of the American Society for Cell Biology (copyright 2005).

not sensitive to the acidic pH environment (the vacuole pH is approximately 6.1), but caution should be exercised in determining whether GFP stability might be affected during a long incubation time under starvation conditions.

4.1.2. Methods

1. *S. cerevisiae* strains harboring the GFP-Atg8-expressing plasmid are grown in YPD or SMD medium (10 ml) containing the appropriate auxotrophic nucleosides and amino acids and lacking those that are necessary to maintain plasmid selection, if any; see section 5 for a transformation protocol. Cells are grown at 30 °C to mid-log phase (O.D.$_{600}$ = 0.8), and then are shifted to SD-N medium to induce starvation.
2. At various time points (e.g., 0, 1, 2, and 4 h) after the shift, 1 O.D.$_{600}$ unit of cells is harvested by centrifugation, and protein extracts are prepared as described in section 2.1.
3. Protein extracts equivalent to O.D$_{600}$ = 0.2 U of yeast cells are subjected to SDS-PAGE (10% gel) and Western blot and probed with anti-GFP monoclonal antibody (Covance, Princeton, NJ) by incubation overnight at 4 °C. Following the primary incubation, HRP-conjugated rabbit anti-mouse IgG (1:10,000 dilutions) is added for 1 h. The signal for GFP is detected using the ECL kit (Pierce, Rockford, IL). The band running just below full-length GFP-Atg8 is a cross-reacting contaminant.

4.2. Atg8 synthesis and lipid modification

4.2.1. Introduction

Atg8 is up-regulated following the induction of autophagy at the transcriptional and translational level (Huang *et al.*, 2000; Kirisako *et al.*, 1999). Atg8 is one of the major Atg proteins that are involved in autophagosome expansion (Abeliovich *et al.*, 2000; Xie *et al.*, 2008). Thus, this process may need a rapid increase in Atg8 synthesis. Although the time frame for increase in the transcription level of Atg8 is faster than translation (Kirisako *et al.*, 1999), Atg8 protein levels are significantly increased during autophagy-inducing conditions. Therefore, the increased level of Atg8 upon starvation can be a useful marker for monitoring autophagy induction. However, significant accumulation of Atg8 is also observed in mutants defective in vacuolar degradation, such as the *pep4*Δ strain that lacks vacuolar proteinase A (Fig. 1.6A). These data indicate that Atg8 is degraded through vacuolar hydrolysis while new Atg8 is being synthesized. Similarly, mutants that are defective in various steps of autophagy, such as fusion of the autophagosome with the vacuole, also accumulate Atg8. Accordingly, it is necessary to determine whether the increase in the Atg8 protein level during starvation is due to induction of autophagy or to a direct or indirect block of vacuolar degradation of Atg8. Thus, to determine whether a particular mutant is autophagy defective, the endogenous Atg8 expression level can be compared by shifting to starvation conditions in the presence and absence of vacuolar protease inhibitors (Klionsky *et al.*, 2007;

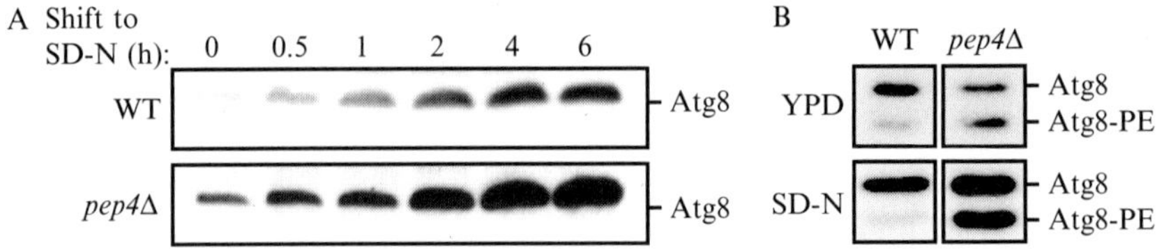

Figure 1.6 Atg8 synthesis and Atg8-PE formation are increased following autophagy induction. (A) Atg8 expression is induced upon starvation, and subsequent turnover depends on vacuolar hydrolysis. Wild-type and *pep4Δ* cells were grown in SMD and subsequently shifted to SD-N medium. Cell aliquots were collected for western blot analysis at the indicated time points. Protein extracts from *pep4Δ* cells were resolved by SDS-PAGE using half the $O.D._{600}$ units of wild-type cells. The increase in the level of Atg8 corresponds to the amount of time incubated in nitrogen starvation medium in both strains. The higher levels of Atg8 seen in the *pep4Δ* strain indicate that Atg8 is being degraded in the vacuole in the wild-type strain, corresponding to normal autophagic flux. This figure was modified from data previously published in Huang *et al.* (2000), which are reproduced by permission of the American Society for Biochemistry and Molecular Biology and Elsevier (copyright 2000). (B) Turnover of Atg8–PE is dependent on vacuolar hydrolysis. Atg8 lipidation was observed in rich or starvation medium in wild-type or *pep4Δ* mutant cells. Cells were grown in YPD to early mid-log phase and then shifted to SD-N for 3 h. Atg8–PE was separated from Atg8 using a 12% SDS-PAGE gel containing 6 *M* urea. Higher levels of Atg8 are seen in starvation conditions, and substantially more Atg8–PE accumulates in the *pep4Δ* mutant, again reflecting autophagic flux in the wild-type strain. This figure was modified from data previously published in Stromhaug *et al.* (2004) and is reproduced by permission of the American Society for Cell Biology (copyright 2004).

Tanida *et al.*, 2005). In wild-type cells, the level of Atg8 should be higher in the presence of the inhibitor (because vacuolar degradation of Atg8 that was delivered to the vacuole will now be blocked), whereas there will be no change in mutants, or following treatments, that block autophagosome formation, fusion with the vacuole, or degradation within the vacuole lumen. In addition, the *pep4Δ* mutation can be introduced as an alternative to using vacuolar protease inhibitors; Pep4 is the major processing protease that is indirectly responsible for the majority of vacuolar proteolytic activity. Finally, some strains, such as those deficient in the Atg12–Atg5-Atg16 conjugation system, and those required for Atg8 conjugation (see Fig. 1.1) show reduced levels or the complete absence of Atg8–PE.

On the basis of fluorescence microscopy and biochemical analyses (Shintani and Klionsky, 2004; Suzuki *et al.*, 2001, 2004), Atg8 exists in two pools: a nonlipidated form and the PE-conjugated species. The nonlipidated form shows a diffuse cytosolic localization based on fluorescence microscopy analysis (see Fig. 4.2 in the chapter by Shintani and Reggiori). The Atg8–PE form is membrane-associated and appears primarily as a punctum at the PAS.

In addition to the increase of Atg8 synthesis during starvation, the PE conjugation level of Atg8 is significantly higher than in vegetative conditions. Thus, the level of lipid modification of Atg8 is an additional criterion that can be used to monitor the induction of nonspecific autophagy. The Atg8–PE form can be separated from nonlipidated Atg8 as a result of the molecular mass shift detected by SDS-PAGE. However, it should be noted that the Atg8–PE form migrates faster than the nonlipidated form (Huang *et al.*, 2000; Kirisako *et al.*, 2000). Furthermore, the Atg8–PE protein is the form that is actually delivered to the vacuole and as a result is degraded in a Pep4-dependent manner. Therefore, the *pep4Δ* strain that is defective in vacuolar degradation generally accumulates substantially more Atg8–PE than the wild-type strain (Fig. 1.6B). Thus, comparison with a *pep4Δ* mutant, or alternatively treatment with vacuolar protease inhibitors, can be used to indicate whether normal autophagy and vacuolar hydrolysis occur in the strain of interest.

4.2.2. Methods

1. *S. cerevisiae* strains are grown in YPD or SMD medium (10 ml) containing the appropriate auxotrophic nucleosides and amino acids and lacking those that are necessary to maintain plasmid selection, if any. Cells are grown at 30 °C to mid-log phase, $O.D._{600} = 0.8$, and then the cultured cells are shifted to SD-N medium for 2–4 h to induce starvation conditions. Protease inhibitor (phenylmethylsulfonyl fluoride (PMSF; Sigma, St Louis, MO)) is added to the cells at a final concentration of 1 m*M*, before shifting to SD-N medium.

   Note: Pefabloc SC (Roche Applied Science) is an alternative to PMSF that is more stable in aqueous solutions and at physiological pH.

2. 1 $O.D._{600}$ unit of cultured cells is harvested at various time points and protein extracts are generated as described previously.
3. Protein extracts equivalent to $O.D._{600} = 0.2$ U of yeast cells are subjected to modified urea-SDS-PAGE for detection of Atg8–PE. The modified urea-SDS-PAGE gel consists of a standard stacking gel and a high percentage (12%–15%) polyacrylamide separating gel that contains the standard concentration of SDS, and 6 *M* urea.
4. After the standard Western blotting procedure, PVDF membranes are probed with antiserum against Atg8 (1:5000 dilution) by incubation overnight at 4 °C. For the secondary antibody, the incubation is performed with HRP-conjugated rabbit anti-mouse IgG with a 1:10,000 dilution for 1 h. The signal for Atg8 is detected using the ECL kit. Antiserum against Atg8 (formerly termed Aut7) is prepared as described previously (Huang *et al.*, 2000).

5. Additional Methods

5.1. High-efficiency yeast transformation

This protocol is based on Gietz and Woods (2002)

1. *S. cerevisiae* strains are grown in DeLong culture flasks overnight at 30 °C in YPD. Grow approximately 1.5 ml of cell culture per transformation. On the next day, the cultured cells are diluted to $O.D_{600} = 0.2$ in fresh YPD and regrown to mid-log phase ($O.D_{600} = 0.8–1$).
2. Sufficient cells for the desired number of transformations (1–1.5 $O.D._{600}$ units per transformation) are harvested by centrifugation at 5,000×*g* for 3 min.
3. Harvested cells are washed in one-tenth volume of 100 m*M* lithium acetate (LiAc) and centrifuged at 5,000×*g* for 3 min. The cells are resuspended in the same volume of 100 m*M* LiAc and the competent cells are aliquoted into 1.7-ml microcentrifuge tubes (dispensing 1–1.5 $O.D._{600}$ units per tube).
4. Salmon sperm (ss)DNA; 2 mg/ml in sterile water or TE (10 m*M* Tris-HCl, pH 7.5, 1 m*M* EDTA) is prepared as carrier DNA, by boiling for 5 min and quickly cooling on ice.

   Note: It is not necessary to boil the carrier DNA every time. Small aliquots of ssDNA can be kept at −20 °C.

5. The following reagents are added to the competent cells in the order listed.
   240 μl of polyethylene glycol-3,350 (50% w/v)
   36 μl 1 *M* lithium acetate (pH 7.5 adjusted with acetic acid)
   25 μl ssDNA (2 mg/ml of sterile water or 1×TE)
   25 μl DNA (any miniprep DNA is sufficient; in the case of plasmid DNA, 1 μg of DNA is adjusted to 25 μl with sterile water)
   34 μl sterile deionized distilled (or MilliQ filtered) water (this volume can be adjusted as necessary)
   360 μl total volume
6. Each tube is mixed vigorously by vortex for 1 min until the cell pellet has been completely mixed.
7. Cells in the microcentrifuge tubes are incubated at room temperature or 30 °C for 30 min and then heat shocked in a 42 °C water bath for 20–30 min.

   Note: If working with temperature-sensitive mutants, the cells can be incubated at room temperature and the temperature of the heat shock can be set to 37 °C.

8. The cells are harvested by centrifugation at 5,000×*g* for 1 min and the supernatant fraction is discarded.

9. The cell pellet is gently resuspended with 200 μl of sterile water and spread on agar plates containing the appropriate auxotrophic nucleosides and amino acids, and lacking the appropriate selective reagent.

   Note: If the cells are allowed to recover after the transformation by adding 1 ml of YPD medium and incubating at 30 °C for 1–2 h before plating, there will be a substantial increase in the number of transformants without a significant increase in cell doubling. The exact times need to be determined empirically.

10. The plates are incubated at 30 °C for 2–3 days to allow growth of the transformants.

5.2. Acid washing for glass beads

1. Glass beads (1 kg of 0.4–0.6-mm size; Thomas Scientific #5663R50) are soaked in 1 liter of 1 N HCl overnight in a glass beaker. The beaker should be covered and placed in a fume hood for safety. This acid washing etches the beads, which makes them better for cell lysis.
2. On the next day, the beads are rinsed 10–15 times with the same volume (1 liter) of tap water.
3. The beads are rinsed with deionized water at least 5 times while mixing them with a spatula and then rinsed with deionized distilled water at least 5 times.
4. After rinsing, the beads are dried in a drying oven overnight. It is recommended that the container for storing the beads is used for this step because the dry beads are difficult to transfer.

REFERENCES

Abeliovich, H., *et al.* (2000). Dissection of autophagosome biogenesis into distinct nucleation and expansion steps. *J. Cell Biol.* **151,** 1025–1034.

Baba, M., *et al.* (1997). Two distinct pathways for targeting proteins from the cytoplasm to the vacuole/lysosome. *J. Cell Biol.* **139,** 1687–1695.

Baba, M., *et al.* (1994). Ultrastructural analysis of the autophagic process in yeast: Detection of autophagosomes and their characterization. *J. Cell Biol.* **124,** 903–913.

Cheong, H., *et al.* (2008). The Atg1 kinase complex is involved in the regulation of protein recruitment to initiate sequestering vesicle formation for nonspecific autophagy in *Saccharomyces cerevisiae*. *Mol. Biol. Cell.* **19,** 668–681.

Cheong, H., *et al.* (2005). Atg17 regulates the magnitude of the autophagic response. *Mol. Biol. Cell.* **16,** 3438–3453.

Dunn, W. A., Jr., *et al.* (2005). Pexophagy: The selective autophagy of peroxisomes. *Autophagy* **1,** 75–83.

Epple, U. D., *et al.* (2001). Aut5/Cvt17p, a putative lipase essential for disintegration of autophagic bodies inside the vacuole. *J. Bacteriol.* **183,** 5942–5955.

George, M. D., *et al.* (2000). Apg5p functions in the sequestration step in the cytoplasm-to-vacuole targeting and macroautophagy pathways. *Mol. Biol. Cell.* **11,** 969–982.
Gietz, R. D., and Woods, R. A. (2002). Transformation of yeast by the LiAc/SS carrier DNA/PEG method. *Methods Enzymol.* **350,** 87–96.
Guan, J., *et al.* (2001). Cvt18/Gsa12 is required for cytoplasm-to-vacuole transport, pexophagy, and autophagy in *Saccharomyces cerevisiae* and *Pichia pastoris. Mol. Biol. Cell.* **12,** 3821–3838.
Gutierrez, M. G., *et al.* (2004). Rab7 is required for the normal progression of the autophagic pathway in mammalian cells. *J. Cell Sci.* **117,** 2687–2697.
Harding, T. M., *et al.* (1996). Genetic and phenotypic overlap between autophagy and the cytoplasm to vacuole protein targeting pathway. *J. Biol. Chem.* **271,** 17621–17624.
Harding, T. M., *et al.* (1995). Isolation and characterization of yeast mutants in the cytoplasm to vacuole protein targeting pathway. *J. Cell Biol.* **131,** 591–602.
He, C., *et al.* (2006). Recruitment of Atg9 to the preautophagosomal structure by Atg11 is essential for selective autophagy in budding yeast. *J. Cell. Biol.* **175,** 925–935.
Huang, W.-P., *et al.* (2000). The itinerary of a vesicle component, Aut7p/Cvt5p, terminates in the yeast vacuole via the autophagy/Cvt pathways. *J. Biol. Chem.* **275,** 5845–5851.
Hutchins, M. U., and Klionsky, D. J. (2001). Vacuolar localization of oligomeric α-mannosidase requires the cytoplasm to vacuole targeting and autophagy pathway components in *Saccharomyces cerevisiae. J. Biol. Chem.* **276,** 20491–20498.
Hutchins, M. U., *et al.* (1999). Peroxisome degradation in *Saccharomyces cerevisiae* is dependent on machinery of macroautophagy and the Cvt pathway. *J. Cell. Sci.* **112** (Pt. 22), 4079–4087.
Ichimura, Y., *et al.* (2000). A ubiquitin-like system mediates protein lipidation. *Nature* **408,** 488–492.
Kamada, Y., *et al.* (2000). Tor-mediated induction of autophagy via an Apg1 protein kinase complex. *J. Cell Biol.* **150,** 1507–1513.
Kihara, A., *et al.* (2001). Two distinct Vps34 phosphatidylinositol 3-kinase complexes function in autophagy and carboxypeptidase Y sorting in *Saccharomyces cerevisiae. J. Cell Biol.* **152,** 519–530.
Kim, J., *et al.* (1999). Apg7p/Cvt2p is required for the cytoplasm-to-vacuole targeting, macroautophagy, and peroxisome degradation pathways. *Mol. Biol. Cell.* **10,** 1337–1351.
Kim, J., *et al.* (2001a). Membrane recruitment of Aut7p in the autophagy and cytoplasm to vacuole targeting pathways requires Aut1p, Aut2p, and the autophagy conjugation complex. *J. Cell Biol.* **152,** 51–64.
Kim, J., *et al.* (2001b). Cvt9/Gsa9 functions in sequestering selective cytosolic cargo destined for the vacuole. *J. Cell Biol.* **153,** 381–396.
Kirisako, T., *et al.* (1999). Formation process of autophagosome is traced with Apg8/Aut7p in yeast. *J. Cell Biol.* **147,** 435–446.
Kirisako, T., *et al.* (2000). The reversible modification regulates the membrane-binding state of Apg8/Aut7 essential for autophagy and the cytoplasm to vacuole targeting pathway. *J. Cell Biol.* **151,** 263–276.
Kiššová, I., *et al.* (2007). Selective and non-selective autophagic degradation of mitochondria in yeast. *Autophagy* **3,** 329–336.
Klionsky, D. J. (2005). The molecular machinery of autophagy: Unanswered questions. *J. Cell. Sci.* **118,** 7–18.
Klionsky, D. J., *et al.* (2007). Methods for monitoring autophagy from yeast to human. *Autophagy* **3,** 181–206.
Klionsky, D. J., *et al.* (1992). Aminopeptidase I of *Saccharomyces cerevisiae* is localized to the vacuole independent of the secretory pathway. *J. Cell Biol.* **119,** 287–299.

Kuma, A., *et al.* (2002). Formation of the approximately 350-kDa Apg12-Apg5·Apg16 multimeric complex, mediated by Apg16 oligomerization, is essential for autophagy in yeast. *J. Biol. Chem.* **277,** 18619–18625.
Levine, B., and Klionsky, D. J. (2004). Development by self-digestion: Molecular mechanisms and biological functions of autophagy. *Dev. Cell.* **6,** 463–477.
Mizushima, N., and Klionsky, D. J. (2007). Protein turnover via autophagy: Implications for metabolism. *Annu. Rev. Nutr.* **27,** 19–40.
Mizushima, N., *et al.* (2008). Autophagy fights disease through cellular self-digestion. *Nature* **451,** 1069–1075.
Mizushima, N., *et al.* (1998). A protein conjugation system essential for autophagy. *Nature* **395,** 395–398.
Monastyrska, I., and Klionsky, D. J. (2006). Autophagy in organelle homeostasis: Peroxisome turnover. *Mol. Aspects Med.* **27,** 483–494.
Oda, M. N., *et al.* (1996). Identification of a cytoplasm to vacuole targeting determinant in aminopeptidase I. *J. Cell Biol.* **132,** 999–1010.
Reggiori, F., and Klionsky, D. J. (2002). Autophagy in the eukaryotic cell. *Eukaryot. Cell.* **1,** 11–21.
Reggiori, F., *et al.* (2005). The actin cytoskeleton is required for selective types of autophagy, but not nonspecific autophagy, in the yeast *Saccharomyces cerevisiae. Mol. Biol. Cell.* **16,** 5843–5856.
Reggiori, F., *et al.* (2004). The Atg1-Atg13 complex regulates Atg9 and Atg23 retrieval transport from the pre-autophagosomal structure. *Dev. Cell.* **6,** 79–90.
Rubinsztein, D. C., *et al.* (2005). Autophagy and its possible roles in nervous system diseases, damage and repair. *Autophagy* **1,** 11–22.
Scott, S. V., *et al.* (2001). Cvt19 is a receptor for the cytoplasm-to-vacuole targeting pathway. *Mol. Cell.* **7,** 1131–1141.
Scott, S. V., *et al.* (2000). Apg13p and Vac8p are part of a complex of phosphoproteins that are required for cytoplasm to vacuole targeting. *J. Biol. Chem.* **275,** 25840–25849.
Shintani, T., *et al.* (2002). Mechanism of cargo selection in the cytoplasm to vacuole targeting pathway. *Dev. Cell.* **3,** 825–837.
Shintani, T., and Klionsky, D. J. (2004). Cargo proteins facilitate the formation of transport vesicles in the cytoplasm to vacuole targeting pathway. *J. Biol. Chem.* **279,** 29889–29894.
Stack, J. H., *et al.* (1993). A membrane-associated complex containing the Vps15 protein kinase and the Vps34 PI 3-kinase is essential for protein sorting to the yeast lysosome-like vacuole. *EMBO J.* **12,** 2195–2204.
Stromhaug, P. E., *et al.* (2004). Atg21 is a phosphoinositide binding protein required for efficient lipidation and localization of Atg8 during uptake of aminopeptidase I by selective autophagy. *Mol. Biol. Cell.* **15,** 3553–3566.
Suzuki, K., *et al.* (2001). The pre-autophagosomal structure organized by concerted functions of *APG* genes is essential for autophagosome formation. *EMBO J.* **20,** 5971–5981.
Suzuki, K., *et al.* (2004). Interrelationships among Atg proteins during autophagy in *Saccharomyces cerevisiae. Yeast* **21,** 1057–1065.
Takeshige, K., *et al.* (1992). Autophagy in yeast demonstrated with proteinase-deficient mutants and conditions for its induction. *J. Cell Biol.* **119,** 301–311.
Tanida, I., *et al.* (2005). Lysosomal turnover, but not a cellular level, of endogenous LC3 is a marker for autophagy. *Autophagy* **1,** 84–91.
Teter, S. A., *et al.* (2001). Degradation of lipid vesicles in the yeast vacuole requires function of Cvt17, a putative lipase. *J. Biol. Chem.* **276,** 2083–2087.
Thumm, M., *et al.* (1994). Isolation of autophagocytosis mutants of *Saccharomyces cerevisiae. FEBS Lett.* **349,** 275–280.
Titorenko, V. I., *et al.* (1995). Isolation and characterization of mutants impaired in the selective degradation of peroxisomes in the yeast *Hansenula polymorpha. J. Bacteriol.* **177,** 357–363.

Tsukada, M., and Ohsumi, Y. (1993). Isolation and characterization of autophagy-defective mutants of *Saccharomyces cerevisiae*. *FEBS Lett.* **333,** 169–174.

Xie, Z., and Klionsky, D. J. (2007). Autophagosome formation: Core machinery and adaptations. *Nat. Cell. Biol.* **9,** 1102–1109.

Xie, Z., *et al.* (2008). Atg8 controls phagophore expansion during autophagosome formation. *Mol. Biol. Cell.* **in press**.

Yang, Z., *et al.* (2006). Atg22 recycles amino acids to link the degradative and recycling functions of autophagy. *Mol. Biol. Cell.* **17,** 5094–5104.

Yen, W.-L., *et al.* (2007). Atg27 is required for autophagy-dependent cycling of Atg9. *Mol. Biol. Cell.* **18,** 581–593.

Zhang, Y., *et al.* (2007). The role of autophagy in mitochondria maintenance: Characterization of mitochondrial functions in autophagy-deficient *S. cerevisiae* strains. *Autophagy* **3,** 337–346.

CHAPTER TWO

Viability Assays to Monitor Yeast Autophagy

Takeshi Noda

Contents

Abstract

In the yeast *Saccharomyces cerevisiae*, autophagy contributes to the sustaining of cell viability under starvation conditions, possibly through the supply of amino acids that is generated as a result of the degradation of cytosolic materials. Therefore, cellular viability is one of the best indexes for monitoring the completion of the entire autophagic process. In this chapter, several assays for monitoring yeast viability are presented. Along with the standard colony-formation assay, assays using the dye phloxine B are introduced.

1. Overview

The physiological roles of autophagy in yeast are diverse and include supplying amino acids for nascent protein synthesis and spore formation (Onodera and Ohsumi, 2005; Tsukada and Ohsumi, 1993). These roles became evident from studies analyzing autophagy-deficient mutants. One of the common phenotypes of autophagy-deficient mutants is the loss of viability that occurs during incubation in starvation medium. The cause of this phenotype is interpreted as attributable to the lack of a necessary supply of amino acids, which normally depends on autophagy; however, the actual cause is still ambiguous. The loss of viability phenotype is shared among

Research Institute for Microbial diseases, Osaka University, Osaka, Japan

Methods in Enzymology, Volume 451
ISSN 0076-6879, DOI: 10.1016/S0076-6879(08)03202-3

mutants that are deficient at any stage of autophagy, including autophagosome formation (George *et al.*, 2000), lysis of the autophagic bodies (the intravacuolar membrane structure derived from the autophagosome) (Teichert *et al.*, 1989), and efflux of amino acids from the vacuole lumen to the cytosol (Yang *et al.*, 2006). Thus, an assay investigating the viability of yeast cells under starvation conditions could provide an index of the completion of the entire autophagic process. In fact, the original autophagy-deficient mutants were isolated on the basis of the loss of viability during starvation (Tsukada and Ohsumi, 1993). Certainly, this phenotype is not specific to an autophagy deficiency, and other mutants such as those of the ubiquitin proteasome system show the same phenotype. Therefore, when analyzing cell viability in relation to autophagy, it must be kept in mind that autophagy is a necessary process, but it is not sufficient by itself to ensure viability. In addition, when attempting to isolate novel autophagy mutants, a viability assay can be the basis of the first screening, but additional and more specific tests appropriate to the question of interest are required.

2. Estimation of Yeast Viability with Phloxine B

Phloxine B is a red dye and a derivative of fluorescein, and it is used as color additive for food, drugs, and cosmetics. When yeast cells are alive, the dye diluted in the growth medium is excluded from the cells. In contrast, once the cell dies, the plasma membrane barrier is disrupted, and the dye penetrates into the cell and stains the cytosolic materials. When the dye is included in agar plate medium, the dye is impregnated into the dying cells and the colonies turn pink. This method was used to isolate *Schizosaccharomyces pombe* mutants of the cell cycle on growth medium (Nurse *et al.*, 1976), and then it was applied later to isolate *Saccharomyces cerevisiae* autophagy-defective mutants on starvation medium (Tsukada and Ohsumi, 1993). Other dyes with similar characteristics may be used with a similar strategy (Kucsera *et al.*, 2000).

2.1. Staining dead colonies under nitrogen starvation

1. Grow the cells of interest as single colonies on an appropriate growth medium plate, such as YPD (1% yeast extract, 2% peptone, and 2% glucose) or SMD (0.67% yeast nitrogen base, 2% glucose and auxotrophic amino acids, nucleosides, and/or vitamins as needed). You will need to use positive and negative control strains derived from the same genetic background, such as wild-type cells and *atg7* mutant cells.
2. Replicate the plate onto nitrogen starvation medium plates containing phloxine B, using sterilized velvet or a substitute. Take care to transfer

the same amount of cells among the strains being compared. Alternatively, the cells can be resuspended in liquid and the optical density determined by spectrophotometry; an equivalent amount of cells can then be aliquoted onto the plates for a higher degree of accuracy. The plates are made of 0.17% (w/v) yeast nitrogen base without ammonium sulfate and amino acids (Difco), 2% glucose, 5 μg/ml phloxine B (Sigma; P2759), and 2% agar. The optimum concentration of phloxine B depends on the cell background and must be determined empirically using the control strains; if you cannot obtain a reasonable difference in pink staining between the negative and positive control cells, you may need to change the concentration. Store the plates at 4 °C, shielded from light, as they are light sensitive.

3. Incubate the plate at 30 °C for 2–5 days until the difference in staining become visible; the negative control (wild-type) colonies should stay white or light pink while the positive control colonies turn deep pink (Fig. 2.1). Note that even wild-type cells will eventually turn pink on phloxine B plates, so it is important to monitor the color changes on a daily basis.

2.2. Estimation of viability by fluorescence microscopy

Yeast viability can also be examined with fluorescence microscopy using phloxine B (Tsukada and Ohsumi, 1993). On the basis of the same principle described in the previous section, phloxine B stains only dead cells, and the dead cells become fluorescent. Sometimes this method estimates more cells as viable than the following direct colony-formation assay. This may reflect the difference of the definition of the dead point, such as loss of membrane

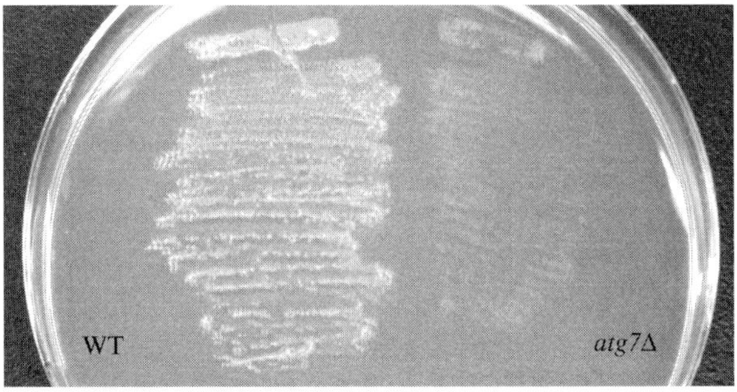

Figure 2.1 Phloxine B staining of yeast colonies under nitrogen starvation. The yeast cells of the wild-type strain (SEY6210) and its isogenic strain deleted for the *ATG7* gene are grown on a YPD plate then replica-plated to a nitrogen-deficient plate containing phloxine B. The plate was incubated at 30 °C for 2 days. The *atg7*Δ strain dies more rapidly under starvation conditions and takes up the phloxine B dye, which stains the cells pink.

integrity versus loss of ability to replicate. Therefore, it is best to take several approaches if subtle differences in viability may affect the conclusions being drawn. As an alternate method, the staining of live cells with a green fluorescent dye, fluorescein diacetate, which is trapped within a cell as a result of an esterase reaction, is a common way to estimate viability (Breeuwer *et al.*, 1995; Calich *et al.*, 1979). There also is a commercial kit available from Molecular Probes (LIVE/DEAD® Yeast Viability Kit; Molecular probes L-7009) that can be used to estimate viability.

1. Grow the cells of interest in an appropriate growth liquid medium such as 50 ml of YPD or SMD to mid log-phase ($O.D._{600} = 1$). You need to use a positive and negative control strain derived from the same genetic background for reference, such as wild-type and *atg7* mutant cells.
2. Wash the cells twice in autophagy-inducing liquid medium, such as nitrogen-depleted SD-(N) medium (0.17% yeast nitrogen base without ammonium sulfate and amino acids, and 2% glucose), by centrifugation at 3,000 rpm for 3 min, then resuspend in the same medium at the initial volume from step 1 and incubate further at the appropriate temperature (typically 30 °C for yeast, unless you are working with temperature-sensitive mutants) for the desired period of time. Be careful not to allow the culture to dry out during the incubation, especially if the volume of the culture is small.
3. After appropriate periods of time, such as 0, 1, 3, 5, and 7 days, remove 1 ml of the culture to a microcentrifuge tube and add 10 μl of a 100x solution of phloxine B (200 μg/ml in water) to a final concentration of 2 μg/ml. Note that it is a good idea to always start with a day 0 culture to set a baseline. Sonicate the tubes just briefly ($\sim$5 s) using a bath-type sonicator to detach the daughter cells from the mother cells that have completed cytokinesis.
4. Count the cell number in a certain area using a hemocytometer and a phase-contrast microscope. Then count the number of cells showing fluorescence in the same field using a fluorescence microscope with a blue filter suitable for FITC (U-MNIB for Olympus). Calculate the number of fluorescent cells among the total cells to determine the percentage of dead cells.

3. Direct Measurement of Cellular Viability by the Colony-Formation Assay

This is the simplest and most classical method to determine cell viability (George *et al.*, 2000). It should be kept in mind that when counting cells on plates, it is possible to count only growing cells, not living cells, because you are counting colonies, not actual cells. Some mutant cells may be able to enter and maintain the resting state (G_0) after nitrogen starvation but unable to exit the state when the nutrient is resupplied. In addition, if the strains of interest tend to flocculate, this method may not be appropriate.

1. Grow the cells of interest in an appropriate growth liquid medium such as 50 ml of YPD or SMD to mid log-phase ($O.D._{600} = 1$). You need to use a positive and negative control strain derived from the same genetic background for reference such as wild-type and *atg7* mutant cells.
2. Wash the cells two times in the appropriate starvation liquid medium, such as nitrogen-depleted SD-(N) medium (0.17% yeast nitrogen base without ammonium sulfate and amino acids [Difco] and 2% glucose), by centrifugation at 3,000 rpm for 3 min, then resuspend in the same medium at the initial volume from step 1 and incubate further at the appropriate temperature. Take care not to allow the culture to dry out during the incubation, especially if the culture volume is small.
3. After an appropriate period, such as 0, 1, 3, 5, and 7 days, remove 1 ml of the culture to a microcentrifuge tube. Note that it is a good idea to always start with a day 0 culture to set a baseline. Sonicate the tube just briefly (~5 s) using a bath-type sonicator to detach the daughter cells from the mother cells that have completed cytokinesis.
4. For the initial time point, count the cell density using a phase-contrast microscope and a hemocytometer. The ratio of cell number to $O.D._{600}$ varies depending on the cell type and the type of spectrophotometer. Dilute the aliquot with distilled water to a concentration of approximately 200 cells/200 μl and plate 200 μl onto YPD plates in triplicate. For the later time points, plate the same volume of culture as at the initial point. Count the colony number after 2 to 3 days of incubation, and calculate the percentage viability by comparing the number to that of day 0.

REFERENCES

Breeuwer, P., Drocourt, J. L., Bunschoten, N., Zwietering, M. H., Rombouts, F. M., and Abee, T. (1995). Characterization of uptake and hydrolysis of fluorescein diacetate and carboxyfluorescein diacetate by intracellular esterases in *Saccharomyces cerevisiae*, which result in accumulation of fluorescent product. *Appl. Environ. Microbiol.* **61,** 1614–1619.

Calich, V. L., Purchio, A., and Paula, C. R. (1979). A new fluorescent viability test for fungi cells. *Mycopathologia* **66,** 175–177.

George, M. D., Baba, M., Scott, S. V., Mizushima, N., Garrison, B. S., Ohsumi, Y., and Klionsky, D. J. (2000). Apg5p functions in the sequestration step in the cytoplasm-to-vacuole targeting and macroautophagy pathways. *Mol. Biol. Cell* **11,** 969–982.

Kucsera, J., Yarita, K., and Takeo, K. (2000). Simple detection method for distinguishing dead and living yeast colonies. *J. Microbiol. Methods* **41,** 19–21.

Nurse, P., Thuriaux, P., and Nasmyth, K. (1976). Genetic control of the cell division cycle in the fission yeast *Schizosaccharomyces pombe*. *Mol. Gen. Genet.* **146,** 167–178.

Onodera, J., and Ohsumi, Y. (2005). Autophagy is required for maintenance of amino acid levels and protein synthesis under nitrogen starvation. *J. Biol. Chem.* **280,** 31582–31586.
Teichert, U., Mechler, B., Müller, H., and Wolf, D. H. (1989). Lysosomal (vacuolar) proteinases of yeast are essential catalysts for protein degradation, differentiation, and cell survival. *J. Biol. Chem.* **264,** 16037–16045.
Tsukada, M., and Ohsumi, Y. (1993). Isolation and characterization of autophagy-defective mutants of *Saccharomyces cerevisiae*. *FEBS Lett.* **333,** 169–174.
Yang, Z., Huang, J., Geng, J., Nair, U., and Klionsky, D. J. (2006). Atg22 recycles amino acids to link the degradative and recycling functions of autophagy. *Mol. Biol. Cell* **17,** 5094–5104.

CHAPTER THREE

The Quantitative Pho8Δ60 Assay of Nonspecific Autophagy

Takeshi Noda* *and* Daniel J. Klionsky†

Contents

Abstract

The measurement of autophagic flux is critical in understanding the regulation of autophagy. The Pho8Δ60 assay employs a very sensitive enzymatic assay that provides a high signal-to-noise ratio and allows for precise quantification of autophagic flow in yeast. Pho8, alkaline phosphatase, is a resident vacuolar enzyme that is delivered to the vacuole membrane through a portion of the secretory pathway. The assay utilizes a genetically engineered version of Pho8 that lacks the N-terminal transmembrane domain that allows for translocation into the endoplasmic reticulum. Accordingly, Pho8Δ60 remains in the cytosol and is delivered to the vacuole only through autophagy. Once in the vacuole lumen, the C-terminal propeptide is proteolytically removed, which results in activation. Thus, the alkaline phosphatase activity reflects the amount of the cytosol delivered to the vacuole through nonspecific autophagy.

1. Overview

Measurement of the activity of autophagy, often referred to as autophagic flux, is an important step in understanding the mechanism and regulation of this process. However, when nonselective protein degradation

* Department of Cellular Regulation, Research Institute for Microbial Diseases, Osaka University, Osaka, Japan
† Life Sciences Institute, University of Michigan, Ann Arbor, Michigan

Methods in Enzymology, Volume 451
ISSN 0076-6879, DOI: 10.1016/S0076-6879(08)03203-5

is monitored by commonly used methods such as western blotting of a particular protein, there are inherent difficulties because the decrease of the protein is generally so small and occurs so slowly that the detection of the changes is hampered by the high background of the remaining protein. In general, it is difficult to quantify the disappearance of a signal (essentially a negative output), whereas it is much easier to monitor the appearance of one (a positive output). Therefore, to overcome this difficulty, the autophagic flow is best translated into the incremental increase of some signal. The method introduced in this chapter, the Pho8Δ60 assay, is a quantitative monitoring system of yeast autophagic flow based on the measurement of the enzyme activity resulting from vacuolar delivery of a zymogen, which reflects the autophagic flow (Noda *et al.*, 1995).

PHO8 is the gene encoding the sole vacuolar alkaline phosphatase in the yeast *Saccharomyces cerevisiae.* The Pho8 protein is a type II transmembrane protein containing an N-terminal cytosolic tail and an integral membrane segment; the active site and the C terminus are located within the vacuolar lumen (Klionsky and Emr, 1989). After synthesis as an inactive form at the endoplasmic reticulum, precursor Pho8 transits to the vacuole via the Golgi complex. After it arrives at the vacuole, the propeptide at the C terminus is cleaved off by vacuolar proteases, which results in the generation of the active form. In order to utilize Pho8 for monitoring autophagy, the amino terminal 60 amino acid residues, including the transmembrane domain, are genetically deleted, and the resultant enzyme is designated as Pho8Δ60 (Noda *et al.*, 1995). As a result of this deletion, the altered enzyme is no longer able to enter the endoplasmic reticulum. Instead, it becomes soluble and is dispersed throughout the cytosol as an inactive zymogen. Once nonselective engulfment of the cytosol by macroautophagy occurs, a portion of the Pho8Δ60 is nonselectively incorporated into autophagosomes and delivered into the vacuolar lumen, where it is activated by vacuolar proteases. Thus, when analyzed in a *pho8Δ* strain, the alkaline phosphatase (ALP) activity in the cell corresponding to Pho8Δ60 is proportional to the amount of cytosol delivered into the vacuole. Among the key aspects of this assay system is that Pho8 is a resident vacuolar protein and is therefore resistant to vacuolar degradation, other than removal of its propeptide. In addition, the physiological importance of the endogenous Pho8-dependent alkaline phosphatase activity seems quite limited under most physiological conditions, and therefore loss of the *PHO8* gene has no apparent effect on cell growth; this allows for the use of Pho8 for the purpose of this assay without obvious indirect effects.

The signal-to-noise (background signal) ratio, namely, the dynamic range, of this method is quite high, and therefore it is sensitive and able to discriminate a small change in autophagic flow. This characteristic provides several opportunities to gain insight into the regulation and mechanism of autophagy through the use of this assay. For example, multicopy suppression by several Atg proteins, such as occurs between Atg1 and Atg13, and between Atg6 and Atg14, was revealed by showing the partial recovery of autophagic activity using this method (Funakoshi *et al.*, 1997; Kametaka *et al.*, 1998). In addition, the size of the autophagosome is proportional to the autophagic flow; accordingly, this assay is suitable for an initial analysis to determine the mechanism involved in controlling the size of the autophagosome (Cheong *et al.*, 2005). Furthermore, this method is applicable to the isolation of mutants involved in autophagy that are either defective in or display hyperactivation of autophagic activity (Noda *et al.*, 2000; Shirahama *et al.*, 1997). Finally, this assay can be modified to monitor the specific degradation of organelles. For example, if Pho8 is fused to a targeting signal that directs it to mitochondria, the Pho8Δ60 assay can be used to follow mitophagy (Campbell and Thorsness, 1998).

2. The Pho8Δ60 Assay

2.1. Construction of a yeast strain expressing Pho8Δ60

To replace the wild-type genomic *PHO8* gene with the *pho8Δ60* construct, the original approach employed a relatively elaborate method of integration of the mutant region at the genomic *PHO8* locus and subsequent recombination to loop out the endogenous gene (Noda *et al.*, 1995). Here, we introduce an easier approach that utilizes the polymerase chain reaction (PCR) and homologous recombination based on the method of Janke *et al.* (2004). It is also possible to generate a strain through conventional genetic methods, by crossing the strain of interest with an existing Pho8Δ60 strain (e.g., TN124; Noda and Ohsumi, 1998) and subsequent tetrad dissection. Another simple method is transformation of the strain of interest with a multicopy plasmid harboring the *pho8Δ60* gene. This strategy is based on the idea that the ALP activity derived from the genome-encoded wild-type Pho8 is treated as background. The result from this latter approach, however, is potentially affected by the change in the wild-type ALP activity that occurs during autophagy induction, and therefore a proper control must be included.

1. Obtain the PCR template plasmid set described by Janke *et al.* (2004) from EUROSCARF (http://www.uni-frankfurt.de/fb15/mikro/euroscarf/index.html).
2. For the PCR, use pYM-N14 or pYM-N15 plasmid as a template, which is designed to amplify a PCR fragment containing the constitutive and strong *GPD1* promoter and yeast selection markers kanMX4

(N14; G418 resistance) or clonNAT (N15; nourseothricin resistance). Using the PCR primer set, pho8S1 is TATCAGCATACGGGACAT-TATTTGAACGCGCATTAGCAGC cgtacgctgcaggtcgac, and pho8S4 is TCACGAAGAATATGACATTCTTCTTCTTGTGTGATGCA-GAcatcga tgaattctctgtcg, which are intended to amplify the pYM-N14 or pYM-N15 template region flanked by 40 nucleotides just upstream of the initiation codon of the *PHO8* gene and 40 nucleotides downstream of amino acid 60; upon recombination in yeast, this construct will replace the endogenous *PHO8* gene with the *pho8Δ60* gene under the control of the *GDP1* promoter (Fig. 3.1). Mix 5 μl of 10x appropriate PCR polymerase buffer, 8.75 μl dNTP-mix (2 m*M* each of dNTP), 3.2 μl of 10 μM pho8S1 primer and 3.2 μl of 10 μM pho8S4 primer, and 1 μl of 100 μg/ml pYM-N14 or pYM-N15 and polymerase for PCR, such as KOD (TOYOBO), KOD-201, for a total reaction mixture volume of 50 μl and hold on ice.

3. Run the PCR reaction using pYM-N14 as a template, at 95 °C for 3 min, 10 cycles of 95 °C for 30 s, 54 °C for 30 s, 68 °C for 2 min 40 s, 20 cycles of 95 °C for 30 s, 54 °C for 30 s, 68 °C for 2 min 40 s, adding 20 s to each subsequent 68 °C cycle; the 12th cycle will be for 3 min, etc. For the reaction

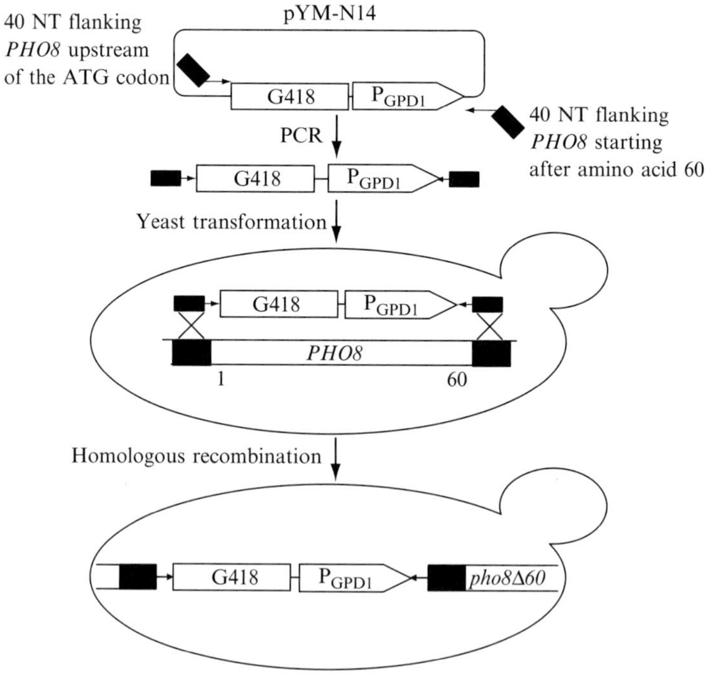

Figure 3.1 Schematic drawing of the method used to integrate the *pho8Δ60* gene at the endogenous *PHO8* locus by homologous recombination in yeast.

using pYM-N15, every 95 °C step is changed to 97 °C, and every incubation at 95 °C for 30 s is changed to 1 min.

4. Yeast strains of interest are transformed with 34 μl of PCR product per 1 transformation using the standard LiAc method (Gietz and Woods, 2002); see Chapter 1 of this volume. To account for the lag time needed for the expression of antibiotic resistance genes, the transformed cells are incubated in liquid YPD medium for 6 h at 30 °C, then spread on the selection plate containing antibiotic. Otherwise, the transformed cells may be spread onto a YPD plate and replica-plated onto a selection plate containing antibiotic after overnight incubation at 30 °C.
5. Among the clones showing antibiotic marker resistance, it is necessary to select the clone with correct replacement of the *PHO8* gene with *pho8Δ60*. This can be verified by genomic PCR of the resultant clones. Correct replacement should yield an approximately 0.6-kb band using the pho8S4 primer and the pho8 check primer: TTGCCAGCAAGTGGCTACATAAACATTTAC.

2.2. Assay for alkaline phosphatase activity

The activity of vacuolar alkaline phosphatase is measured using a modification of a previously published procedure (Nothwehr *et al.*, 1996). There are two methods that can be used to detect ALP activity, which differ in substrate and instrumentation. In the original report, to avoid the influence of Pho13, another alkaline phosphatase that is localized in the cytosol, the *PHO13* gene, is deleted. This method requires a spectrophotometer and is described first. The second method uses a substrate that is primarily utilized by Pho8, so it is not necessary to delete the *PHO13* gene; however, this method requires a fluorometer.

2.2.1. Spectrophotometric assay

1. The *PHO13* gene should be deleted in the *pho8Δ60* strain and replaced by integrative recombination (Noda *et al.*, 1995). The strains of interest are grown in 510 ml of YPD (1% yeast extract, 2% peptone, and 2% glucose) or SMD (0.67% yeast nitrogen base and 2% glucose and auxotrophic amino acids, nucleosides, and/or vitamins as needed) medium at 30 °C in flasks, so that the total culture volume is less than 10% of the flask capacity. Cells are typically grown to log phase ($O.D._{600}$ = 0.8–1.0), then diluted and grown again to log phase. When the cells enter the late-log to stationary phase, autophagy is induced, so it is important to maintain the cells in log phase prior to the desired induction time. If you use an overnight culture at stationary phase for the starting culture, you need to dilute the cells and wait until

they undergo at least two cell divisions prior to use, but it is recommended that the cells be kept in log phase at all times.

Approximately 1.0 O.D.$_{600}$ units of cells are needed for each assay, so the volume can be increased to provide enough cells for a time course, if desired. Care should be taken to harvest cells at the same density to allow comparison among assays performed at different times, although this type of comparison is not recommended; instead, we recommend including positive (*pho8Δ60 pho13Δ*) and negative (*pho8Δ60 pho13Δ* with a deletion of an *ATG* gene or a gene encoding a vacuolar hydrolase, either *PEP4* or *PRB1*) controls in every assay and only comparing samples assayed in the same experiment.

2. Collect the cells by centrifugation at 1,500×*g* (3,000 rpm) for 35 min.
3. The supernatant fraction is discarded and the cell pellet is resuspended and washed in 5 ml of SD-N medium, centrifuged, then resuspended in another 5 ml of SD-N. Alternatively, autophagy may be induced by the addition of rapamycin (0.2 μg/ml final concentration, dissolved in 10% Triton X-100 and 90% ethanol).
4. The cell culture is incubated in SD-N at 30 °C for 4 h to induce autophagy. Autophagy activity starts to increase after approximately 1 h but does not reach an easily detectable level for 3–4 h. Depending on the experiment, it may be appropriate to take time points. If multiple time points are taken, the earlier samples should be cooled on ice or frozen to prevent additional autophagy until all of the samples are ready for processing.
5. At each time point, an aliquot of cells equivalent to 25 O.D.$_{600}$ units is removed and the cells are collected by centrifugation at 1,500×*g* (3,000 rpm) for 5 min. Although approximately 1 O.D.$_{600}$ unit of cells is needed per assay, the larger starting amount allows for some loss of material during sample preparation.
6. The supernatant fraction is discarded and the cell pellet is resuspended and washed in 1 ml of water, followed by centrifugation at 4 °C. The resulting cell pellet is resuspended and washed in 2 ml of ice-cold 0.85% NaCl containing 1 m*M* PMSF and centrifuged to collect the cell pellet.
7. The cell pellet is resuspended in 200–500 μl of ice-cold lysis buffer so that the final cell suspension is at 1.0 O.D.$_{600}$ unit/100 μl. The lysis buffer is 20 m*M* PIPES, 0.5% Triton X-100, 50 m*M* KCl, 100 m*M* potassium acetate, 10 m*M* $MgSO_4$, 10 μ*M* $ZnSO_4$, and 1 m*M* PMSF. The PMSF is dissolved in 95% ethanol and should be added just before the lysis buffer is added to the cells. Only 100 μl of cell extract is used per assay, but it is recommended that a larger volume be prepared to account for loss of sample during glass-bead lysis, to allow for the use of two volumes of sample for the assay, and to calculate protein concentrations.

8. The cell pellet is lysed by adding glass beads equivalent to approximately half the sample volume and mixing vigorously with a vortex for 1–10 min at 4 °C. It is important to keep the glass-bead level below the sample meniscus to prevent excessive bubbling. In addition, this procedure can generate substantial heat, so the vortex should be used for short periods of time (maximum 1 min), followed by cooling on ice. See chapter 1 for additional information on glass-bead lysis, including preparation of acid-washed glass beads.
9. Centrifuge the lysed cells at 14,000×*g* (15,000 rpm) for 5 min at 4 °C to remove unlysed cells and debris. The supernatant fraction is collected for subsequent analysis of alkaline phosphatase activity.
10. The ALP substrate solution is prewarmed to 37 °C. You will need 400 μl of substrate solution per reaction and additional for blank samples. Substrate solution (Mitchell *et al.*, 1981) is prepared from 100 m*M* of stock of *p*-nitrophenyl phosphate (pNPP; Sigma N9389) dissolved in Millipore filtered water. This solution can be kept frozen at 20 °C for a few months or at 4 °C for 1 or 2 days. For the final substrate solution, the pNPP is diluted to 1.25 m*M* in reaction buffer (250 m*M* Tris-HCl, pH 8.5; 0.4% Triton X-100; 10 m*M* $MgSO_4$, and 10 μ*M* $ZnSO_4$).
11. The cleavage of pNPP by Pho8Δ60 generates *p*-nitrophenol. Therefore, a standard curve is generated using the latter. The standard curve uses 1–100 nmoles of *p*-nitrophenol (Sigma, N7660). The appropriate amount of *p*-nitrophenol is placed into 1.7-ml microcentrifuge tubes, and the total volume is brought to 100 μl with reaction buffer. 1.0 μl of the 10 m*M* *p*-nitrophenol standard solution is equivalent to 10 nmoles of *p*-nitrophenol. The 1–100 nmoles of *p*-nitorphenol should result in $O.D._{400}$ values between 0 and 2.0, which is the maximum linear range of most spectrophotometers. It is also recommended that you prepare "blank" samples for enzyme and substrate. For the enzyme blank, 100 μl of reaction buffer is used. For the substrate blank, use 100 μl of cell extract and add 400 μl of reaction buffer without *p*-nitrophenyl phosphate; do not add the alkaline phosphatase substrate solution in step 13 below.
12. Place 50–100 μl of lysed samples into 1.7-ml microcentrifuge tubes and bring the final volume to 100 μl with lysis buffer. We recommend using at least two volumes of sample to allow a determination of linearity; alternatively, two time points can be used for samples prepared in duplicate. Keep the tubes on ice until you are ready to begin the assay.
13. Place the tubes at 37 °C. Add 400 μl of prewarmed alkaline phosphatase substrate solution to the samples, staggering the additions between samples at 15-s intervals (this allows each sample to incubate the same amount of time, allowing for the time needed to add the substrate solution, and later the stop buffer). Incubate the samples at 37 °C for 5–20 min; start timing after the first addition. Add 400 μl of reaction

buffer without *p*-nitrophenyl phosphate to the substrate blank tube, and add substrate solution to the enzyme blank. The length of time for the incubation may need to be determined empirically on the basis of the results with the positive control strain extract; the critical factors are that the substrate is present at saturating amounts (this is why it is important to use duplicate samples with either two volumes of sample or two time points) and that the final spectrophotometric values fall within the linear range of the instrument.

14. Stop the reaction by adding 500 μl of stop buffer (1 *M* glycine/KOH, pH 11.0), again staggering the additions at 15-s intervals.
15. Centrifuge the tubes at maximal speed for 2 min to remove any precipitate or debris.
16. Measure the $O.D._{400}$ using 1.0 ml of the standard curve, blanks, and the samples.
17. Specific enzyme activity can be calculated if protein concentration is determined. We use the BCA (Pierce Chemical Co.) or Bradford assay, with a standard curve having between 0–1.0 mg/ml BSA. You will need approximately 50 μl of sample to measure the protein concentration in the lysates. Subtract the enzyme and substrate blanks from the absorbance readings of the samples, and calculate the concentration in nmol of *p*-nitrophenol in the samples by graphing the adjusted $O.D._{400}$ values relative to the standard curve. The specific activity is calculated as nmol *p*-nitrophenol/min/mg protein.

2.2.2. Spectrofluorometric assay

1. The pho8Δ60 strains of interest are grown in liquid growth medium such as YPD or SMD to mid-log phase ($O.D._{600} = 0.5$–1) as described in step 1 of the preceding section; grow at least 10 ml for one assay time point. The cells are washed twice in nitrogen starvation medium (SD-N) by centrifugation at 1,500×*g* (3,000 rpm) for 3–5 min, then resuspended in the same volume of SD-N as the growth medium. The cells are maintained in starvation conditions at the concentration of $O.D._{600} = 1$ for the desired time periods. Alternatively, autophagy may be induced by the addition of rapamycin as described in step 3 of the preceding section.
2. At each time point, collect 2–4 $O.D._{600}$ units of cells by centrifugation at 1,500×*g* for 3 min and discard the supernatant fraction.
3. For lysis, resuspend the cells in 0.2 ml of ice-cold assay buffer (250 m*M* Tris-HCl, pH 9.0; 10 m*M* $MgSO_4$, and 10 μM $ZnSO_4$) and transfer the cell suspension to a 1.5-ml microcentrifuge tube. Add acid-washed glass beads (425–600 microns), keeping the level of glass beads just below that of the cells. Cool the tube on ice for 5 min, then mix the tubes vigorously with a vortex mixer for 6 × 30 s, stopping for at least 30-s

intervals on ice in between the mixing. Alternatively, the glass-bead lysis can be performed in a cold room using an automatic vortex mixer, although this may still generate enough heat to cause protein denaturation. Add an additional 0.2 ml of assay buffer and mix well. Centrifuge at 14,000×*g* (15,000 rpm) for 1 min to remove unbroken cells and debris, and transfer the supernant fraction to new tubes.

4. For each assay, remove 0.05 ml of the cell lysate solution and place in a new tube. Bring the volume to a total of 0.5 ml with assay buffer.
5. Place the tubes into a 30 °C water bath. Start the reaction by adding 0.05 ml of 55 m*M* α-naphthyl phosphate disodium salt (Sigma N7255) dissolved in assay buffer and mixing well. Stagger the additions as described in step 13 of the preceding section. Incubate at 30 °C for 20 min.
6. Stop the reaction by adding 0.5 ml of 2 *M* glycine-NaOH, pH 11.0, and mixing, and keep the tubes on ice until all of the samples are collected.
7. Measure the fluorescence using a wavelength of 345 nm for excitation and 472 nm for emission.
8. Measure the protein concentration of the cell lysate using either the BCA or Bradford method. The ALP activity is presented as emission per the amout of protein in the reaction (mg) and the reaction time (min).

3. Interpretation of the Results

This method generally yields reproducible results and therefore allows quantitative analysis of nonspecific autophagic activity. A point of special note is that the expression level of the Pho8Δ60 protein potentially affects the activity. The method introduced in this chapter utilizes a constitutive *GPD1* promoter, which normally controls the expression of glycerol-3-phosphate dehydrogenase, but the expression from this promoter may be affected depending on the particular culture conditions. Therefore, if the conditions in which you need to grow your cells might influence expression, you should consider the use of a different promoter, such as glyceraldehyde-3-phosphate dehydrogenase (*TDH3*), which is used in the original method (Noda *et al.*, 1995).

REFERENCES

Campbell, C., and Thorsness, P. (1998). Escape of mitochondrial DNA to the nucleus in *yme1* yeast is mediated by vacuolar-dependent turnover of abnormal mitochondrial compartments. *J. Cell. Sci.* **111,** 2455–2464.

Cheong, H., Yorimitsu, T., Reggiori, F., Legakis, J. E., Wang, C.-W, and Klionsky, D. J. (2005). Atg17 regulates the magnitude of the autophagic response. *Mol. Biol. Cell* **16,** 3438–3453.

Funakoshi, T., Matsuura, A., Noda, T., and Ohsumi, Y. (1997). Analyses of *APG13* gene involved in autophagy in yeast. *Saccharomyces cerevisiae. Gene* **192,** 207–213.

Gietz, R. D., and Woods, R. (2002). Transformtaion of yeast by lithium acetate/single-strand carrier DNA/polyethylene glycol method. *Methods Enzymol* **350,** 87–96.

Janke, C., Magiera, M. M., Rathfelder, N., Taxis, C., Reber, S., Maekawa, H., Moreno-Borchart, A., Doenges, G., Schwob, E., Schiebel, E., and Knop, M. (2004). A versatile toolbox for PCR-based tagging of yeast genes: New fluorescent proteins, more markers and promoter substitution cassettes. *Yeast* **21,** 947–962.

Kametaka, S., Okano, T., Ohsumi, M., and Ohsumi, Y. (1998). Apg14p and Apg6/Vps30p form a protein complex essential for autophagy in the yeast, *Saccharomyces cerevisiae. J. Biol. Chem.* **273,** 22284–22291.

Klionsky, D. J., and Emr, S.D (1989). Membrane protein sorting: Biosynthesis, transport and processing of yeast vacuolar alkaline phosphatase. *EMBO J.* **8,** 2241–2250.

Mitchell, J., Fonzi, W., Wilkerson, J., and Opheim, D (1981). A particulate form of alkaline phosphatase in the yeast, *Saccharomyces cerevisiae. Biochim. Biophys. Acta* **657,** 482–494.

Noda, T., Kim, J., Huang, W.-P., Baba, M., Tokunaga, C., Ohsumi, Y., and Klionsky, D.J (2000). Apg9p/Cvt7p is an integral membrane protein required for transport vesicle formation in the Cvt and autophagy pathways. *J. Cell Biol.* **148,** 465–480.

Noda, T., Matsuura, A., Wada, Y., and Ohsumi, Y. (1995). Novel system for monitoring autophagy in the yeast *Saccharomyces cerevisiae. Biochem. Biophys. Res. Commun.* **210,** 126–132.

Noda, T, and Ohsumi, Y (1998). Tor, a phosphatidylinositol kinase homologue, controls autophagy in yeast. *J. Biol. Chem.* **273,** 3963–3966.

Nothwehr, S. F., Bryant, N. J., and Stevens, T.H (1996). The newly identified yeast *GRD* genes are required for retention of late-Golgi membrane proteins. *Mol. Cell. Biol.* **16,** 2700–2707.

Shirahama, K., Noda, T., and Ohsumi, Y. (1997). Mutational analysis of Csc1/Vps4p: Involvement of endosome in regulation of autophagy in yeast. *Cell Struct. Funct.* **22,** 501–509.

CHAPTER FOUR

Fluorescence Microscopy-Based Assays for Monitoring Yeast Atg Protein Trafficking

Takahiro Shintani* *and* Fulvio Reggiori†

Contents

Abstract

For several years, the yeast *Saccharomyces cerevisiae* has been the leading model organism for the study of autophagy. The amenability of this unicellular eukaryote to genetic and biochemical approaches has allowed the isolation and characterization of most of the genes specifically involved in autophagy, which are known as *ATG* (Reggiori, 2006; Reggiori and Klionsky, 2005). These pioneering studies have been of crucial relevance because most of the yeast *ATG* genes possess orthologs in all eukaryotic organisms. The experimental advantages, all the available reagents, and the established assays still maintain yeast in a prominent position in the study of autophagy and autophagy-related pathways. In this chapter, we describe fluorescent protein-based methodologies that permit one to readily assay the functionality of the autophagic pathway and to assess the trafficking of one of the key protein of this degradative process, Atg9.

* Laboratory of Bioindustrial Genomics, Graduate School of Agricultural Science, Tohoku University, Sendai, Japan
† Department of Cell Biology and Institute of Biomembranes, University Medical Centre Utrecht, Utrecht, The Netherlands

Methods in Enzymology, Volume 451
ISSN 0076-6879, DOI: 10.1016/S0076-6879(08)03204-7

1. Introduction

Autophagy is a complex catabolic process that requires extensive transport and rearrangement of membranes to correctly progress (Reggiori and Klionsky, 2005). These include the delivery of lipid bilayers to the phagophore assembly site or preautophagosomal structure (PAS), the perivacuolar site in yeast that is believed to be where autophagosomes are nucleated and subsequently assemble into these large double-membrane vesicles (Kim *et al.*, 2002; Reggiori and Klionsky, 2005; Suzuki *et al.*, 2001). Furthermore, autophagosomes fuse with vacuoles, releasing their internal autophagic bodies in the interior of this organelle, where, together with their cargo, they are degraded. The factors specifically involved in autophagy, the autophagy-related (Atg) proteins, mediate several of these trafficking events (Reggiori, 2006). As a result, the analysis of the localization and transport of several Atg proteins allow for monitoring the induction, progression, and eventual block of the autophagic pathway.

Fluorescence microscopy-based analyses are a rapid, technically simple way to scrutinize autophagy, and when examining a defect in this transport route, they promptly provide indications about the possible molecular step that is affected. The main disadvantage encountered with this approach is the tedious quantification necessary to characterize partial impairments, which requires time-consuming counting and statistical analyses. There may be other potential disadvantages due to common problems linked to the use of fluorescent protein tags. For example, the tagged protein may not behave as normal. Second, some of the fluorescent proteins tend to oligomerize, which could also be a problem. Finally, the potential cleavage of the fluorescent protein could result in misinterpretation of the results because the object being imaged is the fluorescent tag and not the actual protein. So far, however, there are no reports indicating that the constructs described in this chapter lead to such problems.

2. Ape1 and Atg8 Transport into the Vacuole

The principal role of autophagy is the elimination of cytoplasmic components, and it is possible to assess the functionality of this pathway by following their transport into the vacuole interior. When those cytoplasmic components are linked to a fluorescent protein, their delivery into the vacuole lumen can be monitored by the appearance of the fluorescence signal in the interior of this organelle. Vacuoles are large, circular compartments that can be easily morphologically recognized but also specifically stained with the red fluorescent dye FM 4-64 (Molecular Probes/Invitrogen) (Vida and Emr, 1995).

Transport of cytosolic materials via the autophagic process is generally nonspecific and too slow to use fluorescent protein-based approaches. Therefore, a specific substrate is required to overcome this difficulty. The aminopeptidase Ape1 is an ideal solution. This vacuolar protease is the principal cargo of the cytoplasm to vacuole targeting (Cvt) pathway but also one of the molecules selectively sequestered into autophagosomes (Baba *et al.*, 1997; Kim *et al.*, 1997). After its synthesis as a precursor form, prApe1 forms a dodecamer that further assembles into a higher-ordered structure called the Ape1 complex, which is sequestered in a propeptide-dependent manner into double-membrane vesicles (Kim *et al.*, 1997). When prApe1 is fused with a fluorescent protein and expressed in wild-type cells, most of the fluorescent signal is dispersed uniformly in the vacuole in both growing and starvation conditions, with sometimes one brilliant dot representing the Ape1 complex in the perivacuolar region, presumably at the PAS (Fig. 4.1) (Shintani *et al.*, 2002; Suzuki *et al.*, 2002). The diffused signal in the vacuole is due to the liberation of the fluorescent group from Ape1 through cleavage by resident proteases. It is important to note that the Ape1 chimeras are useful to readily determine whether a specific autophagy step is blocked. Several fluorescent-positive dots are seen moving inside the vacuole in mutant strains, such as *pep4*Δ, which lacks the vacuolar hydrolase proteinase A, which are unable to break down autophagic bodies (see Fig. 4.1) (Shintani *et al.*, 2002; Suzuki *et al.*, 2002). In contrast, in strain backgrounds with mutations that block the fusion of double-membrane vesicles with the vacuole (e.g., the *vam3*Δ strain, which is defective in a SNARE component), several cytoplasmic fluorescent dots representing accumulated completed vesicles can be imaged (see Fig. 4.1). Finally, mutants with a defect in formation of double-membrane vesicles, such as *atg1*Δ, display a single larger fluorescent dot representing the PAS in the perivacuolar region (see Fig. 4.1) (Shintani *et al.*, 2002; Suzuki *et al.*, 2002). Interestingly, this dot is far from the vacuole surface in mutants incapable of recruiting the Ape1 complex to the PAS (Shintani *et al.*, 2002; Suzuki *et al.*, 2002). This latter determination requires staining of the vacuole with FM 4-64 and statistical analysis. It should be noted that the assays relying on prApe1 transport cannot be done in higher eukaryotes because no cargo equivalent to this protease has been identified outside of fungi.

Atg8 initially associates with both the internal and the external membrane of forming double-membrane vesicles (Kirisako *et al.*, 1999). As a result, the internal body and the Atg8 pool present in the interior are delivered and degraded in the lumen of the vacuole after fusion of the double-membrane vesicle outer membrane with this organelle (Kirisako *et al.*, 1999). Thus, fluorescent Atg8 chimeras have a distribution pattern

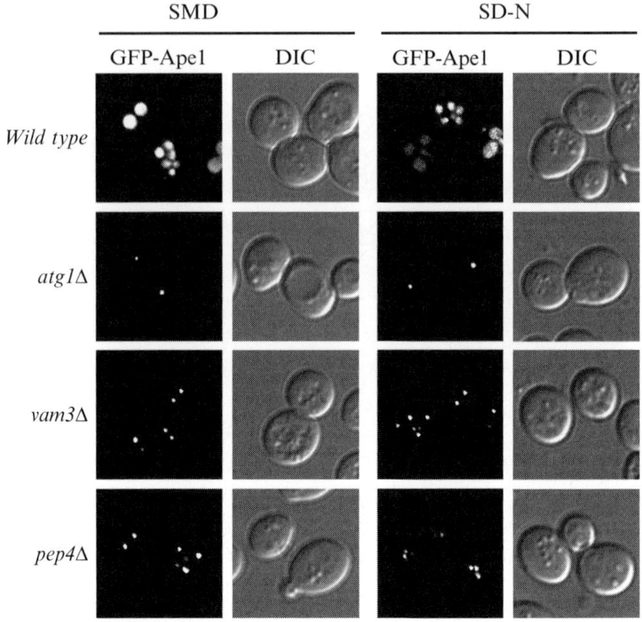

Figure 4.1 Live-cell imaging of cells expressing GFP-Ape1. Wild-type (BY4741, *MATa leu2 Δ0 met15Δ0 ura3Δ0 his3Δ1*), *atg1Δ* (BY4741 *atg1Δ*::KAN), *vam3Δ* (BY4741 *vam3Δ*::KAN), and *pep4Δ* (BY4741 *pep4Δ*::KAN) cells transformed with the plasmid expressing GFP-Ape1 (pTS466) (Shintani *et al.*, 2002) under the control of the authentic promoter were grown in synthetic minimal medium lacking uracil (SMD), or nitrogen starved for 2 h in (SD-N). Images were captured and deconvolved with a DeltaVision RT fluorescence microscope system. In wild-type cells, the GFP-Ape1 signal is primarily seen as diffuse within the vacuole lumen in accordance with its localization as a resident vacuolar hydrolase. In the *atg1Δ* mutant, in contrast, this fluorescent chimera is concentrated in the single spot that represents the PAS. The block of the fusion of double-membrane vesicles with the vacuole and of the autophagic body breakdown with the *vam3Δ* and the *pep4Δ* deletions, respectively, results in the distribution of the GFP-Ape1 signal in numerous puncta. DIC, differential interference contrast.

comparable to that of Ape1, and, similar to this protease, they can be used to assess the functionality of the autophagic pathway and determine whether a specific mutation blocks the formation and/or fusion of double-membrane vesicles and autophagic body breakdown. When a wild-type strain expressing GFP-Atg8 is grown in a rich medium, a single bright punctate structure representing the PAS is observed (Fig. 4.2). When the same cells are starved for at least 1 h, the vacuole lumen is also stained with fluorescence, which reflects the increased number of double-membrane vesicles fusing with the vacuole as a result of the induction of autophagy (see Fig. 4.2). Note that Atg8 also lines the inner membrane of Cvt vesicles that transport prApe1 in

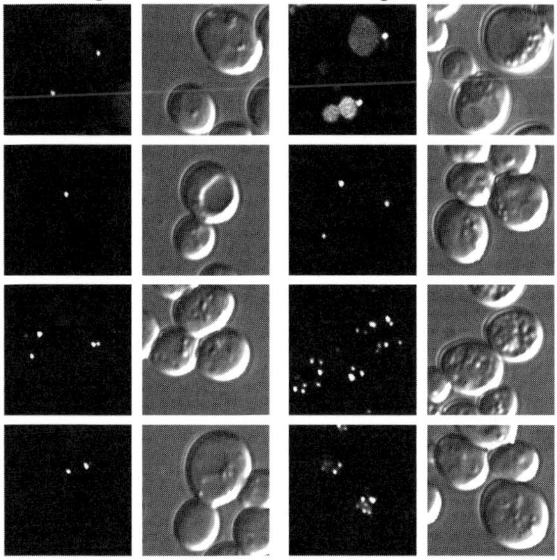

Figure 4.2 Live-cell imaging of cells expressing GFP-Atg8. Wild-type (BY4741, *MATa leu2Δ0 met15Δ0 ura3Δ0 his3Δ1*), *atg1Δ* (BY4741 *atg1Δ*::KAN), *vam3Δ* (BY4741 *vam3Δ*::KAN), and *pep4Δ* (BY4741 *pep4Δ*::KAN) cells transformed with a plasmid expressing GFP-Atg8 [pCuGFPAUT7(416)] (Kim *et al.*, 2001) under the control of the *CUP1* promoter were grown in synthetic minimal medium lacking uracil (SMD), or nitrogen starved for 2 h (SD-N). Images were captured and deconvolved with a DeltaVision RT fluorescence microscope system (Applied Precision Inc., Issaquah, WA). In wild-type cells grown in the nutrient rich medium (SMD), the GFP-Atg8 signal is primarily seen as a single brilliant dot that represents the pool of this fusion localizing to the PAS. In starved wild-type cells (SD-N), however, this fluorescent chimera diffuse within the vacuole lumen in accordance with its higher transport rate into the interior of this organelle. In the *atg1Δ* mutant, in contrast, GFP-Atg8 is concentrated in the single spot that represents the PAS. The block of the fusion of double-membrane vesicles with the vacuole and of the autophagic body breakdown with the *vam3Δ* and the *pep4Δ* deletions, respectively, results in the distribution of the GFP-Atg8 signal in numerous puncta. DIC, differential interference contrast.

vegetative conditions, but the number of these vesicles is quite low relative to autophagosomes that form during starvation. Accordingly, it is difficult to detect the lumenal signal corresponding to GFP-Atg8 under these conditions. The augmentation of the double-membrane vesicle formation rate that occurs during starvation is highlighted in mutants blocking fusion between autophagosomes and vacuoles such as *vam3Δ*, in which a multitude of GFP-Atg8-positive puncta can be seen scattered throughout the cytoplasm, but also in knockout strains such as *pep4Δ*, in which the degradation of autophagic bodies is severely impaired and therefore large clusters of fluorescent dots are visualized in the vacuole lumen (see Fig. 4.2). In contrast, a block in autophagosome biogenesis such as that caused by the deletion of *ATG1*, leads to the formation of a nonfunctional PAS that contains a large part of the membrane-associated Atg8 and appears as the only fluorescent punctate structure present in the cell (see Fig. 4.2).

The Atg8 fusions also have two additional advantages in regard to those of Ape1. Atg proteins are sequentially organized at the PAS, and localization of Atg8 depends on most of the other Atg proteins (Suzuki *et al.*, 2001, 2007). Therefore, the first advantage is that defects in PAS assembly (e.g., the absence of Atg proteins or impairment in their recruitment to this specialized site) can be easily monitored because they lead to a diffuse cytosolic distribution of fluorescent Atg8 (Suzuki *et al.*, 2001, 2007). Atg8 chimeras driven by the authentic promoter generate weak signals when cells carrying this construct are grown in rich media, and basically only the pool at the PAS can be imaged, whereas the one in the vacuole is almost undetectable (Suzuki *et al.*, 2001, 2007). However, Atg8 expression is induced when cells are starved (Kirisako *et al.*, 1999), and the fluorescent staining of the vacuole becomes obvious (see Fig. 4.2). Consequently, the second advantage is that fluorescent Atg8 fusions under the control of the authentic promoter can be used as an assay to monitor the triggering of autophagy.

3. Construction of the prApe1 and Atg8 Fluorescent Fusions

So far, all the generated fluorescent fusions of prApe1 have been put under the control of the authentic promoter and because this aminopeptidase is a relatively abundant protein (Ghaemmaghami *et al.*, 2003), its signal is easy to capture. Precursor Ape1 can be tagged at both termini and retain essentially normal transport characteristics. Strains and plasmids expressing C-terminal fusions of prApe1 with green fluorescent protein (GFP), yellow fluorescent protein (YFP), cyan fluorescent protein (CFP), and monomeric red fluorescent protein (mRFP) have been successfully employed (Meiling-Wesse *et al.*, 2004, 2005; Suzuki *et al.*, 2002). The strains have been created by genomic integration of the fluorescent protein by homologous recombination (Longtine *et al.*, 1998; Suzuki *et al.*, 2002). Plasmids, on the other hand, have been made in two steps by initially cloning the gene encoding for the fluorescent tag and successively *APE1* as well as its promoter (Meiling-Wesse *et al.*, 2004, 2005). Apparently, there are no evident dissimilarities in the results obtained with genomic and plasmid-driven C-terminal Ape1 fusions; however, expression levels of genomic chimeras are more homogeneous throughout the entire cell population.

To construct the plasmids carrying the N-terminal fluorescent fusions, the *APE1* open reading frame including its 5′ and 3′ flanking regions that contain the promoter and the terminator, respectively, was obtained by PCR amplification of genomic DNA and cloned into a centromeric (CEN) vector (Shintani *et al.*, 2002; Sikorski and Hieter, 1989). Successively, a *Bgl*II restriction site was introduced just after the start codon of this protease with a QuikChange Site-Directed Mutagenesis KIT (Stratagene) before ligating the DNA fragment encoding for the *GFP*, *CFP*, *mRFP* or *blue fluorescent protein* (*BFP*) with *Bam*HI sites on both sides into the newly generated *Bgl*II site to make the respective fusion constructs (He *et al.*, 2006; Reggiori *et al.*, 2004; Shintani *et al.*, 2002). The *GFP* and *CFP* coding sequences can be easily acquired from numerous laboratories, whereas that of *mRFP* and BFP can be requested (Campbell *et al.*, 2002) or bought (Quantum Biotechnologies, Canada), respectively. Alternatively, other fluorescent proteins can be considered (Shaner *et al.*, 2004).

In contrast to prApe1, fusions at the two termini of Atg8 are not equivalent. The C-terminal arginine residue of Atg8 is trimmed by Atg4 and the exposed glycine residue is lipidated with phosphatidylethanolamine through the action of Atg7, Atg3, and Atg12–Atg5. This modification is essential for the function of Atg8 in autophagy and any of the other related pathways (Ichimura *et al.*, 2000; Kirisako *et al.*, 2000). Consequently, the C-terminally-tagged fluorescent protein is liberated from Atg8 by Atg4 in the cytoplasm (Kim *et al.*, 2001; Kirisako *et al.*, 2000). Accordingly, C-terminal fusions to Atg8 can be used to monitor the activity of Atg4 but not to follow localization of the chimera. In contrast, N-terminal fusions under the control of both the authentic *ATG8* promoter and the stronger *CUP1* promoter have been successfully and extensively used (Kim *et al.*, 2001, 2002; Suzuki *et al.*, 2001). As mentioned, fluorescent chimeras driven by the authentic promoter generate weak signals when cells carrying this construct are grown in rich media, but those become more prominent when cells are starved, because Atg8 expression is induced under these conditions (Kirisako *et al.*, 1999). The use of a stronger promoter makes it possible to circumvent the difficulty in detecting GFP-Atg8 encountered in growing conditions because it facilitates the imaging of growing cells without obvious artifacts that are often due to protein overexpression (Kim *et al.*, 2001, 2002). However, the use of a strong promoter eliminates the ability to monitor Atg8 expression levels as a criterion for autophagy induction.

Plasmids carrying the Atg8 fusion under the control of the authentic promoter have been generated similarly to the N-terminal fusion of prApe1: the *ATG8* gene plus its promoter and terminator obtained by PCR amplification of genomic DNA has been cloned into a CEN vector before introducing a *Bam*HI site after the *ATG8* start codon, where *GFP*, *YFP*, and *CFP* open reading frames flanked by *Bam*HI sites have been inserted (Sikorski and Hieter, 1989; Suzuki *et al.*, 2001). Plasmids expressing higher levels of Atg8 have been made by cloning the gene encoding this protein into CEN vectors behind *GFP*, *YFP*, or *CFP* preceded by the copper-inducible *CUP1* promoter, which can also be easily obtained by PCR amplification of genomic DNA (Kim *et al.*, 2001, 2002; Labbé and Thiele, 1999). Conventional yeast growth media contain sufficient amounts of copper to guarantee a satisfactory expression of constructs under the control of the *CUP1* promoter without the need to add more of this cation, although the addition of 10 μM of copper will result in even higher levels of expression. High levels of copper are toxic and should be avoided.

4. Visualization of the Fluorescence Signals

The easiest and quickest way to determine the subcellular distribution of fluorescent fusions in yeast is by live-cell microscopy (e.g., Kim *et al.*, 1999; Suzuki *et al.*, 2001).

1. Cells are grown in culture tubes containing 10 ml of medium at 30 °C overnight and then diluted and regrown in the same medium to a logarithmic phase (approximately $OD_{600} = 1$). Rich medium (YPD; 1% yeast extract, 2% peptone, and 2% glucose) is optimal for genomically integrated constructs. Synthetic minimal media (SMD; 6.7% yeast nitrogen base and 2% glucose, amino acids, and vitamins as needed) containing the appropriate components but lacking those needed to maintain plasmid selection, on the other hand, are the obligatory choice when the chimeras are expressed from plasmids.
   a. To image the fluorescent fusions under conditions where the Cvt pathway is active, cells can be directly photographed at this point.
   b. To study the localization of the same constructs during autophagy, cells must be transferred into a synthetic medium lacking nitrogen (SD-N; 0.17% yeast nitrogen base without amino acids and ammonium sulfate and 2% glucose) or carbon (S; 6.7% yeast nitrogen base) for a least 1 h to induce starvation. In this case, cells are centrifuged in 15-ml conical tubes at 3000 ×*g* for 5 min. The supernatant fraction is discarded and the cells are washed by resuspension in 10 ml of sterile water. The cells are again centrifuged and the pellet is resuspended in 10 ml of SD-N in a culture tube, followed by shaking at 30 °C. Alternatively, 200 ng/ml of rapamycin, an inhibitor of the TOR kinase, dissolved in ethanol-Tween 40 (9:1) can be directly added to rich culture media to induce autophagy (Suzuki *et al.*, 2001, 2002, 2007). If rapamycin is used, treatment for 1 h is sufficient to induce autophagy.

   c. Other media can be used if the trafficking of the fluorescent chimeras is analyzed in situations where selective types of autophagy, such as pexophagy (selective degradation of peroxisomes) or mitophagy (selective degradation of mitochondria), are active.
2. In general, 1 ml of culture is sufficient for live-cell imaging. This needs to be centrifuged at 8000 rpm for 1 min and almost all the supernatant removed, leaving 50–100 μl of liquid in the tube to avoid disturbing the cell pellet. Cells must be washed with 1 ml of either SMD medium or water, if they are in YPD, because this medium possesses some autofluorescent components that worsen the quality of the acquired images. In particular, adenine may cause problems, particularly when utilizing red fluorescent fusions. Again, remove the supernatant fraction and leave approximately 50–100 μl for the final resuspension.
3. After vortexing the cell pellet, 1–2 μl of the resuspension can be directly spotted onto a glass slide and the fluorescence image immediately visualized with either a confocal or a fluorescence microscope.

Instead of the live-cell imaging, mild conditions can be used to fix cells without destroying the different fluorescent proteins before being photographed (Reggiori *et al.*, 2005b).

1. Take 2.5 ml of cells grown or starved as described for the live-cell imaging.
2. Harvest the cells by centrifugation at 5000 rpm for 2 min and discard the supernatant.
3. Resupend the pellet in 1 ml of fixation buffer (50 m*M* of KH_2PO_4, pH 8.0; 1 μM of $MgCl_2$; 1.5% formaldehyde) in a 1.5-ml microcentrifuge tube.
4. Incubate the suspension at room temperature for 30 min with occasional mixing.
5. Centrifuge the samples at 5000 rpm for 2 min and discard the supernatant.
6. Resuspend the pellet in 1 ml of 50 m*M* glycine in phosphate buffered saline (PBS; 10 m*M* of Na_2HPO_4; 2 m*M* of KH_2PO_4, pH 7.4; 137 m*M* of NaCl; 2.7 m*M* KCl) with a pipette and let the tube stand at room temperature for 15 min.
7. Centrifuge at 5000 rpm for 2 min and wash the cells twice with 1 ml of fixation buffer without formaldehyde (50 m*M* of KH_2PO_4, pH 8.0; 1 μM of $MgCl_2$).
8. Remove almost all the supernatant, leaving 50–100 μl of liquid in the tube. After vortexing the cell pellet, 1–2 μl of the resuspension can be directly spotted onto a glass slide and the fluorescence image immediately visualized with either a confocal or a fluorescence microscope.

5. Atg9 Trafficking and the TAKA Assay

Atg9 is a transmembrane protein essential for double-membrane vesicle formation and plays a key role in organizing and coordinating the Atg machinery at the PAS (Noda *et al.*, 2000; Suzuki *et al.*, 2007). The exact molecular function of Atg9 is unknown, but because of its association with lipid bilayers, this protein is at least in part likely to be involved in delivering membranes to the forming autophagosomes (Noda *et al.*, 2000; Reggiori *et al.*, 2005b). This idea is supported by the fact that, in yeast, Atg9 cycles between the PAS and several cytoplasmic sites, some of which are in close proximity of mitochondria (Reggiori *et al.*, 2004, 2005b). This almost unique distribution among Atg components can be revealed by fusing Atg9 with a fluorescent protein (Fig. 4.3) (Reggiori *et al.*, 2004). Importantly, Atg9 is almost entirely accumulated at the PAS in the absence of genes such as *ATG1*, *ATG2*, *ATG13*, and *ATG18*, which appear to be essential for its retrograde transport from this structure (Reggiori *et al.*, 2004). Condensation of fluorescent chimeric Atg9 proteins into a single bright dot can be used to easily detect this trafficking defect (see Fig. 4.3) (Reggiori *et al.*, 2004).

It has recently been shown that the Atg9 accumulation in the *atg1*Δ strain can be prevented by the deletion of genes epistatic to *ATG1* and necessary for Atg9 delivery to the PAS (Transport of Atg9 after knocking out Atg1 [TAKA] assay; Cheong *et al.*, 2005; Shintani *et al.*, 2002). In these

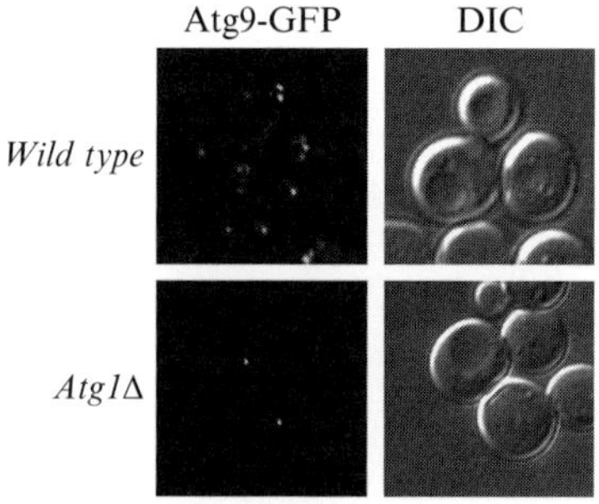

Figure 4.3 Live-cell imaging of strains expressing Atg9-GFP. Wild-type (FRY354, *leu2Δ0 met15Δ0 ura3Δ0 his3Δ1 ATG9-GFP::HIS5 S.p.*) and *atg1*Δ (FRY354, *leu2Δ0 met15Δ0 ura3Δ0 his3Δ1 atg1Δ::KAN ATG9-GFP::HIS5 S.p.*) cells were grown at 30°C in rich medium to an early logarithmic phase before being photographed. The images were captured and deconvolved with a DeltaVision RT fluorescence microscope system. In wild-type cells, Atg9-GFP signal is distributed to several puncta, one of which represents the PAS. In the *Δ* mutant, however, this fluorescent chimera accumulates at this latter location and consequently Atg9-GFP appears as a single brilliant dot. When the TAKA assay is performed and double mutants are analyzed, two results can be expected. If the second mutation is epistatic to *atg1*Δ, the signal will look like the wild type (except that none of the dots will be at the PAS). If the second protein acts at the same time or after Atg1, however, the phenotype will be similar to the *atg1*Δ knockout. DIC, differential interference contrast.

double mutants, fluorescent Atg9 remains disseminated in several punctate structures. When imaged in the presence of a marker such as RFP-Ape1, it is apparent that none of the GFP-Atg9 dots are localized at the PAS, which indicates that the analyzed genes are essential for Atg9 transport to this site. Interestingly, the TAKA assay also can be used to reveal in which situations a certain gene is required for Atg9 transport in the Cvt pathway versus autophagy. To do that, the double mutant expressing fluorescent Atg9 has to be visualized after having been grown in both rich and starved medium (He *et al.*, 2006; Reggiori and Klionsky, 2006; Shintani and Klionsky, 2004). This approach has enabled us to demonstrate, for example, that Atg11 is indispensable for Atg9 delivery to the PAS and successive double-membrane vesicle formation in situations when the Cvt pathway is operational but not when autophagy is active (He *et al.*, 2006; Shintani and Klionsky, 2004). Finally, the TAKA assay can also be used if the gene of interest is essential and a thermosensitive (ts) allele is available (Reggiori and Klionsky, 2006). In this case, it is possible to engineer a strain in such a way that it carries both the ts allele of the gene of interest and an *atg1^{ts}* allele. When this double mutant strain is incubated at restrictive temperatures, the simultaneous inactivation of the proteins encoded by the two alleles allows performing the TAKA assay.

6. Creation of the Strains for the Analysis of Atg9 Trafficking

In the past, plasmids overproducing GFP-tagged Atg9 have been used (Chang and Huang, 2007; Kim *et al.*, 2002; Lang *et al.*, 2000; Noda *et al.*, 2000). The most accurate way to analyze this protein, however, is by PCR-based integration of *GFP* at the 3′ end of *ATG9* to generate strains expressing a C-terminal fusion under the control of its native promoter (Longtine *et al.*, 1998). When Atg9-GFP is overproduced in an *atg1Δ* mutant, additional punctate structures are observed other than the bright one representing the PAS, which makes the expected phenotype less evident (unpublished observations). If co-localization with another protein is necessary, *ATG9* can also be genomically tagged with either *YFP* (He *et al.*, 2006; Legakis *et al.*, 2007; Reggiori and Klionsky, 2006; Reggiori *et al.*, 2004; Shintani and Klionsky, 2004) or *mRFP* (He *et al.*, 2006; Legakis *et al.*, 2007; Reggiori *et al.*, 2005a).

1. Strains are generated by integration of *GFP* (or other fluorescent tags) at the 3′ end of *ATG9* using the PCR-based method described by Longtine *et al.* (1998).
   a. The putative positive clones selected on the appropriate plates have to be first imaged with a fluorescence microscope. Cells revealing fluorescent

puncta distributed throughout the cytoplasm indicate the possible integration at the exact chromosomal locus (Reggiori *et al.*, 2004).

   b. Confirmation of the correctness of the chromosomal fusions can be obtained by either analysis of the *ATG9* locus by PCR (Gueldener *et al.*, 2002; Longtine *et al.*, 1998) or detection of Atg9-GFP by Western blot.
   c. The functionality of the Atg9 chimeras can be determined by analyzing prApe1 transport by Western blot (Kim *et al.*, 1997; Noda *et al.*, 2000).
2. To generate *ATG1* knockout strains, the entire coding regions of this gene can be disrupted by replacement with either an auxotrophic marker (*LEU2*, *TRP1*, or *URA3*) or an antibiotic resistance gene (kanamycin or phleomycin) using PCR primers containing at least $\approx$50 bases of identity to the regions flanking the open reading frame (Gueldener *et al.*, 2002; Longtine *et al.*, 1998).
   a. The deletion of *ATG1* can be assessed in three different ways: (1) analysis of cell extracts by Western blot using specific anti-Atg1 antibodies, (2) PCR analysis examination of the chromosomal *ATG1* locus by PCR (Gueldener *et al.*, 2002; Longtine *et al.*, 1998), and (3) investigation of prApe1 maturation by Western-blot (Kim *et al.*, 1997).
   b. If there is the need to generate a double thermosensitive mutant to perform the TAKA assay, the obtained cells have then to be transformed with the plasmid expressing Atg1^{ts} (Suzuki *et al.*, 2001).
3. The obtained strains are grown and imaged as described above for the visualization of the fluorescence signals. For strains expressing Atg1^{ts}, cells are grown at the permissive temperature of 24 °C and then transferred to 37 °C for at least 1 h before proceeding with the imaging (Reggiori and Klionsky, 2006).

To create the knockout strains required for the analysis of Atg9 trafficking, the entire coding regions can be disrupted by replacement with either an auxotrophic marker (*LEU2*, *TRP1*, or *URA3*) or an antibiotic resistance gene (kanamycin or phleomycin) using PCR primers containing at least $\approx$50 bases of identity to the regions flanking the open reading frame (Gueldener *et al.*, 2002; Longtine *et al.*, 1998).

ACKNOWLEDGMENTS

The authors thank Aniek van der Vaart and Muriel Mari for the critical reading of the manuscript. T.S. is supported by Grants-in-Aid for Scientific Research from the Ministry of Education, Culture, Sports, Science and Technology of Japan (19580391). F.R. is supported by the Netherlands Organization for Health Research and Development (ZonMW-VIDI-917.76.329) and by the Utrecht University (High Potential grant).

REFERENCES

Baba, M., Osumi, M., Scott, S. V., Klionsky, D. J., and Ohsumi, Y. (1997). Two distinct pathways for targeting proteins from the cytoplasm to the vacuole/lysosome. *J. Cell Biol.* **139,** 1687–1695.

Campbell, R. E., Tour, O., Palmer, A. E., Steinbach, P. A., Baird, G. S., Zacharias, D. A., and Tsien, R. Y. (2002). A monomeric red fluorescent protein. *Proc. Natl. Acad. Sci. USA* **99,** 7877–7882.

Chang, C. Y., and Huang, W.-P. (2007). Atg19 mediates a dual interaction cargo sorting mechanism in selective autophagy. *Mol. Biol. Cell* **18,** 919–929.

Cheong, H., Yorimitsu, T., Reggiori, F., Legakis, J. E., Wang, C. W., and Klionsky, D. J. (2005). Atg17 regulates the magnitude of the autophagic response. *Mol. Biol. Cell* **16,** 3438–3453.

Ghaemmaghami, S., Huh, W. K., Bower, K., Howson, R. W., Belle, A., Dephoure, N., O'Shea, E. K., and Weissman, J. S. (2003). Global analysis of protein expression in yeast. *Nature* **425,** 737–741.

Gueldener, U., Heinisch, J., Koehler, G. J., Voss, D., and Hegemann, J. H. (2002). A second set of *loxP* marker cassettes for Cre-mediated multiple gene knockouts in budding yeast. *Nucleic Acids Res.* **30,** e23.

He, C., Song, H., Yorimitsu, T., Monastyrska, I., Yen, W. L., Legakis, J. E., and Klionsky, D. J. (2006). Recruitment of Atg9 to the preautophagosomal structure by Atg11 is essential for selective autophagy in budding yeast. *J. Cell Biol.* **175,** 925–935.

Ichimura, Y., Kirisako, T., Takao, T., Satomi, Y., Shimonishi, Y., Ishihara, N., Mizushima, N., Tanida, I., Kominami, E., Ohsumi, M., Noda, T., and Ohsumi, Y. (2000). A ubiquitin-like system mediates protein lipidation. *Nature* **408,** 488–492.

Kim, J., Dalton, V. M., Eggerton, K. P., Scott, S. V., and Klionsky, D. J. (1999). Apg7p/Cvt2p is required for the cytoplasm-to-vacuole targeting, macroautophagy, and peroxisome degradation pathways. *Mol. Biol. Cell* **10,** 1337–1351.

Kim, J., Huang, W. P., and Klionsky, D. J. (2001). Membrane recruitment of Aut7p in the autophagy and cytoplasm to vacuole targeting pathways requires Aut1p, Aut2p, and the autophagy conjugation complex. *J. Cell Biol.* **152,** 51–64.

Kim, J., Huang, W. P., Stromhaug, P. E., and Klionsky, D. J. (2002). Convergence of multiple autophagy and cytoplasm to vacuole targeting components to a perivacuolar membrane compartment prior to de novo vesicle formation. *J. Biol. Chem.* **277,** 763–773.

Kim, J., Scott, S. V., Oda, M. N., and Klionsky, D. J. (1997). Transport of a large oligomeric protein by the cytoplasm to vacuole protein targeting pathway. *J. Cell Biol.* **137,** 609–618.

Kirisako, T., Baba, M., Ishihara, N., Miyazawa, K., Ohsumi, M., Yoshimori, T., Noda, T., and Ohsumi, Y. (1999). Formation process of autophagosome is traced with Apg8/Aut7p in yeast. *J. Cell Biol.* **147,** 435–446.

Kirisako, T., Ichimura, Y., Okada, H., Kabeya, Y., Mizushima, N., Yoshimori, T., Ohsumi, M., Takao, T., Noda, T., and Ohsumi, Y. (2000). The reversible modification regulates the membrane-binding state of Apg8/Aut7 essential for autophagy and the cytoplasm to vacuole targeting pathway. *J. Cell Biol.* **151,** 263–276.

Labbé, S., and Thiele, D. J. (1999). Copper ion inducible and repressible promoter systems in yeast. *Methods Enzymol.* **306,** 145–153.

Lang, T., Reiche, S., Straub, M., Bredschneider, M., and Thumm, M. (2000). Autophagy and the Cvt pathway both depend on *AUT9*. *J. Bacteriol.* **182,** 2125–2133.

Legakis, J. E., Yen, W. L., and Klionsky, D. J. (2007). A cycling protein complex required for selective autophagy. *Autophagy* **3,** 422–432.

Longtine, M. S., McKenzie, A., III., Demarini, D. J., Shah, N. G., Wach, A., Brachat, A., Philippsen, P., and Pringle, J. R. (1998). Additional modules for versatile and economical PCR-based gene deletion and modification in *Saccharomyces cerevisiae*. *Yeast* **14,** 953–961.

Meiling-Wesse, K., Barth, H., Voss, C., Eskelinen, E. L., Epple, U. D., and Thumm, M. (2004). Atg21 is required for effective recruitment of Atg8 to the preautophagosomal structure during the Cvt pathway. *J. Biol. Chem.* **279,** 37741–37750.

Meiling-Wesse, K., Epple, U. D., Krick, R., Barth, H., Appelles, A., Voss, C., Eskelinen, E. L., and Thumm, M. (2005). Trs85 (Gsg1), a component of the TRAPP complexes is required for the organization of the preautophagosomal structure during selective autophagy via the Cvt pathway. *J. Biol. Chem.* **280,** 33669–33678.

Noda, T., Kim, J., Huang, W. P., Baba, M., Tokunaga, C., Ohsumi, Y., and Klionsky, D. J. (2000). Apg9p/Cvt7p is an integral membrane protein required for transport vesicle formation in the Cvt and autophagy pathways. *J. Cell Biol.* **148,** 465–480.

Reggiori, F. (2006). 1. Membrane origin for autophagy. *Curr. Top. Dev. Biol.* **74,** 1–30.

Reggiori, F., and Klionsky, D. J. (2005). Autophagosomes: Biogenesis from scratch? *Curr. Opin. Cell Biol.* **17,** 415–422.

Reggiori, F., and Klionsky, D. J. (2006). Atg9 sorting from mitochondria is impaired in early secretion and VFT-complex mutants in *Saccharomyces cerevisiae*. *J. Cell Sci.* **119,** 2903–2911.

Reggiori, F., Monastyrska, I., Shintani, T., and Klionsky, D. J. (2005a). The actin cytoskeleton is required for selective types of autophagy, but not nonspecific autophagy, in the yeast *Saccharomyces cerevisiae*. *Mol. Biol. Cell* **16,** 5843–5856.

Reggiori, F., Shintani, T., Nair, U., and Klionsky, D. J. (2005b). Atg9 cycles between mitochondria and the pre-autophagosomal structure in yeasts. *Autophagy* **1,** 101–109.

Reggiori, F., Tucker, K. A., Stromhaug, P. E., and Klionsky, D. J. (2004). The Atg1-Atg13 complex regulates Atg9 and Atg23 retrieval transport from the pre-autophagosomal structure. *Dev. Cell* **6,** 79–90.

Shaner, N. C., Campbell, R. E., Steinbach, P. A., Giepmans, B. N., Palmer, A. E., and Tsien, R. Y. (2004). Improved monomeric red, orange and yellow fluorescent proteins derived from *Discosoma* sp. red fluorescent protein. *Nat. Biotechnol.* **22,** 1567–1572.

Shintani, T., Huang, W. P., Stromhaug, P. E., and Klionsky, D. J. (2002). Mechanism of cargo selection in the cytoplasm to vacuole targeting pathway. *Dev. Cell* **3,** 825–837.

Shintani, T., and Klionsky, D. J. (2004). Cargo proteins facilitate the formation of transport vesicles in the cytoplasm to vacuole targeting pathway. *J. Biol. Chem.* **279,** 29889–29894.

Sikorski, R. S., and Hieter, P. (1989). A system of shuttle vectors and yeast host strains designed for efficient manipulation of DNA in Saccharomyces cerevisiae. *Genetics* **122,** 19–27.

Suzuki, K., Kamada, Y., and Ohsumi, Y. (2002). Studies of cargo delivery to the vacuole mediated by autophagosomes in *Saccharomyces cerevisiae*. *Dev. Cell* **3,** 815–824.

Suzuki, K., Kirisako, T., Kamada, Y., Mizushima, N., Noda, T., and Ohsumi, Y. (2001). The pre-autophagosomal structure organized by concerted functions of *APG* genes is essential for autophagosome formation. *EMBO J.* **20,** 5971–5981.

Suzuki, K., Kubota, Y., Sekito, T., and Ohsumi, Y. (2007). Hierarchy of Atg proteins in pre-autophagosomal structure organization. *Genes Cells* **12,** 209–218.

Vida, T. A., and Emr, S. D. (1995). A new vital stain for visualizing vacuolar membrane dynamics and endocytosis in yeast. *J. Cell Biol.* **128,** 779–792.

CHAPTER FIVE

Measuring Macroautophagy in *S. cerevisiae*: Autophagic Body Accumulation and Total Protein Turnover

Tanja Prick* *and* Michael Thumm*

Contents

Abstract

Macroautophagy has been implicated in various physiological functions and severe human diseases. Accordingly, there is a high interest in determining macroautophagy both qualitatively and quantitatively. In this chapter we discuss how macroautophagy can be followed morphologically in the yeast *Saccharomyces cerevisiae* using light microscopy. To quantitatively measure macroautophagy, we further present two protocols for the determination of total protein turnover.

1. Introduction

During starvation-induced macroautophagy, superfluous cytoplasmic material and organelles are transported for degradation to lysosomes (vacuoles). The identification and characterization of Atg proteins in higher

* Georg-August-University, Center of Biochemistry and Molecular Cell Biology, Goettingen, Germany

Methods in Enzymology, Volume 451
ISSN 0076-6879, DOI: 10.1016/S0076-6879(08)03205-9

and lower eukaryotes significantly broadened the knowledge about the physiological functions of autophagy and its role in diseases (Levine and Kroemer, 2008; Rubinsztein *et al.*, 2007). Autophagy acts as a general host defense mechanism against intracellular bacteria and viruses and is further involved in the presentation of antigens via MHC class II receptors. It is implicated in nonapoptotic programmed cell death and autophagic cell death, and it plays an important role in aging. There is also evidence that autophagy helps clear aggregate-prone mutant proteins and thus protects against neurodegenerative diseases such as Alzheimer's, Huntington's, and Parkinson's. In agreement with a role in tumor suppression, defects in autophagy have further been linked with the development of cancer. As a result of the multiple roles of autophagy, there is growing interest in determining the occurrence of macroautophagy both qualitatively and quantitatively. The autophagic machinery is highly conserved among mammals, plants, and yeasts. In fact, most of the autophagy (Atg) proteins were initially identified in yeast (Suzuki and Ohsumi, 2007; Xie and Klionsky, 2007). Therefore it might be fruitful to analyze the relevance of a protein for autophagy in a model organism such as yeast. Especially in the yeast *Saccharomyces cerevisiae,* a multitude of point or deletion mutants are readily available. A series of methods to measure autophagy in *S. cerevisiae* have been established. Probably the easiest and most rapid way to trace macroautophagy is the direct visualization of intravacuolar autophagic bodies by light microscopy. Macroautophagy uses transport vesicles (autophagosomes), which are limited by two membranes (Suzuki and Ohsumi, 2007; Xie and Klionsky, 2007). Accordingly, the fusion of an autophagosome with the vacuole leads to the release of a membrane-enclosed autophagic body into the vacuole lumen. Autophagic bodies have a diameter between 300 and 900 nm and thus can be monitored with Nomarski (differential interference contrast) optics or by electron microscopy (Takeshige *et al.*, 1992). In some genetic backgrounds of *S. cerevisiae*, the vacuoles are difficult to recognize by Nomarski optics. This problem can be overcome by staining the vacuolar membrane with the red fluorescent lipophilic dye FM 4-64 (also see chapter 7) and by expression of GFP-Atg8. Atg8 is an excellent marker of autophagosomes and autophagic bodies. An autophagy-specific ubiquitin-like conjugation system covalently links Atg8 to the membrane lipid phosphatidylethanolamine (PE) (Nakatogawa *et al.*, 2007). This Atg8–PE conjugate is selectively incorporated into autophagosomes and autophagic bodies. After intravacuolar lysis of the autophagic bodies, their content is exposed to the vacuolar hydrolases. Some proteins are highly resistant against vacuolar proteolysis, and their appearance in the vacuole is a good measure of the autophagic flux. For example, GFP-Atg8 breakdown in the vacuole yields free GFP, whose accumulation can be followed in Western blots and correlated with the autophagic rate (Meiling-Wesse *et al.*, 2002). Another example for this strategy is the use of an enzymatically inactive cytosolic variant of vacuolar alkaline phosphatase (Pho8Δ60), which after maturation inside the vacuole becomes enzymatically active (Noda *et al.*, 1995). Probably the most direct way to monitor macroautophagy is to measure the total protein breakdown after *in vivo* radiolabeling of all intracellular proteins.

This chapter describes common yeast-specific protocols to morphologically study macroautophagy using different microscopy techniques and how to quantitatively determine the autophagic rate by measuring protein turnover of long-lived proteins.

2. Qualitative Measurement of Macroautophagy in *S. cerevisiae*

2.1. Accumulation of autophagic bodies inside the vacuole

In wild-type *S. cerevisiae* cells, autophagic bodies are rapidly degraded after their appearance in the vacuole. It was an important observation that autophagic bodies accumulated during a 4-h starvation period in nitrogen-free medium in the vacuoles of yeast cells either deficient for vacuolar proteinase A or B (Takeshige *et al.*, 1992). The autophagic bodies have diameters ranging from 300 to 900 nm and can be visualized by light microscopy with Nomarski optics. As a result of Brownian movement, the autophagic bodies characteristically dance around within the vacuole. In wild-type cells the degradation of autophagic bodies can easily be hindered by the addition of the proteinase B inhibitor phenylmethylsulfonylfluoride (PMSF) to the starvation medium. The lack of autophagic bodies inside the vacuole of a yeast mutant starved for nitrogen in the presence of PMSF therefore indicates a defect in the autophagic process (Fig. 5.1).

2.1.1. Visualization of autophagic bodies by Nomarski optics

1. Grow the yeast cells logarithmically ($O.D._{600}$ = 0.5–0.8) in YPD medium (1% yeast extract, 2% peptone, and 2% glucose) while shaking at 220 rpm/min at 30 °C. We typically use test tubes and culture volumes of 1–2 ml.
2. Harvest the cells at 1600×*g* for 5 min.
3. Wash the cells twice with SD(-N) medium (0.17% yeast nitrogen base without amino acids and without ammoniumsulfate, and 2% glucose).
4. Resuspend the cells in SD(-N) medium containing 1 m*M* of PMSF (100 m*M* of stock solution in 96% ethanol, always prepared freshly).
5. Incubate the cells for 2 to 4 h at 30 °C while shaking (220 rpm/min).
6. Visualize the cells with Nomarski optics using a light microscope equipped with a 100x objective.

Notes: Always include wild-type cells as a positive control and autophagy-deficient cells as a negative control.

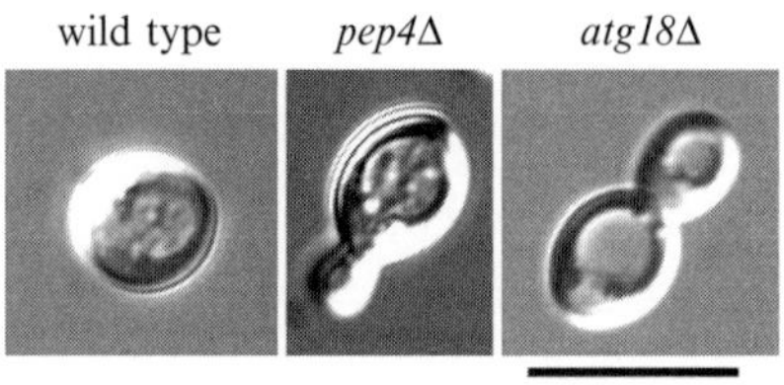

Figure 5.1 Following the accumulation of intravacuolar autophagic bodies in light microscopy. Starvation for 4 h in SD(-N) medium in the presence of the proteinase B inhibitor PMSF leads to the accumulation of autophagic bodies inside the vacuole of wild-type yeast cells. This provides an easy way to monitor autophagy with light microscopy. In yeast, vacuolar proteinases A (encoded by the *PEP4* gene) and B (*PRB1* gene) are required for lysis of autophagic bodies. Therefore *pep4*Δ cells accumulate autophagic bodies inside their vacuoles, even in the absence of PMSF. The *atg18*Δ mutant is defective in autophagy and does not accumulate autophagic bodies, thus serving as a negative control. The bar corresponds to 10 μm.

The rapid movement of the autophagic bodies hampers their documentation. A solution is to take a series of pictures using time-lapse microscopy.

In water, PMSF is stable for only several hours, dependent on the pH. Therefore starvation in SD-(N) medium for more than 4 h is not practical. If extended starvation times are necessary, Pefabloc (4-[2-aminoethyl]-benzenesulfonylfluoride hydrochloride; Sigma-Aldrich) can be used as an alternate proteinase B inhibitor. It is water soluble and more stable against hydrolysis.

2.2. Visualization of autophagic bodies by fluorescence microscopy

In some genetic backgrounds the vacuoles are rather small and hard to detect by light microscopy. In such cases the staining of the vacuolar membrane with the lipophilic styryl dye, *N*-(3-triethylammoniumpropyl)-4-(6-[4-(diethylamino)phenyl]hexatrienyl)pyridinium dibromide (FM 4-64; Invitrogen), is helpful. FM 4-64 is a vital stain (excitation/emission maxima ≈515/640 nm), which in yeast cells follows the endocytic route and finally stains the vacuolar membranes with high selectivity (Vida and Emr, 1995). The visualization of the autophagic bodies can be further improved using the autophagosomal marker GFP-Atg8 (Huang *et al.*, 2000; Kirisako *et al.*, 1999). GFP-Atg8 can be expressed from the centromeric plasmid pGFP-*ATG8*, which encodes a GFP-Atg8 fusion protein under control of the native *ATG8* promoter (Suzuki *et al.*, 2001).

2.2.1. Staining of the vacuolar membrane

1. Grow the yeast cells logarithmically (O.D.$_{600}$ = 0.5–0.8) in 50 ml of YPD or selection medium while shaking (220 rpm/min) at 30 °C.

2. Harvest 20 OD_{600} units of cells by centrifugation at 1500×*g* for 5 min and resuspend in 1 ml of YPD medium in a microcentrifuge tube.
3. Add 2 μl of a 16 m*M* FM 4-64 stock solution (1 mg/100 μl in DMSO) and incubate the cells while shaking for 30 min at 30 °C.
4. Harvest cells at 1500×*g* for 5 min.
5. Wash the cells with SD(-N) medium twice.
6. Resuspend the cells in SD(-N) medium with PMSF (see section 2.1).
7. Incubate the cells for 3 to 5 h at 30 °C while shaking (220 rpm/min).
8. Visualize the cells using a fluorescence microscope equipped with a 100x objective and a CY3 filter set.

To examine the autophagic bodies, which accumulate inside the vacuole under nutrient deprivation conditions more closely, electron microscopy is the preferred method. Autophagic bodies can be seen nicely both in transmission and freeze fracture electron microscopy (Baba *et al.*, 1994; Baba *et al.*, 1995; Lang *et al.*, 2000). Details of these methods will be presented in chapter 10.

3. Quantitative Measurement of Macroautophagy in *S. cerevisiae*

3.1. Proteolysis of long-lived proteins in yeast

Compared to the cytoplasmic ubiquitin-proteasome system, which predominantly degrades short-lived proteins, autophagy is supposed to account for the majority of degradation of long-lived proteins (Takeshige *et al.*, 1992; Teichert *et al.*, 1989). For this reason, measurement of bulk degradation of long-lived proteins is often used to monitor autophagic activity. In principle the cells are incubated with a radioactively labeled amino acid to label all cellular proteins. A subsequent short incubation with an excess of the unlabeled amino acid stops the incorporation of label and allows the degradation of all the labeled, short-lived proteins, which results in the relatively specific labeling of only long-lived proteins. This is the zero time point to determine autophagic protein degradation and therefore an aliquot must be taken to determine the amount of incorporated radioactivity. Then the cells are incubated to allow the degradation of long-lived proteins, and samples are taken at different time points. After washing the cell pellets, the cellular proteins are precipitated using trichloracetic acid (TCA) and separated in pellet and supernatant fractions. The TCA-soluble radioactivity, which corresponds to the liberated amino acid label and small soluble peptides, is determined using a scintillation counter. Here, we present two different protocols to trace the total protein breakdown. Protocol A is specifically adapted for the use of L-4,5-[^{3}H]-leucine for labeling

(Prick *et al.*, 2006). Protocol B uses labeling with L-[^{35}S]-methionine (Straub *et al.*, 1997; Teichert *et al.*, 1989). In our hands, both protocols give reliable results, but for people not familiar with these procedures, protocol A seems to be easier to follow.

3.2. Measurement of total protein breakdown in yeast

3.2.1. Using labeling with L-4,5-[^{3}H]-leucine

3.2.1.1. Day 1: Yeast preculture

- Start a preculture from a single colony in 5 ml of YPD medium at 30 °C with continuous shaking (22S0 rpm/min) for 24 h.

3.2.1.2. Day 2: In vivo *^{3}H-Labeling of yeast proteins*

- Use the preculture to inoculate a 15-ml culture in YPD medium in a 50-ml Erlenmeyer flask and add 1 μCi/ml of L-4,5-[^{3}H]-leucine (164 Ci/mmol) and incubate at 30 °C with continuous shaking (220 rpm/min) until an O.D.$_{600}$ of 0.5–0.8 is reached. The culture should grow for about 16 h until this O.D. is reached. Because yeast mutant strains vary in their growth rates, the required dilution factor of the preculture must be determined in preliminary tests.

3.2.1.3. Day 3: Sampling

- Determine the exact O.D.$_{600}$ by using disposable cuvettes.
- At an O.D.$_{600}$ corresponding to 0.5 to 0.8 transfer an aliquot of 1 O.D.$_{600}$ unit (1 ml of cells at O.D.$_{600}$ = 1.0 is equivalent to 1 O.D. unit) of each culture into a 1.7-ml microcentrifuge tube and harvest the cells by centrifugation at 13,000 rpm for 2 min at room temperature.
- Remove the supernatant fraction and collect it separately in a new tube.
- Supernatant fractions and cell pellets should be stored at −20 °C until the end of sampling.
- Harvest the residual cell suspension (approximately 14 ml) by centrifugation at 3000 rpm for 5 min at room temperature using disposable 50-ml tubes.
- Discard the supernatant, which should be treated as radioactive waste. Add 15 ml of sterile distilled water to the cell pellet, resuspend, and harvest the cells as above twice.
- Resuspend cells in SD(-N) medium with the addition of 100 μM of nonradioactive leucine and incubate at 30 °C up to 6 h with continuous shaking (220 rpm/min).
- Transfer hourly 1 O.D.$_{600}$ unit of cells from each culture into a microcentrifuge tube and harvest cells by centrifugation at 13,000 rpm for 2 min at room temperature.
- Transfer each supernatant into a new tube.

- Supernatants and cell pellets should be stored at −20 °C until the end of sampling.
- To destroy the cell walls, freeze all samples at −80 °C at the end of sampling.

3.2.1.4. Day 4: Sonification, TCA precipitation, and scintillation counting

- Thaw all samples.
- Resuspend each cell pellet in 1 ml of sterile distilled water.
- Sonicate samples (culture supernatants and cell suspensions) for at least 30 s at room temperature using a water-bath sonicator.
- Add TCA to each sample (final concentration 10%), vortex, and place for 10 min on ice.
- Centrifuge for 5 min at 4 °C and 13,000 rpm to pellet TCA-precipitated proteins.
- Transfer supernatant fractions to another microcentrifuge tube.
- Resuspend all TCA pellets in 1 ml of sterile distilled water.
- Measure radioactivity in all four fractions (i.e., the TCA-precipitated, resuspended pellet fraction, the TCA-precipitated supernatant fraction, and the resultant supernatant fractions from both) by scintillation counting with Soluene 350 (Packard) using a TRI-CARB 2900TR counter (Packard). To do this, from each supernatant fraction a 500-μl aliquot is mixed with 4 ml of Soluene 350 in a small scintillation vial. To calculate the complete radioactivity of each supernatant fraction, its specific volume must be taken into account. Each resuspended pellet fraction is put including the microcentrifuge tube into a large scintillation vial and then 10 ml of Soluene 350 is added and thoroughly mixed.
- Using an internal standard, the different quenching effects of the TCA- and NCS-II-containing solutions are corrected.
- Calculate the release of labeled amino acid as the percentage of incorporated radioactivity at time zero after scintillation counting using the following formula: Percentage of released radioactivity = radioactivity in supernatant fractions/sum of radioactivity (all four fractions of the time point zero).

3.2.2. Protocol using labeling with L-[^{35}S]-methionine

- Inoculate a preculture from a single colony in 5 ml of YPD medium at 30 °C for 24 h with continuous shaking (220 rpm/min).
- Inoculate 10 ml of yeast-labeling medium (detailed subsequently) from the preculture in such a way that after 20 h of shaking at 220 rpm at 30 °C a cell density of 5×10^7 cells/ml is reached. Depending on the strain background, we find a dilution of the preculture of 1:250 to 1:1500 to be sufficient.
- Let the culture grow for 6 h at 30 °C while shaking at 220 rpm.

- Add 3.7 MBq of L-[^{35}S]-methionine (Amersham) (1000 Ci/mmol) and grow for an additional 14 h at 30 °C with shaking at 220 rpm.
- Harvest the cells by centrifugation for 5 min at 1200×*g* and wash three times with starvation medium. Then resuspend in 10 ml of starvation medium plus 10 m*M* of nonradioactive methionine and incubate further at 30 °C while shaking at 220 rpm.
- Take a 1-ml sample immediately after the transfer to starvation medium with nonradioactive methionine, place in a 1.7-ml microcentrifuge tube, and add 100 μl of 110% (w/v) TCA and incubate for at least 4 h on ice. Take additional 1-ml samples hourly over 8 h and treat them accordingly.
- To determine the acid-soluble radioactivity, the samples are then centrifuged for 5 min at 14,000×*g*. Then 900 μl of the supernatant fraction is incubated and mixed with 5 ml of liquid scintillator (Rotiszint eco plus, Roth, Karlsruhe, Germany).
- To measure the total incorporated radioactivity, the pellets of the 0-h samples are washed five times with starvation medium containing 10% TCA. Subsequently, the pellets are washed two times with a 1:1 mixture of ethanol and ether. After drying the pellets for 30 min at 40 °C in air, they are dissolved at 40 °C with vigorous shaking in 1 ml of NCS-II tissue solubilizer (90% in water; Amersham, Braunschweig, Germany). Then 900 μl of this solution is mixed for 1 h with 5 ml of liquid scintillator. Radioactivity is afterward measured using a liquid scintillation counter. Using an internal standard, the different quenching effects of the TCA- and NCS-II-containing solutions are corrected.
- To calculate the total protein breakdown rate, the increase of TCA-soluble radioactivity is divided by the total radioactivity of the 0-h sample.

The yeast-labeling medium must be sulfur-free to allow appropriate incorporation of radiolabeled methionine. It consists of 0.17% yeast nitrogen base without amino acids and ammonium sulfate, 2% proline, and 2% glucose and the appropriate auxotrophic nutrients except methionine and cysteine.

As with the first protocol, always include wild-type cells as a positive control and autophagy-deficient cells as a negative control. As an additional negative control *pep4*Δ cells lacking vacuolar proteinase A, and therefore numerous additional hydrolytic activities, are suitable. Fig.5.2 shows a typical outcome using the protocol in section 3.2.1. An example of an experiment using the protocol in section 3.2.2 is shown in Straub *et al.* (1997; see Fig. 4). Under the conditions used in this study, the total protein breakdown rate in wild-type cells corresponds to approximately 2% per hour. In agreement with a block in vacuolar protein breakdown the proteolysis rate in autophagy-deficient mutant cells and in *pep4Δ* cells, lacking vacuolar proteinase A, are similar and are significantly reduced compared to wild-type cells. The residual proteolysis rate of these mutant cells is due to the action of the cytoplasmic proteasome.

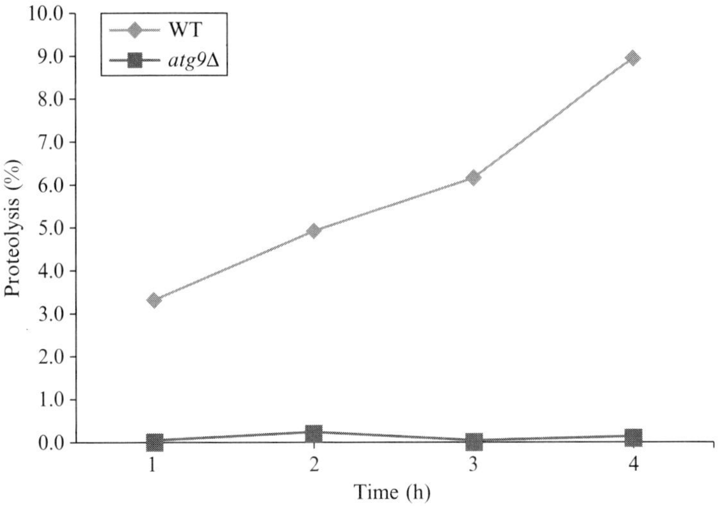

Figure 5.2 Measurement of the total proteolysis rate. The proteolysis rate of yeast cells was determined by the measurement of radioactivity released from yeast cells, which had been labeled for 16 h with 1 μCi/ml of [^{3}H]leucine as outlined in the protocol in section 3.2.1. The cells were starved in nitrogen-free SD(−N) medium, and each hour 1 $O.D._{600}$ unit of cells was harvested. The proteolysis rate was determined by liquid-scintillation counting and was calculated as TCA-soluble radioactivity in the supernatant fractions divided by the total radioactivity. Wild-type (WT) and *atg9Δ* cells, deficient in autophagy, were analyzed.

4. Conclusion

To rapidly check the occurrence of autophagy in *S. cerevisiae* cells, the visualization of autophagic bodies is a reliable method, especially when combined with staining of the vacuolar membrane with FM 4-64 and the labeling of the autophagic bodies by GFP-Atg8. This method requires only several hours and can be applied simultaneously to a series of mutant strains. For the analysis of large numbers of strains, the culture and starvation of yeast cells can easily be done in 24-well microtiter plates. For deeper insights into, for example, the size of the autophagic bodies, electron microscopy can be used. But electron microscopy is a method that requires many skills and much more time.

Following the breakdown of radiolabeled proteins in a pulse-chase experiment is time-consuming and laborious, but it results in an accurate and quantitative determination of the autophagic rate. The use of this method is not limited to *S. cerevisiae*. It has also been adopted for use with mammalian cells and even whole organs such as perfused livers from mice or rats.

REFERENCES

Baba, M., Osumi, M., and Ohsumi, Y. (1995). Analysis of the membrane structures involved in autophagy in yeast by freeze-replica method. *Cell Struct. Func.* **20,** 465–471.

Baba, M., Takeshige, K., Baba, N., and Ohsumi, Y. (1994). Ultrastructural analysis of the autophagic process in yeast: Detection of autophagosomes and their characterization. *J. Cell Biol.* **124,** 903–913.

Huang, W. P., Scott, S. V., Kim, J., and Klionsky, D. J. (2000). The itinerary of a vesicle component, Aut7p/Cvt5p, terminates in the yeast vacuole via the autophagy/Cvt pathways. *J. Biol. Chem.* **275,** 5845–5851.

Kirisako, T., Baba, M., Ishihara, N., Miyazawa, K., Ohsumi, M., Yoshimori, T., Noda, T., and Ohsumi, Y. (1999). Formation process of autophagosome is traced with Apg8/Aut7p in yeast. *J. Cell Biol.* **147,** 435–446.

Lang, T., Reiche, S., Straub, M., Bredschneider, M., and Thumm, M. (2000). Autophagy and the cvt pathway both depend on *AUT9*. *J. Bacteriol.* **182,** 2125–2133.

Levine, B., and Kroemer, G. (2008). Autophagy in the pathogenesis of disease. *Cell* **132,** 27–42.

Meiling-Wesse, K., Barth, H., and Thumm, M. (2002). Ccz1p/Aut11p/Cvt16p is essential for autophagy and the cvt pathway. *FEBS Lett.* **526,** 71–76.

Nakatogawa, H., Ichimura, Y., and Ohsumi, Y. (2007). Atg8, a ubiquitin-like protein required for autophagosome formation, mediates membrane tethering and hemifusion. *Cell* **130,** 165–178.

Noda, T., Matsuura, A., Wada, Y., and Ohsumi, Y. (1995). Novel system for monitoring autophagy in the yeast *Saccharomyces cerevisiae*. *Biochem. Biophys. Res. Commun.* **210,** 126–132.

Prick, T., Thumm, M., Kohrer, K., Haussinger, D., and Vom Dahl, S. (2006). In yeast, loss of Hog1 leads to osmosensitivity of autophagy. *Biochem. J.* **394,** 153–161.

Rubinsztein, D. C., Gestwicki, J. E., Murphy, L. O., and Klionsky, D. J. (2007). Potential therapeutic applications of autophagy. *Nat. Rev. Drug Discov.* **6,** 304–312.

Straub, M., Bredschneider, M., and Thumm, M. (1997). AUT3, a serine/threonine kinase gene, is essential for autophagocytosis in *Saccharomyces cerevisiae*. *J. Bacteriol.* **179,** 3875–3883.

Suzuki, K., Kirisako, T., Kamada, Y., Mizushima, N., Noda, T., and Ohsumi, Y. (2001). The pre-autophagosomal structure organized by concerted functions of APG genes is essential for autophagosome formation. *EMBO J.* **20,** 5971–5981.

Suzuki, K., and Ohsumi, Y. (2007). Molecular machinery of autophagosome formation in yeast, *Saccharomyces cerevisiae*. *FEBS Lett.* **581,** 2156–2161.

Takeshige, K., Baba, M., Tsuboi, S., Noda, T., and Ohsumi, Y. (1992). Autophagy in yeast demonstrated with proteinase-deficient mutants and conditions for its induction. *J. Cell Biol.* **119,** 301–311.

Teichert, U., Mechler, B., Muller, H., and Wolf, D. H. (1989). Lysosomal (vacuolar) proteinases of yeast are essential catalysts for protein degradation, differentiation, and cell survival. *J. Biol. Chem.* **264,** 16037–16045.

Vida, T. A., and Emr, S. D. (1995). A new vital stain for visualizing vacuolar membrane dynamics and endocytosis in yeast. *J. Cell Biol.* **128,** 779–792.

Xie, Z., and Klionsky, D. J. (2007). Autophagosome formation: Core machinery and adaptations. *Nat. Cell Biol.* **9,** 1102–1109.

CHAPTER SIX

Aminopeptidase I Enzymatic Activity

Peter Schu

Contents

Abstract

Aminopeptidase I is the cargo protein of the cytoplasm-to-vacuole targeting (Cvt), autophagy-like protein-targeting pathway of the yeast *Saccharomyces cerevisiae*, the nonclassical vacuolar biosynthetic transport route. The second enzyme following this route to the vacuole, α-mannosidase, is also transported by direct binding to the Atg19 receptor and to aminopeptidase I. Aminopeptidase I forms a homododecameric complex, which is synthesized and assembled in the cytoplasm, packed in double-membrane vesicles, and transported to the vacuole. Only the homododecameric complex of aminopeptidase I has exopeptidase activity directed against amino-terminal leucine residues. Enzymatic activity can be determined spectrofluorometrically in homogenates and semiquantitatively after nondenaturing gel electrophoresis and by yeast colony-overlay assay. This chapter describes the methods to determine aminopeptidase I enzymatic activity used to follow complex assembly and vacuolar transport.

Georg-August-University Göttingen, Zentrum für Biochemie und Molekulare Zellbiologie, Biochemie II, Göttingen, Germany

Methods in Enzymology, Volume 451
ISSN 0076-6879, DOI: 10.1016/S0076-6879(08)03206-0

1. Introduction

Aminopeptidase I is the main cargo protein of the autophagy-like cytoplasm-to-vacuole targeting transport route in the yeast Saccharomyces cerevisiae (Klionsky *et al.*, 1992). It is targeted to this pathway by the Atg19 receptor protein (Scott *et al.*, 2001). The second vacuolar enzyme that follows this pathway to the vacuole, α-mannosidase, binds to the same receptor protein also as an oligomeric complex and to aminopeptidase I (Huang and Klionsky, 2002; Hutchins and Klionsky, 2001; Shintani *et al.*, 2002; Yoshihisa and Anraku, 1990). Thus, the Cvt pathway apparently evolved for the vacuolar transport of only two enzymes, which raises the question of why these enzymes do not follow the endoplasmic reticulum–Golgi pathway like all other vacuolar enzymes. The reason has to do with the unusual biogenesis of aminopeptidase I enzymatic activity. The active enzyme is a homododecameric complex of ~600 kDa, and dissociation into two hexamers leads to loss of enzymatic activity (Marx *et al.*, 1977; Metz *et al.*, 1977). Complex assembly takes place in the cytoplasm, and if this assembly is disturbed, aminopeptidase I is still targeted to the Cvt pathway and will reach the vacuolar lumen. Aminopeptidase I will homododecamerize in the vacuolar milieu; however, these complexes are enzymatically inactive (Andrei-Selmer *et al.*, 2001). Thus, cytoplasmic factors are essential to assemble an enzymatically active complex. Therefore determination of aminopeptidase I activity makes it possible to follow the early steps in the Cvt pathway. Aminopeptidase I releases preferentially leucine residues and the activity is further influenced by downstream residues. Besides aminopeptidase I, *S. cerevisiae* has three more leucine-aminopeptidases (Trumbly and Bradley, 1983). One is found in the cytoplasm, one in the cytoplasm and the nucleus, and one in the cytoplasm and the periplasm. Determination of aminopeptidase I enzymatic activity requires discrimination among these leucine-aminopeptidases. The following sections describe these yeast leucine-aminopeptidases, the biogenesis of aminopeptidase I, and the various assays to discriminate and to determine leucine-aminopeptidase activities.

1.1. Leucine-aminopeptidase activities in *S. cerevisiae*

In cell extracts of the yeast *S. cerevisiae* four aminopeptidase activities with a preference for leucine were described by Trumbly and Bradley (1983).

One of the four leucine-aminopeptidases (LAPs) is the vacuolar aminopeptidase I, Ape1, also referred to as LAP4, yscI, and YKL103C (E.C. 3.4.11.22). Aminopeptidase I is the cargo protein of the autophagy-like Cvt pathway (Klionsky *et al.*, 1992; Scott *et al.*, 2001). Aminopeptidase I is a zinc metalloproteinase that belongs to the peptidase family M18, and zinc is part of the active center polarizing the water molecule, which hydrolyzes the peptide bond.

LAP1 or aminopeptidase 2, also known as yscII, APE2, or YKL157W (E.C. 3.4.11.-) belongs to the peptidase M1 family. LAP1 is found intracellularly and in the periplasm, where the enzyme might be involved in extracellular peptide degradation. A function in the inactivation of the pheromone α-factor could not be demonstrated; however, it may function in the inactivation of killer toxins produced by other yeast strains (Hirsch *et al.*, 1988).

LAP2 or yscIV, which has also been called AP IV or YNL045W, is another member of the peptidase M1 family. It is a zinc-dependent cytoplasmic protein and is also found in the nucleus. It is homologous to the mammalian extracellular leukotriene A4 hydrolase, which converts the epoxid LTA_4 (5*S*-*trans*-5,6-oxido-7,9-*trans*-11,14-*cis*-eicosatetraenoic acid) to the cytokine LTB_4 (5*S*,12*R*-*dihydroxy*-6,14-*cis*-8,10-*trans*-eicosatetraenoic acid). The mammalian enzyme also has alanyl-aminopeptidase activity. Yeast LAP2 does not hydrolyze LTA_4 to LTB_4 but converts LTA_4 to 5*S*,6*S*-dihydroxy-7,9-*trans*-11,14-*cis*-eicosatetraenoic acid, and its leucine-aminopeptidase activity is stimulated tenfold by LTA_4. Moreover, the yeast and mammalian enzymes are inhibited by thioamine and bestatin. Thus, yeast LAP2 is a bifunctional enzyme with a hydrophobic binding site at the active center and appears to be the early ancestral enzyme for the mammalian LTA_4 hydrolases (Kull *et al.*, 1999; Kull *et al.*, 2001).

LAP3 or yscIII, also called YNL239W or BLH1 (E.C. 3.4.22.40), is a cytoplasmic cysteinyl proteinase homologous to the mammalian bleomycin hydrolase (Enenkel and Wolf, 1993). Bleomycin is a glycopeptidic antibiotic used to treat cancers, and antibiotic activity depends on the bleomycin hydrolase expression level of the tumorigenic cells. LAP3 is the only yeast enzyme with bleomycin hydrolase activity as tested with the specific substrate Glu-β-naphthylamide and demonstrated by the bleomycin sensitivity of the Δ*blh1* strain.

None of these four leucine-aminopeptidases is essential for yeast viability under various growth conditions, nor are they required for efficient sporulation. Also, a strain deficient in all four aminopeptidase activities is viable (Trumbly and Bradley, 1983). This indicates that they are not involved in vital cellular processes or that other enzymes can take over their functions. Aminopeptidase I is only one of many vacuolar peptidases, and so its deficiency might be very well compensated by other enzymes, although these will cleave aminopeptidase I substrates at slower rates.

1.2. Biogenesis of aminopeptidase I

Aminopeptidase I is constituitively expressed and its expression is increased upon nutrient starvation, as is true for all vacuolar peptidases. Aminopeptidase I is synthesized in the cytoplasma as a 61 kDa protein with an unusual

45-amino-acid-long N-terminal signal sequence (Chang and Smith, 1989). These 45 amino acids form two amphipathic helices, a helix-turn-helix domain, and the N-terminal helix is essential for aminopeptidase I transport to the vacuole (Martinez *et al.*, 1997; Oda *et al.*, 1996). The N-terminal precursor sequence is sufficient to target proteins to the Cvt pathway (Martinez *et al.*, 1997). Precursor aminopeptidase I (prApe1) is processed upon arrival in the vacuolar lumen by vacuolar endopeptidases to the mature 50 kDa mApe1 form. A maturation intermediate of 55 kDa generated by cleavage in the turn region between the two helices is observed. The cytoplasmic precursor assembles very quickly, with a half-time of two minutes, into the homododecameric complex (Kim *et al.*, 1997). Homododecameric complex assembly probably involves three distinct steps (Fig. 6.1). This is indicated by the stepwise disassembly of the complex induced by increasing proton concentrations. First, stable symmetrical homodimers (6S) are formed, whose dissociation requires a pH < 3.5. These trimerize to form pseudotrimeric rings (12S). Finally, these homohexamers dimerize to form the homododecameric complex (22S). Analysis of negatively stained aminopeptidase I electron microscopy images reveals no aggregates of intermediate size, which also indicates an ordered stepwise

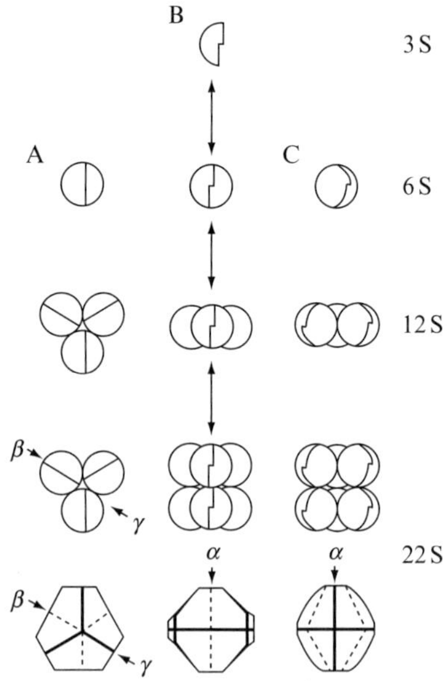

Figure 6.1 Assembly and stability of homododecameric aminopeptidase I complexes. Only the 22S particle has enzymatic activity (Marx *et al.*, 1977; Metz *et al.*, 1977).

assembly and not complex formation by the addition of monomers to existing structures. This demonstrates that the complex is formed by three different types of protein-protein interactions (Marx *et al.*, 1977; Metz *et al.*, 1977). If homododecameric complexes dissociate as a result of high proton concentrations (pH < 5.0), enzymatic activity will be lost. Also monomers and dimers do not have enzymatic activity. Studies to resolve the crystal structure of aminopeptidase I are under way and a description of the crystals has been published. The structure of aminopeptidase I from *Borrelia burgdorferi* (PDB 1y7e) serves as a model (Adachi *et al.*, 2007).

Aminopeptidase I complex assembly in the cytoplasm is essential for formation of an enzymatically active complex. If cytoplasmic homododecamerization of prApe1 is disturbed in mutants, homododecamers of mApe1 will form in the vacuole, but they are enzymatically inactive. The requirement of complex assembly in the cytoplasm explains why this enzyme takes this nonclassical route to the vacuole, bypassing the endoplasmic reticulum–Golgi route and a membrane translocation process (Andrei-Selmer *et al.*, 2001).

Proteolytic enzymes are typically synthesized as precursors, which carry additional either N- or C-terminal domains or even both. Besides targeting information for cellular compartments and chaperone functions, these domains render the precursor proteins enzymatically inactive and prevent unwanted proteolytic activity during protein transport. The proteolytic activity is activated on their arrival in their final cellular compartment. Processing of precursors can be autoproteolytic, initiated by changes in the conformation of the protein (e.g., induced by high proton concentrations). Precursors are also trimmed by other compartment-specific proteolytic enzymes. Aminopeptidase I is also an exception of the rule in this case. The cytoplasmic precursor form is already enzymatically active as can be seen in strains with a deleted *PEP4* gene encoding the endoproteinase proteinase A (Andrei-Selmer *et al.*, 2001). The proteinase A precursor is activated by autoproteolysis induced by the low vacuolar pH. Proteinase A then activates other vacuolar precursor enzymes, including a second vacuolar endoproteinase, proteinase B. Thus, proteinase A-deficient strains are devoid of many vacuolar enzyme activities and are unable to survive starvation conditions (Teichert *et al.*, 1989). In a proteinase A-deficient strain the cytoplasmic precursor form of aminopeptidase I is not processed to the mature mApe1, but the enzyme is enzymatically active (Andrei-Selmer *et al.*, 2001). Aminopeptidase I is enwrapped in Cvt vesicles or autophagosomes under starvation conditions and is thus readily separated from cytoplasmic proteins. An inhibiton of the precursor by the amino terminal extension is therefore not necessary to prevent unwanted aminopeptidase I activity in the cytoplasm.

2. Methods to Measure Leucine-aminopeptidase Activity

Spectrofluorometric enzyme activity assays have been developed that make it possible to discriminate among the four leucine-aminopeptidases in *S. cerevisiae*. However, none of these assays can be used to determine the activity of just one enzyme in crude cell extracts. Determination of specific activities requires either the purification or enrichement of the respective enzyme or, alternatively, deletion of three of the four leucine-aminopeptidase genes. These yeast strains can be readily tested for amino-peptidase activities by yeast colony-overlay assays. Enzyme activities can also be determined in gels after separating proteins in nondenaturing polyacrylamide electrophoresis systems.

The four aminopeptidase activities can be determined using the substrates leucine-β-naphthylamide or leucine-para-nitroanilide; however, the activities determined for each enzyme vary depending on the chromophore (Trumbly and Bradley, 1983). This makes it difficult to determine the contributions of each enzyme to the total leucine-aminopeptidase activity. LAP1 and LAP3 are best measured with leucine-β-naphthylamide, whereas the highest activities for LAP4/aminopeptidase I are obtained with leucine-para-nitroanilide. However, this appears to be a good substrate for many other aminopeptidases as well, because a strain deficient in all four leucine-aminopeptidases still has considerable leucine-para-nitroanilide hydrolyzing activity (Table 6.1) (Trumbly and Bradley, 1983). Only the determination of LAP2 activity does not depend on the substrate that is used. Trumbly and Bradley determined the fraction of each aminopeptidase activity that contributes to the total leucine-aminopeptidase activity using both of these nonphysiological substrates. These data are summarized in Table 6.1.

2.1. Spectrofluorometric assays

1. Cells are grown in 20 ml of rich medium (YPD) or yeast starvation media for highest induction of enzymatic activity to late-exponential (O.D.$_{600}$= 1.4) or early stationary growth phases (O.D.$_{600}$ = 2.0).
2. The cells are harvested by centrifugation at 800*g* for 10 min.
3. The supernatant is discarded and the cell pellet is resuspended by vortexing in 20 ml of 10 m*M* Tris/HCl, pH 7.6. This procedure might be repeated until the yeast cell pellet is white.
4. For the preparation of protein extracts from small cell culture volumes, removal of the cell wall and gentle cell lysis increases the enzyme yield, because of the size of the aminopeptidase I homododecameric complex. If large amounts of cells are broken up, standard yeast fractionation

Table 6.1 : Leucine-aminopeptidase activities in *S. cerevisiae*

LAP present in extracts	Activity in of WTcells (%) Leu-β-NA	Leu-pNA	Activated by 1 m*M*	Inhibited by
LAP1, Ape2	128.8	67.3	Co^{2+}, Fe^{3+}, Mn^{2+}	
LAP2, Ap IV	26.9	16.2	Co^{2+}, Cu^{2+}, Fe^{3+}, Mn^{2+}	Thiamine, Bestatin
LAP3, Blh1	20.4	7.3	Co^{2+}	Iodoacetamide N-Ethylmaleinimide
LAP4, Ape1	9.8	52.8	Zn^{2+}	
None	5.8	11.1	—	—

Cell extracts were prepared from strains mutated in three or all four of the aminopeptidase genes. Cofactor dependence was measured after dialysis of crude protein extracts against 50 m*M* of Tris pH 7.6, 2 mM of EDTA, and reincubation in the presence of the respective cation for 15 min at 30 °C (Trumbly and Bradley, 1983). Leu-β-NA: leucine-β-naphthylamide; Leu-pNA: leucine-para-nitroanilide.

protocols such as sonication, vortexing with glass beads, or shear stress in the French press may be used (please refer to *Methods in Enzymology*, vol. 351). For gentle lysis, small yeast cultures are first treated to convert the cells into spheroplasts.

a. Resuspend cells at a concentration of 10 $O.D._{600}$ units/ml in 0.1 M $Tris/SO_4$, pH 9.4, 10 m*M* DTT, and incubate for 20 min at RT or at 30 °C.
b. Resuspend cells at a concentration of 10 $O.D._{600}$ units/2 ml in 1.2 M sorbitol, 50 m*M* Tris, pH 7.5, 1 m*M* EDTA (remove 50-μl aliquots for cell lysis control).
c. Add 2 μl of Zymolyase 100T (Seikagaku, Cat.-Nr. 120493; stock 10 mg/ml in 1 M sorbitol, 25 m*M* Tris, pH 7.4, 100 m*M* NaCl; 2 μg of zymolyase per O.D.) per 10 $O.D._{600}$ units of cells.
d. Incubate with gentle shaking at 30 °C (i.e., invert the tube to prevent formation of a cell pellet during the incubation).
e. After 15 min, start to monitor spheroplasting by diluting 20 μl of the cell suspension with 980 μl of water and examine cell lysis under the microscope. The aliquot taken before Zymolyase addition serves as prelysis control. After 25 min, approximately 80% of the cells should have turned into spheroplasts. Handle spheroplasts carefully to prevent premature lysis.
f. Spin down the spheroplasts carefully at 500×*g* for 10 s and wash them once with 1.2 M sorbitol, 50 m*M* Tris, pH 7.5.

g. For lysis, resuspend the spheroplasts in 1 to 1.5 ml of water or enzyme assay buffer containing protease inhibitors per 10 $O.D._{600}$ units of starting cells.
   Protease-inhibitor stock-solutions: in water: antipain 2 mg/ml (1000x); leupeptin 2 mg/ml (1000x); α_2-macroglobulin 1 mg/ml (100x); in methanol: chymostatin 2 mg/ml (100x); pepstatin 2 mg/ml (1000x); in isopropanol: PMSF 10 mg/ml (100x) (alternatively, premade inhibitor mixes provided by various vendors might be used).

5. The cell lysates are cleared by centrifugation at 10,000*g* for 10 min. Enzyme activity assays are performed with the supernatant fractions (Trumbly and Bradley, 1983).
6. Prepare stock solutions of substrates. Substrate concentrations in the stock solutions are 20 m*M* leucine-β-naphthylamide (BACHEM, K-1380) dissolved in methanol or 20 m*M* leucine-para-nitroanilide (BACHEM, K-1305) in dimethylsulfoxide (DMSO). The stock solutions may be kept at 4 °C for up to one week or stored frozen at −20 °C for up to one month.
7. Set up 1.7-ml microcentrifuge tubes for the assay. The reaction mixture has a total volume of 1.5 ml and contains 1 ml of 75 m*M* Tris/HCl, pH 7.5; 75 μl of substrate stock solution; and up to 435 μl of protein solution. If less protein solution is used, the volume may be replaced with water.
8. The assay is incubated at 30 °C and spectrophotometer readings are taken at 0, 15, 30 and 45 min. Leucine-β-naphthylamide absorption is highest at 340 nm, and leucine-para-nitroanilide at 405 nm.

2.2. Semiquantitative assays

2.2.1. Enzyme activity after nondenaturing gel-electrophoresis

The four aminopeptidase activities can also be determined in crude cell extracts after separating the proteins in nondenaturing gel electrophoresis systems such as starch or polyacrylamide (Enenkel and Wolf, 1993; Hirsch *et al.*, 1988; Trumbly and Bradley, 1983).

1. A nondenaturing discontinuous polyacrylamide gel is prepared with a 3.5% acrylamide stacking gel and a 7.5% separating gel.
2. The protein concentrations in the supernatant fractions generated as described in section 2.1 are determined using the Bradford or BCA assay. Up to 200 μg of crude cell extract proteins are loaded per lane. Bromophenol blue is added to the loading dye, and the gel is run until the dye reaches the bottom.
3. After the run, the gel is incubated in the assay mixture (70 mg of leucine-β-naphthylamide dissolved in 100 ml of 100 m*M* Tris/HCl, pH 7.5) with or without 0.8% agarose. Agarose prevents dilution of the reaction product into the buffer, thereby increasing assay sensitivity.

Leucine-β-naphthylamide is used as a substrate and the released β-naphthylamide can be visualized directly by activation with UV light at 365 nm; the emitted blue-violet fluorescent light can be recorded with digital camera systems used to document chemifluorescence.

Alternatively, released β-naphthylamide can further react with either Fast Black K salt (TCI, F0092) or Fast Garnet GBC (Sigma, F8761). To couple β-naphthylamide to Fast Black K salt, the same amounts of both Fast Black K salt and leucine-β-naphthylamide are dissolved in 200 m*M* Tris/maleate, pH 6.5 (70 mg each in 100 ml of buffer), in which the gel is immersed. To couple β-naphthylamide to Fast Garnet GBC, 31 mg are dissolved in 100 ml of 200 m*M* potassium phosphate, pH 6.5, and added to the leucine-β-naphthylamide solution (70 mg/100 ml Tris/HCl, pH 7.5).

2.2.2. Yeast colony assay

To screen large numbers of clones and to have a fast and convenient assay, yeast colonies can be tested directly for enzymatic activities in colony-overlay assays. If the enzyme is secreted, yeast colonies on the growth plates can be covered with the assay mixture dissolved in low concentrations of agarose. In most cases the enzyme resides in the cell; therefore, cells must be lysed or permeabilized before they are covered with the assay-agar mixture.

1. Cells are grown in single-colony streaks or small patches on 9-cm Ø agar-plates for 2–4 days at 30 °C. The medium can be rich YPD or a selective medium, if auxotrophic markers are required.
2. Autoclaved chromatography paper (Schleicher & Schüll, 3469), cut to the internal size of the Petri plate (8-cm Ø), is transferred onto the culture plates and left on the plates until wet. If enough cells stick to the paper, the overlay-activity assay can be done directly. If not, the chromatography paper can be transferred, yeast colony side up, onto fresh agar plates and incubated until the colonies grow larger, approximately overnight or even 1 to 2 days. Alternatively, clones can be streaked as patches directly onto the chromatography paper and allowed to grow.
3. Chromatography papers from the plates are placed colony side down onto buffer-saturated $CHCl_3$ for 15 to 20 s (chloroform is mixed with the buffer, which will form a thin water phase on top of the chloroform; this also forms an evaporation barrier for the organic solvent). Be aware that the buffer-saturated chloroform will permeabilize only the upper cell layer.
4. Lift the chromatography paper from the $CHCl_3$, allow excessive $CHCl_3$ to drop off, and air-dry them.
5. Place the chromatography papers in Petri dishes and cover them with the assay-agar mixture as described subsequently.

Aminopeptidase activity assay with leucine-β-naphthylamide:

1. Per chromatography paper in a 9-cm Ø Petri dish, prepare 3 to 5 ml of agar assay solution. Substrate concentration in the stock solution is 200 m*M* leucine-β-naphthylamide dissolved in methanol. This solution should always be prepared fresh. Assay-mixture final concentrations are 75 m*M* Tris/HCl, pH 7.5; 3 m*M* EDTA, 10 m*M* leucine-β-naphthylamide, and 0.8% agarose.
2. Agarose is dissolved in deionized water by heating the mixture in the microwave and added to the buffer solution, with vortexing. Then the substrate solution is added, vortexed, and the solution is poured immediately onto the dried chromatography paper. Mixing the agar solution with buffer and the substrate will cool the solution, preventing heat denaturation of the enzyme. Constant shaking will prevent the agar solution from solidifying. Once poured on the paper, the agar solution will cool down and solidify immediately.
3. Dishes are incubated at room temperature for 10 min and may be incubated as long as 30 min.
4. The released naphthylamide is activated with UV light at 365 nm and will emit a blue-violet fluorescent light, which can be recorded with digital camera systems such as those used to detect chemifluorescence.

Wild-type cells will develop a bright fluorescence within 5–10 min, depending on the growth phase of the cells and their enzyme expression levels, which will not become more intense. Cells with all four leucine aminopeptidases deleted will not emit any blue-violet fluorescent light, even after prolonged incubation times. Mutant activities are determined in comparison to the fluorescence of the wild-type colonies on the same chromatography paper, which allows semi-quantitative assays. If the wild-type activity is set to 100%, mutant strains can be classified in 20% steps, for five categories of aminopeptidase activity (Andrei-Selmer *et al.*, 2001). Transforming strains with aminopeptidase I on various single-copy plasmids makes it possible to correlate expression levels and enzyme activities.

3. Concluding Remarks

The sensitivity of aminopeptidase I enzymatic activity assays is strongly influenced by the method of cell lysis and subsequent fractionation. Enzymatic activity is readily tested by the colony-overlay assay after gentle cell permeabilization. Any subcellular fractionation and even vacuole isolation will result in significant loss of specific activity. This might be caused by complex dissociation or allosteric inhibition of aminopeptidase I. Considering the allosteric regulation that often occurs through homododecamerization, allosteric inhibition might be the most likely mechanism for reduced specific activities upon cell subfractionation procedures.

REFERENCES

Adachi, W., Suzuki, N. N., Fujioka, Y., Suzuki, K., Ohsumi, Y., and Inagaki, F. (2007). Crystallization of *Saccharomyces cerevisiae* aminopeptidase 1, the major cargo protein of the Cvt pathway. *Acta Crystallogr. F* **63,** 200–203.

Andrei-Selmer, C., Knüppel, A., Satyanarayana, C., Heese, C., and Schu, P. V. (2001). A new class of mutants deficient in dodecamerization of aminopeptidase 1 and vacuolar transport. *J. Biol. Chem.* **276,** 11606–11614.

Chang, Y. H., and Smith, J. A. (1989). Molecular cloning and sequencing of genomic DNA encoding aminopeptidase I from *Saccharomyces cerevisiae*. *J. Biol. Chem.* **264,** 6979–6983.

Enenkel, C., and Wolf, D. H. (1993). BLH1 codes for a yeast thiol aminopeptidase, the equivalent of mammalian bleomycin hydrolase. *J. Biol. Chem.* **268,** 7036–7043.

Hirsch, H. H., Suarez Rendueles, P., Achstetter, T., and Wolf, D. H. (1988). Aminopeptidase yscII of yeast: Isolation of mutants and their biochemical and genetic analysis. *Eur. J. Biochem.* **173,** 589–598.

Huang, W.-P., and Klionsky, D. J. (2002). Autophagy in yeast: A review of the molecular machinery. *Cell Struct. Funct.* **27,** 409–420.

Hutchins, M. U., and Klionsky, D. J. (2001). Vacuolar localization of oligomeric a-mannosidase requires the cytoplasm to vacuole targeting and autophagy pathway components in *Saccharomyces cerevisiae*. *J. Biol. Chem.* **276,** 20491–20498.

Kim, J., Scott, S. V., Oda, M. N., and Klionsky, D. J. (1997). Transport of a large oligomeric protein by the cytoplasm to vacuole protein targeting pathway. *J. Cell Biol.* **137,** 609–618.

Klionsky, D. J., Cueva, R., and Yaver, D. S. (1992). Aminopeptidase I of *Saccharomyces cerevisiae* is localized to the vacuole independent of the secretory pathway. *J. Cell Biol.* **119,** 287–299.

Kull, F., Ohlson, E., and Haeggstrom, J. Z. (1999). Cloning and characterization of a bifunctional leukotriene A(4) hydrolase from *Saccharomyces cerevisiae*. *J. Biol. Chem.* **274,** 34683–34690.

Kull, F., Ohlson, E., Lind, B., and Haeggstrom, J. Z. (2001). *Saccharomyces cerevisiae* leukotriene A4 hydrolase: Formation of leukotriene B4 and identification of catalytic residues. *Biochem.* **40,** 12695–12703.

Martinez, E., Jimenez, M. A., Segui-Real, B., Vandekerckhove, J., and Sandoval, I. V. (1997). Folding of the presequence of yeast pAPI into an amphipathic helix determines transport of the protein from the cytosol to the vacuole. *J. Mol. Biol.* **267,** 1124–1138.

Marx, R., Metz, G., and Röhm, K. H. (1977). The quaternary structure of yeast aminopeptidase I. 2. Geometric arrangement of subunits. *Z Naturforsch [C]*. **32,** 938–943.

Metz, G., Marx, R., and Röhm, K. H. (1977). The quaternary structure of yeast aminopeptidase I. 1. Molecular forms and subunit size. *Z Naturforsch [C]*. **32,** 929–937.

Oda, M. N., Scott, S. V., Hefner-Gravink, A., Caffarelli, A. D., and Klionsky, D. J. (1996). Identification of a cytoplasm to vacuole targeting determinant in aminopeptidase I. *J. Cell Biol.* **132,** 999–1010.

Scott, S. V., Guan, J., Hutchins, M. U., Kim, J., and Klionsky, D. J. (2001). Cvt19 is a receptor for the cytoplasm-to-vacuole targeting pathway. *Mol. Cell* **7,** 1131–1141.

Shintani, T., Huang, W.-P., Stromhaug, P. E., and Klionsky, D. J. (2002). Mechanism of cargo selection in the cytoplasm to vacuole targeting pathway. *Dev. Cell.* **3,** 825–837.

Teichert, U., Mechler, B., Müller, H., and Wolf, D. H. (1989). Lysosomal (vacuolar) proteinases of yeast are essential catalysts for protein degradation, differentiation, and cell survival. *J. Biol. Chem.* **264,** 16037–16045.

Trumbly, R. J., and Bradley, G. (1983). Isolation and characterization of aminopeptidase mutants of *Saccharomyces cerevisiae*. *J. Bacteriol.* **156,** 36–48.

Yoshihisa, T., and Anraku, Y. (1990). A novel pathway of import of a-mannosidase, a marker enzyme of vacuolar membrane, in *Saccharomyces cerevisiae*. *J. Biol. Chem.* **265,** 22418–22425.

CHAPTER SEVEN

Monitoring Autophagy in Yeast using FM 4-64 Fluorescence

Dikla Journo, Gal Winter, *and* Hagai Abeliovich

Contents

Abstract

The original observations and experiments dealing with autophagy were purely morphological in nature. Even though more and more molecular techniques have been introduced, experimenters are often asked to provide visual evidence of autophagic processes in order to back up data obtained via other means. In yeast as well, autophagosomes were initially defined morphologically and indirectly, by observing intravacuolar autophagic bodies that accumulate upon starvation. This can be achieved by electron microscopy, which affords very high resolution but is time consuming and costly, or by light microscopy, which is a relatively inaccurate method of scoring autophagy. A third alternative, which we present here, is to use the unique properties of the fluorescent dye FM 4-64 to follow the accumulation of autophagic bodies.

1. Overview

The standard method for using light microscopy in the observation of yeast autophagosomes utilizes the different light refractivity of the yeast vacuole as compared with the cytosolic milieu. Thus, when yeast cells are starved under conditions in which autophagosome breakdown is inhibited,

Department of Biochemistry and Food Science, Hebrew University of Jerusalem, Rehovot, Israel

Methods in Enzymology, Volume 451
ISSN 0076-6879, DOI: 10.1016/S0076-6879(08)03207-2

one can observe intravacuolar bodies that exhibit different optical properties relative to the rest of the vacuolar lumen (Takeshige *et al.*, 1992). Inhibition of vacuolar hydrolase activities is usually achieved by PMSF treatment or by utilizing *pep4Δ* cells, which lack the central activating protease of the vacuolar hydrolase activation cascade. The accumulation of autophagic bodies is then best observed using phase contrast or differential interference microscopy. While this provides a quick and dirty approach for scoring autophagy, it is not universally applicable, such as in species in which the optical properties of the vacuolar lumen are not sufficiently differentiated from those of the cytosol, such as the food-spoilage yeast *Zygosaccharomyces bailii* (Winter *et al.*, 2008). To circumvent this problem, specifically in species in which a large central vacuole is observed (this enables one to define autophagic bodies), we devised a method that does not rely on optical contrast between autophagic bodies and the vacuolar lumen but rather on the physical properties of the FM 4-64 molecule.

2. Molecular Properties and Endosomal-pathway Tracking Behavior of the FM 4-64 Fluorophore

FM 4-64 and its sister dye, FM 1-43, belong to a group of dicationic amphiphilic styryl dyes that fluoresce only when present in a hydrophobic environment (Betz *et al.*, 1992). Because of the charged, bulky nature of their polar head group, they do not cross biological membranes under normal experimental conditions. Rather, they insert their hydrophobic tail into any available membrane leaflet. When added to whole cells in culture, they will attach to the outer leaflet of the cell membrane. Once attached to the membrane, their fluorescence is de-quenched, as the aqueous solution is no longer available to accept the excess energy of the excited state, and this results in plasma membrane fluorescence (Betz *et al.*, 1996). This staining pattern can be stabilized by incubating yeast cells at temperatures in which membrane trafficking is halted (e.g., 4 °C; Vida and Emr, 1995). However under physiological conditions (26 °C–30 °C), endocytic events lead to internalization of the dye, all while maintaining its topology with respect to the inner and outer leaflet of the plasma membrane. In due course, the dye concentrates in the inner leaflet of the vacuole membrane (Vida and Emr, 1995).

2.1. Fluorescent properties of FM 4-64 and endosomal trafficking

2.1.1. Application

The ability of FM dyes to specifically stain the endocytic pathway has enabled numerous applications. The most straightforward is as a marker for the vacuolar membrane. Because FM 4-64 (in contrast with FM 1-43) has no emission in the green range, it is ideal for cofluorescence studies with green-fluorescent protein chimeras (Abeliovich *et al.*, 1998; Sakai *et al.*, 1998). Additional applications are in pulse-chase staining regimes, which in combination with GFP-tagged chimeras allow for the identification of endocytic intermediates (Abeliovich *et al.*, 1998; Burd and Emr, 1998; Vida and Emr, 1995). FM 4-64 staining was also reported as an assay in a screening procedure for FACS-mediated identification of yeast endocytosis mutants (Wendland *et al.*, 1996).

2.1.2. Experimental considerations and design

In all experiments involving FM 4-64 and related compounds, care must be taken to use the lowest possible concentration of dye. Styryl dyes are known to affect membrane behavior at micromolar concentrations, thus dictating a minimal dosage to avoid artifacts (Rodes *et al.*, 1995). We have been routinely using 0.6–0.8 μM FM 4-64, although lower concentrations may be achievable with improved optical detection equipment. Because the dyes concentrate in hydrophobic environments, it is also worthwhile to monitor the cell:dye concentration ratio, as low cell concentrations will lead to high localized dye concentration in the membrane, even when the theoretical dye concentration is low.

Because of their low solubility in aqueous buffers, styryl dyes are usually stored as stocks in dimethylsulfoxide (DMSO). However, at higher concentrations (greater than 0.1%) DMSO also affects membrane behavior. Thus, its final concentration in the presence of living cells must also be kept to a minimum while including appropriate controls where applicable.

2.1.3. Material and methods: Time-resolved versus endpoint measurements

In time-resolved pulse-chase FM 4-64 staining methods, cells are incubated in ice-cold medium containing 0.6 μM FM 4-64 with vigorous shaking, for 30 min. The cells are then collected by centrifugation, washed, and resuspended in growth medium lacking FM 4-64. Cells can then be incubated at 15 °C for 30 min to chase the signal into endocytic intermediates or at 30 °C to chase the signal into the vacuole (Abeliovich *et al.*, 1998; Vida and Emr, 1995). This approach can allow real-time monitoring of the transport of FM 4-64 through the endocytic pathway.

In contrast, the endpoint-measurement method employs a somewhat different protocol. In these experiments, cells are grown to mid-log phase (OD_{600} of 0.4–0.8) and resuspended in 0.6–0.8 μM FM 4-64 at an OD_{600} of 5–10, in a 50-ml conical tube. The cells are incubated for 20 min at 26 °C (or 30 °C, depending on the original growth temperature) and then washed and resuspended at the original optical density. Following a chase period of at least 40 min at 26 °C or 30 min at 30 °C, one can mount the cells on microscope slides for viewing.

2.1.4. Notes on viewing live yeast cells by fluorescence microscopy

Unlike fluorescence studies with fixed cells, observation of cells stained with vital dyes, and to a lesser extent cells expressing fluorescent proteins, requires close attention to the physiological conditions of the cells. A microscope stage with a temperature controller, available from most manufacturers, is recommended, especially for time-lapse photography.

In addition, procedures that are customary for viewing fixed cells, such as sealing the coverslips using acetone-based nail polish, are strongly discouraged, as this has the dual undesirable effects of introducing an organic solvent into contact with the cells as well as restricting gas exchange. The cells should be viewed in the presence of standard growth medium. We have found, however, that in many cases YPD (or YEPD) and other yeast-extract and peptone-based formulations give unacceptable levels of background fluorescence, especially in the red region of the spectrum. This is doubly true if using *ade2* mutants (not recommended), which accumulate a red-fluorescent pigment. One should therefore make an effort to conduct experiments in standard medium based on yeast nitrogen base (YNB; e.g., SD or SMD) to lower background fluorescence. Finally, in many cases, and especially when taking time-lapse photographs, it is necessary to embed the cells in a gel matrix to prevent movement as a result of Brownian motion or surface tension effects that arise from drying at the air-water interface. For this purpose we also include 1%–2% agarose (preferably low-melt agarose) in the mounting solution to hold the cells in place.

2.2. Morphological detection of autophagic bodies

2.2.1. Application

The observation of autophagic bodies in the lumen of yeast vacuoles upon nitrogen starvation was the initial finding that spurred research into autophagy in yeast and allowed for the identification of the first autophagy mutant (*atg1*; Tsukada and Ohsumi, 1993). This protocol may be used as a quick and dirty assay for testing whether autophagy has occurred without recourse to various molecular biological tools and procedures.

2.2.2. Materials and methods: Viewing autophagic bodies in PMSF-treated cells and *pep4Δ* mutants

2.2.2.1. PMSF treatment After growing the cells in minimal dextrose medium (SD; 1x yeast nitrogen base, 2% glucose plus any required amino acid, nucleotide, or vitamin supplement) to mid-log phase under vigorous agitation in 25-mm glass tubes or in flasks, wash 10 ml of cells with 50 m*M* HEPES buffer, pH 7.0 (centrifuge at $2000\times g$ for 3 min in a 15-ml conical polypropylene tube) and then wash 3 times in HEPES buffer containing 2 mM PMSF (centrifuge for 2 min at $6000\times g$ at room temperature in a microcentrifuge).

Note: Because PMSF is highly insoluble, make a 200 m*M* stock dissolved in pure reagent-grade absolute ethanol and dilute the stock solution into the HEPES buffer just before adding to the cells. We add PMSF separately to each batch of HEPES buffer before resuspending the cells, as the half-life of PMSF in aqueous solution is extremely short. An alternative is to use Pefabloc, which is more soluble in water and has a longer half-life but is more expensive. The repeated exposure to PMSF allows high levels of PMSF to accumulate in hydrophobic cell structures, at levels that do not occur during normal exposure in solution. This is compounded by the fact that during centrifugation microcrystals of undissolved PMSF gain access to cell membranes, which leads to additional diffusion of PMSF into the cells without dissolving in the aqueous solution. Thus, the cell is loaded up with PMSF in hydrophobic compartments that allow for a reasonable half-life (low exposure to water) of PMSF for the duration of the experiment.

2.2.2.2. Starvation Once loaded with PMSF, cells are returned to SD medium for 15–30 min, then washed once with sterile double-distilled water, and resuspended in 10 ml of nitrogen starvation medium (1x YNB lacking ammonium sulfate and amino acids, 2% glucose) and containing 1 m*M* PMSF (added fresh from a 200-m*M* stock in absolute ethanol). After a 4-h incubation at 26 °C–30 °C, the cells may be viewed in a microscope equipped with phase contrast or Nomarski (DIC) optics.

2.2.2.3. Modified protocol for* pep4Δ *and* atg15Δ *cells The role of PMSF treatment in the previously mentioned protocol is the inhibition of autophagic body breakdown. An alternative approach is to utilize mutants that accumulate autophagic bodies. Pep4 is a vacuolar aspartyl protease that is required for the posttranslational maturation and activation of many vacuolar hydrolases. Hence, *pep4*Δ mutants are defective in most hydrolytic vacuole functions, although they contain a morphologically intact vacuolar compartment. Likewise, Prb1 is a serine protease apparently required for autophagic body degradation, and Atg15 is a lipase that specifically functions in the degradation of autophagic bodies. In *pep4*Δ, *prb1*Δ, and *atg15*Δ mutants, one observes autophagic body accumulation in the absence of PMSF upon starvation. In these cells, one may simply starve cells by washing in sterile double-distilled water and resuspending in nitrogen starvation medium, followed by incubation and viewing as previously.

2.3. Fluorescent detection of autophagic body accumulation

2.3.1. Application

The observation of autophagic bodies by light microscopy requires that the autophagic bodies have different light-scattering properties relative to the vacuole lumen. In addition, it requires significant experience on the part of the viewer in determining whether the vacuolar content represent autophagic bodies or just "schmutz," given that intravacuolar vesicles occur at a low level in normally growing yeast cells. To circumvent these complications, we have developed an assay that utilizes the fact that FM 4-64 accumulates at the inner leaflet of the vacuolar membrane (topologically identical with the outer leaflet of the plasma membrane). From this location, FM 4-64 can slowly diffuse (as it is soluble in aqueous solution) to any intravacuolar membrane, such as the membranes that delimit autophagic bodies. Hence, the manifold increase in lumenal membrane surface area that occurs when autophagic bodies accumulate in the vacuoles of starved yeast cells can be transformed into a clear shift in the fluorescence pattern of FM 4-64, from an exclusively vacuolar membrane staining ("rim staining") to a diffuse staining of the vacuolar lumen with little or no emphasis on the limiting membrane of the vacuole, in some cases including actual observation of fluorescent intravacuolar bodies (these are often blurred because of the long time scale of the exposures).

2.3.2. Materials and methods: Viewing changes in FM 4-64 fluorescence distribution during the induction of autophagy

In combining the microscopy assay for accumulation of autophagic bodies with FM 4-64 fluorescence, one faces several degrees of freedom in terms of the order of the treatments. Because FM 4-64 is highly stable in membranes (at least up to 8 h) and because its addition may perturb membranes, we chose to stain the cells with FM first, followed by a recovery and chase period, before the rest of the treatments, which allows the membranes time to recover from any perturbations Thus, logarithmically growing cells in SD (OD_{600} of 0.4–0.6) are stained for 20 min with 0.8 μM of FM 4-64 as previously (section 2.1.3), washed in SD, and allowed to recover and to chase the dye into the vacuolar membrane. After a 40-min recovery, the cells can be treated with PMSF and then transferred to various treatments, such as nitrogen starvation. The outcome of such an experiment is illustrated in Fig. 7.1A. The change in fluorescence pattern distribution is due to autophagic trafficking, as it is not observed in *atg8*Δ cells. We have used this protocol to assay autophagy in the food-spoilage yeast *Z. bailii*, to show that autophagy is inhibited by benzoic acid in this yeast species, as also occurs in *Saccharomyces cerevisiae* (Fig. 7.1B). We have found it useful to buffer the starvation medium to pH 4.2–4.3 with 10 m*M* sodium citrate, to better observe changes in FM 4-64 distribution under starvation conditions (Fig. 7.2). Citrate is not consumed by yeast and does not passively cross the cell membrane. In the absence of this buffering agent, less intravacuolar fluorescence is accumulated and autophagic bodies are less obvious, although a change in the FM 4-64 fluorescent pattern is still observed.

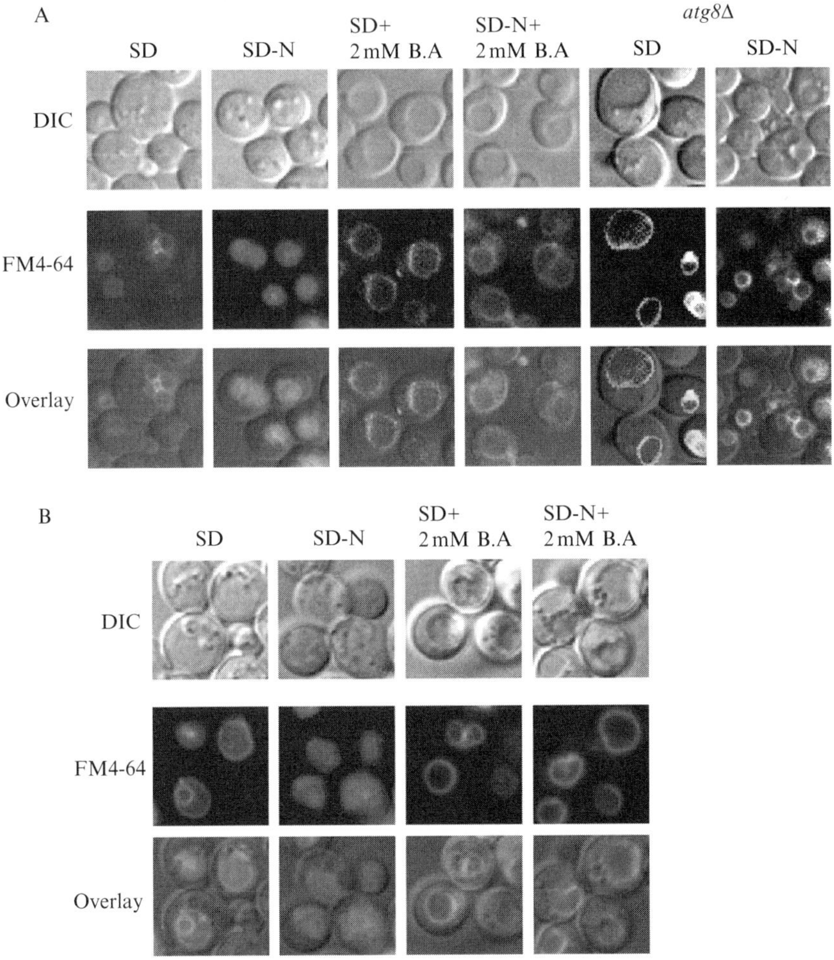

Figure 7.1 Assaying autophagy in *S. cerevisiae* and *Z. bailii* using FM 4-64 fluorescence. (A) Validation of an FM 4-64 fluorescence-based assay for macroautophagy. Wild-type *S. cerevisiae* cells and isogenic *atg8*Δ mutants were grown to mid-log phase. Cells were stained with FM 4-64 (0.8 μM) for 30 min, washed and incubated with or without 2 m*M* benzoic acid in SD medium (pH 4.3) containing 1 m*M* PMSF or in nitrogen starvation medium (SD-N, pH 4.3) containing 1 m*M* PMSF. Cells were viewed by light and fluorescence microscopy after a period of 4 h. Top panel: DIC, differential interference contrast. Middle panel: FM 4-64 fluorescence. Bottom panel: overlay of DIC and FM 4-64 fluorescence images. (B) Visualization of benzoic acid–mediated inhibition of macroautophagy in *Z. bailii*. Cells were grown to mid-log phase and treated as in (A). Top panel: DIC, differential interference contrast. Middle panel: FM 4-64 fluorescence. Bottom panel: overlay of DIC and FM 4-64 fluorescence images. Figure previously published in "Caffeine induces macroautophagy and confers a cytocidal effect on food spoilage yeast in combination with benzoic acid," *Autophagy* 4 (2008): 28–36.

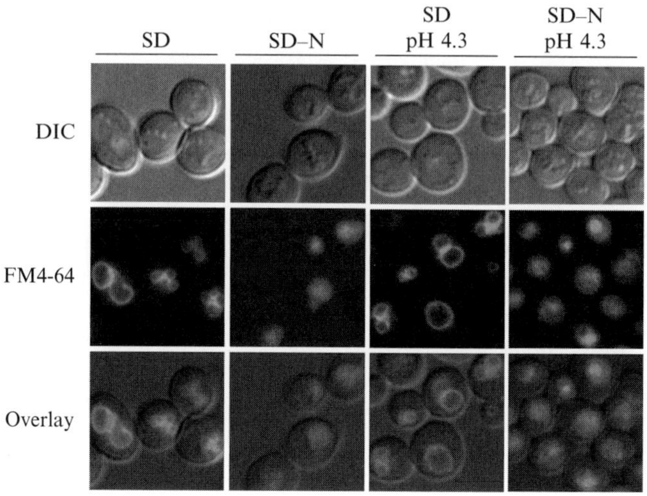

Figure 7.2 Medium acidification enhances the observation of autophagy by FM 4-64 staining. Wild-type *S. cerevisiae* cells were treated as described in Fig. 7.1 except that resuspension was done in unacidified media (pH ~ 6; left panels) versus media buffered at pH 4.3 using 10 m*M* citrate (right panels). (See Color Insert.)

2.3.3. Protocol

1. Grow cells overnight in SD medium: 2% glucose, 1x YNB (without amino acids; Difco), and any required auxotrophic factor, such as amino acids, nucleotides, or vitamins.
2. Spin 10 ml of cells in a tabletop centrifuge (2000×*g* for 3 min) and resuspend in 1–5 ml of SD medium containing 0.8 μM FM 4-64. Final concentration of the cells should result in an approximate OD_{600} of 5–10. The resuspension can be done in a plastic conical 50-ml screw-cap tube (e.g., Falcon) to ensure aeration. The cells are then incubated with vigorous aeration for 20 min at 30 °C.
3. Spin the cells and resuspend in 10 ml of the original growth medium without FM 4-64. Continue incubation for an additional 40 min.
4. Spin the cells in a tabletop centrifuge (2000×*g* for 3 min) and wash 3 times with 10 ml of 50 m*M* HEPES (pH 7.0) containing 1 m*M* PMSF.
5. Resuspend in 10 ml of the original growth medium supplemented with 1 m*M* PMSF and continue shaking for 30 min.
6. Spin the cells in a tabletop centrifuge (2000×*g* for 3 min) and wash once with sterile distilled water. Resuspend in 10 ml of nitrogen starvation medium supplemented with 10 m*M* citrate (pH 4.3) or in control SD medium buffered with 10 m*M* citrate, pH 4.3, both containing 1 m*M* PMSF.
7. Incubate with vigorous shaking for 4 h.

8. Spin 1 ml of cells in a microcentrifuge (6000×*g* for 2 min). Resuspend in 100 μl of starvation medium.
9. Spot 4 μl of cell suspension on a microscope slide.
10. Add 4 μl of nitrogen starvation medium containing 2% low-melting agarose (preequilibrated at 40 °C after melting) to the drop on the slide and rapidly mix by pipetting up and down.
11. Immediately cover the slide with a coverslip. Ensure that you do not trap large air bubbles or allow the agarose to solidify prior to mixing.
12. View in a fluorescence microscope equipped with a long-pass red filter (e.g., Cy3, Texas red, rhodamine).

2.4. Interpreting the results: Limitations and caveats

The use of FM 4-64 staining to detect the accumulation of autophagic bodies adds specificity and coherence to light microscopy–based methods for assessing accumulation of autophagic bodies. As shown in this chapter, it allows the monitoring of autophagy in organisms for which molecular biological tools are not available for this purpose. However, several caveats must also be noted. First, this method, as with any other method that monitors accumulation of autophagic bodies, does not indicate the mechanism by which autophagic bodies accumulate. This is articulated specifically with respect to microautophagy versus macroautophagy. Thus, the change in the fluorescence pattern of FM 4-64 in cells or the observation of autophagic bodies by light microscopy tells us that autophagy has been induced, but it does not indicate whether macroautophagy or microautophagy has taken place (Muller *et al.*, 2000; Roberts *et al.*, 2003). A second caveat is that the observation of the shift in the fluorescence pattern upon starvation is enhanced on acidification of the growth medium to pH 4.3. While it is unclear why this occurs, we speculate that low extracellular pH may affect the vacuolar membrane potential and thus allow for a more obvious distribution of the dye between the limiting membrane and the internal membranes of the vacuole.

REFERENCES

Abeliovich, H., Grote, E., Novick, P., and Ferro-Novick, S. (1998). Tlg2p, a yeast syntaxin homolog that resides on the Golgi and endocytic structures. *J. Biol. Chem.* **273,** 11719–11727.

Betz, W. J., Mao, F., and Bewick, G. S. (1992). Activity-dependent fluorescent staining and destaining of living vertebrate motor nerve terminals. *J. Neurosci.* **12,** 363–375.

Betz, W. J., Mao, F., and Smith, C. B. (1996). Imaging exocytosis and endocytosis. *Curr. Opin. Neurobiol.* **6,** 365–371.

Burd, C. G., and Emr, S. D. (1998). Phosphatidylinositol(3)-phosphate signaling mediated by specific binding to RING FYVE domains. *Mol. Cell.* **2,** 157–162.

Muller, O., Sattler, T., Flotenmeyer, M., Schwarz, H., Plattner, H., and Mayer, A. (2000). Autophagic tubes: Vacuolar invaginations involved in lateral membrane sorting and inverse vesicle budding. *J. Cell Biol.* **151,** 519–528.

Roberts, P., Moshitch-Moshkovitz, S., Kvam, E., O'Toole, E., Winey, M., and Goldfarb, D. S. (2003). Piecemeal microautophagy of nucleus in *Saccharomyces cerevisiae. Mol. Biol. Cell* **14,** 129–141.

Rodes, J. F., Berreur-Bonnenfant, J., Tremolieres, A., and Brown, S. C. (1995). Modulation of membrane fluidity and lipidic metabolism in transformed rat fibroblasts induced by the sesquiterpenic hormone farnesylacetone. *Cytometry* **19,** 217–225.

Sakai, Y., Koller, A., Rangell, L. K., Keller, G. A., and Subramani, S. (1998). Peroxisome degradation by microautophagy in *Pichia pastoris*: Identification of specific steps and morphological intermediates. *J. Cell Biol.* **141,** 625–636.

Takeshige, K., Baba, M., Tsuboi, S., Noda, T., and Ohsumi, Y. (1992). Autophagy in yeast demonstrated with proteinase-deficient mutants and conditions for its induction. *J. Cell Biol.* **119,** 301–311.

Tsukada, M., and Ohsumi, Y. (1993). Isolation and characterization of autophagy-defective mutants of *Saccharomyces cerevisiae. FEBS Lett.* **333,** 169–174.

Vida, T. A., and Emr, S. D. (1995). A new vital stain for visualizing vacuolar membrane dynamics and endocytosis in yeast. *J. Cell Biol.* **128,** 779–792.

Wendland, B., McCaffery, J. M., Xiao, Q., and Emr, S. D. (1996). A novel fluorescence-activated cell sorter-based screen for yeast endocytosis mutants identifies a yeast homologue of mammalian eps15. *J. Cell Biol.* **135,** 1485–1500.

Winter, G., Hazan, R., Bakalinsky, A. T., and Abeliovich, H. (2008). Caffeine induces macroautophagy and confers a cytocidal effect on food spoilage yeast in combination with benzoic acid. *Autophagy* **4,** 28–36.

CHAPTER EIGHT

Monitoring Mitophagy in Yeast

Nadine Camougrand,* Ingrid Kiššová,† Benedicte Salin,* *and* Rodney J. Devenish‡

Contents

* Institut de Biochimie et Génétique Cellulaires, CNRS, Université de Bordeaux2, Bordeaux, France
† Comenius University, Faculty of Natural Sciences, Department of Biochemistry, Bratislava, Slovak Republic
‡ Department of Biochemistry and Molecular Biology, and ARC Centre of Excellence in Structural and Functional Microbial Genomics, Monash University, Clayton Campus, Melbourne, Victoria, Australia

Methods in Enzymology, Volume 451
ISSN 0076-6879, DOI: 10.1016/S0076-6879(08)03208-4

Abstract

Cellular degradative processes including proteasomal and vacuolar/lysosomal autophagic degradation, as well as the activity of proteases (both cytosolic and mitochondrial), provide for a continuous turnover of damaged and obsolete macromolecules and organelles. Mitochondria are essential for oxidative energy production in aerobic eukaryotic cells, where they are also required for multiple biosynthetic pathways to take place. Mitochondrial homeostasis also plays a crucial role in aging and programmed cell death, and recent data have suggested that mitochondrial degradation is a strictly regulated process. A recent study has shown that in yeast cells subjected to nitrogen starvation, degradation of mitochondria by autophagy occurs by both a selective process (termed *mitophagy*) and a nonselective process. This chapter provides an overview of the techniques that enable the study of mitophagy. Fluorescent proteins targeted to mitochondria can be used to follow mitochondrial sequestration within vacuoles. Degradation of mitochondria can be assayed using a mitochondrially targeted alkaline phosphatase (ALP) reporter test in which the delivery of mitochondrial N-terminal truncated Pho8Δ60 to the vacuole results from mitophagy. Degradation of mitochondrial proteins can also be followed by Western immunoblot analyses. Finally, electron microscopy observations permit the discrimination between selective mitophagy and nonselective mitochondrial degradation.

1. Introduction

Autophagy is a highly conserved process in which the cytoplasm, including excess or aberrant organelles, is sequestered into double-membrane vesicles called autophagosomes and delivered to the lysosome/vacuole for breakdown (Yorimitsu and Klionsky, 2005). Although autophagy was originally considered to represent a nonspecific bulk degradation pathway, recent data suggest that there are selective autophagy pathways capable of specifically targeting organelles, such as mitochondria (Kissova *et al.*, 2004, 2007) or ribosomes (Kraft *et al.*, 2008). Because of their central role in many biological processes and pathological issues, mitochondria have been the focus of many investigations aiming to find evidence of selective autophagy of mitochondria, or mitophagy. Experiments on mammalian cells show that mitochondria are actively degraded after induction or blockade of apoptosis (Tolkovsky *et al.*, 2002; Xue *et al.*, 1999, 2001) or following opening of the mitochondrial permeability transition pore (Kim *et al.*, 2007; Rodriguez-Enriquez *et al.*, 2004). Further, the inactivation of mitochondrial catalase induces the autophagic elimination of altered mitochondria (Yu *et al.*, 2006). In yeast, alterations of mitochondrial biogenesis (Priault *et al.*, 2005) or disturbance of the maintenance of ionic balance (Nowikovsky *et al.*, 2007) trigger autophagy. Conversely, recent findings suggest that autophagy-deficient mutants accumulate altered mitochondria (Zhang *et al.*, 2007). The selective degradation of mitochondria through autophagy occurs in yeast in response to starvation or rapamycin (Kissova *et al.*, 2004).

Yeast cells exhibit a versatile metabolism, which depends on the carbon source present in the medium. The fermentable sugar glucose is used in most studies of yeast biology. However, the Crabtree effect prevents the differentiation of fully functional mitochondria, making glucose of poor interest for studies of this organelle in yeast. The Crabtree effect describes the phenomenon whereby in yeasts such as *Saccharomyces cerevisiae* high concentrations of glucose inhibit cellular respiration. Yeast cells can be grown on lactate, a nonfermentable carbon source oxidized to pyruvate by two mitochondrial lactate dehydrogenases that directly transfer electrons to the mitochondrial respiratory chain. This strictly mitochondria-dependent metabolism provides an optimal differentiation of a higher number of mitochondria. The comparison of such lactate-grown cells with glucose-grown cells, after a shift to conditions of nitrogen starvation, reveals a major difference in the type of autophagy that is induced. As expected, the outcome in glucose-grown cells involves the formation of autophagosomes in abundance and their delivery to the vacuole as autophagic bodies, typical of the process of macroautophagy. In lactate-grown cells, on the other hand, these events are rare and cells preferentially undergo microautophagy instead of classical macroautophagy (Kissova *et al.*, 2007).

Extensive electron microscopy studies have shown that autophagy of mitochondria can occur following two distinct processes: one selective and dependent on the mitochondrial protein Uth1, and the other nonselective and Uth1 independent. The selective process was characterized by the early appearance of a large number of physical contacts between mitochondria and vacuoles, followed by the visualization of microautophagic bodies containing almost exclusively mitochondria. The impairment of this process in a Uth1-deficient mutant led to a dramatic decrease in the rate and the efficiency of autophagic mitochondrial degradation (Kissova *et al.*, 2004) and to different phenotypic alterations related to mitochondrial homeostasis (Camougrand *et al.*, 2003; Kissova *et al.*, 2006). This chapter provides an overview of the techniques that enable the study of mitophagy as well as nonselective mitochondrial degradation. Fluorescent proteins targeted to mitochondria can be used to follow mitochondrial sequestration within vacuoles. Degradation of mitochondria can be assayed using an alkaline phosphatase (ALP) reporter in which the delivery of mitochondrially targeted N-terminal truncated Pho8Δ60 to the vacuole results from mitophagy. Degradation of mitochondria can also be followed by Western immunoblot analyses. Finally, electron microscopy observations enable discrimination between selective and nonselective forms of mitophagy.

2. Visualization of Mitochondrial Sequestration within the Vacuole by Fluorescence Microscopy

2.1. Design of constructs

Mitochondria comprise four distinct compartments: outer membrane (OM), inner membrane (IM), intermembrane space (IMS), and matrix. Theoretically, the fluorescent marker can be targeted to any of these compartments; indeed, there are many reports on the use of a fluorescent protein (FP), usually green-fluorescent protein (GFP), to tag proteins specifically targeted to each of the four mitochondrial compartments. To date, however, there are few reports describing the labeling with FPs of mitochondria for the purpose of following their fate during autophagy. In studies carried out in the authors' laboratories, both GFP (Kissova *et al.*, 2004) and Rosella, a fusion of an RFP (DsRed.T3) in tandem with GFP (super ecliptic pHluorin) (Rosado *et al.*, 2008), have been successfully used to follow mitophagy.

2.2. Delivery of the FP to mitochondria

In our studies the FP was targeted to the matrix by use of the N-terminal leader of the *CIT1* gene encoding mitochondrial citrate synthase (CS). It is not yet clear that it is possible to detect mitophagy when an FP is used to label a compartment other than the matrix. For more detailed discussion of this point, see the accompanying chapter by Devenish *et al.* (chapter 9). The precursor protein CS-FP is synthesized in the cytosol and the CS mitochondrial targeting signal (MTS) acts to transport it to the import machinery at the mitochondrial OM. The precursor is translocated across the OM in an unfolded conformation before engaging with the IM components of the import machinery. The precursor is driven across the IM by the combined action of $\Delta\psi$ (the membrane potential) and the PAM (presequence translocase-associated motor) machinery. On entry into the matrix, the MTS is cleaved by the mitochondrial-processing peptidase (MPP), thereby releasing the FP into the matrix (for a review of mitochondrial import machinery, see Baker *et al.*, 2007). Theoretically, the CS sequence, or any other MTS used, could be fused directly to the FP, but typically the MTS sequence fused to the FP will encompass several amino acids downstream of the native MPP cleavage site to preserve the authentic cleavage site. In constructs used in the authors' laboratories, the FP component is targeted to the matrix by use of a 52-amino-acid CS precursor polypeptide derived from the CIT1 gene (Gavin *et al.*, 2002; Okamoto *et al.*, 1998).

Another strategy used to target an FP reporter to a matrix location is to express Rosella as a fusion with Atp5 (subunit OSCP of mitochondrial

F_1F_o-ATP synthase). In this case the targeting capability of the native subunit was used and Rosella was not cleaved from Atp5. The Atp5-Rosella was incorporated into functional mitochondrial ATP synthase complexes, resulting in Rosella being anchored to the matrix face of the IM. In this particular case expression was under control of the *GAL1* promoter in the yeast expression vector pRJ21N-ATP5. The regulated expression was used to establish that delivery of Rosella to the vacuole during starvation did not occur because of redirection to the vacuole without first being imported into the mitochondrial matrix (for details, see Rosado *et al.*, 2008).

Another consideration for labeling mitochondria is the use of single- or multicopy vectors to drive expression. Alternatively, sequences encoding a mitochondrial-FP fusion could be chromosomally integrated such that the reporter protein is expressed under transcriptional control of its native promoter. Use of this form of construct, while successful in other contexts (see Gavin *et al.*, 2002), has not been reported for following mitophagy. Rosella is expressed from a multicopy expression vector under control of the *PGK1* promoter (Rosado *et al.*, 2008). Kissova *et al.* (2004) use a CS-GFP fusion under the control of the *GAL1/10* promoter on the single copy (CEN4-URA3) vector pGAL-CLbGFP (Okamoto *et al.*, 1998).

Westermann and Neupert (2000) have described a set of plasmids for expression of mitochondria-targeted GFP in yeast. Here, the GFP is targeted to mitochondria by the well-characterized MTS of the mitochondrial F_1F_o-ATP synthase subunit 9 from *Neurospora crassa.* The vectors include constructs with strong regulable and constitutive promoters, four different auxotrophic markers for yeast transformation, a green variant (S65T) and a blue-shifted (P4-3) variant of GFP. In general, mitochondria are brightly fluorescent in living yeast cells grown on different carbon sources and at different temperatures, with virtually no background staining. Fluorescence is brighter with expression from a multicopy vector. An example of the visualization of mitochondria using mitochondrial matrix-targeted GFP expressed from such a multicopy vector (pYX232-mtGFP, which incorporates the constitutive *TPI1* [triosephosphate isomerase] promoter), is presented by Nowikovsky *et al.* (2007).

2.3. Transformation of yeast cells with mitochondrial-FP constructs

Yeast transformations were performed by standard procedures (Gietz and Woods, 2002) using only minor modifications. The first phase is to make the cells competent for transformation.

1. Grow the strain to be transformed on plates containing ethanol (e.g., YEPE: 1% yeast extract, 1% peptone, 2% ethanol) or another nonfermentable growth substrate to remove accumulated petite cells (i.e., those having deletions or complete loss of mitochondrial DNA).

2. Inoculate duplicate 2.5-cm diameter glass tubes containing 10 ml of YEPD (1% yeast extract, 1% peptone, 2% glucose, buffered to pH 6.0) with cells taken from the cell stock tested for competence on a non-fermentable carbon source and incubate overnight with shaking at 28 °C.

   Note: Each 10 ml of culture will provide enough cells for approximately two transformations.

3. Determine the cell density and then dilute each overnight culture into 50 ml of prewarmed YEPD (preferably in a fluted flask for optimal aeration) to give 5×10^6 cells/ml ($O.D._{650} = 0.5$).
4. Incubate for 5–6 h at 28 °C with shaking. It is important that 2–4 cell doublings take place (although the transformation efficiency [transformants/pg plasmid/10^8 cells] remains constant for 3–4 cell divisions.] This can be checked by determining the $O.D._{650}$ of the culture.
5. Cells are harvested in sterile 50-ml tubes by centrifugation at $3000 \times g$ (5000 rpm) for 5 min.
6. The cell pellet is washed once in 25 ml of sterile water and the cells are resuspended in 1.0 ml of 100 m*M* LiAc (pH is not adjusted).
7. Following transfer of the cell suspension to a 1.5-ml microcentrifuge tube, cells are pelleted and finally resuspended in 100 m*M* LiAc to achieve a final cell density of 2×10^9 cells/ml (approximately 100 μl for each 10 ml of starting culture).

The second phase is to carry out the transformation with the desired plasmid DNA.

1. Cells prepared as described previously are vigorously mixed and then 50-μl volumes of cell suspension in microcentrifuge tubes are centrifuged at 13,000 rpm for 15 s to pellet the cells.
2. After removal of the LiAc solution, the following components of the transformation mix are carefully added in the order listed: 240 μl of PEG (50% w/v), 36 μl of 1.0 *M* LiAc, 50 μl of carrier DNA (2.0 mg/ml), X μl of plasmid DNA (0.1–10 μg), and 34-X μl sterile water (the final volume will be 360 μl).
   Note: All of the ingredients can be premixed except for the plasmid DNA. If this is done, then 355 μl of the mix is added directly to the cell pellet followed by 5 μl of plasmid DNA, followed by mixing.
3. The contents of each tube are mixed vigorously until the cell pellet has been completely resuspended.
4. Incubate at 30 °C for 30 min and then subject the cell suspension to heat shock at 42 °C for 30 min.

Note: The optimum time can vary for different yeast strains and should be determined empirically to achieve high-efficiency transformation. Note that some strains accumulate petite cells in response to heat shock at 42 °C; in such a case it may be necessary to use an alternative transformation procedure, such as that of Klebe *et al.* (1983).

5. Pellet the cells by centrifugation at 6–8000 rpm for 15 s and remove the supernatant.
6. Add 1.0 ml of sterile water to each tube, and resuspend the pellet by pipetting it up and down very gently. Spread from 2–200 μl of each transformation mix onto appropriate selective minimal medium plates (buffered to pH 5.6). If plating fewer than 200 μl, this is best done by delivering the desired volume of cell suspension into not more than a final volume of 200 μl of sterile water on the plate.

Before undertaking experiments to determine mitochondrial autophagy, it is necessary to select a transformant colony that retains the ability to grow on a nonfermentable substrate and displays strong fluorescence. The latter can be done by taking several transformants for each plasmid-strain combination and viewing the cells using a fluorescence microscope.

2.4. Growth of yeast cells expressing mitochondrial-FP constructs

1. Yeast cells are grown aerobically at 28 °C in a complete medium for 24 h to O.D.$_{650\ nm}$ of 2–4. The Monash laboratory has used 10-ml volumes of *Saccharomyces* Salts medium (SS; comprising $(NH_4)_2SO_4$, 0.12%; KH_2PO_4, 0.10%; $MgCl_2$-$6H_2O$, 0.07%; NaCl, 0.05%; $CaCl_2$, 0.01%; $FeCl_3$, 0.0005%; Proudlock *et al.*, 1971), whereas the Bordeaux laboratory has used YNB medium (yeast nitrogen base without ammonium sulfate and amino acids 0.175%, potassium phosphate 0.1%, ammonium sulfate 5%, Drop-Mix [a mixture of amino acids each at 100 mg/l not including those used for selection] 0.2%, and auxotrophic requirements 0.01%, pH 5.5, supplemented with 2% lactate as a carbon source).
2. Cells are then centrifuged at 13,000 rpm for 60 s and shifted to a nitrogen starvation medium (10 ml) to induce autophagy (Takeshige *et al.*, 1992). Typically, such medium contains yeast nitrogen base (0.17% w/v) without added amino acids and ammonium sulfate (pH 5.5) and containing 2% glucose or 2% ethanol as a carbon source (Rosado *et al.*, 2008). A slightly different formulation of nitrogen starvation medium with 2% lactate as carbon source has been used by Kissova *et al.* (2004, 2007).
3. Cells are harvested from complete medium by centrifugation as previously, and washed 3 times with 1 ml of sterile water.
4. The cells are reinoculated into fresh complete medium (to provide control, nonstarved cells), and into nitrogen starvation medium. For expression of constructs under *GAL* promoter transcription control, the medium must be supplemented with 0.5%–1% galactose to induce transcription. The Monash laboratory starves cells for 6 h and then grows for 6–8 h in galactose supplemented medium (equivalent of 2–4 cell doublings) for good induction. It is important to note that these times

may vary for different strains and optimal times of induction should be determined empirically.

A systematic analysis of the time of starvation for different yeast strains and different carbon sources has not been performed, and therefore this needs to be optimized for each strain and medium combination being used. Thus, a 6-h starvation period was used by Rosado *et al.* (2008) and a 15-h nitrogen starvation period by Kissova *et al.* (2004, 2007).

An alternative to nitrogen starvation for induction of autophagy is rapamycin treatment. Again, such treatment will have to be optimized for strain and medium combinations. In many experiments in different laboratories, rapamycin (Sigma) was added at a final concentration of 0.2 μg/ml (using a stock solution of 20 μg/ml in 90% ethanol and 10% Triton X-100), as first described by Noda and Ohsumi (1998).

2.5. Fluorescence microscopy observations

1. In preparation for microscopy, cells were washed 3 times with sterile water to remove medium and mounted on thin blocks of 0.2% (w/v) low-melting point agarose reconstituted with fresh medium. To form the agarose block, 20 μl of molten agarose is placed on a glass microscope slide. Then 10 μl of cell suspension is immediately added onto the agarose and covered with a coverslip. This serves to restrict movement of cells during imaging.
2. Mounted cells were then immediately imaged using Confocal Laser Scanning Microscopy. The Monash laboratory uses an Olympus FV500 equipped with a 60X water-immersion objective; the Bordeaux laboratory uses a Leica microsystems DM-LB epifluorescence microscope. Note that in the latter case cells were not washed prior to observation of fluorescence.
3. When required, cell vacuoles were stained for imaging by incubation for 30 min at 28 °C with CMAC–Arg (7-amino-4-chloromethylcoumarin, L-arginine amide; Invitrogen, Molecular Probes) at a final concentration of 100 μM, and then washed 3 times with sterile water before mounting. Alternatively, vacuoles can be stained with *N*-(3-triethylammonium-propyl)-4-(*p*-diethylaminophenyl-hexatrienyl) pyridinium dibromide (FM 4-64; Molecular Probes, final concentration of 40 μM) according to a modified method as described previously by Vida and Emr (1995).
4. Delivery of Rosella to the vacuole is quantified by scoring for the accumulation of red fluorescence in the vacuole concomitant with checking for the absence of green fluorescence (see Fig. 8.1). Green fluorescence emission (530 $\pm$ 30 nm) and red fluorescence emission (>590 nm) images are acquired sequentially upon excitation with 488-nm or 543-nm laser light, respectively. Once set, photomultiplier

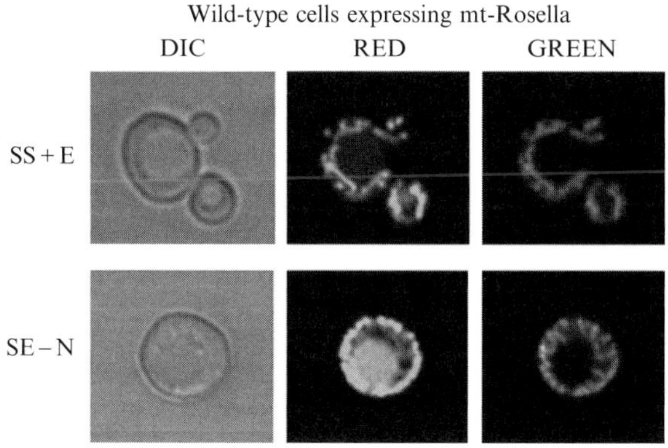

Figure 8.1 The Rosella biosensor targeted to mitochondria (mt-Rosella) is delivered to the vacuole under conditions of nitrogen starvation. DIC and fluorescence images are shown for wild-type cells under growing conditions (SS + E) and after 6 h under nitrogen starvation (SS-N). (See Color Insert.)

tube settings are not altered within individual experiments. In each experiment at least 100–200 cells are scored. ImageJ software (version 1.36b) is employed for image analysis (http://rsb.info.nih.gov/ij/). Other microscope-imaging configurations are used by different laboratories. In the Schweyen laboratory (Vienna) confocal images are captured with a Zeiss Axiovert LSM 510 microscope. The excitation wavelength for GFP and FM 4-64/dsRed is 488 nm and 543 nm, respectively. Images are acquired in multitrack Z stack and projected with LSM5 Image Browser (Nowikovsky *et al.*, 2007). The Bordeaux laboratory uses a Leica microscope with an SIS camera for image acquisition. Images are processed with Corel Draw 9.0 software (Kissova *et al.*, 2004).

3. Mitochondrially Targeted Alkaline Phosphatase

To measure mitochondrial degradation by the vacuole, Campbell and Thorsness (1998) modified an *in vivo* enzymatic assay for the detection of autophagy in yeast that was initially described by Noda *et al.* (1995). *PHO8*, encoding the vacuolar alkaline phosphatase (ALP), is translated as a proenzyme and matured to its active form in the vacuole upon cleavage of the C-terminus by proteinase A (Jones *et al.*, 1982). For this assay, the *PHO8* gene was modified to direct the proenzyme ALP to mitochondria (ALPm) or to the cytoplasm (ALPc) rather than to the vacuole by deleting the first 60 codons bearing the vacuolar targeting signal and a transmembrane domain (*PHO8Δ60*). In a yeast strain bearing a deletion of the genomic alkaline phosphatase loci, *PHO8* and *PHO13*, the only significant source of ALP activity is ALPm or ALPc that has been delivered to the vacuole via mitochondria or autophagosomes and activated by a vacuolar protease. Thus, by assaying ALPm activity in various genetic backgrounds and under different growth conditions, the relative rate of vacuole-dependent mitochondrial turnover can be assessed. Further, measurement and comparison of both ALPm and ALPc activities can be used to determine whether degradation of mitochondria is a selective process or reflects only a general increase in autophagy of bulk cytoplasm (including organelles) under the test conditions.

3.1. Construction of the mitochondrially targeted *Pho*8Δ60 expression vector (Campbell and Thorsness, 1998)

First, the *PHO8* locus was disrupted in strain CCY29 by transformation with an *Eco*RI fragment bearing *pho8Δ::URA3* from pAR2. Next, *PHO*13 was disrupted by transformation of strain CCY28 with an *Eco*RI fragment bearing *pho13Δ::URA3* from pPH13 (Kaneko *et al.*, 1989). Then, strains CCY29 *pho8Δ::URA3* and CCY28 *pho13Δ::URA3* were mated and a Ura+ haploid spore chosen from a tetrad bearing 2:2 Ura+:Ura− segregants, and named strain CCY30. All *pho8* and *pho13* strains were confirmed for phenotype by assaying for endogenous ALP activity using a p-nitrophenylphosphate (Kaneko *et al.*, 1982) or α-naphthylphosphate (Kaneko *et al.*, 1985) colony overlay assay.

The mitochondrially targeted *Pho8Δ60* (ALPm) expression vector was constructed by inserting the 0.81 kb *Eco*RV fragment of *ADH1-COXIV* from *pCOXIV* wt (Pinkham *et al.*, 1994) into the *Eco*RV site of pRS313 (Sikorski and Hieter, 1989), in which the *Xba*I site had been destroyed. Then a PCR fragment corresponding to *PHO8Δ*60 was ligated in frame to the *COXIV* leader at the *Xba*I and *Bam*HI sites, generating the plasmid pCC4. The 5′ *PHO8Δ60* PCR primer was GGTCT AGA TCT GCA TCA CAC AAG AAG AAG AAT GTC ATA TTC TTC GTG and the 3′*PHO8Δ60* PCR primer was GGG GAT CCG GGA GAG TTA GAT AGG ATC AG. A plasmid designed to express cytosolically localized ALP (ALPc), pCC5, was generated by excising the *COXI* leader from pCC4 by *Eco*RI/*Xba*I cleavage and religating with a linker containing an ATG start codon. Two single-stranded oligonucleotides were used for the linker: AATTC ACC ATG G and GTGG TAC CGA TC. Both the pCC4 and pCC5 constructions were confirmed by DNA sequencing. The localization of ALPm (*ADH1-COXIV-PHO8Δ60* expression cassette) and ALPc (*ADH1-PHO8Δ60* expression cassette) was confirmed by cell fractionation and Western immunoblot analysis.

3.2. Alkaline phosphatase assays

ALP activity assays using α-naphthylphosphate (Sigma N7255) as substrate were performed on cells that had been treated with rapamycin (0.2 μg/ml) for 2 h or nitrogen starved for 15 h according to the method described by Nothwehr *et al.* (1996).

1. Triplicate aliquots of 2×10^7 cells are pelleted by centrifugation at 13,000 rpm for 60 s and washed once with 1 ml of assay buffer (250 m*M* Tris, pH 9, 10 m*M* $MgSO_4$, 10 μ*M* $ZnSO_4$, 1 m*M* PMSF, antiprotease cocktail (one Complete EDTA-free Protease Inhibitor Cocktail Tablet from Roche [ref: 11873580001]/50ml buffer).
2. Cells are resuspended in 1 ml of 4% ethanol and 1% toluene in water and vortexed for 10 min.
3. After centrifugation at 10,000×*g* for 3 min, pellets are dried at room temperature for 15 min and then resuspended in 200 μl of assay buffer.
4. The contents of the triplicate tubes are pooled and 100 μl are mixed with 400 μl H_2O in a 1.7-ml microcentrifuge tube and precipitated by adding 50 μl 3 *M* TCA and holding on ice for 20 min for protein assay determined with the Lowry method.
5. Four 100-μl aliquots are transferred to 2-ml microcentrifuge tubes to determine ALP activity.
6. To each 100-μl volume is added 1 ml of assay buffer and 100 μl of 55 m*M* α-naphtylphosphate in assay buffer.
7. The reaction is allowed to proceed at 30 °C for 20 min and then stopped by addition of 1 ml of 2 *M* glycine, pH 11. Finally, 100 μl are used to perform a time zero control by adding to 1 ml of assay buffer, 100 μl of 55 m*M* α-naphthylphosphate in assay buffer and 1 ml 2 *M* glycine, and mixing without incubation.
8. After the reaction is stopped, the cells are pelleted by centrifugation at 13,000 rpm for 60 s and the α-naphthyl product of the reaction in the supernatant is detected by measuring the fluorescence at 472 nm when excited at 345 nm.

4. Degradation of Mitochondrial Proteins

The levels of several mitochondrial proteins localized in the OM (mitochondrial porin, Por1) and IM (cytochrome c oxidase subunit 2, Cox2; or ATP synthase F1 subunit β) can be followed by Western immunoblot analyses in comparison with proteins localized in the cytosol, phosphoglycerate kinase (Pgk1), or actin (Act1), as shown in Fig. 8.2. To verify that the rapamycin-induced or nitrogen starvation–induced protein degradation was really caused by vacuolar proteolysis and not by inhibition of their synthesis, the experiment should be repeated in a *pep4*Δ strain

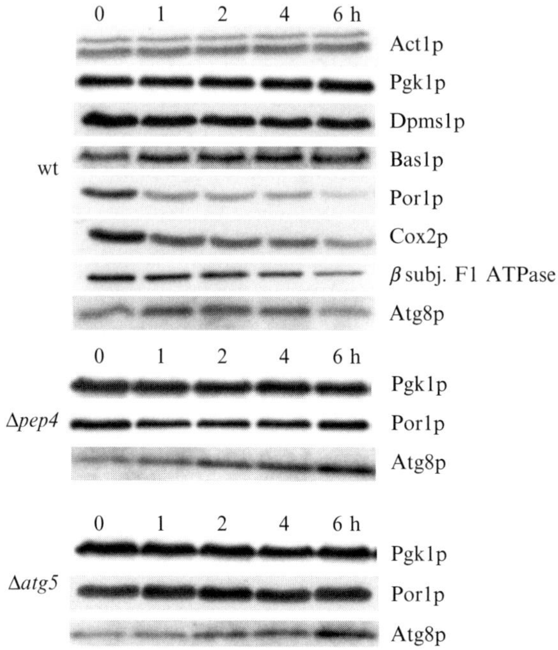

Figure 8.2 Wild-type, (wt) *pep4*Δ, and *atg5*Δ cells were grown in YNB medium supplemented with 2% lactate. Rapamycin (0.2 μg/ml) was added and, at indicated times, whole cell proteins were extracted from the same number of cells, separated by SDS-PAGE and analyzed by Western immunoblot as described in section 4.

(see Fig. 8.2). The lack of degradation in *pep4*Δ cells will confirm that the protein degradation observed in wild-type cells is executed in the vacuoles. Furthermore, by determining the level of proteins from other organelles such as the Golgi apparatus, endoplasmic reticulum, (Dpms1) nucleus (Bas1) or peroxisomes, the selectivity of the process of mitochondrial degradation in different genetic backgrounds and growth conditions can be assayed.

4.1. Protein sample preparation

1. Approximately 2×10^6 cells are harvested, resuspended in 0.5 ml of water, and added to 50 μl of a mixture of 3.5% β-mercaptoethanol in 2 *M* NaOH in a 1.7-ml microcentrifuge tube.
2. After a 15-min incubation on ice, proteins are precipitated by adding 50 μl of 3 *M* trichloroacetic acid for 15 min on ice.
3. After centrifugation for 10 min at $10{,}000 \times g$, the pellet is washed with 0.2 ml of acetone, dried and resolubilized in 8 μl of 5% SDS plus 8 μl of Laemmli buffer (2% β-mercaptoethanol, 2% SDS, 0.1 *M* Tris-HCl pH 8.8, 20% glycerol, 0.02% bromophenol blue).

4.2. SDS-page

Sixteen μl of protein sample are loaded on a polyacrylamide gel in denaturating conditions using a protocol based on Laemmli (1970) and modified by Dr. Jean Velours (Institut de Biochimie et Génétique Cellulaires, CNRS, Université de Bordeaux) in which urea is replaced by glycerol, thus allowing a better separation of membrane proteins. The polyacrylamide solution used is a mixture of acrylamide 30% (w/v) and bis-acrylamide 0.8% (w/v). Ions and organic substances are removed by incubation for 15 min in 0.5% (w/v) vegetable-activated charcoal (Prolabo) and 1.5% (w/v) serdolit M-B2 (Serva), followed by filtration (0.45-μm filter). The polyacrylamide gel is formed in two parts: the concentrating gel (5% polyacrylamide, 125 m*M* Tris-HCl pH 6.7, 0.1% SDS, 0.1% TEMED, 0.1% ammonium persulfate) overlaying the running gel (12.5% polyacrylamide, 375 m*M* Tris-HCl pH 8.8 m*M*, 0.1% SDS, 15% glycerol, 0.1% TEMED, 0.05% ammonium persulfate). A constant current of 7 mA is applied to separate proteins, until the blue marker reaches the bottom of the gel.

4.3. Western immunoblot analysis

After separation, proteins are blotted on polyvinylidene difluoride (PVDF) membranes (Problott; Perkin-Elmer). The primary antibodies used are mouse monoclonal antiyeast Cox2 (1/2000^{e}; Molecular Probes), mouse monoclonal antiyeast porin (1/2000^{e}; Molecular Probes), mouse monoclonal antiyeast phosphoglycerate kinase (1/2000^{e}; Molecular Probes), mouse monoclonal antiyeast dolichol phosphate mannose synthase (dpms) (1/2000^{e}; Molecular Probes), rabbit polyclonal antiyeast actin (1/1000^{e}; Santa Cruz Biotechnology), rabbit polyclonal antiyeast Atp6 and antiyeast β (1/5000^{e}, Dr. Jean Velours), rabbit polyclonal anti-Bas1 (1/5000^{e}; Pinson *et al.*, 2000), goat polyclonal antiyeast Atg8 (1/250^{e}, Santa Cruz Biotechnology). Secondary antimouse and antirabbit antibodies coupled to horseradish peroxidase (Jackson Laboratories) are used at 1/10,000^{e}, secondary antigoat antibodies coupled to horseradish peroxidase (Santa Cruz) are used at 1/5000^{e}. An ECL^{+} kit (Amersham Biosciences) is used for protein detection (see Fig. 8.2). Quantification of protein amounts is done using the ImageJ software (http://rsb.info.nih.gov/ij/).

5. Visualization of Mitophagy by Electron Microscopy

5.1. Yeast cultures

1. Cells from a 3-ml stationary culture are grown aerobically at 28 °C in 50 ml of minimal YNB medium (0.175% yeast nitrogen base without

amino acids and with 0.5% ammonium sulfate, 0.1% potassium phosphate, 0.2% Drop-Mix and 0.01% auxotrophic requirements, pH 5.5) supplemented with 2% lactate as carbon source.
2. Cells are harvested in exponential growth phase ($O.D._{600nm} = 2$) by centrifugation at 13,000 rpm for 60 s.
3. Cells are washed twice with water and once with 1 ml of nitrogen starvation medium. Nitrogen starvation medium (SD-N) contains only 0.175% yeast nitrogen base and 2% lactate, pH 5.5.
4. After the different washes, cells are incubated in nitrogen starvation medium in the presence of 1 m*M* PMSF (phenyl methane sulfonyl fluoride) and harvested at different times.

5.2. Electron microscopy sample preparation (TEM)

1. Harvested cells (10^9 cells) are placed on the surface of Formvar-coated copper grids (HS400 Pelanne instruments 400 mesh).
2. Each grid is quickly submersed in liquid propane (−180 °C) and then transferred to a precooled solution of 4% osmium tetraoxide in dry acetone at −82 °C for 48 h for substitution or fixation.
3. Samples are gradually warmed to room temperature, transferred in a flask, and washed 3 × 10 minutes with 5 ml of dry acetone.
4. Specimens are stained for 1 h with 1% uranyl acetate in acetone at 4 °C.
5. The stained specimens are rinsed 3 × 10 min with 5 ml of dry acetone and infiltrated progressively (one-fourth araldite and three-fourth acetone for 24 h, then one-half araldite and one-half acetone for 24 h, and finally three-fourth araldite and one-fourth acetone for 24 h) with araldite (epoxy resin, Fluka).
6. Ultrathin sections (80 nm) on copper grids (HS400) are stained 1 min by plunging them in 10 μl of 2% lead citrate in water.

Observations were performed on a Philips Tecnai 12 Biotwin (120 kV) electron microscope.

5.3. Immuno-electron microscopy sample preparation (Immuno-EM)

1. For immuno-electron microscopy, the loops are transferred into a precooled solution of 0.1% glutaraldehyde in dry acetone for 3 days at −82 °C.
2. Samples are rinsed with acetone at −20 °C, then embedded progressively (one-fourth LR Gold and three-fourth ethanol for 24 h, then one-half LR Gold and one-half ethanol for 24 h, and finally three-fourth LR Gold and one-fourth ethanol for 24 h) at −20 °C in LR Gold resin (EMS, USA). Resin polymerization was carried out at −20 °C for 7 days under UV illumination.

3. Ultrathin LR Gold sections (80 nm) are cut using a Leica ultracut ultramicrotome and are collected on nickel grids coated with Formvar.
4. Sections are first incubated for 5 min with 1 mg/ml glycine, and 5 min with fetal calf serum (1:20[e]).
5. The grids are incubated 45 min at room temperature with mouse anti-VDAC monoclonal antibody (Molecular Probes, 1/100[e]) or rabbit anti-$F_1\beta$ subunit polyclonal antibody (Dr. Jean Velours, 1/2500[e]).
6. The grids are then rinsed with TBS (Tris-buffered saline 20m*M* Tris, pH 8.2, 150 m*M* NaCl) containing 0.1% BSA and then incubated for 45 min at room temperature with antimouse or antirabbit IgG conjugated to 10-nm gold particles (BioCell).
7. The sections are rinsed with distilled water and contrasted by incubation for 5 min with 10 μl of 2% uranyl acetate in water, followed by incubation for 1 min with 10 μl of 2% lead citrate in water. Observations were performed on a Philips Tecnai 12 Biotwin (120 kV) electron microscope.

5.4. Characteristics of selective and nonselective mitophagy

Within 1 h of the beginning of nitrogen starvation, numerous projections from the vacuolar membrane were observed; many of them were directed toward mitochondria, creating close contacts with these organelles (Fig. 8.3D). These simple contacts appeared to evolve further into more intimate relationships that look like fusions or engulfment of mitochondria by the vacuole. Immuno-EM with an antibody directed against the IM protein $F_1\beta$ and the OM protein porin confirmed the nature of the organelles in tight contact with, and engulfed by, the vacuole. This process was rapid: within the first hour, 45% of the cells exhibited vacuole-mitochondria contacts. This fraction increased to a maximum of 85% after two hours, and fell to 35% after three hours, dropping all the way to 5% at the end of the fourth hour.

In the second stage, whose onset coincided with the ending of the first, mitochondrial microautophagy is observed (Fig. 8.3A–C). Immuno-EM with antibodies directed against F1β (Fig. 8.3I–L) or the OM protein VDAC (not shown) show that whole mitochondria, with their two membranes, are engulfed. After 3 h of nitrogen starvation, about 80% of the cells exhibit vesicles containing mitochondria within the vacuoles. This proportion increases up to 95% at 4 h, at which time the number of mitochondria retained in vacuoles per slice is at its maximum (5–7 units/vacuole). From 6 h to the end of the experiment, this count stands between 1 and 2.

The observation of a large number of microautophagic structures containing mitochondria allows for their classification into two categories on the basis of the presence of mitochondria alone (Fig. 8.3G–H) or that of mitochondria surrounded by cytosol (Fig. 8.3E–F). A visual analysis of the

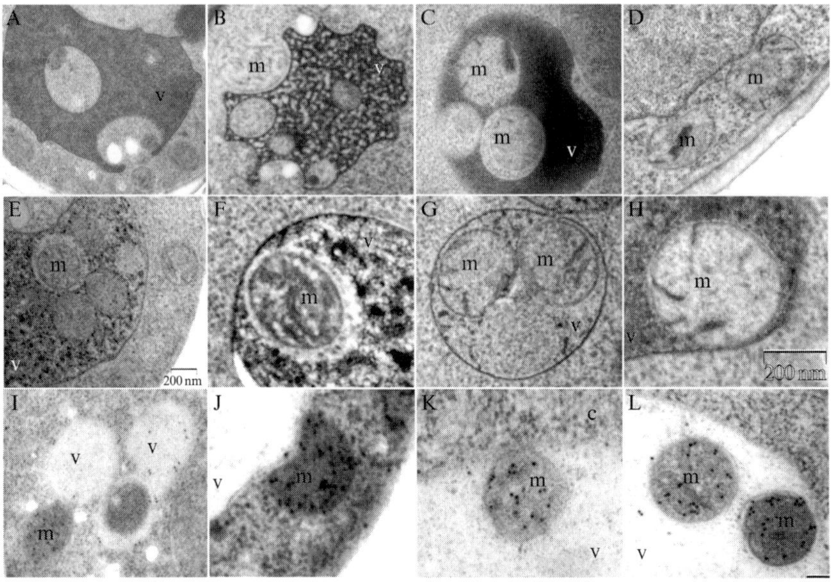

Figure 8.3 Wild-type cells were grown in lactate-supplemented medium and subjected to nitrogen starvation. A, B, C: microautophagy process. D: Contacts between vacuoles and mitochondria observed between 2 h and 3 h after the beginning of starvation. E, F: Vesicles containing mitochondria surrounded by cytosol. G, H: Vesicles containing mitochondria without cytosol. I, J, K, L: Immuno-EM with an antibody directed against $F_1\beta$. c, cytosol; m, mitochondria; v, vacuole. Scale bar in E, H, L: 200 nm.

amount of surrounding cytosol in 45 randomly selected vesicles using an image-processing software reveals a prominent discontinuity in its distribution. The cytosolic content of the first category is in the range of 0%–10% while that in the second category is in the range of 35%–65%. This observation makes it highly unlikely that these two discontinuous categories incidentally result from a random sequestration of different proportions of cytosol, which would instead result in a Gaussian distribution. This finding supports the hypothesis that the two categories reflect different underlying mechanisms: the first, selective (mitophagy), leading to the formation of vesicles containing only mitochondria with some residual cytosol (<10%) and the second, nonselective, leading to the formation of vesicles containing mitochondria randomly engulfed with a significant amount of cytosol (35%–65%).

ACKNOWLEDGMENTS

RJD thanks his Monash colleagues, particularly Dr. Mark Prescott and Mr. Dalibor Mijaljica, who have contributed to studies on mitophagy, for helpful discussions.

NC and IK thank Maika Deffieu, who has contributed to studies on mitophagy, and Dr. Stephen Manon, for helpful discussions. A part of this work was supported by grants from the CNRS, the Université de Bordeaux, and the Conseil Régional d'Aquitaine, a grant from the Science and Technology Assistance Agency (No. APVT-20-012404), and a postdoctoral fellowship from the Fondation pour la Recherche Médicale (to IK). France/ Slovakia collaboration was supported by Egide (EcoNet 10457VF to SM).

REFERENCES

Baker, M. J., Frazier, A. E., Gulbis, J. M., and Ryan, M. T. (2007). Mitochondrial protein-import machinery: Correlating structure with function. *Trends Cell. Biol.* **17,** 456–464.

Camougrand, N., Grelaud-Coq, A., Marza, E., Priault, M., Bessoule, J. J., and Manon, S. (2003). The product of the *UTH1* gene, required for Bax-induced cell death in yeast, is involved in the response to rapamycin. *Mol. Microbiol.* **47,** 495–506.

Campbell, C. L., and Thorsness, P. E. (1998). Escape of mitochondrial DNA to the nucleus in *yme1* yeast is mediated by vacuolar-dependent turnover of abnormal mitochondrial compartments. *J. Cell Sci.* **111,** 2455–2464.

Gavin, P., Devenish, R. J., and Prescott, M. (2002). An approach for reducing unwanted oligomerisation of DsRed fusion proteins. *Biochem. Biophys. Res. Commun.* **298,** 707–713.

Gietz, R. D., and Woods, R. A. (2002). Transformation of yeast by lithium acetate/single-stranded carrier DNA/polyethylene glycol method. *Methods Enzymol.* **350,** 87–96.

Jones, E. W., Zubenko, G. S., and Parker, R. R. (1982). *PEP4* gene function is required for expression of several vacuolar hydrolases in *Saccharomyces cerevisiae*. *Genetics* **102,** 665–677.

Kaneko, Y., Toh-e, A., and Oshima, Y. (1982). Identification of the genetic locus for the structural gene and a new regulatory gene for the synthesis of repressible alkaline phosphatase in *Saccharomyces cerevisiae*. *Mol. Cell Biol.* **2,** 127–137.

Kaneko, Y., Tamai, Y., Toh-e, A., and Oshima, Y. (1985). Transcriptional and post-transcriptional control of *PHO8* expression by *PHO* regulatory genes in *Saccharomyces cerevisiae*. *Mol. Cell. Biol.* **5,** 248–252.

Kaneko, Y., Toh-e, A., Banno, I., and Oshima, Y. (1989). Molecular characterization of a specific *p*-nitrophenylphosphatase gene, PHO*13*, and its mapping by chromosome fragmentation in *Saccharomyces cerevisiae*. *Mol. Gen. Genet.* **220,** 133–139.

Kim, I., Rodriguez-Enriques, J. J., and Lemasters, J. J. (2007). Selective degradation of mitochondria by mitophagy. *Arch. Biochem. Biophys.* **462,** 245–253.

Kissova, I., Deffieu, M., Manon, S., and Camougrand, N. (2004). Uth1p is involved in the autophagic degradation of mitochondria. *J. Biol. Chem.* **279,** 39068–39074.

Kissova, I., Plamondon, L. T., Brisson, L., Priault, M., Renouf, V., Schaeffer, J., Camougrand, N., and Manon, S. (2006). Evaluation of the roles of apoptosis, autophagy and mitophagy in the loss of plating efficiency induced by Bax-expression in yeast. *J. Biol. Chem.* **281,** 36187–36197.

Kissova, I., Salin, B., Schaeffer, J., Bathia, S., Manon, S., and Camougrand, N. (2007). Selective and non-selective autiphagic degradation of mitochondria in Yeast. *Autophagy* **3,** 329–336.

Klebe, R. J., Harriss, J. V., Sharp, Z. D., and Douglas, M. G. (1983). A general method for polyethylene-glycol-induced genetic transformation of bacteria and yeast. *Gene* **25,** 333–341.

Kraft, C., Deplazes, A., Sohrmann, M., and Peter, M. (2008). Mature ribosomes are selectively degraded upon starvation by an autophagy pathway required the Ubp3p/ Bre5p ubiquitin proteasome. *Nature Cell Biol.* **10,** 602–610.

Laemmli, U. K. (1970). Cleavage of structural proteins during the assembly of the head of bacteriophage t4. *Nature* **227,** 680–685.

Noda, T., Matsuura, A., Wada, Y., and Ohsumi, Y. (1995). Novel system for monitoring autophagy in the yeast *Saccharomyces cerevisiae. Biochem. Biophys. Res. Commun.* **210,** 126–132.

Noda, T., and Ohsumi, Y. (1998). Tor, a phosphatidylinositol kinase homologue, controls autophagy in yeast. *J. Biol. Chem.* **273,** 3963–3966.

Nothwehr, S. R., Bryant, N. J., and Stevens, T. H. (1996). The newly identified yeast *GRD* genes are required for retention of late-Golgi membrane proteins. *Mol. Cell. Biol.* **16,** 2700–2707.

Nowikovsky, K., Reipert, S., Devenish, R. J., and Schweyen, R. J. (2007). MDM38 protein depletion causes loss of mitochondrial K^+/H^+ exchange activity, osmotic swelling and mitophagy. *Cell Death Differ.* **14,** 1647–1656.

Okamoto, K., Perlman, P. S., and Butow, R. A. (1998). The sorting of mitochondrial DNA and mitochondrial proteins in zygotes: Preferential transmission of mitochondrial DNA to the medial bud. *J. Cell Biol.* **142,** 613–623.

Pinkham, J. L., Dudley, A. M., and Mason, T. L. (1994). T7 RNA polymerase-dependent expression of COXII in yeast mitochondria. *Mol. Cell Biol.* **14,** 4643–4652.

Pinson, B., Kongsrud, T. L., Ording, E., Johensen, L., Daignan-Fornier, B., and Gabrielsen, O. S. (2000). Signaling through regulated transcription factor interaction mapping of a regulatory interaction domain in the Myb-related Bas1p. *Nucleic Acids Res.* **28,** 4665–4673.

Priault, M., Salin, B., Schaeffer, J., Vallette, F. M., Di Rago, J. P., and Martinou, J. C. (2005). Impairing the bioenergetic status and the biogenesis of mitochondria triggers mitophagy in yeast. *Cell Death Differ.* **12,** 1613–1621.

Proudlock, J. W., Haslam, J. M., and Linnane, A. W. (1971). Biogenesis of mitochondria. 19. The effects of unsaturated fatty acid depletion on the lipid composition and energy metabolism of a fatty acid desaturase mutant of *Saccharomyces cerevisiae. J. Bioenerg. Biomembr.* **2,** 327–349.

Rosado, C. J., Mijaljica, D., Hatzinisiriou, I., Prescott, M., and Devenish, R. J. (2008). Rosella: A fluorescent pH-biosensor for reporting vacuolar turnover of cytosol and organelles in yeast. *Autophagy* **4,** 205–213.

Takeshige, K., Baba, M., Tsuboi, S., Noda, T., and Ohsumi, Y. (1992). Autophagy in yeast demonstrated with proteinase-deficient mutants and conditions for its induction. *J. Cell Biol.* **119,** 301–311.

Rodriguez-Enriquez, S., He, L., and Lemasters, J. J. (2004). Role of mitochondrial permeability transition pores in mitochondrial autophagy. *Int. J. Biochem. Cell Biol.* **36,** 2463–2473.

Sikorski, R. S., and Hieter, P. (1989). A system of shuttle vectors and yeast host strains designed for efficient manipulation of DNA in *Saccharomyces cerevisiae. Genetics* **122,** 19–27.

Tolkovsky, A. M., Xue, L., Fletcher, G. C., and Borutaite, V. (2002). Mitochondrial disappearance from cells: A clue to the role of autophagy in programmed cell death and disease? *Biochimie.* **84,** 233–240.

Vida, T. A., and Emr, S. D. (1995). A new vital stain for visualizing vacuolar membrane dynamics and endocytosis in yeast. *J. Cell Biol.* **128,** 779–792.

Westermann, B., and Neupert, W. (2000). Mitochondria-targeted green fluorescent proteins: Convenient tools for the study of organelle biogenesis in *Saccharomyces cerevisiae. Yeast* **16,** 1421–1427.

Xue, L., Fletcher, G. C., and Tolkovsky, A. M. (1999). Autophagy is activated by apoptotic signalling in sympathetic neurons: An alternative mechanism of death execution. *Mol. Cell. Neurosci.* **14,** 180–198.

Xue, L., Fletcher, G. C., and Tolkovsky, A. M. (2001). Mitochondria are selectively eliminated from eukaryotic cells after bockade of caspases during apoptosis. *Curr. Biol.* **11,** 361–365.

Yorimitsu, T., and Klionsky, D. J. (2005). Autophagy: Molecular machinerie for self-eating. *Cell Death Differ.* **12,** 1542–1552.

Yu, L., Wan, F., Dutta, S., Welsh, S., Liuu, Z., Freundt, E., Baehrecke, E. H., and Lenardo, M. (2006). Autophagic programmed cell death by selective catalase degradation. *Proc. Natl. Acad . Sci. USA* **103,** 4952–4957.

Zhang, Y., Qi, H., Taylor, R., Liu, L. F., and Jin, S. (2007). The role of autophagy in mitochondria maintenance: Characterization of mitochondrial functions in autophagy-deficient *S. cerevisiae* strains. *Autophagy* **3,** 337–346.

CHAPTER NINE

Monitoring Organelle Turnover in Yeast Using Fluorescent Protein Tags

Rodney J. Devenish,* Mark Prescott,* Kristina Turcic,* *and* Dalibor Mijaljica*

Contents

* Department of Biochemistry and Molecular Biology, and ARC Centre of Excellence in Structural and Functional Microbial Genomics, Monash University, Clayton Campus, Melbourne, Victoria, Australia

Methods in Enzymology, Volume 451
ISSN 0076-6879, DOI: 10.1016/S0076-6879(08)03209-6

1. Selective Organelle Degradation by Autophagy in Yeast

Selective organelle degradation (organellophagy) has been best described in yeast. Pexophagy (Dunn *et al.*, 2005), mitophagy (Kiššová *et al.*, 2004, 2007), and reticulophagy (Bernales *et al.*, 2007; Hamasaki *et al.*, 2005) are responsible for specific degradation of peroxisomes, mitochondria and endoplasmic reticulum (ER), respectively. Both macro- and microautophagic processes have been described for vacuolar uptake of peroxisomes and mitochondria. Interestingly, microautophagy of mitochondria (micromitophagy) can apparently occur in two different ways. In the first, mitochondria are directly engulfed by a vacuolar membrane invagination, while in the second numerous projections originating from the vacuolar membrane appear to create close contacts with mitochondria leading to their engulfment (Kiššová *et al.*, 2007). The process of reticulophagy involves the engulfment and degradation of ER fragments during nutrient deprivation (Hamasaki *et al.*, 2005) or when cells are subjected to conditions that place the ER under stress (Bernales *et al.*, 2007). Microautophagic degradation of the nucleus (PMN, *p*iecemeal *m*icroautophagy of the *n*ucleus) has been reported to occur at nucleus-vacuole (NV) junctions that serve as the nucleation sites for protrusion of portions of the nucleus into the vacuole, which are subsequently pinched off and released into the vacuolar lumen (Roberts *et al.*, 2003).

In addition to the characterized autophagy-related (Atg) machinery, it has been suggested that selective organelle turnover uses additional proteins that are required to either confer some degree of specificity (e.g., Uth1 and Aup1 for mitophagy; Kiššová *et al.*, 2004, 2007; Tal *et al.*, 2007) or mediate membrane traffic (e.g., filamentous actin, or F-actin), and the peroxisomal surface protein Pex14 is essential for pexophagy (Dunn *et al.*, 2005). Genetic screens in *Saccharomyces cerevisiae* have led to the identification of 31 genes essential for autophagy and related processes that are termed *ATG*, as well as the steps of autophagosome formation in which they act (Meijer *et al.*, 2007; van der Vaart *et al.*, 2008). Sixteen Atg proteins are essential for the biogenesis of double-membrane vesicles (autophagosomes) in all eukaryotes. These 16 proteins mediate the induction, expansion, and completion steps of autophagosome formation and are reported as being required for both selective and nonselective types of autophagy (van der Vaart *et al.*, 2008). Many of Atg proteins have clear homologs in higher eukaryotes, which in some cases have been shown to function heterologously, indicating that they are true orthologs (Klionsky, 2005; Meijer *et al.*, 2007; Wang and Klionsky, 2003; Yorimitsu and Klionsky, 2005). However, the precise molecular function of most of Atg proteins in selective turnover of organelles remains unclear.

A wide range of methods exists for monitoring autophagy in yeasts and other eukaryotes (see other chapters in this volume, and reviews by Kirkegaard *et al.*, 2004; Klionsky *et al.*, 2007, 2008; Mizushima, 2004). In this chapter we describe the use of fluorescent protein (FP) tags to monitor autophagic degradation of various organelles in the yeast *S. cerevisiae*. The advantages of specific targeting of FPs to different organelles are the ability to follow, under specific conditions: (1) the localization and/or distribution of the labeled (fluorescent) target organelle in relation to the vacuole, and (2) autophagic degradation of the target organelle in a single live cell.

2. Principle of the Assay and General Considerations

The assay relies on correct targeting of FPs to different organelles by use of organelle-specific targeting signals (see section 3; Fig. 9.1A represents an FP construct, Rosella, targeted to mitochondria and designated mt-Rosella), allowing delivery to the vacuole to be followed in a time-dependent manner. Because of their compact β-barrel structures, FPs have enhanced resistance to protease degradation (Tsien, 1998), and those with a suitable pKa can maintain fluorescence emission within the acidic lumen of the vacuole/lysosome (Katayama *et al.*, 2008; Prescott *et al.*, 2006). However, one needs to be aware that different FPs display different characteristics: brightness, maturation, phototoxicity, photostability, and oligomeric state (Olenych *et al.*, 2007). For instance, DsRed is tetrameric and also may self-associate to form higher-order aggregates (Jakobs *et al.*, 2000; Olenych *et al.*, 2007), although it appears that this property does not cause any problems when DsRed is targeted to organelles such as peroxisomes or the ER (see subsequent sections). Of more relevance are differences between FPs in their optical properties. For example, BFP displays limited levels of brightness and rapid photobleaching compared to many other FPs (Olenych *et al.*, 2007). Thus, careful consideration needs to be given to the choice of FP used. However, given the large array of FPs available (Olenych *et al.*, 2007; Prescott *et al.*, 2006; Prescott and Salih, 2008), it should be straightforward to find a suitable FP variant for any application.

Application of FP tagging requires several considerations of construct design to be kept in mind (see section 3). These include (1) selection and fusion to the appropriate targeting signal to achieve organelle-specific targeting; (2) which FP to use (the fluorescence emission of many FPs is pH sensitive, and therefore it is important to use an FP that exhibits significant fluorescence emission down to $\sim$ pH 5.5; Prescott *et al.*, 2006);

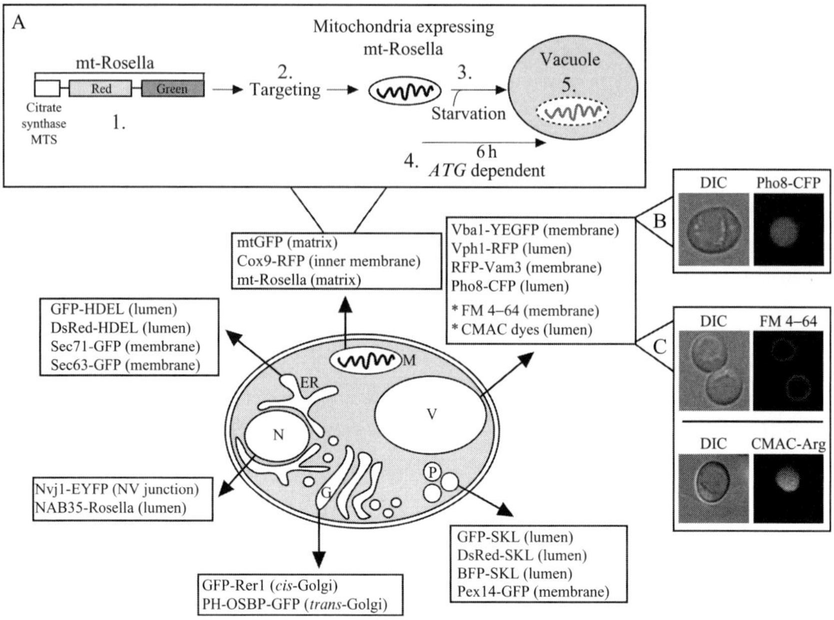

Figure 9.1 Different fluorescent proteins (FPs) are used to label different organelles (for more details, see section 3). An asterisk denotes vacuolar dyes that are commercially available. Abbreviations: ER; endoplasmic reticulum, G; Golgi apparatus, M; mitochondria, N; nucleus, P; peroxisomes, V; vacuole. (A) mt-Rosella is used as an example to explain the concept of the assay: (1) generation of the construct, (2) targeting to the mitochondria, (3) delivery to the vacuole upon starvation, (4) *ATG* and time dependency, and (5) degradation in the vacuole. Labeling of the vacuole (for more details, see section 4.5) can be achieved either by using FP vacuolar protein fusions (B) or commercially available dyes (C).

(3) use of chromosomal integration, or of single- or multicopy vectors for construct expression; and (4) the yeast promoter controlling transcription of the FP fusion (i.e., native or regulable promoter).

Once the construct has been successfully expressed in the chosen yeast host strain a second group of considerations become (1) yeast growth conditions; (2) cell sample preparation for imaging; (3) autophagy induction and time required for the detection of delivery of different organelles to the vacuole; and (4) determining *ATG* dependency of organelle delivery to the vacuole (i.e., the use of appropriate mutants when monitoring turnover of a particular organelle; (see section 5.5). The need to attend to these considerations should not discourage future "fluorophiles" from using FP technology to monitor organellophagy but rather should help them to achieve the greatest chance of success in the application of this approach.

3. Targeting of Fluorescent Proteins (FPs) to Different Organelles

Several research groups have utilized FP fusions (Table 9.1 and Fig. 9.1) to study autophagic degradation of mitochondria, the nucleus, peroxisomes, and endoplasmic reticulum. The possible autophagic degradation of the Golgi apparatus remains to be more fully explored in *S. cerevisiae*.

3.1. Mitochondria

Mitochondria comprise four distinct compartments: outer membrane (OM), inner membrane (IM), intermembrane space (IMS), and matrix. Theoretically, the FP can be targeted to any of these compartments using different mitochondrial targeting signals (MTS). Indeed, there are many reports concerning the use of GFP fusion proteins specifically targeted to each of the four mitochondrial compartments. To date, however, there are few reports describing the labeling with FPs of mitochondria for the purpose of following their fate during autophagy using optical detection. In these reports, mitophagy was successfully followed by targeting FPs (Rosella or GFP) to the mitochondrial matrix (Kiššová *et al.*, 2004, 2007; Nowikovsky *et al.*, 2007; Rosado *et al.*, 2008).

Rosella fusions: In our laboratory we developed and characterized a pH-fluorescent biosensor, Rosella, that is targeted to the mitochondrial matrix by fusion at its N terminus to the MTS of the matrix enzyme citrate synthase. Rosella is expressed from a multicopy expression vector under control of the *PGK1* promoter (Nowikovsky *et al.*, 2007; Rosado *et al.*, 2008). Furthermore, Rosella is also fused to OSCP (Atp5, a subunit of the stator stalk of mitochondrial F_1F_o ATP synthase) and expressed under the control of the *GAL1* promoter (Atp5-Rosella). The Atp5-Rosella protein is incorporated into functional mitochondrial ATP synthase complexes, resulting in Rosella being anchored to the matrix face of the IM. This construct proves a useful tool to show that delivery of mitochondrially targeted Rosella to the vacuole under starvation conditions represents the delivery of Rosella from the mitochondrial matrix rather than redirection of the newly synthesized polypeptide (i.e., mistargeting does not appear to be an issue when using mitochondrially targeted Rosella) (Rosado *et al.*, 2008).

mtGFP/mtDsRed.T3: A citrate synthase MTS fused to GFP was expressed under the control of a *GAL1/10* promoter (for details, see chapter 8 in this volume by Camougrand *et al.*). The same strategy has been used to target other FPs such as DsRed.T3 (a relatively pH-insensitive, fast-maturing variant of the red fluorescent protein, DsRed; Bevis and Glick, 2002) to the mitochondrial matrix. The mtDsRed.T3 was expressed

Table 9.1 FP constructs used to report vacuolar uptake of different organelles[a]

Organelle	Targeting sequence/Protein	FP Identity	Vector		Genomic Integration	Reference
			Copy Number	Promoter		
Mitochondria	Citrate synthase MTS (52 aa)	Rosella (Red-Green fusion)	Multi	Constitutive (*PGK1*)	No	Rosado *et al.*, 2008
		GFP (bGFP - GFP with F99S, M153T and V163A substitutions)	Single	Conditional (*GAL1/10*)	No	Kiššová *et al.*, 2004
	OSCP	Rosella (Red-Green fusion)	Multi	Conditional (*GAL1/10*)	No	Rosado *et al.*, 2008
	Cox9	RFP[b]	ND[c]	Native	Yes	Tal *et al.*, 2007
Endoplasmic reticulum	Kar2 signal presequence and HDEL tetrapeptide retention signal	GFP[b]	ND	Conditional (*MET25*)	No	Hamasaki *et al.*, 2005; Rossanese *et al.*, 2001
		DsRed[b]	ND	N/A	No	Hamasaki *et al.*, 2005; Rossanese *et al.*, 2001
	Sec63	GFP (GFP with S65T and V163A substitutions)	Single	Native	No	Prinz *et al.*, 2000
	Sec71	GFP[b]	Single	Native	No	Sato *et al.*, 2003

Nucleus	Nvj1	EYFP	Multi	Conditional (*GAL1/10*)	No	Roberts *et al.*, 2003
			ND	Conditional (*CUP1*)	No	Pan *et al.*, 2000
			ND	Native	Yes	Roberts *et al.*, 2003
	NAB35 (50 aa region in Nab2)	Rosella (Red-Green fusion)	Multi	Constitutive (*PGK1*)	No	Rosado *et al.*, 2008
Peroxisomes	SKL tripeptide	GFP[b]	ND	Conditional (*CUP1*)	No	Guan *et al.*, 2001
		DsRed[b]	Multi	Constitutive (*PGK1*)	No	Petrova *et al.*, 2004
		BFP	ND	ND	No	Kim *et al.*, 2001[d]
	Pex14	GFP[b]	ND	Native	Yes	Reggiori *et al.*, 2005

[a] The data presented do not necessarily include all constructs reported in the literature.
[b] The precise nature of the FP used is not specified.
[c] Information is not available.
[d] Kim, J., Kamada, Y., Stromhaug, P. E., Guan, J., Hefner-Gravink, A., Baba, M., Scott, S. V., Ohsumi, Y., Dunn, W. A., Jr., and Klionsky, D. J. (2001). Cvt9/Gsa9 functions in sequestering selective cytosolic cargo destined for the vacuole. *J. Cell Biol.* **153,** 381–396.

from a multicopy expression vector under control of the *PGK1* promoter (Mijaljica, D., Rosado, C. J., Prescott, M., and Devenish, R. J., unpublished; Rosado *et al.*, 2008).

It is not yet clear whether mitochondrial compartments other than the matrix, when labeled, are suitable for following mitophagy. Current evidence shows that degradation of OM and IM proteins can involve the outer mitochondrial membrane-associated degradation (OMMAD) pathway, a proteasomal degradation process (Neutzner *et al.*, 2007), and IM-embedded proteases (Arnold and Langer, 2002), respectively. Also, it is possible that IM and OM proteins are more slowly turned over or that their levels of expression are lower than for matrix-targeted proteins. A fusion of the IM cytochrome c oxidase subunit VIIa with RFP (Cox9-RFP) expressed from a chromosomally integrated construct reports apparent vacuolar uptake of mitochondria after a 4-day stationary phase incubation (Tal *et al.*, 2007). However, the suitability of OM and IM proteins for monitoring mitophagy remains to be investigated more fully.

3.2. Nucleus

NAB35-Rosella: To monitor possible autophagic degradation of the nucleus (nucleophagy), we targeted Rosella to the nucleus (Rosado *et al.*, 2008) by fusion to NAB35, a 50-amino-acid segment of Nab2, which encompasses a nuclear localization signal (Siomi *et al.*, 1998). Nab2 is one of the major nuclear proteins associated with nuclear poly(A)$^+$ RNA and a substrate for yeast transportin (Aitchison *et al.*, 1996; Anderson *et al.*, 1993). NAB35-Rosella was expressed from a multicopy expression vector under control of the *PGK1* promoter (Rosado *et al.*, 2008). Fluorescence is distributed throughout the nucleus and nuclear uptake is reported by accumulation of red fluorescence within the vacuole (see section 5.2).

Nvj1-EYFP fusion: PMN has been characterized as occurring at NV junctions through interaction of Vac8 (vacuolar membrane protein) and Nvj1 (outer nuclear membrane protein) (Roberts *et al.*, 2003). Plasmids for expression of *NVJ1* and *NVJ1*-EYFP under *CUP1* promoter control (P_{CUP1}-*NVJ1*-EYFP) and *GAL1* promoter control (P_{GAL1}-*NVJ1*-EYFP) or an integrated c*NVJ1*-EYFP are used in the analysis of NV junctions and PMN structures. Under nitrogen starvation conditions, Nvj1p-EYFP tagged nuclear blebs/vesicles are observed in the vacuole (see chapter 13 in this volume by Millen and Goldfarb; Kvam and Goldfarb, 2007; Pan *et al.*, 2000; Roberts *et al.*, 2003). Thus, Nvj1 fusions allow the movement and accumulation of Nvj1-EYFP tagged blebs and/or vesicles within the vacuolar lumen rather than accumulation of a diffuse fluorescence (NAB35-Rosella; see section 5.2), to be followed as evidence for turnover of nuclear material.

3.3. Peroxisomes

FP (GFP/DsRed/BFP)-SKL fusion: Peroxisomes can be readily labeled with an FP fused at its C terminus to the type 1 peroxisomal targeting signal (PTS) of serine-lysine-leucine (SKL). The SKL sequence is both necessary and sufficient for protein uptake into the peroxisomal matrix (Rachubinski and Subramani, 1995). The SKL sequence has been fused to different FPs, including GFP (Guan *et al.*, 2001; Klionsky *et al.*, 2007), DsRed (Petrova *et al.*, 2004), and BFP (Guan *et al.*, 2001). Each of these FP constructs seemingly reports pexophagy equally well; in our experiments either GFP-SKL or DsRed-SKL, constitutively expressed from the yeast *PGK1* promoter, serve as effective reporters of pexophagy (Mijaljica, D., Prescott, M., and Devenish, R. J., unpublished).

When yeasts cells are grown in media containing oleic acid (*S. cerevisiae*) or methanol (*Hansenula polymorpha* and *Pichia pastoris*) as the sole carbon source peroxisomes proliferate. Upon subsequent shift to glucose or ethanol containing medium, or to starvation conditions, a diffuse fluorescence can be seen within the vacuolar lumen at 2–3 h (for *S. cerevisiae* cells) (Guan *et al.*, 2001). The signal dissipates a few hours later (by 5–6 h), suggesting indicating degradation of peroxisomes (Dunn *et al.*, 2005; Guan *et al.*, 2001; Klionsky *et al.*, 2007; Meijer *et al.*, 2007).

Pex14-GFP: Pex14 is a membrane component of the peroxisomal protein import machinery. Pex14 fused to GFP at its C terminus results in the FP moiety being exposed on the cytosolic face of the peroxisome. Delivery of such tagged peroxisomes to the vacuole results in cleavage of the GFP moiety from its membrane anchor which is detected as free GFP by Western immunoblotting rather than fluorescence microscopy (Klionsky *et al.*, 2007; Reggiori *et al.*, 2005).

3.4. Endoplasmic reticulum

Targeting of FPs to the ER can be achieved either by fusion to lumenal or membrane-anchored proteins.

FP (GFP/DsRed)-HDEL fusion: Targeting of FPs to the ER lumen has been achieved using GFP-HDEL or DsRed-HDEL fusions. Both of these fusions possess a signal sequence at the N terminus (e.g., Kar2-ER lumenal protein presequence joined to the N terminus of GFP-HDEL by a linker peptide; Rossanese *et al.*, 2001). The HDEL sequence (histidine-aspartate-glutamate-leucine) is an ER retrieval signal found at the C-terminal end of ER resident proteins and that is sufficient to prevent secretion from the ER (Hamasaki *et al.*, 2005; Pelham *et al.*, 1988). The plasmids containing GFP-HDEL and DsRed-HDEL were integrated into the chromosomal *TRP1* locus (Rossanese *et al.*, 2001).

ER membrane protein fusions—Sec71-GFP: Sec71 is an ER integral membrane protein that does not normally escape from the ER and it is therefore not found on COPI or COPII trafficking vesicles (Sato *et al.*, 1997). GFP was fused at the C terminus of Sec71 and expressed from a single copy expression vector under its own promoter (Sato *et al.*, 2003).

ER membrane protein fusions—Sec63-GFP: In our experiments we have used the plasmid pJK59 that encodes Sec63-GFP, a fusion of the S65T V163A mutant of GFP to the C terminus of Sec63 under the *SEC63* promoter (Mijaljica *et al.*, 2006; Prinz *et al.*, 2000).

3.5. Golgi apparatus

There are no published reports addressing the possible turnover of the Golgi apparatus by autophagy in yeast.

Golgi FP fusions: Use of two fusion constructs, the *cis*-Golgi reporter GFP-Rer1 (Hamasaki *et al.*, 2005) and PH-OSBP-GFP (Mijaljica, D., Prescott, M., and Devenish R.J., unpublished) has not yet provided any data unequivocally demonstrating autophagy of the Golgi apparatus.

4. Materials and Methods

4.1. Yeast strains and transformation of yeast cells with organelle-specific FP constructs

Depending on the studies that are to be undertaken, different parental yeast strains can be used. In our laboratory we routinely use *S. cerevisiae* strain *BY4741* (*MATa his3*Δ*1 leu2*Δ*0 met15*Δ*0 ura3*Δ*0*) and isogenic deletion strains (*atg* deletion strains) obtained from Research Genetics/Invitrogen.

Transformation of yeast cells with organelle-specific FP constructs can be performed by standard procedures. To introduce FP constructs into wild-type or selected *atg* null mutant cells, we routinely use the EasyComp Transformation Kit (Invitrogen, Australia), a commercially available kit, on the basis of a method described previously (Gietz and Schiestl, 2007; Gietz and Woods, 2002). The procedure includes the following steps:

1. Add 2–2.5 μl ($\sim$100 ng/μl) of DNA encoding the FP fusion to a 1.5-ml tube.
2. Add 5 μl of competent cells prepared according to EasyComp Transformation Kit instructions.
3. Add 50 μl of Solution III (EasyComp Transformation Kit). CAUTION: Add Solution III on the side of the tube not directly onto the cells.
4. Vortex the tubes vigorously for 30 s.

5. Incubate the tubes for 1 h at 30 °C, mixing the contents by inversion every 15 min.
6. After the incubation period is complete, carefully place the cell transformation mixture onto a yeast growth plate (yeast minimal medium [YMM] containing 0.67% yeast nitrogen base without amino acids, 2% glucose, 1.5% agar, and any required auxotrophic supplements [0.01%] but omitting amino acids or nucleosides needed to maintain plasmid selection).
7. Gently distribute the cells around the plate. CAUTION: At this stage the cells are fragile and will not withstand rough treatment.
8. Incubate the plates inverted at 28 °C–30 °C. Yeast colonies should become visible after 3 days.
9. Examine the colonies (put cells on microscope slides and look at individual cells) for fluorescence using a fluorescence microscope (e.g., Nikon IX or Olympus Fluoview FV500).

Note: Before undertaking experiments to determine the delivery of different FP constructs to the vacuole, it is necessary to select a transformant colony that displays strong fluorescence.

10. Patch those colonies showing expression of the FP construct on fresh YMM plates containing appropriate supplements and incubate at 28 °C–30 °C for 2 days.
11. Inoculate the cells into corresponding liquid medium for further experiments (see subsequent sections).

4.2. Initial growth conditions

1. Yeast strains are grown aerobically at 28 °C with shaking to mid-log phase ($O.D._{600}$ = 2.5–4.0) in 2.5-cm diameter glass test tubes containing 10 ml of *Saccharomyces* Salts (SS) medium (comprising $(NH_4)_2SO_4$, 0.12%; KH_2PO_4, 0.10%; $MgCl_2 \cdot 6H_2O$, 0.07%; NaCl, 0.05%; $CaCl_2$, 0.01%; $FeCl_3$, 0.0005%; Proudlock *et al.*, 1971). Auxotrophic supplements (0.01%) are added as required.
2. For mitophagy experiments 2% (v/v) ethanol is added as carbon source (to ensure fully derepressed mitochondria and eliminate petite ($rho^{-/o}$) cells; Linnane and Lukins, 1975). Other nonfermentable carbon sources can be used, such as lactate (Kiššová *et al*, 2004, 2007). For experiments following other organelles, 2% glucose is added as carbon source. For following pexophagy cells are first grown in 10 ml of oleic acid containing medium (YTO; 0.67% yeast nitrogen base without amino acids, 0.1% oleic acid [v/v], and 0.05% Tween 80 [v/v]) for 19 h to induce formation of peroxisomes.
3. After this growth period, 1 ml of cells is harvested by centrifugation (at 13,000 rpm for 1 min), washed three times with sterile water,

and reinoculated into either fresh growth medium or autophagy/pexophagy-induction medium for the required period of time (see subsequent sections).

4.3. Conditions for induction of autophagy and pexophagy

The media used for induction of autophagy or pexophagy are the following.

4.3.1. Autophagy induction medium

Nitrogen-starvation medium contains 0.17% yeast nitrogen base without amino acids and ammonium sulfate (Difco) and either 2% ethanol (v/v) (SE-N) or 2% glucose (SD-N). To induce mitophagy (SE-N) medium was used, whereas to induce nucleophagy and reticulophagy, (SD-N) medium was used.

An alternative to nitrogen starvation for induction of autophagy is rapamycin treatment. Such treatment needs to be optimized for different yeast strains and the growth media used. For example, rapamycin (Sigma) was used to induce mitophagy at a final concentration of 0.2 μg/ml (using a stock solution of rapamycin dissolved in 90% ethanol and 10% Triton X-100; Kiššová *et al.*, 2004), or reticulophagy at a final concentration of 0.5 μg/ml (Hamasaki *et al.*, 2005).

4.3.2. Pexophagy induction medium

To induce pexophagy, three types of medium can be used: nitrogen-starvation SD-N or SE-N (see previous section) or complete starvation medium containing no carbon or nitrogen source (S-CN) (0.17% yeast nitrogen base without amino acids and ammonium sulfate).

4.4. Time required for the delivery of different organelles into the vacuole

Cells are grown at 28 °C with shaking for the required period of time; pexophagy (3 h), mitophagy (6 h); reticulophagy (9 h), and nucleophagy (24 h) (Fig. 9.2). These times represent the minimal time required to observe accumulation of fluorescence in the vacuole and were determined empirically by observing accumulation of the organelle-targeted FP construct in the vacuole under autophagy-induction conditions. These times may vary with different constructs and yeast strains. However, no systematic study has been undertaken, so it remains for each investigator to determine the time course of autophagic turnover under his or her own conditions.

Exactly what is represented by the difference in timing of the ability to detect vacuolar uptake is presently unclear; however, this difference possibly represents, for organelles, the difference in time taken for components of each compartment to accumulate and reach a threshold required to signal

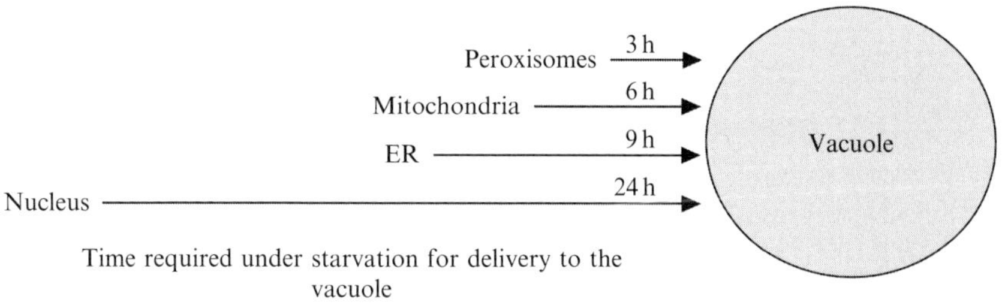

Figure 9.2 Timeline of organellophagy: The delivery of different organelles to the vacuole is a time-dependent process (for more details, see section 4.4).

uptake. The difference in timing might also reflect relative contributions and timing of nonspecific and specific autophagy under particular conditions.

4.5. Labeling of the vacuole

In yeast cells, vacuoles can be readily observed by using transmitted light microscopy (see Fig. 9.1; DIC images). However, it can be useful to confirm delivery of the target organelle to the vacuole using the fluorescence microscope. This can be achieved by labeling of the vacuole with vacuolar FP fusion proteins or chemical dyes. FP fusions that have been successfully used to confirm the delivery of selected organelles to the vacuole under specific conditions include membrane protein fusions such as Vba1-YEGFP (Kiššová *et al.*, 2007) and RFP-Vam3 (Reggiori *et al.*, 2005), or lumenal protein fusions such as Vph1-RFP (Tal *et al.*, 2007) and Pho8-CFP (see Fig. 9.1B; Mijaljica, D., Prescott, M., and Devenish, R. J., unpublished). For vacuolar membrane protein fusions the expected result is to observe fluorescence from the organelle construct delineated by a ring of green or red fluorescence. For vacuolar lumenal protein fusions the expected result is to observe co-localization of fluorescence for the organelle construct with the relevant marker. Use of such fusion constructs requires transformation and expression of a second construct within cells. This may not always be possible depending on the available selectable markers in the host yeast strain.

Alternatively, commercially available dyes can be used to specifically label the vacuolar membrane or the lumen (Fig. 9.1). The most commonly used dye is FM 4-64 (lipophilic styryl dye, *N*-(3-triethylammoniumpropyl)-4-(6-(4-(diethylamino)phenyl)hexatrienyl)pyridinium dibromide) (Fig. 9.1C; Kiššová *et al.*, 2004; Mijaljica *et al.*, 2007a; Vida and Emr, 1995) that specifically labels the outer leaflet of the vacuolar membrane. Less commonly used dyes are CellTracker Blue-CMAC (7-amino-4-chloromethylcoumarin), and the aminopeptidase substrates CMAC-Arg (7-amino-4-chloromethylcoumarin, L-arginine amide) and CMAC-Ala-Pro (7-amino-4-chloromethylcoumarin, L-alanyl-L-proline amide) that selectively stain the lumen of the yeast vacuole. Each of these coumarin-based vacuole markers contains a mildly reactive chloromethyl moiety that reacts with accessible thiols on peptides and proteins to form an aldehyde-fixable fluorescent conjugate. The conjugate formed with CellTracker Blue CMAC is blue fluorescent, whereas CMAC-Arg and CMAC-Ala-Pro substrates require subsequent cleavage by hydrolases in the lumen to activate their blue fluorescence. In our experiments, when required, cell vacuoles (of growing and starved cells) are stained with CMAC-Arg (Molecular Probes-Invitrogen, Australia) at a final concentration of 100 μM for 30 min prior to imaging at 28 °C, and then washed three times with sterile water before mounting (Fig. 9.1C) to successfully label the vacuolar lumen.

4.6. Mounting cells for viewing by fluorescence microscopy

One-ml aliquots of growing and starved cells (at this stage the O.D. of cells is not measured) are harvested by centrifugation and then washed quickly three times with 1 ml of sterile water to remove the medium, and the cell pellet resuspended in 50 μl of corresponding medium (growth or starvation). Cells are mounted on microscope slides either by using low-melting-point agarose (Protocol 1) or concanavalin A (Protocol 2). The second protocol allows a more uniform layer of the cells on the microscope slide to be obtained.

4.6.1. Protocol 1

1. Melt 0.2% (w/v) low-melting-point agarose (Progen Biosciences, Australia) in 1 ml of fresh medium (growth or starvation) at 70°C.

   Note: Mix the contents every 10 min.

2. Place 20 μl of molten agarose onto a glass microscope slide (76 mm × 26 mm) (Menzel-Glaser, Germany).
3. Immediately add 10 μl of cell suspension onto the agarose bed.

   Note: At this stage the O.D. of cells is not measured.

4. Cover with a small square coverslip (22 mm × 22 mm) (Menzel-Glaser, Germany) so that the agarose and cells disperse between the slide and the coverslip. A thin film of nail polish around the edge of the coverslip prevents movement of the coverslip. CAUTION: When and only when the nail polish is completely dry is the slide ready to view. Nail polish can cause damage to microscope objectives.
5. Image the cells under the fluorescence microscope (e.g., Olympus Fluoview FV500).

Note: This procedure allows cells to stay in the medium that was used for their initial growth or starvation treatment. To our knowledge, this procedure does not cause any additional stress (e.g., changes in cell morphology) to the cells for a quite long time period (up to 3 h). However, long-term incubation under such conditions may cause cells to experience anoxia.

4.6.2. Protocol 2

1. Spot 20 μl of concanavalin A (2 mg/ml) (Sigma, Australia) onto the center of a coverslip (either 22 mm × 22 mm or 24 mm × 60 mm) (Menzel-Glaser, Germany).
2. Spread the concanavalin A in the middle of the coverslip using a pipette tip.
3. Leave aside to air dry next to a lit Bunsen burner to aid drying.

   Note: The drying requires at least 20 min (if thin enough).

4. Place the cell suspension (prepared as for Protocol 1) on top of the dried concanavalin A. CAUTION: The concanavalin A on the coverslip has to be completely dry before depositing the cell suspension on it.
5. Leave the cells to settle and bind to concanavalin A for 10–15 min.
6. Wash off excess (unbound) cells by immersing the slide in sterile water 10 times.

   Note: Use fresh water for each slide.

7. Blot off excess water/medium with a tissue applied to the edge of the coverslip.
8. Place coverslip on a microscope slide (76 mm × 26 mm) (Menzel-Glaser, Germany) with cells facing toward the slide.
9. Seal with a thin film of nail polish around the edge of the coverslip. CAUTION: When and only when the nail polish is completely dry is the slide ready to view. Nail polish can cause damage to microscope objectives.
10. Image the cells under the microscope (e.g., Olympus Fluoview FV500).

Note: Leaving the cells on the slides for more than 1 h can cause changes in cell morphology.

4.7. Parameters of fluorescence microscopy used in imaging cells

For imaging purposes any wide-field or confocal fluorescence microscope equipped with a suitable objective can be used. We routinely use Confocal Laser Scanning Microscopy; a Fluoview FV500 mounted on an inverted IX-81 microscope (Olympus, Australia) was used to generate the images

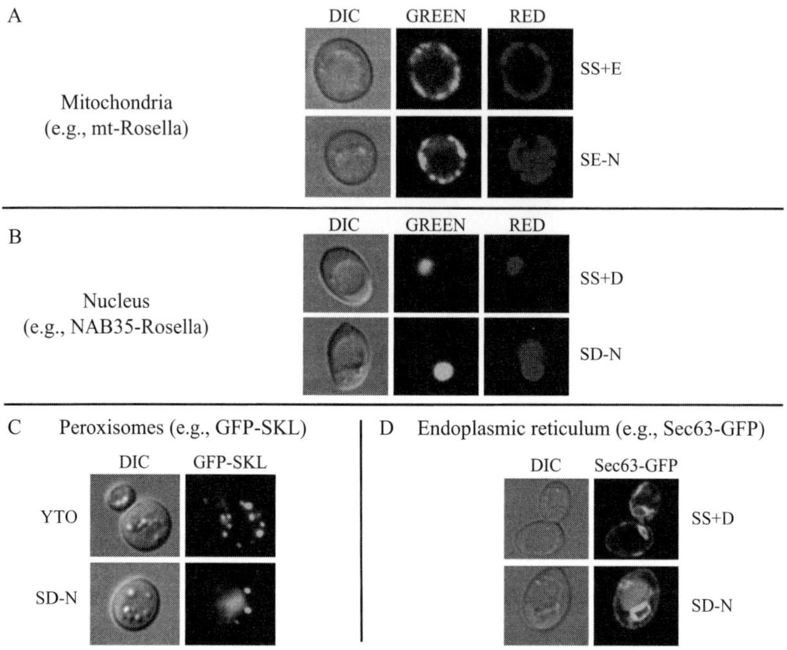

Figure 9.3 Fluorescence images of growing and starved wild-type (*BY4741*) cells expressing FPs targeted to (A) mitochondria, (B) the nucleus, (C) peroxisomes, and (D) endoplasmic reticulum (for more details, see section 5).

shown in Figs. 9.1 and 9.3. This instrument is equipped with multiple laser lines; 405-nm laser, argon (Ar) laser (488 nm), helium neon (HeNe) green laser (543 nm), and HeNe red laser (633 nm). This combination of the laser lines allows us to image four different channels at any point of time either simultaneously or sequentially. In our experiments, green fluorescence emission (530 $\pm$ 30 nm) and red fluorescence emission (>590 nm) are acquired sequentially upon excitation with 488-nm or 543-nm laser light, respectively. Blue fluorescence emission of CMAC dyes (469 nm) is acquired where required prior to obtaining green and red fluorescence and upon excitation with 405-nm laser light. Sequential (as opposed to simultaneous) scanning should be used to eliminate any bleed through from one channel to another. Transmitted light images are acquired to provide additional information on the vacuole and its relative position in the cell. Yeast are among the smallest eukaryotes ($\approx$5–10μm in diameter); therefore, it is necessary to use a high level of magnification in fluorescence microscopy. For visual observation, we typically use a 60X water-immersion (or 100X oil immersion) objective lens with a numerical aperture of 1.2. The confocal pinhole value used in all experiments was 125 μm but it can be increased if required. Changing these values leads to an increased sensitivity but with a loss in the rejection of out of focus light. For example, using the pinhole value of 125 μm allows us to observe the mitochondria as a series of dots localized at the periphery of the cell (Rosado *et al.*, 2008), while using higher values (e.g., 200–250 μm) allows the typical filamentous mitochondrial network distributed throughout the cell to be observed (Mijaljica, D., Rosado, C. J., Prescott, M., and Devenish, R. J., unpublished).

It is important that the parameters selected for fluorescence microscopy (e.g., laser intensity, scan rate, pinhole value, exposure for CCD-camera) are maintained throughout an experiment. This will ensure that comparisons can be made between control (growing cells) and starved samples.

5. Analysis and Typical Images of Organellar Turnover

In this section of the chapter we provide typical images and describe anticipated fluorescence patterns using selected constructs under both growing and starvation-induced conditions. It is expected that similar outcomes would be obtained with other FP constructs.

5.1. Mitophagy

Here, we show typical results obtained using mt-Rosella, expressed in cells as explained previously. The operation of the reporter relies on the pH differences between the vacuole (pH $\sim$ 4.8) and the mitochondrial matrix (pH $\sim$ 8.0). The reporter is a dual color emission biosensor comprising a relatively pH-stable RFP (DsRed.T3) and a pH-sensitive GFP variant (superecliptic pHluorin-SEP; Rosado *et al.*, 2008). Under growing conditions with ethanol as carbon source (SS + E), wild-type cells exhibit a cellular distribution of fluorescence (red and green) typical of mitochondria in yeast cells (Gavin *et al.*, 2002). Red and green fluorescence emission is not detected in the vacuole (Fig. 9.3A). When subjected to nitrogen starvation (SE-N) for 6 h (see Fig. 9.2) then, in addition to red and green fluorescence corresponding to mitochondria, cells also exhibit the accumulation of red fluorescence in the vacuolar lumen but no green fluorescence is observed (Fig. 9.3A). The accumulation of red fluorescence in the vacuolar lumen can be confirmed by labeling the cell vacuole with the blue fluorescent dye CMAC-Arg (see section 4.5; Rosado *et al.*, 2008). To determine whether the delivery of mt-Rosella to the vacuole is dependent on the characterized autophagic machinery, an *atg8* null mutant can be used (Atg8 is required for autophagosome formation). When *atg8* null cells are subjected to nitrogen starvation for the same amount of time, no accumulation of red fluorescence in the vacuole is observed. Such results confirm that mt-Rosella upon starvation is delivered to the vacuole in an autophagy-related process as mt-Rosella should not accumulate in the vacuole of starved *atg8* null cells.

5.2. Nucleophagy

In this section we show results obtained with NAB35-Rosella. Under growing conditions (SS + D), wild-type cells expressing NAB35-Rosella exhibit fluorescent labeling of the entire nuclear lumen, which appears as a single red and green body (Fig. 9.3B). Localization of NAB35-Rosella to the nucleus can be confirmed by labeling with the blue fluorescent dye, Hoechst 33258, which stains DNA (Rosado *et al.*, 2008). When incubated in SD-N medium for 24 h (Fig. 9.2), cells show red and green fluorescence of the nucleus as well as markedly visible accumulation of red fluorescence in the vacuole (Fig. 9.3B; Mijaljica *et al.*, 2007b). NAB35-Rosella labeling allows the morphology of the nucleus to be readily visualized and allows a distinction to be made between cells that have taken NAB35-Rosella into the vacuole (red fluorescence only) and those that show altered nuclear morphology and abrogated delivery of a nuclear bleb. These results emphasize a useful property of Rosella in being able to distinguish different pH environments (Rosado *et al.*, 2008). In addition to fluorescence microscopy, this FP reporter can also be used for analysis of populations of yeast cells by fluorescence-activated cell sorting (Rosado *et al.*, 2008).

5.3. Pexophagy

To follow vacuolar delivery of peroxisomes (pexophagy), we have used two FP constructs; GFP-SKL and DsRed-SKL. Under growing conditions (YTO medium), wild-type cells expressing GFP-SKL exhibit a pattern of fluorescence labeling characteristic of peroxisomes (Guan *et al.*, 2001), which appear as punctate structures in the cytosol (Fig. 9.3C). When shifted from oleic acid–containing medium to medium containing a different carbon source (e.g., glucose, SD-N), in addition to the normal distribution but with reduced number of peroxisomes in the cytosol, an accumulation of green fluorescence in the vacuole is observed (Fig. 9.3C; 3-h pexophagy induction). Similar results are obtained when cells are shifted to either ethanol-containing medium (SE-N) or complete starvation medium (S-CN) (Mijaljica, D., Prescott, M., and Devenish, R. J., unpublished). Equivalent results are observed using the peroxisomal marker DsRed-SKL (Mijaljica, D., Prescott, M., and Devenish, R. J., unpublished).

5.4. Reticulophagy

To follow reticulophagy, we utilize a Sec63-GFP fusion. Under growing conditions (SS+D), wild-type cells exhibit a fluorescence pattern typical of the ER network (cortical and peripheral ER) (Fig. 9.3D). After induction of autophagy by subjecting cells to nitrogen starvation (SD-N) for 9 h (Fig. 9.2), we observe accumulation of green fluorescence in the vacuolar lumen (Fig. 9.3D; Mijaljica *et al.*, 2006).

5.5. The requirement of *ATG* genes for organellophagy

To confirm that the delivery of a specific-organelle FP construct, under particular conditions occurs through an autophagy-related process, it is necessary to examine the outcome in strains lacking different *ATG* genes and grown under the same conditions. As mentioned previously, there are 31 *ATG* genes for autophagy and related processes. In a wide variety of studies typically one (or a small number of) null mutant is chosen for confirmatory studies and a conclusion reached that the process observed is either autophagy dependent or independent on the basis of the outcome. Therefore, it is important to select appropriate *atg* null mutants to investigate their involvement in degradation of each organelle. For example, Atg19 appears to be a specific receptor for the cytoplasm-to-vacuole targeting (Cvt) pathway (Xie and Klionsky, 2007). Therefore, the use of an *atg19* null mutant would not be a good choice to investigate autophagy-dependent degradation in most cases. A second example is the *S. cerevisiae* protein Atg26, which is not involved in the Cvt pathway, macroautophagy (nonspecific autophagy), or pexophagy (Cao and Klionsky, 2007). However, the role of Atg26, if any, in the specific autophagic degradation of other organelles (including mitochondria, endoplasmic reticulum, and the nucleus) remains to be elucidated. Indeed, in this context it is necessary to keep in mind that a complete assessment of the requirement for all *ATG* genes for different forms of specific autophagy has not yet been made and that some early reports may be subject to reevaluation. For example, it was originally reported that PMN is an *ATG7*-independent process (Roberts *et al.*, 2003); however, PMN is now reported to be dependent on Atg proteins (Beau *et al.*, 2008). In our studies, we find that the delivery of nuclear-targeted, NAB35-Rosella to the vacuole occurs in some but not all *atg* null mutants and have concluded that the process is dependent only on some components of the characterized macroautophagic machinery (Mijaljica, D., Prescott, M., and Devenish, R. J., unpublished).

Others have proposed that noncanonical forms of autophagy may exist (Xie and Klionsky, 2007), but without suggesting which, if any, of the characterized components of autophagic machinery may be required. We are of the view that a much more systematic study of the dependence of autophagic processes on *ATG* genes needs to be carried out, not only with regard to the use of FP constructs as reporters of organelle degradation but for other assays of autophagy as well. Only with the outcomes of such studies will it be possible to be more definitive about which *atg* mutants are the appropriate ones to test to confirm organelle-specific forms of autophagy.

6. Summary and Perspectives

Currently, our mechanistic knowledge of organellophagy as a selective form of autophagy is still quite limited. Accordingly, continued efforts are needed to elucidate the detailed mechanism and to uncover the basis of selectivity of organellophagy.

FPs have long been recognized as powerful tools for uncovering dynamic processes in eukaryotic cells. They serve as excellent markers of organelles for studies of autophagic degradation and, as such, have the potential to greatly expand our knowledge of organellophagy in various model systems (from yeast to higher eukaryotes). The use of organelle-specific FPs as a means to follow autophagic degradation of different organelles and its dependency on the characterized autophagic machinery is, relatively speaking, still in its infancy. What is presented here is not a formulaic set of rules, but guidelines for the use of FP technology in the context of autophagy. Further developments and specific uses of different FPs to ask new questions are sure to be forthcoming.

Acknowledgments

Work of the authors' lab described here was supported in part by Australian Research Council funding to the Centre of Excellence in Structural and Functional Microbial Genomics. We thank Dr Pamela Silver (Harvard Medical School) for providing pJK59 and Dr Diane Drescher (Universität des Saarlandes, Saarbrücken, Germany) for providing plasmids encoding GFP-SKL and DsRed-SKL.

References

Aitchison, J. D., Blobel, G., and Rout, M. P. (1996). Kap104p: A karyopherin involved in the nuclear transport of messenger RNA binding proteins. *Science* **274,** 624–627.

Anderson, J. T., Wilson, S. M., Datar, K. V., and Swanson, M. S. (1993). NAB2: A yeast nuclear polyadenylated RNA-binding protein essential for cell viability. *Mol. Cell Biol.* **13,** 2730–2741.

Arnold, I., and Langer, T. (2002). Membrane protein degradation by AAA proteases in mitochondria. *Biochim. Biophys. Acta.* **1592,** 89–96.

Beau, I., Esclatine, A., and Codogno, P. (2008). Lost to translation: When autophagy targets mature ribosomes. *Trends Cell Biol.* **18,** 311–314.

Bernales, S., Schuck, S., and Walter, P. (2007). ER-phagy: Selective autophagy of the endoplasmic reticulum. *Autophagy* **3,** 285–287.

Bevis, B. J., and Glick, B. S. (2002). Rapidly maturing variants of the *Discosoma* red fluorescent protein (DsRed). *Nature Biotech.* **20,** 83–87.

Cao, Y., and Klionsky, D. J. (2007). Atg26 is not involved in autophagy-related pathways in *Saccharomyces cerevisiae. Autophagy* **3,** 17–20.

Dunn, W. A., Jr., Cregg, J. M., Kiel, J. A. K. W., van der Klei, I. J., Oku, M., Sakai, Y., Sibirny, A. A., Stasyk, O. V., and Veenhuis, M. (2005). Pexophagy: The selective autophagy of peroxisomes. *Autophagy* **1,** 75–83.

Gavin, P., Devenish, R. J., and Prescott, M. (2002). An approach for reducing unwanted oligomerisation of DsRed fusion proteins. *Biochem. Biophys. Res. Commun.* **298,** 707–713.

Gietz, R. D., and Schiestl, R. H. (2007). Frozen competent yeast cells that can be transformed with high efficiency using the LiAc/SS carrier DNA/PEG method. *Nat. Protoc.* **2,** 1–4.

Gietz, R. D., and Woods, R. A. (2002). Transformation of yeast by lithium acetate/single-stranded carrier DNA/polyethylene glycol method. *Methods Enzymol.* **350,** 87–96.

Guan, J., Stromhaug, P. E., George, M. D., Habibzadegah-Tari, P., Bevan, A., Dunn, W. A., and Klionsky, D. J. (2001). Cvt18/Gsa12 is required for cytoplasm-to-vacuole transport, pexophagy, and autophagy in *Saccharomyces cerevisiae* and *Pichia pastoris. Mol. Biol. Cell* **12,** 3821–3838.

Hamasaki, M., Noda, T., Baba, M., and Ohsumi, Y. (2005). Starvation triggers the delivery of the endoplasmic reticulum to the vacuole via autophagy in yeast. *Traffic* **6,** 56–65.

Jakobs, S., Subramaniam, V., Schönle, A., Jovin, T. M., and Hell, S. W. (2000). EGFP and DsRed expressing cultures of *Escherichia coli* imaged by confocal, two-photon and fluorescence lifetime microscop. *FEBS Lett.* **479,** 131–135.

Katayama, H., Yamamoto, A., Mizushima, N., Yoshimori, T., and Miyawaki, A. (2008). GFP-like proteins stably accumulate in lysosomes. *Cell Struct. Funct.* **33,** 1–12.

Kirkegaard, K., Taylor, M. P., and Jackson, W. T. (2004). Cellular autophagy: Surrender, avoidance and subversion by microorganisms. *Nat. Rev. Microbiol.* **2,** 301–314.

Kiššová, I., Deffieu, M., Manon, S., and Camougrand, N. (2004). Uth1p is involved in the autophagic degradation of mitochondria. *J. Biol. Chem.* **279,** 39068–39074.

Kiššová, I., Salin, B., Schaeffer, J., Bhatia, S., Manon, S., and Camougrand, N. (2007). Selective and non-selective degradation of mitochondria in yeast. *Autophagy* **3,** 329–336.

Klionsky, D. J. (2005). The correct way to monitor autophagy in higher eukaryotes. *Autophagy* **1,** 65.

Klionsky, D. J., Cuervo, A. M., and Seglen, P. O. (2007). Methods for monitoring autophagy from yeast to human. *Autophagy* **3,** 181–206.

Klionsky, D. J., Abeliovich, H., Agostinis, P., Agrawal, D. K., Aliev, G., Askew, D. S., Baba, M., Baehrecke, E. H., Bahr, B. A., Ballabio, A., *et al.* (2008). Guidelines for the use and interpretation of assays for monitoring autophagy in higher eukaryotes. *Autophagy* **4,** 151–175.

Kvam, E., and Goldfarb, D. S. (2007). Nucleus-vacuole junctions and piecemeal microautophagy of the nucleus in *S. cerevisiae. Autophagy* **3,** 85–92.

Linnane, A. W., and Lukins, H. B. (1975). Isolation of mitochondria and techniques for studying mitochondrial biogenesis in yeasts. *Methods Cell Biol.* **12,** 285–309.

Meijer, W. H., van der Klei, I. J., Veenhuis, M., and Kiel, J. A. K. W. (2007). *ATG* genes involved in non-selective autophagy are conserved from yeast to man, but the selective Cvt and pexophagy pathways also require organism-specific genes. *Autophagy* **3,** 106–116.

Mijaljica, D., Prescott, M., and Devenish, R. J. (2006). Endoplasmic reticulum and Golgi complex: contributions to, and turnover by, autophagy. *Traffic* **7,** 1590–1595.

Mijaljica, D., Prescott, M., Klionsky, D. J., and Devenish, R. J. (2007a). Autophagy and vacuole homeostasis: A case for self-degradation? *Autophagy* **3,** 417–421.

Mijaljica, D., Prescott, M., and Devenish, R. J. (2007b). Nibbling within the nucleus: Turnover of nuclear contents. *Cell. Mol. Life Sci.* **64,** 581–588.

Mizushima, N. (2004). Methods for monitoring autophagy. *Int. J. Biochem. Cell Biol.* **36,** 2491–2502.

Neutzner, A., Youle, R. J., and Karbowski, M. (2007). Outer mitochondrial membrane protein degradation by the proteasome. *Novartis Found. Symp.* **287,** 4–14.

Nowikovsky, K., Reipert, S., Devenish, R. J., and Schweyen, R. J. (2007). Mdm38 protein depletion causes loss of mitochondrial K^+/H^+ exchange activity, osmotic swelling and mitophagy. *Cell Death Differ.* **14,** 1647–1656.

Olenych, S. G., Claxton, N. S., Ottenberg, G. K., and Davidson, M. W. (2007). The fluorescent protein color palette. *In* "Current protocols in cell biology" (J. S. Bonifacino, M. Dasso, J. B. Harford, J. Lippincott-Schwartz, and K. M. Yamada, eds.), UNIT 21.5. John Wiley and Sons, New Jersey.

Pan, X., Roberts, P., Chen, Y., Kvam, E., Shulga, N., Huang, K., Lemmon, S., and Goldfarb, D. S. (2000). Nucleus-vacuole junctions in *Saccharomyces cerevisiae* are formed through the direct interaction of Vac8p with Nvj1p. *Mol. Biol. Cell* **11,** 2445–2457.

Pelham, H. R. B., Hardwick, K. G., and Lewis, M. J. (1988). Sorting of soluble ER proteins in yeast. *EMBO J.* **7,** 1757–1762.

Petrova, V. Y., Drescher, D., Kujumdzieva, A. V., and Schmitt, M. J. (2004). Dual targeting of yeast catalase A to peroxisomes and mitochondria. *Biochem. J.* **380,** 393–400.

Prescott, M., Battad, J., Wilmann, P., Rossjohn, J., and Devenish, R. J. (2006). Recent advances in all-protein chromophore technology. *In* "Biotechnology annual review" (M. R. El-Gewely, ed.), pp. 31–66. Elsevier, Amsterdam.

Prescott, M., and Salih, A. (2008). Genetically encoded fluorescent proteins: Some properties and applications in the life sciences. *In* "Fluorescence applications in biotechnology and the life sciences" (E. Goldys, ed.), pp. 55–86. John Wiley & Sons, New Jersey.

Prinz, W. A., Grzyb, L., Veenhuis, M., Kahana, J. A., Silver, P. A., and Rapoport, T. A. (2000). Mutants affecting the structure of the cortical endoplasmic reticulum in *Saccharomyces cerevisiae*. *J. Cell Biol.* **150,** 461–474.

Proudlock, J. W., Haslam, J. M., and Linnane, A. W. (1971). Biogenesis of mitochondria. 19. The effects of unsaturated fatty acid depletion on the lipid composition and energy metabolism of a fatty acid desaturase mutant of *Saccharomyces cerevisiae*. *J. Bioenerg. Biomembr.* **2,** 327–349.

Rachubinski, R. A., and Subramani, S. (1995). How proteins penetrate peroxisomes. *Cell* **83,** 525–528.

Reggiori, F., Monastyrska, I., Shintani, T., and Klionsky, D. J. (2005). The actin cytoskeleton is required for selective types of autophagy, but not nonspecific autophagy, in the yeast *Saccharomyces cerevisiae*. *Mol. Biol. Cell* **16,** 5843–5856.

Roberts, P., Moshitch-Moshkovitz, S., Kvam, E., O'Toole, E., Winey, M., and Goldfarb, D. S. (2003). Piecemeal microautophagy of nucleus in *Saccharomyces cerevisiae*. *Mol. Biol. Cell* **14,** 129–141.

Rosado, C. J., Mijaljica, D., Hatzinisiriou, I., Prescott, M., and Devenish, R. J. (2008). Rosella: A fluorescent pH-biosensor for reporting vacuolar turnover of cytosol and organelles in yeast. *Autophagy* **4,** 205–213.

Rossanese, O. W., Reinke, C. A., Bevis, B. J., Hammond, A. T., Sears, I. B., O'Connor, J., and Glick, B. S. (2001). A role for actin, Cdc1p, and Myo2p in the inheritance of late Golgi elements in *Saccharomyces cerevisiae*. *J. Cell Biol.* **153,** 47–62.

Sato, K., Sato, M., and Nakano, A. (1997). Rer1p as common machinery for the endoplasmic reticulum localization of membrane proteins. *Proc. Natl. Acad. Sci. USA* **94,** 9693–9698.

Sato, K., Sato, M., and Nakano, A. (2003). Rer1p, a retrieval receptor for ER membrane proteins, recognizes transmembrane domains in multiple modes. *Mol. Biol. Cell* **14,** 3605–3616.

Siomi, M. C., Fromont, M., Rain, J. C., Wan, L., Wang, E., Legrain, P., and Dreyfuss, G. (1998). Functional conservation of the transportin nuclear import pathway in divergent organisms. *Mol. Cell Biol.* **18,** 4141–4148.

Tal, R., Winter, G., Ecker, N., Klionsky, D. J., and Abeliovich, H. (2007). Aup1p, a yeast mitochondrial protein phosphatase homolog, is required for efficient stationary phase mitophagy and cell survival. *J. Cell Biol.* **282,** 5617–5624.

Tsien, R. (1998). The green fluorescent protein. *Annu. Rev. Biochem.* **67,** 509–544.

van der Vaart, A., Mari, M., and Reggiori, F. (2008). A picky eater: Exploring the mechanisms of selective autophagy in human pathologies. *Traffic* **9,** 281–289.

Vida, T. A., and Emr, S. D. (1995). A new vital stain for visualising vacuolar membrane dynamics and endocytosis in yeast. *J. Cell Biol.* **128,** 779–792.

Wang, C. -W., and Klionsky, D. J. (2003). The molecular mechanism of autophagy. *Mol. Med.* **9,** 65–76.

Yorimitsu, T., and Klionsky, D. J. (2005). Autophagy: Molecular machinery for self-eating. *Cell Death Differ.* **12,** 1542–1552.

CHAPTER TEN

ELECTRON MICROSCOPY IN YEAST

Misuzu Baba*

Contents

Abstract

The membrane dynamics of autophagy in yeast were resolved mostly by using various electron microscopy (EM) methods combined with the cryofixation technique. In addition, the similarity of the dynamic process between autophagy and the cytoplasm-to-vacuole targeting (Cvt) pathway was first clarified through EM studies. In this chapter the application of several EM methods to detect the dynamic events involving cellular structures are described, and examples are provided of the typical images obtained in autophagy and the Cvt pathway.

* Department of Chemical and Biological Sciences, Faculty of Science, Japan Women's University, Tokyo, Japan

Methods in Enzymology, Volume 451
ISSN 0076-6879, DOI: 10.1016/S0076-6879(08)03210-2

1. Introduction

In yeast, macroautophagy (hereafter, autophagy) is primarily a nonspecific degradative process induced under starvation conditions. There are also various types of specific autophagy, including the selective degradation of superfluous organelles. Another example of selective autophagy is the cytoplasm-to-vacuolar targeting (Cvt) pathway, which is a biosynthetic route used for the delivery of certain resident vacuolar hydrolases. Although autophagy is degradative, whereas the Cvt pathway is biosynthetic, they largely share a common set of autophagy-related proteins to drive a process of vesicle formation. The hallmark of autophagy is the formation of double-membrane autophagosomes that sequester portions of the cytoplasm for subsequent delivery to the vacuole. Similarly, the most obvious morphological feature of the Cvt pathway is the generation of double-membrane Cvt vesicles. In either case, after fusion with the vacuole-limiting membrane, the inner single-membrane vesicle is released into the organelle lumen; this vesicle is now termed an autophagic or Cvt body, respectively.

To capture accurately one stage of a quickly proceeding process such as protein degradation, protein transport, or the dynamic membrane trafficking events of organelles, physical fixation is a prerequisite. In the autophagic process of yeast, fusion between vacuoles and the outer membrane of the autophagosome was clarified using a rapid-freezing technique (Baba *et al.*, 1995). The best choice for examining intracellular events is to use cryofixation followed by the freeze-substitution fixation method, which keeps cells in a native and frozen-hydrated state. This technique is now widely applied in many fields of morphological analysis. To achieve the state of a vitreous ice in the biological specimen, several types of freezing machines and an automatic substitution device have been developed. Although these freezing machines are expensive, they are essential and provide a consistent basis for producing high-quality electron micrographs (Bernales *et al.*, 2006; Kiššová *et al.*, 2007). Two major techniques are utilized for the preparation of yeast samples. One is to plunge cells into liquid propane, and another is to freeze cells under high pressure (2100 bar). At present, the plunge technique relies on the use of a homemade device for submerging the samples. In this case, extreme caution must be exercised with the use of liquid propane. There are various modified protocols that have been developed for each step of the rapid freezing and freeze-substitution fixation method. The detailed procedures are described by McDonald (2002).

To study yeast autophagy, one approach is to examine starved cells; induction of autophagy can also be achieved with the Tor inhibitor rapamycin, but at least in terms of the regulatory processes, these two treatments are not identical. When yeast cells are starved, typically by nitrogen depletion, they have an increased volume of the vacuolar compartment, and the cell wall becomes thicker than when cells are growing actively. Therefore, the fixation and penetration of the embedding resin is more difficult than with growing cells. Even with the high-pressure freezing technique, autophagic bodies are sometimes broken within the vacuole.

This chapter focuses on imaging techniques for yeast, mainly the freeze-substitution fixation method, followed by histochemical and immunolabeling methods of protein detection, and a simple description of the freeze replica method. The electron microscope produces images that represent only a two-dimensional projection of cells; the information about the third dimension of cells has been lost. To recover this information, numerous applications of electron tomography have been reported (Marsh, 2007; Melloy *et al.*, 2007; O'Tool *et al.*, 2002). In autophagy, a dynamic membrane rearrangement is induced. Electron tomography is a powerful tool to use in tracking a complex cellular event. The outline is described in brief. For all of the methods described, representative examples of the resulting electron micrographs are presented.

2. Morphological Examination of Membrane Dynamics in Autophagy and the Cvt Pathway

The unique point of the method described subsequently is to fracture yeast cells in liquid nitrogen after plunge freezing by using liquid propane. By this process, a part of the cell wall and the outer leaflet of the plasma membrane are stripped away because of the freeze fracture. As a result, the penetration of fixative and various resins for embedding yeast cells becomes easier.

2.1. Freezing method using liquid propane in yeast

This method is one of a number of different rapid-freezing methods.

1. The use of copper disks of two different types is recommended, because you can easily distinguish the copper disk attached to the cells from the other disk that is used as a cover during the fracturing process under liquid nitrogen. You can obtain the copper disks from standard EM vendors (e.g., Cat No. 2475 and Cat No. 247, Nisshin EM or Maxtaform). Alternatively, a copper disk can be made from a sheet of copper (high-purity metal, any company) by punching out the appropriately sized pieces using a punching device (BAL-TEC) or surgical scissors. In this case, the disk will be in the shape of a square (approximately 3 mm). If you use a spacer, EM grids (400 mesh gold or copper, VECO) are needed. You will need two disks per sample and one grid (optional; see Fig. 10.1). Scratch the surface of the copper disk (e.g., Cat No. 2475)

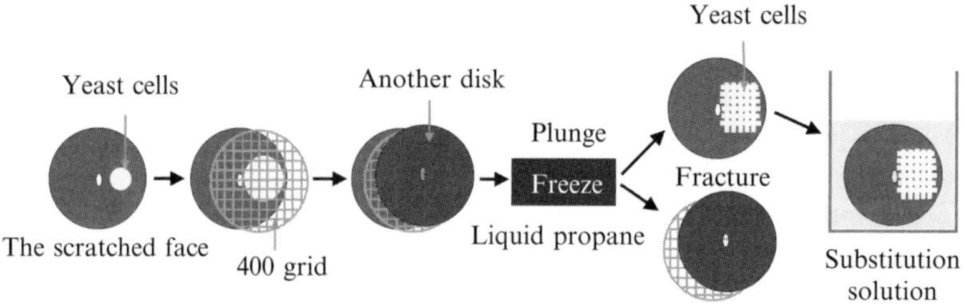

Figure 10.1 The protocol of the plunge method for yeast.

for good attachment of the cells. Be careful to keep the disk flat (i.e., prevent bending the disk) while handling it to facilitate the subsequent sectioning process.

2. Just before use, clean the copper disk and EM grids to remove any oxidation film or grease layer that may prevent good attachment of cells. For cleaning, transfer the copper disk and EM grids to an appropriate tube and add distilled water and several drops of 1N HCl.
3. Vortex for half a minute and check to see whether the surface of the copper disk (and EM grids) has become shiny; the surface should be shiny if the disk is properly cleaned.
4. Wash the disk and EM grids with distilled water several times and rinse with acetone.
5. Keep the copper disk and EM grids in the acetone until they are to be used.
6. Prepare the machine for freezing as described in the instruction manual.
7. Dry the copper and EM grids on filter paper. Line up the disk and grids before use; have two disks and one grid prepared for each sample.
8. Harvest yeast cells by centrifugation (3000 rpm for 1 min in a micro-centrifuge) or filtration. If using centrifugation, aspirate the supernatant carefully.
9. Place a small amount of the yeast pellet on the scratched surface of a copper disk, put the shiny surface of a copper or gold EM grid as a spacer on top of it, and cover with another copper disk. It is possible to omit the spacer grid. In this case cells are sandwiched between two copper disks.
10. Hold the sandwiched disks with a forceps and fix them by the use of a clamp. Then fix the forceps by screwing them to the end of the injector rod of the freezing machine. Plunge the tip of the forceps fixed to the injector rod into liquid propane by pushing the release button. The sandwiched disks are frozen in the liquid propane. This process follows the protocol of the instruction manual that accompanies various types of freezing machines.

11. Transfer the disks to liquid nitrogen. Place the scratched copper disk (e.g., Cat No. 2475) of the sandwiched disks onto the metal block in liquid nitrogen, holding them lightly with tweezers. Remove the upper copper disk (Cat. No. 247, Nisshin EM) with tweezers (a second pair of tweezers) and then carefully remove the 400 mesh grid (if it was included). The unloading device (Cat. No. BU012122-T, BAL-TEC) is useful for this purpose. The cells remain attached to the copper disk (Cat. No. 2475, Nisshin EM) as a thin layer during all procedures.

2.2. Substitution and Embedding

1. Transfer the cells attached to the disk to 2 ml of the substitution solution.
   The particular substitution solution should be chosen on the basis of the aim of the study. In general, 2% osmium tetroxide in anhydrous acetone is used. If necessary, add uranyl acetate (which is recommended to detect cytoskeleton) or other reagents if needed. Note that if the vacuole is stained with certain electron density, it is very difficult to detect the membrane of autophagic bodies. If a deep freezer is used for substitution, a PFA jar (Flon Industry) is convenient for keeping the disks in solution during the freeze substitution, because the area of the bottom is wide, and it has a tight screw cap. Put the jar into a steel container, and then inside a stainless steel container sealed tightly to prevent the spreading of the osmium tetroxide vapors.
2. Allow the substitution in the cells to take place below −80 °C for 2–4 days.
3. Transfer a container in which you have put several PFA jars to −20 °C and keep for 4 h or overnight.
4. In this and the following steps, gently shake the container several times in each step. Transfer the container to 4 °C and let it stand for 2 h.
5. Transfer the container to room temperature and let it stand for 2 h.
6. Remove the substitution solution by aspirating carefully and add 2 ml of anhydrous acetone. Hereafter, all procedures are carried out by aspiration with a polyethylene pipette.
7. Change the anhydrous acetone 3 times (allow 15 min for each change).
8. Change the anhydrous acetone to QY-1 (n-Butyl glycidyl ether; Nisshin EM) by exchanging with the following:

   Mixture of anhydrous acetone:QY-1 = 1:1 (2 ml), 30 min.
   100% QY-1; 30 min.

9. Infiltrate with Quetol 812 mixture as follows (Quetol 812 set, TAAB set; EM vendors):

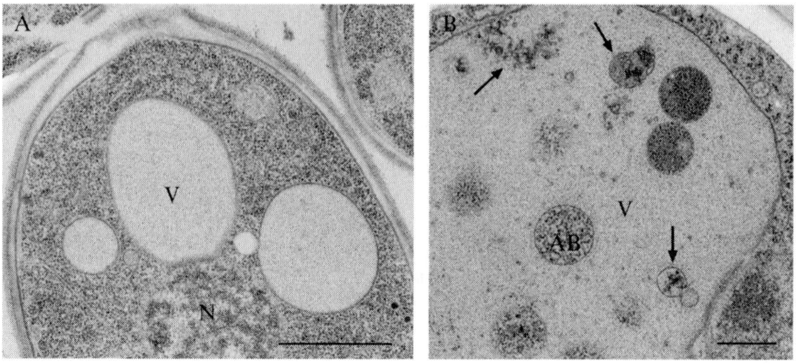

Figure 10.2 The change of vacuolar morphology during autophagy. (A) In wild-type cells grown in rich medium the vacuolar content is in a rather homogeneous state. Scale bars: 1 μm. (B) In cells lacking vacuolar hydrolase activity, the contents of the vacuole are not degraded. Under starvation conditions, the cytoplasmic components are accumulated in the vacuole in the form of an autophagic body. In wild-type cells, the autophagic bodies are typically degraded in the vacuole (arrow) and the degradation products are released back into the cytosol. Scale bars: 500 nm. AB, autophagic body; N, nucleus; V, vacuole.

QY-1:Quetol 812 mixture = 9:1 (2–3 ml); overnight
QY-1:Quetol 812 mixture = 3:1; e.g., ~ 3 h
QY-1:Quetol 812 mixture = 1:1; e.g., for 3 h
QY-1:Quetol 812 mixture = 1:3; e.g., for 3 h or overnight
QY-1:Quetol 812 mixture = 1:9; overnight
100% Quetol 812 mixture twice; e.g., for 4 h each

10. Embedding and polymerization. Transfer a copper disk into a Beem embedding capsule (size "00," Cat. No. #70020-B, Electron Microscopy Sciences) filled with Quetol 812 mixture. The copper disk will sink by itself; the surface with the attached cells should be face up. In general, polymerization is done at 60 °C for 48 h. Cut off the resin surrounding the disk with a razor blade, and take off the disk using tweezers after polymerization. The cells remain in the epoxy resin. (See Figs. 10.2 and 10.3)

3. Characterization of the Autophagosome and Autophagic Body Membrane in Starving Cells

3.1. PATAg method

This method detects polysaccharides on the ultrathin sections. The difference between rough ER and membranes originated from the Golgi can be distinguished. The original method was described by Thièry (1967).

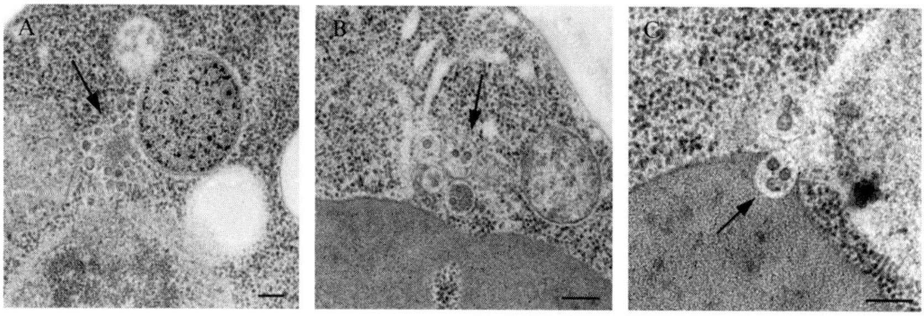

Figure 10.3 The transport mechanism used for aminopeptidase I in growing cells involves a double-membrane vesicle similar to that in autophagy. As the vesicle size is rather small, tracking the membrane is difficult. (A) A Cvt complex (composed of oligomers of precursor aminopeptidase I and unknown particles) not enclosed by a membrane (arrow). (B) Cvt vesicle (arrow) (C) Cvt body (arrow) Scale bars: 200 nm.

Sections were floated on solutions of various reagents and were transferred with a platinum wire loop, moving the sample from drop to drop. In the modified method described here, cells are prepared by using the freeze-substitution fixation method.

1. Prepare Formvar-coated gold grids. Formvar coating is a general method in EM work. Coat the gold grid with 0.5% Neoplate W in toluene solution (EM vendors). Prepare 0.5%–0.6% Formvar in ethylene dichloride (solvent) at room temperature and transfer it to a 50-ml glass jar. Clean a new standard glass slide thoroughly with soap and water to remove the surfactant, and then rinse thoroughly with Milli-Q water. Dry the glass slide and dip it in EtOH, and then polish it with a Kimwipe until the surface of the glass is completely smooth. Note that this is a critical step. Dip the slide approximately 5 cm into the Formvar solution. After 4–5 s, remove it from the jar by using a slide glass lifter (EM vendors). Release the Formvar film from the slide and place the gold grids on the floating film. Avoid any dusty portions of the film. Place a piece of Parafilm over the floating film, and lift it off very slowly from the water surface by using forceps. The detailed protocols are described by Hayat (2000).
2. Pick up the thin sections on the Formvar-supported gold grids.
3. Incubate the gold grids with the section's side on a drop of 1% periodic acid for 20 min. Hereafter, all procedures except washing with distilled water are done using a spot plate made of white porcelain (BAL-TEC). However, it is possible to use a sheet of Parafilm as an alternative. The use of a moist chamber is recommended to prevent evaporation from the drops.
4. Wash the grid with distilled water 3 times (20–25 μl of drop).
5. Incubate the gold grids on a drop of 0.2%–1% thiocarbohydrazide in 20% acetic acid for 60 min.

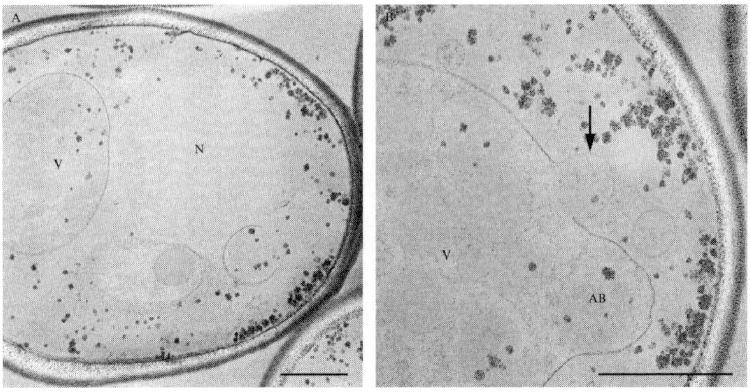

Figure 10.4 PATAg method. (A) The nuclear envelope is not stained by this method. (B) The combination of the freeze-substitution fixation method and the PATAg method reveals the characteristic difference in the membranes between the vacuole (V) and the autophagosome at the fusion site (arrow). Scale bar: 1 μm. AB, autophagic body; N, nucleus. From Baba *et al.* (1994).

6. Wash the gold grids with 10% acetic acid for 5 min (20–25 μl of drop).
7. Wash the gold grids with 5% acetic acid for 5 min.
8. Wash the gold grids with distilled water 3 times.
9. Incubate the gold grids on a drop of 1% silver proteinate for 60 min in the dark.
10. Wash the gold grids with distilled water 3 times.
11. Air-dry the gold grids and examine them without poststaining (see Fig. 10.4).

3.2. Freeze replica method

Although the aldehyde fixation and the use of cryoprotectant induce the aggregation of autophagic bodies in the vacuoles, this method provides good information on the intramembrane particles.

1. For the aldehyde fixation and freezing method harvest cells by centrifugation and add 10 ml of 2.5% glutaraldehyde in 50 m*M* sodium phosphate buffer (pH 7.4). Incubate for 2 h at room temperature. Wash 3 times with 10 ml of sodium phosphate buffer. Resuspend in 10 ml volume of sodium phosphate buffer and add glycerol stepwise to 30%. Harvest cells by centrifugation after standing overnight at 4 °C. Mount on a hat-shaped specimen holder (www.bal-tec.com) and immediately immerse into liquid propane.
2. For rapid-freezing of unfixed cells there are two basic ways to freeze unfixed yeast cells for freeze replica work. One approach is the metal contact method. Mount yeast cells on the specimen holder (Cat. No. LZ02125VN, LZ02124VN, BAL-TEC). The specific specimen holders

and the specimen stage depend on which of various types of freezing machines and freeze replica machines are being used. Place the specimen holder on the plunger tip of the rapid-freezing device (in the case of KF80 or EM CPC, Leica). Release the plunger and let the sample drop onto a gold-coated pure copper block precooled with liquid nitrogen. The depth of the good freezing zone is only 10 μm. The device produced by Heuser (1979, http://www.heuserlab.wustl.edu/) can obtain the vitreous state of the depth to approximately 20 μm by using liquid helium. However, it is extremely difficult to obtain a good fracture of the frozen specimen (10–20 μm) in the fracture apparatus. The other method is high pressure freezing (see McDonald, 2002). In this case, the fracturing process is easy because of the large volume of cells.

3. Prepare the freeze-etching apparatus as described in the manual. Fracture cells in the machine. In yeast, etching is done for approximately 1 min at -100 °C under a vacuum of 10^{-5} pa.
4. Coating. Platinum-carbon is evaporated onto the specimen rotating at 70 rpm at an angle of 25°, followed by carbon evaporation at an angle of 90°. The platinum-carbon and carbon are 1.8 nm and 18 nm in thickness, respectively.
5. Cleaning. Clean the replicas with 5 ml of household bleach overnight at room temperature, and wash five times with 5 ml of distilled water. Mount the replica on copper grids (see Fig. 10.5).

4. Subcellular Localization Analysis of ATG Proteins in Autophagy and the Cvt Pathway

Immunolabeling of yeast cells is performed in many cases by the postembedding method, described here, because the thick cell wall prevents the penetration of antibody (Geng *et al.*, 2008). The detection of antigen on the sections prepared by using the freeze-substitution fixation method provides better results than that of chemical fixation. In the case of proteins expressed at a low level, the labeled protein is difficult to detect. The overexpression of the target protein provides a positive control for detection of the antigen. However, the expression level of the protein should ultimately be based on results from fluorescence microscopy. That is, the protein should first be observed by direct or indirect fluorescence microscopy prior to the start of the immuno-EM analysis. The protein should initially be examined at the endogenous level if possible, and then the localization compared to that of the overexpressed protein. This provides the necessary information on the acceptable level of overexpression, if needed, for the subsequent immuno-EM studies. Co-localization analysis of two proteins is achieved using the double labeling method at the

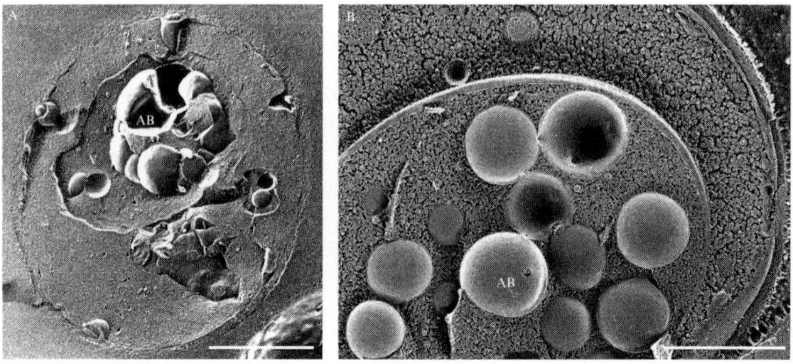

Figure 10.5 Freeze-replica image of nitrogen-starved cells. (A) Glutaraldehyde-fixed cell. Many autophagic bodies (AB) are aggregated in the vacuole (V). (B) Cryofixed cell. Autophagic bodies are dispersed in the vacuole as observed in ultrathin sections. In both (A) and (B), intramembrane particles are barely detected in the autophagic bodies. Scale bars: 1 μm. From Baba *et al.* (1995).

ultrastructural level. A deletion mutant of the target gene is used as a control for the first antibody.

4.1. Preparation of cells

The cryofixation was done according to the procedures described previously, with slight modifications.

1. Prepare the aluminum and copper disks. Scratch the surface of the aluminum disk. In LR White resin the copper starts to dissolve and turn to green; therefore aluminum should be used.
2. Plunge the disks into liquid propane. After freezing, remove the copper disk and then carefully remove the 400 mesh grid using tweezers.
3. Substitute the cells for 2–4 days below −80 °C. Substitution solution; anhydrous acetone containing 0.2%–0.4% formaldehyde, or glutaraldehyde, or 0.01% osmium tetroxide as described previously.
4. Transfer the cells to −20 °C and keep for 2–4 h.
5. Remove the substitution solution and add anhydrous acetone at −20 °C.
6. Change anhydrous acetone 3 times (15 min each) at −20 °C.
7. Change anhydrous acetone to anhydrous EtOH.
8. Change anhydrous EtOH 3 times (15 min each) at −20 °C.
9. Remove the EtOH and add 30% LR White resin in anhydrous EtOH at −20 °C overnight.
10. Infiltrate 50% LR White resin in anhydrous EtOH for 6 h at −20 °C.
11. Infiltrate 70% LR White resin in anhydrous EtOH at −20 °C overnight.
12. Infiltrate 100% LR White resin for 6 h at −20 °C.

13. Change 100% LR White resin at 4 °C overnight.
14. Change 100% LR White resin twice (3 h each) at 4 °C.
15. Dry gelatin capsules (No. 0; Lilly) at 30 °C–45 °C and stand them in a rack made of aluminum (EM Vendors).
16. Transfer the disk into a gelatin capsule half-filled LR White resin.
17. Add LR White resin to fill the gelatin capsule.
18. Cap the gelatin capsule tightly and stand it in a box, which is an attachment of the UV polymerization device (TUV-200 type. Cat. No. 441, D.S.K. EM Co.).
19. Transfer the box into a UV polymerization device (TUV-200 type) in the freezer at −20 °C and keep for 30 min.
20. Start polymerization under UV irradiation at −20 °C for approximately 2–4 days. In general, the polymerization of LR White resin is done from 4 °C to 55 °C. If the required immunoreaction is not subsequently obtained, the use of Lowicryl K4M or Lowicryl HM20 resins are alternate choices for the embedding method. The advantage of LR White is that the toxicity is rather low compared with the other two resins.

4.2. Immuno-staining for ultrathin sections

All incubations are done on a drop placed on a sheet of Parafilm.

1. Prepare Formvar-coated nickel grids.
2. Pick up ultrathin sections on the nickel grids.
3. Prepare a moist chamber.
4. Prepare a polystyrene Petri dish (9-cm diameter) having the cover sheeted with dampened filter paper. Mark freely the name of the required antibody on the back of the bottom of the Petri dish. Put this dish into a larger Petri dish (glass type, 12-cm diameter), the bottom of which is covered with dampened Kimwipes.

4.2.3. Standard immuno-staining method

One of many possible protocols is described subsequently.

1. Block the sections on a drop of the blocking solution for 15 min at room temperature. Use Milli Q water and filter (0.22 μm) all reagents. Blocking solution: PBS containing 2% BSA, Goat IgG (chromatographically purified) 1:50 dilution (as described in the instruction manual (ZYMED Laboratories) or 1%∼2% normal goat serum. The detail of blocking is described by Mulholland and Botstein (2002).
2. Incubate the grids on a drop of primary antibody solution diluted in PBS containing 0.5% BSA at 20 °C for 2 h or an appropriate time based on the affinity of the antibody. It is essential to use affinity-purified antibody or antibody pre-treated with the acetone powder method (the antibody

is incubated in an acetone precipitated powder extract prepared from a yeast strain disrupted for the gene of interest).
3. Wash the grids 6 times on a drop of PBS containing 0.5 % BSA.
4. Incubate the grids on a drop of secondary antibody diluted in PBS containing 0.5% BSA at 20 °C for 1–2 h. Use 5-nm or 10-nm colloidal gold-conjugated goat antirabbit (or antimouse) IgG 1:50 or 1:100 dilution (as described in the instruction manual).
5. Wash the grids 6 times on a drop of PBS.
6. Wash the grids 3 times on drops of Milli-Q water.
7. Fix the grids on a drop of 1% glutaraldehyde in PBS at room temperature for 10–15 min.
8. Wash the grids 5 times with Milli-Q water.
9. Stain the grids with 4%–6% uranyl acetate dissolved in Milli-Q water for 10 min and further stain with 0.04%–0.08% lead citrate dissolved in Milli-Q water for 1 min.

One option is to omit the staining with the lead solution to allow the detection of protein distributed randomly in the cytoplasm. Because the contrast of the stained structures becomes lower on the thin sections without lead staining, the distribution pattern of immunoreactive proteins is easily observed.

4.2.4. For the nanogold staining method

The sensitivity of immunolabeling increases with the use of nanogold.

1. Prepare ultrathin sections as described previously.
2. Incubate the grids on a drop of secondary antibody diluted into PBS containing 0.5% BSA at 20 °C for 1–2 h. Use 1-nm colloidal gold-conjugated goat antirabbit (or antimouse) IgG 1:100 dilution (as described in the instruction manual).
3. Wash the grids 6 times on a drop of PBS.
4. Fix the grids on a drop of 1% glutaraldehyde in PBS at room temperature for 10–15 min.
5. Wash the grids 5 times with Milli-Q water.
6. For silver enhancement, make the enhancement solution before use as indicated in the supplied product information. Enhancement solution (in the case of Aurion R-Gent SE-EM):

   Use a plastic microcentrifuge tube (1.5 ml).
   Place 20 drops of ENHANCER solution in the microcentrifuge tube.
   Add 1 drop of the DEVELOPER solution.
   Mix well by vortex.

7. Incubate the grids on a drop of the enhancer mixture at 20 °C in the dark for 20–40 min.
8. Wash the grids 5 times with Milli-Q water.
9. Stain the grids as described previously (see Fig. 10.6).

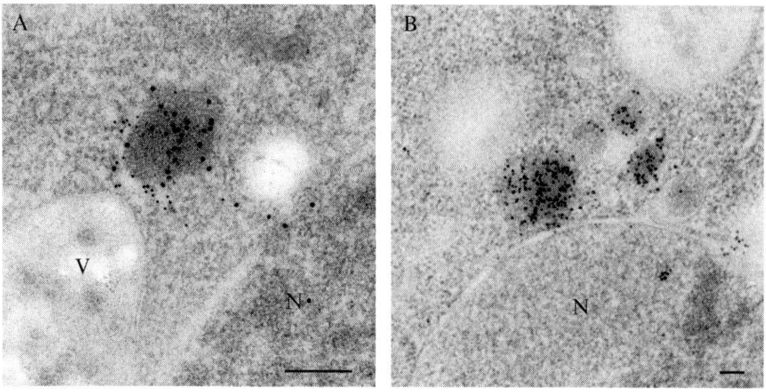

Figure 10.6 (A) Examples of double labeling using monoclonal and polyclonal antibody indicate the object of localization. Atg8 tagged with the HA epitope is detected by using anti-HA monoclonal antibody (5 nm gold) and aminopeptidase I is detected by using anti-aminopeptidase I polyclonal antibody (10 nm gold). (B) Nanogold labeling. Aminopeptidase I is detected by using 1 nm gold labeling and the silver enhancement technique. Scale bars: 200 nm. N, nucleus; V, vacuole.

5. Morphometric Analysis of Autophagy-Related Structures

The mainstream method of the recent three-dimensional (3-D) image analysis is electron tomography. It can be used for the quantitative and qualitative analysis of a cell or organelles as a powerful technique to determine the relationship between their function and structure (Marsh *et al.*, 2007). The approach using serial thin sections requires a whole cell or large area in high resolution (Baba *et al.*, 1994; Hamasaki *et al.*, 2005; Roberts *et al.*, 2003).

5.1. 3-D reconstruction from serial thin sections

Although the resolution along the z-axis is not very good, the image quality of serial thin sections is high. Serial thin sections can be prepared using a standard protocol for EM work. The extrapolation of the real 3-D structure from the EM photographs is made by hand, so this step is critical for reliable results. The detailed theory of 3-D reconstruction is described in Baba (2000).

1. Trace and copy contour lines of objects observed in each EM image with a black pen or a digitizer.
2. Input the coordinates constructing the copied contour line images into a computer in the order of section series aligning the successive section contour line images.

3. Sort the contour lines into groups, each group representing a structural part; in this process contour lines are represented by a set of points.
4. The handmade software or a commercial one (e.g. 3D-DOCTOR, Able Software Corp.) for 3-D reconstruction is utilized as indicated in the manual.

5.2. Electron tomography

In general, an ultrahigh or intermediate voltage electron microscope, or the conventional voltage type attached with a large angled-specimen tilting, is needed. One example applied to yeast cells induced for autophagy is described. It is possible to scrutinize interactions between the vacuolar membrane and autophagic bodies at high resolution. The principles and detailed procedure are described in many reports (Frank, 2006, 1987; Hoog and Antony, 2007). This approach can provide novel insight into the dynamic trafficking events and changes in the cytoskeletal network that occur during autophagy.

5.2.1. Prepare sections

1. Cells should be fixed by using the freeze-substitution fixation methods.
2. Obtain thick sections appropriate to the study. The limits of the thickness observable with an electron microscope relating to the accelerating voltage are approximately 0.15, 0.5, and 2 μm for 100, 300, and 1 M (kV), respectively.
3. The staining of sections is an important step, and it must be uniform with depth.

The staining protocol described by Takaoka *et al.* (2008) is introduced subsequently.

1. Prepare 3% uranyl acetate in 70% methanol solution.
2. Immerge thick sections into the staining solution and place a distilled water vessel in a microwave oven.
3. Heat at 170W for 30 s.
4. Stain further at room temperature for 20 min.
5. Prepare lead solution (Sato, 1968).
6. Place thick sections and a distilled water vessel in a microwave oven.
7. Heat at 170W for 30 s.
8. Stain further at room temperature for 10 min.
9. Coat both surfaces of sections with carbon film.

5.2.2. Collect images of tilt series with a CCD camera

The tilt angle generally used is $\pm 60°$–$70°$ with $1° \sim 3°$ angular increment.

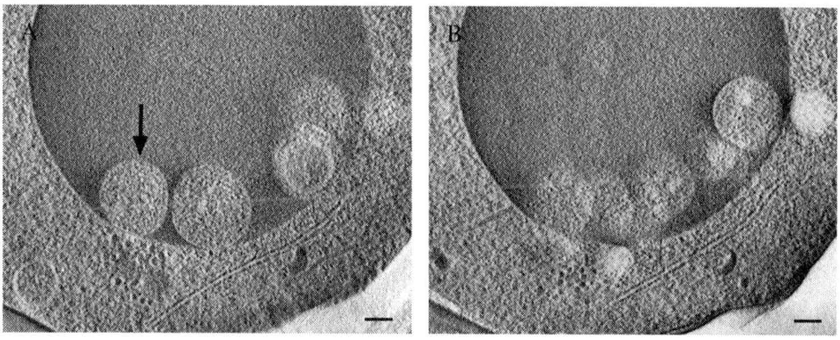

Figure 10.7 Tomographic reconstructed result of a nitrogen starved cell. (A) and (B) Two typical cross-sections parallel to x-y plane. The resolution obtained is ~3 nm. One autophagic body (arrow) includes 0.094 μm^3 cytoplasmic volume, as an example. Scale bars: 200 nm.

5.2.3. Reconstruction and 3-D presentation

In many cases the computer software package IMOD developed by Colorado University and Avizo DEV (Mercury Computer Systems) is utilized; however, there are other software programs available from several companies, such as TEMograpy (JEOL), EMIP-3D (Hitachi) or Inspect 3D (FEI).

5.2.4. 3-D measurement

5.2.4.1. For the usual method

1. Trace contour lines of a 3-D object on a computer display in serially reconstructed (x-y) cross-sections with the aid of some image processing software (e.g., 3D-DOCTOR Able Software Corp.).
2. Fill the object image area surrounded by the contour lines with an identification label-value interactively, with the aid of such similar software, which specifies it as a measurement target.
3. Calculate various shape parameters such as volume and surface area (e.g., using the NIH ImageJ software).

5.2.4.2. For the unusual method Because of the tilt angle limitation, the resolution along the z-axis is very low, which makes it impossible to trace the top parts of round-shaped objects such as a sphere. Even in such a case as this, however, if an object is modeled on a mathematical geometry, an approximate measurement is feasible with the least-squares method as follows. In the present application of yeast cells, the measurement of an autophagic body (AB) is a typical example.

1. Trace approximately 3 contour lines of the AB in (x, y) cross-section images, which are as far apart as possible from each other.
2. Input all 3-D coordinate data points (Xi, Yi, Zi) to a computer.

3. Solve the following least squares error (E) equation numerically for the radius r and the center position (a, b, c).

$$E = \sum[r^2 - (Z_i - c)^2 - (X_i - a)^2 - (Y_i - b)^2]^2 \rightarrow \text{min}.$$

4. Calculate the volume (V) of one AB sphere given by $4\pi r^3/3$.
5. For example, $r = 282$ nm and V = 0.094 μm^3 (see Fig. 10.7).

ACKNOWLEDGMENTS

I am grateful to Dr. D. J. Klionsky for critical reading of the manuscript. I thank Dr. N. Baba for the 3-D analysis in Kogakuin University, Dr. N. Nagata for use of the EM facility in Japan Women's University, and Dr. A. Takaoka and Dr. N. Kajimura for use of the ultrahigh voltage machine at the Research Center for Ultrahigh Voltage Electron Microscopy, Osaka University. A part of this work was supported by the Nanotechnology Network Project of the Ministry of Education, Culture, Sports, Science and Technology (NEXT), Japan, at the Research Center for Ultrahigh Voltage Electron Microscopy, Osaka University (Hitachi multi-functional; Nano-Foundry).

REFERENCES

Baba, M., Takeshige, K., Baba, N., and Ohsumi, Y. (1994). Ultrastructural analysis of the autophagic process in yeast: Detection of autophagosomes and their characterization. *J. Cell Biol.* **124,** 903–913.

Baba, M., Osumi, M., and Ohsumi, Y. (1995). Analysis of the membrane structures involved in autophagy in yeast by freeze-replica method. *Cell Struct. Func.* **20,** 465–471.

Baba, N. (2000). Computer-aided three-dimensional reconstruction from serial section images. *In* "Image analysis: Methods and applications" (D. P. Häder, Ed.), pp. 329–354. CRC Press, LLC.

Bernales, S., McDonald, K. L., and Walter, P. (2006). Autophagy counterbalances endoplasmic reticulum expansion during the unfolded protein response. *PLoS Biol.* **4,** 2311–2324.

Frank, J., McEwen, B. F., Radermacher, M., Turner, J. H., and Rieder, C. L. (1987). Three-dimensional tomographic reconstruction in high voltage electron microsocopy. *J. Electron Microsc. Tech.* **6,** 193–205.

Frank, J. (Ed.), (2006). "Electron tomography," 2nd edn., Springer, New York.

McDonald, K., and Müller-Reichert, T. (2002). Cryomethods for thin section electron microscopy. *Methods. Enzymol.* **351,** 96–123.

Geng, J., Baba, M., Nair, U. and Klionsky, D. J. (2008). Quantitative analysis of autophagy-related protein stoichiometry by fluorescence microscopy. *J. Cell Biol.* **182**, 129–140.

Hamasaki, M., Noda, T., Baba, M., and Ohsumi, Y. (2005). Starvation triggers the delivery of the endoplasmic reticulum to the vacuole via autophagy in yeast. *Traffic* **6,** 56–65.

Hayat, M. A. (2000). Suport film. *In* "Principles and techniques of electron microscopy: Biological applications." 4th edn., pp. 211–220. Cambridge University Press.

Heuser, J. E., Reese, T. S., Dennis, M. J., Jan, Y., Jan, L., and Evans, L. (1979). Synaptic vesicle exocytosis captured by quick freezing and correlated with quantal transmitter release. *J. Cell Biol.* **81,** 275–300.

Hoog, J. L., and Antony, C. (2007). Whole-cell investigation of microtubule cytoskeleton architecture by electron tomography. *Methods Cell Biol.* **79,** 145–167.

Kiššová, I., Salin, B., Schaeffer, J., Bhatia, S., Manon, S., and Camougrand, N. (2007). Selective and non-selective autophagic degradation of mitochondria in yeast. *Autophagy* **4,** 329–336.

Marsh, J. B. (2007). Reconstructing mammalian membrane architecture by large area cellular tomography. *Methods Cell Biol.* **79,** 193–220.

Marsh, B. J., Soden, C., Alarcon, C., Wicksteed, B. L., Yaekura, K., Costin, A. J., Morgan, G. P., and Rhodes, C. J. (2007). Regulated autophagy controls hormone content in sevretory-deficient pancreatic endocrine beta-cells. *Mol. Endocrinol.* **21,** 2255–2269.

Melloy, P., Shen, S., White, E., Mcintosh, J. R., and Rose, M. D. (2007). Nuclear fusion during yeast mating occurs by a three-step pathway. *J. Cell Biol.* **179,** 659–670.

Mulholland, J., and Botstein, D. (2002). Blocking. *Methods Enzymol.* **351,** 69.

O'Toole, E., Winery, M., McIntosh, J. R., and Mastronarde, D. N. (2002). Electron tomography of yeast cells. *Methods Enzymol.* **351,** 81–95.

Roberts, P., Moshkovitz, M. S., Kvam, E., O'Tool, E., Winey, M., and Goldfarb, S. D. (2003). Piecemeal microautophagy of nucleus in *Saccharomyces cerevisiae*. *Mol. Biol. Cell* **14,** 129–141.

Sato, T. (1968). A modified method for lead staining of thin sections. *J. Electron Microsc.* **17,** 158–159.

Takaoka, A., Hasegawa, T., Yoshida, K., and Mori, H. (2008). Microscopic tomography with ultra-HVEM and applications. *Ultracisrosc.* **108,** 230–238.

Thièry, J. P. (1967). Mise en evidence des polysaccharides sur coupes fines en microscopie electronique. *J. Microscopie (Paris)* **6,** 987–1018.

CHAPTER ELEVEN

Cell-Free Reconstitution of Microautophagy in Yeast

Andreas Mayer*

Contents

Abstract

Microautophagy is the direct uptake of soluble or particulate cellular constituents into lysosomes. Here, I describe methods to reconstitute and study this process *in vitro*, using vacuoles (lysosomes) from the yeast *Saccharomyces cerevisiae* as model organelles. Protocols to grow the cells, isolate vacuoles from them, and to induce microautophagy of soluble tracers are presented.

1. Introduction: Microautophagy in Yeast

In the yeast *Saccharomyces cerevisiae*, autophagic degradation happens in the vacuole, the lysosomal compartment of this organism. Vacuoles are the main site for storage, recycling and breakdown of cellular constituents. When yeast cells sense nutrient (e.g., nitrogen) limitation, they stop dividing and enter stationary phase. In stationary phase, yeast vacuoles coalesce into a large single central vacuole. The cells adapt to the new environmental conditions by transferring large quantities of cellular constituents into

* Département de Biochimie, Chemin des Boveresses 155, CH-1066 Epalinges, Switzerland

Methods in Enzymology, Volume 451
ISSN 0076-6879, DOI: 10.1016/S0076-6879(08)03211-4

vacuoles where they are degraded. This phenomenon enables yeast cells to survive long periods of starvation. Cytoplasmic compounds enter vacuoles by macro- or microautophagy (Baba *et al.*, 1994). Vacuoles are an excellent tool for studying microautophagy because they are big (approximately 3 μm diameter), they can easily be stained for *in vivo* fluorescence microscopy, and they can be purified in milligram amounts per day. Microautophagy coincides with invaginations of vacuoles that can be easily seen by light-microscopic inspection of FM 4-64-stained cells. Microautophagy can be reconstituted in a cell-free system from purified vacuoles and cytosolic extracts. This *in vitro* system measures the uptake of a luciferase reporter substrate (Sattler and Mayer, 2000). Using pharmacological substances, the *in vitro* uptake reaction can be dissected into different kinetic stages (Kunz *et al.*, 2004; Uttenweiler *et al.*, 2005, 2007).

Microautophagy in yeast is necessary for the transition from starvation-induced growth arrest to logarithmic growth (Dubouloz *et al.*, 2005). It affects not only soluble cytosolic proteins but also organelles. Piecemeal microautophagy of the nucleus (PMN) transfers parts of the nucleus into vacuoles (Roberts *et al.*, 2003), and micropexophagy leads to the degradation of peroxisomes (Mukaiyama *et al.*, 2002, 2004; Sakai *et al.*, 1998; Tuttle and Dunn, 1995; Tuttle *et al.*, 1993; Veenhuis *et al.*, 1983). Microautophagy of cytosol, the reaction described in this article, is controlled by the TOR and EGO (composed of Ego1, Gtr2 and Ego3) signaling complexes (Dubouloz *et al.*, 2005) and depends on the Vtc proteins and calmodulin (Uttenweiler *et al.*, 2005, 2007). Although microautophagy, similar to macroautophagy, is induced by starvation and Tor kinase signaling, there is no evidence that Atg proteins are directly involved in microautophagic uptake of a soluble reporter protein. However, macroautophagy and the Atg machinery are prerequisites to sustain microautophagy over many rounds of vesicle formation (Sattler and Mayer, 2000). This may reflect an indirect connection via membrane-supply because microautophagy leads to reduction of the vacuolar boundary membrane. Macroautophagic membrane influx into the vacuolar boundary membrane should be necessary to maintain the membrane source for invagination.

Aside from its participation in organelle degradation, a major function of microautophagy may also lie in maintenance of organellar size and membrane composition (Muller *et al.*, 2000). Because microautophagy leads to uptake and degradation of the vacuolar boundary membrane, it can compensate for the influx of membrane caused by macroautophagy. Microautophagic vacuole invagination thus might guarantee maintenance of vacuolar size and membrane composition under nutrient restriction. A function in organelle homeostasis should render microautophagic membrane invagination dependent on membrane influx via macroautophagy, which has been observed (Muller *et al.*, 2000).

2. Why Reconstitute Microautophagy *In Vitro* with Yeast Vacuoles?

Genetic *in vivo* analysis is a powerful tool to identify components involved in a process. Also, most of the components involved in autophagy have been defined by pioneering studies using this approach. However, the possibilities for mechanistic analysis are limited *in vivo*. Here, *in vitro* systems offer a distinct advantage. Reconstituting a complex cell-biological reaction in a cell-free form offers significantly greater possibilities to manipulate the reaction conditions and study the effects. In an *in vitro* system one can use membrane-impermeable inhibitors, purified proteins, or antibodies to selectively inactivate individual proteins on the organellar surface. Furthermore, peripheral proteins can be extracted from the membrane, and cross-link experiments and kinetic analyses are possible. Therefore, *in vitro* reconstitution is a useful and necessary complementation to the *in vivo* approaches.

The vacuole is an excellent model system for reconstitution studies because vacuoles can be prepared in good purity and in milligram amounts per day with reasonable effort, thus facilitating biochemical studies of protein interactions.

3. Methods for Reconstituting Microautophagy with Isolated Vacuoles

3.1. Materials

3.1.1. Buffers and reagents

1. PS buffer (10 m*M* Pipes/KOH, pH 6.8, 200 m*M* sorbitol)
2. Cytosol buffer (40 m*M* Pipes/KOH, pH 6.8, 0.5 m*M* $MgCl_2$, 150 m*M* KCl, 200 m*M* sorbitol, 1 m*M* DTT, 0.2 m*M* PMSF, 0.1 m*M* pefabloc SC, 0.5 μg/ml pepstatin A, 50 μM o-phenanthroline)
3. Spheroplasting buffer for vacuole preparation (50 m*M* potassium phosphate, pH 7.5, 600 m*M* sorbitol in YPD with 0.2% glucose)
4. Washing buffer for lyticase preparation (25 m*M* Tris/HCl, pH 7.4)
5. FM 4-64 (Molecular Probes) is dissolved as a 100 X stock solution (10 m*M*) in DMSO or ethanol and stored at −20 °C.
6. For immobilization of yeast cells use Seaplaque agarose in 10 m*M* PS buffer.

3.1.2. Media

1. YPD (1% yeast extract, 2% Bacto peptone, 2% glucose)
2. YPD containing 200 n*M* Rapamycin (Alexis, dissolve at 100X in DMSO)
3. SD(-N) (0.67% Difco yeast nitrogen base without amino acids and without ammonium sulfate, 2% glucose) for starvation
4. LB medium (2% tryptone, 1% yeast extract, 1% NaCl, pH 7)

3.2. Preparation of vacuoles and cytosolic extracts

3.2.1. Yeast culture

Yeast cells are precultured in YPD for 6–8 h at 30 °C and then diluted for logarithmic overnight growth (14–16 h, 30 °C, 225 rpm; final optical density [$O.D._{600}$] should be approximately 1.0 for harvesting) in 2-liter Erlenmeyer flasks with 1 liter of YPD medium. 1 liter of culture typically yields 250 μg of purified vacuoles (measured by their protein content). For starving cells, overnight cultures are harvested at an $O.D._{600}$ of 2, centrifuged (4 min at 3800*g*), washed with sterile water, resuspended in 1 liter of SD(-N) starvation medium, and incubated (3 h, 30 °C, 225 rpm).

3.2.2. Cytosol preparation

Cytosol is usually prepared from yeast strain K91-1A (kindly provided by Y. Kaneko). This strain is deficient for the alkaline phosphatases Pho8 (vacuolar) and Pho13 (cytosolic). This guarantees a low alkaline phosphatase background in cytosolic preparations and is advantageous when values from microautophagy assays shall be normalized to vacuolar membrane material, which is measured via vacuolar alkaline phosphatase (Pho8) activity.

1. Harvest overnight yeast cultures (1 liter in 2-*l* baffled Erlenmeyer flasks) at $O.D._{600} = 4.5$, by centrifugation (4 min, 3800*g*), wash with sterile water, harvest again, resuspend in 1 liter of SD(-N), and incubate for 3–4 h at 30 °C and 255 rpm. For preparation of cytosol from nonstarved cells, replace SD(-N) by YPD in this incubation.
2. Harvest cells and wash as described previously, first with water, and then with one pellet volume of chilled cytosol buffer. After centrifugation (5 min, 3800*g*, 4 °C), resuspend the pellet in a small volume of cytosol buffer so that a thick slurry results.
3. Freeze the suspension as little nuggets in liquid nitrogen and blend (6–8 times for 30 s) in a Waring blender filled with liquid nitrogen.
4. Thaw cells and centrifuge the lysate (10 min, 12,000*g*, 4 °C). Ultracentrifuge the supernatant fraction (20 min, 125,000*g*, 2 °C), discard the fatty top fraction, and recover the clarified cytosol.

5. Adjust the protein concentration of the cytosolic fraction to 25–30 mg/ml with cytosol buffer (do not use preparations below 10 mg/ml for *in vitro* microautophagy assays). Freeze 50–300 μl aliquots in liquid nitrogen and store at -80 °C.

3.2.3. Production of lyticase for vacuole preparation

Lyticase (β-1,3-glucanase) from Oerskovia xanthineolytica (Scott and Schekman, 1980; Shen *et al.*, 1991) is expressed in *E. coli* strain RSB 805. The protein is purified from the periplasmic space (Reese *et al.*, 2005).

1. Grow 10 l RSB 805 in LB medium with 100 μg/ml ampicillin to $O.D._{600} = 0.6$.
2. Add 0.4 m*M* IPTG and let cells grow for another 5 h at 30 °C.
3. Harvest cells (10 min, room temperature, 4200*g*), wash with 50 ml washing buffer and centrifuge (10 min, room temperature, 4200*g*).
4. Resuspend pellets in 200 ml of washing buffer, supplemented with 2 m*M* EDTA and an equal volume of 40% (w/v) sucrose in washing buffer and gently shake at room temperature for 20 min.
5. Harvest cells (10 min, room temperature, 4200*g*) and completely remove the supernatant fraction. Chill flasks on ice.
6. Resuspend the cells per one liter of culture in 20 ml ice cold 0.5 m*M* $MgSO_4$, shake gently (4 °C, 20 min) and centrifuge (10 min, 4 °C, 4200*g*).
7. The supernatant fraction contains lyticase and some *E. coli* periplasmic proteins. Adjust the protein concentration of the supernatant fraction to 1 mg/ml. The lyticase solution can be stored at -20 °C in 6-ml aliquots without further purification. It should be thawed directly before use for digestion of yeast cell walls.

3.2.4. Vacuole preparation

1. Grow cells (1 l culture in 2-l Erlenmeyer flasks) in YPD overnight to $O.D._{600} = 2.5–3.0$, harvest (2 min, 3800*g*, 4 °C), and resuspend them in 50 ml of 30 m*M* Tris-HCl, pH 8.9, with 10 m*M* DTT. Incubate cells (5 min, 30 °C).
2. Centrifuge cells as above, resuspend in 12 ml of spheroplasting buffer, add 3 ml of a lyticase solution (concentration of stock solution: 1 mg/ml) and transfer into 30-ml Corex tubes. Incubate cells (25 min, 30 °C).
3. Reisolate spheroplasts (4 °C, 1 min, 2500*g*) and resuspend in 2.5 ml 15% Ficoll 400 in PS buffer by gentle stirring with a plastic rod and/or gentle vortexing.
4. Add DEAE-dextran (300 μl) from a 0.4-mg/ml stock in 15% Ficoll 400 in PS buffer. Incubate spheroplasts (2 min at 0 °C, then 90 s at 30 °C), chill again, transfer to an SW41 ultracentrifugation tube, and overlay with 3 ml of 8% Ficoll 400, 3 ml 4% Ficoll 400, and 1.5–2 ml 0% Ficoll

400 in PS buffer. After centrifugation (85 min, 154,000*g*, 2 °C,), harvest vacuoles from the 0%–4% interphase. Note: The optimal $O.D._{600}$ for harvesting cells, the time for spheroplasting and the amount of DEAE-dextran added can vary for different yeast strains. If necessary, vacuoles can be prepared with up to 1 m*M* PMSF in the spheroplasting buffer to reduce protein degradation.

3.2.5. Storage of isolated vacuoles

For storage of vacuoles, add glycerol (final concentration 10% w/v from a 50% stock) to a fresh vacuole suspension. Freeze the suspension as little nuggets in liquid nitrogen and store at −80 °C. The vacuoles maintain their activity for several months. Beware that the freezing procedure may lead to lysis of some vacuoles. Even if this is only a small fraction, proteases released into the buffer may cause proteolytic damage in subsequent incubations, particularly if the organelles stem from fully protease-competent strains. Therefore, vacuoles should be used immediately after thawing. Protease inhibitor cocktail (PIC) can be added from the beginning to avoid proteolytic damage. A 1000x stock solution of PIC is prepared from 500 μl of 200 m*M* pefabloc SC, 100 μl of 500 m*M* o-phenanthroline (in ethanol) and 100 μl of 5 mg/ml pepstatin A (in methanol). Make up the solution to 1 ml, aliquot, and store at −20 °C.

3.3. Assay of *in vitro* microautophagy

3.3.1. Standard reaction conditions and chemical assay of microautophagic activity

The assay of microautophagy *in vitro* is based on the uptake of a soluble reporter protein from the surrounding medium into the vacuolar lumen. After uptake, the reporter is reisolated with the vacuoles in differential centrifugation steps that sediment the vacuoles. The sedimented vacuoles are washed with buffer to remove residual reporter protein included in the pellet. Because reporter protein can also associate nonspecifically with the outer surface of the membranes, the vacuoles are subjected to a protease treatment. The protease digests residual reporter adhering to the vacuole surface, whereas microautophagocytosed reporter enzyme is protected against proteolysis by the surrounding membrane. Protease treatment thus reduces the nonspecific background signal of the assay. Finally, the reisolated vacuoles are lysed in detergent and the activity of the enclosed reporter enzyme is measured. Due to its stability and the high sensitivity of its luminescent detection we chose luciferase as the reporter enzyme. Other reporter enzymes can be used in an analogous fashion. For example, we have also successfully tried β-galactosidase and glucose oxidase.

Detailed procedure:

1. A standard reaction has a volume of 45 μl and contains: Vacuoles (0.2 mg/ml, either freshly prepared or thawed from a −80 °C stock), 3 mg/ml cytosol from starved cells, 105 m*M* KCl, 7 m*M* $MgCl_2$, 2.2 m*M* ATP, 88 m*M* disodium creatine phosphate, 175 U/ml creatine kinase, 17 μg/ml luciferase, 100 μM DTT, 0.1 m*M* pefabloc SC, 0.5 m*M* o-phenanthroline, 0.5 μg/ml pepstatin A, 200 m*M* sorbitol, 10 m*M* PIPES/KOH, pH 6.8. Incubate this mixture for 1 h at 27 °C.
2. For measuring luciferase uptake, chill the samples on ice, dilute with 300 μl of 150 m*M* KCl in PS buffer, centrifuge (6500*g*, 3 min, 2 °C), and wash the pellet once more with 300 μl 150 m*M* KCl in PS buffer by gently pipetting 3–4 times up and down with a 1-ml pipette. Addition of antibodies/pharmaceuticals to the *in vitro* reaction can induce formation of large clusters of vacuoles, which may be difficult to resuspend. In such cases, prolong the resuspension steps carefully as necessary to resuspend the vacuoles without destroying them. Alternatively, prepare, for example, F_{ab} fragments from antibodies to reduce clustering.
3. Centrifuge again and resuspended in 55 μl 150 m*M* KCl in PS buffer by shaking on a shaker (1400 rpm, 8 min, 4 °C).
4. Add proteinase K (0.3 mg/ml from 18x stock) and incubate on ice for 23 min. Stop digestion by adding 55 μl of 1 m*M* PMSF, 150 m*M* KCl in PS buffer.
5. Determine luciferase activity using an assay kit according to the manufacturer's instructions (Berthold Detection Systems, Pforzheim, Germany): mix 25 μl of sample with 25 μl of lysis buffer and add 50 μl of substrate mix (Gaunitz and Papke, 1998) directly before counting light emission in a microplate luminometer (LB 96 V, Berthold Technologies, Bad Wildbad, Germany). Note: Do not add leupeptin to the lysis buffer for cytosol preparation because leupeptin inhibits the firefly luciferase reporter enzyme.
6. Sometimes, addition of pharmacological inhibitors may interfere with the sedimentation of vacuoles, or with the quantitative separation of pellet and supernatants. Therefore, it is recommended to use an internal reference for the quantity of pelleted vacuoles. A practical marker is the activity of the vacuolar membrane-integral alkaline phosphatase (Pho8), which can be assayed in a 10-μl aliquot. Specific uptake activity can then be calculated as the quotient of luciferase activity over alkaline phosphatase activity (counts per second/$O.D._{400}$ per min). When comparing different yeast strain backgrounds, the level of mature alkaline phosphatase in vacuolar preparations has to be equal (beware of mutants affecting Pho8 sorting or maturation).
7. Pho8 activity is measured by withdrawing 10 μl of sample, adding 500 μl of phosphatase assay mix (10 μM p-nitrophenol phosphate, 10 m*M* $MgCl_2$,

0.4% Triton X-100, 250 m*M* Tris/HCl, pH 8.9) and incubating at 27 °C for 5 min. 500 μl of 1 *M* glycine, pH 11.5, is added to stop the reaction. The amount of the p-nitrophenyl phosphate hydrolyzed by Pho8 to p-nitrophenol is determined in a photometer via the $O.D._{400}$. Also see the chapters herein by Noda and Klionsky and Cabrera and Ungermann.

Note: Prepare all samples on ice and choose the order of addition of components wisely. For example, do not expose vacuoles to ATP or high salt concentrations on ice, as this will wash off peripheral vacuolar proteins and/or damage vacuolar enzymes, such as the H^+-pumping V-ATPase. Therefore, add KCl and the ATP immediately before starting the reaction and transferring the samples to 27 °C. Thaw cytosol/vacuole nuggets directly before use and do not freeze again. Direct comparisons of activity should be performed only with vacuoles from the same batch of preparation because there can be variations in absolute activities from batch to batch. These are often related to changes in the composition of media ingredients supplied. Our experience is that the quality of peptones or yeast extracts varies considerably over the year, even when purchased from the same supplier.

3.3.2. Kinetic analysis

Using a pharmacological approach including low molecular weight inhibitors, purified proteins, or antibodies, the *in vitro* uptake reaction can be dissected into different kinetic stages (Kunz *et al.*, 2004). According to their ability to block the reaction at different kinetic stages, these inhibitors have been defined as early acting class A inhibitors (nystatin, GTPγS, aristolochic acid) and late-acting class B inhibitors (W-7, valinomycin/FCCP, K252a and rapamycin). For this kind of analysis, standard uptake reactions are started. Inhibitors are added at different time points after the start of the reaction and the incubation is continued until the end of the normal 60-min reaction period. Inhibitors influencing only early steps (class A inhibitors) are ineffective if added late during the reaction, whereas inhibitors acting on late events of microautophagy (class B inhibitors) remain inhibitory throughout the reaction. As a control for progression of the overall reaction, one sample should be transferred to ice at each time point. Chilling stops the uptake reaction efficiently, except the very last step.

3.3.3. Rapid uptake

Rapid uptake of luciferase occurs after preincubation of vacuoles under standard conditions supporting microautophagic membrane invagination, but in the absence of the reporter enzyme (Kunz *et al.*, 2004). It is assumed that this assays the preparatory reactions for uptake (e.g., to form an invagination) without producing a luciferase signal. The final formation of vesicles from preformed tubes can then be scored by adding luciferase for only a short period of time (Fig. 11.1).

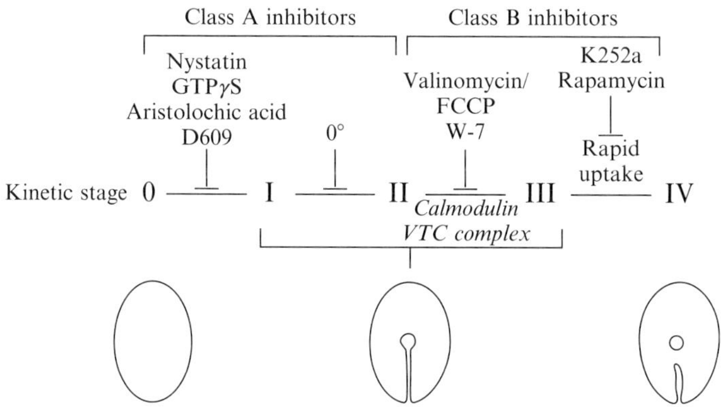

Figure 11.1 Schematic drawing of the stages of microautophagy in yeast and of the inhibitors and proteins influencing the transition between them.

A short incubation of 5 min allows only rapid uptake from preformed invaginations but is too short for the formation of new invaginations, which takes 20 min or more (Sattler and Mayer, 2000). Thus, this criterion can be used to distinguish the preparation for uptake (tube formation) from its completion (vesicle scission). Late-acting inhibitors in microautophagy can be further classified depending on their ability to inhibit this rapid uptake. Rapid uptake reactions are performed as follows:

1. Start standard uptake reactions, but leave out the luciferase reporter enzyme.
2. After 60 min of incubation without luciferase, add the reporter enzyme to an uptake reaction. Incubate the samples for another five minutes to permit reporter enzyme uptake.
3. Terminate uptake by chilling, diluting and centrifuging the reactions and analyse pelleted vacuoles for luciferase uptake as described previously for a standard reaction.

3.3.4. Microscopy assay of microautophagy

Microautophagy *in vitro* can be assayed by fluorescence microscopy. This has the advantage to allow quantification of the formed microautophagic vesicles (e.g., with respect to their number and size). Furthermore, this assay is recommended when using inhibitors that interfere with luciferase activity, precluding the chemical assay described above. The fluorophores should be coupled to high-molecular-weight carriers to exclude entry into vacuoles by diffusion or by vacuolar transporters. For microscopy assays, the procedure of standard uptake assays is modified:

1. Note: Commercial conjugates of fluorophores often contain a fraction of fluorophore that is not coupled. Therefore, the fluorescein dextran must

be purified by ultrafiltration before use. Ten mg of fluorescein dextran 500,000 (Molecular Probes, Leiden) are dissolved in 1 ml of PS buffer and concentrated and rediluted at least 5 times by ultrafiltration in Centricon 30 devices. The retentate can be used for the experiments if low-molecular-weight fluorescein is no longer detectable in the filtrate by fluorescence spectroscopy. Purified concentrated conjugate is rediluted to 20 μM with PS. It can be stored in small aliquots at −80 °C but, after prolonged storage or multiple freeze-thaw cycles, low-molecular weight fluorescein will reappear.

2. Standard uptake reactions are run, but luciferase is replaced by 1 μM purified fluorescein dextran 500,000 (20 μM stock).
3. After 60 min, chill the samples on ice, dilute with 300 μl of 150 mM KCl in PS buffer, centrifuge (6800×g, 4 min, 2 °C, fixed-angle tabletop centrifuge).
4. Wash twice with 300 μl of 150 mM KCl in PS buffer, centrifuging as previously. Resuspend the vacuoles in 30 μl of 150 mM KCl in PS buffer. Ten μl are used to check for complete vacuole recovery (e.g., by determining the activity of the integral vacuolar membrane protein alkaline phosphatase [Pho8] as described previously).
5. The vacuoles are resuspended in PS buffer with 10 μM FM 4-64 to a protein concentration of approximately 1 mg/ml. The suspension is mixed with an equal volume of 0.4% Seaplaque low melting point agarose in PS (kept liquid at 35 °C). 12 μl are transferred to a slide, chilled at 4 °C for 5 min to immobilize the vacuoles. The vacuoles can then be analyzed by a fluorescence microscope with a 100x or 63x objective. We use Zeiss Axiovert 200 microscopes, equipped with a 100 W Hg lamp and high numerical aperture objectives (e.g., 100x Plan-Apochromat, NA = 1.4). Microautophagic bodies in the lumen move very fast. Therefore, sharp photos can only be taken with fluorescence microscopes equipped with sensitive cameras (e.g., Photometrics Coolsnap HQ) that permit very short exposure times (<100 ms). If confocal analysis is desired, we use a spinning disc confocal microscope (Visitron QLC100) which, because of its high scan speed, permits acquisition times of 30 ms per frame (Fig. 11.2).

3.4. Limitations and caveats

As in every reconstitution approach, there are limitations to the reconstituted vacuolar microautophagy reaction. Luciferase as well as other potential reporter proteins adsorb to the vacuolar surface. Depending on the conditions chosen, simple washing of the membranes is thus insufficient to avoid a substantial background of adsorbed reporter, which has not been taken up into the lumen. Therefore, we routinely perform a digestion with proteinase K as part of the separation step.

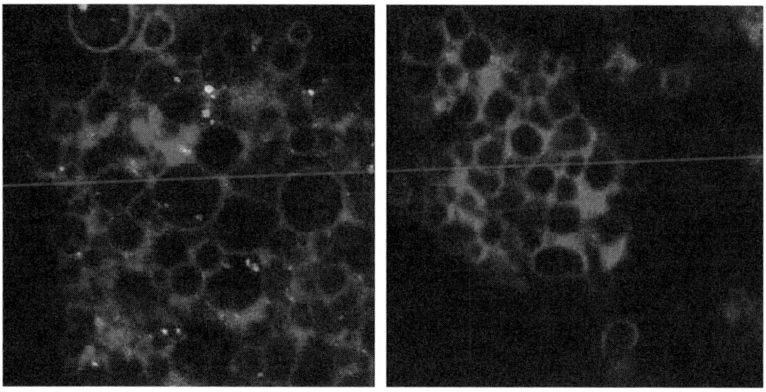

Figure 11.2 Example of uptake of FITC-dextran (green) into vacuoles (red) *in vitro*. Uptake happens in the presence of ATP (left) but not in its absence (right). (See Color Insert.)

Vacuole recovery and integrity are also issues since the assay depends on the retention of soluble indicators in the lumen of the vacuoles. Vacuoles are big membrane vesicles. They are osmotically sensitive and can be broken by shear forces. Therefore, all solutions used to resuspend vacuoles need to be osmotically stabilized with sorbitol (200 m*M*). Extreme changes of salt concentrations should be avoided for the same reasons.

During the assay procedures vacuoles are sedimented and resuspended several times. Resuspension should be performed gently and vacuoles should be pipetted by wide-bore tips whenever possible to avoid excessive mechanical stress. Leakage can be checked if desired using the activities or physical presence of soluble lumenal vacuolar proteins, such as carboxypeptidase Y (Prc1) or proteinase B (Prb1). Proteinase B will cleave almost any commercially available chromogenic substrate for serine proteases, and monoclonal antibodies are available for Prc1 (Molecular Probes). We routinely compare the content of soluble vacuolar material by comparing the content of these markers between freshly prepared vacuoles and the final reisolated and washed organelles. Leakage can be diagnosed as a loss of Prc1 or Prb1 relative to vacuolar membrane markers, such as alkaline phosphatase (Pho8). Antibodies to Pho8 are also available (Molecular Probes).

A further aspect to be considered, especially when using low-molecular weight inhibitors, is the reporter enzyme luciferase. Check whether a compound that is to be used can inhibit the purified luciferase. If so, particular care is necessary and control experiments need to be performed to ensure that the final activity that is read out has not been influenced by nonspecific influences on luciferase.

The vacuolar membrane is impermeable to most inhibitors. Thus, once the reporter has been autophagocytosed, it is usually protected. Dilution of the co-autophagocytosed inhibitor in the final assay step, the determination of luciferase activity in a detergent extract, leads to a >1000-fold dilution of co-autophagocytosed inhibitors. Thus, as long as inhibition is readily reversible, adverse effects on the reporter enzyme are usually not to be expected. If the action of an inhibitor on luciferase is poorly reversible, however, the compound cannot be used.

REFERENCES

Baba, M., Takeshige, K., Baba, N., and Ohsumi, Y. (1994). Ultrastructural analysis of the autophagic process in yeast: Detection of autophagosomes and their characterization. *J. Cell Biol.* **124,** 903–913.

Dubouloz, F., Deloche, O., Wanke, V., Cameroni, E., and De Virgilio, C. (2005). The TOR and EGO protein complexes orchestrate microautophagy in yeast. *Mol. Cell.* **19,** 15–26.

Gaunitz, F., and Papke, M. (1998). Gene transfer and expression. *Methods Mol. Biol.* **107,** 361–370.

Kunz, J. B., Schwarz, H., and Mayer, A. (2004). Determination of four sequential stages during microautophagy *in vitro*. *J. Biol. Chem.* **279,** 9987–9996.

Mukaiyama, H., Baba, M., Osumi, M., Aoyagi, S., Kato, N., Ohsumi, Y., and Sakai, Y. (2004). Modification of a ubiquitin-like protein Paz2 conducted micropexophagy through formation of a novel membrane structure. *Mol. Biol. Cell.* **15,** 58–70.

Mukaiyama, H., Oku, M., Baba, M., Samizo, T., Hammond, A. T., Glick, B. S., Kato, N., and Sakai, Y. (2002). Paz2 and 13 other PAZ gene products regulate vacuolar engulfment of peroxisomes during micropexophagy. *Genes Cells* **7,** 75–90.

Muller, O., Sattler, T., Flotenmeyer, M., Schwarz, H., Plattner, H., and Mayer, A. (2000). Autophagic tubes: Vacuolar invaginations involved in lateral membrane sorting and inverse vesicle budding. *J. Cell Biol.* **151,** 519–528.

Reese, C., Heise, F., and Mayer, A. (2005). Trans-SNARE pairing can precede a hemifusion intermediate in intracellular membrane fusion. *Nature* **436,** 410–414.

Roberts, P., Moshitch-Moshkovitz, S., Kvam, E., O'Toole, E., Winey, M., and Goldfarb, D. S. (2003). Piecemeal microautophagy of nucleus in *Saccharomyces cerevisiae*. *Mol. Biol. Cell* **14,** 129–141.

Sakai, Y., Koller, A., Rangell, L. K., Keller, G. A., and Subramani, S. (1998). Peroxisome degradation by microautophagy in Pichia pastoris: Identification of specific steps and morphological intermediates. *J. Cell Biol.* **141,** 625–636.

Sattler, T., and Mayer, A. (2000). Cell-free reconstitution of microautophagic vacuole invagination and vesicle formation. *J. Cell Biol.* **151,** 529–538.

Scott, J. H., and Schekman, R. (1980). Lyticase: Endoglucanase and protease activities that act together in yeast cell lysis. *J. Bacteriol.* **142,** 414–423.

Shen, S. H., Chretien, P., Bastien, L., and Slilaty, S. N. (1991). Primary sequence of the glucanase gene from *Oerskovia xanthineolytica:* Expression and purification of the enzyme from *Escherichia coli*. *J. Biol. Chem.* **266,** 1058–1063.

Tuttle, D. L., and Dunn, W. A., Jr. (1995). Divergent modes of autophagy in the methylotrophic yeast *Pichia pastoris*. *J. Cell Sci.* **108**(Pt. 1), 25–35.

Tuttle, D. L., Lewin, A. S., and Dunn, W. A., Jr. (1993). Selective autophagy of peroxisomes in methylotrophic yeasts. *Eur. J. Cell Biol.* **60,** 283–290.

Uttenweiler, A., Schwarz, H., and Mayer, A. (2005). Microautophagic vacuole invagination requires calmodulin in a Ca2+-independent function. *J. Biol. Chem.* **280,** 33289–33297.

Uttenweiler, A., Schwarz, H., Neumann, H., and Mayer, A. (2007). The vacuolar transporter chaperone (VTC) complex is required for microautophagy. *Mol. Biol. Cell.* **18,** 166–175.

Veenhuis, M., Douma, A., Harder, W., and Osumi, M. (1983). Degradation and turnover of peroxisomes in the yeast *Hansenula polymorpha* induced by selective inactivation of peroxisomal enzymes. *Arch. Microbiol.* **134,** 193–203.

CHAPTER TWELVE

Autophagy in Wine Making

Eduardo Cebollero,* M. Teresa Rejas,† *and* Ramón González‡

Contents

Abstract

Aging that involves contact with dying yeast cells is one of the differential processes between sparkling and still wine production. The release of the products of autolysis during this aging step is fundamental for the quality of sparkling wines made by the traditional method. These cells undergo an autolysis process characterized by self-digestion of yeast intracellular and cell-wall macromolecules, and the release of the degradation products to the wine. Autolysis is the source of several molecules responsible for the quality of sparkling wines, as well as still wines aged on lees (yeast cells). Autolysis is a slow process under sparkling wine production conditions, and there is interest, from the industrial side, in the design of strategies for rapid development of autolysis. Some years ago our research group hypothesized that, during the

* University Medical Center Utrecht, Department of Cell Biology, Utrecht, The Netherlands
† Servicio de Microscopía Electrónica, Centro de Biología Molecular Severo Ochoa (CSIC-UAM)
‡ Instituto de Ciencias de la Vid y del Vino (CSIC-UR-GR) C. Madre de Dios, 51, 26006 Logroño, Spain

Methods in Enzymology, Volume 451
ISSN 0076-6879, DOI: 10.1016/S0076-6879(08)03212-6

process of sparkling wine production, autophagy would take place. This had important implications for the design of genetic engineering strategies aimed to accelerate autolysis. The relationships between autolysis and autophagy are not completely elucidated, but in case autophagy preceded autolysis during the aging step of sparkling wine production, there were at least two possibilities for accelerating autolysis by targeting genes involved in autophagy. This chapter discusses methods to demonstrate the development of autophagy under enological conditions. This is accomplished by using either laboratory strains defective in autophagy and/or the Cvt pathway, in conditions that mimic sparkling wine production or industrial wine yeast strains under real sparkling wine production conditions.

1. Introduction

Elaboration of sparkling wines through the traditional or Champenoise method involves two fermentation steps followed by an aging period in contact with yeasts that is essential for wine quality. The traditional method is different from other sparkling wine making processes in that the second fermentation, wine aging, and the subsequent removal of yeast sediment from the wine, take place in a bottle, instead of a big vat, and this bottle is the same that will contain the finished sparkling wine on the market (Lallement, 1998). To elaborate on the process, the juice obtained by pressing grape berries is first transferred to tanks, inoculated with selected yeasts, and allowed to ferment until a base wine is obtained. The base wine is then stabilized, clarified, and supplemented with sugar, selected yeasts, and bentonite, a coadjutant that helps to remove the yeasts from the wine at the end of the elaboration process. The base wine is subsequently distributed into thick-walled glass bottles, which are sealed with a crown cap and placed into a cellar for extended periods (at least 9 months in the case of *cava* wines and 12 months for Champagne wines) to allow fermentation and aging to proceed. During the second fermentation, the sugars that have been added to the base wine are fermented to ethanol and CO_2. Because the bottles are hermetically closed, this CO_2 remains dissolved in the wine. This will contribute to the typical effervescence of sparkling wines when the bottles are opened.

The multiple stresses affecting second-fermentation yeasts during sparkling wine aging (starvation, low pH and temperature, high CO_2 pressure, and ethanol concentration), finally lead to cell death and the concomitant yeast autolysis. Yeast autolysis is characterized by self-digestion of yeast intracellular and cell-wall macromolecules, and the release of the degradation products to the extracellular medium (Connew, 1998; Fornairon-Bonnefond *et al.*, 2002). Autolytic products have been shown to contribute to several of the physical and sensorial characteristics desired in wines aged in contact with yeasts, and sparkling wine makers usually employ longer aging times as a way to improve the quality of their wines (Leroy *et al.*, 1990; Martinez-Rodriguez *et al.*, 2002; Moreno-Arribas *et al.*, 1996, 2000; Pueyo *et al.*, 1995). However, autolysis is a slow process under enological conditions, and the longer aging periods are associated with increased production costs. For this reason, several wine biotechnologists have focused their interest on accelerating yeast autolysis, to reduce the time required to obtain premium quality sparkling wines, with consequent advantages for wine producers and consumers. Different approaches based on the modification of the wine-making process or the use of alternative yeast strains have been explored to speed up the autolytic process. For example, yeast autolysis has been accelerated by increasing the temperature of the aging period (Charpentier and Feuillat, 1992). However, the addition of yeast extracts as a way to confer aging-like properties to the final product has also been examined (Charpentier and Feuillat, 1992). These approaches were hampered by the formation of toasty or yeasty off-flavors (Peppler, 1982). A more promising alternative, consisting in the inoculation of a mixture of killer and killer-sensitive yeast strains as second-fermentation starter, has been successfully used but only in laboratory conditions (Todd *et al.*, 2000). In addition, industrial strains with increased autolytic rates, obtained by classical genetics, have been shown to result in improved sparkling wines (Gonzalez *et al.*, 2003; Tini *et al.*, 1995).

Our group is interested in the construction of autolytic industrial yeasts by using genetic engineering techniques. Two different strategies were devised, both based on a hypothetical connection between autophagy and autolysis in wine-making conditions. First, because autolysis takes place after cell death, and autophagy is required for maintenance of cell survival under starvation conditions, we hypothesized that yeast strains impaired in autophagy would die earlier during wine aging and consequently autolyze faster. Second, in addition, autophagy plays a key role in the degradation of intracellular organelles by mediating destructive self-digestion, so we postulated that this pathway could contribute directly to yeast cell autolysis in enological conditions. As a consequence, increased activity of the autophagic pathway would also result in accelerated autolysis. In both cases, demonstration of autophagy under wine-making conditions was required. To demonstrate the induction of autophagy by yeast cells during wine-making, we carried out different biochemical and morphological studies using laboratory and industrial strains.

2. Detection of Autophagy Using Laboratory Yeast Strains Under Enological Conditions

The cytoplasm-to-vacuole targeting (Cvt) pathway is a constitutive and biosynthetic route by which the vacuolar proteins aminopeptidase I (Ape1) and α-mannosidase I (Ams1) are specifically transported from the cytosol to the vacuole, where they carry out their physiological role (Baba *et al.*, 1997; Huang and Klionsky, 2002). Ape1, synthesized in the cytoplasm as a 61-kDa inactive precursor (prApe1), is assembled in a big macromolecular complex and further selectively enwrapped by a double-membrane vesicle, called a Cvt vesicle (Baba *et al.*, 1997; Kim *et al.*, 1997). The Cvt vesicle fuses its external membrane with the vacuole, releasing the inner vesicle, called the Cvt body, into the vacuolar lumen, where the resident hydrolases process the prApe1 into a 50 kDa mature form (mApe1; Scott *et al.*, 1997). As a consequence of the different molecular weights of the cytosolic and vacuolar forms, sorting of Ape1 can be easily monitored by Western blot (Harding *et al.*, 1995). When autophagy is induced, prApe1 is also sequestered by autophagosomes and consequently it is sorted to the vacuole together with the bulk cytoplasmic cargo (Kim *et al.*, 2000; Klionsky and Emr, 2000). Although Cvt and autophagy pathways share most of the molecular machinery (Harding *et al.*, 1996), there are some gene products specifically required for the Cvt pathway. This is the case for Atg19, the prApe1 receptor, which is required for prApe1 transport under vegetative growth conditions (when the Cvt pathway operates) but is not required under autophagy conditions (Leber *et al.*, 2001). Vacuolar import of prApe1 by bulk autophagy is much slower and inefficient, however, and the protein normally utilizes the Atg19 receptor for uptake and vacuolar delivery by either pathway.

To elucidate whether autophagy occurs under wine-making conditions, an *ATG19* knockout strain can be incubated in laboratory enological conditions and the transport of prApe1 to the vacuole can be assayed by Western blot (Cebollero *et al.*, 2005a). Because *atg19Δ* strains are defective in the Cvt pathway but not in nonspecific autophagy, appearance of mApe1 under second fermentation conditions would be indicative of the induction of autophagy. A mutant strain completely defective in both the Cvt and autophagy pathways, such as an *atg1Δ* strain, should be used as a negative control. Because both pathways are defective in the absence of Atg1, mApe1 content should be null or negligible in the extracts of this control strain all through the experiment. This control is also important to rule out one alternative explanation for prApe1 maturation under wine-making conditions (i.e. disruption of the vacuolar membrane and release of vacuolar proteases to the cytoplasm). Because of the genetic complexity of industrial yeasts, which makes it difficult to obtain knockout strains, and the availability of laboratory strains deleted in the previously mentioned genes, this experiment can be tackled using laboratory strains and conditions simulating the industrial wine-making process. Because laboratory strains are more sensitive to ethanol than are industrial strains, a dealcoholized base wine should be used for this approach. Protocols to determine the tolerance to ethanol of laboratory strains and to demonstrate the induction of autophagy as the maturation of prApe1 are described subsequently.

2.1. Determination of ethanol tolerance under second-fermentation conditions for laboratory yeast strains

1. Inoculate 20 ml of YPD (1% yeast extract, 2% peptone, 2% dextrose) with 1 isolated yeast colony and incubate the cells overnight at 30 °C with rotation at 150 rpm.
2. Prepare culture flasks containing 20 ml of synthetic base wine added with 0%–8% ethanol differing in 1% ethanol increments.

Note: Synthetic base wine contains 3 g/l tartaric acid, 6 g/l malic acid, 0.3 g/l citric acid, 1.7 g/l yeast nitrogen base without ammonium sulfate and amino acids, 0.5 g/l ammonium sulphate, 20 g/l sucrose and the specific auxotrophic requirements. The pH is adjusted to 3.5 with KOH. The appropriate amount of ethanol is added into the cool medium after autoclaving.

3. Recover the cells from step 1 by centrifugation (5 min at 3000×*g*), wash twice with a sterile 0.9% NaCl solution, and inoculate the synthetic base wines at 10^6 cells/ml.
4. Close the flasks with a Müller valve previously filled with sulfuric acid or Vaseline oil, record the zero time weight, and incubate the cultures for several days at 17 °C in static conditions.
5. Record the weight loss daily until a constant value is reached (Vaughan-Martini and Martini, 1998). Note: Weight loss is result of the release of CO_2 produced during wine fermentation, so it can be used to monitor the fermentative process.

The highest ethanol concentration that allows for a complete fermentation of the synthetic wine by all of the strains assayed is used for detection of autophagy using the protocol described subsequently.

2.2. Aminopeptidase I maturation under simulated wine-making conditions

1. Evaporate the ethanol contained in 3 liters of a commercial base wine (Cavas Castellblanch, Sant Sadurní D'Anoia, Spain) in a rotavapor during 30 min at 45 °C under vacuum.
2. Measure the remaining ethanol concentration of the dealcoholized base wine in an ebulliometer (Barus, Bordeaux) or by HPLC (Coleman *et al.*, 2007).
3. Reconstitute the wine to its original volume with water and ethanol to get a final content of 3% ethanol.

Note: BY4741, *atg1*Δ and *atg19*Δ strains completely ferment a synthetic base wine containing up to 3% ethanol (section 2.1, step 5).

4. Add sucrose to a final concentration of 2%.

5. Inoculate the dealcoholized base wine at 10^6 cells/ml from an overnight culture. Before inoculation wash the cells twice with a sterile 0.9% NaCl solution.

Note: Wine inoculation should be carried out right after sugar addition to avoid microbial spoilage of the medium.

6. Incubate at 17 °C without agitation.
7. Daily remove samples containing 10^8 cells.
8. Centrifuge the samples at $3000 \times g$ for 5 min to separate the cellular pellet from the supernatant fraction.

Note: Supernatant fractions can be used to monitor wine fermentation by quantifying residual reducing sugars as described by Bernfeld (1955) or residual sugar and ethanol by HPLC (Coleman *et al.*, 2007).

9. Wash the cell pellet twice with a sterile 0.9% NaCl.
10. Mix the cells with 150 μl of cold 50 m*M* Tris-HCl, pH 8, 5 m*M* EDTA and containing protease inhibitors (1 m*M* PMSF, 1 μg/ml aprotinin, 1 μg/ml leupeptin, 1 μg/ml pepstatin).
11. Prepare crude cell extracts by vortexing 10 times with 0.3 g of glass beads for 30 s (see the chapter by Cheong and Klionsky in this volume for additional information on the preparation of glass beads). Chill the samples on ice after each vortex cycle.
12. Eliminate the cell debris by centrifugation at 4 °C for 5 min at $500 \times g$.
13. Separate 10 μg of proteins by SDS-PAGE (Laemmli, 1970).
14. Perform a Western blot (Sambrook *et al.*, 1989) using a polyclonal rabbit antibody against Ape1 (Martínez *et al.*, 1999) and the ECL detection system (Amersham Biosciences Europe, Barcelona, Spain).

An example of this experiment is shown in Fig. 12.1.

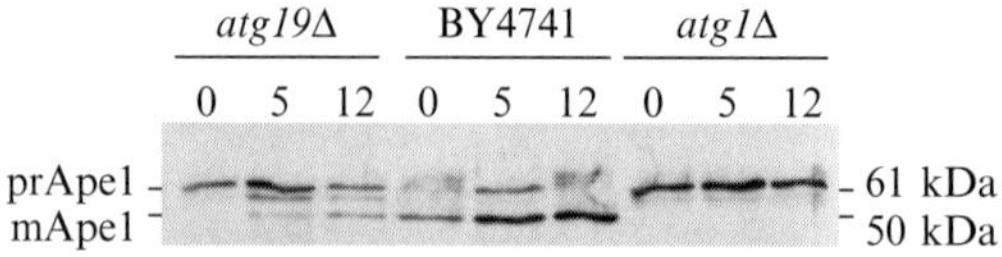

Figure 12.1 Western blot of extracts of the strains BY4741, *atg1*Δ and *atg19*Δ after 0, 5 or 12 days of incubation in second-fermentation conditions in reconstituted base wine. The membranes were probed with rabbit polyclonal anti-Ape1 antibody. mApe1, mature form of aminopeptidase I; prApe1, precursor form of aminopeptidase I. Reprinted with permission from Biotech. Prog. 21: 614–616. Copyright © 2005 American Chemical Society.

3. Detection of Autophagy in Wine-making Using Industrial Yeast Strains

Autophagosomes and autophagic bodies are short-lived intermediate structures during the autophagic process; after completion, autophagosomes quickly fuse with the vacuolar membrane and the resultant autophagic body is rapidly degraded by vacuolar hydrolases. For this reason, inhibitors or specific mutant backgrounds that block the autophagic flux at these levels have been successfully used to induce accumulation of the intermediate vesicles and favor their visualization using microscopy techniques. For example, wild-type cells treated with the proteinase inhibitor PMSF or mutant strains deleted in *PEP4*, coding for the vacuolar proteinase A, show reduced vacuolar hydrolase activity and accumulate autophagic bodies inside the vacuole during the progression of autophagy (Epple *et al.*, 2001; Takeshige *et al.*, 1992).

Autophagy has been morphologically shown to be induced by industrial yeasts during the elaboration of sparkling wines (Cebollero and Gonzalez, 2006). In addition, a biochemical approach different to that based on the maturation of prApe1 has also been carried out to detect autophagy by industrial yeasts (Cebollero and Gonzalez, 2006). It is known that acetaldehyde dehydrogenase (Ald6) is preferentially included into autophagosomes and transported to the vacuole for degradation during autophagy, resulting in a decrease of Ald6 levels (Onodera and Ohsumi, 2006). In this sense, accumulation of autophagic bodies and depletion of Ald6 are autophagic markers that can be used to determine the induction of this pathway in industrial yeasts.

3.1. Preparation of yeast samples to monitor autophagy during wine making

1. Use an overnight culture of the wine yeast strain *Saccharomyces cerevisiae* EC1118 to inoculate 100 ml of YPD with 10^5 cells/ml, and incubate for an additional 48 h at 30 °C shaking at 150 rpm.
2. Transfer the cells to 1 liter of adaptation medium in a 2-liter flask to a final density of 10^6 cells/ml and incubate for 5 days at 30 °C shaking at 150 rpm.

Note: Adaptation medium is 0.08 g/l diammonium phosphate, 2.4 g/l tartaric acid, 20 g/l sucrose, and 482 ml/l base wine.

3. Add 20 g/l of sucrose to 6 liters of base wine and immediately inoculate with 10^6 cells/ml of industrial yeasts adapted to ethanol from step 2.
4. Fill 375 ml thick-walled glass bottles with inoculated base wine, taking care that the cells do not sediment during this process.

5. Seal the bottles with a crown cap and store at 17 °C in a horizontal position for approximately 20 days.
6. Daily remove samples containing 10^8 cells and centrifuge at $3000 \times g$ at 4 °C for 5 min to separate the wine from the cells.

Note: Because hydrostatic pressure increases during fermentation, to maintain the conditions of the cells throughout the experiment, each bottle should be discarded after sample removing. Bottles under pressure are not opened directly, but a hole is made in the crown cap with the aid of a hammer and nail. Excess carbon dioxide is allowed to exit before removing the crown cap.

7. Use the supernatant fraction to monitor the fermentation progress (sugar and ethanol content), and prepare the cells to morphological and biochemically assay autophagy as described subsequently.

3.2. Acetaldehyde dehydrogenase (Ald6) depletion during wine making

1. Wash the cells from step 6 in section 4.1 twice with a 0.9% NaCl sterile solution.
2. Obtain crude cell extracts for Western blotting as described in section 3.1, steps 8-13. Use a rabbit antiacetaldehyde dehydrogenase monoclonal antibody for Ald6 detection (Rockland, Gilbertsville, PA).

Note: This antibody simultaneously reacts with Ald4, providing an internal control for extract concentration and gel loading.

3.3. Visualization of autophagic bodies in wine making using industrial yeast strains

1. Wash the cells from section 3.1 twice with a 0.9% NaCl sterile solution.
2. Fix the cells for 90 min at room temperature with 1 ml of 4% paraformaldehyde and 2% glutaraldehyde in 40 m*M* phosphate buffer, pH 7.2.
3. After fixation, pellet the cells in microcentrifuge, and wash three times at 4 °C for 10 min with 1 ml of 40 m*M* phosphate buffer, pH 7.2.
4. Add 10% gelatin over the cell pellet and mix at 37 °C for 10 min (Merk Gelatin powder food grade 1.04078) in 1x phosphate-buffered saline (PBS). Add an excess of the 10% gelatin, 200–250 μl.
5. Centrifuge the mixture of cells/gelatin at 13,500 rpm for 5 min to increase the cellular concentration, and then solidify the gelatin by incubation on ice for at least 2 h (it can be overnight). Add PBS to help separate the solidified gelatin containing the cells from the microcentrifuge tube walls with a toothpick (you can also cut the tip of the

tube that contains the pellet with a stainless-steel razor blade). Keep the temperature strictly at 4 °C and prepare small blocks from the pellet. The centrifugation results in a high density of cells in the final specimen.

6. Obtain 0.5–1 mm^3 cubes of gelatin-embedded cells. The cubes containing the gelatin embedded cells are kept in PBS.
7. Cryoprotect the cells by progressively adding 30% glycerol in PBS until reaching a concentration of 15% glycerol in 15 min at 4 °C.
8. Take out the 15% glycerol and add 30% glycerol for 15 min.

Note: Cryoprotection is important to prevent ice-crystal formation in the specimen during freezing. Ice-crystal formation can be prevented only when freezing occurs within milliseconds in liquid propane or ethane.

9. Mount the cryoprotected cubes on aluminium specimen holders (those used to mount specimens in a cryoultramicrotome) (Leica) and cryofix the cells by quickly plunging in liquid propane at −180 °C. This was done using a commercial plunge-freezing apparatus (KF80, Leica, Wien).
10. Freeze substitute the cells by incubation at −85 °C for 54 h in anhydrous methanol containing 0.5% uranyl acetate in an Automatic Freeze-Substitution System (AFS, Leica, Wien; see also the chapter by M. Baba in this volume for additional information).
11. Raise the temperature to −45 °C at a rate of 5 °C/h and wash the cells several times with pure methanol (3 × 1 h).
12. Infiltrate the samples at −45 °C with increased concentrations of Lowicryl HM20 in methanol (50% for 2 h, 70% for 2 h, and 100% for 1 h and overnight).
13. Polymerize the resin by UV light irradiation at −45 °C for two days and for an additional day at 22 °C.
14. Cut ultrathin Lowicryl sections in an Ultracut E ultramicrotome (Leica) and collect the sections on Collodion/carbon-coated Cu grids (200 mesh).
15. Stain the sections with 2% aqueous uranyl acetate for 5–7 min and 2% Reynolds lead citrate for 2 min.
16. Examine the samples at 80Kv in an electron microscope (JEOL 1010, Japan).

The protocols described in this section have been designed to follow industrial conditions of wine elaboration. Even though the autophagic bodies are transient structures, it is possible to catch some of them without the use of protease inhibitors or specific mutants, probably because of the relatively low temperature of the experiment. The use of PMSF was ruled out, as opening the bottle for PMSF addition would disturb the experiment, and it could not be added at the beginning of the experiment for two reasons: it would inhibit the onset of the second fermentation, and it would be inactive at sampling time. However, successful experiments have also been carried out using Erlenmeyer flasks closed with cotton plugs instead of closed bottles. In this case, the hydrostatic pressure remains constant during the experiment. The advantage is that PMSF can be added at any time during the progression of the second fermentation. In the case of using Erlenmeyer flasks to monitor the progression of autophagy, we recommend carrying out the following procedure:

1. Follow steps 1–3 of section 3.1.
2. Distribute the inoculated base wine in three Erlenmeyer flasks and incubate at 16 °C for approximately 20 days.

Note: One of the flasks is used to monitor autophagy biochemically. In one of the other flasks PMSF is added during the progression of the second fermentation to allow morphological detection of autophagy, and the other flask is the negative control.

3. Remove samples containing 10^8 cells and centrifuge them at $3000\times g$ at 4 °C for 5 min to separate the wine from the cellular pellet.
4. Use the pellets to monitor depletion of Ald6 (section 3.2) and the supernatant fraction to monitor fermentation progression by HPLC (Coleman *et al.*, 2007).
5. At the beginning of the second fermentation, approximately after 5–10 days, add 1 m*M* PMSF (from a 100 m*M* stock in ethanol) and incubate for an additional 8 h in the presence of the proteinase inhibitor.
6. Remove a sample containing 10^8 cells.
7. Wash the cells twice with a 0.9% NaCl sterile solution.
8. Analyze the cells by TEM as described in section 3.3.

4. Conclusions

The biochemical and morphological methods described in this chapter have been used with success in demonstrating autophagy by yeasts under enological conditions. Sparkling wine elaboration requires the use of selected industrial yeast strains, and these are genetically different from yeasts used for fundamental research. For this reason, monitoring of autophagy by simulating wine-making conditions with genetically more tractable laboratory strains can be advantageous because of the easier genetic manipulation and also the availability of mutant strains. Despite the modifications needed to perform second-fermentation experiments with laboratory strains, including lowering the ethanol content of the wine, this strategy was shown to be a good experimental approach to sparkling wine making. Indeed, autophagy can be detected by the appearance of mApe1 in an *atg19Δ* background at the beginning of the second fermentation (Cebollero *et al.*, 2005a). The timing of the start of autophagy observed in this experiment parallels that observed using actual industrial strains and sparkling wine production conditions, based on recording the depletion of Ald6 (Cebollero and Gonzalez, 2006). It is noteworthy that, in contrast to other winemaking processes that involve the use of big fermentation tanks, second fermentation of base wine and further wine aging take place in bottles. This means that the industrial procedure can by exactly reproduced in the laboratory, and consequently the results should be totally comparable to the actual industrial process.

Visualization of autophagosomes and autophagic bodies is usually facilitated by working with specific mutant backgrounds or by the addition of inhibitors that induce the accumulation of these vesicular intermediates. However, to avoid any disturbance of the real sparkling wine production process, both alternatives were ruled out in the first part of section 3.3. This is why the vacuoles are not completely filled with autophagic bodies. Instead, some isolated vesicles indicating the induction of autophagy are found (Cebollero and Gonzalez, 2006). A reasonable explanation of the ability to observe these vesicles is the low temperatures used during the second fermentation, which would slow down the autophagic process, making these transient structures stable enough to be detected, at least at low rates. Finally, we did not use antibodies targeted against autophagosomes/autophagic bodies marker proteins for the morphological demonstration of autophagy; however, we recommend the use of immuno-electron microscopy methods (described in other chapters of this volume) to facilitate the recognition of these vesicles. Showing autophagy under industrial second fermentation conditions paves the way for the construction of improved wine yeast strains for second fermentation and aging (Cebollero *et al.*, 2005b; Tabera *et al.*, 2006).

REFERENCES

Baba, M., Osumi, M., Scott, S. V., Klionsky, D. J., and Ohsumi, Y. (1997). Two distinct pathways for targeting proteins from the cytoplasm to the vacuole/lysosome. *J. Cell Biol.* **139,** 1687–1695.

Bernfeld, P. (1955). Amylases α and β. *Methods Enzymol.* **1,** 149–158.

Cebollero, E., Carrascosa, A. V., and Gonzalez, R. (2005a). Evidence for yeast autophagy during simulation of sparkling wine aging: A reappraisal of the mechanism of yeast autolysis in wine. *Biotechnol. Prog.* **21,** 614–616.

Cebollero, E., and Gonzalez, R. (2006). Induction of autophagy by second-fermentation yeasts during elaboration of sparkling wines. *Appl. Environ. Microbiol.* **72,** 4121–4127.

Cebollero, E., Martínez-Rodríguez, A. J., Carrascosa, A. V., and González, R. (2005b). Overexpression of *csc1-1*: A plausible strategy to obtain wine yeast strains undergoing accelerated autolysis. *FEMS Microbiol. Lett.* **246,** 1–9.

Charpentier, C., and Feuillat, M. (1992). Yeast autolysis. *In* "Wine Microbiology and Biotechnology" (C. H. Fleet, ed.), pp. 225–242. Harwood Academic Publisher, Chur, Switzerland.
Coleman, M. C., Fish, R., and Block, D. E. (2007). Temperature-dependent kinetic model for nitrogen-limited wine fermentations. *Appl. Environ. Microbiol.* **73,** 5875–5884.
Connew, S. J. (1998). Yeast autolysis: A review of current research. *Aust. New Z. Wine Ind. J.* **13,** 61–64.
Epple, U. D., Suriapranata, I., Eskelinen, E.-L., and Thumm, M. (2001). Aut5/Cvt17p, a putative lipase essential for disintegration of autophagic bodies inside the vacuole. *J. Bacteriol.* **183,** 5942–5955.
Fornairon-Bonnefond, C., Camarasa, C., Moutounet, M., and Salmon, J. M. (2002). New trends on yeast autolysis and wine aging on lees: A bibliographic review. *J. Int. Sci. Vigne Vin.* **36,** 49–69.
González, R., Martínez-Rodríguez, A. J., and Carrascosa, A. V. (2003). Yeast autolytic mutants potentially useful for sparkling wine production. *Int. J. Food Microbiol.* **84,** 21–26.
Harding, T. M., Morano, K. A., Scott, S. V., and Klionsky, D. J. (1995). Isolation and characterization of yeast mutants in the cytoplasm to vacuole protein targeting pathway. *J. Cell Biol.* **131,** 591–602.
Harding, T. M., Hefner-Gravink, A., Thumm, M., and Klionsky, D. J. (1996). Genetic and phenotypic overlap between autophagy and the cytoplasm to vacuole protein targeting pathway. *J. Biol. Chem.* **271,** 17621–17624.
Huang, W.-P., and Klionsky, D. J. (2002). Autophagy in yeast: A review of the molecular machinery. *Cell. Struct. Funct.* **27,** 409–420.
Kim, J., Scott, S. V., and Klionsky, D. J. (2000). Alternative protein sorting pathways. *Int. Rev. Cytol.* **198,** 153–201.
Kim, J., Scott, S. V., Oda, M. N., and Klionsky, D. J. (1997). Transport of a large oligomeric protein by the cytoplasm to vacuole protein targeting pathway. *J. Cell Biol.* **137,** 609–618.
Klionsky, D. J., and Emr, S. D. (2000). Autophagy as a regulated pathway of cellular degradation. *Science* **290,** 1717–1721.
Laemmli, U. K. (1970). Cleavage of structural proteins during the assembly of the head of bacteriophage T4. *Nature* **227,** 680–685.
Lallement, A. (1998). Fermentation and the method champegnoise. *Brewers Guardian* **127,** 31–35.
Leber, R., Silles, E., Sandoval, I. V., and Mazon, M. J. (2001). Yol082p, a novel Cvt protein involved in the selective targeting of aminopeptidase I to the yeast vacuole. *J. Biol. Chem.* **276,** 29210–29217.
Leroy, M. J., Charpentier, M., Duteurtre, B., Feuillat, M., and Charpentier, C. (1990). Yeast autolysis during champagne aging. *Am. J. Enol. Vitic.* **41,** 21–28.
Martinez-Rodriguez, A. J., Carrascosa, A. V., Martin-Alvarez, P. J., Moreno-Arribas, V., and Polo, M. C. (2002). Influence of the yeast strain on the changes of the amino acids, peptides and proteins during sparkling wine production by the traditional method. *J. Ind. Microbiol. Biotechnol.* **29,** 314–322.
Martínez, E., Seguí-Real, B., Silles, E., Mazón, M. J., and Sandoval, I. V. (1999). The prepropeptide of vacuolar aminopeptidase I is necessary and suffcient to target the fluorescent reporter protein GFP to the vacuole of yeast by the Cvt pathway. *Mol. Microbiol.* **33,** 52–62.
Moreno-Arribas, V., Pueyo, E., and Polo, C. (1996). Peptides in musts and wines: Changes during the manufacture of cavas (sparkling wines). *J. Agric. Food Chem.* **44,** 3783–3788.
Moreno-Arribas, V., Pueyo, E., Nieto, F. J., Martin-Alvarez, P. J., and Polo, M. C. (2000). Influence of polysaccharides and the nitrogen compounds on foaming properties of sparkling wines. *Food Chem.* **70,** 309–317.

Onodera, J., and Ohsumi, Y. (2004). Ald6p is a preferred target for autophagy in yeast, *Saccharomyces cerevisiae*. *J. Biol. Chem.* **279,** 16071–16076.

Peppler, H. J. (1982). Yeast extracts. *In* "Economic Microbiology, Fermented Foods" (A. H. Rose, ed.), Vol 7. pp. 293–312. Academic Press, London.

Pueyo, E., Martin-Alvarez, P. J., and Polo, M. C. (1995). Relationship between foam characteristics and chemical composition in wines and cavas (sparkling wines). *Am. J. Enol. Vitic.* **46,** 518–524.

Sambrook, J., Fritsch, E. F., and Maniatis, T. (1989). *In* "Molecular Cloning: A Laboratory Manual." Cold Spring Harbor Laboratory, Cold Spring Harbor, NY.

Scott, S. V., Baba, M., Ohsumi, Y., and Klionsky, D. J. (1997). Aminopeptidase I is targeted to the vacuole by a nonclassical vesicular mechanism. *J. Cell Biol.* **138,** 37–44.

Tabera, L, Muñoz, R., and González, R. (2006). Deletion of *BCY1* from the *Saccharomyces cerevisiae* genome is semidominant and induces autolytic phenotypes suitable for improvement of sparkling wines. *Appl. Environ. Microbiol.* **72,** 2351–2358.

Takeshige, K., Baba, M., Tsuboi, S., Noda, T., and Ohsumi, Y. (1992). Autophagy in yeast demonstrated with proteinase-deficient mutants and conditions for its induction. *J. Cell Biol.* **119,** 301–311.

Tini, V., Zambonelli, C., Benevelli, M., and Castellari, L. (1995). The autolysogenic *Saccharomyces cerevisiae* strains for the sparkling wines production. *Industrie-delle-Bevande.* **24,** 113–118.

Todd, B. E. N., Fleet, G. H., and Henscheke, P. A. (2000). Promotion of autolysis through the interaction of killer and sensitive yeasts: Potential application in sparkling wine production. *Am. J. Enol. Vitic.* **51,** 65–72.

Vaughan-Martini, A., and Martini, A. (1998). Determination of ethanol production. *In* "The Yeasts: A Taxonomic Study" (C. P. Kurtzman and J. W. Fell, eds.), p. 107. Elsevier, Amsterdam.

CHAPTER THIRTEEN

Purification and *In Vitro* Analysis of Yeast Vacuoles

Margarita Cabrera* *and* Christian Ungermann*

Contents

Abstract

The purification of eukaryotic organelles is a prerequisite for the detailed analysis of protein sorting, localization and translocation, membrane fusion and vesicle budding. Yeast vacuoles receive cargo from the exocytic, endocytic, and autophagic pathways and hence represent an excellent model system for the study of organelle biogenesis and protein sorting. Yeast vacuoles undergo fission and fusion *in vivo*, events that can be monitored *in vitro* by an assay that employs purified vacuoles from two tester strains. Here, we describe the methodology of yeast vacuole purification, and provide protocols for the detailed analysis of the fusion reaction. We also include methods to analyze protein dynamics on yeast vacuoles and the controls required to ensure their reliability.

1. Introduction

The secretory and endocytic pathway consist of organelles that are in constant communication via vesicle flux. Vesicular transport of both pathways directs endocytic cargo as well as hydrolases and lipases via the endosome to the lysosome, which has also been termed a terminal organelle.

* Department of Biology, Biochemistry Section, University of Osnabrück, Osnabrück, Germany

Methods in Enzymology, Volume 451
ISSN 0076-6879, DOI: 10.1016/S0076-6879(08)03213-8

Similarly, cytoplasm is delivered to the lysosome through various autophagic pathways. Within the lysosome, the hydrolytic enzymes then digest the delivered lipids, proteins, nucleic acids, and carbohydrates, and the monomers are subsequently exported into the cytoplasm as nutrients for the cell.

In yeast, the vacuole corresponds to the mammalian lysosome. In contrast to the rather small mammalian lysosome, yeast vacuoles are large (>500 nm in diameter) and present in one to five copies per cell. As vacuoles can be easily visualized by fluorescent dyes, vacuole morphology has been taken as a measure to identify proteins involved in protein sorting and recycling between the Golgi complex and the vacuole (leading to the identification of vacuolar protein sorting, *VPS*, genes) (Raymond *et al.*, 1992; Robinson *et al.*, 1988). More recently, the analysis of aminopeptidase I sorting to the vacuole lumen unraveled a class of cytoplasm-to-vacuole targeting (*CVT*) genes that overlap extensively with those involved in autophagy (*APG* and *AUT*); these three sets of genes have now been grouped under the common name of autophagy-related (*ATG*) genes (Klionsky, 2007). In addition, vacuoles can undergo fission and homotypic fusion in response to salt stress and as part of the inheritance of vacuoles from the mother to the daughter cell (Weisman, 2006). All these described reactions depend on a conserved fusion machinery (Ostrowicz *et al.*, 2007). Based on the *in vitro* fusion analyses, homotypic vacuole fusion has been dissected into priming, tethering, docking, and fusion/lipid bilayer mixing. During the priming step NSF/Sec18 and α-SNAP/Sec17 drive disassembly of *cis*-SNARE complexes. Tethering is mediated by the interaction of the HOPS tethering complex and the Rab-GTPase Ypt7. Finally, the pairing of the SNARE proteins Vam3, Vam7, Vti1 and Nyv1 leads to *trans*-SNARE complex formation that drives the fusion of apposed vacuole membranes (Ostrowicz *et al.*, 2007). *In vitro* assays of vacuole fission have not yet been described, but fission has been linked to the dynamin-like Vps1 protein (Peters *et al.*, 2004).

Several assays, which have been used to understand vacuole biogenesis, depend on the efficient and reliable purification of yeast vacuoles. Within this review, we will describe the basic methods of vacuole purification and vacuole fusion. For more detailed information on specific approaches and recently employed activators of the fusion reaction, the reader is referred to the methods section of the respective publications.

2. Methods

2.1. Vacuole purification

Vacuoles are purified from logarithmically grown yeast cell cultures. Two particular strain backgrounds must be used in this assay, *pho8Δ* and *pep4Δ* (see section 2.2 for details). The purification procedure consists of an initial incubation step of the cells in 10 m*M* DTT at pH 9.4 to break disulfide bonds in the cell wall, followed by the digestion of the cell wall with the help of the enzyme lyticase, and the DEAE-dextran mediated gentle lysis of the cells (Bankaitis *et al.*, 1986; Haas, 1995). Purified vacuoles are obtained by flotation in a Ficoll step gradient. The procedure takes approximately 4 h to generate vacuoles that can be used in subsequent experiments. Fresh vacuoles are good for approximately 2 h if used in a fusion experiment and can be frozen if needed later on.

2.1.1. Reagents

Stock solutions

1 *M* Tris, pH 9.4
1 *M* DTT
1 *M* potassium phosphate, pH 7.5
4 *M* sorbitol
1 *M* PIPES/KOH, pH 6.8

YPD medium (per liter)

10 g yeast extract
20 g of bacto peptone
20 g of dextrose
Adjusted to pH 5.5 with HCl and sterilized

0.2x YPD (per liter)

2 g of yeast extract
4 g of bacto peptone
4 g of dextrose
Adjusted to pH 5.5 with HCl and sterilized

DTT solution

0.1 *M* Tris pH 9.4
10 m*M* DTT
Prepare just before use

Spheroplasting buffer

80 ml of 0.2x YPD
15 ml of 4 *M* sorbitol
5 ml of 1 *M* potassium phosphate, pH 7.5

DEAE-dextran solution (10 mg/ml)

10 mg of DEAE-dextran in 1 ml
10 m*M* PIPES/KOH, pH 6.8
200 m*M* sorbitol

PS buffer (10 m*M* PIPES/KOH, pH 6.8, 200 m*M* sorbitol)

4% Ficoll (4% w/v Ficoll in PS buffer)
8% Ficoll (8% w/v Ficoll in PS buffer)
15% Ficoll (15% w/v Ficoll in PS buffer)

20x Protease inhibitor cocktail (PIC)

2 μg/ml of leupeptin
20 m*M* o-phenanthroline
10 μg/ml of pepstatin A
2 m*M* Pefabloc

2.1.2. Preparation of lyticase

Lyticase can be purchased from ICN (from *Arthrobacter luteus*, ICN cat. no. 360944). For our experiments, we and others (Reese *et al.*, 2005; Uttenweiler *et al.*, 2007) purify lyticase (β-1,3-glucanase) from *Oerskovia xanthineolytica* (Shen *et al.*, 1991) using an *Escherichia coli* overproduction strain, RSB 805. Lyticase produced in this strain contains a leader sequence that guides the protein to the periplasmic space. The purification relies on the selective disruption of the outer membrane of *E. coli* by cold shocking, while the plasma membrane stays intact. The crude lysate has sufficient activity for purifying vacuoles from yeast cells.

1. Inoculate a 1-liter flask with LB-amp medium (5 g of yeast extract, 10 g of tryptone, 5 g of NaCl, pH 7.2, per liter) the evening before the purification and grow at 37 °C overnight.
2. Inoculate 4 × 4 L of LB amp (in 6-liter flasks) with equal portions of the overnight culture.
3. Grow to an OD of 0.4, then induce with 0.4 m*M* IPTG for 4–5 h at 37 °C.
4. Collect cells by centrifugation in a Beckman Avanti JA10-Rotor (at least 8 min, at 4400×*g* for each spin).
5. Discard the supernatant fraction and combine the cell pellets into two tubes. The cells can be frozen at −20 °C until further use. If required, repeat the procedure (steps 1–3) three times to obtain 48 liters in total.
6. Thaw pellets, if necessary, and resuspend the cells in 320 mL of 25 m*M* Tris, pH 7.4, which corresponds to one-fiftieth (15L/50 = 320 mL) of the culture volume.
7. Centrifuge the cells at 4400×*g* for 10 min in 500-ml bottles as previously.
8. Resuspend cell pellets completely in the centrifuge bottles in 25 m*M* Tris, pH 7.4, into a total of one-fiftieth of the starting volume, then add EDTA from a 200 m*M* stock solution, pH 8.0, to a final concentration of 2 m*M* and mix. Add one-fiftieth of total initial volume of 25 m*M* Tris, pH 7.4, 40% sucrose.
9. Mix and incubate for exactly 20 min with gentle shaking on a shaker, and centrifuge at 4400×*g* for 10 min.
10. Remove supernatant by aspiration. Be careful, the pellet is very loose at this stage.

11. From now on, all reactions should occur on ice. For the cold shock, resuspend the pellets fast and completely with a glass rod in {1/50} of the initial volume with ice-cold 0.5 m*M* $MgSO_4$.
12. Place the tubes into wet ice and incubate them on a shaker for 20 min, then centrifuge for 10 min at 7250×*g* in a Beckman JLA10.500 in 500-ml centrifuge bottles.
13. Quickly collect the supernatant fraction, which contains the lyticase and discard the pellets, which contain unlysed cells and debris. Be cautious as the pellet is very loose.
14. Determine the protein concentration using the Bio-Rad protein assay. Good preparations contain approximately 1–2 mg/ml. Freeze the lyticase in aliquots in liquid nitrogen and store at −80 °C. For the vacuole preparation, usually 1–2 ml of purified lyticase is necessary per 1000 $O.D._{600}$ units of cells, which is obtained from 1 liter of starting culture. For the lysis of DKY6281 and BY4741 approximately half as much lyticase is needed than for BJ3505 cells.

2.1.3. Yeast culture conditions

1. In the morning, inoculate a 20-ml YPD culture in a 50-ml flask (or Falcon tube) with sufficient cells from a plate with freshly streaked yeast to have a slightly turbid culture. Shake the culture for 6–7 h at 30 °C.
2. In the late afternoon, determine the $O.D._{600}$ of your culture and inoculate one liter YPD in a 2-liter flask with sufficient cells to obtain an $O.D._{600}$ = 0.8–1 the next morning. The amount of cells must be determined for each strain individually. We add 2.5 $O.D._{600}$ units of BJ3505 cells, and 1 $O.D._{600}$ unit of DKY6281 or BY4741. Grow the culture for 16 h at 30 °C on a shaker.

2.1.4. Vacuole isolation by dextran lysis and flotation

1. Harvest the cultures by centrifugation in 500-ml bottles at 4400×*g* (5000 rpm in a Beckman Avanti J-25, JA-10 rotor) for 2 min at room temperature. The $O.D._{600}$ of the culture should be not higher than 1.5. Combine 1 liter of culture into one bottle.
2. Discard the supernatant fractions and resuspend the cell pellets by vortexing in 50 ml of the DTT solution.
3. Incubate the resuspended cells in a water bath at 30 °C for 10 min. Centrifuge as before and discard the supernatant fractions.
4. Prepare spheroplasting buffer (15 ml per each sample corresponding to a 1-liter starting culture is needed).
5. Add 5 ml of spheroplasting buffer and the appropriate amount of lyticase to a 30-ml centrifuge tube. In general, BJ3505 cells lyse with 1.5 ml of lyticase, and DKY6281 or BY4741 cells need 0.8 ml. This needs to be titrated for each batch of lyticase.

6. Resuspend the cell pellets from step 3 in 10 ml of spheroplasting buffer by vortexing. Transfer the suspension to the 30-ml centrifuge tubes containing the lyticase and incubate in a water bath at 30 °C for 20 min with occasional gentle mixing by inversion.
7. Prepare the dextran solution (0.4 mg/ml) from a stock solution (10 mg/ml) in PS buffer.
8. Centrifuge the spheroplasted cells at 4 °C and 5380×*g* (3500 rpm in Beckman Avanti J-25, JA 25.50 rotor) for 3 min. The spheroplast conversion can be tested by adding 10 μl of the cells to 1 ml of PS buffer or to H_2O in a cuvette. Efficient osmotic lysis will lead to a 10-fold reduction in turbidity in the H_2O sample.
9. Carefully discard the supernatant fractions and place the tubes on ice. Add 2.5 ml of 15% Ficoll to each tube and resuspend the spheroplasts gently by pipetting with a 1.0 ml micropipette tip (typically a blue color tip). Avoid any air bubbles at this stage.
10. Add the dextran solution (0.4 mg/ml) to each tube (the appropriate amount must be determined for each strain; we use 120 μl for DKY6281 and 200 μl for BJ3505) and mix gently.
11. Incubate on ice for 5 min. Transfer the tubes to a 30 °C water bath and incubate for 1.5 min. Mix gently by inversion twice during the incubation and return the tubes to ice.
12. Transfer the cell lysates to SW40 or SW41 tubes (5 ml is recommended). Pour the step gradient by adding slowly 2.5–3 ml of 8% Ficoll, then 2.5–3 ml of 4% Ficoll solution, and finally carefully add PS buffer solution to the top of the tube. The gradient fractions should be clearly visible.
13. Place the tubes in SW40 or SW41 buckets and centrifuge at 110,000×*g* (30,000 rpm in a Beckman Optima L-90K) at 4 °C for 90 min.
14. Collect vacuoles from the 0%–4% interface with a 1-ml tip (the volume should be approximately 500 μl) and place in a microcentrifuge tube on ice. Vacuoles are visible as a white fluffy band that lies directly on top of the 4% Ficoll layer. Be careful not to disturb the interface when collecting the vacuoles.

2.1.5. Quantification and storage

Vacuoles contain proteases that also destroy vacuolar surface proteins if the vacuoles break. We therefore add to 500-μl vacuoles 1 μl of 20x PIC prior to the protein determination to stabilize the vacuoles. Determine protein concentration with the Bio-Rad protein assay kit. The protein concentration should be between 0.3 and 1 mg/ml. Dilute the vacuoles to a final concentration of 0.3 mg/ml and add 2.5 μl of 20x PIC to 500 μl vacuoles (final concentration is 0.1x). Fusion reactions therefore contain 0.06x PIC. Vacuoles can be frozen in liquid nitrogen and stored at −80 °C after they have been mixed with glycerol to a final concentration of 10% (from a 50% glycerol stock).

2.1.6. Comments

The protocol works reliably to isolate vacuoles for fusion assays. A failure in purifying vacuoles could occur if the culture is too dense. Also, slow growing strains and strains grown in minimal medium or galactose tend to lyse poorly with lyticase and need longer incubation times or additional lyticase. Some issues should be mentioned at this stage. First, vacuoles isolated via this procedure are only 35-fold enriched and still contain endosomal, ER, and Golgi complex contaminants (Bankaitis *et al.*, 1986). If highly purified organelles are needed, additional steps are advisable (Zinser *et al.*, 1993). Second, freshly isolated vacuoles are labile structures that are sensitive to freezing and storage. For any experimental analysis, they should be used right away. We noticed that vacuoles lose fusion activity significantly within three hours. Vacuoles can be visualized by fluorescence microscopy by adding 3 μM of the lipophilic dye FM 4-64 (Invitrogen). Even though frozen vacuoles can still be used in fusion experiments and immunoprecipitations, they lose their large structure during this procedure. Visual analysis therefore depends on freshly isolated vacuoles. Third, the protocol has been modified to allow for the purification of large amounts of vacuoles, which are used as a prerequisite of purification of the HOPS complex (Starai *et al.*, 2008).

2.2. Applications

2.2.1. Vacuole fusion assay (Pho8-based assay)

Vacuole fusion can be measured in a fast and quantitative manner (Conradt *et al.*, 1992). The most commonly used assay is based on the maturation of precursor alkaline phosphatase, a lumenal membrane protein of the yeast vacuole (Haas, 1995; Haas *et al.*, 1994). Alkaline phosphatase is generated as a proenzyme and sorted to the vacuole via the AP-3 pathway directly from the Golgi complex to the vacuole, bypassing the endosome. Within the vacuole lumen, proPho8 is processed to its active form in a Pep4 protease-dependent manner. For the fusion assay, vacuoles are purified from two different strains. One carries Pep4, but is devoid of Pho8 because of a deletion, whereas the other lacks Pep4 and therefore accumulates inactive proPho8 in the vacuole. Thus, neither strain contains active Pho8. In the fusion assay, vacuoles from the two different strains are incubated in a fusion reaction. Upon fusion of the two vacuole types, the protease Pep4 gains access to proPho8 and converts it into the active form. The amount of active Pho8 enzyme reflects the fusion efficiency. Pho8 activity is determined in an enzymatic reaction that includes lysis and addition of *p*-nitrophenol-phosphate. The yellow color of the dephosphorylated *p*-nitrophenol is easily detected and can be determined spectrophotometrically (Fig. 13.1). In addition, the increase in vacuole size can be used to complement this analysis. (Mayer and Wickner, 1997; Merz

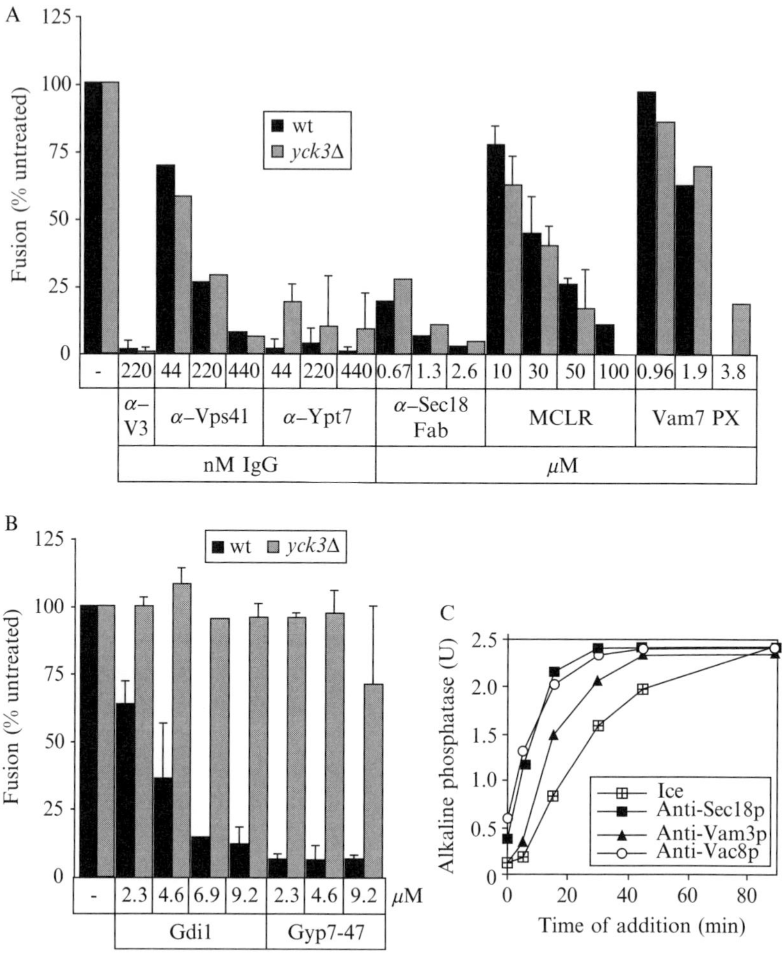

Figure 13.1 Analysis of vacuole fusion. (A) Titration of inhibitors to a fusion reaction. Vacuoles from the two tester strains were incubated with the indicated amounts of antibodies or inhibitors. Inhibitors block the reaction in a dose-dependent manner. In this experiment, wild-type vacuoles are compared to vacuoles lacking the casein kinase Yck3. The fusion values are expressed as a percentage of the fusion activity seen with vacuoles that undergo the normal fusion reaction without inhibitors. In (B) it becomes obvious that *yck3Δ* vacuoles show resistance to the Ypt7-specific inhibitors Gdi1 and Gyp7-47. Parts A and B correspond to Fig. 7A and B in Lagrassa and Ungermann (2005). (C) Time-course inhibition of a fusion reaction. Fusion reactions containing the two tester vacuoles were incubated for the indicated time, and were then set on ice (open boxes), or the reaction received an antibody to Sec18 (filled boxes), Vac8 (open circles), or Vam3 (triangles). Antibody-treated reactions were incubated for the remaining time at 26 °C and then developed together with the ice samples. The figure corresponds to Fig. 7E from Veit *et al.* (2001).

and Wickner, 2004a; Wang *et al.*, 2003). An alternative complementation assay employs the tight interaction between Fos and Jun, which are coupled to two segments of β-lactamase (Jun and Wickner, 2007). Fos and Jun fusion proteins are present in vacuoles from two different strains, and their interaction reconstitutes β-lactamase activity after fusion of these vacuoles. Here, we will exclusively concentrate on the first two assays, which are commonly used in the field.

Strains

BJ3505 (MATa *pep4Δ::HIS3 prb1-Δ1.6R HIS3 lys 2-208 trp1-Δ101 ura3-52 gal2 can*)
DKY6281 (MATα *leu2-3 leu2-112 ura3-52 his3-Δ200 trp1-Δ101 lys2-801 suc2-Δ9 pho8::TRP1*)
BY4741 pep4Δ (MATa *his3Δ1 leu2Δ0 met15Δ0 ura3Δ0 pep4Δ*)
BY4741 pho8Δ (MATa *his3Δ1 leu2Δ0 met15Δ0 ura3Δ0 pho8Δ*)

Proteins that can be added to the fusion reaction

His-tagged Sec18 (purified from *E. coli*) (Haas and Wickner, 1996; Mayer *et al.*, 1996)
His-tagged Sec17 (purified from *E. coli*) (Haas and Wickner, 1996; Mayer *et al.*, 1996)
His-tagged Vam7 (purified from *E. coli*) (Thorngren *et al.*, 2004)
HOPS (purified from yeast vacuoles) (Stroupe *et al.*, 2006)

Reagents

Stock solutions

3 *M* KCl
1 *M* $MnCl_2$
1 *M* $MgCl_2$
4 *M* sorbitol
1 *M* PIPES/KOH, pH 6.8
ATP 0.2M, buffered
1 *M* Tris, pH 8.5
20% Triton X-100
0.1 *M* p-nitrophenol phosphate
1 *M* glycine-KOH, pH 11.5

10x fusion reaction buffer 1

1.25 *M* KCl
50 m*M* $MgCl_2$
200 m*M* sorbitol
10 m*M* PIPES-KOH, pH 6.8

10x ATP regenerating system

5 m*M* ATP
1 mg/ml creatine kinase
400 m*M* creatine phosphate
10 m*M* PIPES-KOH, pH 6.8
200 m*M* sorbitol

Developer

10 m*M* $MgCl_2$
250 m*M* Tris, pH 8.5
0.4% v/v Triton X-100
1.5 m*M* *p*-nitrophenol phosphate

Method

The preparation of fusion reactions requires substantial time for pipetting and therefore should be set up during the centrifugation step. It is recommended to prepare duplicates or triplicates of each reaction. It can also be useful to set up master reactions containing all of the components except the vacuoles and then aliquot from the master mix into the individual tubes. In general, the priming step, which includes cis-SNARE complex disassembly, depends on ATP. This stage can be bypassed by adding purified Vam7 to the fusion reaction (Boeddinghaus *et al.*, 2002; Thorngren *et al.*, 2004). Under these conditions fusion can occur even if ATP is omitted or if priming is blocked by an antibody to Sec17 (Thorngren *et al.*, 2004). The concentration of Vam7 needed for fusion has to be titrated. In addition, purified HOPS can stimulate fusion (Starai *et al.*, 2008; Stroupe *et al.*, 2006) but is not added routinely to fusion reactions.

1. Dilute the isolated vacuoles to 0.3 mg/ml in PS buffer and add PIC to a final concentration of 0.1x from a 20x stock.
2. The 30-μl standard fusion reaction contains 3 μl of 10x fusion reaction buffer 1, 3 μl of 10x ATP regenerating system, 10 μl of vacuoles from the *pep4Δ* strain and 10 μl of vacuoles from the *pho8Δ* strain in PS buffer, and additional PS buffer to bring the reaction to a final volume of 30 μl. We typically use several compounds to enhance vacuole fusion: cytosol (1–3 mg/ml), Sec18 (1 μg/ml) and CoA (10 μ*M*). Prepare the fusion reaction in a microcentrifuge tube on ice adding the vacuoles last. It is generally best to stagger the addition of vacuoles by approximately 15 s to ensure that each tube is incubated the same length of time.
3. We recommend setting up each fusion reaction at least in duplicates or triplicates, and including a background control reaction lacking ATP. As an alternative control, one complete reaction can be kept on ice.
4. Incubate the fusion reaction at 26 °C for 60–90 min.
5. At the end of the reaction, add 470 μl of developer to each tube, again staggering the additions by 15 s, and incubate for 5 min at 30 °C.
6. Add 500 μl of stop solution (1 *M* glycine, pH 11.5).
7. Determine the alkaline phosphatase activity by measuring the absorbance at 400 nm. The blank is 470 μl of developer and 500 μl of stop solution incubated for 5 min at 30 °C and the $O.D._{400}$ reading for the blank is subtracted from each sample. One unit of fusion activity is defined as 1 μmol *p*-nitrophenol developed per min and μg *pep4Δ* vacuoles.

2.2.2. Time course reaction

This assay is very useful if it is necessary to determine the stage, at which an inhibitor interferes with fusion (Conradt *et al.*, 1994; Haas *et al.*, 1994; Mayer *et al.*, 1996; Ungermann *et al.*, 1998a). It can add important information on the order of events taking place during the fusion reaction. Each inhibitor value is plotted against the time axis and thus contributes to a characteristic curve, which is plotted relative to the curve for a sample placed on ice. The priming curve, marked by antibodies to Sec17 or Sec18 precedes the docking curve (Mayer *et al.*, 1996), which is obtained if antibodies to the SNAREs Vam3 or Vti1 are added (Ungermann *et al.*, 1999). The fusion curve is seen in the presence of BAPTA or GTPγS and coincides with the ice curve (Peters and Mayer, 1998).

1. Prepare a master reaction containing all the compounds of a standard fusion reaction. The volume of the master reaction depends on the number of samples to be taken during the assay.
2. For each time point set up two tubes on ice; one contains only buffer, and the other has the inhibitor or reagent to be tested (up to 5 μl). Place these tubes at 26 °C just before they are needed.
3. Start the fusion reaction with the master tube. At each time point, mix the master reaction and transfer a 30 μl aliquot to both tubes, with and without inhibitor. As a control, place one additional aliquot on ice to monitor how far the reaction has proceeded at the time the inhibitor has been added.
4. Incubate at 26 °C for 90 min, develop, and calculate units of fusion as described previously.
5. Plot fusion activity against time to determine the stage where the inhibitor acts. The curves should be clearly separated in the assay (see Figure 13.1.C).

2.2.3. Microscopy assay of vacuole docking and fusion

This assay requires some training and is not as simple as the fusion assay. It has been used initially to determine vacuole docking, and later was also employed to monitor fusion. Controls (incubation on ice, omission of ATP) are absolutely essential to validate the results (Lagrassa and Ungermann, 2005; Mayer and Wickner, 1997; Merz and Wickner, 2004a; Wang *et al.*, 2003). Vacuoles that have not docked remain separated, whereas docked vacuoles form a visible clump.

Materials

10x docking reaction buffer

40 m*M* KCl
5 m*M* $MgCl_2$
10 m*M* PIPES-KOH, pH 6.8
200 m*M* sorbitol

Some commonly used inhibitors	Concentration in reaction	Reference
IgGs	approximately 1 μg/μl	(Price *et al.*, 2000)
BAPTA (30 m*M* stock)	3 m*M*	(Peters and Mayer, 1998)
Gdi1	1–5 μg	(Haas *et al.*, 1995)
PX domain of Vam7	15 μ*M*	(Boeddinghaus *et al.*, 2002)
FYVE domain	10 μ*M*	(Boeddinghaus *et al.*, 2002)
Bromopalmitate (10–30 m*M* stock, DMSO)	100–200 μ*M*	(Veit *et al.*, 2001)
Dethio-PalCoA (1 m*M* stock)	10 μ*M*	(Veit *et al.*, 2001)
Apyrase (1 U/μl)	10–20 U/ml	(Mayer *et al.*, 1996)
Neomycin (5 m*M* stock in PS buffer)	0.3–0.5 m*M*	(Mayer *et al.*, 2000)
Nystatin	50 μg/ml	(Kato and Wickner, 2001)
GTPγS (100 m*M* stock)	3 m*M*, plus $MgCl_2$	(Eitzen *et al.*, 2000)

Method

1. Prepare a 30-μl reaction containing 6 μg of vacuoles, 1x docking buffer, 0.5x ATP regenerating system, 20 μ*M* CoA, 1 μg/ml Sec18 and 2.7 μ*M* FM 4-64 in PS buffer. Alternatively, vacuoles can be labeled with the lipophilic dye MDY-64 at 3 μ*M* (Fratti *et al.*; 2004).
2. The buffer has been optimized to reduce the ATP-independent clustering of vacuoles. Include a reaction without ATP addition.
3. Incubate samples at 27 °C for 30 min.
4. Add 50 μl of 0.6% low-melting-point agarose in PS buffer (kept liquid at 42 °C) and mix by vortexing.
5. Place a 15-μl aliquot on ice-cold glass slides and analyze by fluorescence microscopy.

2.2.4. Vacuole protein dynamics

2.2.4.1. Protein release assay This assay was initially established for monitoring Sec17 release (Mayer *et al.*, 1996). We have employed it successfully to monitor release of Ykt6, Vps41, Vam7 and Ypt7 from vacuoles during the fusion reaction (Boeddinghaus *et al.*, 2002; Dietrich *et al.*, 2005b; Lagrassa and Ungermann, 2005; Ungermann *et al.*, 2000). It is important to thoroughly wash the vacuole pellet to monitor the release.

To obtain a complete picture, both supernatant and pellet fractions should be analyzed (Fig. 13.2).

Materials

10x fusion reaction buffer 2

1.5 *M* KCl
5 m*M* $MnCl_2$
5 m*M* $MgCl_2$
200 m*M* sorbitol
10 m*M* PIPES-KOH, pH 6.8

Method

1. Set up a 150-μl fusion reaction for each time point in a siliconized microcentrifuge tube containing 30 μg of vacuoles isolated from the BJ3505 strain, 1x fusion reaction buffer 2, 1x ATP regenerating system, CoA (10 μM) and PS buffer. Include a reaction without ATP (Sec17 should be not released under these conditions). Sec18 addition (1 μg/ml) is essential to monitor efficient Ykt6 (Figure 13.2.C) release. Gdi1 can be added (64 μg/ml) to analyze protein release that depends on Ypt7 (Figure 13.2.A).
2. Incubate the reactions for 10–30 min at 26 °C, then place them on ice.
3. Dilute each reaction 5-fold with PS buffer at the end of the time course.
4. Reisolate vacuoles by centrifugation for 10 min at 14,000×*g* at 4 °C.
5. Transfer the supernatant fraction to a new tube and precipitate proteins in the tubes containing the supernatant fractions and vacuole pellets with TCA (10% final concentration). Centrifuge at 20,000×*g* for 10 min at 4 °C and discard the supernatant fraction.
6. Wash the TCA pellets by vortexing in PS buffer and repeat the centrifugation step.
7. Analyze proteins from pellets and supernatant fractions by SDS-PAGE and immunoblot.

2.2.4.2. Monitoring Vps41 phosphorylation This method is based on the standard fusion assay and used to monitor Vps41 phosphorylation by mobility shift in SDS-PAGE gels. Vps41 phosphorylation is dependent on the casein kinase Yck3, and consequently is blocked on *yck3*Δ vacuoles (Lagrassa and Ungermann, 2005; (Figure 13.2.A)).

Method

1. Prepare a 150 μl fusion reaction containing 30–40 μg vacuoles isolated from the BJ3505 strain, 1x fusion reaction buffer 2, CoA (10 μM) and with and without 1x ATP regenerating system, in PS buffer.
2. Incubate the reactions for 30 min at 26 °C.
3. Reisolate vacuoles by centrifugation for 10 min at 14,000×g at 4 °C.
4. Discard the supernatant (the increase in Vps41 molecular mass is only seen in the vacuole fraction) and resuspend the pellet in 1x SDS sample buffer.

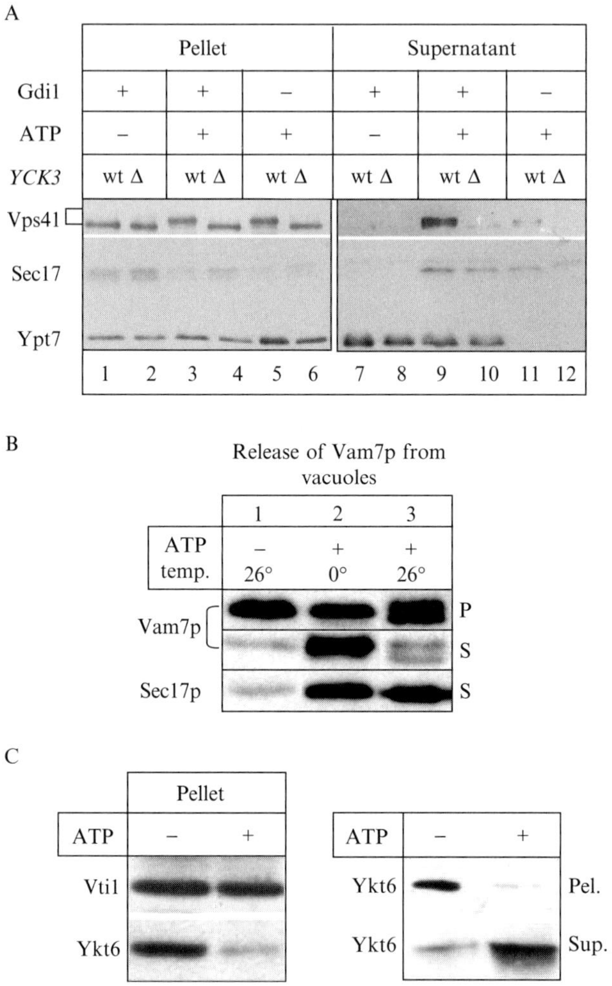

Figure 13.2 Dynamics of vacuole-associated proteins. (A) Vps41 release from yeast vacuoles. Vacuoles were incubated for 30 min in a 60-μl standard reaction containing, where indicated, ATP or Gdi1 (9.2 μM). Vacuoles were reisolated and half of the pellet fraction and the TCA-precipitated supernatant fraction were analyzed by SDS-PAGE and Western blotting. The figure corresponds to Fig. 7B in Lagrassa and Ungermann (2005). (B) Release of Vam7 from yeast vacuoles. Vacuoles were incubated as in (A) for 60 min. Reactions contained His-Sec18 and 10 μM CoA. Pellet and supernatant fractions were analyzed as in (A). The figure is Fig. 1A from Boeddinghaus *et al.* (2002). (C) Ykt6 is released from yeast vacuoles. The incubations and reaction volumes were as (A). The left panel only shows the pellet fraction, whereas the right blot also shows the supernatant fraction. The figure is equivalent to Fig. 4A of Dietrich *et al.* (2005b).

5. Boil the sample for 5 min and analyze the Vps41 mobility shift by SDS-PAGE and inmunoblot. The shift is observed both on 12% and 7.5% gels.

2.2.4.3. Co-immunoprecipitation of vacuolar protein complexes Protein complexes can be isolated by several procedures from yeast vacuoles (Collins *et al.*, 2005; Ungermann *et al.*, 1998a; Ungermann *et al.*, 1999). We present here the general protocol based on the generation of a detergent extract and subsequent immunoprecipitation of the protein complexes. Co-immunoprecipitated proteins are then identified by Western blotting of SDS-PAGE gels.

Method

1. The vacuole pellet (usually from a 150-μl reaction) is washed with 1 ml of PS buffer, 150 m*M* KCl, and then detergent solubilized in 1 ml of solubilization buffer (0.5% TX-100, 20 m*M* HEPES-KOH, pH 7.4, 150 m*M* KCl, 1xPIC, 10 μg/ml α_2-Macroglobulin). Make sure that the pellet is completely resuspended, but do not vortex.
2. Incubate the detergent extract for 10 min on a nutator or rotating wheel at 4 °C.
3. Centrifuge the sample in a microcentrifuge tube for 10 min at 20,000×*g* at 4 °C and transfer the supernatant fraction to a new tube.
4. Remove 1%–10% of the total supernatant and TCA precipitate the proteins ("load" sample).
5. Add the remaining detergent extract to 10–20 μl of protein A-coupled IgG beads. The beads are generated as follows. We incubate 1 ml of antiserum with 500 μl protein A-Sepharose slurry (GE Healthcare) in 10 m*M* Tris, pH 8.0, and then cross-link the antibodies with 20 m*M* Dimethyl pimelimidate (Pierce; added as a solid powder) for 30 min at room temperature to the beads. Beads are then washed with 10 ml of 0.1 *M* glycine, pH 2.6, to remove uncoupled ligand. The beads are then neutralized with 10 ml of 100 m*M* Tris, pH8.0 buffer.
6. Incubate the beads with the detergent extract 1.5 h or overnight at 4 °C on a nutator.
7. Centrifuge briefly (1 min at 1000×*g*) to pellet the beads and remove the supernatant fraction. The supernatant fraction may be saved to monitor the efficiency of the immunoprecipitation reaction (flow-through sample).
8. Add 1 ml of solubilization buffer to the pellet fraction and incubate for 10 min at 4 °C on a nutator.
9. Centrifuge briefly, remove the supernatant fraction, add 1 ml of buffer, and incubate again.
10. Centrifuge briefly, remove the supernatant fraction completely, and add 1 ml of 0.1 *M* glycine, pH 2.6. Incubate by inversion, centrifuge briefly to recover the beads and collect the supernatant fractions in new tubes.

11. TCA precipitate the protein in the supernatant fractions, then boil the pellets in SDS-sample buffer, and analyze the precipitated proteins, together with the load fractions, by SDS-PAGE and Western blotting.

2.2.4.4. Comments Co-immunoprecipitations work best if the antibodies are specific and if the antibodies are cross-linked efficiently to the beads. Any excess antibody that is released from the beads will mask the signal of the specific antibody in the Western blot. Therefore, it is necessary to wash the beads with 0.1 *M* glycine prior to using them in the assay. We always determine the efficiency of the immunoprecipitation by comparison with the load, and >80% of the bait protein should be captured by an efficient antibody. Otherwise, a titration of the antibody might be necessary.

2.2.5. Additional assays

Several additional assays have been applied to monitor reactions at the vacuole. This includes the analysis of *cis*- and *trans*-SNARE complexes by immunoprecipitation (Collins and Wickner, 2007; Dietrich *et al.*, 2005a; Ungermann *et al.*, 1998a,b), the palmitoylation of the fusion factor Vac8 (Dietrich *et al.*, 2004; Subramanian *et al.*, 2006), the release of Ca^{2+} (Merz and Wickner, 2004b; Peters and Mayer, 1998), and the use of the soluble Vam7 protein in specific subassays (Merz and Wickner, 2004b; Thorngren *et al.*, 2004). These assays are not used by all laboratories and are sufficiently described and discussed in the primary literature (see previously) and reviews (Mayer, 2002; Ostrowicz *et al.*, 2007), so that we have not discussed them in this chapter.

3. Discussion

The vacuole purification method described here permits the isolation of intact and functional vacuoles in milligram quantities that can be used for further experimentation. Growth conditions and initial culture density are crucial to guarantee optimal yield and activity.

We employ purified vacuoles for several assays. The most common application is the fusion assay. Within this review, we have described several assays to characterize the proteins that are involved and various subreactions of the fusion process. We would like to emphasize that we have provided only protocols of the basic methodology but did not go into the details of several more specific assays such as the *trans*-SNARE assay (Collins and Wickner, 2007; Dietrich *et al.*, 2005a; Ungermann *et al.*, 1998b), or the detection of palmitoylation on yeast vacuoles (Dietrich *et al.*, 2004; Veit *et al.*, 2001). There are a few issues that should be discussed in the context of the fusion assay. In particular, attention needs to be paid if mutants are employed in this assay. Taking into account that always two types of vacuoles need to be prepared for each mutant, variations in fusion activity can strongly affect the interpretation of the data. Therefore, it is critical that each mutant vacuole is tested against the corresponding wild-type vacuoles to evaluate the effect of the mutation on activity. In addition, it is important to control for the Pep4 and Pho8 content of the purified vacuoles by western blotting, because deficient sorting of these proteins to the vacuole can lead to low fusion *in vitro*. In general, fusion needs to be at least 10-fold greater than the background control for reliable interpretation.

The visual examination of docking and fusion has been employed in several laboratories (Mayer and Wickner, 1997; Merz and Wickner, 2004a; Wang *et al.*, 2002). The major challenge of this assay is the unspecific clustering of vacuoles, which interferes with the analysis. Wickner and colleagues have also developed a visual assay to monitor co-localization of proteins at vacuole docking sites, and observed interdependence of several proteins (Fratti *et al.*, 2004).

Proteins that belong to the fusion machinery are released from vacuole membranes as a part of their functional cycle (Boeddinghaus *et al.*, 2002; Dietrich *et al.*, 2005b; Lagrassa and Ungermann, 2005; Mayer *et al.*, 1996; Price *et al.*, 2000). During *cis*-SNARE complex disassembly, Sec17 is released together with Vam7 and HOPS subunits (Boeddinghaus *et al.*, 2002; Dietrich *et al.*, 2005b; Lagrassa and Ungermann, 2005; Mayer *et al.*, 1996; Price *et al.*, 2000). This release event can be monitored using an easy experiment based on the standard vacuole fusion reaction. Compounds in the reaction should be optimized to facilitate the release of each protein (e.g., Sec18 addition is crucial for the Ykt6 release process; Dietrich *et al.*, 2005b).

How can purified vacuoles be used in future assays? Vacuoles undergo major morphological changes *in vivo* (Weisman, 2003; Weisman, 2006). They can undergo fragmentation in response to osmotic stress and refuse if this stress is relieved (Bonangelino *et al.*, 2002; Lagrassa and Ungermann, 2005). In addition, vacuoles form tubules and fragment in the context of inheritance (Weisman and Wickner, 1988). Among these reactions, assays have been established for vacuole fusion (Conradt *et al.*, 1992; Haas *et al.*, 1994), which subsequently led to mechanistic insights and the identification of the machinery involved (Ostrowicz *et al.*, 2007).

However, purified vacuoles have not been used extensively to address the mechanistic details of membrane alterations during fission or inheritance. In some initial assays, the formation of inheritance structures has been observed if vacuoles are incubated with cytosol (Conradt *et al.*, 1992). It is possible that vacuole-associated actin or tubulin is responsible for these alterations (Eitzen, 2003; Eitzen *et al.*, 2002; Guthrie and Wickner, 1988; Isgandarova *et al.*, 2007). Potentially, purified dynamin-like Vps1 could be a membrane active component that affects vacuole morphology (Peters *et al.*, 2004). In fact, all mutants, which produce a class D phenotype (a very large vacuole) such as *vps1Δ*, do not fragment *in vivo* (Lagrassa and Ungermann, 2005). How the alterations in fission relate to the loss of these proteins is not yet clear. The detailed analysis of purified vacuoles might also help to identify components involved in vacuole fission. In addition to the analysis of vacuole fission, regulation of vacuole fusion will be of major interest. Finally, novel assays are required to dissect the machinery involved in fusion of endosomes, autophagic vesicles or AP-3 vesicles with vacuoles. The available protocols for vacuole purification and analysis will facilitate research in these directions.

ACKNOWLEDGMENTS

This work has been supported by the DFG (CA 806/2-1), the Fundación Ramón Areces (to MC), the SFB 431, and the Hans-Mühlenhoff foundation (to CU).

REFERENCES

Bankaitis, V. A., Johnson, L. M., and Emr, S. D. (1986). Isolation of yeast mutants defective in protein targeting to the vacuole. *Proc. Natl. Acad. Sci. USA* **83,** 9075–9079.

Boeddinghaus, C., Merz, A. J., Laage, R., and Ungermann, C. (2002). A cycle of Vam7p release from and PtdIns 3-P-dependent rebinding to the yeast vacuole is required for homotypic vacuole fusion. *J. Cell Biol.* **157,** 79–90.

Bonangelino, C. J., Nau, J. J., Duex, J. E., Brinkman, M., Wurmser, A. E., Gary, J. D., Emr, S. D., and Weisman, L. S. (2002). Osmotic stress-induced increase of phosphatidylinositol 3,5- bisphosphate requires Vac14p, an activator of the lipid kinase Fab1p. *J. Cell Biol.* **156,** 1015–1028.

Collins, K. M., Thorngren, N. L., Fratti, R. A., and Wickner, W. T. (2005). Sec17p and HOPS, in distinct SNARE complexes, mediate SNARE complex disruption or assembly for fusion. *EMBO J.* **24,** 1775–1786.

Collins, K. M., and Wickner, W. T. (2007). Trans-SNARE complex assembly and yeast vacuole membrane fusion. *Proc. Natl. Acad. Sci. USA* **104,** 8755–8760.

Conradt, B., Haas, A., and Wickner, W. (1994). Determination of four biochemically distinct, sequential stages during vacuole inheritance *in vitro*. *J. Cell Biol.* **126,** 99–110.

Conradt, B., Shaw, J., Vida, T., Emr, S., and Wickner, W. (1992). *In vitro* reactions of vacuole inheritance in *Saccharomyces cerevisiae*. *J. Cell Biol.* **119,** 1469–1479.

Dietrich, L. E., Gurezka, R., Veit, M., and Ungermann, C. (2004). The SNARE Ykt6 mediates protein palmitoylation during an early stage of homotypic vacuole fusion. *EMBO J.* **23,** 45–53.

Dietrich, L. E., Lagrassa, T. J., Rohde, J., Cristodero, M., Meiringer, C. T., and Ungermann, C. (2005a). ATP-independent control of Vac8 palmitoylation by a SNARE subcomplex on yeast vacuoles. *J. Biol. Chem.* **280,** 15348–15355.

Dietrich, L. E., Peplowska, K., Lagrassa, T. J., Hou, H., Rohde, J., and Ungermann, C. (2005b). The SNARE Ykt6 is released from yeast vacuoles during an early stage of fusion. *EMBO Rep.* **6,** 245–250.

Eitzen, G. (2003). Actin remodeling to facilitate membrane fusion. *Biochim. Biophys. Acta* **1641,** 175–181.

Eitzen, G., Wang, L., Thorngren, N., and Wickner, W. (2002). Remodeling of organelle-bound actin is required for yeast vacuole fusion. *J. Cell Biol.* **158,** 669–679.
Eitzen, G., Will, E., Gallwitz, D., Haas, A., and Wickner, W. (2000). Sequential action of two GTPases to promote vacuole docking and fusion. *EMBO J.* **19,** 6713–6720.
Fratti, R. A., Jun, Y., Merz, A. J., Margolis, N., and Wickner, W. (2004). Interdependent assembly of specific regulatory lipids and membrane fusion proteins into the vertex ring domain of docked vacuoles. *J. Cell Biol.* **167,** 1087–1098.
Guthrie, B. A., and Wickner, W. (1988). Yeast vacuoles fragment when microtubules are disrupted. *J. Cell Biol.* **107,** 115–120.
Haas, A. (1995). A quantitative assay to measure homotypic vacuole fusion *in vitro*. *Methods Cell Sci.* **17,** 283–294.
Haas, A., Conradt, B., and Wickner, W. (1994). G-protein ligands inhibit *in vitro* reactions of vacuole inheritance. *J. Cell Biol.* **126,** 87–97.
Haas, A., Scheglmann, D., Lazar, T., Gallwitz, D., and Wickner, W. (1995). The GTPase Ypt7p of *Saccharomyces cerevisiae* is required on both partner vacuoles for the homotypic fusion step of vacuole inheritance. *EMBO J.* **14,** 5258–5270.
Haas, A., and Wickner, W. (1996). Homotypic vacuole fusion requires Sec17p (yeast alpha-SNAP) and Sec18p (yeast NSF). *EMBO J.* **15,** 3296–3305.
Isgandarova, S., Jones, L., Forsberg, D., Loncar, A., Dawson, J., Tedrick, K., and Eitzen, G. (2007). Stimulation of actin polymerization by vacuoles via Cdc42p-dependent signaling. *J. Biol. Chem.* **282,** 30466–30475.
Kato, M., and Wickner, W. (2001). Ergosterol is required for the Sec18/ATP-dependent priming step of homotypic vacuole fusion. *EMBO J.* **20,** 4035–4040.
Klionsky, D. J. (2007). Autophagy: From phenomenology to molecular understanding in less than a decade. *Nat. Rev. Mol. Cell Biol.* **8,** 931–937.
Lagrassa, T. J., and Ungermann, C. (2005). The vacuolar kinase Yck3 maintains organelle fragmentation by regulating the HOPS tethering complex. *J. Cell Biol.* **168,** 401–414.
Mayer, A. (2002). Membrane fusion in eukaryotic cells. *Annu. Rev. Cell Dev. Biol.* **18,** 289–314.
Mayer, A., Scheglmann, D., Dove, S., Glatz, A., Wickner, W., and Haas, A. (2000). Phosphatidylinositol 4,5-bisphosphate regulates two steps of homotypic vacuole fusion. *Mol. Biol. Cell* **11,** 807–817.
Mayer, A., and Wickner, W. (1997). Docking of yeast vacuoles is catalyzed by the Ras-like GTPase Ypt7p after symmetric priming by Sec18p (NSF). *J. Cell Biol.* **136,** 307–317.
Mayer, A., Wickner, W., and Haas, A. (1996). Sec18p (NSF)-driven release of Sec17p (α-SNAP) can precede docking and fusion of yeast vacuoles. *Cell* **85,** 83–94.
Merz, A. J., and Wickner, W. T. (2004a). Resolution of organelle docking and fusion kinetics in a cell-free assay. *Proc. Natl. Acad. Sci. USA* **101,** 11548–11553.
Merz, A. J., and Wickner, W. T. (2004b). Trans-SNARE interactions elicit Ca^{2+} efflux from the yeast vacuole lumen. *J. Cell Biol.* **164,** 195–206.
Ostrowicz, C. W., Meiringer, C. T., and Ungermann, C. (2007). Yeast vacuole fusion: A model system for eukaryotic endomembrane dynamics. *Autophagy* **4,** 1–15.
Peters, C., Baars, T. L., Buhler, S., and Mayer, A. (2004). Mutual control of membrane fission and fusion proteins. *Cell* **119,** 667–678.
Peters, C., and Mayer, A. (1998). Ca^{2+}/calmodulin signals the completion of docking and triggers a late step of vacuole fusion. *Nature* **396,** 575–580.
Price, A., Seals, D., Wickner, W., and Ungermann, C. (2000). The docking stage of yeast vacuole fusion requires the transfer of proteins from a cis-SNARE complex to a Rab/Ypt protein. *J. Cell Biol.* **148,** 1231–1238.
Raymond, C. K., Howald-Stevenson, I., Vater, C. A., and Stevens, T. H. (1992). Morphological classification of the yeast vacuolar protein sorting mutants: Evidence for a prevacuolar compartment in class E *vps* mutants. *Mol. Biol. Cell* **3,** 1389–1402.

Reese, C., Heise, F., and Mayer, A. (2005). Trans-SNARE pairing can precede a hemifusion intermediate in intracellular membrane fusion. *Nature* **436,** 410–414.

Robinson, J. S., Klionsky, D. J., Banta, L. M., and Emr, S. D. (1988). Protein sorting in *Saccharomyces cerevisiae*: Isolation of mutants defective in the delivery and processing of multiple vacuolar hydrolases. *Mol. Cell Biol.* **8,** 4936–4948.

Shen, S. H., Chretien, P., Bastien, L., and Slilaty, S. N. (1991). Primary sequence of the glucanase gene from *Oerskovia xanthineolytica*: Expression and purification of the enzyme from *Escherichia coli*. *J. Biol. Chem.* **266,** 1058–1063.

Starai, V. J., Hickey, C. M., and Wickner, W. (2008). HOPS proofreads the *trans*-SNARE complex for yeast vacuole fusion. *Mol. Biol. Cell* **19,** 2500–2508.

Stroupe, C., Collins, K. M., Fratti, R. A., and Wickner, W. (2006). Purification of active HOPS complex reveals its affinities for phosphoinositides and the SNARE Vam7p. *EMBO J.* **25,** 1579–1589.

Subramanian, K., Dietrich, L. E., Hou, H., Lagrassa, T. J., Meiringer, C. T., and Ungermann, C. (2006). Palmitoylation determines the function of Vac8 at the yeast vacuole. *J. Cell Sci.* **119,** 2477–2485.

Thorngren, N., Collins, K. M., Fratti, R. A., Wickner, W., and Merz, A. J. (2004). A soluble SNARE drives rapid docking, bypassing ATP and Sec17/18p for vacuole fusion. *EMBO J.* **23,** 2765–2776.

Ungermann, C., Nichols, B. J., Pelham, H. R. B., and Wickner, W. (1998a). A vacuolar v-t-SNARE complex, the predominant form *in vivo* and on isolated vacuoles, is disassembled and activated for docking and fusion. *J. Cell Biol.* **140,** 61–69.

Ungermann, C., Price, A., and Wickner, W. (2000). A new role for a SNARE protein as a regulator of the Ypt7/Rab-dependent stage of docking. *Proc. Natl. Acad. Sci. USA* **97,** 8889–8891.

Ungermann, C., Sato, K., and Wickner, W. (1998b). Defining the functions of trans-SNARE pairs. *Nature* **396,** 543–548.

Ungermann, C., von Mollard, G. F., Jensen, O. N., Margolis, N., Stevens, T. H., and Wickner, W. (1999). Three v-SNAREs and two t-SNAREs, present in a pentameric cis-SNARE complex on isolated vacuoles, are essential for homotypic fusion. *J. Cell Biol.* **145,** 1435–1442.

Uttenweiler, A., Schwarz, H., Neumann, H., and Mayer, A. (2007). The vacuolar transporter chaperone (VTC) complex is required for microautophagy. *Mol. Biol. Cell* **18,** 166–175.

Veit, M., Laage, R., Dietrich, L., Wang, L., and Ungermann, C. (2001). Vac8p release from the SNARE complex and its palmitoylation are coupled and essential for vacuole fusion. *EMBO J.* **20,** 3145–3155.

Wang, L., Merz, A. J., Collins, K. M., and Wickner, W. (2003). Hierarchy of protein assembly at the vertex ring domain for yeast vacuole docking and fusion. *J. Cell Biol.* **160,** 365–374.

Wang, L., Seeley, E. S., Wickner, W., and Merz, A. J. (2002). Vacuole fusion at a ring of vertex docking sites leaves membrane fragments within the organelle. *Cell* **108,** 357–369.

Weisman, L. S. (2003). Yeast vacuole inheritance and dynamics. *Annu. Rev. Genet.* **37,** 435–460.

Weisman, L. S. (2006). Organelles on the move: Insights from yeast vacuole inheritance. *Nat. Rev. Mol. Cell Biol.* **7,** 243–252.

Weisman, L. S., and Wickner, W. (1988). Intervacuole exchange in the yeast zygote: A new pathway in organelle communication. *Science* **241,** 589–591.

Zinser, E., Paltauf, F., and Daum, G. (1993). Sterol composition of yeast organelle membranes and subcellular distribution of enzymes involved in sterol metabolism. *J. Bacteriol.* **175,** 2853–2858.

CHAPTER FOURTEEN

PEXOPHAGY IN *HANSENULA POLYMORPHA*

Tim van Zutphen,* Ida J. van der Klei,*,† *and* Jan A. K. W. Kiel*

Contents

Abstract

In the yeast *Hansenula polymorpha* the development and turnover of peroxisomes is readily achieved by manipulation of the cultivation conditions. The organelles massively develop when the cells are incubated in the presence of methanol as the sole source of carbon and energy. However, they are rapidly and selectively degraded when methanol-grown cells are placed at conditions of repression of methanol metabolism (e.g. in glucose or ethanol excess conditions) by a process termed *macropexophagy*. Degradation of peroxisomes is also observed when the cells are placed at nitrogen-depletion conditions (microautophagy). This contribution details the methodologies that are currently in use investigating macropexophagy and microautophagy in *H. polymorpha*. Emphasis is placed on various structural (fluorescence microscopy, electron microscopy) and biochemical (specific enzyme activity measurements, Western blotting) approaches.

* Molecular Cell Biology, University of Groningen, The Netherlands
† Kluyver Centre for Genomics of Industrial Fermentation, Delft, The Netherlands

Methods in Enzymology, Volume 451
ISSN 0076-6879, DOI: 10.1016/S0076-6879(08)03214-X

1. Introduction

Organelle homeostasis is a requisite for optimal functioning of eukaryotic cells. One of the modes to achieve this is that specific cell organelles proliferate when required and are removed when they have become superfluous. In eukaryotes, the two major proteolytic degradation processes are the ubiquitin-proteasome pathway and autophagy. The first process is solely involved in the degradation of single protein molecules (Ciechanover, 2006), whereas the latter is capable of degrading a wide range of intracellular constituents (proteins, lipids, or DNA) in the lysosome/vacuole (Klionsky, 2007). Although the process of autophagy was already described by de Duve 45 years ago (de Duve, 1963; de Duve and Wattiaux, 1966), only recently has this topic gained substantial interest, mainly because of its role in a wide variety of processes, such as cell development, ageing, cell death, and immunity (Mizushima *et al.*, 2008). Hence, autophagy is very important in human health and disease (Shintani and Klionsky, 2004).

The key proteins required for autophagy and autophagy related processes are encoded by *ATG* genes and most are conserved from yeast to man (Meijer *et al.*, 2007). Because of their versatility in handling, yeast species are ideal model organisms for studying the molecular mechanisms of autophagy. Indeed, most *ATG* genes were initially discovered in yeast (Klionsky *et al.*, 2003). In particular, the process of peroxisome degradation is studied in these organisms because the development and turnover of these organelles is readily manipulated by the growth conditions. Here we describe the degradation of peroxisomes by autophagy in the methylotrophic yeast *Hansenula polymorpha* and focus on the experimental approaches that are used to monitor this process.

2. *H. polymorpha* as a Model System for Peroxisome Degradation

A limited number of microorganisms are capable of growing on one-carbon compounds (methylotrophs). In yeast species, methylotrophy is limited to the utilization of methanol as the sole source of carbon and energy. Examples of methylotrophic yeast species are *Candida boidinii*, *H. polymorpha*, and *Pichia pastoris*. In these organisms the initial oxidation of methanol is catalyzed by the enzyme alcohol oxidase (AO), which is a peroxisomal oxidase that generates formaldehyde and hydrogen peroxide from methanol. Formaldehyde can be assimilated via the xylulose-5-phosphate pathway, which involves the peroxisomal enzyme dihydroxyacetone synthase (DHAS), whereas hydrogen peroxide may be decomposed by peroxisomal catalase (CAT). The other enzymes involved in methanol metabolism (e.g., formaldehyde dissimilation enzymes and the other enzymes of the xylulose-5-phosphate pathway) are all localized to the cytosol (reviewed in van der Klei *et al.*, 2006).

When methylotrophic yeast species are grown on media containing glucose as the sole carbon and energy source and ammonium sulfate as nitrogen source, peroxisomal enzymes are not required for primary metabolism. As a consequence, the cells generally contain one or only a few small peroxisomes. However, upon a shift to media containing methanol as the sole carbon source, enzymes involved in methanol metabolism are induced concomitant with an increase in the number and size of peroxisomes. Conversely, placing methanol-grown *H. polymorpha* cells into fresh glucose media leads to rapid degradation of the—now superfluous—organelles. This degradation of peroxisomes (pexophagy), is highly selective and resembles macroautophagy in mammalian cells (Leighton *et al.*, 1975) and hence is designated macropexophagy (Klionsky *et al.*, 2007). During this process individual organelles are consecutively sequestered from the cytoplasm by membranous layers forming autophagosomes. After sequestration is complete, the outer membrane layer of the autophagosome fuses with the vacuolar membrane, resulting in incorporation of the sequestered peroxisome into the vacuole, where the entire organelle becomes degraded by vacuolar hydrolases (depicted in Fig. 14.1A). The molecular mechanisms involved in the formation of autophagosomes and fusion of this organelle with the vacuole overlap with those involved in general macroautophagy, because various Atg proteins are involved in both processes.

Initiating macropexophagy, however, requires the function of two peroxisomal membrane proteins, Pex3 and Pex14. The function of these two peroxins, that are also essential for peroxisome biogenesis, is completely different. Remarkably, one of these, Pex3, has to be removed from the peroxisomal membrane to allow initiation of pexophagy, whereas Pex14 probably acts as the site of recognition of the organelle by the autophagy machinery (Bellu *et al.*, 2001a, 2002). Recognition most likely involves Atg11, a protein also functional in other selective modes of autophagy (Kiel *et al.*, 2003). Atg11 has been localized to the PAS (preautophagosomal structure) and autophagosomal membranes and is thought to be involved in recruiting selective cargo to the autophagosome (Yorimitsu and Klionsky, 2005).

In the related methylotrophic yeast *P. pastoris*, a peroxisomal membrane protein exclusively involved in selective peroxisome degradation was recently uncovered (designated Atg30) and may be involved in this process as well (Farre *et al.*, 2008). A putative Atg30 ortholog is present in *H. polymorpha*, but its function in pexophagy has not yet been confirmed.

In addition to macropexophagy, nitrogen limitation leads to peroxisome degradation in *H. polymorpha* by a mechanism known as microautophagy.

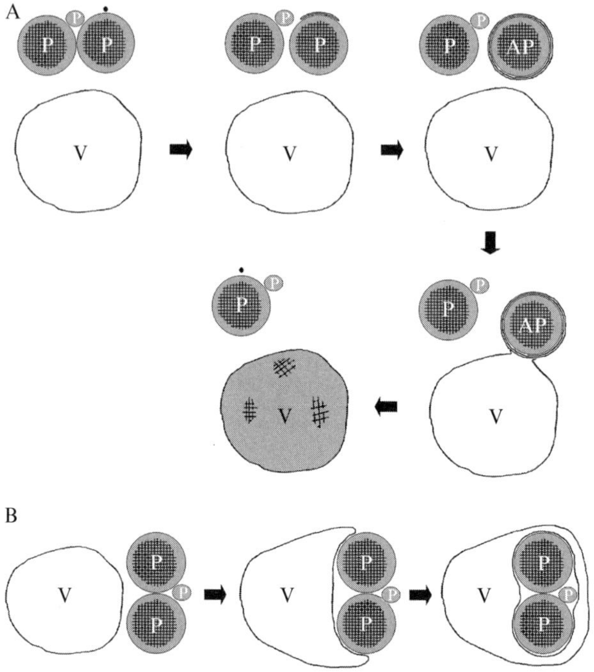

Figure 14.1 Methanol-grown *H. polymorpha* cells contain mature peroxisomes (P) as well as at least one immature organelle. (A) Upon induction of macropexophagy, a single mature organelle is tagged for degradation (marked by the black dot) followed by its sequestration by multimembrane layers forming an autophagosome (AP). After sequestration is completed, the outer membrane layer of the autophagosome fuses with the vacuolar membrane resulting in the uptake of the organelle into the vacuole (V), where it is degraded by vacuolar hydrolases. Next, one by one, other mature peroxisomes are tagged and degraded. Only one (or few) immature peroxisome(s) escape(s) degradation. (B) Upon induction of microautophagy the vacuole engulfs (part of) a cluster of peroxisomes that is subsequently degraded in the vacuole lumen.

However, this process is not considered a selective pathway, as cytosolic components are taken up concomitantly with peroxisomes (Bellu *et al.*, 2001b). Induction of microautophagy by nitrogen starvation results in a direct engulfment and subsequent uptake of peroxisomes and cytoplasmic components by the vacuole (Fig. 14.1B). Microautophagy is therefore morphologically very distinct from macropexophagy. Nevertheless many components of the macroautophagy machinery function also in microautophagy (Sakai *et al.*, 2006). Remarkably, in *H. polymorpha* microautophagic degradation of peroxisomes, but not of other cytoplasmic constituents, requires Atg11, suggesting some mode of selectivity of peroxisome degradation during nitrogen starvation (Komduur *et al.*, 2004).

3. Cultivation of *H. polymorpha* and Induction of Pexophagy

3.1. Cultivation of *H. polymorpha* in media supplemented with methanol as the sole carbon and energy source

For induction of pexophagy, cells from the exponential growth phase on methanol are preferably used. Care should be taken that all nutrients are present in excess and that cultures are optimally aerated.

We generally use (auxotrophic derivatives of) the *H. polymorpha* strain NCYC495 and mutants generated from this strain (Gleeson and Sudbery, 1988). For inoculation of glucose-containing batch cultures, colonies plated on a glucose-containing agar plate (e.g., YPD plate) are used. These plates can be stored at 4 °C for several weeks.

For precultivation, a mineral medium (van Dijken *et al.*, 1976; Table 14.1) is used that contains 0.25% ammonium sulfate as nitrogen source and is supplemented with 0.5% glucose as carbon source. Optimal growth of *H. polymorpha* cells on methanol media and maximal induction of peroxisomes is only obtained when cells are extensively pregrown on

Table 14.1 Composition of mineral medium

Components	g/l	Vishniac stock solution (1000x)	g/l	Vitamin stock solution (1000x)	g/l
$(NH_4)_2SO_4$	2.5	EDTA (Titriplex-III)	10	Biotin	0.1
$MgSO_4$	0.2	$ZnSO_4.7H_2O$	4.4	Thiamin	0.2
K_2HPO_4	0.7	$MnCl_2.4H_2O$	1.01	Riboflavin	0.1
NaH_2PO_4	3.0	$CoCl_2.6H_2O$	0.32	Nicotinic acid	5
yeast extract	0.5	$CuSO_4.5H_2O$	0.315	p-Aminobenzoic acid	0.3
Vishniac stock solution	1.0 ml	$(NH_4)_6Mo_7O_{24}.4H_2O$	0.22	Pyridoxal hydrochloride	0.1
		$CaCl_2.2H_2O$	1.47	Ca-panthothenate	2
Add after autoclaving:		$FeSO_4.7H_2O$	1.0	Inositol	10
Vitamin stock solution	1.0 ml			*Sterilize by filtration*	
Carbon source	0.5 %				

glucose prior to the shift to methanol medium. Therefore, methanol-grown cells should not be used as inoculum for these cultures.

Protocol

1. Normally, cells from fresh glucose plates are used as inoculum. Cells are precultured at 37 °C at 200 rpm in a 100-ml flask with 20 ml of mineral medium containing 0.5% glucose until an optical density (OD, expressed as absorption at 660 nm) of 1.5–1.8 is reached.
2. The culture is diluted to $OD_{660} = 0.1$ in fresh glucose medium and grown again until the mid-exponential growth phase. This procedure is normally repeated 3 times until the cells continuously grow at maximal speed (doubling time of approximately 1 h for wild-type cells).
3. Cells from the mid-exponential growth phase (OD_{660} = 1.5–1.8) are diluted in 100 ml of fresh mineral medium containing 0.5% methanol as the sole source of carbon (starting at $OD_{660} = 0.1$) in a 500-ml flask. After a short lag phase wild-type cells start to grow on methanol (normal doubling time on methanol is approximately 4–4.5 h). Cultures at the late exponential growth phase (OD_{660} = 1.8–2.4) are used to induce peroxisome degradation.

3.2. Glucose/ethanol induced macropexophagy

In *H. polymorpha*, macropexophagy is induced by exposure of methanol-grown cells to excess glucose or ethanol conditions. Both addition of glucose/ethanol to cultures growing on methanol as well as a dilution of methanol-grown cells in fresh glucose/ethanol media have been applied successfully. Although addition of glucose or ethanol to a methanol culture is experimentally easier, it must be noted that prolonged cultivation may ultimately lead to depletion of medium components (e.g., vitamins, amino acids) as cultures with very high densities may be obtained. This carries the risk of induction of nonselective autophagy as a result of starvation as well. Hence, a shift of cells to fresh glucose/ethanol medium is preferred.

Protocol

1. Dilute the methanol culture in fresh, prewarmed medium lacking a carbon source to an $OD_{660} = 0.2$.
2. Immediately and rapidly take a sample (T = 0 h; for OD measurement and biochemical or microscopy analysis, see subsequent sections), followed by addition of inducer (glucose, ethanol) to a final concentration of 0.5%.
3. Continue to incubate the cells at 37 °C at 200 rpm.
4. Take samples of an equal culture volume as performed at step 2, at T = 1, 2 ,3 and 4 h after the addition of the inducer.

3.3. N-starvation induced microautophagy

Upon a shift of *H. polymorpha* cells from media containing excess nitrogen (generally ammonium sulfate) to media lacking any nitrogen source (i.e., also excluding amino acids), nonselective autophagy is induced.

Protocol

1. Cells are grown on methanol (100 ml), as described previously.
2. Cells are collected by centrifugation (5 min at 3000×*g* at 37 °C).
3. The supernatant fraction is discarded and the cells are rapidly resuspended in an identical volume of prewarmed mineral medium lacking any nitrogen source.
4. Samples are taken at 2-h intervals to follow the fate of peroxisomes, and other proteins, over time.

4. Analysis of Peroxisome Degradation

To determine whether degradation of peroxisomes has occurred, several approaches are suitable. A combination of at least two experimental approaches is preferred.

4.1. Biochemical analysis

4.1.1. Preparation of cell extracts for biochemical analysis

Peroxisome degradation can be demonstrated by Western blot analysis of samples taken prior to, and after, the induction of autophagy, using antibodies against peroxisomal marker proteins.

For Western blot analysis, a culture volume corresponding to at least 3 OD_{660} (volume × OD_{660}) units is harvested. Three OD_{660} units correspond to approximately 300 μg of protein. Upon induction of peroxisome degradation, the cells may continue to grow, while peroxisome formation and the synthesis of peroxisomal proteins of methanol metabolism (AO, DHAS, and CAT) is fully repressed. Thus, equal volumes of the culture before and after induction of peroxisome degradation should be studied.

Protocol

1. Cells are precultured in glucose, grown in methanol-containing medium, and then shifted to glucose, ethanol or nitrogen-starvation medium as described in section 3.
2. Cultures are cooled on ice followed by collection of the cells by centrifugation (10,000×*g* for 10 min at 4 °C in case of large volumes or for 1 min at 14,000×*g* in a microcentrifuge when using small volumes).

3. The supernatant fractions are discarded and the cell pellets are resuspended in a solution containing 12.5% trichloroacetic acid (TCA) by vortexing, followed by freezing at −80 °C for at least 30 min.
4. For analysis, the frozen samples are thawed on ice and centrifuged at 14,000×*g* at 4 °C for 5 min to collect the cells.
5. The cell pellet is washed twice (by resuspending thoroughly followed by centrifugation) using 500 μl of 80% ice-cold acetone (v/v) to remove residual TCA.
6. The pellet is air-dried and resuspended in 100 μl of a solution containing 1% SDS and 0.1 N NaOH. The high pH of this solution causes the yeast cell wall to disintegrate whereas SDS dissolves the cellular membranes. The pellet is resuspended (using a vortex) until a homogeneous cell suspension is obtained.
7. The suspension is mixed with an equal volume of concentrated SDS sample buffer (4% SDS, 10% β-mercaptoethanol, 20% glycerol, 0.002% bromophenol blue, 0.1 *M* Tris-HCl, pH 6.8) and boiled for 5 min.
8. Prior to loading onto an SDS-polyacrylamide gel, the sample is centrifuged for 3 min at 10,000×*g* to sediment residual cell debris. The gels are used for Western blotting to analyze the levels of peroxisomal marker proteins. Cytosolic and/or mitochondrial marker proteins should be used as controls to monitor whether or not selective peroxisome degradation has occurred.

4.1.2. Analysis of pexophagy by Western blot analysis

Because the synthesis of the methanol metabolism-related peroxisomal matrix proteins is fully repressed in glucose or ethanol excess conditions, the biochemical analysis of peroxisome degradation has focused on following the amount and/or specific activities of these proteins. A distinction can be made between peroxisomal matrix proteins and membrane proteins. In specific cases in which matrix proteins cannot be used as a marker for peroxisome degradation, for instance when using a mutant in which matrix protein import is strongly impaired (e.g., a *pex* mutant), the fate of peroxisomal membrane proteins should be analyzed.

4.1.3. Analysis of the levels of peroxisomal matrix proteins as a marker for peroxisome degradation

In *H. polymorpha* cells grown on methanol as the sole carbon source, AO is an abundant peroxisomal matrix protein, constituting 5%–30% of the total cellular protein depending on growth conditions. The relatively high abundance of AO renders it an excellent marker to study the fate of peroxisomes after induction of peroxisome degradation, either by following the reduction of the level of AO protein over time (by Western blotting) or by determining the decrease in AO-specific enzyme activities

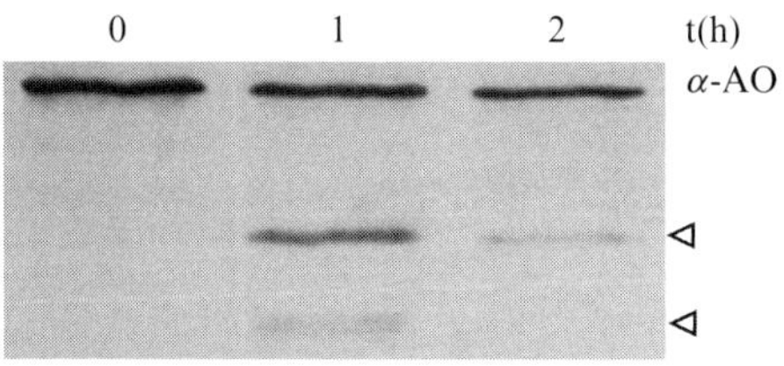

Figure 14.2 Western blot, prepared of crude extracts of methanol-grown *H. polymorpha* cells exposed for the indicated time points to 0.5% glucose, using an α-AO specific antiserum. AO protein levels decrease in time, while also characteristic AO degradation products (arrowheads) are observed.

(described subsequently). For analysis by Western Blotting, successive samples of a time series of cells that were transferred to pexophagy-inducing conditions are loaded on 10% SDS-PAA gels. After Western Blotting by standard protocols, a specific α-AO primary rabbit antiserum is used, followed by an alkaline phosphatase-conjugated secondary antirabbit antibody using NBT-BCIP as substrate for the detection of the immunogenic protein bands. In the protein samples taken after induction of pexophagy, the intensity of the AO band (running at ±70 kDa) should decrease with time, which can be quantified by densitometry scanning of the blots. Additionally, degradation of AO results in the appearance of lower molecular weight degradation bands on the blot (Fig. 14.2). Because AO is inactivated during peroxisome degradation (Bruinenberg *et al.*, 1982), generally the kinetics of the decrease in enzyme activities is faster relative to the decrease of AO protein (see the subsequent section).

4.1.4. Analysis of the levels of peroxisomal membrane proteins as a marker for peroxisome degradation

Next to monitoring the fate of a peroxisomal matrix protein by Western Blotting, the decrease in the level of peroxisomal membrane proteins can also be used. However, proteins containing domains exposed to the cytosol might also be susceptible to other modes of degradation (e.g., via the ubiquitin-proteasome pathway). Before using a novel membrane protein as a tool to study pexophagy, its turnover by alternative pathways should be determined (e.g., by studying the change in levels upon induction of peroxisome degradation in an *atg1* mutant).

H. polymorpha Pex10 is a suitable peroxisomal membrane marker for degradation studies (Veenhuis *et al.*, 1996). However, because of the generally low levels of Pex10, the sensitive chemiluminescent horseradish peroxidase method is preferred (Roche BM Chemiluminescent western Blotting Kit) over the alkaline phosphatase-based method.

In addition, a constructed fusion protein consisting of the first 50 amino acids of the peroxisomal membrane protein Pex3 fused to GFP

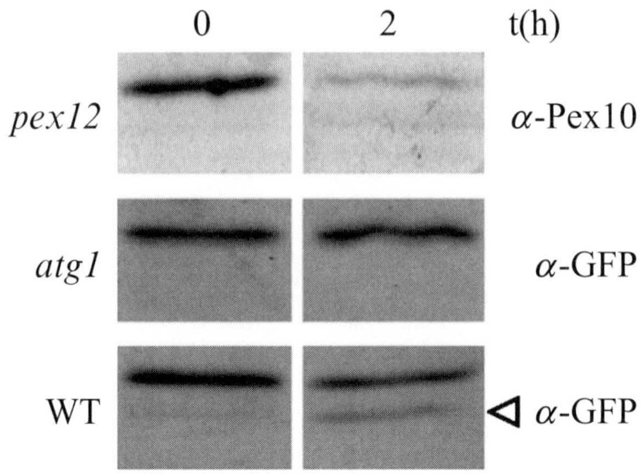

Figure 14.3 Glycerol/methanol-grown *pex12* cells, as well as methanol-grown *atg1* and wild-type cells that produce N50.Pex3.GFP, were exposed to excess glucose conditions to induce macropexophagy. Samples were taken prior to and 2 h after induction and analyzed by Western blotting, using α-Pex10 (for *pex12* cells) and α-GFP antisera (for *atg1* and wild-type cells). In *pex12* cells, which contain peroxisome remnants, and in wild-type cells, but not in *atg1* cells, peroxisome components decline, indicating their proteolytic turnover. In wild-type cells a second α-GFP specific protein band is observed (arrowhead) that represents free GFP accumulating in the vacuole as a result of its relative stability toward vacuolar hydrolases.

(N50.Pex3.GFP), expressed under control of the inducible *AOX* promoter, has been successfully used as a marker (van Zutphen *et al.*, 2008; see Fig. 14.3). The GFP portion of the fusion protein forms a rather stable barrel structure, which is degraded relatively slowly in the vacuole. The formation of cleaved N50.Pex3.GFP can also easily be monitored by Western blotting or fluorescence methods (Shintani and Klionsky, 2004).

4.1.5. Nonperoxisomal control proteins

To determine the specificity of the peroxisome degradation process, turnover of other components in the cell should also be analyzed. This has been performed mainly by Western blot analysis. As markers for mitochondria, porin or malate dehydrogenase may be used, whereas as a representative of the endoplasmic reticulum, the levels of Sec63 may be analyzed (Kiel *et al.*, 1999). As marker for the cytosol translation elongation factor 1-α (eEF1A) and Hsp70 may be applied successfully (Bellu *et al.*, 2001b).

4.1.6. Preparation of cell-free extracts for specific enzyme activity measurements

Specific enzyme activity measurements to follow peroxisome degradation are performed in crude cell free extracts, which are prepared as follows.

Protocol

1. Harvest a culture of cells, grown under the appropriate conditions as described previously to induce pexophagy, with a volume corresponding to approximately 10 OD_{660} units (volume × OD_{660}) and cool on ice.
2. Centrifuge the cells for 5 min at 3000×*g* and wash the pellet twice (by resuspending and centrifugation) using 10 ml of 50 m*M* potassium phosphate buffer, pH 7.2.
3. Resuspend the cells in 500 μl of potassium phosphate buffer and lyse the cells using a vortex or Fastprep (FP120, Bio101/Savant, Qbiogene, Cedex, France) after adding 0.5 volume of acid-washed glass beads for 1 min (see chapter 1), followed by cooling on ice for 1 min. This is repeated until the majority of the cells are broken (checked by light microscopy).
4. Remove unbroken cells and cell debris by centrifugation for 5 min at 14,000 rpm at 4 °C. The resulting supernatant can be used for enzyme activity measurements.

4.1.7. Alcohol oxidase activity assay

Alcohol oxidase catalyzes the oxidation of methanol, thereby producing formaldehyde and hydrogen peroxide. The hydrogen peroxide that is generated can be assayed via the oxidation of reduced ABTS (2,2′-azino-bis (3-ethylbenzthiazoline-6-sulfonic acid)) depicted in formula 1. This reaction is catalyzed by peroxidase. The end product, oxidized ABTS, has a green color with an absorption maximum at 420 nm (Verduyn *et al.*, 1984).

$$\begin{array}{lcl} CH_3OH + O_2 & \rightarrow & H_2CO + H_2O_2 \\ H_2O_2 + 2\ ABTS_{red} & \rightarrow & H_2O + 2\ ABTS_{ox} \end{array} \qquad (1)$$

Protocol

1. Add an appropriate amount of buffer A (0.5 mg/ml ABTS, 10 U/ml horseradish peroxidase, in 50 m*M* potassium phosphate buffer, pH 7.2) to a glass cuvette such that together with the sample the final volume will be 990 μl. Place the cuvette into the thermostated cuvette holder of a spectrophotometer (temperature set to 37 °C).
2. Add the required amount of cell free extract (as prepared in section 4.1.6).
3. Record absorbance at 420 nm (A_{420nm}) and equilibrate until the absorbance remains constant.
4. Add 10 μl of 10 *M* methanol (40% v/v) and mix.
5. Measure the change in absorbance at 420 nm.
6. Perform the assay in triplicate using 3 different sample volumes. All three measurements should result in similar specific activities.

7. Calculate the enzyme activity using the following equation:
   $\text{Units/mg} = \frac{V}{2.\epsilon.d.v.c} \times \frac{\Delta t}{\Delta E}$

   where

   V = total volume (ml),
   v = sample volume (ml),
   ε = extinction coefficient ($cm^2/\mu mol$) = 43,2 $cm^2/\mu mol$,
   d = length of the light path,
   c = protein concentration of the sample in mg/ml,
   ΔE = change in absorbance, and
   Δt = change in time (min).

4.2. Morphological analysis

Uptake of peroxisomes by the vacuole during autophagy can be visualized by electron microscopy or, after introduction of a fluorescent peroxisomal marker, by fluorescence microscopy.

4.2.1. Fluorescence microscopy

To visualize peroxisomes, fluorescent proteins are used that are either sorted to the peroxisomal lumen or to its surrounding membrane. The use of a fluorescent matrix protein is preferred as it is not susceptible to degradation by processes other than autophagy. The most common peroxisomal matrix targeting signal is the C-terminal PTS1 (-SKL). Thus, green fluorescent protein (GFP) fused to SKL localizes to the peroxisomal matrix in *H. polymorpha*. During peroxisome degradation, the matrix contents are released into the vacuolar lumen, as observed in Fig. 14.4.

A suitable peroxisomal membrane marker is the fusion protein N50. Pex3.GFP (see section 4.1.4). This marker can also be used to tag peroxisomal membranes in mutants defective in matrix protein import (see Fig. 14.4; for details, see van der Klei and Veenhuis 2007).

To demonstrate that peroxisomes have indeed been taken up by the vacuole, the vacuolar membrane can be specifically stained with the red fluorescent dye FM 4-64 (N-(3-triethylammoniumpropyl)-4-(6-(4-(diethylamino)phenyl)hexatrienyl)pyridinium dibromide, Molecular Probes Invitrogen). To label with FM 4-64, 0.5 to 1 ml of culture is incubated with 1 μl of FM 4-64 solution (100 μg dissolved in 83 μl DMSO, 1μg/μl) for at least 45 min at 37 °C with shaking at 200 rpm, followed by washing with prewarmed media (also see chapter 7). Because FM 4-64 is taken up by endocytosis, the relatively slower-growing methanol cells need to be incubated longer than cells cultivated on glucose.

4.2.2. Ultrastructural analysis of pexophagy by electron microscopy

Electron microscopy allows for the obtaining of detailed morphological information on peroxisome degradation. Two types of fixation techniques are routinely used prior to embedding the cells for electron microscopy.

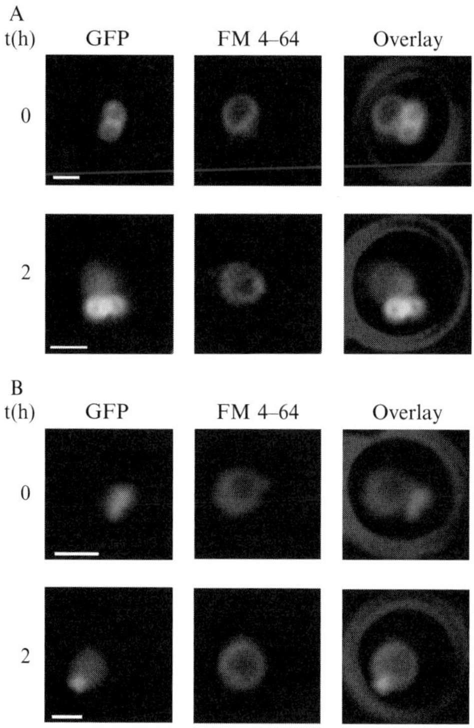

Figure 14.4 Fluorescence microscopy analysis of glucose-induced macropexophagy. (A) In wild-type *H. polymorpha* cells producing GFP.SKL peroxisomes are visualized by GFP fluorescence, the vacuoles by FM 4-64. At T = 0 the vacuoles lack GFP fluorescence, which is evident after 2 h of incubation in the presence of glucose, confirming that macropexophagy has occurred. (B) In *pex13* cells peroxisome remnants are marked by N50.Pex3.GFP. Also these structures are subject to degradation at macropexophagy conditions, as is demonstrated by the presence of GFP fluorescence in the vacuole after 2 h of incubation. The bar represents 1 μm. (See Color Insert.)

Of these, potassium permanganate fixation is suited to visualize overall cell morphology, in particular membranes (Fig. 14.5A), whereas aldehyde fixations are generally used for immunocytochemistry (Fig. 14.5B).

Protocols

$KMnO_4$ fixation

All incubations are performed at room temperature unless otherwise indicated.

1. Grow cells under pexophagy-inducing conditions and harvest a volume of the culture corresponding to 10–20 OD_{660} units.
2. Wash the cells three times with 5 ml of demineralized water by resuspending/centrifugation (5000×*g* 2 min, 10-ml tubes) and subsequently

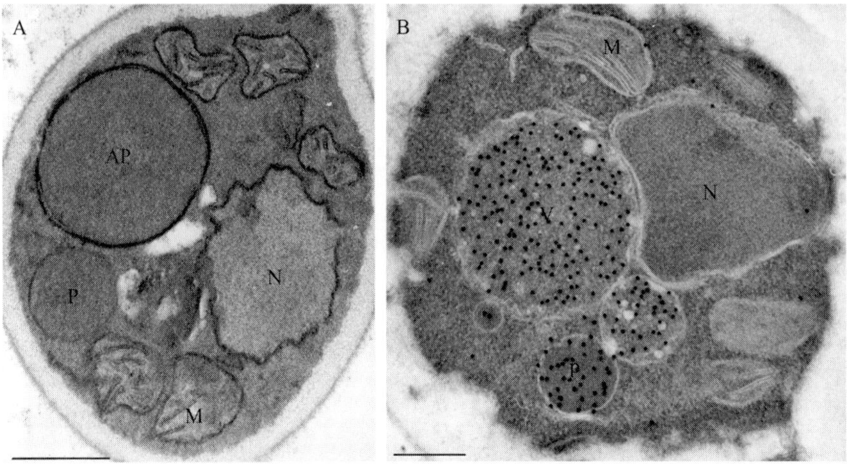

Figure 14.5 (A) Ultrathin section of a $KMnO_4$-fixed *H. polymorpha* cell, 30 min after glucose-induced macropexophagy showing sequestration of the large peroxisome (AP) in the cell leaving the smaller organelle (P) unaffected. (B) Ultrathin section of a glutaraldehyde-fixed *H. polymorpha* cell, labeled with an α-AO antiserum, showing labeling of a peroxisome (P) as well as autophagic vacuoles (V). Key: AP, autophagosome; P, peroxisome; N, nucleus; M, mitochondrion; V, vacuole. The bar represents 0.5 μm.

resuspend the pellet in 5 ml of $KMnO_4$ solution (1.5% $KMnO_4$ in water).

3. Incubate the cell suspension for 20 min at room temperature and shake gently every 5 min.
4. After incubation, collect the cells by centrifugation and wash with demineralized water until the supernatant is colorless (3 times with 5 ml of water each usually suffices).
5. Resuspend the pellet in 5 ml of uranyl acetate solution (0.5% in water) and centrifuge for 15 min at 5000×*g* to obtain a firm pellet. The supernatant should not be discarded but left on top of the pellet for at least 4 h or maximally overnight at room temperature.
6. Decant the uranyl acetate supernatant from the pellet and dehydrate the cells by incubating the pellet with solutions of increasing ethanol concentrations according to the following plan:
   - 15 min in 50% ethanol (without resuspending, the pellet remains intact).
   - 15 min in 70% ethanol (the pellet is broken into small pieces, approximately 1–5 mm^3, using a spatula. Incubation is performed without rotation; therefore, the pieces will be lying at the bottom of the tube and the solution can be poured of directly after incubation).
   - 15 min in 96% ethanol (mix carefully, do not use a vortex, and the small pieces should stay intact).

- 15 min in 100% ethanol (mix carefully).
- Refresh the 100% ethanol solution and incubate for another 30 min.

7. In the subsequent steps, the cells are impregnated with Epon resin. To prepare the Epon embedding resin, mix 100*g* of Epon 812 (glycid ether) with 92*g* of methylnadic anhydride (MNA), then add 2.3*g* of 2,4,6-tri (dimethylaminomethyl) phenol (DMP-30). During the incubations with Epon/ethanol mixtures or pure Epon solutions the tubes are continuously mixed using a slowly rotating incubator. Incubate the samples (i.e., pieces of cell material) with approximately 5 ml of each of the Epon/ethanol mixtures according to the following:
   - 4–8 h in a 1:1 mixture of 100% ethanol and Epon
   - Overnight in a 1:3 mixture of 100% ethanol and Epon
   - 1 h in pure Epon solution
   - Refresh the Epon solution and incubate for another 8 h
8. Fill gelatin capsules {3/4} full with pure Epon and load one piece of sample onto the top of the capsule; it will readily sink.
9. Polymerize the Epon by incubating the capsules for 24 h at 80 °C.
10. Prepare sections using a diamond knife and view the sections in a transmission electron microscope.

Aldehyde fixations

All steps are performed at 4 °C.

1. Harvest at least 20 OD_{660} units of a fresh culture by centrifugation (3 min at 5000×*g*).
2. Wash the cells 3 times with demineralized water and add 5 ml of either one of the fixation solutions:
   3% glutaraldehyde in 0.1 *M* Na-cacodylate, pH 7.2
   3% formaldehyde in 0.1 *M* Na-cacodylate, pH 7.2
   0.5% glutaraldehyde + 2.5% formaldehyde in 0.1 *M* Na-cacodylate, pH 7.2
   3% glutaraldehyde is the preferred fixative. However, glutaraldehyde may affect the antigenicity of specific proteins. In that case formaldehyde or a mixture of formaldehyde and glutaraldehyde can be used.
3. Carefully resuspend the cells (by manual shaking) in the fixative solution and incubate for 2 h on ice. Mix the suspension every 15 min by inversion.
4. Collect the cells by centrifugation (see previous section) and discard the supernatant fraction. Wash the pellet using fresh Na-cacodylate buffer keeping the pellet intact.
   All subsequent steps are performed at room temperature.
5. Wash the intact pellet of fixed yeast cells twice with demineralized water.

6. Add 5 ml of a freshly prepared solution of 0.4% (w/v) Na-periodate in water to the pellet (keep pellet intact).
7. After incubation for 15 min on a slowly rotating incubator, wash the pellet twice with 5 ml of demineralized water.
8. Incubate the pellet in 5 ml of a 1% (w/v) NH_4Cl solution in water for 15 min at room temperature.
9. Decant the NH_4Cl solution and wash the pellet once with demineralized water.
10. Dehydrate the cell material as indicated above for permanganate fixation. The samples are now ready for impregnation with Unicryl, the preferred plastic for immunocytochemical purposes.
11. Incubate the pieces successively as indicated in the following:
    - 3 h in 5 ml of a 1:1 solution of 100% ethanol and Unicryl
    - 1 h in 5 ml of pure Unicryl
    - Overnight in 5 ml of pure Unicryl
    - 6–8 h in 5 ml of pure Unicryl
12. Embed the material in BEEM capsules (Standard polyethylene embedding capsules size 00; 1×1-mm flat bottom) filled {3/4}full with Unicryl. Only use *carefully dried* capsules! (Dry overnight in 37 °C oven).
13. Polymerize the Unicryl for 2 days using ultraviolet light at 4 °C.
14. Fill the capsules completely with Unicryl and incubate for 2 more days at 30 °C.
15. Cut sections using a diamond knife and transfer sections onto nickel grids (400 mesh, Formvar/carbon coated).

Immunocytochemical staining methods are performed on ultrathin sections that are collected on Formvar/carbon coated nickel grids (do not use copper grids). The incubation steps are performed by floating the grids, section side down, on top of small droplets of the solution on a sheet of Parafilm. All steps are performed at room temperature unless stated otherwise.

16. Incubate the grids with the following solutions: 0.5% BSA in PBS-glycine buffer for 5 min as a blocking step (PBS/Glycine/BSA buffer = 2*g*/l sodium chloride, 0.05*g*/l potassium chloride, 0.36*g*/l disodium hydrogen phosphate, 0.055*g*/l sodium dihydrogen phosphate, 0.375*g*/l glycine, 0.025*g*/l sodium azide and 5*g*/l BSA).
    Transfer grids to a droplet of appropriately diluted primary antibody in PBS-glycine buffer containing 0.5% BSA and incubate for 1 h at room temperature (alternatively this step can be performed overnight at 4 °C). The appropriate dilution of the primary antibody is generally 10 times less than that used for Western blotting (e.g., 1:100 if a 1:1000 dilution is the optimal dilution for Western blotting).
17. Rinse the grids with PBS-glycine buffer, 3 times for 5 min each.
18. Incubate the grids with a solution of secondary antibodies conjugated to gold in PBS-glycine buffer containing 0.5% BSA (use dilution as

recommended by the manufacturer). Use the appropriate secondary antibodies (i.e., goat-antirabbit-(GAR)-gold when the primary antibodies were raised in goat or goat-antimouse-gold, when the primary antibodies were raised in mice).
19. Rinse the grids in PBS-glycine buffer, 6 times for 5 min each.
20. Rinse grids in distilled water, 4 times for 5 min each.
21. Remove excess water by carefully tipping one side of the grid (section side up) onto filter paper.
22. Poststain the sections by placing the grid (section side down) onto a droplet of 1% uranyl acetate and 0.2% methylcellulose for 20 s.
23. Remove excess staining solution using filter paper and allow the grid to dry.

5. Concluding Remarks

The success of electron microscopy (EM) studies, much more than biochemical approaches, depends on the availability of fast-growing cells that preferably are in the exponential growth phase. To avoid synthesis of storage products (e.g., glycogen) that strongly interfere with optimal cell architecture, we aim to avoid culturing cells at high concentrations of carbon sources (always <1% glucose). Similar arguments hold for fluorescence microscopy studies aiming at live cell imaging. That is, the cellular processes are optimally visualized when the cells have not been exposed to high concentrations of the carbon source.

It should also be noted that the resolution of fluorescence techniques is still far from the resolution of EM techniques. This may in specific cases lead to overinterpretation of the data. It is therefore recommended that, wherever possible, the two methods are used in parallel.

ACKNOWLEDGMENTS

TvZ is financially supported by a grant from the Netherlands Organisation for Scientific Research/Earth and Life Sciences (NWO/ALW). JAKWK is financially supported by the Netherlands Ministry of Economic Affairs and the B-Basic partner organizations (www.b-basic.nl) through B-Basic, a public-private NWO-ACTS program (ACTS = Advanced Chemical Technologies for Sustainability).

REFERENCES

Bellu, A. R., Komori, M., van der Klei, I. J., Kiel, J. A. K. W., and Veenhuis, M. (2001a). Peroxisome biogenesis and selective degradation converge at Pex14p. *J. Biol. Chem.* **276,** 44570–44574.

Bellu, A. R., Kram, A. M., Kiel, J. A. K. W., Veenhuis, M., and van der Klei, I. J. (2001b). Glucose-induced and nitrogen-starvation induced peroxisome degradation are distinct processes in *Hansenula polymorpha* that involve both common and unique genes. *FEMS Yeast Res.* **1,** 23–31.

Bellu, A. R., Salomons, F. A., Kiel, J. A. K. W., Veenhuis, M., and van der Klei, I. J. (2002). Removal of Pex3p is an important initial stage in selective peroxisome degradation in *Hansenula polymorpha. J. Biol. Chem.* **277,** 42875–42880.

Bruinenberg, P. G., Veenhuis, M., van Dijken, J. P., Duine, J. A., and Harder, W. (1982). A quantitative analysis of selective inactivation of peroxisomal enzymes in the yeast *Hansenula polymorpha* by high-performance liquid chromatography. *FEMS Microbiol. Lett.* **15,** 45–50.

Ciechanover, A. (2006). The ubiquitin proteolytic system: From a vague idea, through basic mechanisms, and onto human diseases and drug targeting. *Neurology* **66,** 7–19.

De Duve, C. (1963). *CIBA Found Symp Lysomomes.* 36.

De Duve, C. (1963). The lysosome concept. *In*: de Reuck, A. V. S. and Cameron, M. P. (eds.). Ciba Foundation Symposium on Lysosomes. J. & A. Churchill, Ltd., London.

Farre, J. C., Manjithaya, R., Mathewson, R. D., and Subramani, S. (2008). PpAtg30 tags peroxisomes for turnover by selective autophagy. *Dev. Cell* **14,** 365–376.

Gleeson, M. A. G., and Sudbery, P. E. (1988). Genetic analysis in the methylotrophic yeast *Hansenula polymorpha. Yeast* **4,** 293–303.

Kiel, J. A. K. W., Rechinger, K. B., van der Klei, I. J., Salomons, F. A., Titorenko, V. I., and Veenhuis, M. (1999). The *Hansenula polymorpha PDD1* gene product, essential for the selective degradation of peroxisomes, is a homologue of *Saccharomyces cerevisiae* Vps34p. *Yeast* **15,** 741–754.

Kiel, J. A. K. W., Komduur, J. A., van der Klei, I. J., and Veenhuis, M. (2003). Macropexophagy in *Hansenula polymorpha*: Facts and views. *FEBS Lett.* **549,** 1–6.

Klionsky, D. J., Cregg, J. M., Dunn, W. A., Jr., Emr, S. D., Sakai, Y., Sandoval, I. V., Sibirny, A., Subramani, S., Thumm, M., Veenhuis, M., and Ohsumi, Y. (2003). A unified nomenclature for yeast autophagy-related genes. *Dev. Cell* **5,** 539–545.

Klionsky, D. J. (2007). Autophagy: From phenomenology to molecular understanding in less than a decade. *Nat. Rev. Mol. Cell Biol.* **8,** 931–937.

Klionsky, D. J., Cuervo, A. M., Dunn, W. A., Jr., Levine, B., van der Klei, I., and Seglen, P. O. (2007). How shall I eat thee? *Autophagy* **3,** 413–416.

Komduur, J. A. (2004). Molecular aspects of peroxisome degradation in *Hansenula polymorpha.* Ph.D. dissertation, University of Groningen, the Netherlands.

Leighton, F., Coloma, L., and Koenig, C. (1975). Structure, composition, physical properties, and turnover of proliferated peroxisomes: A study of the trophic effects of Su-13437 on rat liver. *J. Cell Biol.* **67,** 281–309.

Meijer, W. H., van der Klei, I. J., Veenhuis, M., and Kiel, J. A. K. W. (2007). *ATG* genes involved in non-selective autophagy are conserved from yeast to man, but the selective Cvt and pexophagy pathways also require organism-specific genes. *Autophagy* **3,** 106–116.

Mizushima, N., Levine, B., Cuervo, A. M., and Klionsky, D. J. (2008). Autophagy fights disease through cellular self-digestion. *Nature* **451,** 1069–1075.

Sakai, Y., Oku, M., van der Klei, I. J., and Kiel, J. A. K. W. (2006). Pexophagy: Autophagic degradation of peroxisomes. *Biochim. Biophys. Acta* **1763,** 1767–1775.

Shintani, T., and Klionsky, D. J. (2004). Cargo proteins facilitate the formation of transport vesicles in the cytoplasm to vacuole targeting pathway. *J. Biol. Chem.* **279,** 29889–29894.

Van der Klei, I. J., and Veenhuis, M. (2007). Protein targeting to yeast peroxisomes. *Methods Mol. Biol.* **390,** 373–392.

Van der Klei, I. J., Yorimoto, H., Sakai, Y., and Veenhuis, M. (2006). The significance of peroxisomes in methanol metabolism in methylotrophic yeast. *Biochim. Biophys. Acta* **1763,** 1453–1462.

Van Dijken, J. P., Otto, R., and Harder, W. (1976). Growth of *Hansenula polymorpha* in a methanol-limited chemostat: Physiological responses due to the involvement of methanol oxidase as a key enzyme in methanol metabolism. *Arch. Microbiol.* **111,** 137–144.

Van Zutphen, T., Veenhuis, M., and van der Klei, I. J. (2008). Pex14 is the sole component of the peroxisomal translocon that is required for pexophagy. *Autophagy* **4,** 63–66.

Veenhuis, M., Komori, M., Salomons, F., Hilbrands, R. E., Hut, H., Baerends, R. J. S., Kiel, J. A. K. W., and van der Klei, I. J. (1996). Peroxisome remnants in peroxisome deficient mutants of the yeast *Hansenula polymorpha. FEBS Lett.* **383,** 114–118.

Verduyn, C., van Dijken, J. P., and Scheffers, W. A. (1984). Colorimetric alcohol assays with alcohol oxidase. *J. Microbiol. Methods* **2,** 15–25.

Yorimitsu, T., and Klionsky, D. J. (2005). Atg11 links cargo to the vesicle-forming machinery in the cytoplasm to vacuole targeting pathway. *Mol. Biol. Cell.* **16,** 1593–1605.

CHAPTER FIFTEEN

Pexophagy in *Pichia pastoris*

Masahide Oku[*,†] *and* Yasuyoshi Sakai[*,†]

Contents

Abstract

The peroxisome is an organelle whose quantity is tightly regulated in response to changes in metabolic status, and much knowledge has been accumulated regarding its dynamics. The turnover of peroxisomes through autophagic pathways, termed *pexophagy*, has been especially studied in several methylotrophic yeast strains capable of growth on methanol as a sole carbon source, which led to the identification of factors involved in pexophagy (Dunn *et al.*, 2005; Sakai *et al.*, 2006). In the methylotrophic yeast *Pichia pastoris*, several types of membrane dynamics during pexophagy can be visualized simultaneously under live cell imaging. The decrease of abundant peroxisomal proteins in the cell lysate can be used as a convenient indicator of the completion of pexophagy. In combination, these methods provide basic information for further analysis of pexophagy at the molecular level.

* Division of Applied Life Sciences, Graduate School of Agriculture, Kyoto University, Kyoto, Japan
† CREST, Japan Science and Technology Agency, Japan

Methods in Enzymology, Volume 451
ISSN 0076-6879, DOI: 10.1016/S0076-6879(08)03215-1

1. Introduction: Use of *Pichia pastoris* as a Model Organism to Study Pexophagy

The yeast *Pichia pastoris* is now regarded as a powerful host organism for heterologous protein expression (Cregg, 2007). Gene manipulation of *P. pastoris* has been well established thanks to auxotrophic strains and various plasmids that complement the auxotrophies or confer resistance to certain antibiotics (e.g., Zeocin). In this part of the chapter, we describe several methods that we have carried out using strains derived from one parental auxotrophic strain, PPY12 (*arg4*$^-$, *his4*$^-$)(Gould *et al.*, 1992). All of the methods are applicable to other *P. pastoris* strains, such as GS or JC strains generated by Dr. J. Cregg's group (Keck Graduate Institute).

Because *P. pastoris* massively develops its peroxisomes during growth on methanol, it has made a large contribution as a model organism in revealing the molecular mechanism of peroxisome assembly (van der Klei *et al.*, 2006). In methanol-grown cells, the augmented peroxisomes are clustered and can be readily observed by light microscopy, and a spherical vacuole is also prominent.

Pexophagy is induced when the methanol-grown cells are shifted to media composed of another carbon source whose metabolism does not require peroxisomal enzymes. One remarkable feature of pexophagy in this organism is that two distinct modes of pexophagy are observed depending on the kind or amount of the carbon source in the media (Tuttle and Dunn, 1995) (Fig. 15.1). The switch between the two modes of pexophagy is found to have a close relationship to the intra-cellular ATP level (Ano *et al.*, 2005a), and experimentally we are able to control the mode by choosing ethanol or glucose as the carbon source in the media (Tuttle and Dunn, 1995). A detailed method of the culture conditions is described in the following section.

In one mode of pexophagy, a double-membrane structure that is formed *de novo* in the cytoplasm fully sequesters the targeted peroxisomes one by one, and then its outer membrane fuses with the vacuolar membrane transporting the sequestered peroxisome into the vacuolar lumen (Tuttle and Dunn, 1995) (see Fig. 15.1). Because this process resembles starvation-induced macroautophagy in that a double-membrane structure (similar to an autophagosome) is formed in the cytoplasm, this pathway is termed *macropexophagy* and the double–membrane structure is named a *pexophagosome*. Analogous to the localization of Atg8 to the autophagosome during macroautophagy, PpAtg8, the ortholog of the *S. cerevisiae* Atg8 protein, localizes to the pexophagosome (Mukaiyama *et al.*, 2004).

In the other mode, referred to as *micropexophagy*, the peroxisome cluster as a whole is engulfed by a vacuolar membrane and delivered into the vacuole (Farre and Subramani, 2004; Tuttle and Dunn, 1995)

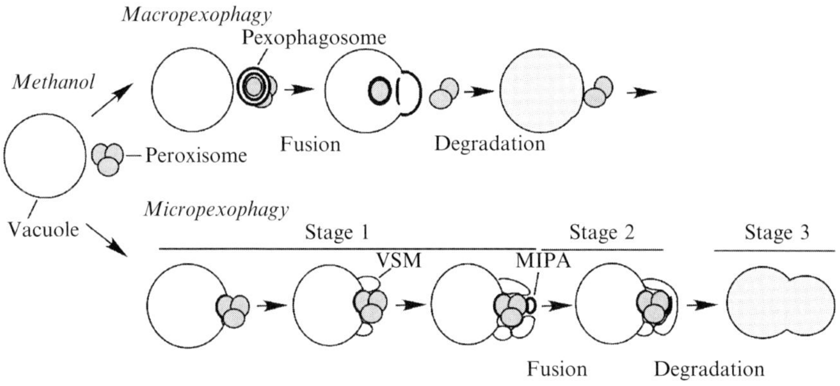

Figure 15.1 Pexophagy in *P. pastoris* has two distinct types, both of which contain formation, fusion and degradation steps of the membrane structures. When methanol-cultured cells (methanol) are transferred to ethanol medium (macropexophagy), the double-membrane structure termed a *pexophagosome* is formed surrounding an individual peroxisome. After completing the sequestration of the peroxisome, the pexophagosome fuses with the vacuolar membrane, releasing the sequestered peroxisome into the vacuolar lumen. The fusion was found to occur at the vertex ring (the edge of the contact plane between the outer membrane of pexophagosome and vacuolar membrane), as judged by a fraction of the vacuolar membrane found in the vacuolar lumen at this stage (Ano *et al.*, 2005b). The sequestered peroxisomes are then lysed in the vacuole. When the cells are transferred to glucose medium (micropexophagy), a portion of the vacuolar membrane extends, termed *vacuolar sequestering membrane* (VSM), along the cluster of the peroxisomes. At the same stage of the vacuolar-membrane extension (stage 1), a flattened membrane structure, termed the *micropexophagy-specific membrane apparatus* (MIPA), forms on the surface of the peroxisome cluster. Both VSM and MIPA participate in the subsequent fusion event that leads to the sequestration (stage 2) and subsequent lysis of the peroxisome cluster (stage 3).

(see Fig. 15.1). The vacuolar-membrane dynamics leading to peroxisome degradation is a complicated process that we previously dissected into three stages (Mukaiyama *et al.*, 2002; Sakai *et al.*, 1998). In stage 1, the vacuole forms an armlike extension (vacuolar sequestering membrane; Fry *et al.*, 2006; Oku *et al.*, 2006) to engulf the peroxisome cluster. The sequestering membrane often comprises several septated parts, as illustrated in Fig. 15.1. After the vacuole completely sequesters the peroxisomes through membrane fusion (stage 2), the membrane portion surrounding the peroxisomes is degraded, releasing the peroxisomes into the vacuolar lumen (stage 3). The peroxisomes are subsequently degraded and the resulting macromolecules are released back into the cytosol for reuse. Fluorescence staining of the vacuolar membrane with a styryl dye, FM 4-64 (Vida and Emr, 1995), in combination with the expression of a peroxisome-targeted GFP chimera, enables live cell imaging of the vacuolar membrane dynamics. This method is described in section 4.1.

The *de novo* formation of a membrane structure also occurs during micropexophagy (Mukaiyama *et al.*, 2004). The formed membrane structure is cup-shaped and associated with a part of the peroxisome cluster (see Fig. 15.1). This peculiar membrane structure is termed MIPA (micropexophagy-specific membrane apparatus). Because PpAtg8 is localized to the MIPA, visualization of PpAtg8 tagged with a fluorescent protein is a useful technique offering detection of not only the pexophagosome but also the MIPA. Section 4.2 describes details of the microscopy observation.

Transport of the peroxisomes to the vacuole leads to degradation of proteins or lipids resident in the target organelle. Meanwhile, the expression of many peroxisomal proteins is induced only in methanol culture and repressed under pexophagy conditions. Thus, measuring the intracellular amount of a peroxisomal protein allows a valid type of chase experiment to monitor the overall progression of pexophagy. Along these lines, the amount of the peroxisomal enzyme alcohol oxidase (Aox) is regarded as the index of pexophagy progression in most studies (Tuttle *et al.*, 1993). Determination of Aox activity in the cell lysate is one method for monitoring pexophagy, which is described elsewhere (See chapter 17). An alternative immunoblot analysis is described in section 3.

2. Culture Methods to Induce Micropexophagy or Macropexophagy in *P. pastoris*

To obtain fresh culture, the cells are cultured to an early-log phase in YPD (1% w/v Bacto Yeast Extract, BD Biosciences; 2% w/v Bacto Peptone, BD Biosciences; and 2% w/v glucose) liquid medium at 28 °C for more than 6 h, harvested by centrifugation (2000 *g*, 5 min) at room temperature, removing the residual YPD medium, and resuspended in YNM medium (0.67% w/v Difco Yeast Nitrogen Base without amino acids, BD Biosciences; 0.05% w/v Bacto Yeast Extract, and 0.5% v/v methanol] at an OD_{600} [optical density] of 0.5 to 0.6). Yeast extract is contained in the YNM medium to provide efficient uptake of FM 4-64 dye (described subsequently) by the cells. To the cultures of the auxotrophic strains, 100 mg/L arginine and/or 100 mg/L histidine is added, depending on the auxotrophies. Additionally, 0.4 mg/L of biotin can be added to the medium to accelerate cell growth.

The cells are cultured in YNM medium for more than 12 h at 28 °C. This culture requires sufficient aeration for the cells to oxidize methanol. For this purpose it is recommended to use a rotary shaker at no less than 180 rpm and an Erlenmeyer flask with at least 10-fold volume relative to the culture. Under suitable conditions, the wild-type *P. pastoris* strains are grown to more than 2 OD_{600} in 15 h.

To induce macropexophagy, the YNM-cultured cells are harvested by centrifugation (2000 *g*, 5 min) and resuspended in SE (synthetic ethanol) medium (0.67% w/v Difco Yeast Nitrogen Base without amino acids, 100 mg/L of the auxotrophic amino acids, and 0.5% v/v ethanol) at an OD_{600} of 1.0. For micropexophagy induction the culture is harvested and resuspended in SD (synthetic dextrose) medium (0.67% w/v Difco Yeast Nitrogen Base without amino acids, 100 mg/L of the auxotrophic amino acids, and 2% w/v glucose) at an OD_{600} of 1.0. Special care should be taken not to cool the culture in carrying out the medium transfer; the media should be prewarmed to 28 °C. Failure in temperature control decreases the frequency of cells undergoing micropexophagy.

3. Immunoblot Analysis to Monitor Pexophagy Progression

Degradation of the peroxisomes can be analyzed by a chase experiment by following the level of peroxisomal Aox protein in the cell lysates harvested at different time points after induction of pexophagy. A standard method for this detection is described subsequently. Because the cells under pexophagy-inducing conditions propagate unaccompanied by the synthesis of Aox, the relative concentration of Aox in the lysate rapidly decreases even when pexophagy is inhibited due to cell division and dilution. Thus, it is strongly recommended to prepare lysates from either the wild-type or a certain pexophagy-deficient strain when a strain of interest is tested for pexophagy competence. Comparison of the kinetics of the Aox decrease among these strains facilitates data interpretation. Alternatively, the protein amount subjected to the immunoblot analysis can be adjusted in proportion to the OD_{600} value at each time point of sampling. It is notable that the degradation kinetics of Aox detected by the immunoblot analysis is slower than the rapid transport of the peroxisomes observed by fluorescence microscopy (see the subsequent section). This discrepancy is thought to arise from the physical properties of Aox, which forms a crystal structure inside the peroxisomes and thus may be rather resistant to proteolytic activities in the vacuole.

In general autophagy systems, measurement of the amount of Atg8/LC3, or its lipidated form can be used as an index of autophagy induction. However, this methodology cannot be applied to the analyses of pexophagy in *P. pastoris*, because PpAtg8 expression as well as its conversion to the lipidated form is induced at the stage of methanol culture (before the pexophagy-inducing conditions). This may reflect the induction of some autophagic activities during the growth on methanol (Mukaiyama *et al.*, 2004).

Immunoblot analysis of Aox

1. Culture the cells as described in section 2.
2. At several time points (0–12 h after the carbon source shift from YNM to SD or SE medium) retrieve 10 OD_{600} units of cells by centrifugation at 2000*g* for 5 min. The retrieved samples should be kept on ice until cell breakage.
3. Resuspend the cells with 500 μL of potassium phosphate buffer (0.1 *M*, pH 7.5) that contains 1 m*M* PMSF and 12.5 μg/mL of leupeptin in a 1.7-ml microcentrifuge tube.
4. Add 500 μL of glass beads and break the cells by 6-time repetition of vigorous mixing at 4 °C for 30 s and cooling for 1 min. It is important to keep the level of glass beads below the level of the liquid to prevent excessive bubble formation.
5. Remove the cell debris by centrifugation (17,000*g* for 5 min) at 4 °C. Retrieve the supernatant and determine the protein concentration of the fraction (lysate).
6. Use 5 μg equivalent of the lysate at time 0 (methanol culture) for the immunoblot analysis. The protein amount of the samples acquired after pexophagy induction is adjusted to Y × 5 μg, where the factor Y represents the fold of the increase in the OD_{600} value (e.g., if the OD_{600} increases by a factor of 2, use 10 μg). We use 1:10,000 dilution of rabbit anti-Aox antiserum (a gift from Dr. J. M. Goodman, University of Texas South Western Medical Center) as a primary antibody and incubate it for 1 h at room temperature, and then incubate the membrane with 1:10,000 dilution of antirabbit IgG horseradish peroxidase conjugate (GE Healthcare Life Sciences) for 1 h at room temperature before detection of the Aox signal with the ECL-detection kit (GE Healthcare Life Sciences). Fig. 15.2 shows the results of the immunoblot analysis.

4. Microscopy Methods to Follow Pexophagy in *P. pastoris*

4.1. Fluorescence microscopy of the peroxisomes and the vacuoles

Pioneer studies on protein import to the peroxisomes identified several peroxisome-targeting signals (PTSs) on the cargo proteins. Tagged with one of the signals (PTS1) consisting of only 3 amino acid residues (SKL) at the C terminus, GFP-PTS1 expressed under the *AOX1* promoter is localized to the peroxisomal lumen and becomes a fluorescent marker for this organelle. One plasmid for the GFP-PTS1 expression (named *pTW51*, a gift

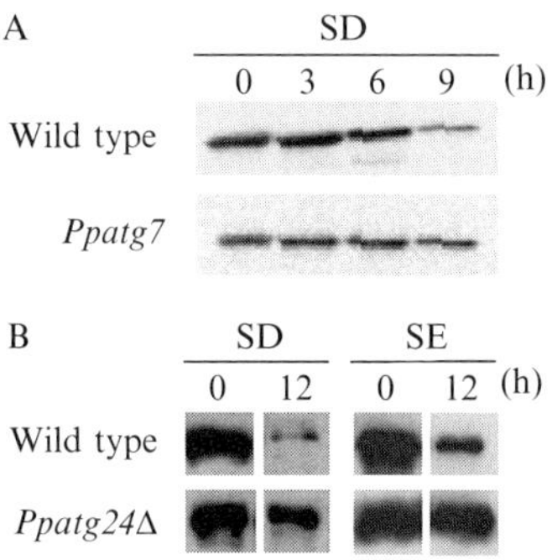

Figure 15.2 Aox-degradation detected by immunoblot analysis. (A) A wild-type strain (PPY12) and its derivative mutant strain of *PpATG7* (strain PZR12, shown as *Ppatg7*; Oku *et al.*, 2003) were methanol-cultured and transferred to glucose medium (SD) for the indicated hours. Lysate fractions were prepared and subjected to immunoblot analysis according to the protocol in section 3. (B) The wild-type (PPY12) and a *Ppatg24Δ* (YAP2401; Ano *et al.*, 2005b) strains were methanol-cultured and transferred to either glucose (SD) or ethanol medium (SE) for the indicated hours, and the lysate samples of the same protein amount were subjected to the immunoblot analysis. Note that the Aox signal decreased even in the samples from the mutant strain.

from Dr. S. Subramani, University of California, San Diego) harbors *PpHIS4*, which is integrated into the chromosome of a *P. pastoris his4*$^-$ strain (Sakai *et al.*, 1998).

FM 4-64 (Invitrogen) is used for fluorescent staining of the vacuolar membrane in a living cell (also see chapter 7). The excitation/emission spectrum of this dye (515/640 nm) is compatible with the fluorescence labeling of other organelles or proteins with BFP, CFP, GFP, or YFP. Because the dye is also fluorescent while being transported through endocytic organelles before reaching the vacuole, exclusive labeling of the vacuolar membrane requires a chase period after incorporation of the dye. For this reason we apply the dye at the beginning of the YNM culture (>12 h before the observation of pexophagy).

Here we describe a standard method for visualizing the dynamics of GFP-labeled peroxisomes and FM 4-64-stained vacuoles in strain STW1 (PPY12, *his4*::pTW51 [*HIS4*]; Sakai *et al.*, 1998). The result of fluorescence microscopy is shown in Fig. 15.3. If the gene product of interest is necessary in any of the steps of the pexophagy process, the mutant will exhibit a pattern of the vacuole/peroxisome blocked at the corresponding stage where the protein normally functions as shown in Fig. 15.1, not undergoing steps further ahead. For instance, in a pexophagy-deficient strain (*Ppatg7*) the vacuolar membrane deeply engulfs the peroxisome after being transferred to the glucose medium, but the diffusion of the GFP signal inside the vacuole is not observed as shown in Fig. 15.3C (Mukaiyama *et al.*, 2002).

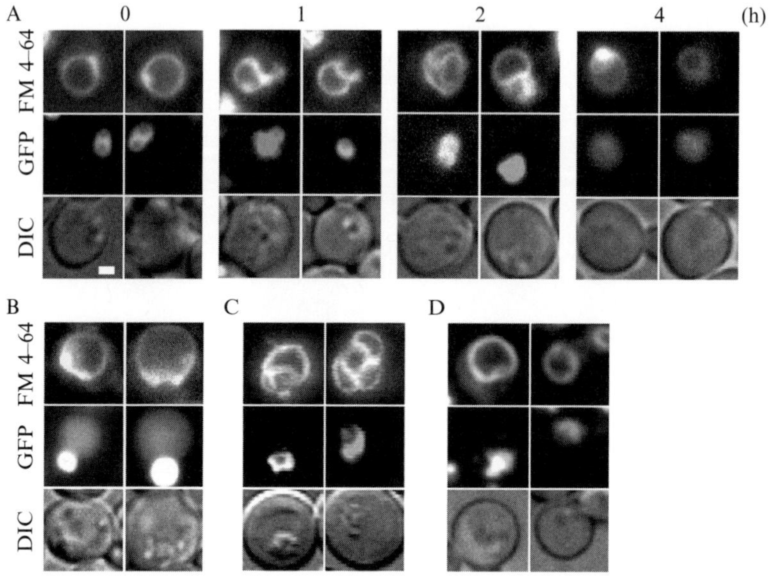

Figure 15.3 Morphological changes of the vacuolar membrane and peroxisomes observed under fluorescence microscopy. (A) A wild-type strain (STW1) harboring peroxisome-targeted GFP was cultured on methanol, labeled with FM 4-64 and transferred to glucose (SD) medium for the indicated hours. FM 4-64, the peroxisome-targeted GFP, and differential interference contrast (DIC) images from two representative cells undergoing morphological changes are shown in each panel. Bar, 1 μm. Note that after completion of peroxisome degradation, the GFP signal is seen as a diffuse pattern inside the vacuole owing to the stability of the GFP moiety. (B) The methanol-cultured STW1 (wild-type) strain labeled with FM 4–64 was transferred to ethanol (SE) medium for 4 h, and the images are shown as in (A). Diffusion of the GFP signal inside the vacuole is detected along with an intense signal outside the vacuole, showing partial sequestration of the peroxisomes. (C) A STW1-derivative strain mutated in *PpATG7* (strain PZR12; Oku *et al.*, 2003) was cultured in methanol and transferred to glucose as in (A) and subjected to fluorescence microscopy after a 4-h incubation in SD medium. (D) The images of the strain PZR12 cultured in methanol and transferred to ethanol medium as described in (B) are shown.

Diffusion of the GFP signal inside the vacuole is also a useful (but qualitative) index for judging the macropexophagy-competence of a strain, as shown in Figure 15.3 (Sakai *et al.*, 1998).

Fluorescence microscopy

1. Culture the cells in 5 mL of YPD medium to OD_{600} of 2, and resuspend in YNM medium as described in section 2.
2. Add one-two hundredth volume of FM 4-64 stock solution (16 m*M* in DMSO) to the culture medium. (The concentration of the dye can be reduced to 0.93 μM depending on the signal-detection sensitivity of the microscopy equipment.)

3. Culture the cells in YNM medium (10 mL) as described in section 2. Retrieve approximately 500 μl of the culture, as a time 0 sample, and keep it on ice until microscopy. Centrifuge the rest of the culture and resuspend it in either SD or SE medium as described in section 2.
4. Harvest 500-μl aliquots of the culture with a time interval of 15–30 min for 3 h. Concentrate the cells by brief centrifugation (2000*g*) at 4 °C, and keep them on ice until microscopy.
5. Observe the samples under fluorescence microscopy. The acquisition of the data should be done by first using a filter set for the FM 4-64 signal (e.g., rhodamine filter set) and then using a filter set for GFP: a reverse use of the filter sets may cause fluorescent peroxisomes even with the filter set for FM 4-64.

4.2. Detection of membrane structures for pexophagy using PpAtg8 tagged with fluorescent proteins

PpAtg8 shows 78% amino-acid sequence identity to *Saccharomyces cerevisiae* Atg8 and was found to be required for both modes of pexophagy of this organism. This protein is localized to the newly formed membrane structures (pexophagosome or MIPA; Ano *et al.*, 2005b; Mukaiyama *et al.*, 2004), similarly to Atg8 localization to autophagosomes during starvation-induced macroautophagy (Kirisako *et al.*, 1999). The expression level of PpAtg8 seems tightly regulated depending on culture conditions: over-expression of PpAtg8 impairs the pexophagy competence (our unpublished data). Thus, we generated a PpAtg8 derivative protein fused with a fluorescent protein expressed under the control of the endogenous *ATG8* promoter. We constructed a plasmid that harbors a DNA fragment encoding YFP-PpAtg8 flanked by the 5′- and 3′-UTR region of *PpATG8*. The transformants of this plasmid (pSAP115 possessing *PpARG4*) are suitable for the chase analysis of membrane dynamics (Mukaiyama *et al.*, 2004). As shown in Fig. 15.4, the same culture and microscopy method as described in section 4.1 enables visualization of YFP-PpAtg8 and vacuolar-membrane dynamics in this strain.

Analysis by fluorescence microscopy facilitates separation of the membrane-formation process into two steps: nucleation and elongation. Nucleation means the concentration of multiple factors for MIPA formation at a dot; a similar phenomenon is reported as for the so-called PAS (preautophagosomal structure) in the Cvt pathway or macroautophagy of *S. cerevisiae* (Noda *et al.*, 2002). *Pp*Atg8 itself is recruited to the concentration site depending on the factors of a ubiquitin-like conjugation system (Ichimura *et al.*, 2000). For instance, mutation of PpAtg7, the E1-like factor in the conjugation system (Yuan *et al.*, 1999), abolishes the YFP-PpAtg8 concentration as shown in Fig. 15.4 (Mukaiyama *et al.*, 2004).

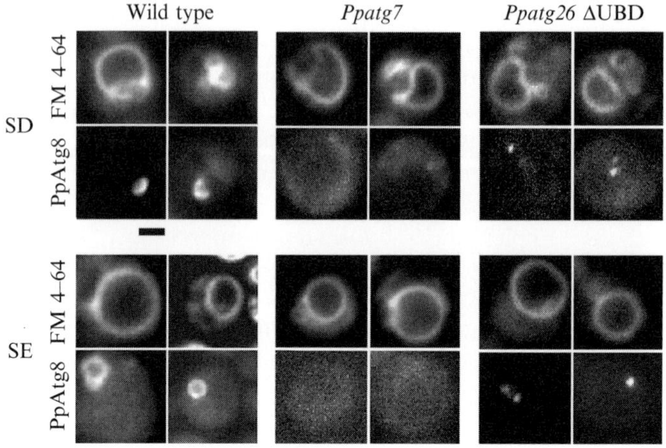

Figure 15.4 *De novo* formation of the membrane structure visualized by fluorescence microscopy. A wild-type strain SA1017 harboring YFP-PpAtg8 (Mukaiyama *et al.*, 2004), along with its derivative mutant strains of *PpATG7* (*Ppatg7*) and *PpATG26* (*Ppatg26*ΔUBD), were cultured on methanol, labeled with FM 4-64, and transferred to glucose for 30 min (SD, upper panels) or ethanol for 1 h (SE, lower panels) as described in section 4.1. The FM 4-64 and YFP signals from two representative cells are shown. Bar, 1 μm.

Elongation of the membrane structure is subsequently observed, giving rise to the cuplike pattern of *Pp*Atg8, a representative morphology of the MIPA. Our studies identified PpAtg26, a sterol glucosyltransferase, as a specific factor for the elongation step (Oku *et al.*, 2003; Yamashita *et al.*, 2006). In the cells harboring PpAtg26 without its enzymatic activity (*Ppatg26*ΔUBD), the YFP-PpAtg8 signal exhibited a dot pattern, which did not elongate to the cup-shaped MIPA or the pexophagosome, as shown in Fig. 15.4. These gene disruptants will contribute to the molecular analysis of specific steps in membrane-structure formation during pexophagy.

5. Concluding Remarks

The pexophagy processes include several types of membrane dynamics: vacuolar-membrane deformation and *de novo* synthesis of the two membrane structures (the pexophagosome and the MIPA). All these dynamics can be examined by chase experiments using fluorescence microscopy, which highlights their intermediate steps. Once the immunoblot analysis of Aox for a certain mutant strain indicates involvement of the mutated gene in pexophagy, the subsequent fluorescence microscopy specifies the step for which the gene product is required. This information is particularly useful for surveying the physical and functional interactions of the analyzed proteins, because there is a substantial possibility that the molecules acting on the same step may exert their functions as a molecular complex. A better understanding of the pexophagy mechanism will rely on the finding of the molecular links required for each step of the membrane dynamics.

ACKNOWLEDGMENTS

We thank Dr. S. Subramani for the gifts of the PPY12 strain and the pTW51 plasmid. We also thank Dr. J. M. Goodman for providing the antiserum for Aox. This work is supported by (1) Grant-in-Aid for Scientific Research on Priority Areas 18076002 of the Ministry of Education, Culture, Sports, Science and Technology, Japan and (2) CREST, Japan Science and Technology Agency.

REFERENCES

Ano, Y., Hattori, T., Kato, N., and Sakai, Y (2005a). Intracellular ATP correlates with mode of pexophagy in *Pichia pastoris*. *Biosci. Biotechnol. Biochem.* **69,** 1527–1533.

Ano, Y., Hattori, T., Oku, M., Mukaiyama, H., Baba, M., Ohsumi, Y., Kato, N., and Sakai, Y. (2005b). A sorting nexin PpAtg24 regulates vacuolar membrane dynamics during pexophagy via binding to phosphatidylinositol-3-phosphate. *Mol. Biol. Cell* **16,** 446–457.

Cregg, J. M. (2007). Introduction: Distinctions between *Pichia pastoris* and other expression systems. *Methods Mol. Biol.* **389,** 1–10.

Dunn, W. A., Jr., Cregg, J. M., Kiel, J. A. K. W., van der Kiel, I. J., Oku, M., Sakai, Y., Sibirny, A., Stasyk, O. V., and Veenhuis, M. (2005). Pexophagy: The selective autophagy of peroxisomes. *Autophagy* **1,** 75–83.

Farre, J. C., and Subramani, S. (2004). Peroxisome turnover by micropexophagy: An autophagy-related process. *Trends Cell Biol.* **14,** 515–523.

Fry, M. R., Thomson, J. M., Tomasini, A. J., and Dunn, W. A., Jr. (2006). Early and late molecular events of glucose-induced pexophagy in *Pichia pastoris* require Vac8. *Autophagy* **2,** 280–288.

Gould, S. J., McCollum, D., Spong, A. P., Heyman, J. A., and Subramani, S. (1992). Development of the yeast *Pichia pastoris* as a model organism for a genetic and molecular analysis of peroxisome assembly. *Yeast* **8,** 613–628.

Ichimura, Y., Kirisako, T., Takao, T., Satomi, Y., Shimonishi, Y., Ishihara, N., Mizushima, N., Tanida, I., Kominami, E., Ohsumi, M., Noda, T., and Ohsumi, Y. (2000). A ubiquitin-like system mediates protein lipidation. *Nature* **408,** 488–492.

Kirisako, T., Baba, M., Ishihara, N., Miyazawa, K., Ohsumi, M., Yoshimori, T., Noda, T., and Ohsumi, Y. (1999). Formation process of autophagosome is traced with Apg8/Aut7p in yeast. *J. Cell Biol.* **147,** 435–446.

Mukaiyama, H., Baba, M., Osumi, M., Aoyagi, S., Kato, N., Ohsumi, Y., and Sakai, Y. (2004). Modification of a ubiquitin-like protein Paz2 conducted micropexophagy through formation of a novel membrane structure. *Mol. Biol. Cell* **15,** 58–70.

Mukaiyama, H., Oku, M., Baba, M., Samizo, T., Hammond, A. T., Glick, B. S., Kato, N., and Sakai, Y. (2002). Paz2 and 13 other PAZ gene products regulate vacuolar engulfment of peroxisomes during micropexophagy. *Genes Cells* **7,** 75–90.

Noda, T., Suzuki, K., and Ohsumi, Y. (2002). Yeast autophagosomes: *De novo* formation of a membrane structure. *Trends Cell Biol.* **12,** 231–235.

Oku, M., Nishimura, T., Hattori, T., Ano, Y., Yamashita, S., and Sakai, Y. (2006). Role of Vac8 in formation of the vacuolar sequestering membrane during micropexophagy. *Autophagy* **2,** 272–279.

Oku, M., Warnecke, D., Noda, T., Muller, F., Heinz, E., Mukaiyama, H., Kato, N., and Sakai, Y. (2003). Peroxisome degradation requires catalytically active sterol glucosyltransferase with a GRAM domain. *EMBO J.* **22,** 3231–3241.

Sakai, Y., Koller, A., Rangell, L. K., Keller, G. A., and Subramani, S. (1998). Peroxisome degradation by microautophagy in *Pichia pastoris*: Identification of specific steps and morphological intermediates. *J. Cell Biol.* **141,** 625–636.

Sakai, Y., Oku, M., van der Klei, I. J., and Kiel, J. A. (2006). Pexophagy: Autophagic degradation of peroxisomes. *Biochim. Biophys. Acta* **1763,** 1767–1775.

Tuttle, D. L., and Dunn, W. A., Jr. (1995). Divergent modes of autophagy in the methylotrophic yeast *Pichia pastoris*. *J. Cell Sci.* **108** (Pt 1), 25–35.

Tuttle, D. L., Lewin, A. S., and Dunn, W. A., Jr. (1993). Selective autophagy of peroxisomes in methylotrophic yeasts. *Eur. J. Cell Biol.* **60,** 283–290.

van der Klei, I. J., Yurimoto, H., Sakai, Y., and Veenhuis, M. (2006). The significance of peroxisomes in methanol metabolism in methylotrophic yeast. *Biochim. Biophys. Acta* **1763,** 1453–1462.

Vida, T. A., and Emr, S. D. (1995). A new vital stain for visualizing vacuolar membrane dynamics and endocytosis in yeast. *J. Cell Biol.* **128,** 779–792.

Yamashita, S., Oku, M., Wasada, Y., Ano, Y., and Sakai, Y (2006). PI4P-signaling pathway for the synthesis of a nascent membrane structure in selective autophagy. *J. Cell Biol.* **173,** 709–717.

Yuan, W., Stromhaug, P. E., and Dunn, W. A., Jr. (1999). Glucose-induced autophagy of peroxisomes in *Pichia pastoris* requires a unique E1-like protein. *Mol. Biol. Cell* **10,** 1353–1366.

CHAPTER SIXTEEN

Methods of Plate Pexophagy Monitoring and Positive Selection for *ATG* Gene Cloning in Yeasts

Oleh V. Stasyk,* Taras Y. Nazarko,*,† *and* Andriy A. Sibirny*,‡

Contents

Abstract

Methods for colony assay of peroxisomal oxidases in yeasts provide a convenient and fast approach for monitoring peroxisome status. They have been used in several laboratories for the isolation of yeast mutants deficient in selective autophagic peroxisome degradation (pexophagy), catabolite repression of peroxisomal enzymes or mutants deficient in oxidases themselves. In this chapter,

* Institute of Cell Biology, NAS of Ukraine, Lviv, Ukraine
† Section of Molecular Biology, University of California, San Diego, La Jolla USA
‡ Department of Biotechnology and Microbiology, Rzeszów University, Rzeszów, Poland

Methods in Enzymology, Volume 451
ISSN 0076-6879, DOI: 10.1016/S0076-6879(08)03216-3

protocols for monitoring peroxisomal alcohol oxidase and amine oxidase directly in yeast colonies and examples of their application for mutant isolation are described. These methods were successfully utilized in several methylotrophic yeasts and the alkane-utilizing yeast *Yarrowia lipolytica*.

1. Introduction

Peroxisomes are ubiquitous in eukaryotic cells. They are specialized organelles that harbor certain oxidases and hydrogen peroxide decomposing enzymes and are involved in various catabolic and biosynthetic processes (van den Bosch *et al.*, 1992). Methods of plate assays for peroxisomal enzymes in yeasts have been proved to be a convenient approach to quickly monitor peroxisome status: organelle biogenesis, repression, or degradation. This especially concerns commonly used methylotrophic yeasts (i.e., *Hansenula polymorpha* [syn. *Pichia angusta*], *Pichia pastoris*, *Pichia methanolica* [formerly *Pichia pinus*] and *Candida boidinii*). Peroxisome proliferation in these species is highly induced in cells exposed to methanol but is strongly repressed in cells grown on such carbon sources as hexoses (glucose, fructose), disaccharides (sucrose, maltose), and ethanol (Leao and Kiel, 2003; Stasyk *et al.*, 2004). Peroxisomal alcohol (or methanol) oxidase is unique to the methylotrophs' enzyme that, upon methanol induction, constitutes a large portion of total cellular protein and can be used as a convenient peroxisomal marker activity. Significant peroxisome proliferation in methylotrophs as well as other yeasts can also be induced by oleic acid and primary amines (Veenhuis *et al.*, 2003). In the latter case, peroxisomal amine oxidase can be used as a reference enzymatic activity for the organelle.

When yeast cells grown on peroxisome proliferator substrates are shifted in liquid culture or replica plated on solid media supplemented with glucose (as well as ethanol in methylotrophs), rapid autophagic peroxisome degradation (pexophagy) that occurs in vacuoles is triggered (Dunn *et al.*, 2005; Veenhuis *et al.*, 1981). To monitor pexophagy in yeast colonies, alcohol oxidase and amine oxidase assays have been successfully used (Nazarko *et al.*, 2002; Stasyk *et al.*, 1999; Titorenko *et al.*, 1995). In this chapter, we first describe methods for a plate assay for peroxisomal oxidases and then provide several examples of their utilization in yeast mutant selection, with the main emphasis on pexophagy. In addition, we describe the method of positive selection for pexophagy (*ATG*, autophagy-related) gene cloning by functional complementation in yeasts based on the alcohol oxidase substrate, allyl alcohol (Stasyk *et al.*, 1999).

2. Methods of Plate Assays for Peroxisomal Enzymes in Yeast Colonies

2.1. Alcohol oxidase

Here we provide a protocol for visualization of peroxisomal alcohol oxidase (AO) activity in yeast colonies as described by us previously (Sibirny and Titorenko, 1986), with some minor modifications. This protocol can be applied to all methylotrophic yeasts.

1. Plate yeast cells on YPD (1% yeast extract, 1% peptone, 1% glucose, and 2% agar) plates as small patches and incubate for 1–2 days at 30 °C (37 °C for *H. polymorpha*).
2. Replica-plate the patches onto YNM solid medium (6.7 g/l YNB without amino acids, 0.05% (w/v) yeast extract, 1% (v/v) methanol and 2% agar) and incubate for 24 h. Keep the YPD master plate at 4 °C if necessary for subsequent colony analyses.
3. Prepare agarized AO assay reaction mixture (100 mM K-phosphate buffer, pH 7.5, 0.3% (w/v) agarose, 0.05% (w/v) chromogen o-dianisidine, 0.25% (w/v) cetyltrimethylammonium bromide (CTAB) as a permeabilizing agent, 1% (v/v) methanol and 3 U/ml of horseradish peroxidase).
   To prepare the mixture, the first three components of the reaction mixture are briefly boiled until the agarose and chromogen are fully dissolved. Then CTAB is added and the solution is vigorously mixed. Upon cooling to approximately 40 °C, the mixture is sequentially supplemented with methanol and peroxidase. Peroxidase is added as a water solution of appropriate concentration. As alternative components of the reaction mixture, 2,2′-Azino-bis(3-ethylbenzothiazoline-6-sulfonic acid) (ABTS) can be utilized as a peroxidase substrate (0.05%, w/v), and digitonin as a permeabilizing agent (0.1%, w/v) instead of o-dianisidine and CTAB, both added upon dissolving agarose (Titorenko *et al.*, 1995).

   Note that ABTS does not work efficiently in combination with CTAB as a permeabilizing agent, whereas o-dianisidine/digitonin combination can be utilized (Sibirny and Titorenko, 1986). CTAB is also less expensive than digitonin and provides better permeabilization in cases of extensive cell biomass on plates (Nazarko, T., unpublished observation).

4. Carefully overlay the plates with 7–9 ml of AO assay reaction mixture, let the mixture solidify at room temperature (10–15 min), cover the plates, and incubate them with the medium side up for 1–2 h at 30 °C. Patches with high AO activity will be stained red (with o-dianisidine) or green (with ABTS).

2.2. Amine oxidase

Here we provide the protocol for the peroxisomal amine oxidase (AMO) plate assay optimized for the colonies of *Yarrowia lipolytica* (Nazarko *et al.*, 2002). It was developed based on the plate colony assays of alcohol oxidase in methylotrophic yeasts (Sibirny and Titorenko, 1986; Titorenko *et al.*, 1995).

1. Plate the *Y. lipolytica* strains on YPD (1% yeast extract, 1% peptone, 1% dextrose, and 2% agar) medium as small patches and incubate for 1–2 days at 28 °C.
2. Replica-plate the patches to YEE (1.7 g/l YNB without amino acids and ammonium sulfate, 0.05% yeast extract, 0.5% ethanol, 0.2% ethylamine-HCl, and 2% agar) medium and incubate for 18 h at 28 °C. Keep the YPD master plate at 4 °C.
3. Prepare the AMO assay mixture. First, boil 0.3% agar in 50 m*M* phosphate buffer, pH 7.0, with 0.05% o-dianisidine (serves as a chromogen). Then add 0.5% cetyltrimethylammonium bromide (CTAB), as a cell-permeabilizing agent, mix thoroughly and cool down to 37 °C. Finally, add 2.3 U/ml of peroxidase and 2 m*M* ethylamine.
4. Carefully overlay the plates with 7–9 ml of AMO assay mixture, let the mixture solidify, cover the plates, and incubate them with the mixture side up for 1–2 h at 28 °C. Patches with high residual activity of AMO will be stained red.
5. Identify the patches that have a high activity of AMO on a master YPD plate.

2.3. Catalase

Peroxisomal catalase (CAT) of methylotrophic yeasts can be visualized in colonies as well, essentially as described and used by us previously (Gonchar *et al.*, 1990; Sibirny and Titorenko, 1986). However, this enzyme is not as strictly regulated by carbon sources as peroxisomal oxidases.

1. Follow steps 1 and 2 of the protocol for visualization of AO activity in yeast colonies.
2. Prepare agarized CAT assay reaction mixture (50 m*M* Tris-HCl buffer, pH 8.0, 0.3% (w/v) agarose, 0.05% (w/v) chromogen o-dianisidine, 0.1% (w/v) digitonin, and 0.015% (v/v) hydrogen peroxide). H_2O_2 is added to the reaction mixture after the agarose is fully melted in the buffer by brief repeated rounds of boiling, and the reaction mixture is cooled to 40 °C.
3. Carefully overlay the plates with 7–9 ml of CAT assay reaction mixture, let the mixture solidify at room temperature (10–15 min), and incubate plates at 30 °C. Colonies with high CAT activity will be stained red. Alternatively, yeast colonies can be overlaid with reaction mixture containing 50 mM K-phosphate buffer, pH 7.0, 0.3% (w/v) agarose, and 2%

(v/v) hydrogen peroxide. Colonies with high CAT activity will produce visible bubbles of oxygen resulting from decomposition of hydrogen peroxide by peroxisomal catalase, as originally described (Rytka *et al.*, 1976).

3. Enzyme Plate Assays in Selection of Yeast Mutants Defective in Pexophagy or Catabolite Repression

3.1. *P. pastoris*

The AO plate assay can be utilized in the selection of *P. pastoris* mutants deficient in pexophagy. Mutations can be induced by means of UV-light or random insertional mutagenesis (REMI) as described elsewhere (Schroder *et al.*, 2007). To identify *P. pastoris* mutants deficient in glucose-induced micropexophagy or ethanol-induced macropexophagy (Dunn *et al.*, 2005), or both, follow the subsequent protocol:

1. Plate the mutagenized *P. pastoris* cells on YND solid medium (6.7 g/l yeast nitrogen base (YNB) without amino acids, 1% glucose, and 2% agar) with a dilution that provides about 200 yeast colonies per plate, and incubate for 2–3 days at 30 °C.
2. Replica-plate colonies onto YNM solid medium (6.7 g/l YNB without amino acids, 0.05% (w/v) yeast extract, 1% (v/v) methanol and 2% agar) and incubate for 24 h. Keep the YND master plate at 4 °C for subsequent colony analyses.
3. Replica-plate colonies grown on YNM plates onto YND and YNE solid media (the latter supplemented with 1% v/v ethanol instead of glucose). Incubate plates overnight at 30 °C.
4. Overlay YNM and YND plates with AO reaction mixture as described in paragraph 2.1 (steps 3–4). Mutant colonies deficient in AO degradation via pexophagy will exhibit high residual AO activity. These colonies are picked up from the initial master YND plate and the analysis through steps 1–4 has to be repeated to verify the phenotype. At this stage, cells of the nonmutagenized parental (wild-type) strain have to be used as a negative control. The scheme and an experimental example of the AO plate analysis described in this paragraph are shown in Fig. 16.1.

3.2. *H. polymorpha*

In *H. polymorpha*, mutants deficient in macropexophagy upon adaptation to glucose can be isolated essentially as described in paragraph 3.1, except that cells are incubated on all media at 37 °C (Titorenko *et al.*, 1995). In

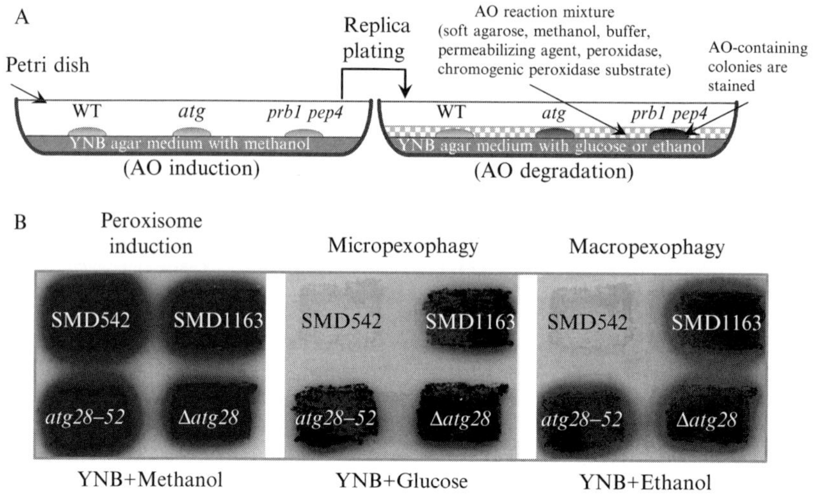

Figure 16.1 AO plate assay in yeast colonies. **(A)** Scheme depicting AO visualization for isolation of *P. pastoris* mutants deficient in pexophagy. WT, wild-type strain; *atg*, autophagy (pexophagy)-deficient mutant; *prb1 pep4*, vacuolar protease-deficient mutant that serves as a positive control in the analysis. **(B)** An example of an AO colony qualitative assay with UV-induced *atg28–52* and constructed Δ*atg28* deletion mutants under pexophagy-triggering conditions (from Stasyk *et al.*, 2006). Cells were grown for 2 days on mineral agar medium with methanol (1% v/v) to induce peroxisomes and peroxisomal AO, then replica-plated onto YND (1% w/v glucose) or YNE (1% v/v ethanol) plates to induce pexophagy. After incubation for 14 h, AO activity was visualized by overlaying cells with the AO activity reaction mixture. Wild-type strain SMD542 and vacuolar protease-deficient mutant SMD1163 (*pep4 prb1*, impaired in selective and nonselective autophagy) served as negative and positive controls, respectively.

addition, the AO plate assay can be utilized in a search for *H. polymorpha* mutants deficient in glucose (catabolite) repression of AO (Stasyk *et al.*, 1997, 2004). For this, negative or positive selection can be used based on UV, chemical or insertional mutagenesis (van Dijk *et al.*, 2001).

1. For negative selection, plate the mutagenized *H. polymorpha* cells on YND solid medium with a dilution that provides about 200 yeast colonies per plate and incubate for 2–3 days at 37 °C. For positive selection, mutagenized cells are plated with 10 times higher density on YNM medium supplemented with a nonmetabolizable glucose analogue, 2-deoxyglucose (2-DG) at 100–150 mg/l. 2-DG exerts a repression effect on peroxisomal enzymes, and wild-type cells do not grow in its presence on YNM plates.
2. Replica-plate growing colonies on YND plates or YND plates supplemented with a second carbon source, 1% v/v methanol (YNDM), and incubate for 24–36 h.

3. Overlay the YND and YNDM plates with AO reaction mixture as described in paragraph 2.1 (steps 3–4). Mutant colonies deficient in glucose repression will exhibit pronounced AO activity. These colonies are picked up from the master plate and the analysis through steps 1–4 has to be repeated to verify the phenotype. At this stage, cells of the non-mutagenized parental (wild-type) strain have to be used as a negative control. Mutants exhibiting high AO activity on YND plates represent a constitutive phenotype (i.e., they do not require methanol for AO induction in the presence of glucose). Mutant strains that display AO activity only on YNDM plates belong to the so-called inducible phenotype (i.e., they are deficient in glucose repression of AO synthesis only when the inducer methanol is present).

3.3. *Y. lipolytica*

Y. lipolytica mutants affected in autophagic peroxisome degradation (pexophagy) can be isolated as mutants deficient in inactivation of peroxisomal amine oxidase (AMO) after exposure of ethanol/ethylamine-grown cells to excess glucose/ammonia conditions. Here we provide the modified AMO plate assay that was used in the selection of *Y. lipolytica* pexophagy mutants (Nazarko *et al.*, 2002, 2005).

1. Plate the mutagenized *Y. lipolytica* cells on YND (6.7 g/l yeast nitrogen base (YNB) without amino acids, 1% dextrose and 2% agar) medium with a dilution that provides about 200 yeast colonies per plate and incubate for 2 days at 28 °C.
2. Replica-plate the yeast colonies to YEE (1.7 g/l YNB without amino acids and ammonium sulfate, 0.05% yeast extract, 0.5% ethanol, 0.2% ethylamine-HCl, and 2% agar) medium and incubate for 18 h at 28 °C. Keep the YND master plate at 4 °C.
3. Prepare the AMO inactivation mixture containing 3% glucose (to induce pexophagy), 1% ammonium sulfate (to repress *de novo* AMO synthesis), and 0.3% agar. Boil the mixture and cool it down to 37 °C.
4. Carefully overlay the plates with 7–9 ml of AMO inactivation mixture, let the mixture solidify, cover the plates and incubate them with the mixture side up for 10 h at 28 °C.
5. Prepare the AMO assay mixture. First, boil 0.3% agar (or agarose) in 100 m*M* phosphate buffer pH 7.0 with 0.05% o-dianisidine (serves as a chromogen). Then add 0.5% cetyltrimethylammonium bromide (cell-permeabilizing agent), mix thoroughly, and cool to 37 °C. Finally, add 2.3 U/ml of peroxidase and 4 mM ethylamine.
6. Carefully overlay the plates with 7–9 ml of the AMO assay mixture, let the mixture solidify, cover the plates, and incubate them with the

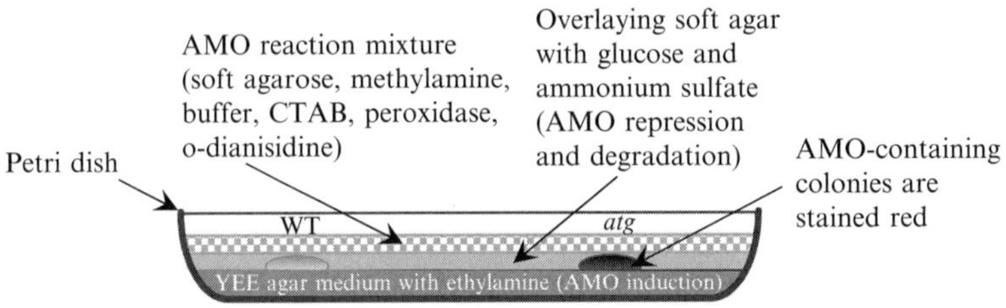

Figure 16.2 Scheme depicting AMO visualization in yeast colonies for the isolation of *Y. lipolytica* mutants deficient in pexophagy. WT, wild-type strain; atg, autophagy (pexophagy)-deficient mutant. Note that for the AMO plate assay sandwich-like repeated overlays with inactivation solution followed by AMO reaction mixture are applied. See main text for details.

mixture side up for 14 h at 28 °C. Colonies with a high residual activity of AMO will be stained red.

7. Identify the colonies with high residual activity of AMO on a master YND plate.
8. Plate the identified positive colonies and nonmutagenized parental strain as a negative control on YPD (1% yeast extract, 1% peptone, 1% dextrose, and 2% agar) medium as small patches and repeat steps 2–7 to verify the phenotype.

The scheme of AMO plate analysis described in this paragraph is shown in Fig. 16.2.

4. Positive Selection of Pexophagy Mutants and Cloning of *ATG* Genes using Allyl Alcohol as a Selective Agent

4.1. Isolation of the mutants deficient in the peroxisomal enzyme alcohol oxidase in *H. polymorpha* and *Pichia methanolica*

Mutants of methylotrophic yeasts deficient in AO synthesis and/or enzymatic activity can be isolated via negative selection as strains unable to utilize methanol as a single carbon source and subsequent analysis of such candidates by the AO plate assay on YNM solid medium, essentially as described in section 2.1 (Lahtchev *et al.*, 2002). Alternatively, positive selection can be applied based on allyl alcohol as an AO substrate. Allyl alcohol is efficiently converted by AO to a much more toxic aldehyde product, acroleine. Therefore, mutants deficient in AO synthesis and/or activity exhibit higher resistance to exogenous allyl alcohol (Motruk *et al.*, 1989; Sibirny *et al.*,

1989). Allyl alcohol-based selection can also be applied to isolate mutants deficient in peroxisome biogenesis in *P. pastoris* (Johnson *et al.*, 1999) or mutants with altered AO kinetic properties, as described for *H. polymorpha* (Dmytruk *et al.*, 2007).

4.2. Allyl alcohol-based positive selection of *ATG* genes from *P. pastoris* genomic libraries

Allyl alcohol-based selection can be utilized to clone *ATG* genes from genomic libraries by functional complementation. This specifically concerns UV- or chemically-induced mutants deficient in pexophagy, for which the defective genes cannot be identified by other means. Here we provide a protocol successfully utilized in *P. pastoris* (Stasyk *et al.*, 1999, 2003, 2006). An analogous protocol has been also developed for *H. polymorpha* (Komduur *et al.*, 2002).

1. Transform a *P. pastoris* mutant that is deficient in pexophagy with a *P. pastoris* genomic library (usually a *his4* recipient strain and a replicative genomic library based on the *S. cerevisiae HIS4* gene are used (Lin Cereghino *et al.*, 2001). Plate cells on YND (6.7 g/l yeast nitrogen base (YNB) without amino acids, 1% glucose, and 2% agar) minimal plates and select for prototrophic transformants (3–4 days at 30 °C).
2. Replica-plate colonies onto YNM solid medium (6.7 g/l YNB without amino acids, 0.05% (w/v) yeast extract, 1% (v/v) methanol, and 2% agar) and incubate for 24–36 h. Keep the YND master plates at 4 °C.
3. Replica-plate colonies grown on YNM plates onto YND and YNE (with 1% v/v ethanol instead of glucose) solid media supplemented with 0.1 m*M* and 0.15 m*M* allyl alcohol, respectively. Incubate plates for 2–3 days at 30 °C.

Colonies functionally complemented to the wild-type phenotype (with restored pexophagy) degrade peroxisomal AOX rapidly after a shift from methanol to glucose or ethanol and are, therefore, less sensitive to allyl alcohol than the initial pexophagy mutant. Thus, they exhibit relatively faster growth in the presence of allyl alcohol. Candidate fast growers are picked up from the YND master plate and analyzed for restored pexophagy by the AO plate assay as described in paragraph 3.1 (steps 2–4). Usually, allyl alcohol screening provides 100 times enrichment for colonies with restored pexophagy. Finally, functionally complemented colonies from the YND master plate with a confirmed wild-type pexophagy rate are utilized for isolation of total DNA, with its subsequent transformation into *E. coli* and identification of the complementing plasmid according to standard procedures (Sambrook *et al.*, 1989).

5. Concluding Remarks

In this chapter we describe methods of monitoring peroxisomal oxidases (alcohol oxidase in methylotrophic yeasts and amine oxidase in the alkane-utilizing yeast *Y. lipolytica*) directly in yeast colonies and provide several examples of their application in mutant isolation and analysis. These methods are reliable, fast, and relatively simple and, in our opinion, can be further useful for a variety of applications in yeast research.

ACKNOWLEDGMENTS

This publication was supported in part by Award No. UB1-2447-LV-02 of the U.S. Civilian Research and Development Foundation (CRDF), FIRCA grant 2R03TW00547-04A2, and Polish National Research Grant N N302 1385 33 from the Polish Ministry for Science and Higher Education (to AAS).

REFERENCES

Dmytruk, K. V., Smutok, O. V., Ryabova, O. B., Gayda, G. Z., Sibirny, V. A., Schuhmann, W., Gonchar, M. V., and Sibirny, A. A. (2007). Isolation and characterization of mutated alcohol oxidases from the yeast *Hansenula polymorpha* with decreased affinity toward substrates and their use as selective elements of an amperometric biosensor. *BMC Biotechnol.* **7,** 33.

Dunn, W. A., Jr., Cregg, J. M., Kiel, J. A. K. W., van der Klei, I. J., Oku, M., Oku, M., Sakai, Y., Sibirny, A. A., Stasyk, O. V., and Veenhuis, M. (2005). Pexophagy: The selective autophagy of peroxisomes. *Autophagy* **2,** 75–83.

Gonchar, M. V., Ksheminska, G. P., Hladarevska, N. M., and Sibirny, A. A. (1990). Catalase-minus mutants of methylotrophic yeast *Hansenula polymorpha* impaired in regulation of alcohol oxidase synthesis. *In* Proc. Intern. Conf. 'Genetics of Respiratory Enzymes in Yeasts', (July 29–August 3, 1990, Karpacz, Poland, and T. M. Lachowicz, eds.) pp. 222–228. Wroclaw University Press.

Johnson, M. A., Waterham, H. R., Ksheminska, G. P., Fayura, L. R., Cereghino, J. L., Stasyk, O. V., Veenhuis, M., Kulachkovsky, A. R., Sibirny, A. A., and Cregg, J. M. (1999). Positive selection of novel peroxisome biogenesis-defective mutants of the yeast *Pichia pastoris*. *Genetics* **151,** 1379–1391.

Komduur, J. A., Leao, A. N., Monastyrska, I., Veenhuis, M., and Kiel, J. A. K. W. (2002). Old yellow enzyme confers resistance of *Hansenula polymorpha* towards allyl alcohol. *Curr. Genet.* **41,** 401–406.

Lahtchev, K. L., Semenova, V. D., Tolstorukov, I. I., van der Klei, I., and Veenhuis, M. (2002). Isolation and properties of genetically defined strains of the methylotrophic yeast *Hansenula polymorpha* CBS4732. *Arch. Microbiol.* **177,** 150–158.

Leao, A. N., and Kiel, J. A. K. W. (2003). Peroxisome homeostasis in *Hansenula polymorpha*. *FEMS Yeast Res.* **4,** 131–139.

Lin Cereghino, G. P.., Lin Cereghino, J., Sunga, A. J., Johnson, M. A., Lim, M., Gleeson, M. A., and Cregg, J. M. (2001). New selectable marker/auxotrophic host strain combinations for molecular genetic manipulation of *Pichia pastoris*. *Gene* **263,** 159–169.

Motruk, O. M., Tolstorukov, I. I., and Sibirny, A. A. (1989). Selection of alcohol oxidase deficient mutants in methylotrophic yeast *Pichia pinus*. *Biotekhnologiya (Moscow)* **5,** 692–698 (in Russian).

Nazarko, T. Y., Huang, J., Nicaud, J. M., Klionsky, D. J., and Sibirny, A. A. (2005). Trs85 is required for macroautophagy, pexophagy and cytoplasm to vacuole targeting in *Yarrowia lipolytica* and *Saccharomyces cerevisiae*. *Autophagy* **1,** 37–45.

Nazarko, T., Mala, M. J., and Sibirny, A. A. (2002). Development of the plate assay screening procedure for isolation of the mutants deficient in inactivation of peroxisomal enzymes in the yeast *Yarrowia lipolytica*. *Biopolymers Cell (Kiev)* **18,** 135–138.

Rytka, J., Sledzewski, A., Litwinsky, J., and Bilinski, T. (1976). Haemoprotein formation in yeast: Isolation of catalase regulatory mutants. *Mol. Gen. Genet.* **145,** 37–42.

Sambrook, J., Fritsch, E. F., and Maniatis, T. (1989). Molecular cloning: A laboratory manual 2nd ed. Cold Spring Harbor, NY, Cold Spring Harbor Laboratory.

Schroder, L. A., Glick, B. S., and Dunn, W. A. (2007). Identification of pexophagy genes by restriction enzyme-mediated integration. *Methods Mol. Biol.* **389,** 203–218.

Sibirny, A. A., and Titorenko, V. I. (1986). A method of quantitative determination of alcohol oxidase and catalase in yeast colonies. *Ukr. Biokhim. Zh.* **58,** 65–68 (in Russian).

Sibirny, A. A., Vitvitskaya, O. P., Kulachkovsky, A. R., and Ubiyvovk, V. M. (1989). Selection and properties of the mutants of *Hansenula polymorpha* yeast deficient in alcohol oxidase. *Mikrobiologiya (Moscow)* **58,** 751–759 (in Russian).

Stasyk, O. V., Ksheminskaya, G. P., Kulachkovsky, A. R., and Sibirny, A. A. (1997). Mutants of the methylotrophic yeast *Hansenula polymorpha* with impaired catabolite repression. *Microbiologiya (Moscow)* **66,** 755–760 (in Russian).

Stasyk, O. V., Monastyrska, I. L., Veenhuis, M., Cregg, J. M., and Sibirny, A. A. (1999). *In* New genes involved in autophagic peroxisome degradation in themethylotrophic yeast *Pichia pastoris* p. 456. Abstract of XIX International Conference on Yeast Genetics and Molecular Biology, Rimini (Italy).

Stasyk, O. V., Nazarko, T. Y., Stasyk, O. G., Krasovska, O. S., Warnecke, D., Nicaud, J. M., Cregg, J. M., and Sibirny, A. A. (2003). Sterol glucosyltransferases have different functional roles in *Pichia pastoris* and *Yarrowia lipolytica*. *Cell Biol. Int.* **27,** 947–952.

Stasyk, O. V., Stasyk, O. G., Komduur, J., Veenhuis, M., Cregg, J. M., and Sibirny, A. A. (2004). A hexose transporter homologue controls glucose repression in the methylotrophic yeast *Hansenula polymorpha*. *J. Biol. Chem.* **279,** 8116–8125.

Stasyk, O. V., Stasyk, O. G., Mathewson, R. D., Farré, J.-C., Nazarko, V. Y., Krasovska, O. S., Subramani, S., Cregg, J. M., and Sibirny, A. A. (2006). Atg28, a novel coiled-coil protein involved in autophagic degradation of peroxisomes in the methylotrophic yeast *Pichia pastoris*. *Autophagy* **2,** 30–38.

Titorenko, V. I., Keizer, I., Harder, W., and Veenhuis, M. (1995). Isolation and characterization of mutants impaired in the selective degradation of peroxisomes in the yeast *Hansenula polymorpha*. *J. Bacteriol.* **177,** 357–363.

van den Bosch, H., Schutgens, R. B., Wanders, R. J., and Tager, J. M. (1992). Biochemistry of peroxisomes. *Annu. Rev. Biochem.* **61,** 57–97.

van Dijk, R., Faber, K. N., Hammond, A. T., Glick, B. S., Veenhuis, M., and Kiel, J. A. K. W. (2001). Tagging *Hansenula polymorpha* genes by random integration of linear DNA fragments (RALF). *Mol. Genet. Genomics* **266,** 646–656.

Veenhuis, M., Kiel, J. A. K. W., and van der Klei, I. (2003). Peroxisome assembly in yeast. *Micr. Res. Tech.* **61,** 139–150.

Veenhuis, M., Zwart, K. B., and Harder, W. (1981). Biogenesis and turnover of peroxisomes involved in the concurrent oxidation of methanol and methylamine in *Hansenula polymorpha*. *Arch. Microbiol.* **129,** 35–41.

CHAPTER SEVENTEEN

Autophagy in the Filamentous Fungus *Aspergillus fumigatus*

Daryl L. Richie* *and* David S. Askew*

Contents

Abstract

Many breakthroughs in our understanding of the function and molecular basis of autophagy have been achieved in mammalian and yeast systems. However, we still know very little about the contribution of autophagy to the biology of filamentous fungi. A comparative analysis of autophagy between genera will expand our knowledge of the autophagy machinery and has the potential to identify novel functions that are relevant to multiple biological systems. This chapter will discuss methods that have been employed for studying autophagy in the opportunistic mold pathogen *Aspergillus fumigatus*. Understanding how autophagy influences the growth of this important human pathogen could lead to the development of novel antifungal drugs that restrict the growth of the fungus by manipulating the autophagy pathway.

* Department of Pathology and Laboratory Medicine, University of Cincinnati College of Medicine, Cincinnati, USA

Methods in Enzymology, Volume 451
ISSN 0076-6879, DOI: 10.1016/S0076-6879(08)03217-5

1. Introduction

Aspergillus fumigatus is a filamentous fungus that resides in compost and decaying vegetation. Human encounters with this mold are practically unavoidable because the organism is ubiquitous and releases high concentrations of conidia (spores) into the atmosphere after physical disturbance of its environment (Goodley *et al.*, 1994; Hospenthal *et al.*, 1998; Woodcock *et al.*, 2006). The inhalation of these conidia is generally innocuous in a healthy individual but can be life-threatening in patients with depressed immunity (Maschmeyer *et al.*, 2007). Because current antifungal drugs have been unable to prevent the high mortality rate associated with invasive aspergillosis (Upton *et al.*, 2007), there is a need for increased understanding of fungal pathways that could be targeted with novel therapy. One possible target that has been largely unexplored in this fungus is the autophagy machinery. As the major pathway for bulk turnover of organelles and other cytoplasmic constituents, active autophagy is likely to be incompatible with rapid fungal growth and would therefore be a promising drug target if further research was able to identify ways to selectively induce fungal autophagy through pharmacological intervention (Mizushima *et al.*, 2008).

In *A. fumigatus*, autophagy is required for conidiation and hyphal foraging, both of which are adaptive responses to nutrient deficiency that are important to the survival of the organism in its native environment (Richie *et al.*, 2007). Recent data have also suggested that autophagy contributes to metal ion homeostasis in *A. fumigatus*, but the mechanism by which this is accomplished requires further investigation (Richie *et al.*, 2007). Similarly, questions remain about the importance of autophagy to fungal virulence. For example, although autophagy is required for the virulence of some eukaryotic pathogens (Besteiro *et al.*, 2006; Hu *et al.*, 2008; Liu *et al.*, 2007; Veneault-Fourrey *et al.*, 2006), it is dispensable for the virulence of others (Palmer *et al.*, 2007), including *A. fumigatus* (Richie *et al.*, 2007). Thus, a comprehensive understanding of how autophagy influences the virulence of some, but not all, eukaryotic pathogens, may provide insight into mechanisms of eukaryotic pathogenesis. Because autophagy in eukaryotic pathogens is an emerging field of investigation, it becomes important to adapt current methodologies to the unique physiology of these organisms. This chapter outlines protocols that have been used to examine autophagy and autophagy-dependent processes in *A. fumigatus*.

2. Analysis of Autophagy-Dependent Processes in *A. fumigatus*

Analysis of autophagy in *A. fumigatus* has identified three quantifiable processes that depend on autophagy for optimal activity: starvation foraging, conidiation, and growth under metal ion–deficient conditions. Assays for these processes can therefore be used as a qualitative indicator of autophagy function when evaluating autophagy mutants, with the caveat that autophagy is not the only pathway that can influence these functions.

2.1. Starvation foraging

The mycelium of a filamentous fungus comprises a network of interconnected hyphae that are separated by perforated septa. When the mycelium exhausts its nutrient supply, the organism is able to take advantage of these interconnections to support a limited amount of growth at the colony periphery, even though the rest of the colony is starved (Prosser and Tough, 1991; Robson, 1999). This foraging-like behavior allows the colony to expand into regions of unexplored substrate and has been hypothesized to involve the recycling of cytoplasmic contents by autophagy (Shoji *et al.*, 2006). To test this, an assay was developed to monitor the foraging capacity of *A. fumigatus* (Fig. 17.1). Hyphal plugs containing portions of mycelium are transferred to water/agarose (WA) starvation medium and colony diameter is monitored daily as a measure of starvation foraging. The ability of *A. fumigatus* to grow in this assay requires the Atg1 kinase (Richie *et al.*, 2007), which suggests that autophagy-controlled recycling of endogenous nutrients supports the foraging of the hyphal tips under acute starvation conditions. The autophagy dependence of this assay, and its ease-of-use, makes it a convenient adjunct to more sophisticated analyses of autophagic activity.

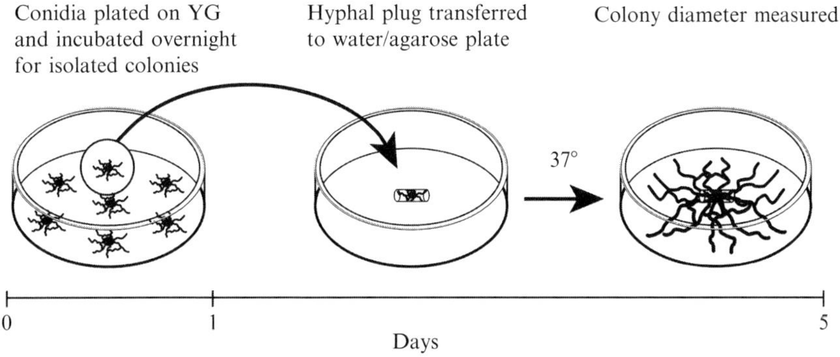

Figure 17.1 Schematic representation of the procedure used to monitor starvation-associated foraging in *A. fumigatus*. Hyphal plugs are transferred from rich medium onto WA lacking any nutrients, thereby forcing the organism to use autophagy to fuel any further growth. Colony diameter is measured daily as a relative indicator of active autophagy.

2.1.1. Media

YG plates: 2% glucose, 0.5% yeast extract, 1.5% Bacto agar. Combine ingredients in sterile distilled water, autoclave, and cool in a 50 °C water bath prior to pouring plates.

WA plates: Add 1g of agarose (genetic analysis grade) to 100 ml of Millipore ultrapure water in a clean 250 ml Erlenmeyer flask, autoclave, cool to 50 °C in a water bath, and pour 10 ml into 60-mm Petri dishes. Note: Bacto agar cannot be used in these plates because it contains sufficient nutrients to support the growth of an autophagy mutant.

2.1.2. Assay

1. *A. fumigatus* lacks a known sexual stage, but produces abundant asexual conidia (spores) on most types of laboratory medium. To harvest the conidia, flood the plate with 20 ml of sterile distilled water and dislodge the conidia by gently rubbing the top of the colony with a sterile Q-tip.
2. Aspirate the liquid with a 25-ml pipette, and remove the hyphal debris by filtration with sterile Miracloth into a 50-ml conical centrifuge tube (Calbiochem cat. 475855). Miracloth is a rayon-polyester material with a typical pore size of 22–25 μm.
3. Pellet the conidia by centrifuging at 4200*g* for 5 min at 4 °C. Wash twice with 50 ml of sterile distilled water and resuspend in 10 ml of sterile distilled water.
4. The spherical conidia are 2–3 μm in diameter and can be counted microscopically using a standard hemacytometer. Adjust the concentration to 1×10^4 conidia/ml.
5. Spread 20 μl of the 10^4 conidia/ml conidial stock solution (200 conidia) onto the surface of a 60-mm YG plate using a glass rod or hockey stick to evenly distribute the conidia.
6. Incubate at 37 °C for 24 h.
7. Using a five and three-fourth inch Pasteur pipette plugged with cotton wool, remove a hyphal plug from the edge of an individual colony on the YG plate and expel the plug onto the center of a WA plate using a small bulb.
8. Incubate at 37 °C for 4–6 days.
9. Measure the diameter of the colony starting with the first measurement after 24 h of incubation. The extent of growth after 24 h provides a baseline for further measurement because it is the same in both wild type (wt) *A. fumigatus* and an autophagy mutant (probably as a result of the carryover of nutrients from the original plug of YG agar). However, subsequent growth is autophagy dependent, which results in radial extension of very thin hyphal filaments. Because hyphal density is much lower on WA medium than rich medium, colony diameter is best monitored with a dissecting scope (Fig. 17.2). This is accomplished by drawing a horizontal line through the colony on the back of the plate

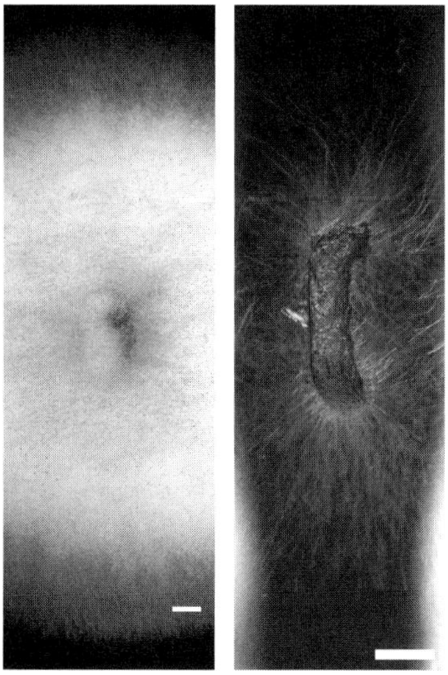

Figure 17.2 Hyphal density is reduced on WA starvation medium. An agar plug containing wt *A. fumigatus* was isolated from a plate of rich medium (YG) using a Pasteur pipette and transferred to a plate of YG or WA and incubated at 37 °C for 24 h. Colony size and hyphal density are greatly reduced on WA relative to YG, although the hyphae continue to expand radially for up to 5 days on WA. Colony morphology was photographed using a Nikon SMZ1000 dissecting microscope. Scale bar represents 10 mm.

on day 1 and marking the subsequent migration of the hyphal tips along this line with continued incubation.

Note: The ability to grow under these conditions is strain dependent. The wt strains H237 and CEA17 work well, but Af293 is less effective. However, this can be alleviated by transferring more Af293 biomass to the WA plates. The reason for this difference is not known.

2.2. Conidiation

In filamentous fungi, the production of conidia requires the construction of new morphological structures that are responsible for packaging nuclei into spores. Current evidence suggests that autophagy is essential for this process, presumably by providing the building blocks that support the necessary developmental changes (Richie *et al.*, 2007).

2.2.1. Media

Potato-dextrose agar (PDA): 1% glucose, 2% potato flakes, and 1.5% Bacto agar. Combine all ingredients in sterile distilled water, autoclave, and cool in a 50 °C water bath prior to pouring into 60-mm Petri plates.

2.2.2. Assay

1. Harvest conidia as described in section 2.1.2 and resuspend in sterile distilled water at a concentration of 1 × 10^6 conidia/ml.
2. Spot 5 μl of the 1 × 10^6 conidia/ml stock (5000 conidia) onto the center of triplicate PDA plates and incubate at 37 °C for 3 days.

Note: wt *A. fumigatus* conidiates extensively on PDA medium, but an autophagy mutant requires supplementation of the medium with increased nitrogen to support the same level of conidiation. Ammonium tartrate, ammonium chloride, ammonium sulfate, or sodium nitrate can be added to autoclaved PDA medium to a final concentration of 40 m*M* to restore conidiation to wild-type levels.

3. To harvest conidia, add 10 ml of sterile distilled water to each plate and gently rub the surface of the plate with a sterile Q-tip. Aspirate the water from the plate with a 10-ml pipette, and filter through Miracloth to remove hyphal fragments.
4. Repeat step 2 three times to maximize conidial recovery.
5. Count the total number of conidia recovered from each of the triplicate plates using a hemacytometer. For the wt strain a typical dilution of 1:100 will provide a countable number of conidia; autophagy mutants will usually require a lower dilution.

2.3. Growth in metal-depleted medium

The hypersensitivity of an *A. fumigatus* autophagy mutant to ethylene diamine tetraacetic acid (EDTA) suggests that the autophagy machinery contributes to the maintenance of metal-ion homeostasis in this fungus, possibly by recycling metals from preexisting metal-associated proteins (Richie *et al.*, 2007). This hypersensitivity can be quantified by inoculating conidia into liquid cultures containing EDTA and measuring percentage germination.

1. Place a sterile coverslip into a 60-mm Petri dish and overlay with 5 ml of liquid YG medium in the presence or absence of 0.5–0.75 m*M* EDTA.
2. Harvest conidia as described in section 2.1.2 and resuspend in sterile distilled water at a concentration of 1 × 10^6 conidia/ml.
3. Inoculate the YG-EDTA medium with 200 μl of the 1 × 10^6 conidia/ml stock solution and incubate at 37 °C for 3–6 days.
4. Remove the coverslip, and count the number of germinated conidia microscopically. The germination of *A. fumigatus* conidia involves a series of morphological changes, beginning with a short period of isotropic growth (swelling) that is followed by the establishment of an axis of polarity and the extension of the first germ tube. In this study, a conidium was scored as germinated if it had extended a germ tube that was equal to, or greater than the length of the conidium.

3. Analysis of Autophagosome Accumulation

After the induction of autophagy and the fusion of the autophagosome to the vacuolar membrane, a single membrane vesicle termed an autophagic body is delivered into the vacuole lumen (Mizushima *et al.*, 2008). Under normal circumstances, autophagic bodies are rapidly degraded by vacuolar hydrolases, making them difficult to visualize. However, their degradation can be delayed by incorporating the serine protease inhibitor phenylmethanesulfonyl fluoride (PMSF) into the medium. This allows autophagic bodies to accumulate within the vacuole, and their appearance by light microscopy can be used as a qualitative marker of active autophagy (Klionsky *et al.*, 2007). To differentiate autophagic vesicles from other material that may accumulate in the vacuole, a second assay is recommended that involves fluorescent tagging of autophagosomes. This approach involves the creation of a strain of *A. fumigatus* that expresses a green fluorescent protein (GFP)-tagged Atg8 protein. Because a fraction of the Atg8 produced in a cell is continually associated with the autophagosome until it is finally degraded in the vacuole, the presence of GFP-Atg8 fluorescence can be used as a marker of autophagosome delivery into the vacuole (Fig. 17.3; Klionsky *et al.*, 2007).

3.1. GFP-tagging Atg8

1. Link the N terminus of the *A. fumigatus atg8* gene (GenBAnk accession No. Q4WJ27), or the corresponding cDNA, to GFP by standard cloning procedures in a plasmid vector. A 6-His linker can be used to separate GFP from the Atg8 protein (Richie *et al.*, 2007), but it is not known whether the linker is absolutely necessary. The plant-adapted GFP works

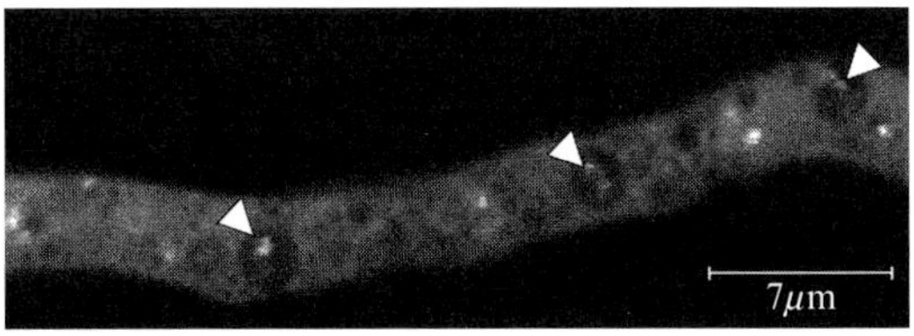

Figure 17.3 Analysis of autophagic body accumulation in starved cultures of *A. fumigatus*. Conidia from a strain expressing GFP-AfAtg8 were incubated in *Aspergillus* minimal medium for 16 h at 37 °C. The hyphae were washed and transferred to sterile distilled water (starvation conditions) containing 2 m*M* PMSF (to impair vacuolar protease activity). After incubating at 37 °C for 4 h, autophagic body accumulation in vacuoles was visualized by laser confocal microscopy. The image shown is a single optical section through a hypha. Arrowheads denote autophagic bodies in vacuoles. Scale bar represents 7 μm.

well in *A. fumigatus* (Bhabhra *et al.*, 2004), but other fluorescent tags can also be used, including ECFP and DsRed-Monomer (unpublished observations). The strong constitutive promoter *gpdA* has been successfully used to drive GFP-*atg8* from an ectopically integrated transgene in *A. fumigatus* (Richie *et al.*, 2007). However, techniques for tagging chromosomal genes with GFP have also been reported in *Aspergillus spp*, allowing expression levels to be under the control of the native promoter (Szewczyk *et al.*, 2006).

2. Digest 10 μg of the GFP-atg8 plasmid to linearize and introduce into the desired strain as an ectopic transgene by standard protoplast transformation as previously described (Bhabhra *et al.*, 2004).
3. The accumulation of GFP-labeled autophagic bodies within vacuoles can be visualized by fluorescence microscopy using the PMSF technique described subsequently.

3.2. Assay

1. Place a sterile glass coverslip into a 35-mm Petri dish
2. Harvest conidia as described in section 2.1.2, and inoculate 10,000–50,000 conidia into 5 ml of *Aspergillus* minimal medium. Incubate at 30–37 °C for 16 h.

Note: The incubation time and temperature can be adjusted to ensure that the culture has completely germinated. However, it is important not to overgrow to maintain synchrony between hyphal compartments (certain compartments become starved and begin to vacuolate prematurely when overgrown). In addition, overgrown cultures are more resistant to PMSF.

3. Remove the medium gently so as not to dislodge the mycelium from the coverslip.
4. Wash gently 3 times with sterile distilled water.
5. Remove the last of the distilled water and replace with 5 ml of starvation medium supplemented with 2 m*M* PMSF. The starvation medium can be water lacking any nutrients, or minimal medium lacking a nitrogen source. Note: PMSF precipitates in *Aspergillus* minimal medium at concentrations above 2 m*M*.
6. Incubate at 37 °C for 2–4 h. Extended incubation times cannot be used because the PMSF is toxic to the organism and results in cytoplasmic breakdown.
7. Aspirate the medium, wash the coverslip once with sterile distilled water, and mount the coverslip on a glass slide with the mycelium facing down.
8. Visualize autophagic bodies by fluorescence microcopy.

4. Conclusion

These methods have provided the first insight into the contribution of the autophagy kinase Atg1 to the growth and virulence of *A. fumigatus*. However, autophagy in filamentous fungi is still an emerging field that awaits adaptation of the many sophisticated techniques that have been developed in other systems. The high level of conservation of the autophagy machinery between genera underscores the importance of this pathway to eukaryotic cell biology, which makes continued investment in vigorous methods of analysis a high priority for future research.

REFERENCES

Besteiro, S., Williams, R. A., Morrison, L. S., Coombs, G. H., and Mottram, J. C. (2006). Endosome sorting and autophagy are essential for differentiation and virulence of *Leishmania major*. *J. Biol. Chem.* **281,** 11384–11396.

Bhabhra, R., Miley, M. D., Mylonakis, E., Boettner, D., Fortwendel, J., Panepinto, J. C., Postow, M., Rhodes, J. C., and Askew, D. S. (2004). Disruption of the *Aspergillus fumigatus* gene encoding nucleolar protein CgrA impairs thermotolerant growth and reduces virulence. *Infect. Immun.* **72,** 4731–4740.

Goodley, J. M., Clayton, Y. M., and Hay, R. J. (1994). Environmental sampling for aspergilli during building construction on a hospital site. *J. Hosp. Infect.* **26,** 27–35.

Hospenthal, D. R., Kwon-Chung, K. J., and Bennett, J. E. (1998). Concentrations of airborne *Aspergillus* compared to the incidence of invasive aspergillosis: Lack of correlation. *Med. Mycol.* **36,** 165–168.

Hu, G., Hacham, M., Waterman, S. R., Panepinto, J., Shin, S., Liu, X., Gibbons, J., Valyi-Nagy, T., Obara, K., Jaffe, H. A., Ohsumi, Y., and Williamson, P. R. (2008). PI3K signaling of autophagy is required for starvation tolerance and virulence of *Cryptococcus neoformans*. *J. Clin. Invest.* **118,** 1186–1197.

Klionsky, D. J., Cuervo, A. M., and Seglen, P. O. (2007). Methods for monitoring autophagy from yeast to human. *Autophagy* **3,** 181–206.

Liu, X. H., Lu, J. P., Zhang, L., Dong, B., Min, H., and Lin, F. C. (2007). Involvement of a *Magnaporthe grisea* serine/threonine kinase gene, MgATG1, in appressorium turgor and pathogenesis. *Eukaryot. Cell* **6,** 997–1005.

Maschmeyer, G., Haas, A., and Cornely, O. A. (2007). Invasive aspergillosis: epidemiology, diagnosis and management in immunocompromised patients. *Drugs* **67,** 1567–1601.

Mizushima, N., Levine, B., Cuervo, A. M., and Klionsky, D. J. (2008). Autophagy fights disease through cellular self-digestion. *Nature* **451,** 1069–1075.

Palmer, G. E., Kelly, M. N., and Sturtevant, J. E. (2007). Autophagy in the pathogen *Candida albicans*. *Microbiology* **153,** 51–58.

Prosser, J. I., and Tough, A. J. (1991). Growth mechanisms and growth kinetics of filamentous microorganisms. *Crit. Rev. Biotechnol.* **10,** 253–274.

Richie, D. L., Fuller, K. K., Fortwendel, J., Miley, M. D., McCarthy, J. W., Feldmesser, M., Rhodes, J. C., and Askew, D. S. (2007). Unexpected link between metal ion deficiency and autophagy in *Aspergillus fumigatus*. *Eukaryot. Cell* **6,** 2437–2447.

Robson, G. (1999). Hyphal Cell Biology. *In* "Molecular Fungal Biology" (R. Oliver and M. Schweizer, eds.), pp. 164–184. Cambridge University Press, Cambridge, UK.

Shoji, J. Y., Arioka, M., and Kitamoto, K. (2006). Possible involvement of pleiomorphic vacuolar networks in nutrient recycling in filamentous fungi. *Autophagy* **2,** 226–227.

Szewczyk, E., Nayak, T., Oakley, C. E., Edgerton, H., Xiong, Y., Taheri-Talesh, N., Osmani, S. A., and Oakley, B. R. (2006). Fusion PCR and gene targeting in *Aspergillus nidulans*. *Nat. Protoc.* **1,** 3111–3120.

Upton, A., Kirby, K. A., Carpenter, P., Boeckh, M., and Marr, K. A. (2007). Invasive aspergillosis following hematopoietic cell transplantation: outcomes and prognostic factors associated with mortality. *Clin. Infect. Dis.* **44,** 531–540.

Veneault-Fourrey, C., Barooah, M., Egan, M., Wakley, G., and Talbot, N. J. (2006). Autophagic fungal cell death is necessary for infection by the rice blast fungus. *Science* **312,** 580–583.

Woodcock, A. A., Steel, N., Moore, C. B., Howard, S. J., Custovic, A., and Denning, D. W. (2006). Fungal contamination of bedding. *Allergy* **61,** 140–142.

CHAPTER EIGHTEEN

Monitoring Autophagy in the Filamentous Fungus *Podospora anserina*

Bérangère Pinan-Lucarré* *and* Corinne Clavé[†]

Contents

Abstract

Autophagy has been monitored in the filamentous fungus *Podospora anserina* using electron, light, and fluorescence microscopy. In this organism autophagy can be induced either by starvation or rapamycin treatment or by *het* gene incompatibility. Incompatible HET products signal a cell death reaction referred to as cell death by incompatibility. In *het-R het-V* strain bearing the two incompatible *het-R* and *het-V* genes, cell death is induced by a simple shift in

* Laboratoire de Génétique Moléculaire des Champignons, Institut de Biochimie et de Génétique Cellulaires, Université de Bordeaux 2 et CNRS, Bordeaux, France
† Department of Molecular Biology and Biochemistry, Rutgers, The State University of New Jersey, Piscataway, New Jersey, USA

Methods in Enzymology, Volume 451
ISSN 0076-6879, DOI: 10.1016/S0076-6879(08)03218-7

growth temperature, as incompatibility is thermosensitive. In this strain large autophagosomes are formed as revealed by electron microscopy or using the GFP-PaATG8 marker. This strain constitutes an alternative model to study autophagy. Analysis of the three autophagy mutants, Δ*PaATG*1, Δ*PaATG*8, and Δ*pspA*, reveals that autophagy is essential for aerial hyphae and female organ differentiation and involved in spore germination. During the incompatibility reaction, autophagy might protect cells from cell death as suggested by accelerated cell death observed in autophagy mutants.

1. Introduction

The filamentous fungus *Podospora anserina* is a valuable model system for the study of senescence, prion protein structure and inheritance, cytoplasmic heredity, non-self-recognition, sexual reproduction, and vacuolar cell death (Bonnet *et al.*, 2006; Coppin *et al.*, 2005; Golstein *et al.*, 2003; Kicka *et al.*, 2006; Lorin *et al.*, 2006; Maddelein *et al.*, 2002; Malagnac *et al.*, 2007; Paoletti *et al.*, 2007; Pinan-Lucarré *et al.*, 2007; van Diepeningen *et al.*, 2008; Wasmer *et al.*, 2008). Autophagy and its relevance in biological processes such as differentiation or cell survival have been analyzed in this fungus (Pinan-Lucarré *et al.*, 2003, 2005). *P. anserina* has a small (36 Mb), compact, and haploid genome. This genome has been sequenced, which makes it easier to identify gene function (Espagne *et al.*, 2008). In this article we will describe strategies and conditions to monitor and explore autophagy in filamentous fungi.

2. *Podospora anserina*

The vegetative apparatus, or mycelium, of *P. anserina* is a network of interconnected filaments, or hyphae, forming a syncitium: hyphae are divided into plurinucleated cells. Within the filaments, cytoplasmic continuity is allowed by perforated cross-walls termed *septa* (Pinan-Lucarré *et al.*, 2007). Fungi are heterotrophic organisms that absorb nutrients from the growth medium. The mycelium represents an appropriate feeding system as it displays extensive interaction surface with the external medium. *P. anserina* is an ascomycete and its life cycle includes sexual but not asexual reproduction (Coppin *et al.*, 1997). Specialized structures involved in sexual reproduction can differentiate from the mycelium. Under starvation conditions and in the presence of light, female reproductive structures termed *protoperithecia* differentiate. Inside each protoperithecium the female gamete, termed an *ascogonium*, is embedded. Their fertilization occurs by a male gamete of the opposite mating type that is a small uninucleated cell termed a *microconidium*. Fertilization initiates the development of the fruiting body or perithecium, in which asci are formed through a complex developmental process. After fertilization, the male and female haploid nuclei proliferate before formation of dikaryotic cells bearing each parental nucleus. Karyogamy and meiosis then proceed in this ascus mother cell giving rise to the nuclei of ascospores. After a postmeiotic mitosis, four large ascospores, each containing two nonbrother nuclei of opposite mating type, are delineated inside the ascus. A low percentage of asci contain five spores, in which one large binucleated ascospore is substituted by two uninucleated ascospores of each mating type. Germination of uninucleated spores results in homokaryotic colonies that are preferentially used in genetic analysis. Spore germination yields short hyphae, which soon branch and fuse to form the mycelium. This life cycle allows production of approximately a hundred asci from a single fertilization event.

3. Vegetative Incompatibility as an Alternative Way of Autophagy Induction

In filamentous fungi, somatic fusions can occur between hyphae from the same or different individuals. Fusion between individuals leads to the formation of heterokaryotic cells. Most often, the heterokaryotic cells are nonviable as a result of genetic differences at specific loci termed *het* loci and non-self-recognition in these mixed cells. This is referred to as *vegetative* or *heterokaryon incompatibility*. Heterokaryotic fusion cells are destroyed through a degenerative and lytic process termed *cell death by incompatibility*. Genetic and molecular analyses of *het* genes have been conducted both in *P. anserina* and *Neurospora crassa* (for a review, see Glass and Dementhon, 2006). The incompatibility reaction has been characterized in *Podospora* using the self-incompatible (SI) *het-R het-V* strain The *het-R het-V* strain was obtained from the progeny of the cross between the *het-R het-V1* strain and the *het-r het-V* strain. The *het-r het-V* strain corresponds to the *s* wild-type isolate from the G. Rizet and J. Bernet collections. The *het-R het-V1* strain (Bernet, 1967) is isogenic to the *s* strain except for the *het-r* and *het-v* loci. The *het-R* and *het-V1* genes were present in the M wild-type isolate from the same collection. The SI strain bears the two incompatible *het-R* and *het-V* genes in all nuclei; their coexpression triggers cell death in all the cells and hence in the entire mycelium. This SI strain was particularly convenient for cell death characterization, as cell death triggering is thermosensitive (Labarère, 1973). The strain can grow at 32 °C. The transfer to 26 °C triggers cell death by incompatibility. The SI *het-R het-V* strain has been extensively used to characterize biochemical and molecular aspects of the incompatibility reaction (for a review, see Pinan-Lucarré *et al.*, 2007). In particular this reaction is associated with transcriptional up-regulation of a specific set of genes termed *idi* genes (*i*nduced *d*uring *i*ncompatibility). Two of them at least (*idi-6/pspA* and *idi-7*) are involved in the autophagic process, induced upon starvation and by rapamycin treatment, and essential for differentiation in *Podospora* (Dementhon *et al.*, 2003; Pinan-Lucarré *et al.*, 2003) as in most eukaryotes (Levine and Klionsky, 2004). *idi-6/pspA* is the ortholog of the yeast *PRB1* gene encoding the vacuolar protease B and *idi-7* is the ortholog of the yeast *ATG8* gene and therefore termed *PaATG8*. Up-regulation of these genes during cell death by incompatibility suggests the induction of the autophagic process.

The SI *het-R het-V* strain has been also used to characterize cytological aspects of the incompatibility reaction. Using optical microscopy, the development of this cell death reaction has been recorded during 14 h after transfer to restrictive temperature (Pinan-Lucarré *et al.*, 2005). The death reaction occurs in nearly all cells but in an asynchronous manner. The main feature of dying cells is vacuolization occurring through a rapid morphological change of the vacuolar compartment from a preexisting tubular network to round-shaped structures (Pinan-Lucarré *et al.*, 2003). These spherical vacuoles invade most of the cell volume and fuse, and finally cell lysis occurs. Cell collapse could result either from vacuolar or cytoplasmic membrane rupture. This cell death reaction is also associated with the increased deposition of septa, accumulation of lipid droplets, and abnormal deposition of cell wall material (Dementhon *et al.*, 2003). Using electron and fluorescence microscopy, we observe large autophagosomes and autophagic bodies in an SI strain as soon as 15 min after cell death triggering (Pinan-Lucarré *et al.*, 2003, 2005, 2007). This induction occurs while cells are grown on rich medium. Vegetative incompatibility constitutes an alternative to starvation for induction of autophagy. Interestingly, rapamycin treatment mimics the incompatibility reaction suggesting that *het* gene coexpression would lead to TOR kinase inactivation (Dementhon *et al.*, 2003; Pinan-Lucarré *et al.*, 2006). To determine the role of autophagy in vacuolization and cell death by incompatibility the *PaATG1* gene encoding the ortholog of the yeast kinase Atg1 has been inactivated. Neither vacuolization nor cell death is suppressed in the Δ*PaATG1* autophagy mutant, indicating that autophagy is not responsible for the noticeable enlargement of the vacuolar compartment and does not play a causal role in this cell death reaction (Pinan-Lucarré *et al.*, 2005). Rather, autophagy participates in cell survival, as both *PaATG1* and *PaATG8* gene deletions lead to accelerated cell death in SI mutant strains.

In natural conditions, cell death by incompatibility is observed mainly after somatic fusion between individuals. This cell death reaction is limited to the heterokaryotic cell resulting from the fusion and never spreads throughout the parental strains despite their syncitial structure. To survive an incompatible interaction, fungal strains have to carefully control cell death. Soon after the fusion, the syncitial structure is lost as perforated cross-walls bordering the heterokaryotic cell are sealed. Septal pore occlusion occurs probably through plugging by Woronin bodies, which are fungal-specific specialized peroxisomes (Jedd and Chua, 2000). The induction of autophagy in cells surrounding the heterokaryotic cell could participate in the protection of these neighboring cells from cell death. Autophagy might be required to eliminate prodeath signal(s) or damaged organelles moving out of the heterokaryotic cell into adjacent homokaryotic cells. Both autophagy and cross-wall sealing could be responsible for the confinement of cell death by incompatibility to the fusion cell. When programmed cell death (PCD) occurs as a host defense in the hypersensitive response (HR) in plants, or in apoptosis triggered by viral infection, the cell death reaction has to be precisely spatially restricted (Everett and McFadden, 1999; Lam, 2004). Liu *et al.* have shown that autophagy negatively regulates PCD and functions to restrict HR PCD to the site of pathogen infection (Liu *et al.*, 2005). Similar to the HR, cell death by incompatibility is precisely spatially restricted.

4. Autophagosome and Autophagic Body Examination

Autophagosomes can be examined either by electron microscopy or by fluorescence microscopy using GFP-tagged PaATG8 protein. The 121 amino acid long PaATG8 protein is the ortholog of the yeast Atg8 protein (90% similarity, 79% identity) (Pinan-Lucarré *et al.*, 2003). The yeast full-length protein is cytosolic under favorable growth conditions. Upon starvation, this protein relocalizes from the cytosol to the forming autophagosomal membrane and last to the vacuolar lumen (Huang *et al.*, 2000; Kim *et al.*, 2001; Kirisako *et al.*, 1999). The Atg8 protein is thus either soluble or tightly bound to the autophagosomal membrane as a result of a two-step C-terminal end processing: the last amino-acid is removed by the Atg4 protease and the glycine residue exposed as the new C-terminal end is covalently conjugated to a phosphatidylethanolamine (Ichimura *et al.*, 2000; Kim *et al.*, 2001; Kirisako *et al.*, 2000). This phospholipid allows anchorage of Atg8 on both external and internal membranes of the forming autophagosome. Before autophagosome fusion with the vacuole, external membrane Atg8 deconjugation occurs. Internal membrane-associated Atg8 is delivered to the vacuolar lumen after fusion. The processed glycine residue is strongly conserved among Atg8 orthologs and present in the PaATG8 polypeptide. As a result of this processing, Atg8 can only be GFP-tagged at the N-terminal end to follow its localization throughout the autophagic process. The corresponding *Podospora* fusion protein was termed GFP-PaATG8.

4.1. The GFP-PaATG8 marker

The fusion protein GFP-PaATG8 has been constructed on the model of the yeast GFP-Atg8 protein described by Kim *et al.* (2001). In these constructs, GFP is fused to the third NH_2-terminal amino acid of PaATG8 and Atg8, through the same two-amino-acid linker. The *gfp-PaATG8* fusion gene is under the control of the *PaATG8* promoter.

4.1.1. Construct design, fungal transformation, and fluorescence microscopy analysis

The pCB*gfp-PaATG8* plasmid bears the *egfp-PaATG8* fusion gene and bacterial and fungal selection markers. To construct the pCB*gfp-PaATG8* plasmid, a *Xho*I-*Cla*I fragment containing 2.6 kb of the *PaATG8* locus was first cloned into the pBluescript SKII+ vector. Unique *Nco*I and *Bsr*GI sites were then introduced by PCR at the 5′ end of the *PaATG8* ORF, and a *Nco*I-*Bsr*GI fragment containing the *egfp* ORF was excised from the pEGFP vector and cloned into these unique sites. The resulting construct contains 1.7 kb of *PaATG8* upstream sequences containing the promoter region, followed by the *egfp* ORF, the *PaATG8* ORF, and 355 bp of *PaATG8* downstream sequences. This construct was then subcloned into the pCB1004 vector bearing hygromycin and chloramphenicol resistance genes by an *Xho*I-*Cla*I digest resulting in the pCB*gfp-PaATG8* plasmid. The sequence of the fusion region between the two ORFs is given in Fig. 18.1A.

The pCB*gfp-PaATG8* plasmid was used in transformation experiments of hygromycin-sensitive protoplasts prepared from either the wild-type strain or the SI *het-R het-V* strain.

4.1.2. Preparation and transformation of protoplasts

Protoplasts are prepared for transformation according to the following procedure. Optimal temperatures for growth are respectively 26 °C for the wild-type and 32 °C for the *het-R het-V* strain. All steps are performed in sterile conditions:

1. Roux flasks are inoculated with mycelia grown on solid G4 medium and fragmented either with a scalpel or in a Waring blender for 15 s. G4 medium corresponds to D0 medium supplemented with ammonium acetate 4g/l. Solid D0 medium is a corn meal agar medium containing corn flour 25 g/l and corn cream 15 g/l, dissolved with H_2O, mixed and incubated overnight at 58 °C before filtering the mix through standard filter paper (cat. no. JO33107, Prat-Dumas), volume adjustment with H_2O, bacto agar addition (20 g/l) and sterilization by autoclaving for 30 min at 110 °C (Esser, 1974).

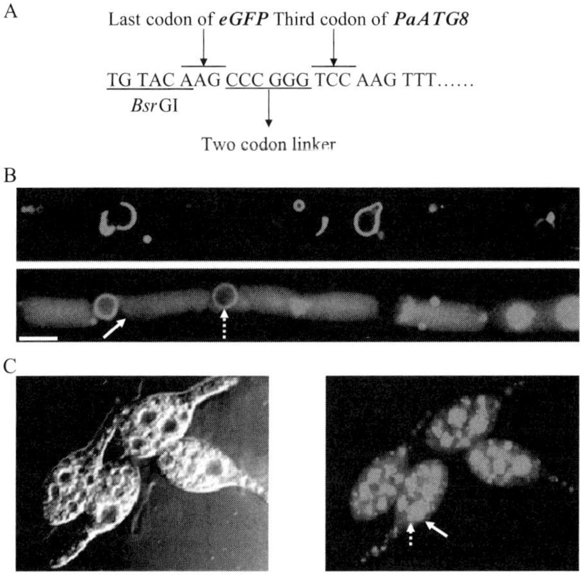

Figure 18.1 The GFP-PaATG8 autophagy marker. (A) Fusion region between *egfp* and *PaATG8* ORFs in the *egfp-PaATG8* fusion gene. (B) Autophagosomes and vacuoles labeled by GFP-PaATG8 during cell death by incompatibility: the SI *egfp-PaATG8* strain was observed after transfer to 26 °C for 4 h. Adapted from Pinan-Lucarré *et al.* (2007). (C) Autophagosomes and vacuoles labelled by GFP-PaATG8 upon ascospore formation: ascospores were obtained in the progeny of an homozygous cross of a compatible *egfp-PaATG8* strain and observed before melanization of the cell wall as fluorescence was not observable in mature black ascospores. The solid arrows point to a vacuole; the dotted arrows indicate an autophagosome. Scale bar: 3 μm. (See Color Insert.)

2. Cultures are grown for 30 h at optimal temperature (26 °C or 32 °C) in the dark in liquid FRIES medium (100ml/roux flask). FRIES medium is 2.5 g/l yeast extract, 5g/l peptone, 5 g/l malt extract, 5 g/l glucose, 5g/l sucrose, 5 μg/l biotin, and 0.1 ml/l of trace element concentrate solution. The trace-element concentrate solution contains 50 g/l citric acid, 50 g/l $ZnSO_4$ heptahydrate, 10 g/l $Fe(NH_4)_2(SO_4)_2$ hexahydrate, 2.5 g/l $CuSO_4$ pentahydrate, 0.5 g/l $MnSO_4$ monohydrate, 0.5 g/l boric acid and 0.5 g/l $MoNa_2SO_4$ dihydrate.
3. The mycelium is collected by filtration on sterile gauze and washed with TPS1 buffer (0.6 *M* sucrose, 5 m*M* Na_2HPO_4, 45 m*M* KH_2PO_4).
4. Weigh the mycelium. X g of wet mycelium is put in X ml of TPS1 containing 40 mg/ml of Glucanex (Novo Nordisk Ferment AG) and digested for 2–3 h at 37 °C.
5. The protoplasts are separated from mycelial debris by filtration through sterile gauze.

6. The filtrate is centrifuged 10 min at 3200 rpm, and the protoplasts are washed twice with 0.1 X ml TPS1 and once with 0.1 X ml TPC buffer (0.6 *M* sucrose, 10 m*M* $CaCl_2$, 10 m*M* Tris, pH 7.5).
7. Protoplast concentration is determined in TPC suspension before the last centrifugation by counting under a microscope with a hemacytometer. The final pellet is resuspended in TPC buffer in a volume such that the protoplast concentration is 10^8/ml. Protoplasts can be transformed immediately or stored at −70 °C.
8. Before transformation, the protoplasts are subjected to a 5-min heat shock at 48 °C and transferred to ice.
9. The DNA is added (5 μg of DNA/0.2 ml protoplasts) and protoplasts are incubated for 10 min at room temperature.
10. Two ml of a PEG solution (60% polyethylene glycol 4000, 10 m*M* $CaCl_2$, 10 m*M* Tris, pH 7.5) are added and carefully mixed.
11. After a 15-min incubation, the protoplasts are distributed in tubes containing top agar placed in a 42 °C water bath and plated on minimal selective medium containing 0.8 *M* sucrose as an osmotic stabilizer (top agar contains 0.2 *M* sorbitol in addition to sucrose). Solid minimal medium is 5 g/l dextrin (starch gum). [23277.364, VWR], 0.25 g/l KH_2PO_4, 0.3 g/l K_2HPO_4, 0.25 g/l $MgSO_4$ (if $MgSO_4{\cdot}7H_2O$), 0.5 g/l urea, 0.05 mg/l thiamine, 0.25 μg/l biotin, 0.1 ml/l of trace element concentrate solution, and agar 12.5 g/l. Routinely, 10–50 transformants are obtained per μg of plasmid DNA. The *hph* gene allows selection of transformants resistant to hygromycin (100 μg/ml).

4.1.3. Fluorescence microscopy analysis

To be analyzed by fluorescence microscopy, the mycelium is observed directly on its growth medium solidified with agarose.

1. 2% agarose solid SA medium is used as growth medium. SA medium is 20 g/l dextrin, 4 g/l ammonium acetate, 0.05 g/l NaCl, 7.5 mg/l $CaCl_2$, 0.17 g/l K_2HPO_4, 0.1 g/l $MgSO_4$, 5 μg/l biotin, 100 μg/l thiamine, and 0.1 ml/l of trace-element concentrate solution. 2% agarose SA medium is SA medium solidified with 2% agarose. 20 ml of 2% agarose SA medium is prepared in Falcon 100-mm-diameter Petri dishes (cat. no. 391.2202, VWR) by pouring two successive 10-ml layers of medium.
2. One explant (approximately 10 mm^3 colonized agar piece) of approximately 9 freshly cultured hygromycin-resistant transformants are inoculated per plate containing SA agarose medium and grown for 16–24 h at optimal temperature (26 °C or 32 °C).
3. The whole mycelium (approximately 1–1.5 cm in diameter) grown on the top layer of the agarose SA medium was cut out (approximately 2–3 cm^2 agar piece) and either examined directly by light and fluorescence

microscopy or transferred to another SA agarose plate for further incubation at different temperature or rapamycin treatment before examination. To observe the GFP fluorescence, excitation was at 450–490 nm and fluorescence was detected at 515–560 nm.

4.1.4. Use of GFP-PaATG8 as an autophagy marker

In *Podospora*, autophagy induction can be observed upon nutritional starvation, rapamycin treatment, or along the vegetative incompatibility reaction.

1. Transformants bearing the *gfp-PaATG8* fusion gene can be analyzed by fluorescence microscopy to localize the GFP fusion protein following several different regimen:
   a. After 16 h growth at 26 °C on rich medium (SA agarose medium) or on low nitrogen medium (SA agarose lacking ammonium acetate).
   b. After 16 h growth on rich medium and further incubation for 1 h at 26 °C after transfer of the whole colony to SA agarose supplemented or not with rapamycin (200 ng/ml).
   c. The *gfp-PaATG8 het-R het-V* transformants can be analyzed after 16 h growth on rich medium (SA agarose medium) at 32 °C or after various times (between 15 min and 4 h) following transfer to nonpermissive temperature (26 °C, on preheated SA agarose medium).

      Under nutrient-rich growth conditions and permissive temperature, the GFP-PaATG8 fusion protein has a diffuse cytoplasmic distribution and is associated with dotlike structures distinct from the vacuolar compartment that can be revealed by the fluorescent FM 4-64 dye.
2. For FM 4-64 labeling (Molecular Probes), 10 μl of a 4 μg/ml solution was applied to the mycelium during 20 min and then the mycelium was washed in 150 m*M* NaCl. FM 4-64 excitation was at 530–595 nm and fluorescence was detected at >615 nm (see also the chapters by H. Abeliovich, van Zutphen *et al.*, and Oku and Sakai in this volume).

In noninducing conditions, vacuoles formed a tubular network and were often located near the cytoplasmic membrane. When autophagy was induced by nitrogen starvation or rapamycin treatment (wild-type strain) or shift to nonpermissive temperature (SI strain), spherical vacuoles are observed. The GFP-PaATG8 fusion protein localizes to large punctate perivacuolar structures and to the lumen of the vacuoles (Dementhon *et al.*, 2003; Pinan-Lucarré *et al.*, 2003, 2005). As expected, the PaATG8 protein relocalizes from the cytoplasm to the vacuoles upon autophagy induction. Perivacuolar structures can be clearly identified as autophagosomes or preautophagosomal structures in SI strains using an inverted microscope (Zeiss 200*M*). Autophagosomes appear as fluorescent circles of various size (largest size observed is 2 μm; Figure 18.1B and Pinan-Lucarré *et al.*, 2005). GFP-PaATG8 relocalization specificity can be tested by analysis of the distribution of a cytosolic GFP protein that is not affected by autophagy induction. In particular, no accumulation of the GFP protein in the vacuole is observed during starvation (Pinan-Lucarré *et al.*, 2003).

4.1.5. Use of GFP-PaATG8 to monitor the loss of autophagy

To analyze autophagy function in *Podospora*, inactivation of *PaATG1* and *PaATG8* was achieved by gene replacement both in the wild-type strain and the self-incompatible *het-R het-V* strain. The construction of the self-incompatible *het-R het-V* Δ*PaATG* strains by a cross between the *het-R het-V* strain and the Δ*PaATG* mutant strain could not be performed because of the low fertility of crosses between strains bearing incompatible *het-R* and *het-V* genes and to the female sterility conferred by the Δ*PaATG* mutations (Labarère *et al.*, 1974; Pinan-Lucarré *et al.*, 2003). Loss of autophagy in the Δ*PaATG1* strain can be verified using the GFP-PaATG8 fusion protein.

1. Protoplasts of the SI Δ*PaATG1* strain are generated as described in section 4.1.1.
2. The protoplasts are transformed with the pCB*gfp-PaATG8* plasmid as described in section 4.1.1.
3. Hygromycin-resistant transformants are selected and examined by fluorescence microscopy as described previously.

After transfer to 26 °C, no autophagosomes are observed inside the cytoplasm and no fluorescence accumulates inside the vacuole compared to the control, indicating, as expected, that neither autophagosomes nor autophagic bodies are formed in this strain (Pinan-Lucarré *et al.*, 2005). The perivacuolar fluorescent dots likely correspond to preautophagosomal structures as observed in the yeast *atg1*Δ mutant (Suzuki *et al.*, 2004).

4.2. Limiting proteolytic activities to detect autophagic bodies

Autophagy induction can be also revealed by the presence of autophagic bodies inside the vacuoles (see the chapter by M. Thumm in this volume). When autophagy is induced in yeast, a few autophagic bodies are observed transiently 1–2 h after starvation (Takeshige *et al.*, 1992). Further accumulation of autophagic bodies occurs either in the presence of the proteinase inhibitor PMSF or in a vacuolar protease B-deficient strain. Autophagic bodies are degraded rapidly by vacuolar enzymes, but their observation can be greatly enhanced using limiting proteolysis conditions.

4.2.1. Using PMSF to monitor autophagic body accumulation

The effect of PMSF in preventing the vacuolar degradation of autophagic bodies has been reproduced in *P. anserina*.

1. Approximately 4 explants of the self-incompatible *het-R het-V* strain are inoculated onto 100-mm-diameter plates containing SA agarose solid medium and incubated at 32 °C for 16 h.
2. For PMSF treatment, the whole mycelium of a single colony grown on the top layer of the SA agarose medium is cut out (approximately 2–3 cm^2 agar piece) and transferred for further incubation at 32 °C for 1 h on another 100-mm-diameter plate containing SA agarose medium supplemented with 4 m*M* PMSF.

   Note: PMSF must be dissolved in ethanol and added to the medium just before use because of its low stability in aqueous solutions.

3. Each colony (on agar piece) is then transferred to 26 °C for 4 h on individual preheated plates containing the same SA agarose medium supplemented with 4 m*M* PMSF before observation by light microscopy.

In the presence of PMSF, the vacuoles of the self-incompatible strain are full of spherical bodies that show ceaseless Brownian movement and corresponding to autophagic bodies (Pinan-Lucarré *et al.*, 2003).

4.2.2. The accumulation of autophagic bodies in specific genetic backgrounds

The *Podospora pspA* gene encodes a major vacuolar protease orthologous to the yeast protease B. Loss of function mutation of *pspA* has been obtained by gene replacement using a two-marker strategy (Pinan-Lucarré *et al.*, 2003). The accumulation of autophagic bodies inside the vacuole can be evidenced by comparison of the wild-type and the Δ*pspA* mutant strains using various growth conditions and light microscopy. The two strains are analyzed after 16 h growth at 26 °C on rich nitrogen medium (SA agarose medium) or on low nitrogen medium (SA agarose medium lacking ammonium acetate), or after 16 h growth at 26 °C on SA agarose medium and further incubation at 26 °C for 1 h after transfer of the whole colony (see 4.1.1) to SA medium supplemented with rapamycin (200 ng/ml). On rich medium, at the microscopic level, wild-type and Δ*pspA* strains are indistinguishable. Upon starvation conditions or rapamycin treatment, large spherical vacuoles are observed in both strains. Vacuoles of the Δ*pspA* mutant cells have a granular aspect that results from the accumulation of autophagic bodies (Pinan-Lucarré *et al.*, 2003).

4.3. Electron microscopy to detect autophagic bodies and autophagosomes

4.3.1. Autophagic body examination

To image autophagic bodies, the vacuoles of the Δ*pspA* mutant strain grown on low nitrogen medium can be analyzed. A modification of the chemical fixation technique described by Chang and Tanaka (1970) for electron microscopy is performed.

1. 100-mm-diameter plates containing SA solid medium are covered by a sterile cellophane sheet (cat. no. 1650963, Bio-Rad). Approximately 4 explants of the Δ*pspA* strain are inoculated onto the cellophane sheet and the plates are incubated at 26 °C for 16 h.
2. Cellophane sheets on which fungal colonies grew are transferred using sterile forceps onto preheated plates containing low nitrogen SA medium before further incubation for 4 h at 26 °C.
3. A round piece of cellophane on which *Podospora* was grown is cut and stripped from the growth medium.
4. These samples, placed in a petri dish, are fixed for 4 hours in 10 ml of 2.5% glutaraldehyde solution in 0.1 *M* phosphate buffer, pH 7.2
5. After several washes in 0.1 *M* phosphate buffer, pH 7.2, samples are postfixed overnight at 0 °C with 10 ml of 1% OsO_4 in 0.1 *M* phosphate buffer, pH 7.2.
6. After rinsing in water, the samples are dehydrated in 10 ml of 50% acetone for 15 min and three times in 10 ml of anhydrous acetone for 15 min at room temperature.
7. The fixed mycelia are stained in 10 ml of 2% uranyl acetate in acetone for 2 hours at 4 °C in a darkroom.
8. Samples are then embedded at room temperature in araldite resin progressively (24 h in 10 ml of araldite:acetone (1:3, v/v), 24 h in 10 ml of araldite:acetone (1:1, v/v) and 24 h in 10 ml of araldite:acetone (3:1, v/v) (Araldite epoxy resin, cat. no. A3183, Sigma-Aldrich).
9. Ultrathin sections (80 nm) on copper grids (HS400 Pelanne Instruments) are contrasted 1 min by plunging them in 10 μl of 2% lead citrate aqueous solution and examined in a Philips Tecnai 12 Biotwin (120 kV) electron microscope.

Autophagic bodies appear inside the vacuoles of the Δ*pspA* mutant as single-membrane vesicles with a content that is morphologically similar to the cytosol (Fig. 18.2A). Vacuoles of the mutant contain numerous autophagic bodies, whereas those of the wild type are less dense to electrons and appear empty as a result of full proteolytic activities.

4.3.2. Autophagosome observation

To observe autophagosomes by electron microscopy, the SI strain was analyzed at different times during the course of the cell death reaction. Cryofixation was required for sample preparation and autophagosome observation.

1. Cultures are performed as indicated in 4.3.1. Approximately 4 explants of the *het-R het-V* strain are inoculated onto the cellophane sheet and the plates are incubated at 32 °C for 16 h.

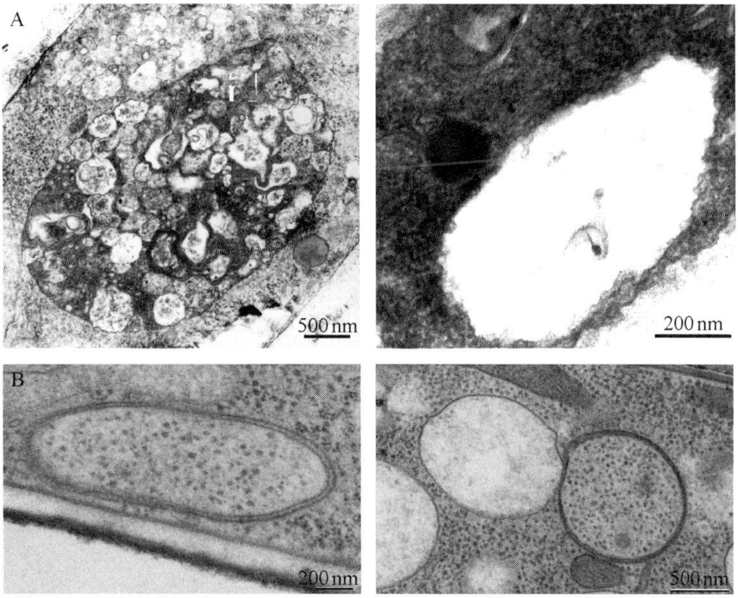

Figure 18.2 Electron microscopy examination of autophagic bodies and autophagosomes. (A) Autophagic bodies examination. The vacuoles of the wild type (*right*) and the Δ*pspA* mutant (*left*) grown under nitrogen starvation conditions are observed. The two strains were grown on SA medium and transferred onto SA medium lacking ammonium acetate for 4 h. (B) Autophagosome examination. The *het-R het-V* strain is observed 1 h after cell death triggering. *Left panel:* large autophagosome (i.e. double-membrane vesicle with cytoplasmic content; diameter between 0.5 μm and 1.6 μm). *Right panel:* Fusion between a vacuole and an autophagosome of similar size (diameter: 1.2 μm). Note in right panels of A and B, the low electron density of the vacuolar lumen rich in hydrolytic activities. Adapted from Pinan-Lucarré *et al.* (2003, 2007).

2. Fungal colonies are transferred (see 4.3.1) for cell death triggering onto preheated plates containing SA medium and further incubated for various times (30 min to 4h) at 26 °C.
3. Pieces of mycelium are collected as follow: the explant in the center of the colony is pulled up and pieces of mycelium carried along are used immediately for cryofixation.
4. Mycelium samples are quickly submersed in liquid propane (−180 °C) and then transferred to a precooled solution of 4% osmium tetraoxide in dry acetone at −82 °C for 48 h for cryosubstitution.
5. Samples are gradually warmed to room temperature, transferred in a flask and washed 3×10 minutes with 5 ml of dry acetone.
6. Stainings, resin embedment, and observation are performed as indicated in 4.3.1 (steps 7–9).

Upon cell death by incompatibility in the SI strain, autophagosomes of various sizes are observed and revealed by their two specific features that are a double membrane and a cytoplasmic content (Fig. 18.2B).

5. Phenotypic Traits of *Podospora* Autophagy Mutants

Podospora Δ*PaATG1*, Δ*PaATG8* and Δ*pspA* autophagy mutants displayed similar growth phenotypes. Their linear growth rate is comparable to that of the wild type, but the aerial hyphae density is decreased, the differentiation of female reproductive organs is suppressed and spore germination is affected. When introduced into the SI genetic background, the Δ*PaATG* mutations confer acceleration of cell death by incompatibility.

5.1. Mycelium without aerial hyphae

To compare aerial hyphae differentiation ability, wild-type and mutant strains are grown on synthetic SU medium (20 g/l dextrin, 0.5 g/l KH_2PO_4, 0.6 g/l K_2HPO_4, 0.5 g/l $MgSO_4$ heptahydrate, 5 μg/l biotin, 100 μg/l thiamine, 1 g/l urea, 0.1 ml/l of trace-element concentrate solution [see section 4.1.1], and 25 g/l bacto-agar). After growth on SU medium at 26 °C for 7 days, the mutants display fewer aerial hyphae than the wild type or no aerial hyphae in the case of the Δ*pspA* mutant. Aerial hyphae appear as white rings around the explants of the wild type (Fig. 18.3 and Pinan-Lucarré *et al.*, 2003).

The three autophagy mutants differentiate fewer aerial hyphae or none. Feeding of aerial hyphae might occur via a specialized mechanism, as no direct contact to the growth medium exists. Nutrients might be transported from hyphae growing at the surface of the medium to the cytoplasm of aerial hyphae. Transport of nutrients could be ensured by vacuoles as proposed (Shoji *et al.*, 2006). Aerial hyphae and/or foot cells during aerial hyphae differentiation might continuously undergo autophagy to enrich vacuoles with nutrients and support growth. When autophagy is affected, insufficient feeding of aerial hyphae would occur leading to a defect in aerial hyphae differentiation.

5.2. Female sterility

The three mutants are male fertile and female sterile. Fertilization experiments are performed to control the production of male and female reproductive structures.

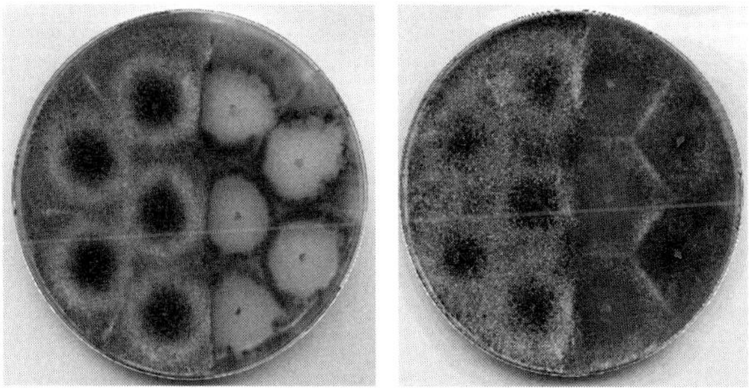

Figure 18.3 Differentiation defects of the Δ*pspA* and Δ*PaATG8* mutant strains. Each 100-mm-diameter plate containing solid SU medium was inoculated with five wild-type (*left*) or mutant (*right*) explants and incubated for 7 days at 26 °C. Mycelium from the Δ*pspA* and Δ*PaATG8* strains was less dense and less pigmented than the wild-type mycelium (WT). Aerial hyphae forming the white rings around the explants of the wild type are few for the Δ*PaATG8* strain or absent for the Δ*pspA* strain. Adapted from Pinan-Lucarré *et al.* (2003).

1. Wild-type and mutant strains are used either as male or as female. The strains are grown separately in the dark for 5 days at 26 °C on solid D0 medium (see 4.1.1).
2. The plates are then transferred to the light on a bench top to trigger differentiation of sexual organs and further incubated at 26 °C.
3. Three days later, cultures are overlaid at room temperature with a suspension containing microconidia (male cells) of the opposite mating type (fertilization step) for 15 min. The remaining liquid, not absorbed by the medium, is discarded. To collect the male cells, approximately 2 ml of sterile water is added to the plates containing the wild-type or mutant strain grown in the same conditions as the female.
4. The plates are checked after a few days for the development of fruiting bodies triggered by the fertilization of female organs by the microconidia. The presence of female organs before fertilization and of male cells in the suspension can be directly observed using a binocular (for female organs) or a microscope (for microconidia). In fertilization experiments their presence is indirectly revealed by the development of fruiting bodies but, importantly, functional organs are detected.

Fertilization of wild-type protoperithecia by wild-type, Δ*pspA*, Δ*PaATG1*, or Δ*PaATG8* mutant microconidia leads to the differentiation of normal fruiting bodies. The three mutants are male fertile similar to the wild type. The Δ*pspA* mutant produces very few protoperithecia whose fertilization by wild-type microconidia gives no perithecia (fruiting bodies).

The protoperithecia become larger and roughly double their size but do not produce asci. This suggests that fertilization occurs but that further maturation of the fertilized organ is blocked. The Δ*PaATG1* and Δ*PaATG8* mutants produce no protoperithecia at all. These three mutants are female sterile (Pinan-Lucarré *et al.*, 2003, 2005).

Development of female reproductive structures in filamentous fungi is an intricate developmental process, which occurs when nutrients allowing vegetative growth are exhausted. Therefore, differentiation of the female organ has to rely on internal sources for the nutrient supply. Autophagy provides new amino acids pools from the degradation of preexisting proteins. If autophagy is affected, this internal nutrient supply providing the new building blocks for differentiation of female tissue is abolished. This readily could explain why inactivation of autophagy in Δ*pspA* and Δ*PaATG* mutants leads to female sterility in Podospora. Autophagy thus plays an essential role in fungal development.

5.3. Spore germination defects

In addition to its requirement for female organ differentiation, autophagy is involved in another step of sexual development, as Δ*PaATG* and Δ*pspA* autophagy mutants display defects in ascospore germination. As autophagy mutants are female sterile, spore germination can be analyzed only with the progeny of heterozygous crosses.

1. Δ*PaATG* and Δ*pspA* mutant strains are crossed as male parents with wild-type females as indicated in section 5.2 with one modification. The strains and importantly the females are grown on D0 medium supplemented with dihydrostreptomycin 4g/l. Addition of dihydrostreptomycin to the growth medium increases the percentage of 5-spore asci.
2. Perithecia are opened with a sterile needle and then the rosette of asci is transferred to a solid medium (agar 3%) to facilitate dissection of the asci.
3. Two small spores of a 5-spore asci corresponding to the two uninucleated spores are picked and inoculated onto G4 germination medium (see 4.1.1). Because of a germination delay, the spores are inoculated on separate small G4 medium plates (5.5-cm diameter) and incubated at optimal temperature (26 °C).

In the progeny of the heterozygous cross between the Δ*pspA* strain and the wild type, only 64% (112/176) of the analyzed uninucleated spores could germinate while 100% germinated in a wild-type cross. Among the 64%, 84% (94/112) germinated normally within the first 48 h after inoculation and 16% (18/112) showed up to 7 days of germination delay. Similar results are obtained with Δ*PaATG* mutants. The segregation of the mutant alleles can be followed on the basis of the absence of aerial hyphae. Overall, approximately 52% of the analyzed spores produced wild-type strains (92/176) and 11% (20/176) produced mutant strains. Thus, all wild-type spores germinated, while less than a quarter of the mutant spores did, indicating that autophagy is involved in spore germination. Induction of autophagy occurs during spore formation as shown by the labeling of autophagosomes and of the vacuolar lumen by the GFP-PaATG8 marker in young spores (Fig. 18.1C). Autophagy would be involved in internal nutrient supply required for optimal germination.

5.4. Accelerated cell death

To compare the time course of cell death by incompatibility in the wild-type and the Δ*PaATG* SI strains, video microscopy and quantification of dead cells by light microscopy can be performed. To measure the percentage of dead cells, the mycelia are labeled with Evans blue (cat. no. 720-1179, VWR) and observed at different times before and after transfer of agar pieces containing whole mycelia to 26 °C (see 4.1.1). Cells are defined as part of hyphae delimited by two consecutive septa. For each time point approximately 1000 fungal cells are analyzed. The percentage of cell death corresponds to the percentage of stained cells. For Evans blue staining, the whole colony (approximately 1 cm^2) is covered with 10 μl of a 1% (w/v) solution in sterile distilled water for 20 min and then the mycelium is washed in 150 m*M* NaCl. Evans blue can be both observed as blue staining by light microscopy, or as red fluorescent staining (excitation at 530–595 nm and fluorescence detected at >615 nm).

Globally, the rate of cell death is increased in the Δ*PaATG* mutants (Pinan-Lucarré *et al.*, 2005). Six hours after the SI strain is transferred to 26 °C, approximately 50% of the cells are dead in the Δ*PaATG* mutants, whereas the cell death rate only reaches 10% in the control. Approximately 80% of cell death occurs 9 h after transfer for the SI Δ*PaATG* strains, whereas in the control the same percentage is only observed 24 h after transfer. No cell death is observed for the three strains maintained at 32 °C. Cell death is accelerated in the Podospora autophagy mutants indicating that autophagy negatively regulates cell death in this fungal model as in the plant hypersensitive response.

6. Concluding Remarks

In fungi, autophagy induction occurs upon starvation or rapamycin treatment; in the *Podospora* SI strain induction occurs in nutrient rich conditions. Investigation of this additional mechanism of induction will be of interest to fully understand the control of autophagy and its physiological functions. A simple change in temperature is able to trigger massive autophagy in this SI strain. Numerous autophagosomes are observed as soon as 15 min after temperature shift and their size can reach several micrometers. This strain could constitute an alternative model to investigate still debated or unsolved points of autophagosome biogenesis such as membrane origin or elongation. In particular the size and abundance of autophagosomes that are rapidly formed during the incompatibility reaction might make ultrastructural characterization particularly favorable.

ACKNOWLEDGMENTS

This work was supported by the European Commission (Transdeath: Contract #511983) and by ANR SexDevMycol NT05-1-41707

REFERENCES

Bernet, J. (1967). Systems of incompatibility in *Podospora anserina*. *C. R. Acad. Sci. Hebd. Seances. Acad. Sci. D* **265,** 1330–1333.

Bonnet, C., Espagne, E., Zickler, D., Boisnard, S., Bourdais, A., and Berteaux-Lecellier, V. (2006). The peroxisomal import proteins PEX2, PEX5 and PEX7 are differently involved in *Podospora anserina* sexual cycle. *Mol. Microbiol.* **62,** 157–169.

Chang, S. T., and Tanaka, K. (1970). Culturing and embedding of filamentous fungi for electron micrography in the plane of the filaments. *Stain Technol.* **45,** 109–113.

Coppin, E., de Renty, C., and Debuchy, R. (2005). The function of the coding sequences for the putative pheromone precursors in *Podospora anserina* is restricted to fertilization. *Eukaryot. Cell* **4,** 407–420.

Coppin, E., Debuchy, R., Arnaise, S., and Picard, M. (1997). Mating types and sexual development in filamentous ascomycetes. *Microbiol. Mol. Biol. Rev.* **61,** 411–428.

Dementhon, K., Paoletti, M., Pinan-Lucarré, B., Loubradou-Bourges, N., Sabourin, M., Saupe, S. J., and Clavé, C. (2003). Rapamycin mimics the incompatibility reaction in the fungus *Podospora anserina*. *Eukaryot. Cell* **2,** 238–246.

Espagne, E., Lespinet, O., Malagnac, F., Da Silva, C., Jaillon, O., Porcel, B. M., Couloux, A., Aury, J. M., Segurens, B., Poulain, J., Anthouard, V., Grossetete, S., *et al.* (2008). The genome sequence of the model ascomycete fungus *Podospora anserina*. *Genome Biol.* **9,** R77.

Esser, K. (1974). Podospora anserina. *In* "Handbook of genetics," (R. C. King, ed.), pp. 531–551. Plenum Press, New York.

Everett, H., and McFadden, G. (1999). Apoptosis: An innate immune response to virus infection. *Trends Microbiol.* **7,** 160–165.

Glass, N. L., and Dementhon, K. (2006). Non-self recognition and programmed cell death in filamentous fungi. *Curr. Opin. Microbiol.* **9,** 553–558.

Golstein, P., Aubry, L., and Levraud, J. P. (2003). Cell-death alternative model organisms: Why and which? *Nat. Rev. Mol. Cell Biol.* **4,** 798–807.

Huang, W.-P., Scott, S. V., Kim, J., and Klionsky, D. J. (2000). The itinerary of a vesicle component, Aut7p/Cvt5p, terminates in the yeast vacuole via the autophagy/Cvt pathways. *J. Biol. Chem.* **275,** 5845–5851.

Ichimura, Y., Kirisako, T., Takao, T., Satomi, Y., Shimonishi, Y., Ishihara, N., Mizushima, N., Tanida, I., Kominami, E., Ohsumi, M., Noda, T., and Ohsumi, Y. (2000). A ubiquitin-like system mediates protein lipidation. *Nature* **408,** 488–492.

Jedd, G., and Chua, N. H. (2000). A new self-assembled peroxisomal vesicle required for efficient resealing of the plasma membrane. *Nat. Cell Biol.* **2,** 226–231.

Kicka, S., Bonnet, C., Sobering, A. K., Ganesan, L. P., and Silar, P. (2006). A mitotically inheritable unit containing a MAP kinase module. *Proc. Natl. Acad. Sci. USA* **103,** 13445–13450.

Kim, J., Huang, W.-P., and Klionsky, D. J. (2001). Membrane recruitment of Aut7p in the autophagy and cytoplasm to vacuole targeting pathways requires Aut1p, Aut2p, and the autophagy conjugation complex. *J. Cell Biol.* **152,** 51–64.

Kirisako, T., Baba, M., Ishihara, N., Miyazawa, K., Ohsumi, M., Yoshimori, T., Noda, T., and Ohsumi, Y. (1999). Formation process of autophagosome is traced with Apg8/Aut7p in yeast. *J. Cell Biol.* **147,** 435–446.

Kirisako, T., Ichimura, Y., Okada, H., Kabeya, Y., Mizushima, N., Yoshimori, T., Ohsumi, M., Takao, T., Noda, T., and Ohsumi, Y. (2000). The reversible modification regulates the membrane-binding state of Apg8/Aut7 essential for autophagy and the cytoplasm to vacuole targeting pathway. *J. Cell Biol.* **151,** 263–276.

Labarère, J. (1973). [Properties of an incompatibility system in Podospora anserina fungus and value of this system for the study of incompatibility]. *C. R. Acad. Sci. Hebd. Seances. Acad. Sci. D* **276,** 1301–1304.

Labarère, J., Bégueret, J., and Bernet, J. (1974). Incompatibility in *Podospora anserina*: Comparative properties of the antagonistic cytoplasmic factors of a nonallelic system. *J. Bacteriol.* **120,** 854–860.

Lam, E. (2004). Controlled cell death, plant survival and development. *Nat. Rev. Mol. Cell Biol.* **5,** 305–315.

Levine, B., and Klionsky, D. J. (2004). Development by self-digestion: Molecular mechanisms and biological functions of autophagy. *Dev. Cell* **6,** 463–477.

Liu, Y., Schiff, M., Czymmek, K., Tallóczy, Z., Levine, B., and Dinesh-Kumar, S. P. (2005). Autophagy regulates programmed cell death during the plant innate immune response. *Cell* **121,** 567–577.

Lorin, S., Dufour, E., and Sainsard-Chanet, A. (2006). Mitochondrial metabolism and aging in the filamentous fungus *Podospora anserina*. *Biochim. Biophys. Acta* **1757,** 604–610.

Maddelein, M. L., Dos Reis, S., Duvezin-Caubet, S., Coulary-Salin, B., and Saupe, S. J. (2002). Amyloid aggregates of the HET-s prion protein are infectious. *Proc. Natl. Acad. Sci. USA* **99,** 7402–7407.

Malagnac, F., Klapholz, B., and Silar, P. (2007). PaTrx1 and PaTrx3, two cytosolic thioredoxins of the filamentous ascomycete *Podospora anserina* involved in sexual development and cell degeneration. *Eukaryot. Cell* **6,** 2323–2331.

Paoletti, M., Saupe, S. J., and Clavé, C. (2007). Genesis of a fungal non-self recognition repertoire. *PLoS ONE* **2,** e283.

Pinan-Lucarré, B., Balguerie, A., and Clavé, C. (2005). Accelerated cell death in Podospora autophagy mutants. *Eukaryot. Cell* **4,** 1765–1774.

Pinan-Lucarré, B., Iraqui, I., and Clavé, C. (2006). *Podospora anserina* target of rapamycin. *Curr. Genet.* **50,** 23–31.

Pinan-Lucarré, B., Paoletti, M., and Clavé, C. (2007). Cell death by incompatibility in the fungus Podospora. *Semin. Cancer Biol.* **17,** 101–111.

Pinan-Lucarré, B., Paoletti, M., Dementhon, K., Coulary-Salin, B., and Clavé, C. (2003). Autophagy is induced during cell death by incompatibility and is essential for differentiation in the filamentous fungus *Podospora anserina*. *Mol. Microbiol.* **47,** 321–333.

Saupe, S. J., Clavé, C., and Bégueret, J. (2000). Vegetative incompatibility in filamentous fungi: Podospora and neurospora provide some clues. *Curr. Opin. Microbiol.* **3,** 608–612.

Shoji, J. Y., Arioka, M., and Kitamoto, K. (2006). Possible involvement of pleiomorphic vacuolar networks in nutrient recycling in filamentous fungi. *Autophagy* **2,** 226–227.

Suzuki, K., Noda, T., and Ohsumi, Y. (2004). Interrelationships among Atg proteins during autophagy in Saccharomyces cerevisiae. *Yeast* **21,** 1057–1065.

Takeshige, K., Baba, M., Tsuboi, S., Noda, T., and Ohsumi, Y. (1992). Autophagy in yeast demonstrated with proteinase-deficient mutants and conditions for its induction. *J. Cell Biol.* **119,** 301–311.

van Diepeningen, A. D., Debets, A. J., Slakhorst, S. M., and Hoekstra, R. F. (2008). Mitochondrial pAL2-1 plasmid homologs are senescence factors in *Podospora anserina* independent of intrinsic senescence. *Biotechnol J.* In press.

Wasmer, C., Lange, A., Van Melckebeke, H., Siemer, A. B., Riek, R., and Meier, B. H. (2008). Amyloid fibrils of the HET-s(218–289) prion form a β solenoid with a triangular hydrophobic core. *Science* **319,** 1523–1526.

CHAPTER NINETEEN

Monitoring Autophagy in *Magnaporthe oryzae*

Xiao-Hong Liu,* Tong-Bao Liu,* *and* Fu-Cheng Lin*[,1]

Contents

* State Key Laboratory for Rice Biology, Institute of Biotechnology, Zhejiang University, Huajiachi Campus, Hangzhou, Zhejiang, China
[1] Corresponding author

Methods in Enzymology, Volume 451
ISSN 0076-6879, DOI: 10.1016/S0076-6879(08)03219-9

Abstract

Autophagy is a ubiquitous degradative pathway for the bulk degradation of eukaryotic macromolecules and organelles in eukaryotic cells (Klionsky, 2005; Levine and Klionsky, 2004). Previously, the role of autophagy in turgor generation in plant pathogenic fungi was unknown. Currently, autophagy is confirmed as an important pathway for turgor accumulation in the appressorium (the tips of the invasive hyphae; Liu *et al.*, 2007b) using a technique of targeted gene replacement, deleting the genes that code for *Magnaporthe oryzae* homologs of yeast autophagy-related (*ATG*) genes *ATG*2, *ATG*4, *ATG*5, *ATG*8, *ATG*9, and *ATG*18 (Liu *et al.*, 2007a). All of these null mutants fail to breach the cuticle of the host. This chapter will first look at some methodologies to analyze the functions of autophagy-related gene products at the biological, cellular, and molecular level in this model plant pathogenic fungi, and then provide some research evidence of the role of autophagy in the promotion of the formation of the infection structure and pathogenicity to point out some significant areas for further research in this field.

1. Introduction

As the cause of one of the most widespread and devastating plant diseases in the world, rice blast, the pathogenic fungus *Magnaporthe oryzae* has emerged as a model system for dissecting fungal-plant interactions (Dean *et al.*, 2005; Ebbole, 2007; Talbot, 2003). Considerable progress has been made in identifying genes necessary for the regulation of pre- and post-penetration events in the development and pathogenicity of the fungus, such as the appressorium (a dome-shaped infection structure) mediated penetration of the host and invasive growth of the pathogen. However, the detailed molecular mechanisms of these developmental stages, especially appressorium morphogenesis (Lu *et al.*, 2005a,b), are still relatively poorly understood. Currently, we isolated 24 autophagy genes from the genome of this fungus, which are highly conserved among other eukaryotes, including humans and plants, and that control the capability of surviving starvation, conidiation, conidial germination, lipid turnover, and appressorium turgor generation. Identification, monitoring, and clarification of the functions of these autophagy genes and their interaction networks will facilitate our understanding of the role of autophagy genes in fungal pathogenesis and help new-comers join this hot and exciting research field.

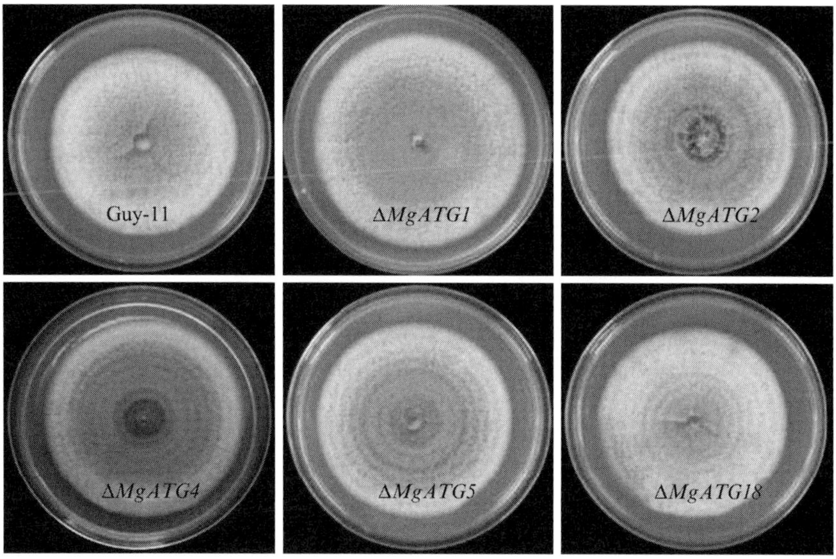

Figure 19.1 Colonies of *atg* null mutants. The *atg* null mutants have sparse aerial hyphae that are different from the dense mycelium seen in the wild-type Guy-11 strain, when cultured on CM plates at room temperature for 10 days.

2. Targeted Gene Replacement of Autophagy-Related Genes in *M. oryzae* Leading to Nonpathogenicity

To determine the genes essential for the target fungus to influence autophagy, we take a more direct approach by specifically disrupting a predetermined gene that is perceived as essential for autophagy, thereby resolutely disproving/proving its role in autophagy. Gene replacement comprises the direct substitution of the wild-type gene with its mutant allele, which has been disrupted by the insertion of a selectable marker (e.g., antibiotic resistance) within its coding region, via homologous recombination mediated by transformation. By generating a precise mutation in the studied genome, it can provide a powerful means for assessing gene function. We get many *atg* null mutants (Fig. 19.1) to study the effects on autophagy.

3. Construction of Gene Replacement Vectors

To examine the effect of deleting *ATG* genes, it is necessary to generate appropriate gene replacement vectors. Construction of the targeted gene (e.g., *MgATG1*) disruption vector pBSATG1 was carried out by inserting two flanking sequences into the pBS-HPH1 vector, a modified pBluescriptII SK (+) vector with the hygromycin expression cassette cloned from the pCB1003 vector. A 1.6-kb *Xho*I-*Sal*I upstream flanking sequence fragment, amplified from genomic DNA, was cloned into the pBS-HPH1 vector to generate pBS-ATG1up. After inserting the *Hin*dIII-*Xba*I downstream flanking sequence fragment amplified using ATG1low primers (p1: 5′-TTCGAG-GATGCCCCGAAGCGAAGCTT-3′) and (p2: 5′-CCtctagaCAGGGCG TGCGACAAACCAAAGAGTA-3′) into pBS-ATG1up, the gene targeted disruption vector pBSATG1 was generated. Next, the 4.6-kb *Xho*I-*Cla*I digested targeted disruption fragment of pBSATG1 was transformed into the protoplasts of strain Guy11. Fungal transformation of *M. oryzae* requires the generation of protoplasts (Balhadere *et al.*, 1999) as described in a later section. Hybridization and washes were conducted under high stringency following the manufacturer's instructions for the DIG high-prime DNA labeling and detection starter kit I (Roche, Germany). Identification of the gene disruption mutants was first performed by PCR and subsequently confirmed by Southern blot.

4. Complementation of the Δ*MgATG1* Mutant

To confirm that the phenotypic differences observed in the *ATG* mutants were all associated with the gene replacement event, complementation assay was carried out. The 6-kb PCR product containing 2 kb of the 5′ upstream sequence, the full-length *M. oryzae MgATG1* gene coding region and 1-kb 3′ downstream sequence, were amplified from genomic DNA using primers hb1 (5′-GTCGTTCATCAGGCGTTCTATTTG-3′) and hb2 (5′-TACTCACTCCTGCTCCTGGGCCTG-3′) and cloned into the pCR-XL-TOPO vector (Invitrogen, USA) to produce the pTOPO-ATG1 vector. Then the *Eco*RI-digested PCR fragment was ligated into the *Eco*RI sites in pBarKS1 to obtain pBar-ATG1. Resistance to glufosinate ammonium (150 μg/ml) was used as a fungal selectable marker during transformation (see section 11.3) with XhoI-linearized pBar-ATG1 (and pBarKS1 as control), and DNA gel blot analysis was performed to confirm successful single-copy genomic integration.

5. Analysis of Genomic DNA

5.1. Mycelia are grown in CM liquid medium

1. Aerial mycelia were harvested by removing them with a sterilized brush from 10-day-old CM agar cultures.
2. The harvested mycelia were inoculated into 100 mL of CM liquid medium and grown for 4 days on a rotary shaker at 150 rpm.

3. Remove the mycelium from the flask by filtration through Miracloth (Calbiochem). Blot the mycelium dry.
   Complete Media (CM): 10 g/L of glucose, 2 g/L of peptone, 1 g/L of yeast extract, 1 g/L of casamino acids, 0.1% (v/v) trace element, 0.1% (v/v) vitamin supplement, 6 g/L of $NaNO_3$, 0.52 g/L of KCl, 0.52 g/L of $MgSO_4$, 1.52 g/L of KH_2PO_4, pH 6.5.

5.2. Fungal genomic DNA was extracted using the CTAB (cetyltrimethyl ammoniumbromide) protocol

1. Grind the sample (about 5 g of wet mycelia) to a fine powder with a mortar and pestle by using liquid nitrogen.
2. Transfer the powdered mycelium to a tube and add 5 ml of CTAB lysis buffer (2% CTAB, 100 m*M* Tris, 10 m*M* EDTA, 0.7 *M* NaCl, store at room temperature) that has been preheated to 65 °C. Gently disperse the powdered mycelium by inverting the tube.
3. Incubate for 30 min at 65 °C, inverting the tube every 10 min to ensure adequate mixing.
4. Add 5 ml of CIA (chloroform/isoamyl alcohol, 24:1, v/v) to the tube and incubate tubes on a shaker for 15 min at 60 rpm.
5. Extract by centrifugation at 10,000 rpm for 10 min at 4 °C.
6. Remove the aqueous (top) phase carefully and be careful not to remove denatured proteins and debris at the interface. Repeat steps 4–6 twice more.
7. Remove the aqueous phase and add to 0.5 volumes of isopropanol in a fresh tube to precipitate the DNA.
8. Leave the tube on ice for 10 min and then centrifuge at 10,000 rpm for 20 min at 4 °C. Wash the pellet in 10 ml 70% ethanol and centrifuge at 10,000 rpm for 10 min.
9. Dry the pellet on the bench by inverting the tube on paper. Leave the samples for 30 min. Resuspend the pellet in 100 μl of TE buffer, pH 8.0, and store at 4 °C.

5.3. DNA gel blot analysis

Genomic DNA was digested with the proper restriction endonucleases, separated on a 0.7% agarose gel in 1 × TAE buffer, and transferred to a positively charged nylon transfer membrane. Select the proper labeled probe to hybridize. DNA gel blot analyze in detail using the DIG high-prime DNA labeling and detection starter kit I (Roche, Germany) following the manufacturer's instructions.

6. Plant Infection Assays

To assay the effect of deleting *ATG* genes, we can monitor the size of the lesions that result from fungal infection (Fig. 19.2).

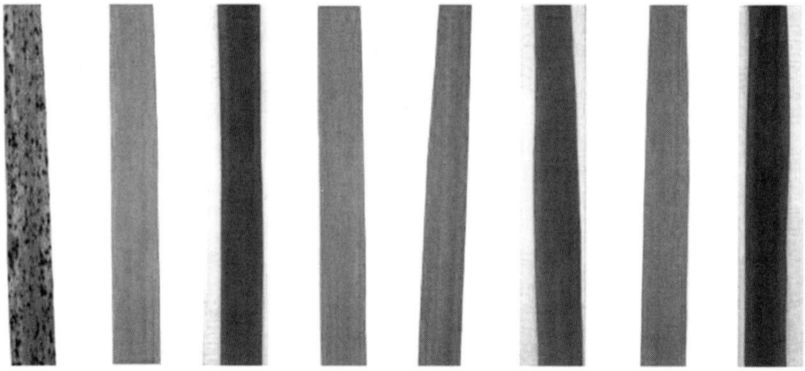

Guy-11 Δ*MgATG1* Δ*MgATG2* Δ*MgATG4* Δ*MgATG5* Δ*MgATG9* Δ*MgATG18* Gelatin

Figure 19.2 atg mutants lose the ability to penetrate the host plants. Leaves from CO-39 were spray-inoculated individually with conidia from the wild-type Guy-11 strain, or the Δ*MgATG1*, Δ*MgATG2*, Δ*MgATG4*, Δ*MgATG5*, Δ*MgATG9*, Δ*MgATG18* mutants. Disease symptoms were allowed to develop for 7 days. (See Color Insert.)

1. Two-week-old seedlings of the rice cultivars CO-39 and 8-day-old barley ZJ-8 were used for spray assays.
2. *M. oryzae* are grown on CM plates for 10 days.
3. Conidia were harvested by removing aerial mycelia with a sterilized brush from 10-day-old CM agar cultures, and filtered through three layers of lens papers (Hangzhou, China). The conidial suspension was concentrated in sterile distilled water.
4. The conidia counted with a hemocytometer under a microscope and were resuspended to 1×10^5 conidia/ml in a 0.2% (w/v) gelatin solution (gelatin was dissolved in sterile distilled water).
5. The suspension was sprayed evenly onto the rice leaves using an artist's airbrush (Badger Co., Franklin Park, IL).
6. Inoculated plants were placed in a dew chamber at 25 °C for 48 h (barley for 24 h) in the dark, and then transferred to a growth chamber with a photoperiod of 12 h using fluorescent lights.
7. Lesion formation was examined 7 days after inoculation in rice and 4 days after inoculation in barley. Disease severity on rice was rated according to Bonman *et al.* (1986) (see the following paragraph) and disease lesion densities were recorded from 20–30 infected leaves using a 5-cm section of each leaf. Infection assays were carried out three times using 45 plants per assay.

Bonman *et al.* (1986) classification standard: 0 = no evidence of infection; 1 = brown specks smaller than 0.5 mm in diameter; 2 = brown specks smaller than 0.5 mm in diameter, no sporulaton; 3 = roundish to elliptical lesions about 1–3 mm in diameter with gray center surrounded by brown margins, lesions capable of sporulation; 4 = typical spindle-shaped blast lesions capable of sporulation, 3 mm or longer with necrotic gray centers

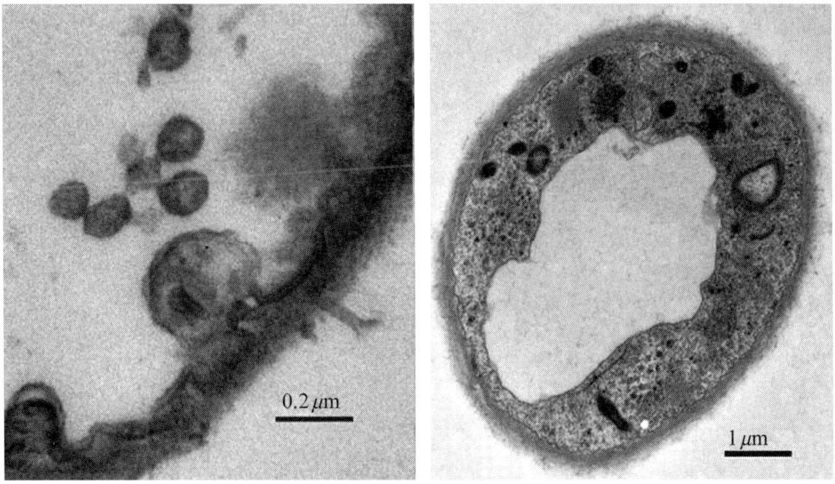

Figure 19.3 Autophagic vesicles in the vacuole. Accumulation of autophigic vesicles in the vacuolar lumen was blocked in the Δ*MgATG1* mutant. After 4 h of nitrogen starvation in the presence of 1 mM PMSF, the vacuole of the wild-type Guy-11 strain appears to contain vesicles (left). Under the same conditions, no vesicles are evident in the vacuole or the cytoplasm of the Δ*MgATG1* mutant (right).

and water-soaked or reddish brown margins, little or no coalescence of lesions; and 5 = lesions as in 4 but about half of one or two leaf blades killed by coalescence of lesions. Scores of 0–3 are considered resistant reactions, and scores of 4 and 5 are considered susceptible reactions.

7. Electron Microscopy

Another method for analyzing the effect of the *ATG* gene mutations is to examine the cells for the presence of autophagic structures (such as autophagosomes) ultrastructurally by electron microscopy (Fig. 19.3).

1. *M. oryzae* are grown on CM plates as described previously.
2. Conidia collected from 10-day-old mycelia were cultured at 28 °C for 24 h in 100 ml of CM liquid medium in flasks continuously shaken at 150 rpm.
3. The mycelial growth was collected by filtration through three layers of lens paper, thoroughly washed in distilled water, transferred to MM–N liquid medium (10 g of glucose, 0.52 g of KCl, 0.52 g of $MgSO_4$, 1.52 g of KH_2PO_4, 0.5% Biotin dissolved in 1 liter of distilled water, pH 6.5) with 2 m*M* PMSF, and incubated at 28 °C for 4 h on a shaker at 150 rpm.
4. The fungal mass was then collected by filtration through three layers of lens paper, and fixed overnight at 4 °C in modified Karmovsky's fixative

containing 2% paraformaldehyde and 2.5% glutaraldehyde (v/v) in 0.1 *M* phosphate buffer, pH 7.2.

5. The fixed samples were washed three times, for 10 min each time, with 0.1 *M* phosphate buffer (pH 7.2).
6. The samples were postfixed in 1% OsO_4 for 2 h at 25 °C, washed three times with phosphate buffer as before, and dehydrated in a graded ethanol series. The samples was first dehydrated by a graded series of ethanol (50%, 70%, 80%, 90%, 95%, and 100%) for about 15–20 minutes at each step, then transferred to absolute acetone for 20 min.
7. The samples were placed in a 1:1 mixture of absolute acetone and the final Spurr resin mixture for 1 h at room temperature, and then transferred to a 1:3 mixture of absolute acetone and the final resin mixture for 3 h and to the final Spurr resin mixture for overnight.
8. The embedding samples was placed in capsules contained embedding medium and heated at 70 °C for about 9 h. The sample sections were stained by uranyl acetate and alkaline lead citrate for 15 min respectively and observed in TEM of Model JEM-1230 electron microscope (JEOL, Tokyo, Japan) operating at 70 kV.

8. Subcellular Localization of GFP-tagged MgATG1

A method to analyze the induction of autophagy relies on monitoring the localization of an Atg protein fused to a fluorescent tag. For example, to investigate the localization of MgATG1 in *M. oryzae* (Fig. 19.4), we constructed an MgATG1-green fluorescent protein gene fusion expression cassette (pGFP-ATG1) by the following steps:

1. Place a GFP gene (without stop codon) under control of the promoter of *NAR* (Lu *et al.* 2007) and the full-length coding domain of *MgATG1* (without initiation codon) into the vector pBS-HPH1 to generate fusion cassette pHNATG1.
2. The plasmid pHNATG1 linearized with *Sma*I was introduced to the protoplasts of Guy-11 (see section 11.3).
3. Transformant NGA7 was selected for further studies after Southern blot analysis, which confirmed that it contains a single-copy integration of pGFP-ATG1.
4. Expression of GFP-ATG1 was uniformly detectable in the cytoplasm of conidia, mycelia, and appressoria of the transformant NGA7 (see Fig. 19.4).
5. The transformant was observed using an Olympus-BX51 microscope (Japan) with UV epifluorescence.

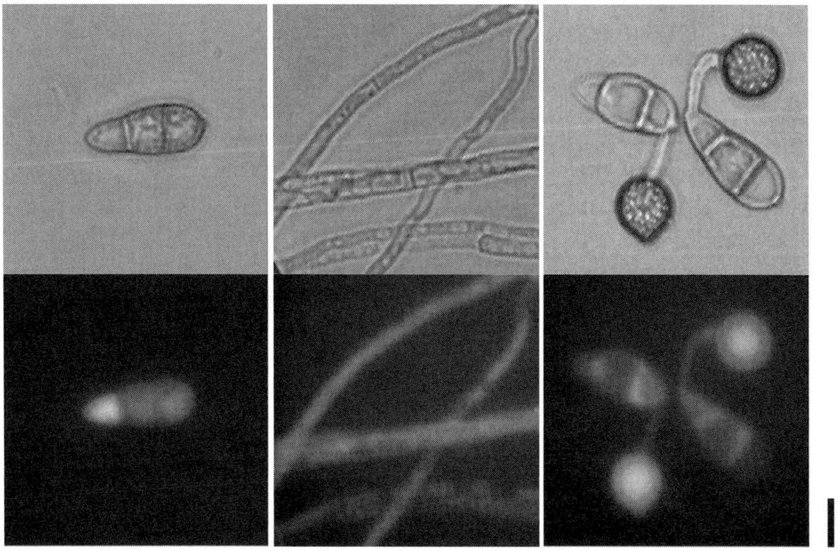

Figure 19.4 Localization of MgATG1 in *M. oryzae*. The MgATG1-GFP protein appeared in the cytoplasm of *M. oryzae* strain NGA7. Bright-field (upper) and fluorescence (bottom) are shown. Bar = 10 μm. (See Color Insert.)

9. Modified TAKA Assay in *M. oryzae*

The transport of Atg9 after knocking-out ATG1 (TAKA) assay allows us to analyze the transport of this protein from peripheral sites to the phagophore assembly site (PAS), the presumed site of autophagosome nucleation and/or formation. Atg9 participates in the formation of double-membrane vesicles. Once the vesicle is completed, Atg9 is retrieved; it is not found associated with the final cytoplasm to vacuole targeting pathway vesicle or autophagosome (Cheong *et al.*, 2005; Klionsky *et al.*, 2007; Noda *et al.*, 1995). In wild-type cells, Atg9 is detected in multiple puncta, one of which corresponds to the PAS. In certain mutants such as *atg1*Δ, Atg9 accumulates at the PAS. The TAKA assay is an epistasis assay in which a second mutation is introduced into the *atg1*Δ background, and the resulting Atg9-GFP phenotype is examined. To look at the effect of various *atg* mutations on Atg9 localization, we utilize the following assay.

Materials and reagents

Agrobacterium tumefaciens AGL1 stock culture
DNA for transformation
Glycerol (10% [v/v])
Sterile, ice-cold LB liquid medium (without antibiotics)

LB liquid medium (the standard medium for *Escherichia coli*) and agar plates containing appropriate antibiotics
Liquid nitrogen
Agrobacterium induction medium (AIM) (see the subsequent sections)

Equipment
Cuvettes for electroporation, chilled
Centrifuge with rotor prechilled to 4 °C
Electroporator (Bio-Rad) incubator, preset to 28 °C
Cellulose nitrate membrane (Whatman, 47 mm Dia, 0.45-μm pore size, WCN type, cat no. 7141114)
Pipettes
Spectrophotometer
Tubes and microcentrifuge
Water bath, ice cold

Construction of plasmids expressing MgATG9-GFP and GFP-MgATG9
Restriction digestion, gel electrophoresis, and ligation reactions were all carried out using standard procedures (Fig. 19.5). In our experience, it does not matter whether MgATG9 is tagged at the N or C terminus in *M. oryzae*.
Preparation of electrocompetent Agrobacterium cells

1. Grow an overnight culture of wild-type *Agrobacterium* at 28 °C in 5ml of LB.
2. Inoculate 100 ml of LB with 2.5 ml of the overnight culture at 28 °C with vigorous agitation until the OD_{600} reaches 0.5–0.7.
3. Chill the culture in an ice-water bath for 10 min.
4. Pellet the cells by centrifuging at 8000 rpm for 10 min at 4 °C in a prechilled rotor.
5. Discard the supernatant and resuspend the cells in 10 ml of ice-cold sterile 10% glycerol. Pipette the cells gently up and down until no clumps remain. Repeat steps 4 and 5 four times.
6. Resuspend the cells in 1mL of ice cold sterile 10% glycerol by pipetting vigorously.
7. Dispense 50-μl aliquots of the cells into microcentrifuge tubes and freeze in liquid nitrogen. Store cells at −70 °C until use.

Electrotransformation of agrobacterium

1. Thaw competent cells on ice (50 μl per transformation).
2. Add 1 μl of *E. coli* miniprep plasmid DNA to the cells, and mix them together on ice.
3. Transfer the mixture to a prechilled electroporation 2-mm cuvette. Electroporate at 25 μF (apacitance), 1.8 kV (voltage), and 300 Ω (resistance).

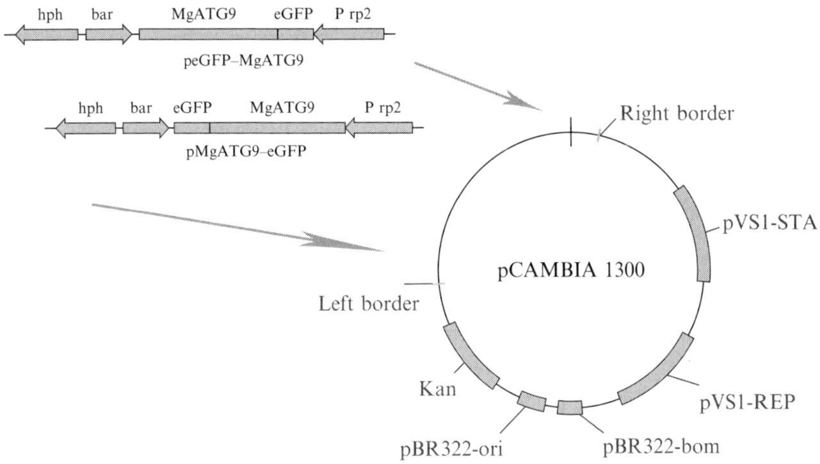

Figure 19.5 Schematic representation of the modified TAKA assay vector.

4. Immediately, add 1 ml of ice-cold LB to the cuvette, mix and transfer the suspension to a 1.5-ml culture tube.
5. Incubate for 2 h at 28 °C with gentle agitation.
6. Collect the cells by centrifuging briefly, and spread them on an LB agar plate containing the appropriate antibiotic.
7. Incubate the cells for 2 days at 28 °C to allow the colonies to appear.
8. Grow small liquid cultures of the colonies, and carry out minipreps of the DNA and/or PCR to verify the presence of plasmid DNA. Make glycerol stocks of the appropriate clones, and store them at −20 °C.

Agrobacterium-mediated transformation of *M. oryzae*

1. Grow an overnight culture of the transformed *Agrobacterium* strain in 2 mL of LB + antibiotics at 28 °C (kanamycin at 50–100 μg/mL).
2. Collect cells and wash with *Agrobacterium* induction medium (AIM) + antibiotics.
3. Resuspend cells in 10 ml of AIM + antibiotics + 200 *μM* of acetosyringone and grow for 6 h at 28 °C.
4. Prepare *M. oryzae* conidia suspension in ddH_2O at 10^5–10^6 conidia/ml.
5. Place the cellulose nitrate membrane on the AIM plate with 200 *μM* acetosyringone. Mix 100 μl of the *Agrobacterium* culture with the 100 μl conidia suspension, and spread onto the cellulose nitrate plate. Incubate the plate at 22 °C for 48 h.
6. Lift off the membrane and place on selection plates: CM + 200 μg/mL Hygromycin B/bialophos + 400 μg/ml cefotaxime + 60 μg/ml streptomycin. Incubate at 28 °C until colonies appear (Hellens, *et al.*, 2000; Mullins, *et al.*, 2001; Rho *et al.*, 2001; Vijn and Govers, 2003; Zwiers and De Waard, 2001).

Agrobacterium Induction Medium (AIM)

0.8 ml of 1.25 K-phosphate-buffer, pH 4.8 (make stocks of KH_2PO_4 and K_2HPO_4; add one to the other until pH 4.8 is reached)
20 ml of MN-buffer (30 g/l $MgSO_4 \cdot 7H_2O$, 15 g/l NaCl)
1 ml of 1% $CaCl_2 \cdot 2H_2O$ (w/v)
10 ml of 0.01 % $FeSO_4$ (w/v)
5 ml of spore elements (100 mg/l $ZnSO_4 \cdot 7H_2O$, 100 mg/l $CuSO_4 \cdot 5H_2O$, 100 mg/l H_3BO_3, 100 mg/l $Na_2MoO_4 \cdot 2H_2O$), filter sterilized
2.5 ml of 20% NH_4NO_3 (w/v)
10 ml of 50% glycerol (v/v)
40 ml of 1 *M* MES, pH 5.5 (adjust pH with NaOH, stocked solution)
20% glucose (w/v): 10 ml for liquid medium
5 ml for solid medium
200 μM acetosyringone (always use fresh stock 0.2 *M* in DMSO)
Add H_20 to 1000 ml and 15 g/l bactoagar for plates

Fluorescence microscopy analysis of Atg9-GFP

1. Conidia were harvested from 10-day-old growth on CM plates and the concentration of the conidial suspension, in sterile distilled water or CM liquid medium, was adjusted to 1×10^5 conidia/ml.
2. Droplets (20 μl) of the suspension were placed on glass coverslips and incubated under a humid environment at 25 °C. The frequency of conidial germination, mycelium and appressorium formation was determined at different times (2 h, 4 h, 6 h, 18h, 24 h, and 32 h after inoculation).
3. Fluorescence of the transformants was observed using an inverted confocal laser scanning microscopy.

10. Western Blot Analyses of *M. oryzae* Autophagic Proteins expressed in *Pichia pastoris*

Protein blotting is an analytical method that involves the immobilization of proteins on membranes before detection using monoclonal or polyclonal antibodies. There are different blotting protocols (e.g., dot blot, 2-D blot); one of the most powerful is Western blotting; the name *Western blotting* was given to the technique by W. Neal Burnette (1981).

In Western blotting, prior to protein immobilization on the PVDF or nitrocellulose membranes, sample proteins are separated using SDS polyacrylamide gel electrophoresis (SDS-PAGE) providing information about molecular weight and the potential existence of different isoforms of the proteins under study. In yeast, Atg4 is a unique cysteine protease responsible for the cleavage of the carboxyl terminus of Atg8, which is essential for Atg8 lipidation during the formation of autophagosomes. However, it is still unclear whether an Atg4 homolog cleaves the carboxyl termini of the *Magnaporthe* Atg8 homolog. Here we use Western blot analyses to detect the *Magnaporthe* autophagic protein MgATG4, which is expressed in *Pichia pastoris*.

10.1. Sample preparation

Before you run a Western blot, analyze what cell lines or tissues you are using. This will affect how you prepare your samples. Also take precautions to prevent degradation of proteins by proteases. For example, if you are analyzing phosphoproteins or phosphorylation, you need to inhibit phosphatases. Autophagy-related protein MgATG4 and MgATG8 were expressed in *P. pastoris* strains GS115 by cloning into the pPICZαA vector (Invitrogen, USA).

Preparation of supernatant (secreted expression only)

1. Using a single colony, inoculate 25 ml of BMGY(1% yeast extract, 2% peptone, 100 m*M* potassium phosphate, 1.34% YNB, 4×10^{-5}% biotin, 1% glycerol) in a 250 ml baffled flask. Grow at 28–30 °C in a shaking incubator (250–300 rpm) until the culture reaches an $O.D._{600} = 2$–6 (approximately 16–18 h).
2. Harvest the cells by centrifuging at 1500–3000 × *g* for 5 minutes at room temperature. Decant the supernatant fraction and resuspend the cell pellet to an $O.D._{600}$ of 1.0 in BMMY (1% yeast extract, 2% peptone, 100 m*M* potassium phosphate, 1.34% YNB, 4×10^{-5}% biotin, 0.5% methanol) medium to induce expression (approximately 100–200 ml).
3. Place the culture in a 1-liter baffled flask. Cover the flask with 2 layers of sterile gauze or cheesecloth and return to the incubator to continue growth.
4. Add 100% methanol to a final concentration of 0.5% methanol every 24 h to maintain induction.
5. At each of the times indicated below, transfer 1 ml of the expression culture to a 1.5-ml microcentrifuge tube. Centrifuge at maximum speed in a tabletop microcentrifuge for 2–3 min at room temperature. Transfer the supernatant to a separate tube. Time points (h): 0, 6, 12, 24 (1 day), 36, 48 (2 days), 60, 72 (3 days), 84, and 96 (4 days).
6. Store the supernatant and the cell pellets at −80 °C until ready to assay. Freeze quickly in liquid nitrogen or a dry ice/alcohol bath.
7. Thaw supernatants and place on ice.
8. Mix 50 μl of the supernatant with 50 μl of SDS-PAGE gel loading buffer (see subsequent sections).

9. Boil 10 min, then load 10–30 μl onto the gel. Remaining sample may be stored at −20 °C for Western blots, if necessary. Supernatants may be stored at −80 °C for further analysis.

Bradford assay

1. Dilute Bradford dye concentrate with dH_2O (see bottle for specific; however, 1ml of dye + 4 ml of dH_2O is common).
2. Filter the dye solution (gravity filtration); the diluted reagent may be used for approximately 2 weeks.
3. A standard curve is needed to determine the concentration of the unknowns (i.e., samples).
   a. Using 6 cuvettes (1-ml capacity), add the following amounts of BSA: 0 (baseline), 2, 5, 8, 10, 15 μgs.
   b. Fill with Bradford dye (to a volume of 1 ml).
   c. Carefully vortex the cuvettes and allow them to incubate for approximately 5 min at room temperature.
   Note: Absorbance will increase with time; therefore, be consistent with the amount of time that the samples are allowed to incubate before spectrophotometric measurement.
   d. Using a spectrophotometer, measure the OD at 595 nm.
4. Measure the OD of samples; measure the OD of the lysis buffer and subtract it as background to remove its contribution to the total protein measurements.
5. Plot the results for the BSA and make a standard curve.
6. Using the curve, determine the protein concentration of your sample(s).

10.2. Sodium dodecyl sulfate polyacrylamide gel electrophoresis: SDS-PAGE

SDS-PAGE is a commonly used molecular biology protocol that allows for the separation of molecules (i.e., the proteins for Western blotting). Proteins are separated according to weight and electrical properties as they migrate through a polyacrylamide gel matrix. Acrylamide gel preparation is a process that involves the cross-linking of acrylamide monomers with the use of catalysts. Once the gel has set, protein samples can be loaded into the gel and separated along an electrical field.

Acrylamide gel preparation

1. Clean two plates by spraying with 70% ethanol and drying with a towel.
2. Make a plate sandwich, separating the two plates with spacers and align in a clamping device. The bottom of the plates must be temporarily sealed so the gel can be poured, but the bottom can later be exposed to the running buffer.

3. Make resolving gel cocktail containing the proper amounts of the following:

30% Bis/Acrylamide	5 ml
1.5 *M* Tris-HCl, pH8.8	5 ml
10% SDS (pH 7.2)	0.2 ml
nanopure H_2O	9.7 ml
10% Ammonium persulfate (APS)	100 μl
TEMED	10 μl

4. Pour into the plate sandwich and overlay with 100 μl of 70% butanol.
5. Allow to polymerize at least 20 min. If the gel will not be used for some time, wash off the butanol, rinse with water, and overlay with the resolving gel buffer (omit Bis/acrylamide, APS and TEMED).
6. Aspirate the butanol and wash 5 times with water.
7. Make the 5% stacking gel:

30% Bis/Acrylamide	1.3 ml
1 *M* Tris-HCl, pH 6.8	6.81 ml
10% SDS (pH 7.2)	80 μl
nanopure H_2O	5.5ml
10% APS	80 μl
TEMED	5 μl

8. Take 1 ml to wash the top of each of the gels.
9. Place the combs (15-well or 10-well, depending on the number of samples) in the gel at an angle. Alternatively, the combs can be placed after the stacking gel is added.
10. Add 10 μl of TEMED.
11. Pour 1 ml (or enough to fill) at the corner under the comb.
12. Push the comb down at an angle making sure not to trap any bubbles.
13. Allow the gel to polymerize for 20 min.

Electrophoretic separation (SDS-PAGE)1

1. Pour 1X SDS-PAGE Buffer (0.025 *M* Tris-HCl, 0.192 *M* Glycine, 0.1% SDS, pH 8.3) into the Western blot tank.
2. Position the acrylamide gels in the gel holder assembly and immerse into the tank.
3. Fill the upper compartment with 1×SDS-PAGE buffer.
4. Carefully load the samples in the wells (using a fine-tipped pipette).
5. Place the lid on the tank and plug it into the power source. (Note: this procedure uses electric current to separate proteins; use caution when working with tank electrophoresis to avoid injury).
6. Run the apparatus at 100V until the samples have passed the stacking gel.

7. Turn the voltage up to 160V and allow the samples time to separate; use a prestained molecular weight marker to determine the end point of the electrophoresis. Alternatively, the loading buffer can contain a dye such as bromophenol blue, which will run ahead of the proteins.

10.3. Coomassie Blue staining

Coomassie Brilliant Blue R250 binds nonspecifically to virtually all proteins. Coomassie Blue staining is less sensitive than silver staining but effective nonetheless, and easy to perform. After staining the PVDF membrane, wash the membrane in destaining solution to remove excess dye.

Procedure

1. Mix the protein stain components listed herein; filter the solution using a Whatman No. 1 filter.
2. Soak the gel in the dye solution for 30 min.
3. Excess Coomassie stain will be washed out during destaining (30 min).
4. The destaining solution is prepared as indicated in the recipe for the staining solution without the Coomassie R250 dye powder.

10.4. Electrophoretic transfer to PVDF membranes

The electrophoretic transfer of sample proteins to membranes for immunoblotting occurs after protein separation by SDS-PAGE. Care must be used when working with PVDF (polyvinylidene fluoride) membranes: use forceps and handle membranes by their edges so as not to damage the surface.

Transfer protocol

1. Cut filter paper (typical chromatography paper) in approximately 7 cm × 20 cm pieces.
2. Cut PVDF membrane to 7 cm × 20 cm.
3. Prewet the PVDF membrane using 100% methanol for 10 s and immerse in dH_2O.
4. Soak the filter pads for the Western blot apparatus in PVDF transfer buffer (see subsequent section).
5. Assemble the membrane sandwich according to the following order:
   1 fiber pad
   1 Whatman 3MM filter paper
   Gel
   PVDF membrane
   1 Whatman 3MM filter paper
   1 fiber pad
6. Place the sandwich into the transfer assembly.
7. Fill the transfer tank with PVDF transfer buffer.

8. Remove the cooling block from storage at −20 °C; place into transfer apparatus.
9. Place the tank in a refrigerator at 4 °C.
10. Run the transfer protocol at 0.1 A (constant amperage) overnight. Note: higher current and voltage settings may be used; however, use caution as the transfer buffer may heat up substantially, damaging proteins or equipment.

10.5. Western blot protocol

Used daily in medicine, biology and other disciplines, the western blot method allows for the study of protein expression. The Western blot protocol below should be optimized for specific experiments because of differences in antibody specificity, protein copy number, and other parameters.

Western blot protocol

1. After completing SDS-PAGE and electrophoretic transfer, block the membrane in 50 ml of 5% nonfat dry milk (NFDM) in TBS containing 0.05% Tween-20 for 2 h.
2. Incubate the membrane with the primary antibody for 16 h at 4 °C. Dilute the antibody with a solution of 2.5% NFDM in TBST (see subsequent sections). A total antibody and diluent solution of 5 ml will coat a small membrane in a rotating tube very well.
3. Remove the antibody and perform 3 room temperature washes (10 min of rotation each) with TBST.
4. Reblock the membrane in 10% NFDM in TBST for 10 min at room temperature.
5. Incubate the membrane with the secondary antibody for 30 min at room temperature. Dilute the antibody with a solution of 2.5% NFDM in TBST. You can also add blocking serum from the same species in which the secondary antibody was produced.
6. Remove the antibody and perform 3 room temperature washes (10 min of rotation each) with TBST.
7. Add 50 ml of diluted alkaline-phosphatase conjugated secondary antibody and incubate for 30 min at room temperature with shaking (anti-mouse IgG conjugate is diluted 1:5000 in 5% milk-TBST).
8. Wash 4 times for 5 min per wash with TBST at room temperature.
9. Add 10 ml of BCIP/NBT substrate (0.02% BCIP (Sangon, China) and 0.03% NBT (Sangon, China) in 0.1 *M* TBS, pH 9.5). Wait approximately 15 min for bands to appear. Watch carefully because reaction times will vary. Rinse with water to stop developing.
10. Photograph the membranes (7ypical results are shown in Fig. 19.6).

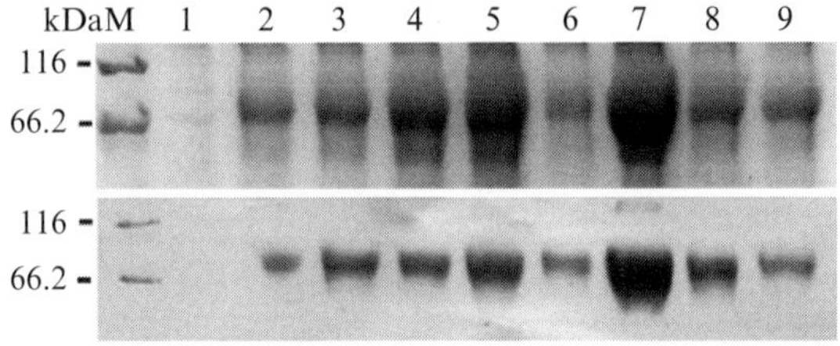

Figure 19.6 Expression of MgATG4 in Pichia pastoris. The supernatant of the transformants was prepared as described above (see section 8D). Proteins in the supernatant were separated by SDS-PAGE, and assayed by immunoblotting with an anti-his antibody. Lane 1: The negative control, transformant transformed with the empty vector pPICZαA; Lane 2-9: The supernatants of eight transformants transformed with pPICZαA-ATG4 expressing 70KD MgATG4 protein glycosylied. (See Color Insert.)

10.6. Recipes

5% NFDM

Prepare using Carnation Nonfat Dry Milk in TBST.

10% Ammonium Persulfate (APS) Solution

0.12 g APS

1.2 ml H_2O

Keep for no more than one week in the refrigerator.

Coomassie Brilliant Blue Stain

0.25 g Coomassie Brilliant Blue R250

90 ml MetOH:H2O (1:1 v/v)

10 ml Glacial acetic acid

TBST (10 m*M* Tris-HCl, 0.15 *M* NaCl, 8 m*M* sodium azide, 0.05% Tween-20, pH 8.0)

Chemical for 4 liters

Tris: 4.84 g

NaCl: 35.06 g

NaN3: 2.0 g

Tween-20: 2.0 ml

Adjust pH to 8.0 with HCl. Bring volume up to 4 liters with water and store at 4 °C.

Tank Buffer (25 m*M* Tris-HCl, 0.2 *M* glycine, 0.35% SDS)

Chemical for 4 liters and for 500 ml

Tris (121.114): 12.11g and 1.51 g Tris

Glycine (75.07): 57.6 g and 7.2 g

SDS (288.38): 4.0 g and 17.5 ml 10% SDS

Bring volume up to 4 liters or 500 ml with H_2O. The pH should be 8.3 but adjusting is not needed. Store at 4 °C.

Transfer Buffers

Wet blot transfer buffer (3 L) (25 mM Tris-HCl, 0.2 M glycine, 20% methanol)
9.08 g of Tris
43.24 g of glycine
600 ml of methanol
Bring volume up to 3 liters with water.

Lower Gel Buffer (1.5 M Tris-HCl, 0.4% SDS, pH 8.8)
For 100 ml: 18.17 g Tris and 4 ml 10% SDS
Adjust pH to 8.8 with HCl. Bring volume up to 100 ml with water. Store at room temperature.

Upper Gel Buffer 0.5 M Tris-HCl, 0.4% SDS, pH 6.8)
For 100 ml: 6.06 g Tris and 4 ml 10% SDS
Adjust pH to 6.8 with HCl. Bring volume up to 100 ml with water. Store at room temperature.

2x SDS loading Buffer (Hoeffer Scientific)
Prepare in fume hood!
2.5 ml of 0.5 *M* Tris-HCl, pH 6.8
4.0 ml of 10% SDS
2.0 ml of glycerol
1.0 *ml concentrated β-mercaptoethanol*
0.4 mg bromophenol blue
Bring volume up to 10 ml with water. Transfer to 1-ml aliquots and store at −20 °C.

11. Visualization of Autophagic Proteins Interactions in *M. oryzae* Using Bimolecular Fluorescence Complementation

In living cells, many proteins interact with others and form a regulatory network that controls many cellular functions. To analyze protein-protein interactions, many methods have been developed for the identification and characterization of protein-protein interactions *in vitro* and *in vivo*, each with its advantages and limitations. For example, a widely used screening technique for interacting proteins, yeast two-hybrid system (Fields and Song, 1989), has been developed to facilitate large-scale screening and detection of protein-protein interactions *in vivo*. Although very useful, this system suffers from the high occurrence of false-positives and the requirement that the interacting proteins accumulate in the cell nucleus (Golemis *et al.*, 1999).

Recently, the bimolecular fluorescence complementation (BiFC) approach, a visualization-based method, for the simple and direct visualization of protein interactions in living cells has been developed by Hu *et al.* (2002). This approach is based on the ability of fusions of interacting proteins to non-fluorescent fragments of the yellow fluorescence protein (YFP) to form a fluorescent complex and reconstruct a functional YFP when two fragments of a fluorescent protein are brought together by an interaction between interacting proteins. Now this method has been implemented for the detection of protein-protein interactions in mammalian, bacterial, plant, and fungal cells (Grinberg *et al.*, 2004; Hoff and Kück, 2005; Hu and Kerppola, 2003; Lacroix *et al.*, 2005; Stolpe *et al.*, 2005; Tsuchisaka and Theologis, 2004; Zhao and Xu, 2007).

Plasmid vectors for expression of the proteins of interest fused to N- and C-terminal fragments of YFP must be designed. In most cases, fusions to both the N- and C-terminal ends of the proteins under investigation should be analyzed. The following describes the approach used for the construction of expression vectors for BiFC analysis and fungal transformation in *M. oryzae*.

11.1. Design of expression vectors for fusion proteins

1. To place the *GFP* reporter gene under control of the *MgATG4* promoter, an XbaI–BamHI fragment containing *MgATG4* upstream sequences was amplified with the primers ATG4 PROP1 (5′-ggtctagacgcccaccgtgtcgagcataaata-3′) and ATG4 PROP2 (5′- ccggatccccagccacggccgaatccat-3′) .
2. The PCR product was digested at the introduced XbaI and BamHI sites and ligated to the 4.5-kb XbaI–BamHI-digested pEGFP-HPH containing an HPH gene expression cassette to generate pMgATG4(p)-GFP.
3. The coding domain of the MgATG4 fragment without the stop codon (TAG) was amplified with primers (5′-ggggtacccatggattcggccgtggctggtg-3′ and 5′- ccggtaccccatccaaaacggtatcgccat-3′) from *MgATG4* cDNA and digested at the introduced KpnI sites and inserted into the KpnI sites of pMgATG4(p)-GFP to generate pATG4-GFP.
4. The N-terminal region of EYFP (1–154) was amplified with primers YFPNF (5′-ggaccggtgcgctccatcgccacgatggtgagcaagggcgaggagctgt-3′) and YFPNR (5′-gggcggccgcttaggccatgatatagacgttgtggctgt -3′) from pEYFP (Clontech, 2005) and digested at the introduced *Age*I and *Not*I sites (underlined) and cloned into pATG4-GFP as pATG4-YFPN-HPH.
5. To generate a truncated ATG4-YFPN fusion construct, the plasmid pATG4-YFPN-HPH was digested with *Nco*I and self-ligated, resulting a negative control plasmid pATG4Δ-YFPN-HPH in which a 570-bp fragment of *MgATG4* was cut off.

6. The C-terminal region of EYFP (155–238) without the stop codon (TAA) was amplified from pEYFP with primers YFPCF (5′-ggggatcccgccgacaagcagaagaacggca-3′) and YFPCR (5′-ccaccggtgtggttcatgaccttctgtttcaggtggttcgggatcttgcaggccgggcgcttgtacagctcgtccatgccg-3′) and digested at the introduced *Bam*HI and *Age*I sites (underlined) and cloned into the same sites of pATG4-YFPN-HPH as pYFPC-HPH.
7. The coding domain of the MgATG8 fragment with the stop codon (TGA) was amplified with primers ATG8 P1 (5′-ggaccggtatgcgctccaagttcaaggacg-3) and ATG8 P2 (5′-gggcggccgctcactcgacttcctcaaacagg-3′) and digested at the introduced *Age*I and *Not*I sites and cloned into the same sites of pYFPC-HPH as pYFPC-ATG8-HPH.
8. A *BAR* fragment was amplified with primers (5′-aagcggccgcagaagatgatattgaagga-3′ and 5′- aaactagtctaaatctcggtgacgggc -3′) from pBarKS1 (Pall and Brunelli, 1993) and digested at the introduced *Not*I and *Spe*I sites (underlined) and inserted into NotI-SpeI sites of pYFPC-ATG8-HPH to generate pYFPC-ATG8-BAR.

11.2. Protoplast preparation and fungal transformation

1. A 6-day old colony of *M. oryzae* strain Guy11 mycelium was cut from the surface of a complete medium (CM) plate, smashed by a triturator (Supor, China) in 200 ml of CM liquid media, and incubated 2 days at 28 °C on a rotary shaker at 125 rpm.
2. The culture was harvested by filtration with 4 layers of sterile gauze, and protoplasts produced by Glucanax (Sigma) digestion in 0.7 *M* NaCl buffer. For each gram of mycelium, 2 ml of 0.7 *M* NaCl buffer with 15 mg of Glucanax was added.
3. Protoplasts were collected at the bottom of a 50-ml tube with 0.7 *M* NaCl buffer after centrifugation at 3000 g.
4. The protoplasts were washed twice in STC buffer (1.2 *M* sorbitol, 10 m*M* Tris-HCl, pH 7.5, 20 m*M* $CaCl_2$).
5. Transformation was carried out using 1×10^8 protoplasts, with 2 μg DNA, in the presence of 1 ml of PTC buffer (60% polyethylene glycol 4000, 10 m*M* Tris-HCI, pH 7.5, 10 m*M* $CaCl_2$).
6. The pATG4-YFPN-HPH and pYFPC-ATG8-BAR vectors were cotransformed as linear *Eco*RV fragments. For the negative control, pATG4Δ-YFPN-HPH and pYFPC-ATG8-BAR vectors were cotransformed as linear *Eco*RV fragments.
7. Protoplasts were incubated with PTC for 25 min and then added to 5 ml of OCM liquid media (CM stabilized osmotically by adding 20% (w/v) sucrose), and protoplasts were allowed to regenerate after incubation overnight.
8. The morning of the second day, the regenerated protoplasts were mixed with 45 ml of 1% OCM agar containing 200 μg/ml Hygromycin B and 300 μg/ml glufosinate ammonium and poured into 90-mm plates.

9. The plates were incubated for 7 days at 30 °C in the dark.
10. Monoconidial isolations were conducted on all hygromycin and glufosinate ammonium-resistant transformants.

11.3. Fluorescence microscopy

Fluorescence was detected using a Leica TCS SP5 inverted confocal laser scanning microscope equipped with a Leica Axiovert 100 microscope. We use primarily filters with excitation at 500/520 nm and emission at 535/530 nm for YFP fragments. Observation was performed with a 63× immersion oil objective. The conidia, germinated conidia, appressoria, and mycelia were observed under fluorescence microscopy. If we detect a YFP fluorescence signal in any of the preceding, we can confirm the direct interaction between MgATG4 and MgATG8. Here we use pATG4Δ-YFPN-HPH and pYFPC-ATG8-BAR cotransformed transformants resistant to both hygromycin and glufosinate ammonium as a negative control.

ACKNOWLEDGMENTS

This study was supported by grants (No. 30270049, 30470064 and 30671351) from the National Natural Science Foundation of China to FC Lin.

REFERENCES

Balhadere, P. V., Foster, A. J., and Talbot, N. J. (1999). Identification of pathogenicity mutants of the rice blast fungus *Magnaporthe grisea* by insertional mutagenesis. *Mol. Plant-Microbe Interactions* **12,** 129–142.

Bonman, J. M., Vergel de Dios, T. I., and Khin, M. M. (1986). Physiologic specialization of *Pyricularia oryzae* in the Philippines. *Plant Disease* **70,** 767–769.

Bracha-Drori, K., Shichrur, K., Katz, A., Oliva, M., Angelovici, R., Yalovsky, S., and Ohad, N. (2004). Detection of protein-protein interactions in plants using bimolecular fluorescence complementation. *Plant J.* **40,** 419–427.

Burnette, W. N. (1981). Western blotting: Electrophoretic transfer of proteins from sodium dodecyl sulfate-polyacrylamide gels to unmodified nitrocellulose and radiographic detection with antibody and radioiodinated protein A. *Analytical Biochem.* **112,** 195–203.

Cheong, H., Yorimitsu, T., Reggiori, F., Legakis, J. E., Wang, C. W., and Klionsky, D. J. (2005). Atg17 regulates the magnitude of the autophagic response. *Mol. Biol. Cell.* **16,** 3438–3453.

Dean, R. A., Talbot, N. J., Ebbole, D. J., Farman, M. L., Mitchell, T. K., Orbach, M. J., Thon, M., Kulkarni, R., Xu, J. R., Pan, H., Read, N. D., Lee, Y. H., *et al.* (2005). The genome sequence of the rice blast fungus *Magnaporthe grisea. Nature* **434,** 980–986.

Deppmann, C. D., Thornton, T. M., Utama, F. E., and Taparowsky, E. J. (2003). Phosphorylation of BATF regulates DNA binding: A novel mechanism for AP-1 (activator protein-1) regulation. *Biochem. J.* **374,** 423–431.

Ebbole, D. J. (2007). Magnaporthe as a model for understanding host-pathogen interactions. *Ann. Rev. Phytopathol.* **45,** 437–456.

Fields, S., and Song, O. (1989). A novel genetic system to detect protein-protein interactions. *Nature* **340,** 245–246.
Gershoni, J. (1988). Protein blotting. *Meth. Biochem. Anal.* **33,** 1–58.
Gershoni, J. M., and Palade, G. E. (1983). Protein blotting: Principles and applications. *Anal. Biochem.* **131,** 1–15.
Gershoni, J. M., and Palade, G. E. (1982). Electrophoretic transfer of proteins from sodium dodecyl sulfate-polyacrylamide gels to a positively charged membrane filter. *Anal. Biochem.* **124,** 396–405.
Golemis, E. A., Serebriiskii, I., and Law, S. F. (1999). The yeast two-hybrid system: criteria for detecting physiologically significant protein-protein interactions. *Curr. Issues Mol. Biol.* **1,** 31–45.
Grinberg, A. V., Hu, C. D., and Kerppola, T. K. (2004). Visualization of Myc/Max/Mad family dimers and the competition for dimerization in living cells. *Mol. Cell. Biol.* **24,** 4294–4308.
Hellens, R., Mullineaux, P., and Klee, H. (2000). Technical focus: A guide to Agrobacterium binary Ti vectors. *Trends in Plant Science* **5,** 446–451.
Hoff, B., and Kück, U. (2005). Use of bimolecular fluorescence complementation to demonstrate transcription factor interaction in nuclei of living cells from the filamentous fungus *Acremonium chrysogenum*. *Curr. Genet.* **47,** 132–138.
Hu, C. D., and Kerppola, T. K. (2003). Simultaneous visualization of multiple protein interactions in living cells using multicolor fluorescence complementation analysis. *Nature Biotechnol.* **21,** 539–545.
Hu, C. D., Chinenov, Y., and Kerppola, T. K. (2002). Visualization of interactions among bZIP and Rel family proteins in living cells using bimolecular fluorescence complementation. *Mol. Cell* **9,** 789–798.
Klionsky, D. J. (2005). The molecular machinery of autophagy: unanswered questions. *J. Cell Sci.* **118,** 7–18.
Klionsky, D. J., Cuervo, A. M., and Seglen, P. O. (2007). Methods for monitoring autophagy from yeast to human. *Autophagy* **3,** 181–206.
Lacroix, B., Vaidya, M., Tzfira, T., and Citovsky, V. (2005). The VirE3 protein of *Agrobacterium* mimics a host cell function required for plant genetic transformation. *EMBO J.* **24,** 428–437.
Levine, B., and Klionsky, D. J. (2004). Development by self-digestion: Molecular mechanisms and biological functions of autophagy. *Dev. Cell* **6,** 463–477.
Liu, X. H., Lu, J. P., and Lin, F. C. (2007a). Autophagy during conidiation, conidial germination and turgor generation in *Magnaporthe grisea. Autophagy* **3,** 472–473.
Liu, X. H., Lu, J. P., Zhang, L., Dong, B., Min, H., and Lin, F. C. (2007b). Involvement of a *Magnaporthe grisea* serine/threonine kinase gene, MgATG1, in appressorium turgor and pathogenesis. *Eukaryot Cell.* **6,** 997–1005.
Lu, J. P., Duan, Z. B., Liu, T. b., and Lin, F. C. (2007). Cloning, sequencing and expression analysis of the NAR promoter activated during hyphal stage of Magnaporthe grisea. *J. Zhejiang Univ. Sci.* **8,** 661–665.
Lu, J. P., Liu, T. B., and Lin, F. C. (2005a). Identification of mature appressorium-enriched transcripts in *Magnaporthe grisea*, the rice blast fungus, using suppression subtractive hybridization. *FEMS Microbiol. Lett.* **245,** 131–137.
Lu, J. P., Liu, T. B., Yu, X. Y., and Lin, F. C. (2005b). Representative appressorium stage cDNA library of *Magnaporthe grisea. J. Zhejiang Univ. Sci. B.* **6,** 132–136.
Mullins, E. D., Chen, X., Romaine, P., Raina, R., Geiser, D. M., and Kang, S. (2001). Agrobacterium-mediated transformation of *Fusarium oxysporum*: An efficient tool for insertional mutagenesis and gene transfer. *Phytopathol.* **91,** 173–180.

Noda, T., Matsuura, A., Wada, Y., and Ohsumi, Y. (1995). Novel system for monitoring autophagy in the yeast *Saccharomyces cerevisiae. Biochem. Biophys. Res. Commun.* **210,** 126–132.

Pall, M. C., and Brunelli, P. (1993). A series of six compact fungal transformation vectors containing polylinkers with multiple unique restriction sites. *Fungal Genet. Newsl.* **40,** 59–62.

Rackham, O., and Brown, C. M. (2004). Visualization of RNA-protein interactions in living cells: FMRP and IMP1 interact on mRNAs. *EMBO J.* **23,** 3346–3355.

Rajaram, N., and Kerppola, T. K. (2004). Synergistic transcription activation by Maf and Sox, and their subnuclear localization are disrupted by a mutation in Maf that causes cataract. *Mol. Cell. Biol.* **24,** 5694–5709.

Ramlau, J. (1987). Use of secondary antibodies for visualization of bound primary reagents in blotting procedures. *Electrophor.* **8,** 398–402.

Rho, H. S., Kang, S., and Lee, Y. H. (2001). *Agrobacterium tumefaciens*-mediated transformation of the plant pathogenic fungus, *Magnaporthe grisea. Mol. Cells* **12,** 407–411.

Stolpe, T., Süßlin, C., Marrocco, K., Nick, P., Kretsch, T., and Kircher, S. (2005). In plant analysis of protein-protein interactions related to light signaling by bimolecular fluorescence complementation. *Protoplasma* **226,** 137–146.

Talbot, N. J. (2003). On the trail of a cereal killer: Exploring the biology of *Magnaporthe grisea. Ann. Rev. Microbiol.* **57,** 177–202.

Tsuchisaka, A., and Theologis, A. (2004). Heterodimeric interactions among the 1-aminocyclopropane-1-carboxylate synthase polypeptides encoded by the Arabidopsis gene family. *Proc. Natl Acad. Sci. USA* **101,** 2275–2280.

Vijn, I., and Govers, F. (2003). *Agrobacterium tumefaciens* mediated transformation of the oomycete plant pathogen *Phytophthora infestans. Mol. Plant Path.* **4**.

Yu, H., West, M., Keon, B. H., Bilter, G. K., Owens, S., Lamerdin, J., and Westwick, J. K. (2003). Measuring drug action in the cellular context using protein-fragment complementation assays. *Assay Drug Dev. Technol.* **1,** 811–822.

Zhang, S., MA Charles, M., and Martin, C. (2004). Combinatorial marking of cells and organelles with reconstituted fluorescent proteins. *Cell* **119,** 137–144.

Zhao, X., and Xu, J. R. (2007). A highly conserved MAPK-docking site in Mst7 is essential for Pmk1 activation in *Magnaporthe grisea. Mol. Microbiol.* **63,** 881–894.

Zwiers, L. H., and De Waard, M. A. (2001). Efficient *Agrobacterium tumefaciens*-mediated gene disruption in the phytopathogen *Mycosphaerella graminicola. Curr. Genet.* **39,** 388–393.

CHAPTER TWENTY

Methods for Functional Analysis of Macroautophagy in Filamentous Fungi

Yi Zhen Deng,* Marilou Ramos-Pamplona,* *and* Naweed I. Naqvi*

Contents

Abstract

Autophagy is a bulk degradative process responsible for the turnover of membranes, organelles, and proteins in eukaryotic cells. Genetic and molecular regulation of autophagy has been independently elucidated in budding yeast and mammalian cells. In filamentous fungi, autophagy is required for several important physiological functions, such as asexual and sexual differentiation, pathogenic development, starvation stress and programmed cell death during heteroincompatibility. Here, we detail biochemical and microscopy methods useful for measuring the rate of induction of autophagy in filamentous fungi,

* Fungal Patho-Biology Group, Temasek Life Sciences Laboratory and Department of Biological Sciences, National University of Singapore, Singapore

Methods in Enzymology, Volume 451
ISSN 0076-6879, DOI: 10.1016/S0076-6879(08)03220-5

and we summarize the methods that have been routinely used for monitoring macroautophagy in both yeast and filamentous fungi. The role of autophagy in carbohydrate catabolism and cell survival is discussed along with the specific functions of macroautophagy in fungal development and pathogenesis.

1. Introduction

Autophagy is a highly conserved catabolic process in eukaryotes that is responsible for organellar turnover, membrane recycling, and protein degradation in vacuoles/lysosomes. Autophagy is induced in response to environmental stress or developmental signals during cellular differentiation (Besteiro *et al.*, 2006; Noda and Ohsumi, 1998; Pinan-Lucarre *et al.*, 2003). Autophagy can act as a prosurvival signal or participate in programmed cell death, depending on the particular physiological conditions (Codogno and Meijer, 2005).

There are three distinct classes of autophagy: macroautophagy, microautophagy, and chaperone-mediated autophagy (CMA); the latter is selectively used to degrade cytosolic proteins containing a specific pentapeptide consensus motif (Majeski and Dice, 2004; Salvador *et al.*, 2000). Macroautophagy and microautophagy (Reggiori and Klionsky, 2002) are considered nonselective and thus have more degradative capacity. The major difference between macroautophagy and microautophagy is whether the double-membrane vesicles, autophasogosomes, sequester cytoplasmic proteins or organelles (macroautophagy) for delivery to the lysosome/vacuole for degradation (Suzuki *et al.*, 2001), or whether the cytoplasm is directly engulfed into the vacuoles (microautophagy) (Mortimore *et al.*, 1989). Besides general autophagy, some specific types of autophagy exist, such as crinophagy (the activity of lysosomes related to the secretory pathway and endocrine functions; Glaumann, 1989), reticulophagy (degradation of ER; Bernales *et al.*, 2007; Bolender and Weibel, 1973) and pexophagy (degradation of peroxisomes; Sakai and Subramani, 2000).

Thus far, 31 *ATG* genes (autophagy-related) have been characterized in *Saccharomyces cerevisiae*, which has led to a better understanding of the genetic and molecular regulation of autophagy (Kabeya *et al.*, 2007; Klionsky *et al.*, 2003), particularly the formation of autophagy-associated vesicular compartments, such as preautophagosomal structures (PAS), autophagosomes (cytosolic), and autophagic bodies (vacuolar) (Suzuki *et al.*, 2001).

1.1. Cellular functions of autophagy in filamentous fungi

Autophagy is reported to play a crucial role during differentiation of several filamentous fungi such as *Podospora*, *Aspergillus*, *Colletotrichum*, and *Magnaporthe*. Autophagy-deficient mutants of *Magnaporthe oryzae* are nonpathogenic and show highly reduced asexual development (Liu *et al.*, 2007; Veneault-Fourrey *et al.*, 2006). Loss of autophagy-assisted programmed cell death in *atg8*Δ appressorium is proposed to be responsible for the failure of penetrating the host cuticle (Veneault-Fourrey *et al.*, 2006). Furthermore, autophagy is involved in lipid body turnover and thus is essential for turgor generation and appressorium-mediated penetration (Liu *et al.*, 2007). However, *Colletotrichum gloeosporioides*, with a related infection strategy as *M. oryzae*, does not require autophagic cell death for successful infection (Nesher *et al.*, 2008). Surprisingly, infection structures/appressoria from a *CLK1*-deletion (an ortholog of *MgATG1*) mutant in *Colletotrichum lindemuthianum*, are unable to penetrate the host cuticle (Dufresne *et al.*, 1998), similar to the result from *M. oryzae*. In *S. cerevisiae*, loss of autophagy leads to failure of sporulation, sensitivity to nitrogen starvation, and increased pseudohyphal growth (Cutler *et al.*, 2001; Ma *et al.*, 2007; Tsukada and Ohsumi, 1993). In *A. oryzae*, autophagy is required for the differentiation of aerial hyphae and in conidial germination (Kikuma *et al.*, 2006). In contrast to its function in fungi mentioned previously, autophagy plays little or no role in the differentiation of the dimorphic yeast *Candida albicans* within the host tissue (Palmer *et al.*, 2007). The *atg9*Δ mutant in *C. albicans* remains unaffected for yeast-hypha or chlamydospore differentiation, though it shows specific defects in autophagy and the cytoplasm-to-vacuole targeting (Cvt) pathway.

In *P. anserina*, autophagy is essential for sexual differentiation and cell death by incompatibility. It remains controversial whether autophagy executes a programmed cell death function or acts as a prosurvival response in *Podospora* (Dementhon *et al.*, 2003, 2004; Pinan-Lucarre *et al.*, 2003; Pinan-Lucarre *et al.*, 2005). It was initially thought that autophagy acts as the cause of cell death during incompatible interactions for it is induced when cells of unlike genotypes fuse in *P. anserina* (Dementhon *et al.*, 2004; Pinan-Lucarre *et al.*, 2003). A recent study suggests that autophagy serves a prosurvival role during incompatibility, as loss of autophagy results in accelerated cell death (Pinan-Lucarre *et al.*, 2005).

In this chapter, we present a technical review of the most frequently used methods to study autophagy in yeast and fungal species and focus on the methods that are useful to monitor the induction and rate of autophagy in filamentous fungi. The following protocols and methods have been validated in the model fungus *M. oryzae* and can be easily adapted and optimized for use in other filamentous fungi of interest.

2. Methods for the Functional Analysis of Autophagy in Filamentous Fungi

2.1. Gene-deletion analyses to assess macroautophagy in filamentous fungi

Generally, the one-step gene deletion strategy (schematized in Fig. 20.1) is used for disruption of requisite gene function in filamentous fungi. For gene disruption of *Magnaporthe ATG1* (MGG_06393.5), genomic DNA fragments

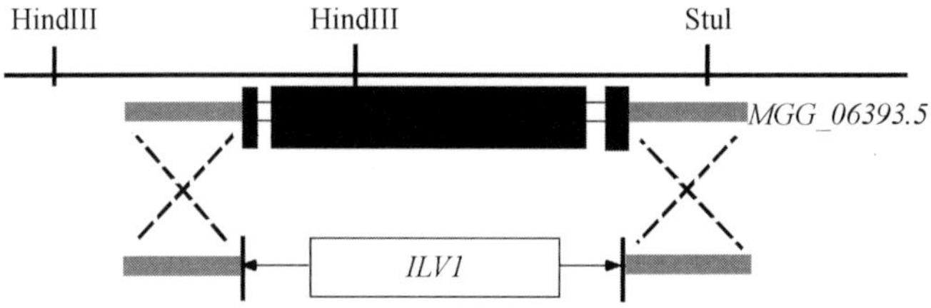

Figure 20.1 Schematic representation of a one-step gene replacement strategy for the *ATG1* locus (MGG.06393.5) in *Magnaporthe*. Solid bars and short open boxes represent coding regions and introns, respectively, whereas grey bars indicate the genomic flanks used as regions of homology for gene targeting. Relevant restriction enzyme sites have been depicted and *ILV1* refers to the sulfonyl urea-resistance cassette used to replace *ATG1* to create the *atg1*Δ strain.

(about 1 kb each) representing the 5′ and 3′ flanks of the *ATG1* open reading frame were amplified by PCR, and ligated sequentially so as to flank the *ILV1* cassette (confers resistance to sulfonyl urea) in pFGL385 to obtain plasmid vector pFGLatg1KO. The pFGLatg1KO plasmid was transformed into wild-type *M. oryzae* using *Agrobacterium* T-DNA-mediated transformation for homology-dependent replacement of the *ATG1* gene. Gene-disruption constructs can also be delivered into the fungal species of choice using electroporation of spheroplasts (Talbot *et al.*, 1993; Vollmer and Yanofsky, 1986; Xoconostle-Cazares *et al.*, 1996; Yelton *et al.*, 1984). Transformants were selected for resistance to chlorimuron ethyl (100 mg/mL; Cluzeau Labo, France) and correct gene-replacement confirmed by PCR analysis and Southern blotting. The primers used for amplifying the 1-Kb region at the 5′- and 3′-flank of the *ATG1* gene were as follows: ATG1-5F (5′- GAGTGA-GAATTCGCGGGACTAAGCAGGCCCAGGA-3′), ATG1-5R (5′- GA GTGAGAATTCTGCACTTAGAAACACTCGGGCT-3′), ATG1-3F (5′- GAGACTGTTCTGCAGCCTGGCAGTGGTTATCGGTTCG-3′), and ATG1-3R (5′- GAGAGTGTTAAGCTTGGACGTACAGTAGG-TAATTGGT-3′). The preceding protocol is validated for *Magnaporthe* but can be easily optimized for other fungal species of choice, provided the requisite sequence information is available. Other selectable marker cassettes for transformation in fungi include hygromycin resistance (*HPH1*) or ammonium-gluphosinate resistance (*BAR*). Gene targeting of the marker cassette can also be achieved by providing homology within the coding sequence per se and need not be restricted to the flanking sequences as described previously.

2.2. Use of chemical inhibitors to investigate autophagy in fungi

There are several chemical inhibitors of autophagy that are commercially available and routinely used in mammalian cells. Although these inhibitors block autophagy, their effects are not entirely specific. Wortmannin (WM) is an inhibitor of PI3-kinase and blocks the induction of autophagy (Blommaart *et al.*, 1997; Petiot *et al.*, 2000). 3-methlyadenine (3-MA) is also a classical inhibitor of the autophagic pathway (Seglen and Gordon, 1982). N-ethylmaleimide (NEM) inhibits several vesicular transport events and thus blocks the formation of autophagic vacuoles (Woodman, 1997). Such inhibitors can be potentially useful in studying autophagy in fungal systems provided special caution is exercised in analyzing the results. WM-treatment of vegetative mycelia (Fig. 20.2) of *Magnaporthe* wild-type strain mimics the phenotype of an *atg8*Δ mutant, in which starvation fails to induce autophagy (Deng and Naqvi, unpublished data).

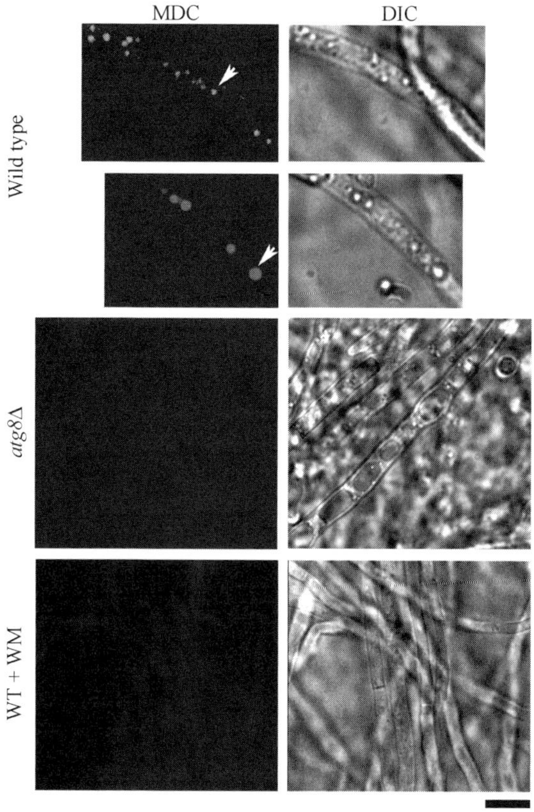

Figure 20.2 Epifluorescence microscopy-based assessment of autophagosomes. Mycelia from the wild type or *atg8*Δ mutant were stained with MDC and imaged using epifluorescence microscopy. MDC-stained wild-type mycelia pretreated with the autophagic-inhibitor Wortmannin (WM) prior to starvation, served as a negative control. Scale bar denotes 5 μm. Micrographs were pseudo colored using Photoshop Version 7.

Procedure

1. Small amount of vegetative mycelia (approximately 50 mg of wet weight; scraped from a colony surface with inoculation loop) or conidia (ca 4×10^3) of wild-type and *atg8*Δ strains are cultured in 20 ml of complete medium (CM; yeast extract 0.6%, casein hydrolysate 0.6%, sucrose 1%) for 48 h at 28 °C with gentle shaking (150 rpm) to obtain sufficient biomass.
2. Wortmannin stock (1 m*M*, in DMSO) is added into the CM, to a final concentration of 200 n*M*. The mycelia are treated with Wortmannin for 3 h at 28 °C with gentle shaking.
3. Mycelia are harvested by filtration through sterile Miracloth (Calbiochem, USA) and washed twice by filtration with sterile distilled water.
4. A small amount of the freshly cultured mycelia is then inoculated into 20 ml of minimal medium lacking nitrogen [MM-N: 0.5 g/L KCl, 0.5 g/L $MgSO_4$, 1.5 g/L KH_2PO_4, 0.1% (v/v) trace elements, 10 g/L glucose, pH6.5; (Talbot *et al.*, 1993)] containing 2 m*M* PMSF, and grown for 16 h with gentle shaking at 28 °C. Please note that this step is carried out to induce autophagy and there may not be a visible change in fungal biomass.
5. The pretreated and starved wild-type mycelia are stained with MDC and observed using epifluorescence microscopy, as described subsequently.

2.3. Microscopy methods to detect autophagy-associated membrane structures

Activation or induction of autophagy can be visualized by differential interference contrast (DIC; or Nomarski optics) microscopy. In *S. cerevisiae* (Lang *et al.*, 1998), *C. albicans* (Palmer *et al.*, 2007), and *P. anserina* (Dementhon *et al.*, 2004), starvation stress leads to enlarged vacuoles, which can be observed by DIC optics. Large punctuate perivacuolar structures or vesicles inside the vacuolar lumen can also be visualized, indicating the formation of autophagosomes or accumulation of autophagic bodies. In the *Podospora* Δ*pspA* mutant, which lacks the PSPA vacuolar protease, the accumulation of autophagic bodies in the vacuolar lumen is even more striking and easily detectable by simple microscopy observations (Dementhon *et al.*, 2003).

Fluorescence microscopy is another method to monitor the induction of autophagy. N-terminal tagging of Atg8 with a fluorescent protein such as GFP or RFP helps in epifluorescent detection of autophagosomes and has been used in several filamentous fungi such as *A. oryzae* (Kikuma *et al.*, 2006), *P. anserina* (Pinan-Lucarre *et al.*, 2005), and *M. oryzae* (Deng and Naqvi, unpublished results). An advantage of utilizing GFP- or RFP-tagged Atg8 is that the extent of autophagosome formation correlates well with the amount of GFP/RFP-Atg8PE (Kabeya *et al.*, 2000), so that the induction of autophagy can be easily quantified by Western blotting using commercially available anti-GFP/RFP antibodies.

Transmission electron microscopy (TEM) is the gold standard for ultrastructural investigation of autophagy-associated membrane compartments in filamentous fungi such as *P. anserina* (Pinan-Lucarre *et al.*, 2003), *A. oryzae* (Kikuma *et al.*, 2006), and *M. oryzae* (Liu *et al.*, 2007; Veneault-Fourrey *et al.*, 2006). Fig. 20.3 depicts the TEM analysis of the vacuolar lumen in the wild type and an autophagy-deficient mutant (*atg8Δ*) of *Magnaporthe*.

Procedure

1. Fresh conidia (4 × 10^3) or small amounts of mycelia (scraped from a colony using inoculation loop; about 50 mg) from the wild-type or *atg8Δ* strain in *Magnaporthe* are grown in 20 ml of CM for 48 h at 28 °C with gentle shaking (150 rpm).
2. Mycelia are harvested by filtration through Miracloth and washed thoroughly using sterile distilled water.
3. Washed mycelia from individual strains are grown in liquid MM-N medium [0.5 g/L KCl, 0.5 g/L $MgSO_4$, 1.5 g/L KH_2PO_4, 0.1% (v/v) trace elements, 10 g/L glucose, pH 6.5; Talbot *et al.*, 1993] containing 2 m*M* PMSF, for 16 h with gentle shaking, at 28 °C.
4. A small amount of the fungal biomass harvested by filtration through Miracloth, is placed in a microfuge tube and resuspended in 200 μl of fixation reagent (2.5% glutaraldehyde in 0.1 *M* phosphate buffer, v/v, pH 7.2) or sufficient to cover the mycelial sample. Initially the fixation is carried out under vacuum for 15 min at room temperature and subsequently at 4 °C overnight.

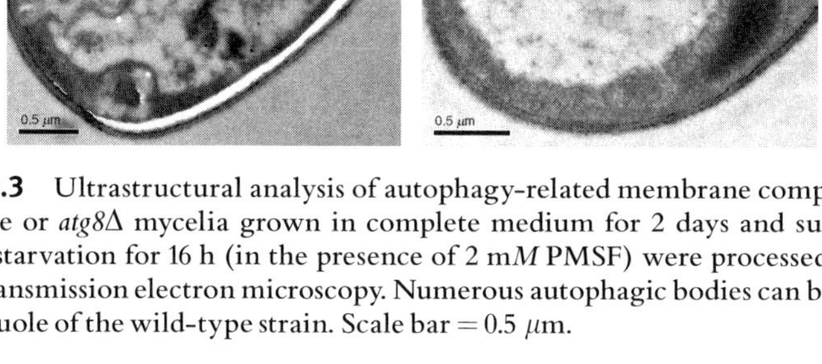

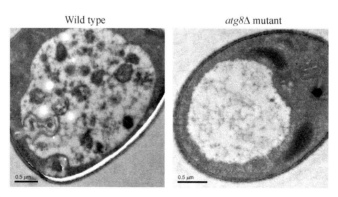

Figure 20.3 Ultrastructural analysis of autophagy-related membrane compartments. Wild-type or *atg8Δ* mycelia grown in complete medium for 2 days and subjected to nitrogen starvation for 16 h (in the presence of 2 m*M* PMSF) were processed for thin-section transmission electron microscopy. Numerous autophagic bodies can be detected in the vacuole of the wild-type strain. Scale bar = 0.5 μm.

5. Fixed mycelia are washed 3 times (10 min each) with 0.1 *M* phosphate buffer, pH 7.2.
6. The washed mycelial samples are postfixed for 3 h in 250 μl of osmium tetraoxide (1%, w/v).
7. Fixed mycelia are again washed 3 times (10 min each) in 0.1 *M* phosphate buffer, pH 7.2.
8. Samples are dehydrated in a graded ethanol series (25%, 50%, 75%, 100%; 10 min in 500 μl each).
9. The samples are then washed twice, for 15 min each, in 250 μl of propylene oxide.
10. Samples are infiltrated in 500 μl of propylene oxide-Spurr's resin (1:1) for 2 h, and then infiltrated overnight in 100% Spurr's resin.
11. Next, samples are embedded in Spurr's resin and polymerized overnight at 70 °C in EMS embedding molds (Electron Microscopy Sciences, USA).
12. Ultrathin (80 nm) sections are generated using Leica Ultracut UCT and mounted on 200 mesh copper grids.
13. Mounted sections are stained for 15 min at room temperature with a mixture of 2% uranyl acetate and 2% Reynolds' lead citrate (10 μl for each grid) and examined using a JEM1230 transmission electron microscope (Jeol, Tokyo, Japan) at 120 kV.

2.4. Monodansylcadaverine (MDC) staining of autophagic vesicles

MDC is an acidotropic dye that labels late stage autophagosomes or autophagic vesicles (Niemann *et al.*, 2001). MDC staining was successfully used for monitoring the increased autophagic activity in nitrogen-starved *Magnaporthe* mycelia (see Fig. 20.2), and the incorporation of MDC into late-stage autophagosomes or autophagic vesicles was inhibited by pretreatment with WM, the chemical inhibitor of autophagic sequestration. Furthermore, MDC staining with the conidiating cultures of *Magnaporthe* at different stages likely reflects the natural induction of autophagy (either basal levels or developmentally-induced), without starvation. MDC-incorporated compartments were copious in the conidiation-specific cell types, including aerial hyphae, and both young and mature conidia. An important limitation is that MDC staining fails to differentiate between late stage autophagosomes and autophagic (acidified) vacuoles.

Procedure

1. *Magnaporthe* wild-type or *atg8Δ* strains are cultured on Prune agar medium (PA; per liter: prune juice 40 ml, lactose 5 g, yeast extract 1 g, agar 20 g) in the dark at 28 °C for 2 days.

2. The wild-type and *atg8*Δ strains are then subjected to constant illumination (using overhead room lighting) to induce conidiation at room temperature.
3. At 6, 12, and 48 h after photoinduction, the conidiating cultures of the wild-type and *atg8*Δ strain are harvested by scraping with an inoculation loop and stained with 0.05 m*M* MDC solution (stock solution: 5 m*M* in normal phosphate buffered saline, pH 7.0) at 37 °C for 15 min. The MDC is then washed out with PBS before microscopic observation.
4. MDC-stained mycelia are observed using an epifluorescence microscope equipped with the following filter sets: excitation wavelength 350 nm, emission 320 to 520 nm.

2.5. LysoTracker-based visualization of vacuoles and vesicular compartments

LysoTracker Green DND-26 and LysoTracker Red DND-99 (Invitrogen-Molecular Probes, USA) are commonly used to stain and visualize acidic compartments, including autophagic compartments (Liu *et al.*, 2005; Scott *et al.*, 2004). LysoTracker dyes differ slightly from MDC and label acidified autophagic vacuoles (Fig. 20.4) but fail to incorporate into late-stage autophagosomes. In conidiating aerial hyphae of *Magnaporthe*, MDC staining was prominent, while very rare staining of LysoTracker Green DND-26 or LysoTracker Red DND-99 was detected. Both the LysoTracker- and MDC-labeled spherical compartments were evident in conidia, indicating that aerial hyphae are devoid of vacuoles that are mostly formed in conidia. One major drawback of the use of LysoTrackers is the inability to perform co-localization studies with RFP-Atg8 labeled vesicles (Deng and Naqvi, unpublished data).

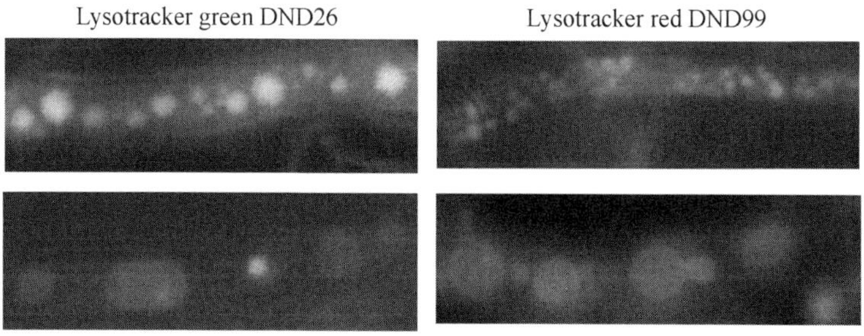

Figure 20.4 Mycelia from the wild-type B157 strain of *Magnaporthe* was stained with LysoTracker Green DND-26 or LysoTracker Red DND99 and subjected to the requisite epifluorescence microscopy to visualize acidified autophagic vacuoles. Bar = 5 μm. Different morphological variants of the fungal vacuoles (numerous small vesicles, top, or big round vesicles, bottom) are detected through LysoTracker DND staining.

Procedure

1. *Magnaporthe* wild-type and *atg8*Δ strains are grown on PA medium in the dark at 28 °C in an incubator for 2–3 days. The diameter of the resulting fungal colonies is about 1–2 cm at this stage.
2. Conidiation is induced in the wild-type and *atg8*Δ colonies by subjecting them to constant illumination at room temperature.
3. At 6, 12, and 48 h after photoinduction, the wild-type and *atg8*Δ conidiating cultures are harvested by scraping with an inoculation loop in sterile distilled water and collected through filtration using Miracloth. The harvested biomass is then stained with 50 n*M* LysoTracker Green DND26 solution (Invitrogen-Molecular Probes, USA; the stock is 1 *M* in DMSO) for 10 min at 37 °C. Staining with LysoTracker Red DND99 (1 *M* in DMSO; Invitrogen-Molecular Probes, USA) followed the same protocol as described previously.
4. The stained mycelia are observed with epifluorescence microscopy using the following filter sets: excitation wavelength 488 nm, emission 505 to 530 nm. LysoTracker Red DND99 staining is visualized using 543 nm excitation and a 560 nm emission filter.

2.6. Analysis of glycogen sequestration and estimation of glycogen content

In most eukaryotic cells, glycogen is stored as a carbohydrate reserve. Autophagy is involved in glycogen catabolism, which occurs in response to depletion of nutrients or to particular growth/differentiation conditions. In *S. cerevisiae*, glycogen can be degraded in the cytoplasm or inside vacuolar compartments. In the cytoplasm, Gph1 mediates glycogen breakdown resulting in the release of glucose 1-phosphate (G1P) (Hwang *et al.*, 1989). In vacuoles, glycogen degradation is catalyzed by the vacuolar glucoamylase (Colonna and Magee, 1978), which produces glucose 6-phosphate (G6P). The vacuolar degradation of glycogen is sporulation-specific and probably relies on autophagy for sequestration and delivery of glycogen (Fonzi *et al.*, 1979; Francois and Parrou, 2001; Wang *et al.*, 2001; Yamashita and Fukui, 1985).

To monitor the sequestration of glycogen in budding yeast, KI-I_2 staining or iodine vapor exposure is widely used (Hwang *et al.*, 1989; Wang *et al.*, 2001). In *M. oryzae*, detection of glycogen by KI-I_2 staining in differentiating conidia (Park *et al.*, 2004; Thines *et al.*, 2000) or by PAS (Periodic acid-Schiff) staining with sectioned fungal materials (Clergeot *et al.*, 2001) has been reported. Total glycogen content in the yeast or fungal cells can be determined enzymatically by hydrolyzing the extracted glycogen with α-amylase and amyloglucosidase, and then measuring the released glucose in a colorimetric assay containing glucose oxidase, peroxidase and *o*-dianisidine (Gascon and Lampen, 1968; Lillie and Pringle, 1980). The kits for total starch (Megazyme, UK) estimations are also commercially available and have been successfully used in *Magnaporthe* samples (Deng and Naqvi, unpublished data).

Procedures

Iodine vapor staining:

1. The fungal strain of interest is subcultured on PA medium and allowed to grow in the dark at 28 °C for 2 days.
2. The dark-grown cultures are then subjected to constant illumination at room temperature to induce conidiation.
3. At 0, 2, and 4 d after photoinduction, the culture dishes containing the colonies are inverted directly (with the lids removed) over a glass beaker containing iodine crystals for approximately 15 min. Caution: This step has to be carried out in a proper fume hood using appropriate safety measures.
4. The iodine-stained colonies (Fig. 20.5) are photographed immediately.

Estimation of total glycogen in fungal tissue(s)

1. The fungal strains (wild type or the autophagy-deficient mutant of interest) are grown on prune agar medium in the dark at 28 °C for 2 days.
2. Conidiation is induced in the fungal strains by subjecting them to constant illumination at room temperature.
3. At 0, 2, and 4 d after photoinduction, colonies from the respective strains are harvested by scraping the colony surface with inoculation loops in approximately 10–15 ml of sterile distilled water and then filtered using sterile Miracloth (Calbiochem, USA).
4. The harvested biomass is ground to a powder in liquid nitrogen using a mortar and pestle.
5. The total glycogen content is estimated using the Megazyme Total Starch Kit (Megazyme, UK) as instructed. Total glycogen content is normalized to the wet weight of the fungal biomass used for the estimation.

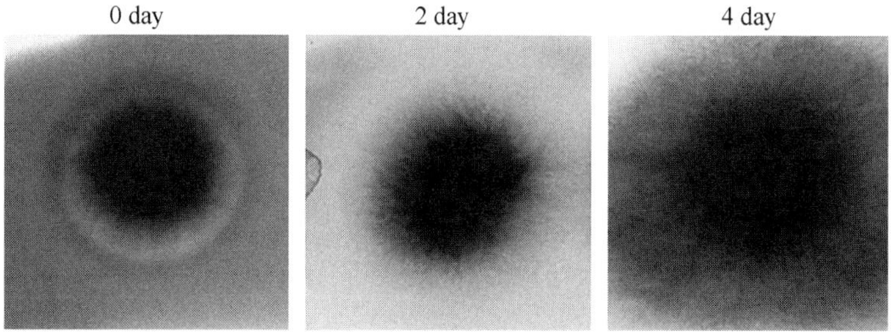

Figure 20.5 Analysis of glycogen accumulation during *Magnaporthe* conidiation. A wild-type strain grown in the dark for 2 days was subjected to constant illumination for the specified time intervals, and finally exposed to iodine vapor for 15 min and quickly photographed.

2.7. Comparative proteomics for identifying the targets of autophagic degradation

Because autophagy is a catabolic process responsible for lysosomal (vacuolar) degradation of proteins, a block of the autophagy pathway will lead to accumulation of proteins that are destined for autophagic degradation. To identify proteins that are regulated by autophagic degradation during conidiogenesis, we performed an SDS-PAGE fractionation of total lysates from 4-day-old conidiating cultures of the wild-type, the *atg8*Δ mutant, and the complemented strains. Mass spectrometry was then used to identify the proteins that showed differential accumulation. A vacuolar α-mannosidase (MGG_04464) in the *Magnaporthe* genome was present in the *atg8*Δ mutant but absent in the wild-type or the complemented strains. In yeast, Ams1 is delivered to the vacuoles via Cvt and autophagy pathway and acts as a vacuolar hydrolase for free oligosaccharide degradation (Chantret *et al.*, 2003; Hutchins and Klionsky, 2001). Another protein identified in this assay is Gph1 (MGG_01819), which is responsible for cytoplasmic glycogen degradation (Hwang *et al.*, 1989). The third protein that accumulates in *atg8*Δ mutant but not the wild type was a methionine synthase (MGG_06712). In *S. cerevisiae*, *Pichia pastoris*, *Cryptococcus neoformans*, and *C. albicans*, the cobalamin-independent methionine synthase is required for the growth of yeast or mycelial cells on methionine-free minimal media (Burt *et al.*, 1999; Csaikl and Csaikl, 1986; Huang *et al.*, 2005; Pascon *et al.*, 2004). In the human pathogenic fungi *C. neoformans* and *C. albicans*, methionine synthase acts as a virulence factor (Burt *et al.*, 1999; Pascon *et al.*, 2004). An aconitase (MGG_03521) function showed reduced expression on loss of autophagy. All the identified proteins are involved in fungal metabolism and potentially subjected to regulation by autophagy, thus validating the importance of such a comparative proteomics approach in identifying the potential targets of the autophagic degradation pathway for proteins.

Procedure

1. Wild type and the *atg8*Δ mutant are cultured on PA medium in the dark at 28 °C for 2 days.
2. The dark-grown cultures are subjected to constant illumination to induce conidiation.
3. At 4 d after photoinduction, the conidiating cultures are harvested by scraping the colony with inoculation loops in 10–15 ml of sterile distilled water. The fungal biomass is collected by filtration through sterile Miracloth (Calbiochem, USA).
4. The harvested fungal biomass is ground to a fine powder in liquid nitrogen, using an autoclaved mortar and pestle, and resuspended in 300 ml of extraction buffer (10 m*M* Na_2HPO_4, pH7.0, 0.5% SDS, 1 m*M* DTT and 1 m*M* EDTA). Lysates were cleared by centrifugation

at 12,000 *g* for 30 min at 4 °C. Protein concentration in the supernatant fraction is determined by Nanodrop ND-1000 spectrophotometer (Thermo Scientific, USA).

5. Normalized protein sample from each extract is fractionated by SDS-PAGE (10%), mixing with equal volume of loading buffer (Bio-Rad, USA).
6. Protein bands of interest (showing differential expression levels between the wild-type and *atg* mutant) are excised from the gel.
7. In-gel digestion and peptide extraction are performed using the following protocol: http://www.millipore.com/userguides.nsf/dda0cb48c91c0fb6852567430063b5d6/2265a5645f93475785256cbe00558a60/$FILE/P36505D.pdf.
8. Proteins of interest are identified by peptide sequencing or MALDI-TOF mass spectrometry.

3. Concluding Remarks

A recent review (Klionsky *et al.*, 2007) discusses methodologies to study the dynamics of autophagy in yeast and mammalian cells. The methods described previously measure the steady-state levels of autophagy, covering the induction, vesicle formation, and vesicle fusion. Very few assays that monitor the complete autophagic flux have been reported in filamentous fungi. One such assay relates to measuring the total protein breakdown in starved cells in wild-type versus an autophagy-deficient mutant in yeast (Tsukada and Ohsumi, 1993). An alternate method requires quantification of the total alkaline phosphatase activity in a yeast strain expressing a truncated form of Pho8, Pho8Δ60, which exists as an inactive precursor in the cytosol and is entirely dependent on autophagy to be delivered to the vacuole for its activity during starvation. Pho8Δ60 gains its phosphatase activity on specific processing by vacuolar enzymes so that the measurement of its activity can be an indicator of the fusion of autophagosomes with the vacuole and the subsequent processing therein (Kirisako *et al.*, 1999; Wang *et al.*, 2001). See the chapters by Noda and Klionsky, Cabrera and Ungermann, and Mayer in this volume for information on the Pho8Δ60 assay.

ACKNOWLEDGMENTS

We thank Yang Ming for generating the *atg1* deletion mutant, and the Fungal Patho-biology group for discussions and suggestions. We are grateful to S. Naqvi for critical reading of the manuscript. Intramural research support from the Temasek Life Sciences Laboratory, Singapore is gratefully acknowledged.

REFERENCES

Bernales, S., Schuck, S., and Walter, P. (2007). ER-phagy: Selective autophagy of the endoplasmic reticulum. *Autophagy* **3,** 285–287.

Besteiro, S., Williams, R. A., Morrison, L. S., Coombs, G. H., and Mottram, J. C. (2006). Endosome sorting and autophagy are essential for differentiation and virulence of *Leishmania major. J. Biol. Chem.* **281,** 11384–11396.

Blommaart, E. F., Krause, U., Schellens, J. P., Vreeling-Sindelarova, H., and Meijer, A. J. (1997). The phosphatidylinositol 3-kinase inhibitors wortmannin and LY294002 inhibit autophagy in isolated rat hepatocytes. *Eur. J. Biochem.* **243,** 240–246.

Bolender, R. P., and Weibel, E. R. (1973). A morphometric study of the removal of phenobarbital-induced membranes from hepatocytes after cessation of threatment. *J. Cell Biol.* **56,** 746–761.

Burt, E. T., O'Connor, C., and Larsen, B. (1999). Isolation and identification of a 92-kDa stress induced protein from *Candida albicans. Mycopathologia* **147,** 13–20.

Chantret, I., Frenoy, J. P., and Moore, S. E. (2003). Free-oligosaccharide control in the yeast *Saccharomyces cerevisiae*: Roles for peptide:N-glycanase (Png1p) and vacuolar mannosidase (Ams1p). *Biochem. J.* **373,** 901–908.

Clergeot, P. H., Gourgues, M., Cots, J., Laurans, F., Latorse, M. P., Pepin, R., Tharreau, D., Notteghem, J. L., and Lebrun, M. H. (2001). PLS1, a gene encoding a tetraspanin-like protein, is required for penetration of rice leaf by the fungal pathogen *Magnaporthe grisea. Proc. Natl. Acad. Sci. USA* **98,** 6963–6968.

Codogno, P., and Meijer, A. J. (2005). Autophagy and signaling: Their role in cell survival and cell death. *Cell Death Differ.* **12,** 1509–1518.

Colonna, W. J., and Magee, P. T. (1978). Glycogenolytic enzymes in sporulating yeast. *J. Bacteriol.* **134,** 844–853.

Csaikl, U., and Csaikl, F. (1986). Molecular cloning and characterization of the *MET6* gene of *Saccharomyces cerevisiae. Gene* **46,** 207–214.

Cutler, N. S., Pan, X., Heitman, J., and Cardenas, M. E. (2001). The TOR signal transduction cascade controls cellular differentiation in response to nutrients. *Mol. Biol. Cell* **12,** 4103–4113.

Dementhon, K., Paoletti, M., Pinan-Lucarre, B., Loubradou-Bourges, N., Sabourin, M., Saupe, S. J., and Clave, C. (2003). Rapamycin mimics the incompatibility reaction in the fungus *Podospora anserina. Eukaryot. Cell* **2,** 238–246.

Dementhon, K., Saupe, S. J., and Clave, C. (2004). Characterization of IDI-4, a bZIP transcription factor inducing autophagy and cell death in the fungus *Podospora anserina. Mol. Microbiol.* **53,** 1625–1640.

Dufresne, M., Bailey, J. A., Dron, M., and Langin, T. (1998). clk1, a serine/threonine protein kinase-encoding gene, is involved in pathogenicity of *Colletotrichum lindemuthianum* on common bean. *Mol. Plant Microbe Interact.* **11,** 99–108.

Fonzi, W. A., Shanley, M., and Opheim, D. J. (1979). Relationship of glycolytic intermediates, glycolytic enzymes, and ammonia to glycogen metabolism during sporulation in the yeast *Saccharomyces cerevisiae. J. Bacteriol.* **137,** 285–294.

Francois, J., and Parrou, J. L. (2001). Reserve carbohydrates metabolism in the yeast *Saccharomyces cerevisiae. FEMS Microbiol. Rev.* **25,** 125–145.

Gascon, S., and Lampen, J. O. (1968). Purification of the internal invertase of yeast. *J. Biol. Chem.* **243,** 1567–1572.

Glaumann, H. (1989). Crinophagy as a means for degrading excess secretory proteins in rat liver. *Revis. Biol. Celular* **20,** 97–110.

Huang, L., Li, D. Y., Wang, S. X., Zhang, S. M., Chen, J. H., and Wu, X. F. (2005). Cloning and identification of methionine synthase gene from *Pichia pastoris. Acta Biochim. Biophys. Sin. (Shanghai)* **37,** 371–378.

Hutchins, M. U., and Klionsky, D. J. (2001). Vacuolar localization of oligomeric alpha-mannosidase requires the cytoplasm to vacuole targeting and autophagy pathway components in *Saccharomyces cerevisiae*. *J. Biol. Chem.* **276,** 20491–20498.
Hwang, P. K., Tugendreich, S., and Fletterick, R. J. (1989). Molecular analysis of GPH1, the gene encoding glycogen phosphorylase in *Saccharomyces cerevisiae*. *Mol. Cell. Biol.* **9,** 1659–1666.
Kabeya, Y., Mizushima, N., Ueno, T., Yamamoto, A., Kirisako, T., Noda, T., Kominami, E., Ohsumi, Y., and Yoshimori, T. (2000). LC3, a mammalian homologue of yeast Apg8p, is localized in autophagosome membranes after processing. *EMBO J.* **19,** 5720–5728.
Kabeya, Y., Kawamata, T., Suzuki, K., and Ohsumi, Y. (2007). Cis1/Atg31 is required for autophagosome formation in *Saccharomyces cerevisiae*. *Biochem. Biophys. Res. Commun.* **356,** 405–410.
Kikuma, T., Ohneda, M., Arioka, M., and Kitamoto, K. (2006). Functional analysis of the ATG8 homologue Aoatg8 and role of autophagy in differentiation and germination in *Aspergillus oryzae*. *Eukaryot. Cell* **5,** 1328–1336.
Kirisako, T., Baba, M., Ishihara, N., Miyazawa, K., Ohsumi, M., Yoshimori, T., Noda, T., and Ohsumi, Y. (1999). Formation process of autophagosome is traced with Apg8/Aut7p in yeast. *J. Cell. Biol.* **147,** 435–446.
Klionsky, D. J., Cregg, J. M., Dunn, W. A., Jr., Emr, S. D., Sakai, Y., Sandoval, I. V., Sibirny, A., Subramani, S., Thumm, M., Veenhuis, M., and Ohsumi, Y. (2003). A unified nomenclature for yeast autophagy-related genes. *Dev. Cell* **5,** 539–545.
Klionsky, D. J., Cuervo, A. M., and Seglen, P. O. (2007). Methods for monitoring autophagy from yeast to human. *Autophagy* **3,** 181–206.
Lang, T., Schaeffeler, E., Bernreuther, D., Bredschneider, M., Wolf, D. H., and Thumm, M. (1998). Aut2p and Aut7p, two novel microtubule-associated proteins are essential for delivery of autophagic vesicles to the vacuole. *EMBO J.* **17,** 3597–3607.
Lillie, S. H., and Pringle, J. R. (1980). Reserve carbohydrate metabolism in *Saccharomyces cerevisiae*: Responses to nutrient limitation. *J. Bacteriol.* **143,** 1384–1394.
Liu, X. H., Lu, J. P., Zhang, L., Dong, B., Min, H., and Lin, F. C. (2007). Involvement of a *Magnaporthe grisea* serine/threonine kinase gene, MgATG1, in appressorium turgor and pathogenesis. *Eukaryot. Cell* **6,** 997–1005.
Liu, Y., Schiff, M., Czymmek, K., Talloczy, Z., Levine, B., and Dinesh-Kumar, S. P. (2005). Autophagy regulates programmed cell death during the plant innate immune response. *Cell* **121,** 567–577.
Ma, J., Jin, R., Jia, X., Dobry, C. J., Wang, L., Reggiori, F., Zhu, J., and Kumar, A. (2007). An interrelationship between autophagy and filamentous growth in budding yeast. *Genetics* **177,** 205–214.
Majeski, A. E., and Dice, J. F. (2004). Mechanisms of chaperone-mediated autophagy. *Int. J. Biochem. Cell Biol.* **36,** 2435–2444.
Mortimore, G. E., Poso, A. R., and Lardeux, B. R. (1989). Mechanism and regulation of protein degradation in liver. *Diabetes Metab. Rev.* **5,** 49–70.
Nesher, I., Barhoom, S., and Sharon, A. (2008). Cell cycle and cell death are not necessary for appressorium formation and plant infection in the fungal plant pathogen *Colletotrichum gloeosporioides*. *BMC Biol.* **6,** 9.
Niemann, A., Baltes, J., and Elsasser, H. P. (2001). Fluorescence properties and staining behavior of monodansylpentane, a structural homologue of the lysosomotropic agent monodansylcadaverine. *J. Histochem. Cytochem.* **49,** 177–185.
Noda, T., and Ohsumi, Y. (1998). Tor, a phosphatidylinositol kinase homologue, controls autophagy in yeast. *J. Biol. Chem.* **273,** 3963–3966.
Palmer, G. E., Kelly, M. N., and Sturtevant, J. E. (2007). Autophagy in the pathogen *Candida albicans*. *Microbiol.* **153,** 51–58.
Park, G., Bruno, K. S., Staiger, C. J., Talbot, N. J., and Xu, J. R. (2004). Independent genetic mechanisms mediate turgor generation and penetration peg formation during plant infection in the rice blast fungus. *Mol. Microbiol.* **53,** 1695–1707.

Pascon, R. C., Ganous, T. M., Kingsbury, J. M., Cox, G. M., and McCusker, J. H. (2004). *Cryptococcus neoformans* methionine synthase: Expression analysis and requirement for virulence. *Microbiol.* **150,** 3013–3023.

Petiot, A., Ogier-Denis, E., Blommaart, E. F., Meijer, A. J., and Codogno, P. (2000). Distinct classes of phosphatidylinositol 3′-kinases are involved in signaling pathways that control macroautophagy in HT-29 cells. *J. Biol. Chem.* **275,** 992–998.

Pinan-Lucarre, B., Paoletti, M., Dementhon, K., Coulary-Salin, B., and Clave, C. (2003). Autophagy is induced during cell death by incompatibility and is essential for differentiation in the filamentous fungus *Podospora anserina. Mol. Microbiol.* **47,** 321–333.

Pinan-Lucarre, B., Balguerie, A., and Clave, C. (2005). Accelerated cell death in *Podospora* autophagy mutants. *Eukaryot. Cell* **4,** 1765–1774.

Reggiori, F., and Klionsky, D. J. (2002). Autophagy in the eukaryotic cell. *Eukaryot. Cell* **1,** 11–21.

Sakai, Y., and Subramani, S. (2000). Environmental response of yeast peroxisomes: Aspects of organelle assembly and degradation. *Cell. Biochem. Biophys.* **32,** 51–61.

Salvador, N., Aguado, C., Horst, M., and Knecht, E. (2000). Import of a cytosolic protein into lysosomes by chaperone-mediated autophagy depends on its folding state. *J. Biol. Chem.* **275,** 27447–27456.

Scott, R. C., Schuldiner, O., and Neufeld, T. P. (2004). Role and regulation of starvation-induced autophagy in the *Drosophila* fat body. *Dev. Cell* **7,** 167–178.

Seglen, P. O., and Gordon, P. B. (1982). 3-Methyladenine: specific inhibitor of autophagic/lysosomal protein degradation in isolated rat hepatocytes. *Proc. Natl. Acad. Sci. USA* **79,** 1889–1892.

Suzuki, K., Kirisako, T., Kamada, Y., Mizushima, N., Noda, T., and Ohsumi, Y. (2001). The pre-autophagosomal structure organized by concerted functions of *APG* genes is essential for autophagosome formation. *EMBO J.* **20,** 5971–5981.

Talbot, N. J., Ebbole, D. J., and Hamer, J. E. (1993). Identification and characterization of MPG1, a gene involved in pathogenicity from the rice blast fungus *Magnaporthe grisea. Plant Cell.* **5,** 1575–1590.

Thines, E., Weber, R. W., and Talbot, N. J. (2000). MAP kinase and protein kinase A-dependent mobilization of triacylglycerol and glycogen during appressorium turgor generation by *Magnaporthe grisea. Plant Cell.* **12,** 1703–1718.

Tsukada, M., and Ohsumi, Y. (1993). Isolation and characterization of autophagy-defective mutants of *Saccharomyces cerevisiae. FEBS Lett.* **333,** 169–174.

Veneault-Fourrey, C., Barooah, M., Egan, M., Wakley, G., and Talbot, N. J. (2006). Autophagic fungal cell death is necessary for infection by the rice blast fungus. *Science* **312,** 580–583.

Vollmer, S. J., and Yanofsky, C. (1986). Efficient cloning of genes of *Neurospora crassa. Proc. Natl. Acad. Sci. USA* **83,** 4869–4873.

Wang, Z., Wilson, W. A., Fujino, M. A., and Roach, P. J. (2001). Antagonistic controls of autophagy and glycogen accumulation by Snf1p, the yeast homolog of AMP-activated protein kinase, and the cyclin-dependent kinase Pho85p. *Mol. Cell. Biol.* **21,** 5742–5752.

Woodman, P. G. (1997). The roles of NSF, SNAPs and SNAREs during membrane fusion. *Biochim. Biophys. Acta* **1357,** 155–172.

Xoconostle-Cazares, B., Leon-Ramirez, C., and Ruiz-Herrera, J. (1996). Two chitin synthase genes from *Ustilago maydis. Microbiol.* **142,** 377–387.

Yamashita, I., and Fukui, S. (1985). Transcriptional control of the sporulation-specific glucoamylase gene in the yeast *Saccharomyces cerevisiae. Mol. Cell. Biol.* **5,** 3069–3073.

Yelton, M. M., Hamer, J. E., and Timberlake, W. E. (1984). Transformation of *Aspergillus nidulans* by using a trpC plasmid. *Proc. Natl. Acad. Sci. USA* **81,** 1470–1474.

CHAPTER TWENTY-ONE

AUTOPHAGY IN *CANDIDA ALBICANS*

Glen E. Palmer*

Contents

Abstract

Candida albicans is considered a commensal organism of humans, colonizing the oral cavity, gastrointestinal, and reproductive tracts. However, when host defenses are compromised *C. albicans* can transform into a tissue invasive pathogen. Infections fall into two broad categories, those of mucosal tissue and serious disseminated disease, which involves transport through the bloodstream and invasion of the deeper organs. The ability of *C. albicans* to colonize diverse host microenvironments and switch from benign commensal to invasive pathogen suggests that *C. albicans* is able to undergo rapid and highly specialized adaptive responses *in vivo*. To date the role played by autophagy in facilitating asymptomatic host colonization, persistence, and transition of *C. albicans* into its pathogenic form have not been fully explored. Also the therapeutic potential of manipulating autophagic degradation within an invading fungus is yet to be established. In this chapter we describe several methods that have been adapted to detect autophagy within the opportunistic pathogen *C. albicans*.

* Department of Oral and Craniofacial Biology, LSUHSC School of Dentistry, New Orleans, Louisiana, USA

Methods in Enzymology, Volume 451

ISSN 0076-6879, DOI: 10.1016/S0076-6879(08)03221-7

1. Introduction

Candida albicans is considered a commensal organism of humans, colonizing the oral cavity, gastrointestinal and reproductive tracts. Under conditions where the host defenses are suppressed, overgrowth of *C. albicans* can lead to tissue invasion and symptomatic disease. This includes a diverse array of mucosal manifestations including oral and vaginal thrush (de Repentigny *et al.*, 2004; Nyirjesy and Sobel, 2003). Furthermore, in seriously ill patients *C. albicans* has the ability to disseminate through the bloodstream and invade the deeper organs. These systemic infections are associated with significant patient mortality (Fraser *et al.*, 1992; Wey *et al.*, 1988). The diverse host niches occupied by *C. albicans* necessitate the ability to readily adapt to new environmental challenges.

Autophagy provides a means for the cell to degrade cytoplasmic material and organelles in a nonspecific manner, either to recycle cellular nutrients during times of starvation or to facilitate cellular remodeling (Levine and Klionsky, 2004) when wholesale changes in protein expression are required. It has been demonstrated that autophagy plays a crucial role in growth, development, differentiation, and programmed cell death (PCD) in higher eukaryotes (Mizushima, 2005); however, the functional significance of autophagy in *C. albicans* and other pathogenic fungi has yet to be established. An initial study revealed that autophagy is apparently conserved in *C. albicans* (Palmer *et al.*, 2007). In addition, the closely related cytoplasm to vacuole targeting (Cvt) pathway described for *S. cerevisiae* (Wang and Klionsky, 2003) is also seemingly conserved in *C. albicans*. A *C. albicans atg9Δ* mutant has been described and is blocked in both autophagic and Cvt trafficking (Palmer *et al.*, 2007). The *atg9Δ* strain is sensitive to nitrogen restriction, indicating that autophagy is important during periods of starvation. However, the mutant is unaffected in yeast-hypha or chlamydospore differentiation. Moreover, the *atg9Δ* mutant is fully virulent in a phagocytic cell challenge model and a mouse model of hematogenously disseminated infection (Palmer *et al.*, 2007). Thus, the biological significance of autophagy in *C. albicans* is yet to be established. The role of autophagy in *C. albicans* PCD (Phillips *et al.*, 2003) is currently under investigation, as are the consequences of autophagic hyperstimulation. Furthermore, the role that autophagy plays in commensal persistence, biofilm formation, and mucosal disease have yet to be addressed. In this chapter we describe methods that have been successfully applied to monitor autophagy in *C. albicans*.

2. Resistance to Nitrogen Starvation

One of the primary roles of autophagy is nutrient recycling during periods of starvation. This seems to be especially important in lower eukaryotes, including fungi (Tsukada and Ohsumi, 1993). Therefore, resistance to nitrogen starvation can provide an indirect assay of autophagic activity. This can simply be determined using a method similar to that described for *S. cerevisiae* (Noda *et al.*, 2000). This involves propagating cells in a rich growth medium before transferring to nitrogen-free growth medium. Viability may then be followed as colony-forming units (CFUs) from samples removed from the starvation culture. Interestingly, *C. albicans* wild-type strains are able to undergo two or three cell divisions before final growth arrest in nitrogen-free medium (Palmer *et al.*, 2007) (Fig. 21.1). The cells then maintain viability for a prolonged period of time (>30 days), demonstrating that *C. albicans* is highly resistant to nitrogen starvation. However, the *C. albicans atg9Δ* mutant, blocked in autophagy is sensitive to nitrogen starvation (see Fig. 21.1). A second *C. albicans* mutant, *ypt72Δ*, is blocked in membrane fusion events at the vacuole, and is predicted to be deficient at a late stage of autophagy. This mutant is also sensitive to nitrogen starvation (see Fig. 21.1). The assay described subsequently provides a simple method for measuring resistance to nitrogen starvation. However, it should be noted that sensitivity to nitrogen starvation can result from defects in multiple cellular activities. Thus, sensitivity to nitrogen starvation alone does not necessarily infer defects in autophagy. Additional methods are required to confirm defects in autophagic trafficking.

2.1. Method

1. Inoculate *C. albicans* strains into 10 ml of a rich growth medium such as yeast extract peptone dextrose (YEPD) (1% yeast extract, 2% peptone, 2% glucose), in a 100-ml flask. Culture at 30 °C overnight with shaking.

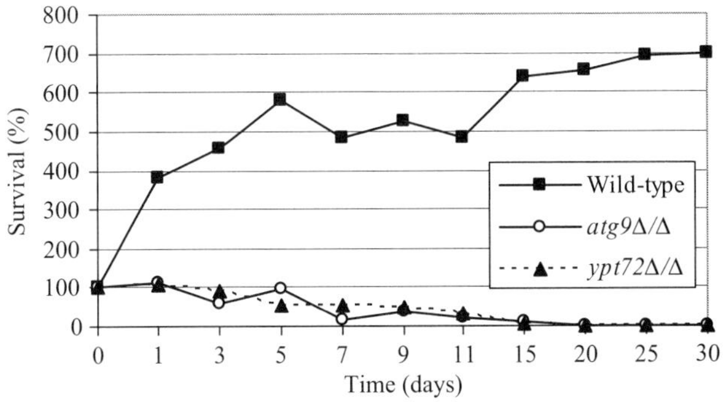

Figure 21.1 Resistance to nitrogen starvation. *C. albicans* strains were grown in YEPD overnight at 30 °C, switched to nitrogen free SD-N medium, and viability measured as CFUs over a 30-day period. Viability is expressed as a percentage of the CFU at t = 0. *C. albicans* mutant strains: *atg9Δ/Δ* = defective in autophagosome formation; *ypt72Δ/Δ* = blocked in vacuolar fusion.

2. Determine cell density using a hemocytometer. Note: if more than one strain is to be compared, cultures should be at similar cell densities.
3. Pellet 10^8 cells of each strain in a microcentrifuge tube at 13,000 rpm for 1 min, and wash twice in 1 ml of SD-N medium (0.17% yeast nitrogen base without ammonium sulfate or amino acids, 2 % glucose; Noda *et al.*, 2000).
4. Resuspend the cells in 1 ml of SD-N at 10^8 cells/ml.
5. Add 100 μl of the cell suspension to 4.9 ml SD-N media in a sealable culture tube (e.g., a sterile Biologix 50-ml polypropylene tube, #10-0501, or similar, which is important to avoid excessive evaporation of culture volume during prolonged culturing periods), to yield a cell density of 2×10^6/ml.
6. Immediately remove 100 μl of each culture and dilute 1:1000 by serial dilution in sterile distilled water.
7. Plate 100 μl onto a YEPD agar plate to determine viability at time 0. This should be done in triplicate to obtain reliable colony counts. After 2 days at 30 °C count the number of colonies formed. (This should yield approximately 200 colonies per plate). The average number of CFUs should then be calculated for each strain.
8. Maintain starvation cultures at 30 °C with shaking.
9. At the desired time intervals repeat steps 6 and 7 to determine viability as CFU. (We have typically followed viability up to 30 days).
10. Viability at each time point should be expressed as a percentage of the viability at time 0 (i.e., CFU/CFU_{t0} x 100; see Fig. 21.1).

3. Cytological Methods for Monitoring Autophagy

To date, two methods have been successfully adapted for observing autophagic/Cvt trafficking in *C. albicans*. The first is based on that described for *S. cerevisiae* (Noda *et al.*, 2000) and induces autophagosome formation through starvation. Inhibition of the vacuolar serine proteases blocks breakdown of autophagic bodies (AB) within the vacuole lumen (Takeshige *et al.*, 1992). The resulting accumulation of ABs confers a granular appearance upon the *C. albicans* cells (Fig. 21.2). The second method utilizes a cargo protein, aminopeptidase I (Ape1), which transits to the vacuole through Cvt and autophagic trafficking (Palmer *et al.*, 2007; Suzuki *et al.*, 2002). A simple genomic tagging method is employed to tag the C-terminus of Ape1 with green fluorescent protein (GFP; Gerami-Nejad *et al.*, 2001). The localization of the Ape1-GFP fusion in the vacuole lumen is dependent upon a reflection of Cvt/autophagic trafficking activity. Both of these methods provide a simple, albeit qualitative assay of autophagic trafficking in *C. albicans*.

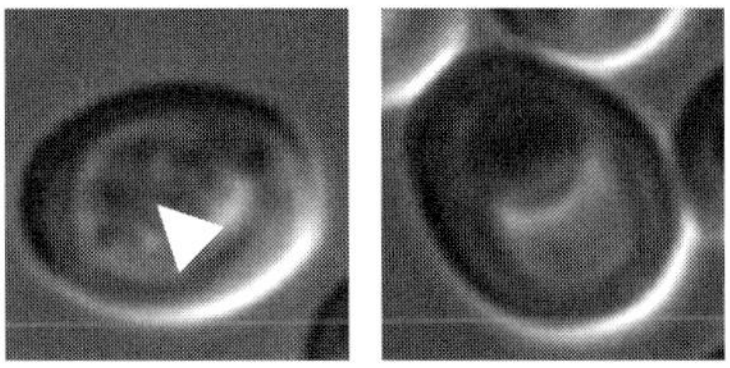

Figure 21.2 Autophagosome formation. *C. albicans* strains were incubated in nitrogen free SD-N medium + 1.5 m*M* PMSF to inhibit vacuolar serine proteases, and examined by DIC microscopy. Accumulated autophagic bodies (arrowhead) within the vacuole lumen give the wild-type cell a granular appearance not observed in the *atg9Δ* mutant, which is blocked in autophagosome formation.

3.1. Autophagosome formation

This protocol induces autophagy in nitrogen-free medium. However, following delivery to, and fusion of the autophagosome with, the vacuole, breakdown of the AB is blocked in the presence of the serine protease inhibitor PMSF (phenylmethylsulphonyl fluoride). Visible accumulation of ABs can be observed after 16–20 h (see Fig. 21.2).

3.1.2. Method

1. Inoculate *C. albicans* into 2 ml of a rich growth medium such as YEPD and culture at 30 °C overnight.
2. Determine cell density using a haemocytometer. Note: if more than one strain is to be compared, cultures should be at similar cell densities.
3. Pellet 10^8 cells in a microfuge tube at 13,000 rpm for 1 min.
4. Wash cells twice in 1 ml of SD-N medium, to remove traces of YEPD.
5. Resuspend cells in 5 ml of SD-N medium containing 1.5 m*M* of the serine protease inhibitor PMSF. (PMSF can be prepared as a 100 m*M* stock in isopropanol, stored in the dark at −20 °C. This must be warmed to get into solution. Note that the PMSF cannot be added to the medium until just before use because of its low stability in water).
6. Culture yeast cells at 30 °C with shaking for 16–48 h.
7. Apply cell culture to a poly-L-lysine coated slide [to make these apply 15 μl of 0.1% w/v poly-L-lysine in water (Sigma-Aldrich #8920) to each of the wells of an 8-well slide (Erie Scientific #30-14 or similar). Let stand at room temperature for 15 min. Rinse slides twice with sterile distilled water. Dry the slides at room temperature.] and apply a coverslip. Cells can then be inspected for AB accumulation by differential interference contrast (DIC) microscopy. Accumulated ABs should confer a granular appearance upon the cells (see Fig. 21.2).

Note: Important controls should include cultures in which PMSF has been omitted. In the absence of PMSF, the ABs should not accumulate.

3.2. Aminopeptidase I–GFP trafficking

In *S. cerevisiae*, Ape1 transport to the vacuole is dependent on Cvt/autophagy trafficking (Wang and Klionsky, 2003). Searches of the *C. albicans* genome database have revealed two homologs of *S. cerevisiae* Ape1, encoded by the *LAP4* and *LAP41* genes. The *LAP41* encoded protein most closely resembles *S. cerevisiae* Ape1 (57% identity). The method described subsequently involves the construction of a strain harboring a *LAP41-GFP* fusion and localization analysis by fluorescence microscopy.

3.2.1. Method

Strain construction. The *LAP41* gene of *C. albicans* can be GFP tagged using the genomic tagging method described by Gerami-Nejad *et al.* (2001). This involves PCR amplification of a *C. albicans*-optimized *GFP-URA3* cassette with flanking sequences to direct in frame integration at the 3′ end of the *LAP41* ORF to generate a *LAP41-GFP* fusion (Fig. 21.3A). Importantly, transcription of the gene fusion is dependent on the endogenous *LAP41* 5′ UTR sequences. This should ensure an appropriate transcription level of the fusion product.

1. Obtain PAGE purified oligonucleotides for production of the *LAP41-GFP-URA3* cassette. Megaprimers LAP41GFPF 5′-ATTAGGTATTAAATTCTTCTATGGTTTCTTCAAGAATTGGAGAGATGTCTATGATAATTTTGTTGATTTA*GGTGGTGGT*TCTAAAGGTGAAGAATTATT-3′ (forward), and LAP41GFPR 5′-TAAACTAACAAATAACTAATACTTGCAAATCAACTTGCAAATCAACTTTTAATGATTCCTTTTCCTGTTGTTCAATGTTCGATCTAGAAGGACCACCTTTGATTG-3′ (reverse; see Fig. 21.3A). These consist of 70 nucleotides of *LAP41* homologous targeting sequence, a glycine linker sequence (italics), and sequences for amplification of the *GFP-URA3* cassette (underlined).
2. Amplify the *GFP-URA3* cassette by PCR using the above primers and plasmid pGFP-URA3 as template (Gerami-Nejad *et al.*, 2001).
3. Transform the *GFP-URA3* cassette into a $ura3^{-}$ recipient strain of *C. albicans*, such as CAI4 (Fonzi and Irwin, 1993) or BWP17 (Wilson *et al.*, 1999) using the lithium acetate transformation method described by Gietz *et al.* (1992).
4. Select Ura^{+} transformants on selective yeast nitrogen base (YNB) agar (lacking uridine), and grow at 30 °C for 2–4 days.
5. Patch transformant colonies to selective agar plates.
6. Inoculate each transformant to 2 ml of YEPD liquid medium, grow overnight, and prepare genomic DNA using the glass-bead breakage or other method (Amberg *et al.*, 2005).

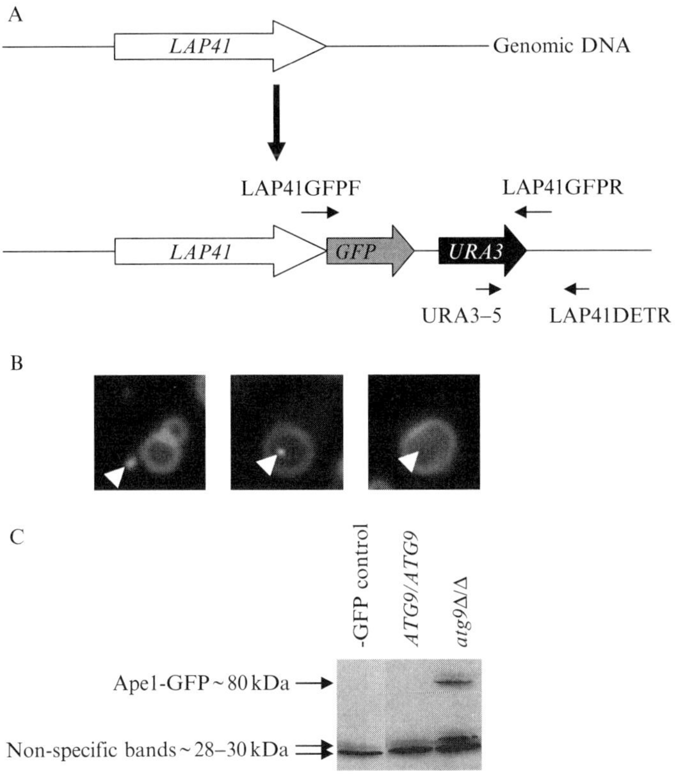

Figure 21.3 Localization and processing of Ape1-GFP. *LAP41* encodes *C. albicans* Ape1. (A) Construction of *LAP41-GFP* fusion. Megaprimers LAP41GFPF and LAP41GFPR are used to amplify a *GFP-URA3* cassette with flanking sequences to direct in-frame integration at the 3′ end of the *LAP41* ORF. *C. albicans* transformants are selected using the *URA3* marker, and correct integration confirmed by PCR detection with the primer set URA3-5 and LAP41DETR. (B) Ape1-GFP localization in wild-type *C. albicans*. Cells harboring the *LAP41-GFP* were pulse-labeled with FM 4-64 to label the vacuolar membrane. Cells were observed using an epifluorescence microscope with FITC-C and TRIT-C filter sets and 100x objective. GFP (green) and FM 4-64 (red) images were merged. Three distinct localizations are observed, often within the same cell: (1) intense GFP spot outside of the vacuole (*left*); (2) intense spot in the vacuole lumen (*center*); and (3) diffuse staining of the vacuole lumen (*right*). (C) Western blot analysis of Ape1-GFP. Protein extracts from $ATG9^+$ and *atg9Δ* *C. albicans* strains harboring the *LAP41-GFP* fusion and grown in YEPD media were probed with anti-GFP. A strain without the fusion was used as a negative control (−GFP control). (See Color Insert.)

7. Confirm integration of the *GFP-URA3* cassette into the 3′ region of *LAP41* by PCR using primers URA3-5 (5′-CCTATGAATCCACTATT-GAACC-3′) and LAP41DETR (5′-TTAATGTACAAGCCTTGCG-GC-3′) (Fig. 21.3A), with genomic DNA as template. Correct integration should yield a product of 470 b.p.

3.2.2. Ape1-GFP localization

Once the *LAP41-GFP* fusion has been constructed in the desired strain background, Ape1-GFP localization can be determined under a range of conditions. In general, it is expected that in rich growth media such as YEPD, autophagy is largely suppressed and Ape1-GFP trafficking is dependent on the Cvt pathway (Khalfan and Klionsky, 2002). However, in starvation media such as SD-N, autophagy will predominate. Ape1-GFP localization is assessed by staining with a vacuolar marker dye such as FM 4-64 (Invitrogen; Vida and Emr, 1995; see also the chapters by Journo *et al.*, van Zutphen *et al.*, and Oku and Sakai in this volume). In wild-type cells grown in rich YEPD media Ape1-GFP is present as three distinct patterns: (1) intense cytoplasmic spots, (2) intense spot within the vacuole lumen, and (3) diffuse fluorescence throughout the vacuole lumen (Fig. 21.3B). This likely corresponds to the stages of Cvt/autophagy trafficking: (1) Ape1-GFP oligomer formation and cytoplasmic sequestration within the Cvt vesicle/autophagosome, (2) release of the Cvt/autophagic body into the lumen of the vacuole following vacuolar delivery, and (3) maturation and release of Ape1-GFP into the vacuole lumen following degradation of the Cvt/autophagic body (Suzuki *et al.*, 2002). However, in the autophagy-defective *atg9*Δ mutant, Ape1-GFP accumulates as intense cytoplasmic spots (Palmer *et al.*, 2007). The following method describes the localization procedure in a rich growth medium (YEPD) when Cvt trafficking predominates and utilizes FM 4-64 to visualize the vacuolar compartment.

1. Inoculate *LAP41-GFP* tagged strain to 2–5 ml YEPD and culture overnight at 30 °C with shaking.
2. Subculture 1:50 into 2 ml of fresh YEPD and culture for a further 3–5 h at 30 °C with shaking.
3. Transfer 1 ml of culture to a microcentrifuge tube and pellet cells at 13,000 rpm for 1 min. Remove supernatant fraction and resuspend pellet in 50 μl of fresh YEPD.
4. Add 1 μl of FM 4-64 [FM 4-64 should be prepared as a 1 μg/μl stock in DMSO], and incubate for 30 min at 30 °C with agitation.
5. Pellet the cells at 13,000 rpm for 1 min, and resuspend in 2 ml of YEPD, transfer back to a suitable culture tube, and incubate for a further 1.5–2 h at 30 °C with shaking.
6. Pellet 1 ml of cells in a microcentrifuge tube and wash twice with 1 ml of sterile distilled water. Resuspend in 200 μl distilled water, and apply to a poly-L lysine coated microscope slide.
7. Cover with a suitable coverslip and inspect cells using either an epifluorescence or confocal microscope with a 100X objective lens. GFP excitation 470–490 nm, emission 520–580 nm; FM 4-64 excitation maximum 558 nm, emission maximum 734 (Invitrogen). Both GFP and FM 4-64 images should be combined to enable the relative distribution of FM 4-64 and GFP to be analyzed.

4. Tracking Autophagy Through Western Blot Analysis

In *S. cerevisiae* Ape1 is synthesized with a propeptide that is proteolytically cleaved on delivery to the vacuole (Oda *et al.*, 1996). This involves the removal of an N-terminal fragment, which can be followed as a mobility shift on a western blot. Thus, defects in the maturation of Ape1 are indicative of a block in Cvt/autophagic trafficking. The method described subsequently examines Ape1-GFP processing through western blot analysis as a marker of autophagy/Cvt trafficking. On the basis of the *Candida* genome database, the predicted mass of Ape1 (Lap41) is 56.7 kDa, and the mass of GFP is 26.9 kDa. Thus the Ape1-GFP fusion has a predicted mass of 83.6 kDa. In wild-type cells the ~80-kDa Ape1-GFP fusion was not detected by western blot using anti-GFP, presumably because of cleavage of the GFP upon vacuolar delivery. However, the Ape1-GFP fusion could be detected in the autophagy/Cvt-defective *atg9*Δ mutant (Fig. 21.3C), presumably because of accumulation of the unprocessed cytoplasmic form (Palmer *et al.*, 2007). Thus, the accumulation of the ~80-kDa Ape1-GFP fusion is indicative of defects in autophagy. Unfortunately, many antibodies against GFP recognize a *C. albicans* protein in the 25–30 kDa size range nonspecifically (see minus GFP control, Fig. 21.3C). Thus, appearance of the presumptively cleaved GFP moiety on vacuolar delivery of Ape1-GFP cannot be utilized as an indicator of Ape1-GFP processing and hence Cvt/autophagic trafficking. The use of monoclonal antibodies may overcome this problem.

The following method necessitates the construction of *LAP41-GFP* tagged strains as described in section 3.2.

4.1. Method

1. Construct *LAP41-GFP* fusions as described in section 3.2 in the desired strain background.
2. Inoculate the tagged strains to 5 ml of YEPD medium and grow overnight at 30 °C with shaking. Subculture 1 ml to 50 ml of fresh YEPD, and grow at 30 °C with shaking for 4 h. Note: In this rich growth medium one would expect autophagy to be suppressed and Ape1 trafficking to depend on the Cvt pathway (Wang and Klionsky, 2003).
3. Make cell extracts using glass-bead breakage or another suitable method (Amberg *et al.*, 2005).
4. Run 20–40 μg of each protein extract on a 10% SDS-PAGE gel.
5. Transfer protein to a nitrocellulose membrane and incubate the membrane in blocking buffer (5% powdered milk in wash buffer [150 m*M* NaCl, 50 m*M* Tris-HCL, pH 8, 5 m*M* EDTA]) at room temperature for 1 h with gentle rocking.

6. Pour off the blocking buffer from the membrane. Dilute the anti-GFP (Rabbit polyclonal anti-GFP, Ana Spec #29779) in blocking buffer, 1:1000, and pour over membrane. Incubate at 4 °C overnight with gentle rocking.
7. Wash membrane using enough wash buffer to cover the membrane, for 5 min at room temperature with rocking. Pour off the wash buffer and repeat three times.
8. Incubate membrane with goat antirabbit IgG conjugated to horseradish peroxidase (HRP; Rockland #611-1302), diluted 1:20,000 in blocking buffer, at room temperature for 1 h with gentle rocking.
9. Repeat step 7.
10. Detect bound HRP conjugate using a chemiluminescent substrate (e.g., Pierce #34080), per the manufacturers' instructions.
11. An increase in the abundance of the ~80-kDa Ape1-GFP species relative to the wild-type control indicates an accumulation of unprocessed Ape1-GFP and a block in Cvt/autophagy trafficking.

5. Summary

The biological significance of autophagy in the pathogenesis of *C. albicans* has yet to be established. However, autophagy is a process of profound importance to eukaryotic cell biology, which makes it unlikely that autophagy is completely superfluous in this prevalent pathogen. Interestingly, not all methods used to monitor autophagy in other species have been readily adapted to *C. albicans*. For example, Atg8/LC3 expression levels and post-translational processing have been utilized as a means to quantify autophagic activity in a variety of species (Klionsky *et al.*, 2007). An *ATG8* (*AUT7*) homolog has been identified in the *Candida* genome database, the product of which shares 81% identity to *S. cerevisiae* Atg8. However, it has not been possible to confirm expression or processing of the gene product. Thus, the utility of Atg8 analysis is uncertain. Taken together with the fact that autophagy is apparently dispensable for *C. albicans* cellular differentiation events (Palmer *et al.*, 2007), this could point to divergence in the mechanism and/or biological function of autophagy in this fungus. To date only a limited number of methods have been applied in the analysis of autophagy in *C. albicans*. To unravel the biological role that autophagy plays in *C. albicans*, additional methods of detection and quantification will need to be developed.

Acknowledgments

This publication was made possible by Grant No. P20RR020160 from the National Center for Research Resources (NCRR), a component of the National Institutes of Health (NIH). Its contents are solely the responsibility of the author and do not necessarily represent the official view of NCRR or NIH.

REFERENCES

Amberg, D. C., Burke, D. J., and Strathern, J. N. (2005). Methods in Yeast Genetics: A Cold Spring Harbor Laboratory Course Manual. Cold Spring Harbor Laboratory Press, Cold Spring Harbor, NY, USA, 2005. pp 230.

de Repentigny, L., Lewandowski, D., and Jolicoeur, P. (2004). Immunopathogenesis of oropharyngeal candidiasis in human immunodeficiency virus infection. *Clin. Microbiol. Rev.* **17,** 729–759.

Fonzi, W. A., and Irwin, M. Y. (1993). Isogenic strain construction and gene mapping in *Candida albicans. Genetics* **134,** 717–728.

Fraser, V. J., Jones, M., Dunkel, J., Storfer, S., Medoff, G., and Dunagan, W. C. (1992). Candidemia in a tertiary care hospital: Epidemiology, risk factors, and predictors of mortality. *Clin. Infect. Dis.* **15,** 414–421.

Gerami-Nejad, M., Berman, J., and Gale, C. A. (2001). Cassettes for PCR-mediated construction of green, yellow, and cyan fluorescent protein fusions in *Candida albicans. Yeast* **18,** 859–864.

Gietz, D., St Jean, A., Woods, R. A., and Schiestl, R. H. (1992). Improved method for high efficiency transformation of intact yeast cells. *Nucleic Acids Res.* **20,** 1425.

Khalfan, W. A., and Klionsky, D. J. (2002). Molecular machinery required for autophagy and the cytoplasm to vacuole targeting (Cvt) pathway in *S. cerevisiae. Curr. Opin. Cell Biol.* **14,** 468–475.

Klionsky, D. J., Cuervo, A. M., and Seglen, P. O. (2007). Methods for monitoring autophagy from yeast to human. *Autophagy* **3,** 181–206.

Levine, B., and Klionsky, D. J. (2004). Development by self-digestion: Molecular mechanisms and biological functions of autophagy. *Dev. Cell* **6,** 463–477.

Mizushima, N. (2005). The pleiotropic role of autophagy: From protein metabolism to bactericide. *Cell Death Differ.* **12**(Suppl. 2), 1535–1541.

Noda, T., *et al.* (2000). Apg9p/Cvt7p is an integral membrane protein required for transport vesicle formation in the Cvt and autophagy pathways. *J. Cell Biol.* **148,** 465–480.

Nyirjesy, P., and Sobel, J. D. (2003). Vulvovaginal candidiasis. *Obstet. Gynecol. Clin. North Am.* **30,** 671–684.

Oda, M. N., Scott, S. V., Hefner-Gravink, A., Caffarelli, A. D., and Klionsky, D. J. (1996). Identification of a cytoplasm to vacuole targeting determinant in aminopeptidase. *I. J. Cell Biol.* **132,** 999–1010.

Palmer, G. E. (2007). Autophagy in the invading pathogen. *Autophagy* **3,** 251–253.

Palmer, G. E., Kelly, M. N., and Sturtevant, J. E. (2007). Autophagy in the pathogen *Candida albicans. Microbiology* **153,** 51–58.

Phillips, A. J., Sudbery, I., and Ramsdale, M. (2003). Apoptosis induced by environmental stresses and amphotericin B in *Candida albicans. Proc. Natl. Acad. Sci. USA* **100,** 14327–14332.

Suzuki, K., Kamada, Y., and Ohsumi, Y. (2002). Studies of cargo delivery to the vacuole mediated by autophagosomes in *Saccharomyces cerevisiae. Dev. Cell* **3,** 815–824.

Takeshige, K., Baba, M., Tsuboi, S., Noda, T., and Ohsumi, Y. (1992). Autophagy in yeast demonstrated with proteinase-deficient mutants and conditions for its induction. *J. Cell Biol.* **119,** 301–311.

Tsukada, M., and Ohsumi, Y. (1993). Isolation and characterization of autophagy-defective mutants of *Saccharomyces cerevisiae. FEBS Lett.* **333,** 169–174.

Vida, T. A., and Emr, S. D. (1995). A new vital stain for visualizing vacuolar membrane dynamics and endocytosis in yeast. *J. Cell Biol.* **128,** 779–792.

Wang, C. W., and Klionsky, D. J. (2003). The molecular mechanism of autophagy. *Mol. Med.* **9,** 65–76.

Wey, S. B., Mori, M., Pfaller, M. A., Woolson, R. F., and Wenzel, R. P. (1988). Hospital-acquired candidemia. The attributable mortality and excess length of stay. *Arch. Intern. Med.* **148,** 2642–2645.

Wilson, R. B., Davis, D., and Mitchell, A. P. (1999). Rapid hypothesis testing with *Candida albicans* through gene disruption with short homology regions. *J. Bacteriol.* **181,** 1868–1874.

CHAPTER TWENTY-TWO

Analysis of Autophagy During Infections of *Cryptococcus neoformans*

Guowu Hu,* Jack Gibbons,† *and* Peter R. Williamson*,‡

Contents

* Section of Infectious Diseases, Department of Medicine, University of Illinois at Chicago, Chicago, Illinois, USA
† Division of Biological Sciences, University of Illinois at Chicago, Chicago, Illinois, USA
‡ Jesse Brown VA Medical Center, Chicago, Illinois, USA

Methods in Enzymology, Volume 451
ISSN 0076-6879, DOI: 10.1016/S0076-6879(08)03222-9

Abstract

Cryptococcus neoformans is a yeastlike fungus that causes a lethal meningoencephalitis in a broad spectrum of immunocompromised patients and has become the most common cause of meningitis due to AIDS-related infections in Africa. Key to the development of new agents to control and prevent this infection is the identification of cellular mechanisms required for pathogenesis. Survival of the fungus within the hostile and nutrient-deprived environments of the host has recently been shown to depend on the induction of autophagy, whereby the cell recycles nutrients by slowly digesting itself in a regulated fashion. Further study of the role of autophagy during infection by *C. neoformans* requires the use of markers of autophagy that are specially adapted to the fungus within the mammalian host.

1. Introduction

Cryptococcus neoformans is a yeastlike basidiomycetous fungus that causes meningoencephalitis in a wide variety of patients including those with AIDS. As antiretroviral therapy has reduced the number of opportunistic infections such as cryptococcosis in the developed world, the number of cases of this disease in the developing world has exploded, resulting in an estimated 13%–44% of AIDS-related deaths. The fungus has become the leading cause of meningitis in most areas of the developing world and, if estimates of USAID about the number of deaths from AIDS are accurate, approximately 2–4 million deaths may have occurred in the past decade from cryptococcosis in Africa alone (Bicanic and Harrison, 2004). Poor responses to standard therapies such as Amphotericin B and fluconazole have also been reported in these African infections (Bicanic *et al.*, 2007), suggesting an urgent need for new approaches to treat and prevent the disease. In addition, emergence of an outbreak of cryptococcosis in persons with apparently normal immune systems in a multistate region of the northwestern United States suggests how deficiencies in the understanding of mechanisms of pathogenesis in this organism may have global impact, afflicting those in the developed world as well (Nicol *et al.*, 2008).

Virulence factors of the fungus have classically focused on interactions of the fungus with the host immune system leading to a good understanding of the function of these factors. For example, an extensive polysaccharide capsule prevents phagocytosis of the fungus and prevents complement deposition on the cell wall (Janbon, 2004). A second virulence factor is a cell wall laccase that produces highly reactive *o*-quinones from neurological catecholamines of the host brain, resulting in the formation of a variety of immunomodulatory end products (Huffnagle *et al.*, 1995; Nosanchuk *et al.*, 2000), and its ferroxidase activity can prevent fenton product formation in host macrophages (Zhu and Williamson, 2004). More recently, the appreciation of the role of a number of metabolic pathways during infection has led to the identification of genes involved in pathogenic fitness, a trait that allows the invading pathogen to survive and effect virulence within the host (Panepinto and Williamson, 2006). For example, growth at 37 °C is an obvious requirement of growth in the mammalian host, and is dependent on calcineurin (Brown *et al.*, 2007). More recently, nutrient deprivation of the invading fungus is increasingly being recognized as a host-defense trait in cryptococcosis (Panepinto and Williamson, 2006). This has led to the identification of cellular pathways such as gluconeogenesis (Panepinto *et al.*, 2005), cell cycle arrest (Liu *et al.*, 2006), and most recently autophagy (Hu *et al.*, 2008) as important for fungal survival during residence in the mammalian host. However, there remain many questions regarding the roles of autophagy during cryptococcal infection. For example, the fungus occupies numerous host tissues during infection including lung, spleen and brain and can survive both intracellularly and extracellularly (Feldmesser *et al.*, 2000), but it is not known to what degree autophagy is required in each of these host environments. In addition, it is not known during which stage of infection autophagy may be most important—initial infection, latency, or reactivation. Finally, if the ability to induce autophagy by the pathogen is so essential to successful infection, could there be host mechanisms to suppress fungal autophagy? These and other questions will undoubtedly require methods for the analysis of cellular processes such as autophagy that can be adapted to the fungus as it traverses and damages multiple host environments.

2. Mouse Models of Infection by *C. neoformans*

Understanding cellular processes such as autophagy during infection of the mammalian host requires the adaptation of biochemical methods typically applied to yeast cells in culture. Fortunately, a number of mouse models exist that mimic the human response to the fungus, and allow modeling of the infectious process. The principal mouse models for cryptococcosis are acute models in that the mice develop progressive disease over time, normally resulting in mortality in several weeks to months. The first is an intravenous model that best mimics dissemination of fungi within the bloodstream to the central nervous system and results in almost simultaneous inoculation of lung, spleen, and brain tissue. In contrast, the intranasal pulmonary model allows study of the lung infection stage, which results in progressive lung disease followed by death.

2.1. Intravenous model of disseminated cryptococcosis

For this mouse model, we normally order mice to arrive 10 days prior to the experiment so that the mice become acclimated to the environment. We use only female mice, as they are more docile during the observation period.

Method

1. Fungal cells for virulence studies are plated out on YPD (2% glucose, 2% bactopeptone, 1% yeast extract; Difco Laboratory) agar for 2 days, followed by a second inoculation on a new agar plate 2 days prior to inoculation. Note: Cells should not be serially plated more than 5 times in the laboratory before being recultured from frozen stock. Many pathogenic strains such as the serotype A strain, H99, lose virulence traits with serial plating. (A reference H99 strain can be obtained from ATCC, strain 208821, in the event of questions of genetic alterations).
2. Fungal cells are collected by scraping the plate, washing once in sterile phosphate-buffered saline (PBS, Mediatech) and resuspending again in sterile PBS.
3. NIH Swiss Albino mice, 18–20 g (Jackson Labs) are inoculated in the lateral tail vein using a 25 gauge needle with 100 μl of sterile PBS containing 1×10^6 fungal cells. A heat lamp can be used to effect vasodilatation, but this is usually not required for Swiss Albino mice that have larger tail veins.
4. Mouse health is monitored twice daily, and moribund mice, defined as mice that are lethargic (arousable but relapse into slumber within 1 min or have repetitive neurological signs such as circling) or are not able to reach food or water, are sacrificed by CO_2 narcosis (for further details on humane methods of euthanasia, see www.avma.org/resources/euthanasia.pdf).
5. Statistics: The Kaplan-Meier test followed by the log rank test is used to compare between two groups of mice and ANOVA is used to determine significant differences in mortality between any of three groups using Graph Pad software (Prism 4; see Fig. 22.1 for an example).
6. Necropsies are performed to isolate infected brains, which can be either fixed in 4% phosphate-buffered formalin (10 m*M* sodium phosphate, pH 7.4) for histology or whole brains can be homogenized in PBS and plating directly on YPD agar (after 1:10, 1:100 and 1:1,000 serial dilutions in sterile distilled water) to determine fungal burden.
7. Separation of fungal cells from brain material. Alternatively, after necropsy, fungal cells can be separated from brain material after homogenization as in step 6. These fungal cells can then be used to prepare protein, mRNA or for microscopy, using methods described below. For this method, the homogenized brain material is centrifuged on a 10 ml 60 to 80% discontinuous sucrose gradient at 3000 rpm for 30 min in a

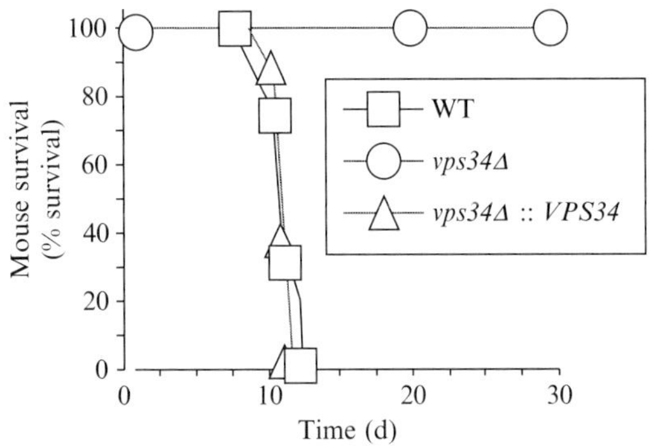

Figure 22.1 Demonstration of the role of *VPS34* in an intravenous model of cryptococcosis. Indicated strains were inoculated intravenously into 10 Swiss Albino mice (10^6 cfu) and sacrificed when moribund. All figures are from Hu *et al.*, (Hu *et al.*, 2008).

swinging bucket rotor to isolate purified yeast cells at the 60%–80% interface, leaving the brain lysate at the top of the gradient. If only smaller quantities are available, approximately 100–200 mg of brain homogenate, the same sucrose density centrifugation (1.5 ml total volume) can be done in a microcentrifuge tube, centrifuging at 13,000g × 1 min.

2.2. Pulmonary model of cryptococcosis

We use female CBA/J mice (Harlan) because it is an immunologically intact strain. Again, typically, 3 groups of 10 mice each are infected with 10^4 cfu of wild-type, mutant strain, or the mutant strain complemented with the wild-type gene. Complementation of mutant strains ensures that any defect in virulence of the mutant is not due to attributes acquired during the transformation and recovery process.

Method

1. Nasal inhalations are performed by first anesthetizing the mice with phenobarbital via intraperitoneal injections (1.2 mg diluted in sterile PBS) and then suspending the mice via the incisors on a silk thread so that the necks are fully extended.
2. Yeast inocula are made up in 25 μl volumes (1 × 10^4 cells in 25 μl of sterile normal saline, prepared in the same way as for the intravenous model) and slowly pipetted using a standard sterile disposable pipette tip, directly into the nostrils of each animal.
3. The mice are suspended for 10 min after infection, followed by recovery.
4. The mice are fed ad libitum and monitored with twice-daily inspections. Mice that appear moribund are sacrificed using CO_2 inhalation as previously.
5. Survival data are analyzed in the same way as the intravenous model (see Fig. 22.2 for an example).

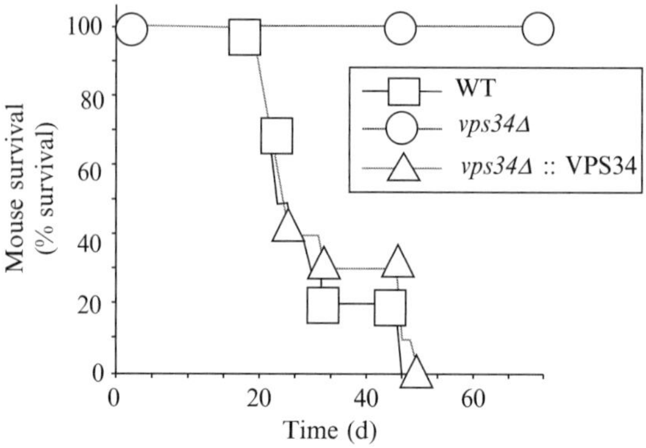

Figure 22.2 Demonstration of the role of *VPS34* in a pulmonary model of cryptococcosis. Indicated strains were inoculated intranasally into 10 CBA/J mice (10^5 cfu) and sacrificed when moribund.

3. Suppression Plasmids for Autophagy-Related Genes

To study regulation of autophagy, RNA interference techniques (Liu *et al.*, 2002) can be successfully employed to identify autophagy-related gene function. Expression plasmids are useful tools for performance of these molecular biological techniques.

3.1. Use of shuttle plasmid pORA-SK for cryptococcal expression

Because of limitations regarding size, and the absence of unique restriction sites in available vectors, a *C. neoformans* shuttle vector was constructed for episomal retention and expression, based on the pPM8 shuttle vector derived previously (Mondon *et al.*, 2000). As shown in Fig. 22.3, the shuttle vector pORA-2XK contains an *E. coli* origin of replication, a cryptococcal stabilization sequence, Stab1, a kanamycin transformation marker and two telomere-like sequences that result in stable replication as an episome in the fungus. An expression plasmid, pKUTAP (5880 bp), has been made from pORA-2XK and consists of (1) an actin promoter inserted 5′–3′ at the *Bgl*II-*Eco*RI sites, (2) an EF1alpha terminator inserted 5′–3′ at the *Eco*RI-*Age*I sites, and (3) a *URA5* transformation marker at the *Kpn*I site as described previously (Liu *et al.*, 2006). The plasmid, pKUTAP, is freely available from the authors.

3.2. Methods for effective RNAi in *C. neoformans*

We will demonstrate the usefulness of the pKUTAP expression plasmid in RNAi suppression of *ATG8* similar to that described previously. Atg8 is a ubiquitin-like protein that is a required component of autophagic bodies

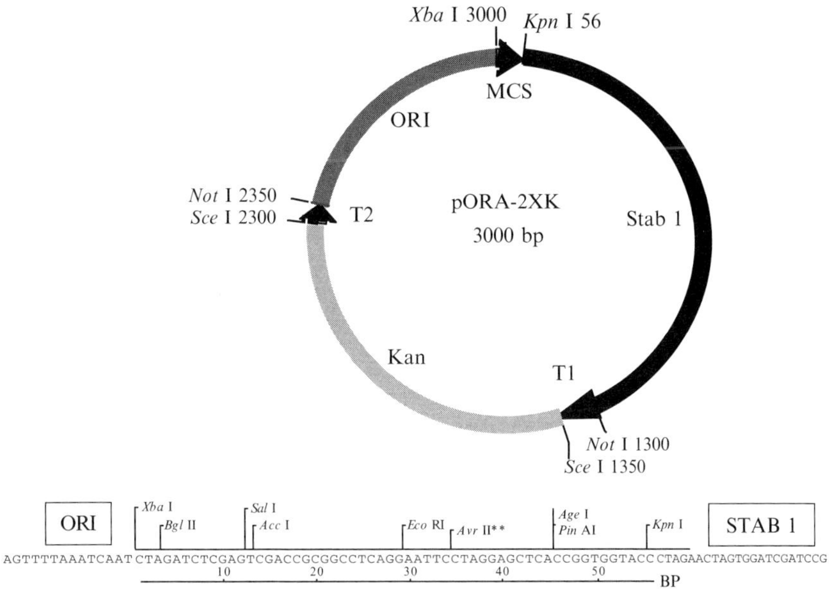

Figure 22.3 Scheme of cryptococcal shuttle vector, pORA-2XK. *Top panel*: The circularized plasmid consists of an *E. coli* origin of replication (ORI), a cryptococcal stabilization sequence (Stab 1) and an *E. coli* kanamycin-resistance cassette (Kan) and two telomeric sequences, T1 and T2 required for replication in *C. neoformans*. *Bottom panel*: Sequence of multiple cloning sites between ORI and Stab1 sequences with restriction sites.

and is essential for autophagy (Lang *et al.*, 1998). RNAi suppression requires amplification of a 500 bp fragment of the *ATG8* coding region as well as a 500-bp fragment of intron I of the laccase gene from *C. neoformans* strain B-3501. Introduction of an intron between the sense and anti-sense fragment produces a hairpin ds-RNA transcript after splicing in *C. neoformans* but allows recovery in *E. coli* by preventing the formation of hairpin RNA in the heterologous organism and the strategy is based on methods developed for other eukaryotes (Smith *et al.*, 2000). Fig. 22.4 shows a scheme of the i*ATG8* construct inserted within the *Bgl*II and *Age*I restriction sites of the pKUTAP expression vector.

Method

1. The *ATG8* gene is PCR amplified from an H99 cDNA mass-excised library (Stratagene) using primers ATG8-2238S-RI (5′-GCCGCGAATTCTGTGAGAAGGCTGAGAAGA) and ATG8-2859A-XhoI (5′-GCCGCCCTCGAGTTAAACATAGTTGCTTT GCTTA).
2. The fragment is purified using the Qiagen PCR purification kit and digested with *Eco*RI and *Xho*I overnight at 37 °C followed by repeat PCR purification.

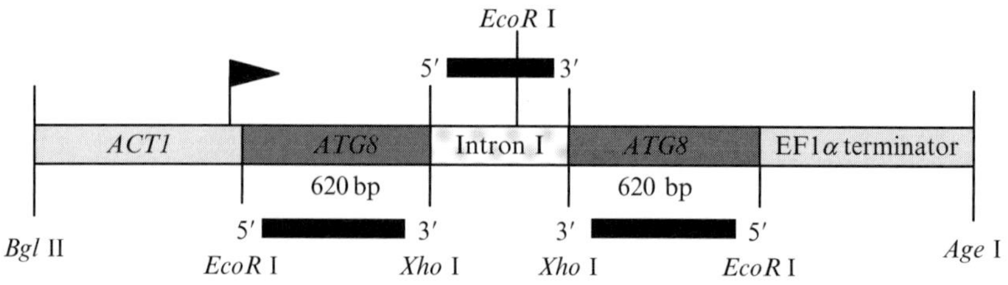

Figure 22.4 Scheme of *ATG8* RNAi expression construct. The RNAi construct is expressed using the *ACT1* cryptococcal promoter, and is inserted between the *Bgl* II and *Age* I sites of pORA-2XK plasmid described in Fig. 22.3. Putative transcription start site and direction of transcription in *C. neoformans* are indicated by the flag symbol.

3. A 500-bp PCR-amplified fragment of intron I of *LAC1* is generated from a *LAC1* genomic clone using primers INTRON-XhoS having sequence 5′-GCCGCTCGAGATCCTAATCGGTAAATATTTCT TTC) and INTRONXhoA (5′-GCCGCTCGA GCGTCGGTATA GCTAAATTG).
4. The intron sequence is PCR purified using the Qiagen PCR purification kit and digested with *Xho*I in appropriate buffers overnight at 37 °C followed by repeat PCR purification.
5. The plasmid pKUTAP is digested with *Eco*RI for 3 h at 37 °C and gel purified using the Qiagen gel purification kit.
6. Plasmid, intron and *ATG8* fragments are ligated using T4 ligase (Invitrogen) at room temperature for 4 h, followed by electroporation into Electromax *E. coli* cells (Invitrogen) and recovery on LB agar plates containing 50 μg/ml of kanamycin according to the manufacturer's directions. Screening of colonies is facilitated by the presence of an *Eco*RI site within the intron (bp 277) which, after digestion, results in a 5.8-kb plasmid band and a doublet of approximately 800 bp and 900 bp in appropriately constructed clones, corresponding to the different sized *ATG8*-intron fragments. Sequencing of the plasmid is then required to establish that the Intron is in the correct 5′–3′ orientation. This orientation of the intron is required for splicing of the intron in the fungus (but not in *E. coli*), resulting in formation of the *ATG8* transcript hairpin.
7. Suitable plasmids are linearized by digestion with *Sce*I, which exposes the telomeric sequences.
8. Linearized plasmids are then transformed into *C. neoformans* strain H99Matα *ura5* using electroporation by standard methods (Varma *et al.*, 1992) and recovered on yeast nitrogen base (YNB, Difco) + 2% glucose.
9. Control strains are produced by transforming the H99 Matα *ura5* strain with an identical pORA-KUTAP plasmid without the *ATG8* RNAi

construct. Whereas *URA* transformation markers have been problematic for virulence studies in some fungi such as *Candida* due to gene dosing of the inserted uracil marker (52), *C. neoformans* virulence does not appear to be as sensitive to levels of uracil synthesis, although the requirement is sufficient to retain the plasmid during infection. Nevertheless, to control for *URA5* expression between strains, independent transformants carrying control pORA-KUTAP plasmids and strains containing the same plasmid with the above i*ATG8* RNA interference construct (both plasmids contain a single copy of the *URA5* locus) are then selected based on equivalent copy number.

10. Copy number is determined by hybridizing Southern blots of uncut DNA from RNAi and control strains with a fragment of the *C. neoformans URA5* gene and ratios of the genomic copy to the smaller episomal copy are matched. The hybridization signal from the genomic copy is intact, since the H99Matα, *ura5* strain (Hu *et al.*, 2008) contains only a partial deletion of the gene. In addition, *URA5* expression is matched within 20% for equivalent expression of *URA5* (ratio of *URA5* to rRNA by densitometry) by Northern blot analysis. Such a strategy typically results in statistically similar mortality of strains containing equivalent plasmids using mouse models (Hu *et al.*, 2008).
11. Strains should also be matched for growth rate at 37 °C prior to virulence studies. This is performed by inoculating a fresh overnight culture in YNB +2% glucose at an A_{600} of 0.10 and recording absorbance every 2 h for two doubling times (approximately 8–10 h). Doubling times should be statistically indistinguishable over this time interval.
12. Strains can then be used for biochemical work as described subsequently or for the study of virulence using either of the two mouse models described previously.

4. Biochemical and Microscopy Methods for Study of Autophagy in *C. neoformans*

The study of cellular processes such as autophagy during infection is facilitated by methods using microscopy. However, the study of autophagy by the use of microscopy techniques can have drawbacks. For example, measurement of autophagic flux is not measured, in contrast to as methods such as protein turnover measurements (Takeshige *et al.*, 1992) are difficult to determine during infection. Nevertheless, identification of essential autophagy-related structural elements or demonstration of autophagy-related genes as described subsequently can provide evidence for autophagy during infectious processes.

4.1. Methods useful for *C. neoformans* cells grown in culture

4.1.1. Autophagic body formation by DIC

DIC microscopy is a convenient method for assessment of autophagic body (AB) formation during autophagy. Autophagic bodies are precursor vesicles, transporting proteins to be degraded in the vacuole required for autophagic degradation (Klionsky *et al.*, 2008). The method for *C. neoformans* is an adaptation of yeast protocols (see the chapter by Prick and Thumm in this volume), but with the addition of the microtubule inhibitor and vacuole-modifying agent, nocodazole, that results in larger vacuole formation in *C. neoformans*, more similar to that of ascomycete yeast. In addition, the visualization of ABs does not appear to be as dependent on protease inhibition, though slightly more ABs are visualized with its use.

Method

1. Mid-log phase cells (overnight culture in 40 ml of YPD to an A_{600} between 0.3-0.6) are washed 1x in YNB without amino acids and ammonium sulfate by centrifuging at 3000*g* × 10 min and inoculated in the same glucose-free YNB, containing 1 m*M* PMSF and 10 μg/ml nocodazole for 3 h at 30 °C.
2. Pellet the cells by centrifugation at 3000*g* × 5 min, followed by mixing 5 μl of a thick suspension (5×10^8 cells/ml; $A_{600} = 1.0$ corresponds to approximately 1×10^7 cells) with an equal volume (5 μl) of 2% low-melting agarose (NuSieve, GTG agarose) on a prewarmed microscope slide (37 °C).
3. Apply a coverslip and let settle at 37 °C for 12 min, then cool to 10 °C briefly. (a PCR machine works well to do this). Exact cell density must be determined empirically, but should result in a monolayer of cells immersed in the agarose in the same plane as determined by microscopy. This method of soft agarose also works well for live cell microscopy of fluorescently-labeled proteins.
4. Cells can be observed for AB accumulation using differential interference contrast (DIC) microscopy for the presence of small vacuolar bodies.

4.1.2. Immunoprecipitation of autophagic complexes

Conjugation of autophagy-related proteins is required for the formation of autophagic bodies, which can be conveniently measured *in vitro* by coimmunoprecipation techniques. A convenient measure of the initiation of autophagy is Vps34-Atg6 condensation, which occurs in yeast, forming the autophagy-specific PI(3)K complex, localizing to the initial phagophore assembly site (PAS) (Kihara *et al.*, 2001). Adaptation for use in *C. neoformans* requires consideration of its highly cross-linked cell wall, which makes cell lysis more difficult and the presence of a large number of proteases. In this protocol, coimmunoprecipitation is performed by the Seize X Protein G immunoprecipitation kit (Pierce Biotechnology, Prod. 45210).

Method

1. Antibody is first bound to immobilized protein G (50% slurry) in a Handee spin cup column (Pierce, Cat. 69702) and the gel is washed twice with 400 μl of Binding/Wash Buffer (Pierce Biotechnology, BupH modified Dulbecco's PBS Pack, Prod. 28374) each time (3000*g*–5000*g*, × 1 min). Apply 3 μl of anti-Atg6 antibody (a kind gift of K. Obara and Y. Ohsumi) prepared in 300 μl Binding/Wash Buffer to the prepared gel. Place the tube on a rocker for a 30-min incubation at room temperature to allow the antibody to bind to the Protein G. Wash the gel three times with Binding/Wash Buffer. Transfer the column into a new microcentrifuge tube and add 400 μl of Binding/Wash Buffer.
2. Cross-linking the bound antibody: Add 12.5 μl of Disuccinimidyl Suberate (DSS; Pierce, No-Weigh DSS, Prod. 21658) solution to the column containing the bound antibody support. Place the tube on a roller for 60-min incubation a room temperature. Wash the gel 5 times with 500 μl of Elution Buffer (ImmunoPure IgG Elution Buffer, Pierce, Prod. #1856202) each time.
3. Generation of a cryptococcal cell lysate: Mid–log phase cryptococcal cells are incubated under starvation conditions in YNB without amino acids and ammonium sulfate for 3 h at 30 °C. Cells are pelleted by centrifugation (3000*g* × 5 min), and the remaining supernatant fraction is removed by suction, followed by addition of an equal volume of dry glass beads (0.5 mm, cat. 9831, Research Products International Corp; see the chapter by Cheong and Klionsky in this volume for the preparation of acid-washed glass beads), as well as 1:500 volume of a 1 *M* solution of PMSF in DMSO + 1:100 volume of protease inhibitor cocktail (Sigma, cat. P-8215). The cell/glass bead slurry is then vortexed 5 times for 1 min, maintaining cells on ice between vortexing. The material is next extracted with 1 cell volume of phosphate-buffered saline. Estimated protein concentration should be in the 5–9 mg/ml concentration. Incomplete disruption is normally the result of incomplete removal of all liquid prior to addition of the dry glass beads.
4. Dilute 200 μl of cell lysate (0.35 mg/ml) 1:1 with Binding/Wash Buffer and add the sample into the column containing the prepared gel. Mix the tube gently and incubate on a roller at 4 °C overnight. Wash the gel 3 times with Binding/Wash Buffer (3000*g*–5000*g* × 1 min). Place the column into a new collection tube and wash one additional time with Binding/Wash Buffer.
5. Elution of immunoprecipitate: Elute the gel twice with 100 μl of Elution Buffer each time and pool the two fractions.

6. Add 5 μl of sample buffer (Pierce, Lane Marker Non-Reducing Sample Buffer (5x) Prod. 39001) and 3 μl of 1 *M* dithiothreitol (DTT) (Sigma-Aldrich, Prod. D9779) solution into 20 μl of eluted sample. Mix gently and thoroughly. Boil for 5 min and allow the sample to cool to room temperature. Now the sample is ready for SDS-PAGE.
7. Western blot can be conducted using standard protocols. Our antibodies are the kind gift of K. Obara and Y. Ohsumi, National Institute for Basic Biology, Okazaki, Japan. Primary antibody concentrations: Anti-Vps34 1:2,000, Anti-Atg8 1: 5,000, Anti-Atg6 1:1,000. Secondary antibody concentrations: Anti-Rabbit IgG 1:10,000 (Invitrogen). Immunodetection of Vps34 after immunoprecipitation using anti-Atg6 antibodies is demonstrated in Fig. 22.5.

4.1.3. Antibody labeling of autophagic bodies from cells grown in culture

Homologs of Atg8 have been used widely as markers of autophagic bodies (Klionsky *et al.*, 2008) and therefore add specificity to the detection of these vesicles over light microscopy methods such as DIC. The methanol fixation method of Tucker and Casadevall (54) was adapted as follows:

Method

1. Cells are grown to mid-log phase in 40 ml of YPD ($A_{600} = 0.5–0.7$) and either harvested by centrifugation at 3000*g* for 10 min or washed twice in

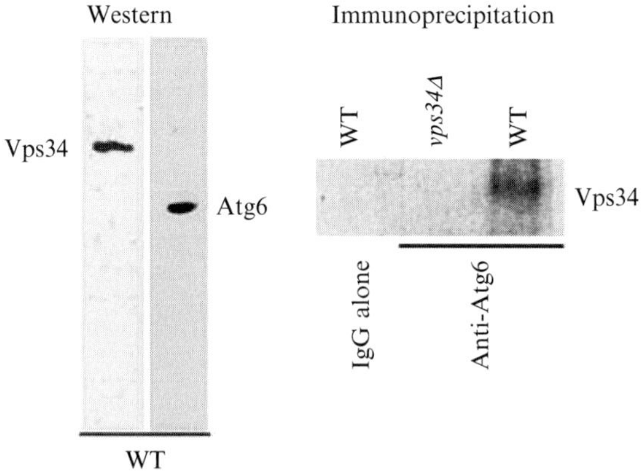

Figure 22.5 Immunoprecipitation of autophagy-related protein complexes from *C. neoformans* cell lysates. Indicated cells were subjected to starvation conditions and whole cell lysates (*left panel*) were subjected to Western blot using anti-Vps34 antibody or anti-Atg6 antibody or immunoprecipitated (*right panel*) with anti-Atg6 antibody or rabbit IgG, washed and eluted and subjected to western blot using the indicated antibodies.

sterile water and incubated for 3 h in YNB without amino acids and ammonium sulfate at 37 °C.

2. Cells are washed in 50 ml of PBS and fixed in 10 ml of 3% formaldehyde for 1 h at 4 °C, washed extensively in sterile distilled water, and then subjected to spheroplasting using 40 mg/ml of lysing enzyme from *Trichoderma harzianum* (Sigma-Aldrich, Prod. L1412) in 4 ml of 1 *M* sorbitol, 10 m*M* sodium citrate, pH 5.8, as previously described (58) for 4 h at 30 °C.
3. Cells are washed in 1 *M* sorbitol and 10 m*M* sodium citrate buffer, diluted 1:8 in PBS, dried on microscope slides, fixed in 100% anhydrous methanol at –60 °C for 10 min, and then dried again. Mild heat fixation (70 °C × 30 min in an oven) can be used to bind cells to the microscope slide more strongly to prevent loss during antibody treatment. We use microscope slides with wells, (Bellco Glass, cat. # 2951W006T), as it helps in the localization of labeled cells and allows up to 8 sets of reactions to be performed on the same slide.
4. To quench autofluorescence, cells are incubated in 0.1 *M* glycine for 10 min.
5. Cells are incubated with a solution of a 1:500 dilution of rabbit anti-yeast Atg8 (Abcam) or an isotype control (IgG purified from rabbit serum, Sigma-Aldrich) in PBS containing 1 mg/ml BSA (Sigma-Aldrich) at 4 °C for 1 h, followed by extensive washing in PBS, then incubation for 1 h at 4 °C with 1:1,000 Alexa Fluor 594 chicken anti-rabbit antibody (Invitrogen) followed by 3 washes with PBS.
6. Immunofluorescence is examined using an Olympus IX-70 microscope with Slide Book 4 deconvolution software (Intelligent Imaging Innovations Inc.). Punctate immunostaining autophagic bodies should be visualized.

5. Light and Electron Microscopy to Study Autophagy During Macrophage Infection

Important to the study of cellular processes such as autophagy is the demonstration of salient features during residence within the mammalian host. The pathogen *C. neoformans* is predominately a facultative intracellular pathogen, residing within host macrophages (Feldmesser *et al.*, 2000). Several macrophage cell lines have been useful for the simulation of the infective environment of *C. neoformans*. For example, the J774.16 macrophage-like cell line (ATCC) has been widely used to evaluate the ability of cryptococcal cells to survive under the host macrophage environment and is described below. For example, an autophagic defective *vps34*Δ cryptococcal mutant strain was found to be rapidly killed by J774.16 cells

(54). In addition, primary mouse brochoalveolar cells can also be used for these purposes, protocols of which are described elsewhere (Liu *et al.*, 1999).

5.1. Phagocytic uptake of *C. neoformans* yeast by J774.16 cells

This method uses the J884.16 macrophage cell line, and requires activation of the phagocytic activity of the cells using IFN-γ and lipopolysaccharide (LPS), and opsonization of the fungal cells using an anticapsular monoclonal antibody.

Method

1. Macrophage cells are allowed to grow for 5–7 days in DMEM (Cellgro) supplemented with 10% fetal calf serum, 100 μg/ml Cellgro penicillin streptomycin at 37 °C in the presence of 5% CO_2, and then harvested from a monolayer using 0.25% trypsin, and the number of cells counted with a hemocytometer. The macrophage concentration is adjusted to 10^6 cells/ ml, and 100 μl of the macrophage suspension is added to each well of a 12-well plate. The cells are then primed with murine IFN-γ (Sigma-Aldrich) at a concentration of 50 U/ml and incubated at 37 °C, 5% CO_2 overnight.
2. Yeast cell suspensions (10^7/ml) of fungal cells are prepared from mid-log phase cells ($A_{600} = 0.5–0.7$) grown in 40 ml of YPD and antibody 18B7 (IgG1) (a generous gift of A. Casadevall), added (10 μg/ml), and incubated at 37 °C for 30 min. Viability of the inoculum is determined by plating serial dilutions on YPD agar and should be >80% by CFU measurement in each case. To each well in a 96-well plate, 10^6 antibody-treated cryptococcal cells are then added plus 50 units of IFN-γ (Sigma-Aldrich) and 1 μg LPS (cat. L4774, Sigma-Aldrich) and incubated at 37 °C 5% CO_2.
3. The macrophage and yeast mixtures are incubated for 30 min, and extracellular yeast cells removed by washing with PBS 3 times, followed by gentle aspiration. Successful phagocytosis can be assayed by incubation of the macrophage-fungal co-culture with the macrophage impermeant dye uvitex 2B (Polysciences, cat. 19517-10) for 3 min followed by washing and should show greater than 90% of adherent cells to be intracellular (lack of epifluorescence in microscopy) by the method of Levitz (55).
4. At various times after addition, macrophages can be lysed with 0.1% SDS in water and the suspension plated on YPD agar to measure live fungal cells. Incubation in 0.1% SDS results in no loss of viability of cryptococcal strains. Alternatively, macrophages can be gently removed by mechanical disruption, followed by fixation in 4% paraformaldehyde and preparation for sectioning as described subsequently.

5.2. Tissue and macrophage sectioning and staining for light microscopy

Tissue staining allows demonstration of autophagic bodies and expression of autophagy-related proteins in the fungus and the host during infection.

Method

1. Fix the tissue in 10% formalin for less than 24 h.
2. Wash fixed tissue in water or buffer.
3. Dehydrate tissue in a graded series of ethanol 35%, 50%, 70%, 95% for 1–2 h each, followed by two changes of 100% ethanol (over molecular sieves) for 1 h each.
4. Place tissue in intermediate solvent (xylene).
5. Infiltrate tissue in Paraplast Plus/xylene solution (cat. 502004, manufacturer McCormick Scientific, St. Louis, MO 63134) (56–58 °C) eventually replacing the solution with 100% Paraplast Plus (temperature as previously). Repeat immersion of tissue in pure Paraplast Plus (4 h at a time) until xylene is not detected (based on odor/smell).
6. Embed tissue in appropriate molds for microtome chucks.
7. Section Paraplast Plus blocks, float sections on deionized water drop.
8. Deposit on clean glass slides, warm slides to 40 °C–45 °C to relax sections.
9. Continue warming the slides until dry.
10. Clear the Paraplast Plus with xylene (volume 100%) for ten minutes; move slides to a fresh dish of xylene for an additional 10 min. Rinse the slides twice for 2 min in 100% alcohol mixture (18:1:1 100% ethanol: 100% methanol: 100% isopropanol).
11. Rinse the slides twice for 2 min in a 95% solution of the 100% alcohol mixture.
12. Place slides in an 80% solution of the 100% alcohol mixture for 2 min, followed by deionized water for 5 min.
13. Rinse slides several times with fresh deionized water followed by another 5-min wash using fresh water.
14. Place slides faceup in incubation tray and cover each section with 1% SDS in TBS (100 m*M* Tris-HCL, pH 7.4, 138 m*M* NaCl, 27 m*M* KCl).
15. Incubate for 5 min at room temperature, followed by three 5-min washes with TBS.
16. Immerse slides in a dish containing blocking buffer (serum from host species of secondary antibody to be used, diluted 1:10 in TBS).
17. Incubate at 37 °C for 1 h.
18. Cover the tissue sections with primary antibody diluted in blocking buffer. An initial antibody concentration of 1.0–10 μg/ml is recommended.
19. Incubate for 2 h at 37 °C.

20. Blot excess liquid from slides and rinse 3 times in TBS for 5 min each wash.
21. Cover the tissue sections with secondary antibody diluted in blocking buffer according to manufacturer's instructions. An affinity purified donkey anti-mouse IgG conjugated to Cy5 (Jackson ImmunoResearch Laboratories) can be used.
22. Incubate at 37 °C for 1 h.
23. Blot excess liquid and rinse twice in TBS for 5 min each wash.
24. Use of an Anti-Ig HRP Detection kit may facilitate the detection of primary antibody. For fig. 22.5, a 1:500 dilution of anti-Atg8 (We have three kits available, anti-mouse IgG HRP, anti-rat IgG HRP, and anti-hamster HRP. These kits contain DAB substrate/chromogen solution that allows for visualization of antibody staining by light microscopy. Please see the kit instructions for the detailed protocol.

5.3. Special considerations for electron microscopy and immuno-electron microscopy of *C. neoformans*

The method of Zhu *et al.* is used for immunoelectron microscopy (57). It is important to be aware that *C. neoformans* presents several obstacles for good electron microscopy, principally due to the extensively cross-linked cell wall that reduces permeability to reagents and creates artifacts during cutting. Electron microscopy provides fine detail of the autophagic body membrane structure and immunolabeling with an AB marker such as Atg8 provides confirmation of the identity of the AB.

Method

1. Fungal cells are taken up by phagocytosis as described previously and are incubated 3 h after phagocytosis by J774.1 cells. Cells are then removed by gentle agitation using a cell scraper for 30 s, transferred to microcentrifuge tubes, and fixed in 4% (vol/vol) paraformaldehyde, 0.1% (vol/vol) glutaraldehyde in 100 m*M* sodium phosphate buffer, pH 7.2, for 16 h at 4 °C. Cells are then washed 3 times for 10 min per wash in 100 m*M* sodium phosphate buffer, pH 7.4, then dehydrated in a graded series of ethanol concentrations (15%, 30%, 50%, 75%, and 100% vol/vol) with 2 changes each for 45 min.
2. Infiltration is continued with 2 parts ethanol to 1 part LR White resin, then 1:1, and 1:2 each for 2 days. Pure LR white infiltration is completed over 24 h with 3 changes, and the samples are then polymerized in 1 ml of pure LR White resin in a vacuum oven at 50 °C for 3 d.
3. Cured blocks are trimmed and thin-sectioned with a DiATOME (Fort Washington, PA) diamond knife on a Reichert-Jung Ultracut E

Ultramicrotome (Leica Microsystems, Bannockburn, IL). The sections are picked up on 200-hex-mesh parlodion carbon coated Ni grids.

4. The thin sections on Ni grids are incubated at room temperature in blocking solution (0.8% [wt/vol] BSA, 0.1% [wt/vol] immunogold-silver stain-quality gelatin, 5% [wt/vol] normal goat serum in PBS, pH 7.4) for 45 min, washed twice in washing solution (0.8% (wt/vol) BSA, 0.1% (wt/vol) gelatin, 0.025% (vol/vol) Tween 20 in PBS), and incubated overnight with a 1:200 dilution of a rabbit anti-Atg8 antibody in 0.8% (wt/vol) BSA, 0.1% (wt/vol) gelatin, 1% (vol/vol) normal goat serum in PBS. The sections are then washed four times in washing solution, and incubated with immunogold-labeled secondary antibody (goat antirabbit IgG; Amersham) diluted 1:100 for 4 h.
5. Grids are washed 6 times with the washing solution for 5 min per wash, then twice with PBS, and the samples fixed with 2% (vol/vol) glutaraldehyde in PBS for 10 min.
6. Grids are then washed twice in PBS and 3 times in deionized water, dried, stained with 2% (wt/vol) aqueous uranyl acetate for 5 min, washed in deionized water and dried. Grids were also stained with Reynolds lead citrate for 2 min, washed with deionized water, dried, and photographed with a JEOL 1200 EX (JEOL USA, Peabody, MA) transmission electron microscope.

6. Detection of Autophagy-Related Gene Products During Human Infection

Genes such as Atg8 are repressed during nutrient rich conditions and are expressed during starvation conditions leading to autophagy (Kirisako *et al.*, 1999). Thus, while not a definitive proof of autophagy, demonstration of Atg8 expression is a required condition for autophagic processes.

6.1. Sources of human tissue and issues of informed consent

Validation of key aspects of mouse modeling experiments in humans is key to the application of biological principals to the disease state. However, three considerations are necessary when considering such experiments: (1) the ability to generate useful data is influenced by the tissue quality and the stage at which the infection is sampled, (2) a larger amount of variability is inherent to the study of an outbred population of humans, and (3) it is essential that the utmost care be taken regarding ethical issues in obtaining and utilizing tissues. A facility's institutional review board should be consulted and all protocols reviewed and approved before use of such tissue. For example, the following protocol is based on work approved by

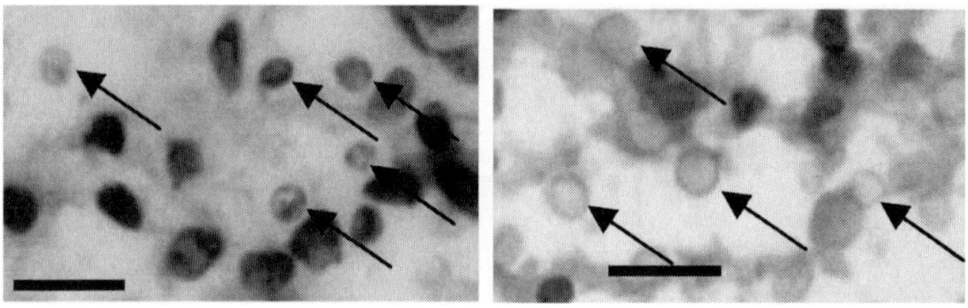

Figure 22.6 Identification of Atg8 expression in *C. neoformans* during human infection. Antibody staining of *C. neoformans*-infected human brain tissue incubated with an anti-Atg8 antibody (+anti-Atg8) or rabbit IgG (IgG control), antirabbit horseradish peroxidase secondary antibody, and developed with diaminobenzidine and counterstained with hematoxylin.

the University of Illinois at Chicago Office for the Protection of Research Subjects and the Institutional Review Board (protocol #2004-0600).

6.2. Detection of Atg8 protein in human tissue by immunohistochemistry

Atg8 expression in yeast is induced during autophagy and thus serves as a proxy for the potential for autophagy of the fungus during infection. However, because the brain material was obtained from an HIV+ individual during autopsy, the quality of fixation does not allow the demonstration of punctate autophagic bodies, because brains are normally subjected to whole brain fixation, to reduce the potential for viral transmission. Whole brain fixation, rather than fixation of small tissue sections, results in solvent fronts that diminish the fine detail of immunostaining.

Method

1. After completion of whole-brain fixation in 3% formaldehyde, 1-cm blocks of tissue are prepared and subjected to microsectioning as previously.
   These data demonstrate the expression of Atg8 during infection of the human brain by *C. neoformans* as shown in Fig. 20.6

7. Concluding Remarks

The ability of the pathogen to utilize cellular stress-response pathways to survive and promote virulence in the mammalian host is an important, though underappreciated, principal of pathogenesis. For example, the identification of the important role of autophagy during cryptococcal pathogenesis suggests a mechanism whereby the pathogen can lie dormant in macrophages or other cellular compartments during long latent phases before reactivation. Thus, the study of mechanisms of latent infections requires the adaptation of methods developed for yeast in cell culture to the environment of the mammalian host. Methods are presently available that will detect the formation of autophagic bodies that are an essential component of autophagic competency. Further development of methods that will enable measurement of protein flux similar to aminopeptidase 1 (Ape1) processing in *C. albicans* (Palmer *et al.*, 2007) will enable a more complete description of the autophagic process. Unfortunately, *C. neoformans* dose not have a clear Ape1 homolog, so further study will be required to identify appropriate surrogate markers. Adaptation of key attributes to studies during human infection will also help to validate the mouse models of infection.

REFERENCES

Bicanic, T., and Harrison, T. S. (2004). Cryptococcal meningitis. *Br. Med. Bull.* **72,** 99–118.

Bicanic, T., Meintjes, G., Wood, R., Hayes, M., Rebe, K., Bekker, L., and Harrison, T. (2007). Fungal burden, early fungicidal activity, and outcome in cryptococcal meningitis in antiretroviral-naive or antiretroviral-experienced patients treated with amphotericin B or fluconazole. *Clin. Infect Dis.* **45,** 76–80.

Brown, S. M., Campbell, L. T., and Lodge, J. K. (2007). *Cryptococcus neoformans*, a fungus under stress. *Curr. Opin. Microbiol.* **10,** 320–325.

Feldmesser, M., Kress, Y., Novikoff, P., and Casadevall, A. (2000). *Cryptococcus neoformans* is a facultative intracellular pathogen in murine pulmonary infection. *Infect. Immun.* **68,** 4225–4237.

Hu, G., Hacham, M., Waterman, S. R., Panepinto, J., Shin, S., Liu, X., Gibbons, J., Valyi-Nagy, T., Obara, K., Jaffe, H. A., Ohsumi, Y., and Williamson, P. R. (2008). PI3K signaling of autophagy is required for starvation tolerance and virulence of *Cryptococcus neoformans. J. Clin. Invest.* **118,** 1186–1197.

Huffnagle, G. B., Chen, G. H., Curtis, J. L., McDonald, R. A., Strieter, R. M., and Toews, G. B. (1995). Down-regulation of the afferent phase of T cell-mediated pulmonary inflammation and immunity by a high melanin-producing strain of *Cryptococcus neoformans. J. Immunol.* **155,** 3507–3516.

Janbon, G. (2004). *Cryptococcus neoformans* capsule biosynthesis and regulation. *FEMS Yeast Res.* **4,** 765–771.

Kihara, A., Noda, T., Ishihara, N., and Ohsumi, Y. (2001). Two distinct vps34 phohatidylinositol 3-kinase complexes function in autophagy and carboxyeptidase Y sorting in *Sacharomyces cerevisiae. J. Cell Biol.* **152,** 519–530.

Kirisako, T., Baba, M., Ishihara, N., Miyazawa, K., Ohsumi, M., Yoshimori, T., Noda, T., and Ohsumi, Y. (1999). Formation process of autophagosome is traced with Apg8/Aut7p in yeast. *J. Cell Biol.* **147,** 435–446.

Klionsky, D. J., Abeliovich, H., Agostinis, P., Agrawal, D. K., Aliev, G., Askew, D. S., Baba, M., Baehrecke, E. H., Bahr, B. A., Ballabio, A., Bamber, B. A., Bassham, D. C., *et al.* (2008). Guidelines for the use and interpretation of assays for monitoring autophagy in higher eukaryotes. *Autophagy* **4,** 151–175.

Lang, T., Schaeffeler, E., Bernreuther, D., Bredschneider, M., Wolf, D., and Thumm, M. (1998). Aut2p and Aut7p, two novel microtubule-associated proteins are essential for delivery of autophagic vesicles to the vacuole. *EMBO J.* **17,** 3597–3607.

Liu, L., Tewari, R. P., and Williamson, P. R. (1999). Laccase protects *Cryptococcus neoformans* from antifungal activity of alveolar macrophages. *Infect. Immun.* **67,** 6034–6039.

Liu, H., Cottrell, T., Pierini, L., Goldman, W., and Doering, T. (2002). RNA interference in the pathogenic fungus *Cryptococcus neoformans*. *Genetics* **160,** 463–470.

Liu, X., Hu, G., Panepinto, J., and Williamson, P. (2006). Role of a *VPS41* homolog in starvation response and virulence of *Cryptococcus neoformans*. *Mol. Microbiol.* **61,** 1132–1146.

Nicol, A. M., Hurrell, C., McDowall, W., Bartlett, K., and Elmieh, N. (2008). Communicating the risks of a new, emerging pathogen: the case of *Cryptococcus gattii*. *Risk Anal.* **28,** 373–386.

Nosanchuk, J. D., Rosas, A. L., Lee, S. C., and Casadevall, A. (2000). Melanisation of *Cryptococcus neoformans* in human brain tissue. *Lancet* **355,** 2049–2050.

Palmer, G., Kelly, M., and Sturtevant, J. (2007). Autophagy in the pathogen *Candida albicans*. *Microbiology* **153,** 51–58.

Panepinto, J., Liu, L., Ramos, J., Zhu, X., Valyi-Nagy, T., Eksi, S., Fu, J., Jaffe, H., Wickes, B., and Williamson, P. (2005). The DEAD-box RNA helicase Vad1 regulates multiple virulence-associated genes in *Cryptococcus neoformans*. *J. Clin. Invest.* **115,** 632–641.

Panepinto, J., and Williamson, P. (2006). Intersection of fungal fitness and virulence in *Cryptococcus neoformans*. *FEMS Yeast Res.* **6,** 489–498.

Smith, N. A., Singh, S. P., Wang, M. B., Stoutjesdijk, P. A., Green, A. G., and Waterhouse, P. M. (2000). Total silencing by intron-spliced hairpin RNAs. *Nature* **407,** 319–320.

Takeshige, K., Baba, M., Tsuboi, S., Noda, T., and Ohsumi, Y. (1992). Autophagy in yeast demonstrated with proteinase-deficient mutants and conditions for its induction. *J. Cell Biol.* **119,** 301–311.

Varma, A., Edman, J. C., and Kwon-Chung, K. J. (1992). Molecular and genetic analysis of *URA5* transformants of *Cryptococcus neoformans*. *Infect. Immun.* **60,** 1101–1108.

Zhu, X., and Williamson, P. (2004). Role of laccase in the biology and virulence of *Cryptococcus neoformans*. *FEMS Yeast Res.* **5,** 1–10.

CHAPTER TWENTY-THREE

Autophagy and Autophagic Cell Death in *Dictyostelium*

Emilie Tresse,* Corinne Giusti,* Artemis Kosta,* Marie-Françoise Luciani,* *and* Pierre Golstein*

Contents

Abstract

Autophagic cell death can be conveniently studied in *Dictyostelium discoideum*, an exceptionally favorable model not only because of its well-known genetic and experimental advantages but also because in *Dictyostelium* there is no apoptosis machinery that could interfere with nonapoptotic cell death. Moreover, autophagic cell death in *Dictyostelium* can be dissociated into a starvation-induced sensitization stage, during which autophagy is induced, and a death induction stage. We show here how to demonstrate, assess and analyze this autophagic cell death. This can be studied *in vivo* during the development of *Dictyostelium*, and *in vitro*, using modifications of the

* Centre d'Immunologie de Marseille-Luminy, Marseille, France

Methods in Enzymology, Volume 451
ISSN 0076-6879, DOI: 10.1016/S0076-6879(08)03223-0

monolayer technique of Rob Kay *et al.* Methods to follow this autophagic cell death qualitatively and quantitatively are reported.

1. Introduction

Dictyostelium discoideum is a eukaryote, a protist, normally found in the soil of deciduous forests, where it feeds on a wide variety of microbes. Axenic strains were derived and are used in most laboratories. *Dictyostelium* is unicellular in the presence of enough nutrients (e.g., in HL5 medium; see subsequent sections) and undergoes multicellular development upon starvation (e.g., in buffer alone). This development ultimately leads to a fruiting body, which includes a mass of spores and a stalk. *Dictyostelium* stalk cells are highly vacuolated and dead, as they do not regrow in culture medium (Whittingham and Raper, 1960), demonstrating the existence in this model of developmental vacuolar cell death. Because vacuolization can be suppressed (Kosta *et al.*, 2004) by targeted mutagenesis of the autophagy gene atg1 (Otto *et al.*, 2004), a link between vacuolization and autophagy exists and the corresponding cell death is referred to as autophagic cell death (ACD).

Dictyostelium brings considerable advantages to the study of autophagic cell death. First, autophagy and ACD can be induced separately, as follows. Starting from a *Dictyostelium* cell line, under *in vitro* monolayer conditions (Kay, 1987) mimicking development, starvation and cAMP induce autophagy but are not sufficient to lead to ACD. Addition of the differentiation-inducing factor DIF-1 to these starved cAMP-treated cells is required to promote their differentiation into stalk cells (Morris *et al.*, 1987; Sobolewski *et al.*, 1983) through ACD (Cornillon *et al.*, 1994; Levraud *et al.*, 2003): cells undergo *atg1*-dependent (Kosta *et al.*, 2004) vacuolar cell death with easily visible morphological changes, namely paddle cell formation, vacuolization, and death (Cornillon *et al.*, 1994; Levraud *et al.*, 2003). An *iplA*-mutation (affecting the IP3R, thus decreasing the Ca^{2+} flux from the endoplasmic reticulum into the cytoplasm) marks this DIF-1 pathway, resulting in no vacuolization and no cell death (Lam and Golstein, 2008; Lam *et al.*, 2008).

Second, in *Dictyostelium* there are no main members of the apoptosis machinery that could interfere with nonapoptotic cell death: there are no caspase family members (except one paracaspase gene that is not involved in autophagic or necrotic cell death), no Bcl-2 family member, and no BH3-(Bcl-2 family domain)-bearing molecule (Lam *et al.*, 2007; Roisin-Bouffay *et al.*, 2004).

Third, *Dictyostelium* shows marked genetic tractability, with in particular haploidy facilitating the search by random mutagenesis for molecules required for ACD.

In this chapter, we show how in *Dictyostelium* to induce ACD, how to identify genes involved in autophagy and/or in ACD, and how to study some of the characteristics of the corresponding mutants.

2. Induction of Autophagic Cell Death

Programmed cell death in *Dictyostelium* is the outcome of terminal differentiation of stalk cells. As indicated previously, this can be obtained in two different ways: by inducing development (dead stalk cells will constitute stalks and basal disks of the resulting fruiting bodies, which corresponds to approximately 15% of the cells) or by mimicking this differentiation *in vitro*, in monolayers, which is more convenient for most applications.

2.1. *Dictyostelium* culture conditions

Dictyostelium cells are grown in HL5 medium (Sussman, 1987) with the following modifications: bactopeptone (Oxoid, Basingstoke, Hampshire, England) 14.3 g/l; yeast extract (Difco Laboratories, Detroit, MI) 7.15 g/l; maltose (Sigma Chemical Co., St Louis, MO) 18 g/l; Na_2HPO_4, 3.6 m*M*; KH_2PO_4, 3.6 m*M*; source water. Cultures are in 75- or 175-cm^2 flasks (Falcon, Becton Dickinson Labware, Lincoln Park, NJ) incubated without shaking at 22–23 °C in a water-saturated atmosphere with no added CO_2. To keep cells in log phase growth (up to about 2×10^6 cells/cm^2; doubling time is around 10 h; cells are counted using a hemocytometer), when necessary cells are diluted 25 times after detachment by mere shaking.

2.2. Development on filters

1. Development can occur at an air-wet solid interface. Filters are the most common surface used. For development on filters, 10^7 *in vitro* grown cells are harvested by shaking and centrifuged at 700*g*.
2. The harvested cells are washed twice with SB 1X (Soerensen buffer 50 X: 100 m*M* Na_2HPO_4, 735 m*M* KH_2PO_4, pH 6.0) and resuspended by pipetting in 2 ml of SB.
3. Cells are starved on nitrocellulose filters (0.8 mm, 47 mm Ø, AABP04700, Millipore, Billerca, MA) and kept on 3 MM paper (Whatman, Maidstone, UK) soaked with SB. Fifty μl of SB containing 2×10^6 cells are spotted on the filters. Development can also be obtained by spotting cells similarly on SM/5 Petri dishes (2g/l bactopeptone, 2 g/l glucose, 0.2 g/l yeast extract, 1g/l $MgSO_4 \cdot 7H_2O$, 2.2 g/l KH_2PO_4, 1g/l

K_2HPO_4, 12 g/l agar, 35 ml SM/5 per Petri dish). On both filters and Petri dishes, upon incubation at 23 °C without shaking fruiting bodies are obtained in about 24 h.

2.3. Autophagic cell death induction in monolayer

Because microscopy observation of cells in a stalk is not easy, cell morphology is more difficult to assess because cells are enclosed in the cellulose sheath tube and tightly packed. Furthermore, regrowth of stalk cells exclusively is difficult to score, so stalk cell differentiation is most often studied in monolayers *in vitro*. As indicated earlier, we use a protocol derived from that described by R. R. Kay (1987), with, under starvation, a first incubation with cAMP and a second one with DIF-1. This protocol works for most of the usual strains of *Dictyostelium*. However, the percentage of cells differentiating into stalk cells is strongly strain dependent: cells of V12M2 origin (e.g., HMX44A) differentiate more efficiently than cells of NC4 origin (e.g., AX-2). This is largely because of a difference in sensitivity of inhibition of the DIF-1-dependent step by cAMP (Berks and Kay, 1988), implying an additional washing step for some strains. The protocol described here is optimized for HMX44A cells (see derivation in Levraud *et al.*, 2003).

1. Vegetative cells are collected in log phase from cells grown in Falcon flasks as indicated earlier and washed twice with SB 1X.
2. The cells are resuspended in SB 1X containing 3mM cAMP (Sigma A6885, stock solution at 60 m*M* in demineralized autoclaved water, filter sterilized and kept at −20 °C in 1-ml aliquots).
3. We then distribute 3–5 × 10^5 cells in 1ml of (SB + 3 m*M* cAMP) in each well of 2-well-Lab-Tek chambers (155380, Nalge Nunc, coverglass #1 German borosilicate sterile), which are very convenient for subsequent microscopy examination.
4. Cells are incubated for 8 h at 22 °C. Most (>80%) cells should adhere firmly to the bottom slide. Because of the high concentration of cAMP saturating corresponding receptors, almost no aggregation should be seen at this stage, and the cells should be randomly scattered.
5. Carefully remove the liquid by aspiration and wash once with 1 ml of SB per well.
6. Replace with 1 ml of SB containing 100 n*M* DIF-1 (Differentiation Inducing Factor-1, 1-(3,5-dichloro-2,6-dihydroxy-4-methoxyphenyl)-hexan-1-one; DN1000, Affiniti Research Products, Exeter, UK; make stock solution 10 m*M* in absolute ethanol; working stocks are diluted to 0.1 m*M* in absolute ethanol in 0.1 ml aliquots. Store at −20 °C.). We generally use as a control a chamber without DIF-1 (ethanol as solvent alone has no obvious effect at the corresponding concentration). Incubate at 22 °C.

Twenty-four hours after addition of DIF-1, in the DIF-1 containing chamber most cells should have differentiated to stalk cells (Fig. 23.1): highly vacuolated, cellulose-encased, nonrefringent by phase contrast microscopy. Almost no vacuolated cells should be seen in the control chamber for cells either isolated or in clumps, if a cell line producing little endogenous DIF-1 (e.g. HMX44A) is used. Further incubation will increase the proportion of dead cells, up to 50% at 30–36 h as assessed by propidium iodide staining (see below and Cornillon *et al.*, 1994), and vacuolization will become progressively more prominent as the cytoplasm of dying/dead cells continues to shrink.

3. Mutagenesis to Obtain Autophagy and Autophagic Cell Death Mutants

A major advantage of *Dictyostelium* as a model organism is the relative ease with which it can be genetically manipulated. Transfection is easy, homologous recombination efficient, and insertional mutagenesis routinely performed. Targeted and random insertional mutagenesis involve the same basic protocol to introduce DNA into *Dictyostelium*, which we describe first, after which we detail the specifics of each procedure, especially the screening strategies to isolate cell death mutants.

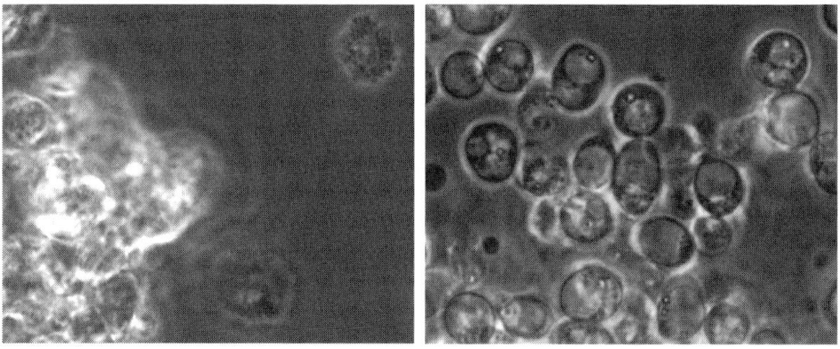

Figure 23.1 *Observation of vacuolization under phase contrast microscopy.* HMX44A cells were incubated for 8 h in SB + cAMP 3 m*M*, then in SB without cAMP and with or without DIF-1 for 24 h, and were observed under phase contrast microscopy. (*Left*) SB, cell clumps, and isolated flat cells. (*Right*) SB + DIF-1, markedly vacuolated cells. Cell clumps appear on removal of the excess cAMP (which allows for normal cAMP chemotactism) in both control and DIF-1 groups. However, in the prolonged presence of DIF-1 cells tend to progressively leave the clumps.

3.1. Transfection and selectable markers

3.1.1. Selectable markers

Several laboratories have used a selection system based on complementation of the uracil pathway through transfection with a plasmid bearing the *Dictyostelium* pyr5-6 gene (Kuspa and Loomis, 1992). However, strains to be transformed must carry a mutated pyr5-6 gene to not grow in minimal medium in the absence of uracil.

Dictyostelium is more sensitive to protein synthesis inhibitors than many other eukaryotic cells, a characteristic that is used to select transfected cells. Dictyostelium cells are transfected with plasmids containing the desired DNA and a gene conferring resistance to a given inhibitor. Whenever possible, blasticidin (from Invitrogen, Groningen, The Netherlands) is used as a selection drug because it is efficient and reliable, usually at a final concentration of 10 μg/ml. One copy of the blasticidin-resistance gene is sufficient to confer resistance, making this suitable for homologous recombination or insertional mutagenesis (Sutoh, 1993). When it is not possible to use blasticidin (e.g., when transfecting an already blasticidin-resistant strain), neomycin selection (G418-sulfate, Gibco-BRL) is often used, at a final concentration of 10–15 μg/ml (Nellen *et al.*, 1984). It is wise to check the sensitivity of the different strains and the activity of different drug batches. Neomycin-resistant cells appear to usually contain multiple copies of the plasmid (Knecht *et al.*, 1986), although homologous recombination has been successfully achieved using this marker (Manstein *et al.*, 1989).

3.1.2. Electrotransfection of *Dictyostelium* cells

To transfect cells, we use the following protocol:

1. Grow enough cells (check by inverted microscopy, count in hemocytometer) to collect 10^7 vegetative cells in log phase per transfection.
2. The cell suspension in HL5 growth medium is chilled on ice for 30 min. Cuvettes (0.4-cm gap, EP-104 Cell Projects) are also cooled on ice.
3. Cells are collected by centrifuging at 700*g* for 5 min at room temperature.
4. Discard the supernatant and wash cells by centrifugation in 5 ml of ice-cold sterile electroporation buffer (10 m*M* $NaPO_4$, pH 6.1, 50 m*M* sucrose).
5. Cells are resuspended at 12.5 × 10^6 cells/ml in electroporation buffer and 0.8 ml of cell suspension with 10 μg of transforming DNA are distributed per ice-cold cuvette.
6. Cells are then electroporated in a Bio-Rad Gene Pulser (1 kV and 3 μF).
7. Immediately after electroporation the content of each cuvette is added to 50 ml of growth medium in a 175-cm^2 flask, then incubated at 22 °C without shaking.

8. After at least 24 h, blasticidin is added at a final concentration of 10 μg/ml. Most cells die, and colonies of blasticidin-resistant cells become visible 5–6 days after addition of blasticidin. There is no medium change during this period.

3.2. Random mutagenesis and screening strategies

3.2.1. Random insertional mutagenesis

For random insertional mutagenesis we use the REMI approach (Restriction Enzyme Mediated Integration; Kuspa and Loomis, 1992). This technique increases the efficiency of integration into the genome of *Dictyostelium*. Plasmid integration leads to gene disruption and tagging of the disrupted gene, allowing its subsequent identification through the analysis of the plasmid-flanking sequences. We use the original transfection protocol (Kuspa and Loomis, 1992) with minor modifications. The pUCBsrΔBamHI plasmid, which bears a blasticidin resistance gene, is BamHI-linearized for transfection. To facilitate insertion of the plasmid we add to the mix of electroporation buffer and plasmid DNA, 1.2 U/ml of the endonuclease DpnII. This enzyme cuts the genome every 600 bp on average, and digested ends can recombine with the BamHI-digested ends of the plasmid. Blasticidin thus selects cells that have incorporated a plasmid in their genome. Among these cells, one can further select for individual cells in which plasmid insertion has disrupted a gene involved in the function of interest.

3.2.2. Development-based screening

Random-mutagenized blasticidin-resistant *Dictyostelium* cell populations (of strains that are capable of development; e.g., DHI and JH10 strains) are first enriched for mutants resistant to cell death by two rounds of DIF-1-induced cell death in monolayers, as follows.

1. 12×10^6 transformants are subjected to cell death induction as described in section 2.3 in monolayer in 75-cm^2 flasks (Falcon; BD Biosciences Discovery Labware, Bedford, MA) in 20 ml of final volume.
2. 48 h after addition of DIF-1, cells are harvested by centrifugation (700*g*, 5 min) and resuspended in 5 ml of HL5 to enable growth of surviving cells, in 25-cm^2 flasks.
3. Around 48 h after resuspension in HL5, a second similar round of cell death induction and selection is performed.
4. After 48 h of incubation in DIF-1, the resulting surviving cells are counted and transferred to a suspension in SB containing *Klebsiella aerogenes* dense enough to obtain a white nontransparent suspension. We then plate on each of SM/5 Petri dishes (see previous sections)

250 μl of the solution containing *K. aerogenes* and 75 *Dictyostelium* cells to facilitate subsequent screening.

5. Development of each clone as a separate plaque on the bacterial lawns is examined with a binocular photomicroscope (Carl Zeiss). This allows us to visually screen these clones, reject those where development did not take place at all, and select those where development took place, but the stalks were abnormal thus possibly made of death-resistant mutant cells.
6. Cells at the edge of plaques with abnormal stalks (Fig. 23.2A) are recovered by scraping and grown in 20 ml HL5 medium supplemented

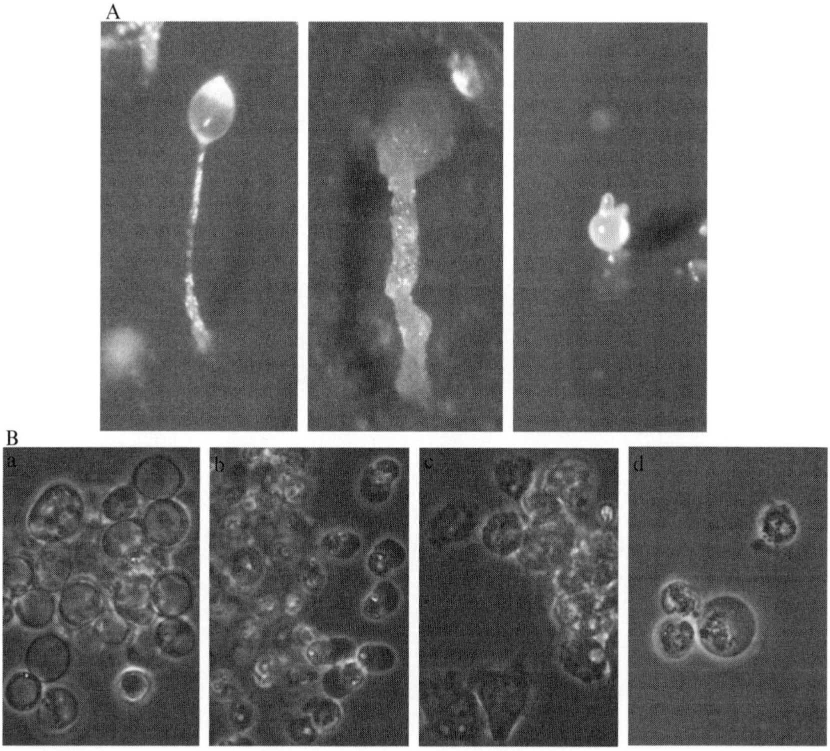

Figure 23.2 *Screening for cell death mutants.* (A) Wild-type or mutant fruiting bodies with aberrant stalks. Cell death mutants were obtained by random mutagenesis and screened for abnormalities of stalk morphology (see Fig. 23.4A) after 26 h of development on filters. (*Left*) Wild-type fruiting body, (*middle*) mutant fruiting body with broad irregular stalk, (*right*) mutant fruiting body with very small stalk. All fruiting bodies are observed under the same magnification. (B) Wild-type or mutant cells with abnormal vacuolization. After 8 h of incubation in starvation medium plus cAMP, autophagic vacuolar cell death was induced by addition of 10^{-7} *M* DIF-1 for 40 h. Unfixed cells were directly visualized in Lab-Tek chambers by phase contrast microscopy, x100. (a) Normal vacuolization of wild-type cells. (b–d) Examples of mutant cells without large vacuoles, due to still-undefined independent mutations.

with antibiotics (100 μg/ml ampicillin and 300 μg/ml streptomycin; Sigma-Aldrich) to remove remaining bacteria. Several such mutants have been obtained and are currently under study (Lam *et al.*, 2008; Tresse *et al.*, 2008).

3.2.3. Vacuolization-based screening

Another way to detect mutants is based on the occurrence or not of vacuolization on induction of ACD in monolayer culture *in vitro*. Initial steps are as above, namely random insertional mutagenesis and two rounds of induction of cell death and regrowth.

1. After the second induction of death, surviving cells are grown for 6 days, and a suspension of 10^6 cells per mL of HL5 *plus* antibiotics (100 μg/ml ampicillin and 300 μg/ml streptomycin; Sigma-Aldrich) is used to distribute by cell sorting (BD-FACS Vantage) one cell per HL5-containing well of 96-well plates (Falcon), using PBS as sheath solution, with the following parameters: pressure, 9 PSI; nozzle, 100 μ; frequency, 21.2 kHZ. It is important to avoid centrifugation just before cell sorting.
2. Cells were then left to grow for a week at 22 °C without shaking.
3. When they were close to confluence in most wells, cell death was induced as described previously.
4. Microscopy observation (phase contrast, x32, inverted microscope) was at 20–40 h after DIF-1 addition. In some wells, cells do not vacuolize. These are allowed to regrow by addition of rich medium and tested again (Fig. 23.2B).

4. Study of Autophagy and Autophagic Cell Death Mutants

4.1. Regrowth assay

This is a clonogenic assay to quantify surviving cells after ACD.

1. Vegetative cells are induced to die as described previously. Cell suspensions are distributed in Lab-Tek chambers, or in 12-well plates (331143, BD Falcon) using the same volume and number of cells as in Lab-Tek chambers. Most cells in DIF-1-containing wells will die but will not in control wells without DIF-1.
2. When by microscopy most cells show a vacuole occupying most of their surface area, that is 24–48 h after addition of DIF-1, 0.5 ml of medium is removed and 1 ml of HL5 culture medium is added to initiate regrowth of surviving cells.

3. After incubation at 22 °C for 40–72 h, growing vegetative cells are detached by vigorous pipetting, and counted in an hemocytometer. Before counting, it should be checked that all vegetative cells are detached. Remaining stalk cells will adhere because of cellulose synthesis, which is not a problem because they are not to be counted: in the hemocytometer the rare heavily vacuolated, nonrefringent stalk cells are easily distinguished and excluded.

Results are expressed as the ratio of the number of regrowing cells in DIF-1 wells to the number of cells in control wells. Because the cells are collected after a period of exponential growth, slight variations in culture conditions may significantly affect the results. This ratio expresses the percentage of surviving cells after DIF-1-induced cell death. Sixteen hours after addition of DIF-1, 50% of the cells have lost their ability to multiply; 48 h after DIF-1, about 80% of the cells do not multiply, while at that time only 50% of the cells show membrane permeability to propidium iodide.

4.2. Fluorescence labeling in monolayer

Among fluorescent reagents we use mainly propidium iodide, fluorescein diacetate, calcofluor, and LysoSensor Blue to study autophagy and autophagic cell death in *Dictyostelium*. Reagents for more specific molecular markers of autophagy are lacking as yet in *Dictyostelium* or are only poorly efficient. We tried protein degradation assays or inhibition by 3MA with only minor differences between *Dictyostelium* wild type or *atg1* mutant cells, either starved or not. Antibodies against autophagy proteins are not yet available in *Dictyostelium*. In our hands GFP-ATG8 (Otto *et al.*, 2003, 2004) gave only weak fluorescence even in starved cells.

4.2.1. Use of propidium iodide

Propidium iodide (PI) is a DNA intercalating dye that cannot cross cell membranes freely; thus, cells will fluoresce only if their membrane become permeable, a late sign of cell death irrespective of its nature. This implies that cells cannot be fixed for this test. Propidium iodide (P4170, stock solution 10 mg/ml in water kept at 4 °C; Sigma Aldrich) is used at a final concentration of 10 μg/ml. An incubation for 10 min at room temperature away from light is necessary. This labeling can be observed with a classical fluorescence microscope or under confocal microscopy. Typically, this red fluorescence appears as a cytoplasmic crescent between the vacuole and the cellulose shell of dead cells.

4.2.2. Use of fluorescein diacetate

In contrast to PI, fluorescein diacetate (FDA) stains living cells. The nonfluorescent, hydrophobic compound freely enters the cell where it is cleaved by cytoplasmic lipases of metabolically active cells into a green fluorophore unable to leave the cell if the membrane is intact. Again, cells must not have been fixed. Fluorescein diacetate (F7378, stock solution 10 mg/ml in acetone kept at 4 °C; Sigma-Aldrich) is used at a final concentration of 50 μg/ml. An incubation for 10 min at room temperature away from light is necessary. This labeling can be observed under classical fluorescence microscopy or under confocal microscopy. Cells alive by this criterium appear uniformly bright green.

4.2.3. Use of calcofluor

As they differentiate, stalk cells encase themselves in a cellulose shell that may be labeled with calcofluor. While this is a useful differentiation marker, positive staining does not constitute evidence of cell death, and conversely mutants with alteration of cellulose synthesis can differentiate normally to cell death without producing a cellulose shell (Levraud *et al.*, 2003) (Tresse *et al.*, submitted). Cells may be fixed before staining with calcofluor. We use calcofluor (Fluorescent brightener 28, F3543, stock solution 1% in water kept at 4 °C; Sigma-Aldrich) at a final concentration of 0.01%. Cells are incubated for 10 min and observed under a fluorescence microscope. Cellulose shells appear bright blue around dying or dead cells. FDA, PI, and calcofluor can be used for triple staining by directly mixing the dyes.

4.2.4. Use of LysoSensor Blue

LysoSensor Blue is an acidotrophic dye, which is able to label acidic structures including autolysosome during autophagic cell death in *Dictyostelium* (Fig. 23.3). However, LysoSensor Blue, like other acidotrophic dyes, is not able to label early autophagosomes. Moreover, it is able to stain other acidic compartments such as lysosomes (Giusti *et al.*, submitted).

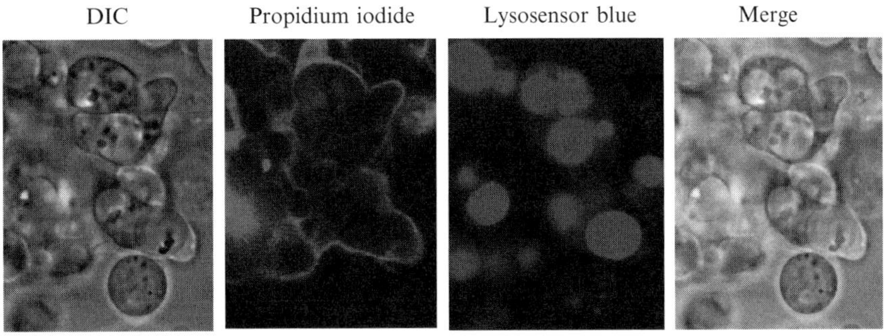

Figure 23.3 Dying cells with autophagy present a large autophagic vacuole that can be labeled with LysoSensor Blue. HMX44A cells observed 24 h after DIF-1 induction, under confocal microscopy. (*Left*) DIC (*second panel from left*, cells with permeabilized membrane), fluorescence with propidium iodide, and (*third panel from left*, acidic compartments) LysoSensor Blue, were all merged in the right panel.

We use LysoSensor Blue (L7533, Invitrogen, Carlsbad, CA) at a final concentration of 5 μM. Cells are incubated for 10 min and can be observed under classical fluorescence or confocal microscopy.

4.3. Fluorescence labeling of stalks

Vacuolar cell death can be observed directly in stalks (Fig. 23.4). Cells are seeded on nitrocellulose filters as described previously and allowed to develop for 48 h. We use a procedure modified from Levraud *et al.* (2003)

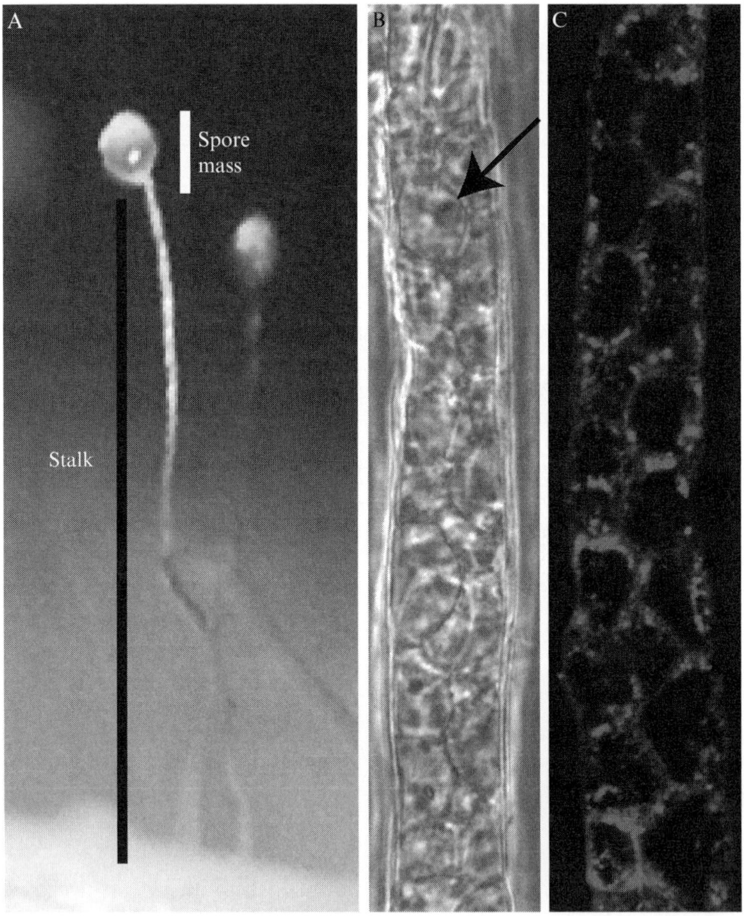

Figure 23.4 Vacuolar cell death in stalks. Wild-type fruiting bodies are composed of a mass of spores on a stalk (as indicated). Images were taken after 5 days of development on a bacterial lawn. Fruiting bodies are 1–2 mm high. Stalks are made of vacuolized cells (*arrow*). Stalk observed under phase contrast microscopy, at 48 h of development on filters. Stalks are composed of dead cells. Stalk labeled with propidium iodide at 48 h of development and observed under confocal microscopy. Propidium-iodide-labeled cytoplasm of cells with a permeabilized membrane. Vacuoles are not labeled.

to examine stalks. Each filter is lightly pressed manually against a coverslip (18 mm × 18 mm No. 1, ESCO, Oakridge, NJ), allowing adhesion of some mature fruiting bodies. A 15-μl volume of solution of fluorochromes (FDA, PI, calcofluor can be used) in SB (final concentrations as previously) is deposited on a glass slide (LLR2BL SuperfrostR, CML, Nemours, France), and the coverslip with the fruiting bodies is mounted over it. Microscopy examination is done coverslip down with a classical fluorescence or a confocal microscope.

4.4. Observation of autophagy and autophagic cell death by electron microscopy

Electron microscopy is a valuable tool that allows the structure of organelles and cellular subsystems to be characterized with unprecedented detail and reliability. Cells are processed as described earlier (Kosta *et al.*, 2006).

1. Cells (10^6 cells/ml, 20 ml) under starvation for 24 h with or without DIF-1 are prefixed at 22 °C by adding an equal volume of 2% glutaraldehyde in PBS buffer, pH 7.2 (grade I; Sigma-Aldrich) to the medium.
2. After 20 min the cells are scraped carefully from the culture flask and collected by centrifugation for 3 min at 3000 rpm.
3. The supernatant is replaced by 2 ml of 1% glutaraldehyde in PBS for 1 h at 4 °C.
4. Cells are washed in 5 ml of PBS, and concentrated in agarose (LMP Agarose, Sigma A9414) by removing the excess of liquid and resuspending the cells in a drop of 2% agarose prepared in PBS. Quickly centrifuge them in an microcentrifuge tube for 3 min at 3000 rpm and cool the pellet on ice. Cut off the tip of the tube and remove the agarose-embedded pellet. On ice, slice the pellet into blocks of the desired size and put them in a glass vial. These blocks will stay solid at room temperature (RT).
5. The agar blocksare washed again in PBS and postfixed in 1% aqueous osmium tetroxide (EMS 19150), not reduced, for 1–2 h at 4 °C.
6. Samples are washed again in distilled water, and treated with 1% uranyl acetate (EMS 22400) for 1 h at 4 °C in the dark (*en bloc* staining).
7. The samples are washed with distilled water, then dehydrated in a graded series of acetone (50%, 70%, 90%, 100% 15 min and two times each) and embedded in Epon (Polysciences 8792) as follows: 1 part resin:3 parts acetone, 1 h, RT; 1 part resin:1 part acetone, 1 h, RT; 3 parts resin:1 part acetone, 1 h, RT (or overnight at 4 °C); pure resin, 2 h, RT; pure resin, 1 h, 37 °C. In the meantime prepare paper labels (use a pencil, as printer ink or pen will smear) for the flat molds. Transfer the cell pellets in the molds with fresh resin, orientate them for best sectioning position, remove bubbles, and polymerize at 60 °C for 48 h.

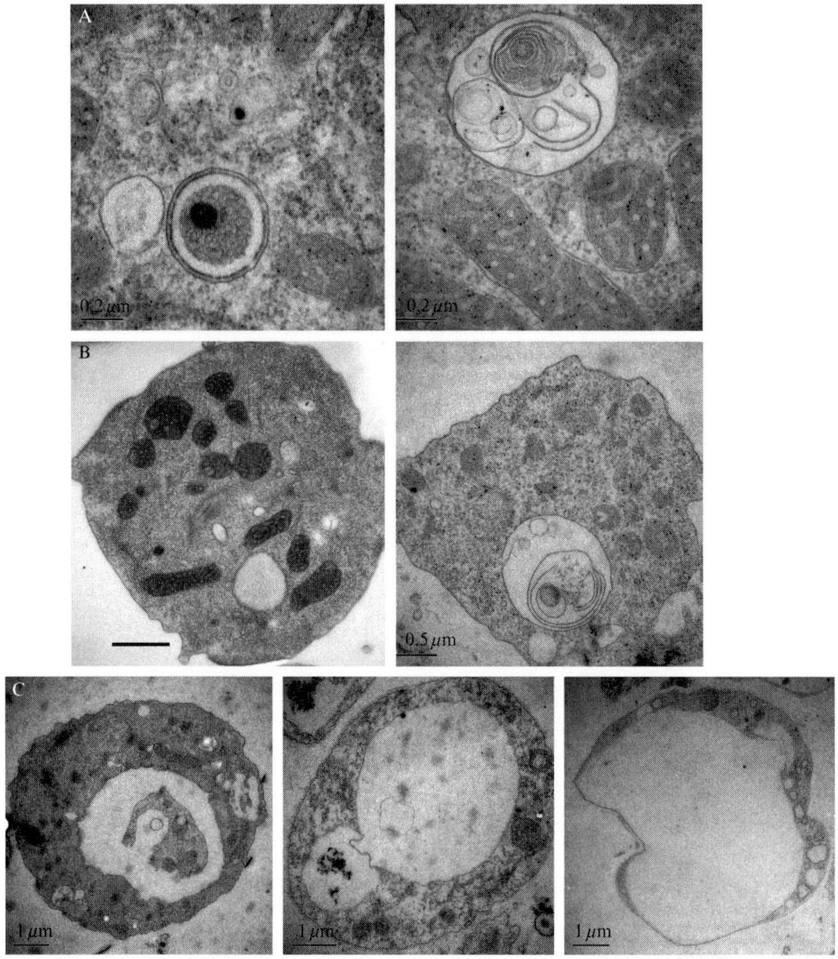

Figure 23.5 Observation of autophagy and autophagic cell death using electron microscopy. Autophagy observed by electron microscopy on HMX44A cells starved for 32 h. (*Left*) a probable autophagosome with a double membrane; (*right*) autophagolysosome with partially degraded cell components. Scale bar: 200 nm. Cells with or without autophagy. (*Left*) an HMX44A atg1 mutant cell obtained by targeted mutagenesis (Kosta *et al.*, 2004) and starved for 17 h. No obvious autophagic vesicles can be observed on this section (and on others, not shown). Scale bar: 1 μm. (*Right*) An HMX44A wild-type cell starved for 32 h with a probable autophagolysosome containing partially degraded cell components. Scale bar: 500 nm. Cells dying by autophagic cell death. Cells were observed at 24 h post-DIF. Autophagolysosomes may fuse to form only one large vacuole which occupies most of the cell at the end of the autophagic cell death process. Scale bars: 500 nm.

8. Ultrathin sections (60–90 nm) are collected on copper grids, stained with 2% aqueous uranyl acetate for 5 min in the dark, washed on 5 drops of distilled water and dried on filter paper. Subsequently, the grids are floated on 1% aqueous lead citrate for 2 min and then washed on 10 drops of distilled water and dried on filter paper and examined using a Zeiss EM 912 Electron Microscope. This technique allows the observation of different steps of autophagy and autophagic cell death (Fig. 23.5). While electron microscopy is a static, not a dynamic method, it takes snapshots of what happens in the fixed cells, and there are morphological (or immunoEM) criteria to characterize an early lysosome and early or late autophagosomes.

ACKNOWLEDGMENTS

We thank Pierre Grenot for help with cell sorting. This work was supported by institutional grants from INSERM and CNRS, and by Agence Nationale pour la Recherche (Dicty-Death ANR-05-BLAN-0333-01), the European Community (FP6 STREP TransDeath LSHG-CT-2004-511983), the Ministère pour la Recherche (ACI BCMS174) and Association pour la Recherche sur le Cancer.

REFERENCES

Cornillon, S., Foa, C., Davoust, J., Buonavista, N., Gross, J. D., and Golstein, P. (1994). Programmed cell death in *Dictyostelium*. *J. Cell Sci.* **107,** 2691–2704.

Kay, R. R. (1987). Cell differentiation in monolayers and the investigation of slime mold morphogens. *Methods Cell Biol.* **28,** 433–448.

Knecht, D. A., Cohen, S. M., Loomis, W. F., and Lodish, H. F. (1986). Developmental regulation of *Dictyostelium discoideum* actin gene fusions carried on low-copy and high-copy transformation vectors. *Mol. Cell Biol.* **6,** 3973–3983.

Kosta, A., Laporte, C., Lam, D., Tresse, E., Luciani, M. F., and Golstein, P. (2006). How to assess and study cell death in *Dictyostelium discoideum*. *Methods Mol. Biol.* **346,** 535–550.

Kosta, A., Roisin-Bouffay, C., Luciani, M. F., Otto, G. P., Kessin, R. H., and Golstein, P. (2004). Autophagy gene disruption reveals a non-vacuolar cell death pathway in *Dictyostelium*. *J. Biol. Chem.* **279,** 48404–48409.

Kuspa, A., and Loomis, W. F. (1992). Tagging developmental genes in *Dictyostelium* by restriction enzyme-mediated integration of plasmid DNA. *Proc. Natl. Acad. Sci. USA* **89,** 8803–8807.

Lam, D., and Golstein, P. (2008). A specific pathway inducing autophagic cell death is marked by an IP3R mutation. *Autophagy* **4,** 349–350.

Lam, D., Kosta, A., Luciani, M. F., and Golstein, P. (2008). The IP3 receptor is required to signal autophagic cell death. *Mol. Biol. Cell* **19,** 691–700.

Lam, D., Levraud, J. P., Luciani, M. F., and Golstein, P. (2007). Autophagic or necrotic cell death in the absence of caspase and bcl-2 family members. *Biochem. Biophys. Res. Commun.* **363,** 536–541.

Levraud, J.-P., Adam, M., Luciani, M.-F., De Chastellier, C., Blanton, R. L., and Golstein, P. (2003). *Dictyostelium* cell death: Early emergence and demise of highly polarized paddle cells. *J. Cell Biol.* **160,** 1105–1114.

Manstein, D. J., Titus, M. A., De Lozanne, A., and Spudich, J. A. (1989). Gene replacement in *Dictyostelium*: Generation of myosin null mutants. *EMBO J.* **8,** 923–932.

Morris, H. R., Taylor, G. W., Masento, M. S., Jermyn, K. A., and Kay, R. R. (1987). Chemical structure of the morphogen differentiation inducing factor from *Dictyostelium discoideum. Nature* **328,** 811–814.

Nellen, W., Silan, C., and Firtel, R. A. (1984). DNA-mediated transformation in *Dictyostelium. In* "Molecular Biology of Development" (E. H. Davidson and R. A. Firtel, eds.), A.R. Liss, New York, 1984, pp. 633–645.

Otto, G. P., Wu, M. Y., Kazgan, N., Anderson, O. R., and Kessin, R. H. (2003). Macroautophagy is required for multicellular development of the social amoeba *Dictyostelium discoideum. J. Biol. Chem.* **278,** 17636–17645.

Otto, G. P., Wu, M. Y., Kazgan, N., Anderson, O. R., and Kessin, R. H. (2004). *Dictyostelium* macroautophagy mutants vary in the severity of their developmental defects. *J. Biol. Chem.* **279,** 15621–15629.

Roisin-Bouffay, C., Luciani, M. F., Klein, G., Levraud, J. P., Adam, M., and Golstein, P. (2004). Developmental cell death in *Dictyostelium* does not require paracaspase. *J. Biol. Chem.* **279,** 11489–11494.

Sobolewski, A., Neave, N., and Weeks, G. (1983). The induction of stalk cell differentiation in submerged monolayers of *Dictyostelium discoïdeum*. Characterization of the temporal sequence for the molecular requirements. *Differentiation* **25,** 93–100.

Sussman, M. (1987). Cultivation and synchronous morphogenesis of *Dictyostelium* under controlled experimental conditions. *Methods Cell Biol.* **28,** 9–29.

Sutoh, K. (1993). A transformation vector for *Dictyostelium discoideum* with a new selectable marker *bsr. Plasmid* **30,** 150–154.

Tresse, E., Kosta, A., Giusti, C., Luciani, M. F., and Golstein, P. (2008). A UDP-glucose derivative is required for vacuolar autophagic cell death. *Autophagy* **4**, in press.

Whittingham, W. F., and Raper, K. B. (1960). Non-viability of stalk cells in *Dictyostelium*. *Proc. Natl. Acad. Sci. USA* **46,** 642–649.

CHAPTER TWENTY-FOUR

Analysis of Autophagy in the Enteric Protozoan Parasite *Entamoeba*

Karina Picazarri,* Kumiko Nakada-Tsukui,*,† Dan Sato,*,‡,§ *and* Tomoyoshi Nozaki*,†

Contents

Abstract

Entamoeba histolytica is the enteric protozoan parasite that causes human amoebiasis. We have previously shown that autophagy is involved in proliferation and differentiation in the related species *Entamoeba invadens*, which infects reptiles and develops similar clinical manifestations. Because this group of protists possesses only a limited set of genes known to participate in autophagy in other eukaryotes, it potentially represents a useful model for

* Department of Parasitology, Gunma University Graduate School of Medicine, Maebashi, Japan
† Department of Parasitology, National Institute of Infectious Diseases, Tokyo, Japan
‡ Institute for Advanced Biosciences, Keio University, Yamagata, Japan
§ Center for Integrated Medical Research, School of Medicine, Keio University, Tokyo, Japan

Methods in Enzymology, Volume 451
ISSN 0076-6879, DOI: 10.1016/S0076-6879(08)03224-2

studying the core system of autophagy and provides tools to elucidate the evolution of eukaryotes and their organelles. Here we describe the methods to study autophagy in *Entamoeba*.

1. Introduction

1.1. Organisms

Entamoeba histolytica is the enteric protozoan parasite that causes human amoebiasis (Petri *et al.*, 2002; Stanley *et al.*, 2003). This anaerobic or microaerophilic eukaryote has a simple life cycle consisting of two forms: the motile, proliferative trophozoite (the active, feeding stage) (Fig. 24.1), which is responsible for the pathology of amoebiasis, and the dormant, infective cyst, which is essential for transmission. This organism lacks organelles commonly observed in other eukaryotes, such as the mitochondria, the peroxisome, and the Golgi apparatus, and is considered one of the early branching eukaryotes (Hasegawa *et al.*, 1993; Loftus *et al.*, 2005). Recent discovery of the mitochondria-related genes (e.g., Cpn60) and the mitochondrial-related remnant organelle, named the *mitosome* (Tovar *et al.*, 1999), led to the presumption that this organism secondarily lost the mitochondria (Clark, 2000). Thus, this organism potentially helps in the elucidation of important questions on the evolution of eukaryotes and organelles.

The developmental transition of the trophozoite to the cyst stage, called *encystation*, is the essential process for transmission and reinfection of the organism. Therefore, its interruption is potentially exploitable to interfere with dissemination of this organism. However, inability to induce encystation *in vitro* hampers molecular understanding of the process in *E. histolytica* (Eichinger *et al.*, 2001) . *Entamoeba invadens*, a related *Entamoeba* species that

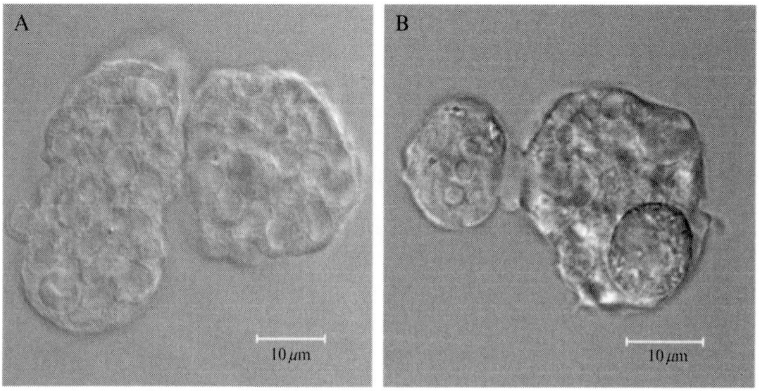

Figure 24.1 Differential interference contrast images of *E. histolytica* trophozoites. (A) two trophozoites containing numerous vacuoles. (B) A trophozoite (*right*) ingesting a Chinese hamster ovary cell (*left*).

infects reptiles and causes similar clinical manifestations, is instead used as a model of encystation because of the ease of inducing encystation using an artificial low-osmolarity, glucose-deprived encystation medium (Sanchez *et al.*, 1994).

1.2. Disease and clinical manifestations

There have been an estimated 40–50 million cases of amoebic colitis and liver abscess, which cause 40,000 to 110,000 deaths worldwide each year (Clark *et al.*, 2000; WHO, 1997). Infection of human and other mammalian hosts occurs upon ingestion of water or food contaminated with cysts. *E. histolytica* cysts are round, usually 10–15 μm in diameter, and protected by a chitin-containing wall. After ingestion, the cyst excysts in the small intestine to release the polymorphic trophozoite, which varies in size from 10–50 μm in diameter. Highly motile trophozoites colonize the large intestine. Epidemiological studies suggested that only 3% of infected individuals develop symptoms such as amoebic colitis and dysentery (Haque *et al.*, 2001; Stanley *et al.*, 2003), while the rest of the infected individuals remain asymptomatic and are able to clear the infection without developing disease (Haghighi *et al.*, 2003; Haque *et al.*, 2006; Stanley *et al.*, 2001). However, asymptomatic carriers represent a risk of contagion, and up to 10% of them develop disease within a year after infection (Stanley *et al.*, 2003). Children (Haque *et al.*, 2001; Warunee *et al.*, 2007), immunocompromised individuals (Hung *et al.*, 2008), men who have sex with men, and mentally handicapped persons (Nozaki *et al.*, 2006) are often more susceptible to infection. The trophozoite is responsible for all clinical manifestations, including abdominal pain, tenderness, and bloody diarrhea. *E. histolytica* can also colonize organs other than the intestine via hematogenous (i.e., originating in the blood) spread of trophozoites from the colon in 5%–10% of the diarrheal/dysenteric patients. Liver abscess is the most common extraintestinal form of amoebic infection. Patients who develop liver abscess present fever, right-upper-quadrant pain, hepatic tenderness, cough, anorexia, and weight loss.

2. Unique Features of Autophagy in *Entamoeba*

2.1. Genome-based identification of genes involved in autophagy in *Entamoeba*

In *Saccharomyces cerevisiae* approximately 30 genes have been identified as involved in autophagy (Klionsky *et al.*, 2003; Suzuki and Ohsumi, 2007; Xie and Klionsky, 2007), of which 17 genes encode proteins composing the core machinery of autophagy (Suzuki and Ohsumi, 2007). Most of them are also conserved in higher eukaryotes including mammals (Mizushima *et al.*,

2002; Xie and Klionsky, 2007). Autophagy genes have been categorized into four functional groups: regulation of autophagy induction, vesicle nucleation, vesicle expansion and completion (consisting two ubiquitin-like conjugation systems), and retrieval (Levine and Yuan, 2005). Among them, *Entamoeba* lacks genes involved in the regulation of the autophagy induction except TOR (target of rapamycin), the entire Atg12–Atg5 conjugation system, and the retrieval system, whereas genes involved in the nucleation of the isolation membrane forming the phosphatidylinositide complex (Vps15, Vps34, and Atg6/Beclin1) and the Atg8 conjugation system (Atg7, Atg3, Atg8, and Atg4) are conserved (Picazarri *et al.*, 2008).

2.2. Uniqueness of autophagy in *Entamoeba*

In both *E. histolytica* and *E. invadens*, the Atg8-positive structures have been identified by confocal microscopy as punctate particulate (or dotlike), vesicular, vacuolar (1–5 μm in diameter), linear, or aggregate-like structures (Fig. 24.2). The size and shape of autophagosomes in this organism is rather unique, and among the largest similar to ones containing intracellular pathogens, Group A *Streptococci* and *Mycobacterium tuberculosis* (Andrade *et al.*, 2006; Gutierrez *et al.*, 2004; Nakagawa *et al.*, 2004).

Autophagy is induced under particular conditions such as starvation, differentiation, defense, and antigen presentation in other organisms (Andrade *et al.*, 2006; Besteiro *et al.*, 2006; Nakagawa *et al.*, 2004; Nimmerjahn *et al.*, 2003). In *Entamoeba*, neither nutrient (e.g., glucose and serum) deprivation nor stress (e.g., heat and oxidative stress) induces the formation Atg8 structures in *E. histolytica*. Autophagosomes are constitutively present in the proliferative trophozoite. In *E. invadens*, where encystation can be induced *in vitro*, autophagosome formation is up-regulated at the mid-to-late logarithmic growth phase and at the early phase of encystation. In *E. invadens*, phosphatidylinositol 3-kinase inhibitors simultaneously inhibit the formation of the Atg8-postive structures and encystation in a dose-dependent manner. This observation suggests a close correlation between autophagy and encystation via phosphatidylinositol 3-kinase-mediated signaling. These data are consistent with the premise that autophagy plays a housekeeping role in *Entamoeba*, as seen in neurons where autophagy was suggested to be involved in the constant turnover of undesirable polyubiquitinated proteins (Komatsu *et al.*, 2006). This chapter describes some essential protocols to understand the function of autophagy in *Entamoeba*: immunoblot and immunofluorescence assays, as well as the creation and analysis of *E. histolytica* transformants expressing an epitope-tagged protein of interest.

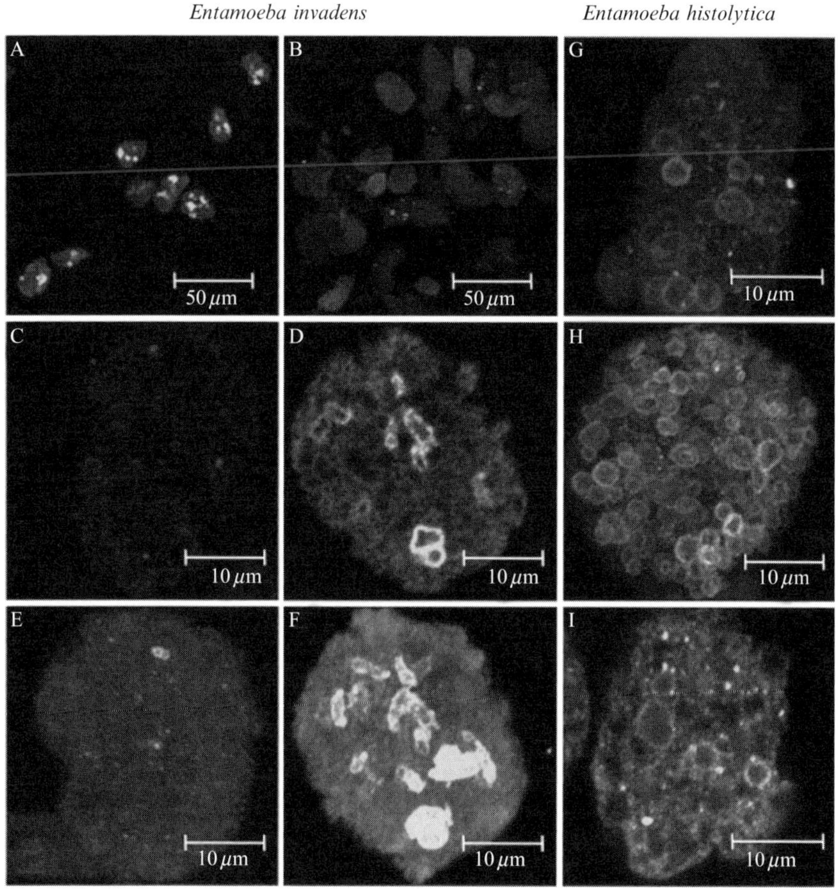

Figure 24.2 Immunofluorescence images of autophagy in *E. invadens* and *E. histolytica* on a confocal microscope. (A, B) Autophagosomes in the proliferating *E. invadens* trophozoites at 1 (logarithmic phase, A) and 2 weeks after the initiation of the culture (stationary phase, B). (C–F) Autophagosomes in the encysting *E. invadens* trophozoites at 0 (C, E) and 24 h postencystation induction (D, F). Single slices (C, E) and maximum projections of 20 slices taken at 1-μm intervals on the z-axis (E, F) are shown. (G, H) Autophagosomes in the proliferating *E. histolytica* trophozoites at days 1 (logarithmic phase, G) and 5 (stationary phase, H) (maximum projection). (I) Colocalization of Atg8 (green) and the lysosome marker, LysoTracker Red (red) in an *E. histolytica* trophozoite (day 3). (See Color Insert.)

3. Analysis of Autophagy in *Entamoeba*

3.1. Production of recombinant *E. histolytica* Atg8 and its antibody

Oligonucleotide primers and conditions of PCR amplification of *EhAtg8a* cDNA have been described elsewhere (Picazarri *et al.*, 2008). Cloning of the *EhAtg8a* cDNA into the pGEX-6P-2 (GE Healthcare Bioscience,

27-4598-01) expression vector to make pGST-EhAtg8a as well as antibody production have also previously been described in detail (Picazarri *et al.*, 2008).

3.1.1. Production of recombinant EhAtg8

1. Transform *E. coli* BL21(DE3) competent cells with pGST-EhAtg8 plasmid.
2. Select transformants on ampicillin plates. Grow a transformant overnight in 20 ml of LB containing 50 μg/ml ampicillin and use to inoculate 200 ml of the same medium. Shake at 37 °C until the O.D.$_{600}$ reaches 0.5.
3. Add 1 m*M* IPTG to the culture to induce expression of GST-EhAtg8 recombinant protein and continue shaking for 3 h at 37 °C.
4. Harvest bacteria at 6000×*g* for 10 min, wash the pellet twice with phosphate-buffered saline (PBS), pH 7.4, and resuspend in 5–20 ml of the lysis buffer (50 m*M* NaH_2PO_4, pH 8.0, 300 m*M* NaCl, 1 m*M* DTT, and complete mini protease inhibitor cocktail).
5. Sonicate the suspension using a VP-15S UltraS Homogenizer (TAITEC) or equivalent in an ice-water bucket. Occasionally examine the lysate under a light microscope with a 40x phase contrast objective to confirm the disruption of bacteria.
6. Centrifuge at 12,000×*g* for 20 min at 4 °C to remove debris.

3.1.2. Purification of recombinant EhAtg8 using glutathione-sepharose affinity chromatography

1. Incubate 1 volume (5 ml) of the clarified lysate with two-thirds volume of glutatione-sepharose (GE Healthcare Bioscience, 17-5279-01) slurry (50%) for 0.5 h on a rocking platform at 4 °C. Transfer the resin to a column, and wash the resin three times with 10 ml of PBS.
2. Elute GST-EhAtg8 recombinant protein with 5 ml of elution buffer (PBS containing 10 m*M* reduced glutathione) twice at room temperature.
3. Filter the eluted fractions with a 0.45-μm syringe filter and dialyze it against protease cleavage buffer (50 m*M* Tris-HCl, pH 7.0, 150 m*M* NaCl, 1 m*M* EDTA, 1 m*M* DTT) using Slide-A-Lyzer (PIERCE, #66110; molecular weight cut off, 3500) overnight at 4 °C.
4. To remove the GST tag from the recombinant protein, mix 10 ml of the purified protein with 80 μg of PreScission protease (GE Healthcare Bioscience) and incubate at 4 °C for 4 h.
5. Add 2 ml of glutathione-Sepharose (50% slurry) to the mixture and rotate it at room temperature for 30 min. Briefly centrifuge the mixture and filter the supernatant with a 0.45-μm syringe filter.

6. Dialyze the filtrate against 2 L of PBS for 2 h and then 3 L of PBS overnight at 4 °C. Finally, remove remaining GST tag and PreScission protease by passing the dialyzed solution through GSTrap (GE Healthcare Bioscience).

3.1.3. Purification of recombinant EhAtg8 using Mono Q anion exchange chromatography

1. Dilute the eluate from GSTrap by 5-fold with the starting buffer (50 m*M* 2, 2′-iminodiethanol, pH 8.4).
2. Preequilibrate a Mono Q HR 5/5 anion-exchange column (GE Healthcare Bioscience) with the starting buffer on an AKTA Explorer.
3. Apply the affinity-purified recombinant EhAtg8 protein onto the column with the starting buffer, and wash the column with 5 volumes of the starting buffer to remove unbound proteins.
4. Elute the proteins with a linear gradient of NaCl (0–1 *M* in 20 ml) with a flow rate of 1.0 ml/min. The recombinant EhAtg8 protein elutes at circa 150 m*M* NaCl.
5. Dialyze the eluted EhAtg8 protein against PBS at 4 °C for 4–5 h.

3.2. Quantitative analysis of kinetics and modification of EhAtg8 by immunoblot analysis

3.2.1. Preparation of amoeba lysates and separation of membrane and soluble fractions by centrifugation

1. Chill a 36-ml semiconfluent amoeba culture ($\sim 3 \times 10^6$ cells) cultivated in a 25-cm^2-plastic flask (Nunc Brand Products, Denmark) on ice for 5 min. Cultivate *E. histolytica* and *E. invadens* trophozoites in BI-S-33 medium [3.4% BLL Biosate Peptone (Becton, Dickinson, France; 211862), 63 m*M* D-glucose, 38.9 m*M* NaCl, 5 m*M* KH_2PO_4, 6.52 m*M* K_2HPO_4, 9.37 m*M* L-cysteine hydrochloride, 1.3 m*M* L-ascorbic acid, 0.1 m*M* Ferric ammonium citrate] (Diamond *et al.*, 1978) at 35.5 °C and 26 °C, respectively, as described subsequently (see section 3.4.1). To obtain *E. invadens* cysts, transfer the 1-week-old trophozoite culture to 47% LG medium [1.5% BLL Biosate Peptone, 17 m*M* NaCl, 2.195 m*M* KH_2PO_4, 2.86 m*M* K_2HPO_4, 3.159 m*M* L-cysteine hydrochloride, 0.565 m*M* L-ascorbic acid, 0.0428 m*M* Ferric ammonium citrate] (Sanchez *et al.*, 1994) at $\sim 6 \times 10^5$ cells/ml.
2. To monitor encystation, examine resistance to 0.05% Sarkosyl (Sanchez *et al.*, 1994) as follows. Centrifuge 1.2×10^5 cells at $200 \times g$ for 5 min, discard supernatant, and add 200 μl of PBS containing 0.05% Sarkosyl. After incubation for 25 min, add 200 μl of PBS containing 0.44%

Trypan blue. After 5 min, count unstained live cysts and stained dead trophozoites to calculate the percentage of cysts.

3. Harvest cells by centrifugation at 400×g for 5 min at 4 °C.
4. Resuspend the pellet with 10 ml of PBS containing 2% glucose, centrifuge at 400×g for 5 min, and carefully discard the supernatant by decanting.
5. Resuspend the pellet in 1 ml of homogenization buffer (50 m*M* Tris, pH 7.5, 250 m*M* sucrose, 50 m*M* NaCl, 200 μ*M* trans-epoxysuccinyl-L-leucylamido-[4-guanidino butane] (E64)).
6. Transfer the suspension to a Dounce glass homogenizer and mechanically homogenize cells with 50–300 strokes depending on homogenizers and cell types. Occasionally examine the lysate under a light microscope with a 10–40x phase contrast objective to verify completeness of homogenization.
7. Transfer the lysate to a 1.5-ml tube, centrifuge at 700×g for 2 min, recover the supernatant, and discard unbroken cells in the pellet.
8. Centrifuge the supernatant at 100,000×g in an ultracentrifuge tube at 4 °C for 1 h.
9. Recover the supernatant and the pellet fractions separately. Carefully resuspend the pellet by pipetting with 1 ml of PBS and centrifuge to wash the pellet to minimize carryover from the supernatant.
10. Resuspend the pellet in the original volume (step 4) of the lysis buffer (50 m*M* Tris, pH 7.5, 1% Triton X-100, 1.34 m*M* E64).
11. Electrophorese 5 μg of each sample by SDS-PAGE on a 13.5% polyacrylamide gel containing 6M urea.
12. Carry out immunoblot analysis using anti-EhAtg8a antibody (Picazarri *et al.*, 2008) (1:1000 dilution in Tris-buffered saline containing 0.05% Tween 20). Develop the membrane using chemiluminescence with the Immobilon Western Substrate (Millipore Corporation) according to the manufacturer's instructions.

3.2.2. Delipidation of EhAtg8 by phospholipase D and quantitation of unmodified and phosphatidylethanolamine-modified EhAtg8

1. Harvest amoeba cells from a 6-ml culture (~6×10^5 cells) as described above and resuspend and lyse the cell pellet in 50 μl of lysis buffer (50 m*M* Tris, pH 7.5, 1% Triton X-100, 1.34 m*M* E64) on ice. Determine the protein concentration of the lysate, after centrifugation to remove unbroken cells.
2. Mix 10 μg of the crude lysate with 2 μl of 20U/μl phospholipase D (Sigma-Aldrich, P8023) dissolved in 50 m*M* Tris, pH 7.5. Incubate the mixture at 37 °C for 1 h.

3. Perform SDS-PAGE on a 13.5% polyacrylamide gel containing 6 *M* urea and transfer proteins to a nitrocellulose membrane. Conduct immunoblot analysis as described previously in section 3.2.1, step 12.
4. Visualize unmodified and phosphatidylethanolamine-modified Atg8 on a Lumi-Imager F1 workstation (Roche Applied Science) and capture the images with Lumianalyst Software. Quantify the bands with ImageJ. Unmodified Atg8 is present as a single band in *E. invadens* and doublets in *E. histolytica* around 15.0 kDa, while phosphatidylethanolamine-modified Atg8 is observed as a 14.5-kDa band in *E. invadens* and two bottom bands in *E. histolytica*.

3.3. Visualization of autophagy by indirect immunofluorescence assay

3.3.1. Sample preparation

1. Harvest and wash amoebae from a 6-ml semiconfluent culture (~6×10^5 cells) at 4 °C as described previously.
2. Resuspend the pellet with 50 μl of PBS containing 3.7% paraformaldehyde. Incubate the mixture for 10 min at room temperature.
3. Centrifuge the mixture at 800×*g* for 3 min at room temperature, remove the supernatant fraction, and carefully resuspend the pellet by pipetting with 1 ml of PBS .
4. Centrifuge again, discard the supernatant, resuspend the pellet in 50 μl of PBS containing 0.2% saponin (PBSS), and incubate for 10 min at room temperature.
5. Wash the amoebae with PBS as described previously.
6. After discarding the supernatant, add to the pellet 100 μl of PBS containing anti-EhAtg8a antiserum (1:1000 dilution), resuspend, and incubate at room temperature for 1 h.
7. Wash the amoebae once with PBS as described previously.
8. Add 100 μl of anti-rabbit IgG conjugated with Alexa Flour 488 (1:1000 dilution) to the washed amoebae and incubate 1 h at room temperature.
9. Wash the amoebae once with PBS as described previously.
10. Resuspend the amoebae in ~10 μl of the mounting medium (PBS, pH 8.0, containing 5.5 m*M* O-phenylenediamine dihydrochloride and 90% glycerol).

3.3.2. Confocal microscopy

Perform confocal microscopy on a Zeiss LSM 510 microscope or equivalent. Typically, we use a Plan-Apochromat 63x/1.4 Oil objective with an argon laser (488 nm) with appropriate configurations for Alexa Flour 488. It is important to optimize detector gain and amplifier offset with preimmune serum to eliminate background. Typically, images of 5–30 slices on the z-axis at 1–3 μm intervals should be captured. Obtained images can be also analyzed after creating a 3-D projection.

3.4. Construction of *E. histolytica* transformants expressing an HA-tagged or Myc-fused protein of interest

Construct a plasmid of interest using epitope (hemagglutinin or myc) tag-containing vector [e.g., pKT-3M (Saito-Nakano *et al.*, 2004) and pEhExHA (Nakada-Tsukui *et al.*, unpublished)], or a GFP- or RFP-fusion vector [e.g., pKT-MR (Nakada-Tsukui *et al.*, unpublished)], according to general recombinant DNA techniques.

3.4.1. Cell preparation

1. Inoculate an appropriate number of trophozoites of *E. histolytica* HM1: IMSS cl6 strain (Diamond *et al.*, 1972), harvested in the mid-late logarithmic growth, phase into fresh BI-S-33 medium in 25-cm^2-ml plastic flasks and cultivate for 24–30 h until the culture reaches semiconfluence (1.5–2 $\times$ 10^6 cells per flask).
2. After decanting the culture medium, add 5 ml of the cold fresh medium to each flask and resuspend cells by chilling the flask on ice for 5 min. Mix suspensions and adjust the cell density to 7–9 $\times$ 10^4 amoebae per ml.
3. Transfer 5 ml of the suspension to a well of a 12-well flat bottom plate (Corning).
4. Seal the plate in a plastic bag containing Anaerocult A (Merck, 64271 Darmstadt, Germany). Incubate the plate at 35.5 °C for at least 30 min.

3.4.2. Lipofection

1. Right before transfection, prepare the transfection medium [TM, Opti-MEM I (Invitrogen, 31985-070), containing 56.7 m*M* L-ascorbic acid and 317 m*M* L-cysteine, pH 6.8, filter-sterilized].
2. Prepare DNA-liposome mixture as follows. Mix ~3–5 μg of DNA, sterilized with ethanol precipitation and dissolved in 30 μl of double-distilled sterile water, with 10 μl of Plus Reagent (Invitrogen, 11514015) and 10 μl of TM (mixture #1). Incubate the mixture #1 at room temperature for 15 min. Plasmids prepared using any spin column-based kits commercially available usually give comparable results.
3. Mix 20 μl Lipofectamine (Invitrogen, 183224-012) reagent (40 μg of Lipofectamine) with 30 μl of TM (mixture #2).
4. Add mixture #2 to mixture #1 (mixture #3) and incubate the mixture at room temperature for 15 min.
5. Add 400 μl of TM to the mixture #3.
6. Carefully remove BI-S-33 medium from the wells of the 12-well plate described in section 3.4.1.

7. Add 500 μl of the DNA-Lipofectamine mixture #3 to the well. Then, incubate the plate under anaerobic conditions as described above at 35.5 °C for 5 h.
8. Place the plate on ice to detach transfected cells for 5 min, and then transfer the whole suspension to a 6-ml glass tube containing 5.5 ml of fresh BI-S-33 medium.
9. Incubate the tube overnight at 35.5 °C.

3.4.3. Drug selection

1. Remove the medium from the tube by aspiration and add 6 ml of a fresh prewarmed medium containing 1 μg/ml G418.
2. Incubate the tube at 35.5 °C for another 24 h.
3. Every 24 h replace the medium containing 1 μg/ml of G418 for 5–10 days. After 3–10 days G418-sensitive cells start to die and the cell number dramatically decreases. Keep replacing the medium containing 1 μg/ml G418 every day until G418-resistant cells become visible. It usually takes an additional 2–5 days. Once G418-resistant cells start to rapidly grow, increase G418 concentrations in a step-wise fashion with an increment of 1 μg/ml per day or per passage until it reaches 10 μg/ml G418. Adjustment of G418 concentration may be necessary depending on the plasmids used. Dead cells may not be apparent after 24 h but usually become visible after 48–72 h (dead cells often accumulate at the bottom of the tube).
4. After growth of the transfectants is established (typically at 1–2 weeks postlipofections), confirm the expression of the gene of interest by immunoblot and immunofluorescence assays. If the expression level is not sufficient, G418 concentrations may be further gradually increased to 20–50 μg/ml.

4. Conclusion

Repression (knockdown) of gene expression by siRNA or gene silencing has gradually become available in *E. histolytica* (Boettner *et al.*, 2008; Bracha *et al.*, 2006; Solis and Guillen, 2008). Gene silencing of *Atg8* and other genes involved in autophagy has recently been accomplished in *E. histolytica* (unpublished). Such reverse genetic tools should help in unequivocal assignment of a role of individual gene products. Because *Entamoeba* is a primitive eukaryote, which apparently possesses only a limited, if not minimal, set of autophagy genes identified in both mammals and yeasts, an understanding of the molecular mechanisms of autophagy in this group of organisms should contribute to the elucidation of the origin and evolution of this important cellular mechanism for protein degradation.

ACKNOWLEDGMENTS

This work was supported by Grant-in-Aids for Scientific Research from the Ministry of Education, Culture, Sports, Science and Technology of Japan to TN (18GS0314, 18050006, 18073001), a grant for research on emerging and re-emerging infectious diseases from the Ministry of Health, Labour and Welfare of Japan to TN, and a grant for research to promote the development of anti-AIDS pharmaceuticals from the Japan Health Sciences Foundation to TN. We thank Roderick Williams for his suggestions on Atg8 delipidation and Yoko Yamada for her help in the recombinant EhAtg8a production.

REFERENCES

Andrade, R. M., Wessendarp, M., Gubbels, M. J., Striepen, B., and Subauste, C. S. (2006). CD40 induces macrophage anti-*Toxoplasma gondii* activity by triggering autophagy-dependent fusion of pathogen-containing vacuoles and lysosomes. *J. Clin. Invest.* **116,** 2366–2377.

Bakatselou, C., Beste, D., Kadri, A. O., Somanath, S., and Clark, C. G. (2003). Analysis of genes of mitochondrial origin in the genus *Entamoeba. J. Eukaryot. Microbiol.* **50,** 210–214.

Bakatselou, C., and Clark, C. G. (2000). A mitochondrial-type hsp70 gene of *Entamoeba histolytica. Arch. Med. Res.* **31,** S176–S177.

Besteiro, S., Williams, R. A., Morrison, L. S., Coombs, G. H., and Mottram, J. C. (2006). Endosome sorting and autophagy are essential for differentiation and virulence of *Leishmania major. J. Biol. Chem.* **281,** 11384–11396.

Boettner, D. R., Huston, C. D., Linford, A. S., Buss, S. N., Houpt, E., Sherman, N. E., and Petri, W. A., Jr. (2008). *Entamoeba histolytica* phagocytosis of human erythrocytes involves PATMK, a member of the transmembrane kinase family. *PLoS Pathog.* **4,** e8.

Bracha, R., Nuchamowitz, Y., Anbar, M., and Mirelman, D. (2006). Transcriptional silencing of multiple genes in trophozoites of *Entamoeba histolytica. PLoS Pathog.* **2,** e48.

Clark, C. G. (2000). The evolution of *Entamoeba*, a cautionary tale. *Res. Microbiol.* **151,** 599–603.

Diamond, L. S., and Cunnick, C. C. (1978). A new medium for axenic cultivation of *Entamoeba histolytica* and other *Entamoeba. Trans. Roy. Soc. Trop. Med. Hyg.* **72,** 431–432.

Diamond, L. S., M., Cunniick, C. C., and Bartgis, I. L. (1972). Viruses of *Entamoeba histolyitica. J. Virol.* **9,** 326–341.

Eichinger, D. (2001). A role for a galactose lectin and its ligands during encystment of *Entamoeba. J. Eukaryot. Microbiol.* **48,** 17–21.

Gutierrez, M. G., Master, S. S., Singh, S. B., Taylor, G. A., Colombo, M. I., and Deretic, V. (2004). Autophagy is a defense mechanism inhibiting BCG and *Mycobacterium tuberculosis* survival in infected macrophages. *Cell* **119,** 753–766.

Haghighi, A., Kobayashi, S., Takeuchi, T., Thammapalerd, N., and Nozaki, T. (2003). Geographic diversity among genotypes of *Entamoeba histolytica* field isolates. *J. Clin. Microbiol.* **41,** 3748–3756.

Haque, R., Ali, I. M., Sack, R. B., Farr, B. M., Ramakrishnan, G., and Petri, W. A., Jr. (2001). Amebiasis and mucosal IgA antibody against the *Entamoeba histolytica* adherence lectin in Bangladeshi children. *J. Infect. Dis.* **183,** 1787–1793.

Haque, R., Mondal, D., Duggal, P., Kabir, M., Roy, S., Farr, B. M., Sack, R. B., and Petri, W. A., Jr. (2006). *Entamoeba histolytica* infection in children and protection from subsequent amebiasis. *Infect. Immun.* **74,** 904–909.

Hasegawa, M., Hashimoto, T., Adachi, J., Iwabe, N., and Miyata, T. (1993). Early branchings in the evolution of eukaryotes: Ancient divergence of *Entamoeba* that lacks mitochondria revealed by protein sequence data. *J. Mol. Evol.* **36,** 380–388.

Hung, C. C., Ji, D. D., Sun, H. Y., Lee, Y. T., Hsu, S. Y., Chang, S. Y., Wu, C. H., Chan, Y. H., Hsiao, C. F., Liu, W. C., and Colebunders, R. (2008). Increased risk for *Entamoeba histolytica* infection and invasive amebiasis in HIV seropositive men who have sex with men in Taiwan. *PLoS Negl. Trop. Dis.* **2,** e175.

Klionsky, D. J., Cregg, J. M., Dunn, W. A., Jr., Emr, S. D., Sakai, Y., Sandoval, I. V., Sibirny, A., Subramani, S., Thumm, M., Veenhuis, M., and Ohsumi, Y. (2003). A unified nomenclature for yeast autophagy-related genes. *Dev. Cell* **5,** 539–545.

Komatsu, M., Waguri, S., Chiba, T., Murata, S., Iwata, J., Tanida, I., Ueno, T., Koike, M., Uchiyama, Y., Kominami, E., and Tanaka, K. (2006). Loss of autophagy in the central nervous system causes neurodegeneration in mice. *Nature* **441,** 880–884.

Levine, B., and Yuan, J. (2005). Autophagy in cell death: An innocent convict? *J. Clin. Invest.* **115,** 2679–2688.

Loftus, B., Anderson, I., Davies, R., Alsmark, U. C., Samuelson, J., Amedeo, P., Roncaglia, P., Berriman, M., Hirt, R. P., Mann, B. J., Nozaki, T., Suh, B., *et al.* (2005). The genome of the protist parasite *Entamoeba histolytica. Nature* **433,** 865–868.

Mizushima, N., Ohsumi, Y., and Yoshimori, T. (2002). Autophagosome formation in mammalian cells. *Cell Struct. Funct.* **27,** 421–429.

Nakagawa, I., Amano, A., Mizushima, N., Yamamoto, A., Yamaguchi, H., Kamimoto, T., Nara, A., Funao, J., Nakata, M., Tsuda, K., Hamada, S., and Yoshimori, T. (2004). Autophagy defends cells against invading group A *Streptococcus. Science.* **306,** 1037–1040.

Nimmerjahn, F., Milosevic, S., Behrends, U., Jaffee, E. M., Pardoll, D. M., Bornkamm, G. W., and Mautner, J. (2003). Major histocompatibility complex class II-restricted presentation of a cytosolic antigen by autophagy. *Eur. J. Immunol.* **33,** 1250–1259.

Nozaki, T., Kobayashi, S., Takeuchi, T., and Haghighi, A. (2006). Diversity of clinical isolates of *Entamoeba histolytica* in Japan. *Arch. Med. Res.* **37,** 277–279.

Petri, W. A., Jr. (2002). Pathogenesis of amebiasis. *Curr. Opin. Microbiol.* **5,** 443–447.

Picazarri, K., Nakada-Tsukui, K., and Nozaki, T. (2008). Autophagy during proliferation and encystation in the protozoan parasite *Entamoeba invadens. Infect. Immun.* **76,** 278–288.

Saito-Nakano, Y., Yasuda, T., Nakada-Tsukui, K., Leippe, M., and Nozaki, T. (2004). Rab5-associated vacuoles play a unique role in phagocytosis of the enteric protozoan parasite *Entamoeba histolytica. J. Biol. Chem.* **279,** 49497–49507.

Sanchez, L., Enea, V., and Eichinger, D. (1994). Identification of a developmentally regulated transcript expressed during encystation of *Entamoeba invadens. Mol. Biochem. Parasitol.* **67,** 125–135.

Solis, C. F., and Guillen, N. (2008). Silencing genes by RNA interference in the protozoan parasite *Entamoeba histolytica. Methods. Mol. Biol.* **442,** 113–128.

Stanley, S. L. (2001). Pathophysiology of amoebiasis. *Trends. Parasitol.* **17,** 280–285.

Stanley, S. L., Jr. (2003). Amoebiasis. *Lancet* **361,** 1025–1034.

Suzuki, K., and Ohsumi, Y. (2007). Molecular machinery of autophagosome formation in yeast, *Saccharomyces cerevisiae. FEBS. Lett.* **581,** 2156–2161.

Tovar, J., Fischer, A., and Clark, C. G. (1999). The mitosome, a novel organelle related to mitochondria in the amitochondrial parasite *Entamoeba histolytica. Mol. Microbiol.* **32,** 1013–1021.

Warunee, N., Choomanee, L., Sataporn, P., Rapeeporn, Y., Nuttapong, W., Sompong, S., Thongdee, S., Bang-On, S., and Rachada, K. (2007). Intestinal parasitic infections among school children in Thailand. *Trop. Biomed.* **24,** 83–88.

WHO/PAHO/UNESCO (1997). A consultation with experts on amoebiasis. *Epidemiol Bull.* **46,** 105–109.

Xie, Z., and Klionsky, D. J. (2007). Autophagosome formation: Core machinery and adaptations. *Nature Cell. Biol. Rev.* **9,** 1102–1109.

CHAPTER TWENTY-FIVE

Kinetoplastida: Model Organisms for Simple Autophagic Pathways?

Viola Denninger,* Rudolf Koopmann,* Khalid Muhammad,* Torsten Barth,* Bjoern Bassarak,* Caroline Schönfeld,* Bruno Kubata Kilunga,† *and* Michael Duszenko*

Contents

Abstract

Phylogenetic analyses based on defined proteins or different RNA species have revealed that the order kinetoplastida belongs to the early-branching

* Interfaculty Institute of Biochemistry, University of Tübingen, Tübingen, Germany
† NEPAD/Biosciences Eastern and Central Africa, ILRI Campus, Nairobi, Kenya

Methods in Enzymology, Volume 451
ISSN 0076-6879, DOI: 10.1016/S0076-6879(08)03225-4

eukaryotes and may thus contain organisms in which complex cellular events are easier to analyze. This view was further supported by results from a bioinformatic survey that suggested that nearly half of the autophagy-related proteins existent in yeast are missing in trypanosomatids. On the other hand, these organisms have evolved a highly sophisticated machinery to escape from the different host immune-response strategies and have learned to cope with extremely variable environmental conditions by morphological and functional reorganization of the cell. For both the stress response and the differentiation processes, autophagy seems to be an indispensable prerequisite. So far autophagy has not been systematically investigated in trypanosomatids. Here we present technical information on how to handle the different parasites belonging to this order and give an overview of the current status of autophagy research in these organisms.

1. Introduction

The invention of the lysosome was certainly a major breakthrough in the development of eukaryotic cells, as it enabled them to get rid of any kind of cellular material without losing the basic chemical components, such as amino acids, bases, sugars, or lipids. In the most primitive way, a bunch of acid hydrolases, sequestered in a membrane-surrounded vesicle, are sufficient to do the job. Here, the vesicle meanders through the cell, thereby engulfing and grabbing small parts of the cytosol or cytoplasm, respectively, which are degraded to building blocks of macromolecules, finally reused by the cell (Marzella *et al.*, 1981). This process is highly unspecific and only marginally controlled but sufficient for a steady turnover of biomacromolecules and individual damaged organelles. From here more sophisticated functions for lysosomal digestion and nutrient supply have evolved, which include fusion of lysosomes with phagosomes on phagocytosis or endosomes on receptor-mediated endocytosis (Morgan *et al.*, 2002a,b). In addition, lysosomes serve as an intracellular defense machinery to destroy bacteria or viruses on infection of the cell (Kirkegaard *et al.*, 2004). Most specific and controlled, however, are the various forms of autophagy. In its simplest form, autophagy is a limited form of self-digestion for the benefit of cell survival, such as during starvation. In this case, bulk cytosol or disposable organelles are surrounded by specifically formed membranes of unknown origin called *phagophores*. These are thought to originate at phagophore assembly sites or preautophagosomal structures (PAS). Upon completion, the phagophore forms a double-membrane vesicle, termed an *autophagosome*, that fuses with lysosomes forming so-called autophagolysosomes (or autolysosomes). If essential nutrients reappear in time, the cell will recover and replace missing organelles and molecules. Otherwise, affected cells may enter a controlled cell death pathway and dissolve, altruistically leaving nutrients for surrounding cells. This form of autophagy has been called programmed cell death (PCD) type II (Levine and Yuan, 2005; Maiuri *et al.*, 2007).

Even more interesting is autophagy as a mechanism to maintain the functional integrity of a cell during changing environmental conditions. For example, at a certain time hepatocytes may have to produce bulk quantities of serum proteins and thus contain abundant ER and Golgi membranes. If, under changing conditions, these organelles become obsolete and should be removed, they will be specifically labeled for digestion, engulfed by phagophores and delivered to lysosomes for degradation. Thus, removal of defined organelles during the life span of a cell is a common and regular process and needs a precise and controlled targeting. There is even a specific denotation such as pexophagy for the removal of peroxisomes or mitophagy for the removal of mitochondria (Dunn *et al.*, 2005; Mijaljica *et al.*, 2007). Likewise, during differentiation a cell adjusts to the organism's needs and has to remodel or replace some organelles to fulfill its duties. Autophagy is thus for an average cell a continuous and indispensable mechanism to cope with cellular needs (e.g. turnover of housekeeping enzymes) and changing life cycle (e.g. differentiation) or environmental (e.g. starvation or stress response) conditions. These different functional elements make it also a process of considerable complexity with a plethora of different factors and the involvement of a variety of protein-protein, protein-organelle, and membrane-membrane interactions (Codogno and Meijer, 2005; Yorimitsu and Klionsky, 2005). As described throughout this volume and elsewhere, some of the interaction partners and molecular events have already been dissected, while others are still missing or of mysterious function.

Using bioinformatic approaches, only a limited number of genes seem to be involved in autophagy in trypanosomes as compared with yeast and higher eukaryotes (Herman *et al.*, 2006; Rigden et *al.*, 2005). This may reflect the specific situation for the order kinetoplastida, which branched off the evolutionary development at a very early time point (Baldauf *et al.*, 2000) and offers the hope of finding a less complex network of molecular partners in these model organisms (Klionsky, 2006). On the other hand, several members of this order are very important pathogens (Table 25.1) such as *Trypanosoma brucei* (sleeping sickness), *Trypanosoma cruzi* (Chagas' disease) and *Leishmania donovani* (kala-azar). A precise understanding of this elementary cellular process in these cells may provide new targets for the development of effective and safe drugs to treat the respective parasitic diseases. Here we present a comprehensive and largely complete picture of what is currently known about autophagy in the trypanosomatids.

Table 25.1 Host specificity and morphological stages of species belonging to the suborder trypanosomatina

		Stages in		Stages in	
Genus	Stages exclusively in invertebrates	Invertebrates	Plants	Invertebrates	Vertebrates
Phytomonas	—	a- and promastigotes	promastigotes	—	—
Herpetomonas	a-, pro-, epi-, and ophistomastigoters	—	—	—	—
Leishmania	—	—	—	promastigotes	amastigotes
Leptomonas	a- and promastigotes	—	—	—	—
Crithidia	a-, pro-, or epimastigotes	—	—	—	—
Blastocrithidia	pro- and epimastigotes	—	—	—	—
T. brucei group	—	—	—	epi- and trypomastigotes	trypomastigotes
T. cruzi	—	—	—	epi- and trypomastigotes	a- and trypomastigotes

2. Experimental Procedures to Handle the Different Species of the Order Kinetoplastida

The order kinetoplastida was named for the presence of a Feulgen-positive structure in a distinct region of the single mitochondrion in these organisms (Feulgen stain is used to detect chromosomal, or in this case mitochondrial, DNA). This compartment contains the mtDNA, which comprises up to several thousand minicircles and some dozen maxicircles intercalated to a cuboidlike structure easily detected in electron micrographs (Shapiro, 1993). Whereas only maxicircles contain genetic information, maxi- and minicircles encode so-called guide RNAs necessary for the excessive RNA editing observed in these organisms (Smith *et al.*, 1997). The kinetoplast is always closely associated with the basal body of the flagellum. Another characteristic feature of these cells is the existence of glycosomes. These organelles resemble peroxisomes and contain most of the glycolytic enzymes. Thus, cells of the order kinetoplastida are the only organisms known so far where glycolysis is compartmentalized (Parsons *et al.*, 2001).

The different species of the order kinetoplastida are heterotrophic and feed as saprozoic or parasitic organisms. In general, they are flagellated. The flagellum arises from a depression of their surface (flagellar pocket) and includes a paraxial rod in addition to the axoneme (De Souza and Souto-Padrón, 1980). The flagellar pocket is the only place within the cell where endo- or exocytosis occurs. Characteristic of some groups of the kinetoplastida is their facultative or obligate assumption of different shapes (polymorphism), so that they may appear as a-, pro-, epi-, or trypomastigotes. Reproduction occurs by an asexual longitudinal binary fission; sexual processes were described in cloned strains, indicating that some sort of genetic exchange must occur at a certain stage (Tait, 1980, 1983). Transmission of the endoparasites belonging to this group mainly occurs via bloodsucking vectors such as biting flies (*Glossina*, *T. brucei*), reduviidae bugs (predatory bloodsucking insects that typically feed on other insects, but in some cases feed on humans; *Triatoma*, *T. cruzi*), or sand flies (*Phlebotoma*, *L. donovani*). Kinetoplastida are divided into two suborders named *bodonina* (containing two flagella) and *trypanosomatina* (containing one flagellum). Table 25.1 shows the members of the latter suborder, of which *Crithidia*, *Leishmania*, and *Trypanosoma* are subjects of this survey. It should be noted that *Crithidia* infects only insects and, though showing polymorphism in nature, grows exclusively in the promastigote (choanomastigote) form in culture (trypanosomatids have a variety of forms during their life cycle, and these vary in the position of the flagellum; in the promastigote it is located anterior of the nucleus and is free from the cell body). Autophagy induced by differentiation is thus negligible under this condition. For handling, all these parasites can be conveniently frozen and stored indefinitely in liquid nitrogen. For freezing, a cell suspension (2×10^8 parasites per ml) is mixed with freezing medium (18 m*M* Na_2HPO_4, 2 m*M* NaH_2PO_4, 5 m*M* KCl, 80 m*M* NaCl, 1 m*M* $MgSO_4$, 20 m*M* glucose, pH 7.7; 20% glycerol) at a ratio of 1:1. Glycerol is used as a cryoprotectant; no other precautions are needed to ensure survival. Cells are frozen in 1 ml aliquots at −20 °C for 1 h, −70 °C overnight and finally in liquid nitrogen. This leads to a cooling down rate of about 1 °C per minute, which is optimal for these parasites.

When needed for infection or cultivation, cells can be treated as follows:

1. After rapid thawing under warm water, cells are transferred to a 15 ml centrifuge tube and washed twice in sterile MEM containing 10 m*M* glucose (5 min at $1500\times g$ each).
2. For cultivation the cell pellet is resuspended in sterile culture medium and cells are immediately seeded in culture vessels at a cell density of 2×10^5 cells per ml. For further cultivation details see 2.2.2.
3. For animal infection the cell pellet is resuspended in sterile CGA (100 m*M* sodium citrate, 40 m*M* glucose, pH 7.7) and (with the exception of *Crithidia* that parasitize arthropods only) parasites are injected *i.p.* into rodent laboratory animals.
4. Cell growth will subsequently be calculated from a hemocytometer count using aliquots of a thoroughly resuspended (i.e., up-and-down pipetting) culture or tail blood conveniently diluted with CGA. Because the parasites are moving, we use a Neubauer chamber with a special depth of 0.02 mm instead of 0.1 mm to avoid continuous focusing. We usually dilute the sample in a way that up to 100 cells are to be counted per square (i.e., 1 mm^2 or 20 nl, respectively.)

2.1. Crithidia

2.1.1. Characteristics of Crithidia

These parasites exclusively infect arthropods, mainly insects, and are passed from host to host as cysts in infective feces. They develop in the digestive tract and may interact with the intestinal epithelium using their flagellum. Host specificity is low and a single parasite can infect a large range of invertebrate hosts. At different stages during their life cycle they appear with amastigote, promastigote, or epimastigote morphology (the flagellum in the amastigote is reduced in size or absent, whereas in the epimastigote it is anterior of the nucleus but connected to the cell body by a short membrane). From the several Crithidia species, *C. fasciculata* is often used as a less advanced parasite in comparison with other trypanosomatids, mainly *T. brucei*.

2.1.2. Cultivation technique

C. fasciculata is very easy to handle *in vitro* as it is rather undemanding and grows axenically similar to bacteria or yeast in submersed culture. Stabilates (i.e., the organisms that were frozen alive) are taken from frozen stocks and diluted into the same medium as used for procyclic trypanosomes (Table 25.2). Following a brief lag phase, the parasite grows logarithmically up to 10^8 cells per ml and can be continuously kept in culture by dilution with fresh medium. Culturing is performed using a volume of 100 ml in 250 m*M* culture flasks at 27 °C without shaking. Generation doubling time is about 6 h.

Table 25.2 Media composition to grow the different forms of *Crithidia*, *T. brucei*, *T. cruzi*, and *Leishmania*.

Basic medium, pH 7.4: (all values in m*M*)
Earle salts: $CaCl_2$ (1.8); KCl (5.4); $MgSO_4$ (0.8); NaCl (116); NaH_2PO_4 (1)
MEM non essential amino acids (1% of stock solution; A, D, N, G, P), MEM essential amino acids (1% of stock solution; R, C, H, I, L, K, M, F, T, W, Y, V)
MEM vitamins (1% of stock solution)
HEPES (30)
adenosine (0.05)
phenol red (0.03)
penicillin/streptomycin (1% of stock solution)

Bloodstream-form medium used for *T. brucei* bloodstream forms	**Additions to procyclic medium used for transformation from *T. brucei* bloodstream to procyclic forms**	**Procyclic medium used for *T. brucei* procyclics, *C. fasciculata*, *L. major* promastigote**
glucose (33)	glucose (10)	proline (5.2)
$NaHCO_3$ (37)	cis-aconitate (3)	glutamic acid (1.2)
cysteine (0.2)	Na-citrate (14)	pyruvate (2)
glutamic acid (1.2)		hemin (0.04)
bathocuproinsulfonate (0.002)	incubation at 27 °C without CO_2 gassing	fetal calf serum (10%)
fetal calf serum (15%)		incubation at 27 °C without CO_2 gassing
incubation at 37 °C in a CO_2 atmosphere (5%)		

2.2. African trypanosomes

2.2.1. Characteristics of trypanosomes

African trypanosomes belong to the so-called salivarian type of parasites, as they are transmitted by saliva, which is released by the tsetse fly during a blood meal to avoid blood clogging. Because of their host specificity, these parasites are divided into several species, of which the human infective *T. brucei* family is the most important. This family contains three members, *T.b. gambiense, T.b. rhodesiense*, and *T.b. brucei*; the former two subspecies cause the chronic or the acute form of sleeping sickness, whereas the latter is not human infective and thus the favored candidate for laboratory use. Here we will deal exclusively with the *Brucei* group. Similar to *Crithidia*, African trypanosomes are extracellular parasites that develop within blood vessels and the lymphatic system and, at a certain time of infection, enter the cerebrospinal fluid (Fig. 25.1). It should thus be easily possible to grow these cells axenically in submersed culture. However, this is true only for the procyclic insect form, not for bloodstream forms.

Considering bloodstream forms, we have to differentiate between pleomorphic and monomorphic strains, where the latter develop by continuous syringe infection from mouse to mouse. These cells adjust to the mammalian host, grow to higher cell densities and gradually lose their ability to develop within the insect vector. Monomorphic strains grow much better in axenic culture and are thus preferred in many laboratories for experimental work.

2.2.2. Growth of trypanosomes

Alternatively to axenic culture, co-cultivation with fibroblast cells is possible and most mono- as well as pleomorphic strains grow nicely under these conditions (Brun *et al.*, 1981; Hirumi *et al.*, 1977). Several media are in use (Hamm *et al.*, 1990; Hirumi and Hirumi, 1989), but we have put some emphasis on the use of the same basic medium (MEM) for all forms of trypanosomes (and the other parasites we work with) and add defined constituents if needed. Thus we use the Earle salt solution (plus HEPES for buffering) as the basic component and invariably add commercially available vitamins and essential as well as nonessential amino acids to obtain a basic medium suitable for all forms of the parasite (see Table 25.2).

To cultivate bloodstream-form parasites, shaking of the culture vessels must be avoided, as these cells do not grow in submersed culture. Instead, to ensure sufficient oxygen supply a surface to volume ratio of about 4 cm^2 per ml is ideal (i.e., a T-25 flask with a surface of 28 cm^2 should be filled with about 7 ml of medium). Bloodstream-form parasites are grown in a CO_2 incubator at 37 °C and 5% CO_2. The main problem in cultivation of bloodstream forms is, however, their inability to take up cystine (Duszenko *et al.*, 1985), a regular constituent of commercially available media, as the reduced form cysteine is not stable and readily oxidized to

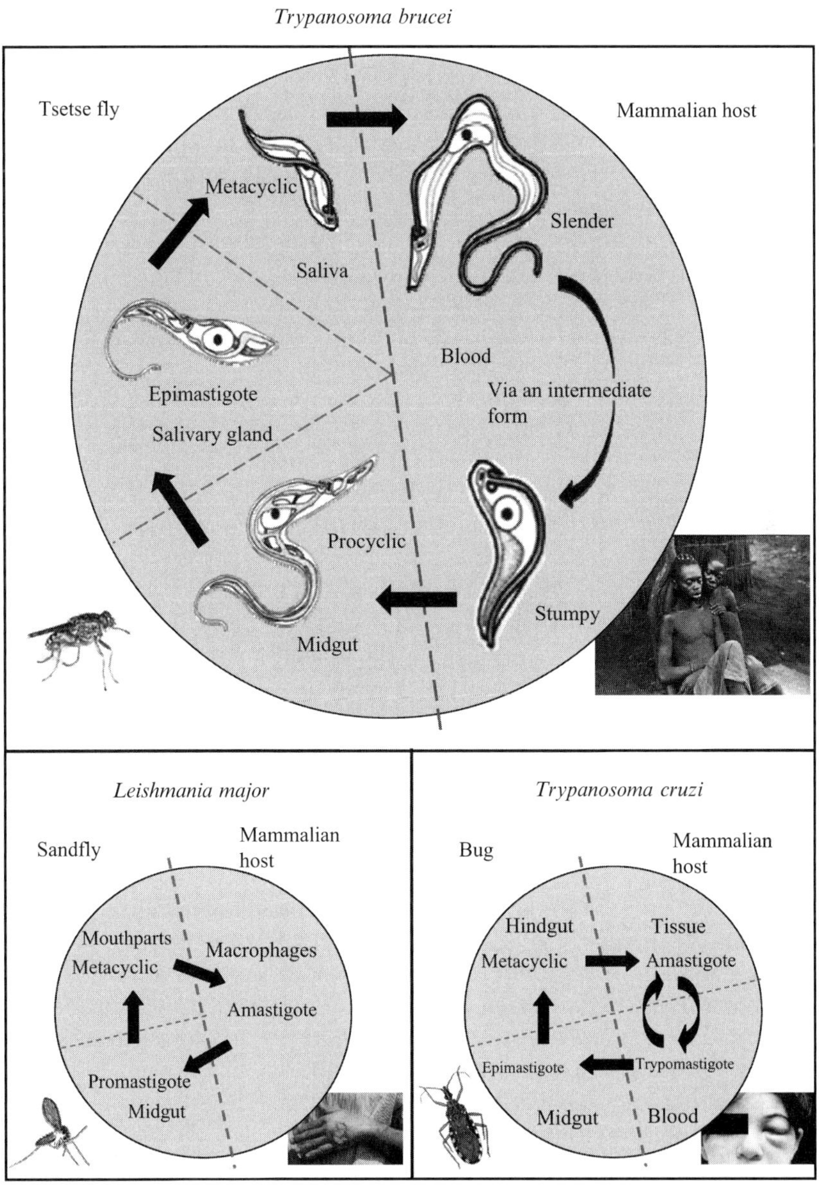

Figure 25.1 Life cycles of *T. brucei*, *T. cruzi* and *Leishmania.*

cystine anyway. Because cysteine is an essential amino acid for bloodstream forms, it must be added one way or another to support cell growth. Straight addition of this amino acid, however, is problematic, as oxidation is catalyzed by ubiquitously dispersed copper ions (e.g., from glassware) and leads to formation of H_2O_2, which kills the parasites easily (Higuchi, 1963). One way to add cysteine to media is the use of feeder cells, which take up cystine and slowly release cysteine (Bannai and Ishii, 1982). Another way is to use bathocuproine sulfonate, a chelating agent for copper ions, which renders cysteine stable (Duszenko *et al.*, 1992). Finally, reducing agents such as β-mercaptoethanol may be used, which are more readily oxidized and thus stabilize cysteine. Likewise to the supplementation with essential amino acids, trypanosomes have a need for purines, which they are unable to synthesize *de novo* (Landfear, 2001). We thus add adenosine to all media, because these parasites have efficient nucleoside transporters and purine salvage pathways are well expressed. It should be noted that bloodstream forms adopt stumpy-like behavior and morphology during the stationary phase in culture, which is due to a trypanosomal-derived differentiation factor (Hamm *et al.*, 1990; Reuner *et al.*, 1997). Growth is thus limited to a maximal density of 5×10^6 trypanosomes per ml which can only be increased by replacement of total medium at regular intervals (Hesse *et al.*, 1995). Continuous culturing of bloodstream forms is only possible if dilution occurs at a cell density below 10^6 cells per ml.

So far, it is possible to cultivate bloodstream forms and procyclic insect forms. For the latter, procyclic medium is used as given in Table 25.2. These cells grow in submersed culture and can be treated as stated for *Crithidia* cells (see previously). In addition, transformation from bloodstream to procyclic forms (but not vice versa) is possible *in vitro*. For this purpose bloodstream form parasites are transferred to 27 °C into a transformation medium containing cis-aconitate and citrate (see Table 25.2), whereby both citric acid metabolites and the temperature shift are necessary to obtain a homogeneous transformation (Brun and Schonenberger, 1981; Overath *et al.*, 1986). This transformation includes an extensive remodeling of cells, especially the activation of the single mitochondrion, which is morphologically and functionally largely reduced in bloodstream forms (Priest and Hajduk, 1994; Vickerman, 1965) and of glycosomes, which obtain additional metabolic functions (Parsons *et al.*, 2001). Autophagic processes are readily observed during transformation and this process is thus well suited to study the involved mechanisms.

Instead of using *in vitro* cultivation, parasites may also be grown *in vivo* in laboratory animals as follows:

1. Rodents can be infected *i.p.* with trypanosomes taken from thawed stabilates or from a bloodstream-form culture. For cloning experiments it is noteworthy that even one single parasite is able to start an infection; however, usually infection rates of about 1×10^5 parasites per animal are more suitable.
2. Parasite counts are determined from tail blood as described in section 2. With a monomorphic strain (e.g., MIT at 1.2) a very high infection rate

of near to 1×10^9 parasites per ml of blood will be obtained within 3–4 days, at which time the animal should be sacrificed.

3. The thorax will be opened and filled with an appropriate volume (about 0.5 ml in mice or 2 ml in rats) of ice-cold citrate-glucose-anticoagulant (CGA).
4. The vena cava inferior is cut and collected blood will be removed and transferred to 4 °C into a centrifuge tube containing additional 5 ml of CGA. The blood system of the animal may be washed by injecting 1 ml (mice) or 5 ml (rats) of ice-cold CGA into the left ventricle of the heart.
5. The collected blood is then centrifuged for 10 min at $1500 \times g$ and the buffy coat on top of the erythrocytes resuspended gently in ~20 ml of separation buffer (57 m*M* Na_2HPO_4, 8.7 m*M* KCl, 3 m*M* KH_2PO_4, 120 m*M* NaCl, pH 8.0).
6. The resuspended solution is then run through an anion exchange column (DEAE cellulose). We use DEAE Sephacel (GE Healthcare), which is spherical and allows a higher flow-through rate. Because of the surface coat of bloodstream forms (Borst and Ulbert, 2001; Richards, 1984), which is uncharged at pH 8, trypanosomes appear in the flow through while all contaminating blood cells will be bound to the column material. Isolated trypanosomes (up to 1×10^9 (mice) or 1×10^{10} (rats) parasites) can than be used for further experiments.

2.3. American trypanosomes

2.3.1. Characteristics of *Trypanosoma cruzi*

Here, we talk solely about *Trypanosoma cruzi*, the causative agent of Chagas' disease. This species causes trypanosomiasis diseases in humans and animals in America. Transmission occurs when the reduviid bug deposits feces on the skin surface and subsequently bites; the human host then scratches the bite area and facilitates penetration of the infected feces (Fig. 25.1). Chagas' disease is a potentially fatal disease of humans. It has two forms, a trypomastigote (the flagellum is posterior of the nucleus and connected to the cell body by a long membrane) found in blood and an amastigote found in tissues. The acute form usually goes unnoticed and often appears as a localized swelling where the parasites enter the skin. The chronic form may develop many years after infection. This form affects internal organs (e.g., heart, esophagus, colon, peripheral nervous system). Affected people often die from heart failure.

2.3.2. Growth of *T. cruzi*

Amastigote and trypomastigote forms are grown by co-cultivation with irradiated muscle cells (L_6E_9 myoblasts) as described by Moreno *et al.* (1992). The following procedure may be used for transformation of trypomastigotes into amastigotes:

1. 10^8 cells are washed once with DMEM (Dulbecco's minimal essential medium, pH 7.5), supplemented with 0.4% BSA and spun for 5 min at $1500\times g$ in a 10-ml centrifuge tube.
2. After adjusting the cell density to 5×10^7 cells per ml in DMEM, the pH is lowered to 5 by buffering with 20 m*M* MES (morpholinoethanesulfonic acid).
3. The parasites are incubated at 35 °C in a nonaerated incubator until differentiation is completed (Furuya *et al.*, 2000; Tomlinson *et al.*, 1995).

The epimastigote form of *T. cruzi* is cultivated in liver tryptose medium (0.5% NaCl, 0.75% Na_2HPO_4, 0.5% tryptose, 0.3% yeast extract, 0.3% liver infusion broth, 0.04% KCl, pH 7.2), containing 5% fetal calf serum at a temperature of 28 °C (Fang *et al.*, 2007).

2.4. Leishmania

2.4.1. Characteristics of *Leishmania*

In general, *Leishmania* ssp. can cause visceral and cutaneous infections in their mammalian hosts. In contrast to trypanosomiasis, leishmaniasis is spread over many countries all over the world, especially in South America, Africa and Asia. The parasite is transmitted by the sand fly, belonging to the family of the *Psychodidae*. After a blood meal the parasite transforms in the digestive tract into the promastigote form. Entering the mouthparts of the insect, promastigotes differentiate into the infectious but nondividing metacyclic form. When this form infects the bloodstream of the mammal, they infiltrate macrophages. The emerging amastigote parasite multiplies and eventually disrupts its host cell infecting new ones (Fig. 25.1). At the best, this process ends up in ulcerating cutaneous lesions. However, in the case of the visceral form, the disease leads to anemia and severe damage of some organs, especially the liver and spleen.

2.4.2. Growth of *Leishmania*

Promastigotes can be cultivated as described for the procyclic form of *T. brucei* (Table 25.2). They can be grown axenically if they are diluted every 2–3 days to a cell density of about 2×10^5 cells per ml; in this way the parasite keeps on multiplying logarithmically with a generation doubling time of ~12 h. Promastigotes reach the stationary phase at a cell density of 2×10^7 cells per ml and then exhibit all the markers of the metacyclic form (Berberich *et al.*, 1998; Rogers *et al.*, 2002; Sacks and da Silva, 1987). Additionally, metacyclogenesis can be initiated by lowering the pH (Bates and Tetley, 1993). Metacyclics are transformed into the amastigote form by transferring them into UM54 medium and incubation for four days at 35 °C at pH 5.5, thus simulating the conditions within the parasitophorous vacuole, as described in Schaible *et al.* (1999). For axenic cultivation and characterization of the amastigote form of the respective *Leishmania* ssp., see P.A. Bates (1993).

3. Autophagy in Protozoa

3.1. Microautophagy

Microautophagy seems to be a continuous process in any eukaryotic cell to have a constant turnover of cytosol/cytoplasm. The most obvious morphological difference between macro- and microautophagy is the form of sequestration. While macroautophagy involves formation of a cytosolic double-membrane vesicle that picks up cellular materials and eventually fuses with the lysosome, microautophagy describes the direct invagination or embracing protrusions of the lysosome to take up parts of the cytosol or even cytoplasm. At least invagination of the lysosome occurs also in protozoa. As known so far, this process seems to be nonspecific, though the molecular details of the process have yet to be explored.

3.2. Starvation

One of the basic functions of autophagy is obviously to cope with periods of starvation by degrading disposable cellular elements such as biomacromolecules or organelles. As published recently, a rather limited number of orthologs of autophagy-related proteins exist in *Trypanosoma* and *Leishmania*, compared to the some 31 *ATG*-genes in yeast (Herman *et al.*, 2006; Rigden *et al.*, 2005). Nevertheless, the respective genomes contain the key proteins involved in autophagy including regulation and signaling pathways. One of the very early proteins is the phosphatidylinositol 3-kinase TOR (target of rapamycin), which is normally activated by nitrogen and leads to an induction of cell growth, cell cycle progression, and protein biosynthesis in higher eukaryotes. Rapamycin, a macrolide isolated from *Streptomyces hygroscopicus*, can form a complex with FKBP (FK506-binding protein), which then binds to TOR. As a result, the kinase is inhibited and autophagy is initiated. As determined in our laboratory, the IC_{50} value of rapamycin in the *T. brucei* bloodstream form is 6.5 μM (data not published). At this concentration, rapamycin induced cell death without any signs of typical apoptotic or necrotic markers (Fig. 25.2). Instead, autophagic vacuoles were observed by fluorescence microscopy using monodansylcadaverine staining. In these cells electron microscopy confirmed the formation of autophagic organelles, from the stage of phagophores onward, including autophagosomes with the typical double bilayer, up to autolysosomes (Figs. 25.3–25.6). In addition, densely packed membrane clusters (DPMC) occurred associated with the mitochondrion near the flagellar pocket (Figs. 25.3–25.6). Formation of DPMC was also observed in cells starved for 3–4 h in amino acid-free TDB (starvation medium), but not under several other stress conditions or during the differentiation processes (Figs. 25.3–25.6). Similar structures have been described for *T. cruzi* epimastigotes (Alvarez *et al.*, 2008) after starvation for 16 h in PBS. Here, this autophagic organelle showed a high concentration of the ubiquitin-like protein ATG8.1 as judged by immunofluorescence. Mutation of the glycine residue at the C terminus abolished accumulation, thus indicating that processing of this protein by the cysteine peptidase ATG4 and coupling of the respective glycine residue to phosphatidylethanolamine (PE) of defined membranes not only occurs in higher eukaryotes but also is part of the autophagic pathway in kinetoplastids. It should be kept in mind, however, that two classical inhibitors of autophagy, wortmannin and 3-methyladenine, could not abolish autophagosome formation at concentrations used in yeast. In Leishmania a 1000-fold higher concentration was needed to observe inhibition of autophagy, while trypanosomes died because of the toxicity of the drugs before effects on autophagy could be observed (Williams *et al.*, 2006).

3.3. Stress response

The differences between the known autophagy-inducing triggers and their molecular consequences are not well defined and overlapping. The concentration of H_2O_2 is somehow of central importance, as it rises because of nutritional stress conditions (Scherz-Shouval *et al.*, 2007), in case of apoptotic events, and after appearance of stress-inducing compounds. With regard to

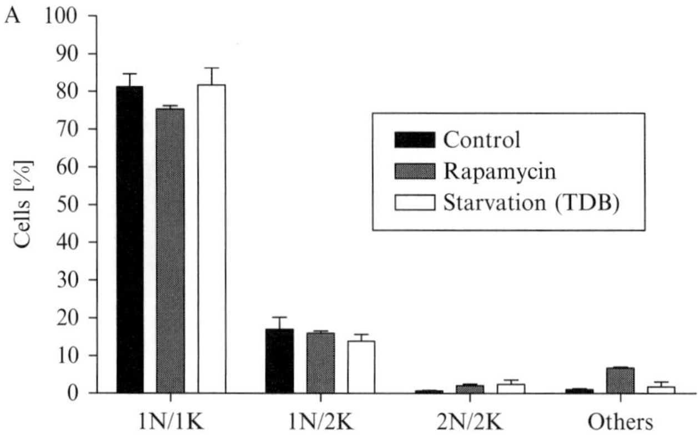

Figure 25.2 (Continued)

B

Control

Sub	10.91
G1	28.20
S	12.90
G2	41.83
Poly	10.08

Rapamycin

Sub	11.50
G1	22.97
S	11.51
G2	37.43
Poly	20.25

Starvation

Sub	5.82
G1	33.05
S	16.64
G2	38.83
Poly	11.37

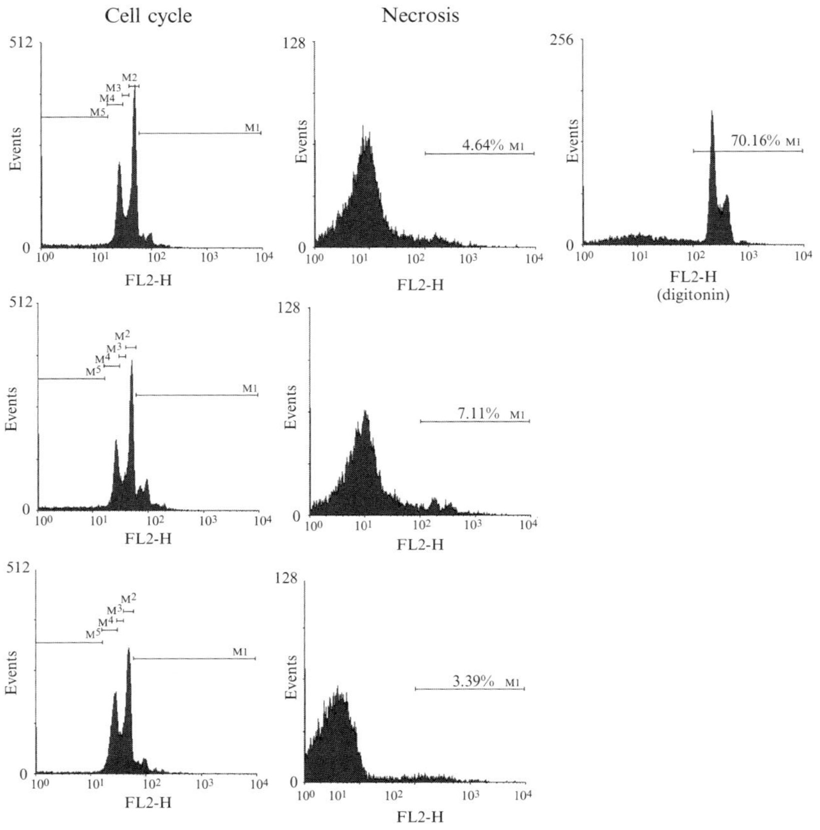

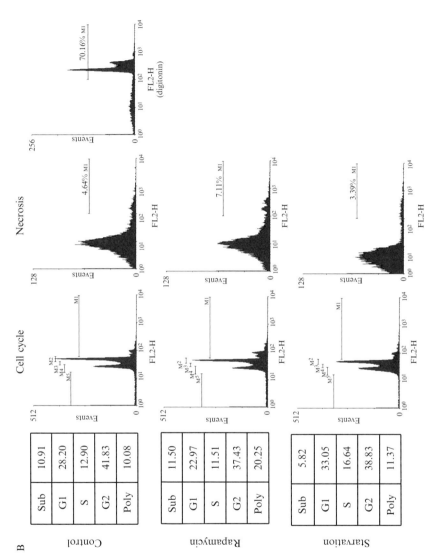

Figure 25.2 (Continued)

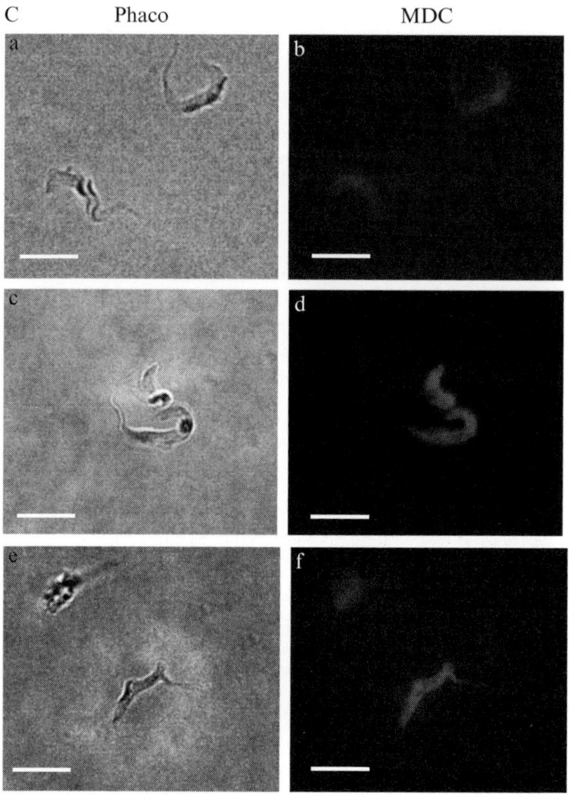

Figure 25.2 Light microscopy and FACS analysis. (A) Indicated is the proportion of cells in different cell cycle phases out of three independent experiments after bisbenzimide staining of their nucleus and kinetoplastid. N = nucleus, K = kinetoplast. 1N/1K corresponds to G1-phase, 1N/2K and 2N/2K correspond to cells at different stages of cell division. (B) FACS analyses. Measurement of the cells' DNA-content with propidium iodide revealed no cell cycle arrest or DNA-degradation. Rapamycin induced a slight increase of polynucleated cells, which could be verified by electron microscopy. Under all conditions no necrosis could be observed by propidium iodide staining of the whole cells. Digitonin treatment served as a positive control. (C) Monodansylcadaverine (MDC) staining of autophagic vacuoles. (A, B) control cells (C, D) rapamycin 6.5 μM (E, F) starved cells. Bar = 10 μm. Phaco, phase contrast. (See Color Insert.)

T. brucei, hydrogen peroxide is produced in bloodstream forms during prostaglandin-induced apoptosis (Figarella *et al.*, 2005, 2006). However, adding H_2O_2 to culture media at IC_{50}-concentrations ($\sim$35 μM), other events predominate. Most obvious is the damage of intracellular structures, which eventually leads to necrosis. Nevertheless, as visualized by transmission electron microscopy (Figs. 25.3–25.6), parasites try as long as possible to dismantle impaired structures by autophagy to survive.

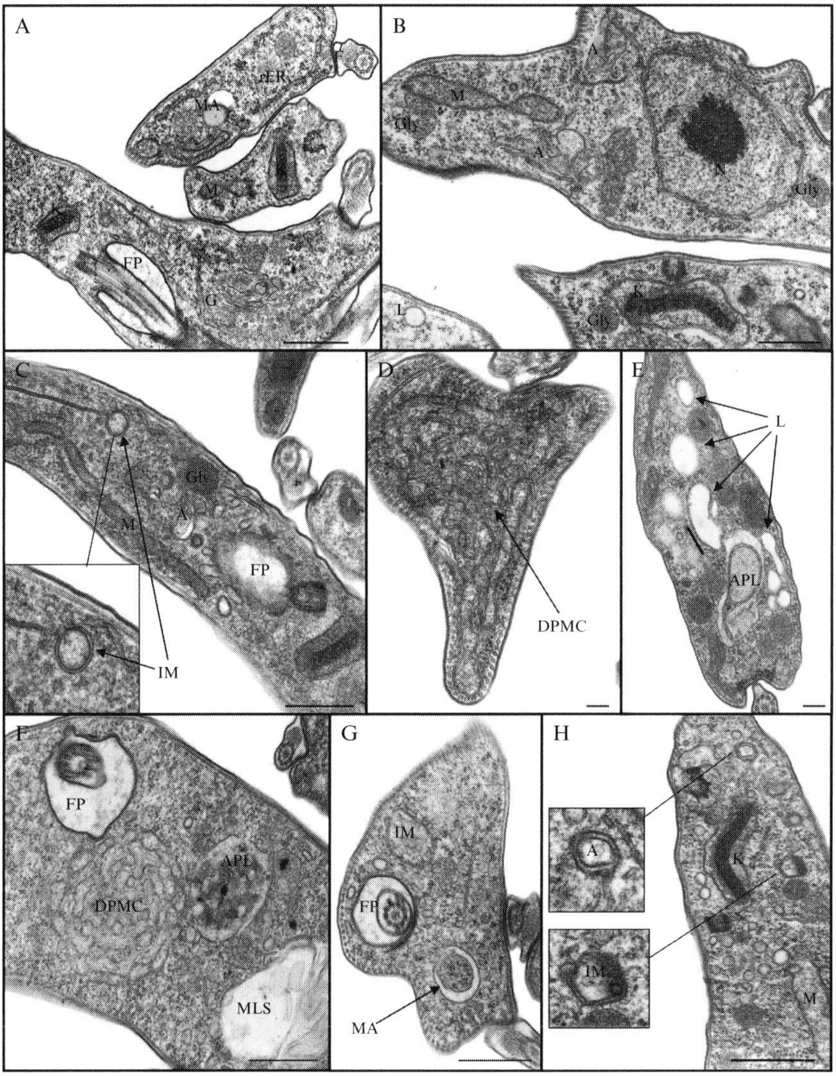

Figure 25.3 Electron micrographs of *T. brucei* bloodstream forms. (A, B) Untreated control cells from a 24-h-old culture; (C, D, E) cells from a 24-h-old culture, starved for the last 3 h in TDB; (F, G) cells from a 24-h-old culture treated with 6.5 μM rapamycin. Please refer to the text for further information. Abbreviations used to label electron micrographs: Figs. 25.3–25.6: A: autophagosome; AC: acidocalcisome; APL: autophagolysome; DL: digestive lysosome; DPMC: densely packed membrane cluster; F: flagellum; FP: flagellar pocket; G: Golgi apparatus; Gly: glycosome; IM: isolation membrane (phagophore); K: kinetoplast; L: lysosome; LD: lipid drop; M: mitochondrion; MA: microautophagy; MLS: multilamellar structure; N: nucleus; rER: rough endoplasmic reticulum.

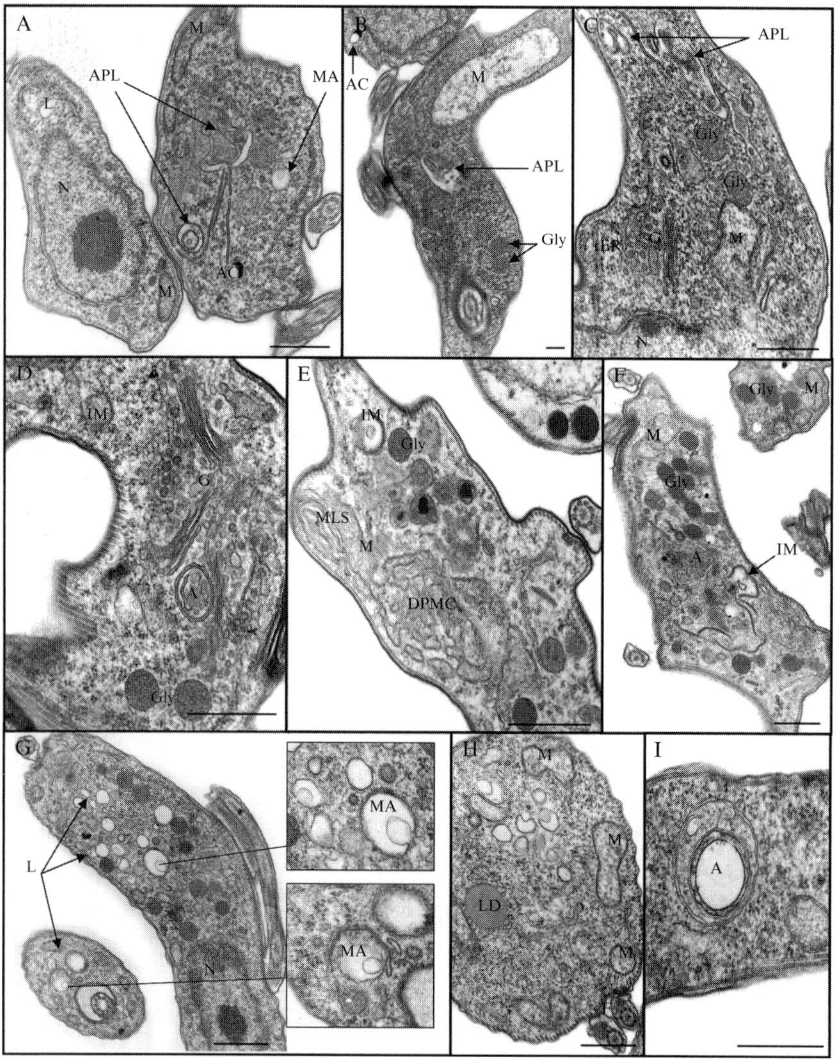

Figure 25.4 Electron micrographs of *T. brucei* bloodstream forms under stress conditions. (A, B, C) cells from a 24-h-old culture treated with 100 μM H_2O_2 for the last 3 h; (D, E, F) cells from a 24-h-old culture treated with 3 mM dihydroxyacetone: (G, H, I) cells were kept for 24 h in transformation medium. Please refer to the text for further information.

Besides reactive oxygen species, other compounds are shown to induce autophagy in trypanosomes. For example, dihydroxyacetone (DHA) is taken up very well by the parasite's aquaglyceroporins (Uzcategui *et al.*, 2004). Because DHA phosphate is an intermediate of glycolysis, DHA can be used as a carbon source at low concentrations due to a side activity of glycerol

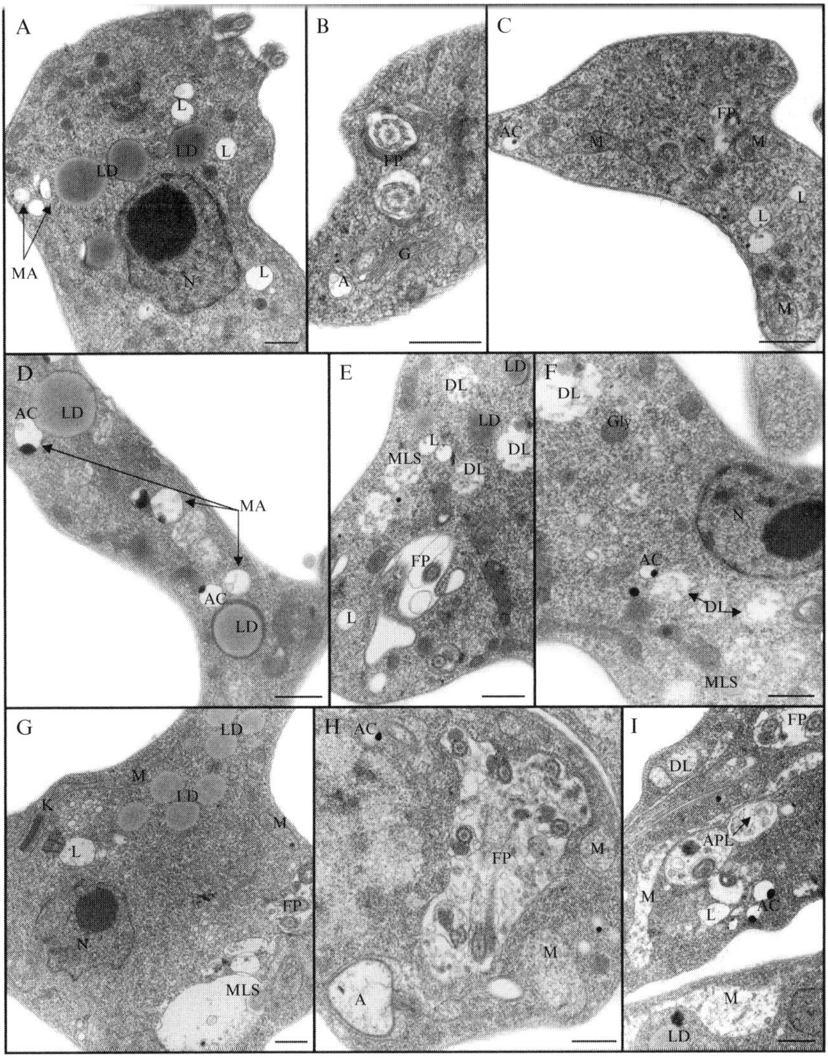

Figure 25.5 Electron micrographs of *T. brucei* procyclic forms. (A, B, C) untreated control cells from a 24-h-old culture; (D, E, F) cells from a 24-h-old culture, starved for the last 3 h in TDB; (G, H, I) cells from a 24-h-old culture treated with 100 μM H_2O_2 for the last 3 h. Please refer to the text for further information.

kinase. However, increasing amounts of DHA lead to a cell cycle arrest in the G_2-phase and formation of autophagic and multilamellar structures. FACS-analyses performed in the presence of different DHA-concentrations (1–4 mM) reveal that in this case autophagic events go along with an increase in cell membrane permeability, formation of ROS and phosphatidylserine-exposure (Uzcategui *et al.*, 2007a,b). This supports the observation by us and

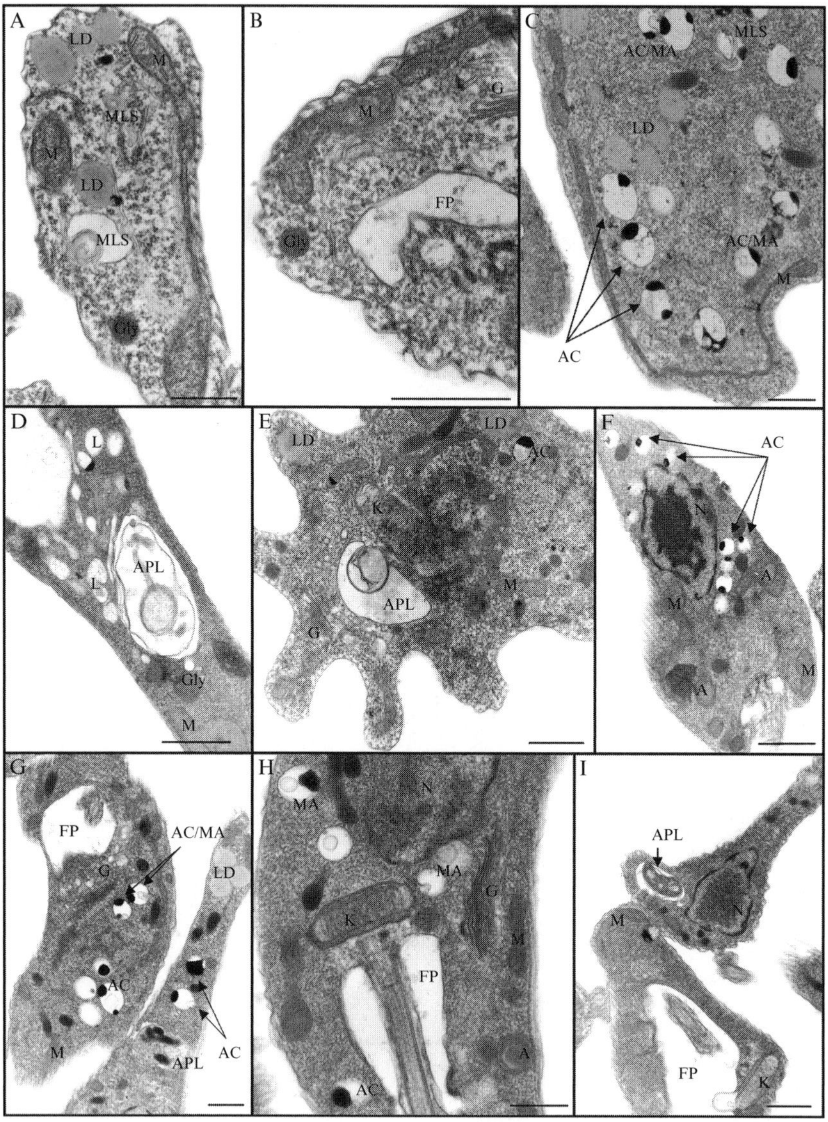

Figure 25.6 Electron micrographs of *Crithidia fasciculata*. (A, B, C) untreated control cells from a 24-h-old culture; (D, E, F) cells from a 24-h-old culture, starved for the last 3 h in TDB; (G, H, I) cells from a 24-h-old culture treated with 100 μM H_2O_2 for the last 3 h. Please refer to the text for further information.

by others that autophagy is not a single, isolated and restricted event but is interconnected with necrosis and apoptosis.

Induction of autophagy is also observed in *Leishmania donovani*, the causative agent of visceral leishmaniasis, by treatment with antimicrobial peptides in the low pM range (Bera *et al.*, 2003). Using monodansylcadaverine staining and transmission electron microscopy, autophagic bodies are readily detected. Indeed, quite impressive ultrastructural changes appear in both, amastigotic and promastigotic forms of *Leishmania amazoniensis* after addition of 0.1–10 μM azasterol (Rodrigues *et al.*, 2002). This compound interferes with endogenous sterol biosynthesis, thus leading to a change in the phospholipid composition of the parasite's membranes. Interestingly, this treatment leads to various autophagic structures, multilamellar bodies, an increasing number of acidocalcisomes, and invaginations of swollen mitochondria.

In conclusion, there are numerous substances probably interfering with the homeostasis of certain metabolic and catabolic pathways of the parasites and leading to cellular stress. The cellular answer involves autophagy for the generation of necessary substances, a very simple but effective way to ensure a longer survival.

3.4. Cell differentiation and homeostasis of organelles

Autophagy during cell differentiation is a common and physiological event. Especially in kinetoplastids, each differentiation step goes along with morphological and metabolic rearrangements to adapt to the diverse environmental conditions concerning the mammalian host and the invertebrate. First hints that autophagy could be necessary for these changes were provided by transmission EM-pictures of *T. congolense*, published by Vickerman in 1969.

T. brucei bloodstream forms possess a single mitochondrion that contains only a few christae and none of the classical metabolic pathways such as β-oxidation, the citric acid cycle, or a respiratory chain. Their energy metabolism depends exclusively on glycolysis, which, as a singularity for the order kinetoplastida, is located within peroxisome-like organelles called *glycosomes*. After being taken up by the tsetse fly, trypanosomes are exposed to a lower temperature and have to cope with low and variable glucose-concentrations. They thus develop a fully equipped mitochondrion and change the enzymatic composition of their glycosomes (Vertommen *et al.*, 2008). Among several possibilities for switching from one glycosome population to another, enzymes could be specifically degraded and replaced by others imported into the organelles, or old glycosomes could be completely disintegrated while new ones are built, resulting in a gradual shift. Current work suggests the latter, the so-called glycophagy (Herman *et al.*, 2008; Rigden *et al.*, 2005) during the differentiation from slender to stumpy bloodstream forms, and during transition from bloodstream to the procyclic insect form. Using electron and fluorescence microscopy, the authors observed a certain population of these organelles to be closely associated with lysosomes, implicating microautophagic events. Apart from glycophagy, transmission EM of stationary phase trypanosomes (development of stumpy forms) and parasites during *in vitro* bloodstream to procyclic transition at least confirms a role of autophagy in differentiation.

It has been shown for both *T. brucei* and *T. cruzi*, that high cell density (Hamm *et al.*, 1990; Hesse *et al.*, 1995; Reuner *et al.*, 1997) and nutritional stress (Figueiredo *et al.*, 2000) cause differentiation. Even if the latter trigger has no physiological relevance, stress response and differentiation obviously share common pathways. Within the insect vector of *T. cruzi*, the reduviidae bugs, parasites develop from the epimastigote to the infective metacyclic form. This metacyclogenesis was demonstrated to be triggered by nutritional stress (Figueiredo *et al.*, 2000) and involves autophagy, as demonstrated by ATG8.1 fluorescence staining of autophagosomes (Alvarez *et al.*, 2008). Interestingly, the mammalian life cycle stages (amastigotes and trypomastigotes) as well as the noninfective epimastigote form express ATG8.1 but not metacyclics. In contrast to ATG8.1, the other Tc-ortholog (ATG8.2) cannot substitute for the yeast Atg8, so that the function of this protein is still unclear. On the other hand, both orthologs of the peptidase ATG4 are expressed during all life cycle stages and are both able to complement an *atg4* Δ strain of *Saccharomyces cerevisiae*.

As a perfect marker for autophagic events and quite easy to monitor, Besteiro *et al.* (2006) and Williams *et al.* (2006) used the ATG8 pathway for their investigations on the relationship of autophagy and differentiation in different *Leishmania* species. By fusing GFP to the N terminus of uncleaved ATG8 and following its usual track, they could show that the respective orthologs are indeed involved in formation of autophagosomes. Whereas the green fluorescence was equally distributed throughout the cytosol in logarithmic phase promastigotes, GFP-ATG8 accumulated to intensive green spots during differentiation to metacyclics. These conglomerates disappeared after successful transformation to the infective metacyclic form and ended up in the MVT (multivesicular tubule)-lysosome, a specific organelle of this cell-cycle-arrested stage. Likewise, Western blot analyses revealed that the uncleaved ATG8 was predominant in promastigotes but changed to the PE-coupled form during differentiation. In the resulting metacyclic parasites, the cysteine peptidase ATG4 cleaved ATG8 once more, leaving behind the soluble ATG8 (free at the C-terminal glycine-residue) form and the PE-linked membrane-bound form. Obviously, autophagy is involved not only in differentiation of the insect forms but also in transformation into the intracellular amastigote form in macrophages. After 18 h differentiation from metacyclic promastigotes into amastigotes, cells showed an increase of fluorescent organelles with a high amount of lipidated ATG8, declining after 72 h (Williams *et al.*, 2006). Additionally, Besteiro *et al.* (2006) could demonstrate the importance of LmATG4.2, one of the *L. major* ATG4-orthologs, for cleavage of ATG8 and its linkage to PE. Knockout of the respective gene resulted in accumulation of GFP-ATG8-labeled autophagosomes, because ATG8 remained lipidated and thus bound to the organelle. This mutant was also unable to undergo metacyclogenesis, as was true for *Leishmania mexicana* knockout mutants of two other cysteine peptidases, CPA and CPB. The latter are potentially involved in degradation of the engulfed proteins within the autolysosomes.

In general, accumulation of autophagosomes can be induced in several ways but always leads to impairment of autophagy, cell growth, and differentiation. Mutation of VPS4 to an ATP hydrolysis-defective protein or use of the ATPase inhibiting compound bafilomycin A_1 leads to a defect in escorting autophagosomes to lysosomes and fusion of both organelles. In addition, inhibition of microtubule formation or degradation using vinblastine or paclitaxel, respectively, resulted in cell death, because this network is obviously necessary for movement of these organelles (Williams *et al.*, 2006).

4. Analysis of Autophagy in Kinetoplastida

4.1. Light microscopy and FACS analysis

4.1.1. Fluorescence microscopy

Rapamycin treatment of various cells such as yeast (Chung *et al.*, 1992), myoblasts (Jayaraman and Marks, 1993), and T-lymphocytes (Morice *et al.*, 1993) leads to a cell cycle arrest especially in G_1/S-phase. Because of the cell cycle progression by binary fission, in trypanosomatids the number of nuclei and kinetoplasts within one cell are indicative of the respective cell cycle phase (Woodward and Gull, 1990); therefore, inhibition of proliferation can be easily visualized.

1. Cells are subjected to nitrogen starvation for 3 h in TDB (18 m*M* Na_2HPO_4, 2 m*M* NaH_2PO_4, 5 m*M* KCl, 80 m*M* NaCl, 1 m*M* $MgSO_4$, 20 m*M* glucose, pH 7.7) or treated with 6.5 μM rapamycin for 24 h in culture medium.
2. 10^6 cells of each sample are spun down in a 1.5 ml microcentrifuge tube at $10{,}000\times g$ for 20 s, and washed twice with 1 ml of PBS (phosphate buffered saline, pH 7.4) each.
3. Cells are resuspended in 200 μl of PBS and fixed by adding 300 μl of 4% paraformaldehyde/0.1% glutaraldehyde at 4 °C overnight.
4. Cells are stained with 10 μg/ml bisbenzimide for 10 min.
5. To monitor the number of nuclei and kinetoplasts in each cell, at least 100 cells have to be counted.

Fig. 25.2 shows the results of three independent experiments. In contrast to published data of other organisms, in *T. brucei* bloodstream forms no cell cycle arrest was observed in the G_1-phase, neither after rapamycin treatment nor after 3-h starvation in TDB. Rapamycin-treated cells showed a slight increase of multinucleated cells, which obviously had some problems in cell division. Nevertheless, electron microscopy revealed a significant increase in autophagy under both conditions (Fig. 25.3).

4.1.2. FACS analyses

Autophagy can be linked to apoptotic as well as necrotic events (Maiuri *et al.*, 2007), and though each type of cell death has its own and specific markers, they often overlap. For this reason the DNA-content of the nuclei is measured by FACS analysis using propidium iodide, which stains the nuclei of dead or necrotic cells due to a plasma membrane disruption.

1. 10^6 cells are collected by centrifugation at 10,000×*g* for 20 s and the supernatant fraction is discarded.
2. Hypotonic lysis of the cell pellet is performed in 100 μl of phosphate buffer (10 m*M*, pH 7.4, containing 100 ng/ml RNase and 64 μ*M* digitonin) and monitored by light microscopy.
3. After addition of the same volume of propidium iodide dissolved in PBS (20 μg/ml), samples are transferred immediately into FACS tubes for measurements (red channel FLH-2; 20,000 counts).

The resulting histograms showed no significant changes of the cell population in the respective cell cycle phases and the number of polynucleated cells increased only slightly after rapamycin treatment. In contrast to apoptosis, no DNA degradation could be observed.

For detection of necrosis, cellular permeability is assayed using propidium iodide. Rapamycin treated, starved and control cells are incubated for 10 min each at room temperature in PBS containing 0.5 μg/ml propidium iodide and transferred into FACS-tubes for measurements. Digitonin (6 μ*M*) is used as a positive control to induce necrosis. In this way, necrotic events, that could accompany autophagy in *T. brucei* under starvation, can be excluded.

4.1.3. Staining of autophagic vacuoles

Studies of Biederbick *et al.* (1995) have shown that the autofluorescent dye monodansylcadaverine selectively stains late autophagic vacuoles but not endocytic acidic compartments or autophagosomes before fusion with the lysosome.

1. 10^6 cells are spun down at 10,000×*g* for 20 s in 1.5 ml microcentrifuge tubes.
2. The cell pellet is resuspended in 100 μl of TDB (pH 7.7).
3. After addition of 5 μ*M* monodansylcadaverine (dissolved in DMSO), samples were incubated at 27 °C for 30 min.
4. The cells are then pelleted and washed twice with TDB before they were subjected to fluorescence microscopy. As in other cells, a slight blue background staining can be observed, but the increase of fluorescent autophagolysosomes in treated and starved cells should be striking.

4.2. Analysis by electron microscopy

Characteristic autophagic structures are easily recognized in electron micrographs, so that transmission EM is a very well established tool to investigate autophagy on the subcellular level.

1. 10^8 parasites are harvested by centrifugation (1500 × g for 5 min) after the respective treatment and washed twice with TDB (pH 7.7).
2. The cell pellet is resuspended in 500 μl of cacodylate buffer (100 m*M* sodium cycodylate, 120 m*M* sucrose; pH 7.4) and fixed by adding 500 μl of glutaraldehyde (2% (v/v) diluted in the same buffer) for 1 h at 4 °C.
3. After washing 4 times in 1 ml cacodylate buffer (10 min each) and storage overnight, cells are postfixed in osmium tetroxide (1.5% (wt/v) in cacodylate buffer), washed once in buffer, and twice in distilled water. Staining is done with uranyl acetate (5% (wt/v) in distilled water).
4. After washing once, cells are dehydrated in ethanol (1 × 50%, 1 × 70%, 1 × 95% and 3 × 100%), cleared in propylene oxide and embedded in Agar 100 according to standard procedures (Hirsch and Fedorko, 1968).
5. Polymerization is performed in Beem cups for 12 h at 45 °C and 24 h at 60 °C. Ultrathin sections are transferred onto grids, stained this time with lead citrate (1mg/ml in distilled water) for 45 s and finally washed with distilled water.

The results are shown in Figs. 25.3–25.6. Please refer to the figure legends for details. We here present *T. brucei* bloodstream and procyclic insect forms as well as *C. fasciculata* cells. All these organisms have been starved and stressed, and bloodstream-form trypanosomes also differentiated to procyclic forms. Under all these circumstances autophagy was readily detected and characteristic examples are shown in Figs. 25.3–25.6. It should be noted that besides the well-known structures of phagophores (isolation membranes), microautophagy, autophagic bodies, autophagolysosomes and multilamellar structures, the trypanosomatids frequently but not under all autophagy-inducing circumstances show densely packed membrane clusters (DPMC), which obviously originate from the mitochondrion (Fig. 25.4E). These structures have also been observed in *Leishmania* and *T. cruzi* (Rodrigues *et al.*, 2002; Santa-Rita *et al.*, 2005), but so far no function has been assigned. Indeed, we often see morphological changes in the mitochondria, usually a dilation of the organelle, but how in trypanosomes this structure is involved in autophagy has yet to be analyzed.

4.3. Bioinformatic analysis of involved proteins

Because more and more genomic data of several kinetoplastida became available during the past few years, one way to identify potential protozoan autophagy-related (ATG) genes is to use a sequence alignment approach.

In this way it was shown that only about half of the autophagy-related proteins found in yeast have orthologs in trypanosomatids (Herman *et al.,* 2006). According to this work it seems that in trypanosomatids some autophagy-related processes are missing completely, such as the cytoplasm-to-vacuole targeting (Cvt) system, or partially, as is the case for most of the components used for the ATG12 pathway. These results suggest that either autophagy is only rudimentarily developed in these cells or they use proteins that are largely different from their yeast counterparts.

Fig. 25.7 lists trypanosomatid proteins that exist in all three species, *T. brucei, T. cruzi* and *L. major* as clear orthologs (predicted by jaccard cog clustering at the Sanger GeneDB) and show significant homology to respective yeast proteins (OmniBLAST at www.genedb.org). For the sake of clarity, only the GeneDB IDs of *T. brucei* are shown.

As an example of a relatively well conserved autophagy-related protein, the yeast Atg8 and its homologs in trypanosomatids are shown in a multiple sequence alignment (Fig. 25.8). *T. brucei* has three homologs, ATG8A, ATG8B, and ATG8.2, of which the first two are nearly identical in sequence. Interestingly ATG8A and ATG8.2 end with a well-conserved glycine residue, which usually becomes the C terminal amino acid only after maturation of this protein by the cysteine peptidase ATG4. This latter step may be omitted here and membrane association depends only on ATG3 and ATG7. On the other hand, ATG8B (which is the closest homolog to the yeast protein) contains one additional amino-acid residue at its C terminus and could thus be the only expressed form of ATG8 in trypanosomes. This would correspond to the results obtained by Alvarez *et al.* (2007), who showed that the respective ortholog in *T. cruzi* is the major functional gene in this species.

According to Herman *et al.* (2006), the ATG12 pathway seems to be completely absent in trypanosomes. However, there is some indication that at least some proteins of this pathway may exist. For example there is a *T. brucei* protein annotated as ATG16 (Tb11.02.4790) in the GeneDB, which shares homology with the autophagy-related protein 16–1 in *Mus musculus* (SwissProt ID Q8C0J2) but not with the respective yeast protein. Furthermore a protein in GeneDB of *T. brucei* exists annotated as ATG5 (Tb927.6.2430), which shows weak homology to the yeast counterpart and is said to have some possible involvement in autophagy. Interestingly, the yeast ATG12 shows a weak homology to ATG8A (Tb927.7.5900) of *T. brucei*.

Because the genomic data for *Crithidia* is still in the sequencing and annotation process we did not include this trypanosomatid in this analysis. But according to personal communication with Dr. Andrew Jackson (Wellcome Trust Sanger Institute), the fact that most of the autophagy-related genes are conserved in the same genomic position in *T. brucei, T. cruzi* and *L. major* makes it highly likely that *C. deanei* or any other *Crithidia* will have homologs to these genes too.

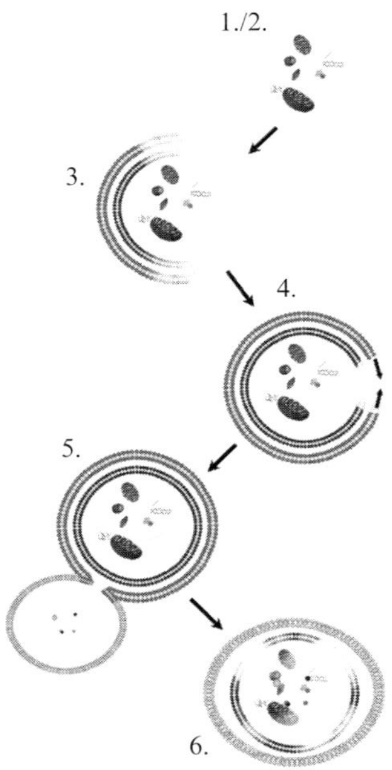

Process	Protein	*T.brucei geneDB ID*	Annotation	Yeast homolog (GI)
1. Induction/initiation	TOR1	Tb10.6k15.2060	PI-3 kinase	6322525
	TOR2	Tb927.4.420	PI-3 kinase	42759850
	ATG1	Many homolog	Serine/threonine kinase	6321258
2. PAS formation	ATG9	Tb11.03.0130	Hypothetical protein	6320052
	ATG18	Tb927.3.4150	Hypothetical protein	16740527
3. Vesicle nucleation	ATG6	Tb927.5.3270	PI3-kinase	6325137
	VPS15	Tb11.01.0930	Protein kinase, putative	536462
	VPS34	Tb927.8.6210	PI3-kinase	6323269
4. Vesicle expansion and completion	ATG3	Tb927.8.6210	E2-like ubiquitin-conjugation enzyme	6324334
	ATG4.1 ATG4.2	Tb11.01.7970 Tb927.6.1690 }	Cysteine peptidase	37362688
	ATG7	Tb10.26.0840	E1-like ubiquitin-activating enzyme	458899
	ATG8B ATG8A ATG8.2	Tb927.7.5910 Tb927.7.5900 Tb92.7.7.3320 }	Ubiquitin-like protein	6319393
5. Docking and fusion		No specific autophagy-proteins		
6. Vesicle breakdown	ATG15	Tb927.4.3170	Lipase	10383802

Figure 25.7 Macroautophagy/autophagy processes and corresponding *T. brucei* orthologs of yeast autophagy-related proteins showing significant homology to respective yeast proteins (Edgar, 2004).

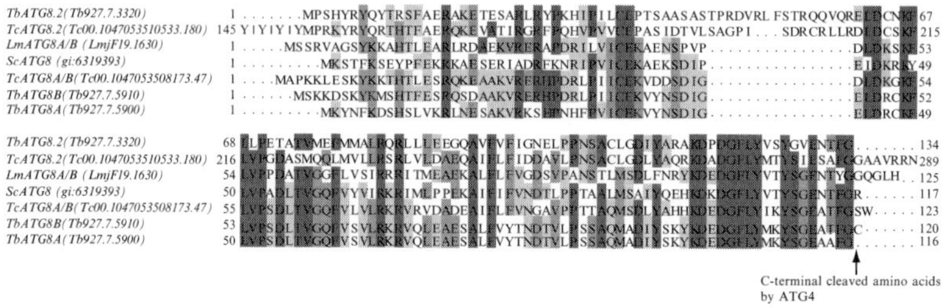

Figure 25.8 MUSCLE-Alignment of yeast Atg8 together with trypanosomastid orthologs (www.ebi.ac.uk) visualized with JalView, Shading: percentage identity. TcATG8.2 is shown from amino acid 145 on (Clamp *et al.*, 2004).

At present, one can definitely argue that there is a core of conserved autophagy-related proteins in trypanosomatids (see Fig. 25.7), but for many of the yeast proteins no true homologs could be convincingly identified, and it requires more experimental data to verify the different pathways.

4.4. Detection and intracellular localization of involved proteins and mRNAs

Inhibition of the phosphatidylinositol 3-kinase TOR is pivotal for autophagy-induction in all cells investigated so far. Its inhibition either by amino-acid starvation or by binding of the rapamycin/FKBP-complex usually leads to induction of autophagy, cell cycle arrest, and a block in protein biosynthesis. To evaluate whether transcription of TbTOR is affected because of its inhibition, Northern blot analyses were performed. For this purpose total RNA was isolated from 3×10^8 cells after the manufacturer's instructions (QIAGEN) and Northern blots were performed as described earlier (Jungwirth *et al.*, 2001). Samples were separated on a 1.2% agarose gel containing 220 m*M* formaldehyde. Large rRNAs (2250, 1850, 1350 nt) were used as loading controls. RNA bands were transferred to a nylon membrane (Hybond-N+, Amersham) and hybridized with $[^{32}P]$-labeled DNA probes from either TbTOR-transcripts or the β-tubulin gene as a control. Northern blots were visualized by autoradiography, using an X-ray developer (SRX-101A, Konica). The analyses revealed no transcriptional regulation of TbTOR in the respective samples (unpublished observation). However, this result was no surprise, as posttranscriptional regulation is rather uncommon in kinetoplastids (Clayton, 2002) and was not observed in other eukaryotes.

Because regulation of TOR would thus be more likely on the protein level, we performed Western blot analyses using rapamycin-treated and untreated trypanosomes. For these very preliminary experiments commercially available antibodies against the human TOR and phospho-TOR have been used, as significant homology exists between the parasite and the human protein. Interestingly, positive signals have been observed in control and rapamycin-treated cells using the phospho-TOR antibody but not with the TOR antibody. This would suggest that TOR is always in its activated form (unpublished observations).

Alvarez *et al.* (2008) have successfully developed polyclonal antibodies against the two orthologs of *T. cruzi*, TcATG8.1 and TcATG8.2. Only the first one is able to complement for Atg8 in yeast, and the corresponding protein is visualized in epimastigote, amastigote, and trypomastigote forms in Western blot analyses. Additionally, immunofluorescence studies of starved cells show an accumulation of the protein in contrast to an equal distribution in control parasites. Williams *et al.* (2006) chose another way to track ATG8 in *L. mexicana*. Using a GFP-fusion protein, they detected ATG8 in all life-cycle stages via anti-GFP antibodies. Using band shift analysis on SDS-PAGE and Western blot analysis, uncleaved, coupled, and cleaved forms could be distinguished.

4.5. Analysis of protein biosynthesis inhibition

Amino-acid starvation leads to autophagy because of declined environmental conditions. To gain enough energy to survive, cells start to degrade present proteins and reduce their protein biosynthesis to a basic level. For this reason the rate of protein biosynthesis was measured after induction of autophagy in *T. brucei* by different triggers. Additionally, parasites were incubated for 1 h in the presence of 10 μg/ml cycloheximide (CHX) as a positive control, as this antibiotic from *Streptomyces griseus* inhibits translation by blocking the peptidyl transferase.

1. For starvation, 1×10^7 bloodstream-form trypanosomes from the logarithmic phase are collected by centrifugation, washed twice, and incubated in TDB at a cell density of 1×10^6 cells per ml for 3 h at 37 °C/5% CO_2. Mimicking these conditions, cells are grown for 24 h after addition of 6.5 μM rapamycin (IC_{50}-concentration in *T. brucei*), which inhibits the phosphatidylinositol 3-kinase TOR, generally leading to the same signal transduction.
2. Following the respective treatments, cells are spun down, adjusted to a cell density of 1×10^6 cells per ml in TDB and preincubated for 5 min at 37 °C.
3. Measurement starts after addition of 5 μCi/ml [^{35}S]-methionine.
4. After 1.5 ml incubation at 37 °C, aliquots of 200 μl of cell suspension are taken and together with 600 μl of acetone added to 20 μl of a BSA solution (10 mg/ml). Protein-precipitation lasts for at least 20 min at −20 °C before centrifugation for 15 min at 13,800×*g* and 4 °C.

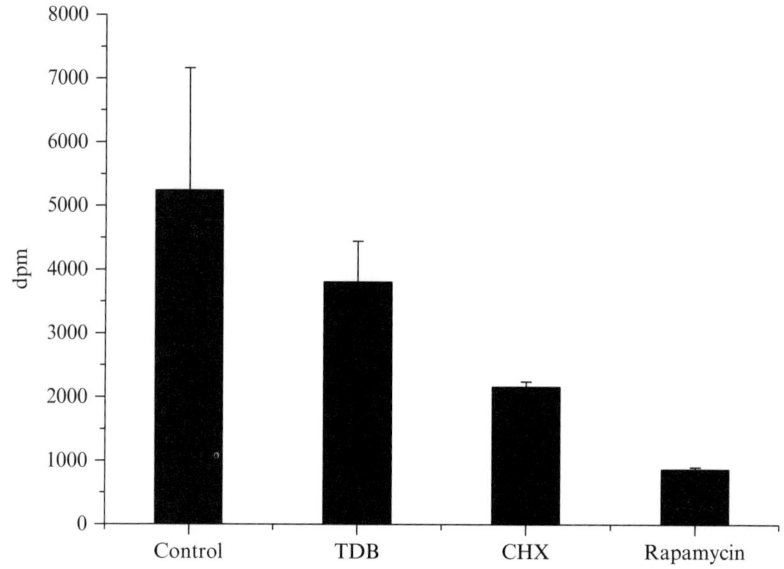

Figure 25.9 Protein biosynthesis of bloodstream-form trypanosomes. Control: 3 h incubation in culture medium; TDB: 3 h incubation in TDB; CHX: incubation with 10 μg/ml cycloheximide for 1 h in culture medium; Rapamycin: 24 h incubation with 6.5 μM rapamycin in culture medium. All samples were washed with TDB and incubated for 15 min with 5 μCi/ml [^{35}S]-methionine.

5. The supernatant has to be discarded and the pellet resuspended in 1 ml of 5% TCA containing 2 mM methionine.
6. Samples are added to filter papers and washed three times with 3 ml TCA/methionine and once with the same volume of 1% acidic acid.
7. Scintillation vials are filled with the dried papers and 5 ml of Ultima Gold (Packard, Frankfurt, Germany). Measurement of implemented radioactivity, representative for the rate of protein biosynthesis, can be performed the next day using a liquid scintillation counter.

The data revealed that protein biosynthesis was blocked in cells treated with cycloheximide as compared to untreated control cells of the logarithmic phase. Starvation and addition of rapamycin led also to a significant reduction of protein biosynthesis, leading to the conclusion that both triggers do induce autophagic processes (Fig. 25.9).

4.6. Knockdown of autophagic pathways by RNAi

In many eukaryotic cells, including trypanosomes, double-stranded RNA triggers RNA interference (RNAi) and has thus become a valuable tool for functional characterization. It works exceptionally well in trypanosomes

(Ngo *et al.*, 1998) and is often used as a fast method to generate conditional knockdown mutants in *T. brucei*.

We use an inheritable inducible knockdown system (Shi *et al.*, 2000) to study the function of essential genes in *T. brucei*. Initial dsRNA is cleaved into short interfering RNAs which are 24–26 nucleotides long and eventually lead to the specific degradation of mRNA of homologous sequences (Djikeng *et al.*, 2001). We generally use DNA fragments of 100 to 1000 bp from the respective gene for RNAi. For trypanosomes a helpful application (RNAit) is available online (Redmond *et al.*, 2003). Our RNA interference vector for trypanosomes is p2T7TAblue, which allows TA cloning of unmodified PCR products and blue/white selection (Alibu *et al.*, 2005). The complementing cell lines, blood form and procyclic, express both *tet* repressor and T7 RNA polymerase. The procyclic 427 cell line is PC1313-514; instead of the corresponding blood form cell line (BF1313-514), we use the single-marker line (SMB), which is under the control of only one selection marker (Alibu *et al.*, 2005). For application, the appropriate gene fragment is amplified and cloned into the respective RNAi vector. The linearized plasmid (20μg) is then transferred into the cells by electroporation. We use 5×10^7 cells in 400 μl of transfection medium in a 2 mm cuvette with a BioRad GenePulser XCell at 1.5 kV, 25 μF, and ∞ resistance. The transfection medium is a modified Cytomix (120 m*M* KCl, 0.15 m*M* $CaCl_2$, 5 m*M* $MgCl_2$, 10 m*M* K_2HPO_4/KH_2PO_4, 25 m*M* HEPES, 2 m*M* EGTA, 100 μg/ml BSA, 0.5% glucose, 1 m*M* hypoxanthine) adjusted to pH 7.6 with KOH and filter sterilized. Hereby the plasmid integrates by homologous recombination into a region of the genome that is usually transcriptionally silent, such as the ribosomal RNA spacer. Those cells survive the following application of the selection marker and diluting the cells leads to the desired clonal cell lines. Experiments are in progress in our lab to apply RNAi in *T. brucei* for silencing the ATG8 genes individually and all together to determine the respective phenotypes under normal and under starvation and stress conditions. Once results have been established, the other autophagy-related genes will be treated the same way.

5. Concluding Remarks

As judged from an *in silico* survey, members of the order kinetoplastida seem to perform autophagy in a rather primitive way, as only half of the respective proteins expressed in yeast have homologs in trypanosomes and Leishmania (Rigden *et al.*, 2005). Moreover, they represent an early evolutionary branch point. This led to the idea that trypanosomes may serve as model organisms to investigate autophagy in a simple form. However, it should be kept in mind that the presence or absence of gene orthologs in a different organism is a rather simplified way of analyzing complex cellular events. Proteins with the same defined function may have developed independently in different species or may have orthologs of the respective gene in one but not in other organisms (as shown here for TbATG16; see previous sections). In addition, if a specific protein is missing, the respective function of this protein may be performed by another more unspecific enzyme (e.g., the yeast proteases Prb1 and Pep4 are replaced by the non-orthologous LmCPA and LmCPB in *Leishmania*; Williams *et al.*, 2006).

We are just in the beginning and more or less in the descriptive phase to understand autophagy in protozoa and to develop a concise picture of how this process is regulated and controlled in kinetoplastida. Because autophagy is obviously involved in the parasite's struggle to escape from the host's immune reactions, it may turn out to be far more complicated than expected and perhaps even as complicated as in some higher eukaryotes. Indeed, electron micrographs, as presented here and elsewhere, clearly indicate that autophagy is involved in the discussed protozoa during the same different cellular events as described for higher eukaryotes, such as microautophagy, starvation, stress, and differentiation. Although the total number of involved proteins has yet to be established, it remains to be analyzed if the number of autophagy-related proteins in trypanosomes is indeed much less than in yeast and/or higher eukaryotes. As mentioned earlier, at least for some orthologous proteins (especially ATG4 and ATG8) intracellular localization and complementation studies could reveal their pivotal role in autophagy for *Leishmania* and *T. cruzi*. In this respect, conditional mutants showing significant down regulation of one or several of the other autophagy related proteins will lead to additional information in case they show a clear phenotype.

REFERENCES

Alibu, V. P., Storm, L., Haile, S., Clayton, C., and Horn, D. (2005). A doubly inducible system for RNA interference and rapid RNAi plasmid construction in *Trypanosoma brucei*. *Mol. Biochem. Parasitol.* **139,** 75–82.

Alvarez, V. E., Kosec, G., Sant'Anna, C., Turk, V., Cazzulo, J. J., and Turk, B. (2008). Autophagy is involved in nutritional stress response and differentiation in *Trypanosoma cruzi*. *J. Biol. Chem.* **283,** 3454–3464.

Baldauf, S. L., Roger, A. J., Wenk-Siefert, I., and Doolittle, W. F. (2000). A kingdom-level phylogeny of eukaryotes based on combined protein data. *Science* **290,** 972–977.

Bannai, S., and Ishii, T. (1982). Transport of cystine and cysteine and cell growth in cultured human diploid fibroblasts: Effect of glutamate and homocysteate. *J. Cell Physiol.* **112,** 265–272.

Bates, P. A. (1993a). Axenic culture of Leishmania amastigotes. *Parasitol. Today* **9,** 143–146.

Bates, P. A., and Tetley, L. (1993b). *Leishmania mexicana*: Induction of metacyclogenesis by cultivation of promastigotes at acidic pH. *Exp. Parasitol.* **76,** 412–423.

Bera, A., Singh, S., Nagaraj, R., and Vaidya, T. (2003). Induction of autophagic cell death in *Leishmania donovani* by antimicrobial peptides. *Mol. Biochem. Parasitol.* **127,** 23–35.

Berberich, C., Marin, M., Ramirez, J. R., Muskus, C., and Velez, I. D. (1998). The metacyclic stage-expressed meta-1 gene is conserved between Old and New World Leishmania species. *Mem. Inst. Oswaldo Cruz* **93,** 819–821.

Bergmann, A. (2007). Autophagy and cell death: No longer at odds. *Cell* **131,** 1032–1034.

Besteiro, S., Williams, R. A., Morrison, L. S., Coombs, G. H., and Mottram, J. C. (2006). Endosome sorting and autophagy are essential for differentiation and virulence of *Leishmania major*. *J. Biol. Chem.* **281,** 11384–11396.

Besteiro, S., Williams, R. A., Coombs, G. H., and Mottram, J. C. (2007). Protein turnover and differentiation in Leishmania. *Int. J. Parasitol.* **37,** 1063–1075.

Biederbick, A., Kern, H. F., and Elsasser, H. P. (1995). Monodansylcadaverine (MDC) is a specific *in vivo* marker for autophagic vacuoles. *Eur. J. Cell Biol.* **66,** 3–14.

Borst, P., and Ulbert, S. (2001). Control of VSG gene expression sites. *Mol. Biochem. Parasitol.* **114,** 17–27.

Brun, R., Jenni, L., Schönenberger, M., and Schell, K. F. (1981a). *In vitro* cultivation of bloodstream forms of *Trypanosoma brucei*, *T. rhodesiense*, and *T. gambiense*. *J. Protozool.* **28,** 470–479.

Brun, R., and Schonenberger, M. (1981b). Stimulating effect of citrate and cis-Aconitate on the transformation of *Trypanosoma brucei* bloodstream forms to procyclic forms *in vitro*. *Z. Parasitenkd.* **66,** 17–24.

Chung, J., Kuo, C. J., Crabtree, G. R., and Blenis, J. (1992). Rapamycin-FKBP specifically blocks growth-dependent activation of and signaling by the 70 kd S6 protein kinases. *Cell* **69,** 1227–1236.

Clamp, M., Cuff, J., Searle, S. M., and Barton, G. J. (2004). The Jalview Java alignment editor. *Bioinformatics* **20,** 426–427.

Clayton, C. E. (2002). Life without transcriptional control? From fly to man and back again. *EMBO J.* **21,** 1881–1888.

Codogno, P., and Meijer, A. J. (2005). Autophagy and signaling: Their role in cell survival and cell death. *Cell Death Differ.* **12,** 1509–1518.

de Souza, W., and Souto-Padrón, T. (1980). The paraxial structure of the flagellum of trypanosomatidae. *J. Parasitol.* **66,** 229–236.

Djikeng, A., Shi, H., Tschudi, C., and Ullu, E. (2001). RNA interference in *Trypanosoma brucei*: Cloning of small interfering RNAs provides evidence for retroposon-derived 24-26-nucleotide RNAs. *RNA* **7,** 1522–1530.

Dunn, W. A. J., Cregg, J. M., Kiel, J. A., van der Klei, I. J., Oku, M., Sakai, Y., Sibirny, A. A., Stasyk, O. V., and Veenhuis, M. (2005). Pexophagy: The selective autophagy of peroxisomes. *Autophagy* **1,** 75–83.

Duszenko, M., Ferguson, M. A., Lamont, G. S., Rifkin, M. R., and Cross, G. A. (1985). Cysteine eliminates the feeder cell requirement for cultivation of *Trypanosoma brucei* bloodstream forms *in vitro*. *J. Exp. Med.* **162,** 1256–1263.

Duszenko, M., Mühlstädt, K., and Broder, A. (1992). Cysteine is an essential growth factor for *Trypanosoma brucei* bloodstream forms. *Mol. Biochem. Parasitol.* **50,** 269–273.

Edgar, R. C. (2004). MUSCLE: A multiple sequence alignment method with reduced time and space complexity. *BMC Bioinformatics* **5,** 1–19.

Fang, J., Ruiz, F. A., Docampo, M., Luo, S., Rodrigues, J. C., Motta, L. S., Rohloff, P., and Docampo, R. (2007). Overexpression of a Zn^{2+}-sensitive soluble exopolyphosphatase from *Trypanosoma cruzi* depletes polyphosphate and affects osmoregulation. *J. Biol. Chem.* **282,** 32501–32510.

Figarella, K., Rawer, M., Uzcategui, N. L., Kubata, B. K., Lauber, K., Madeo, F., Wesselborg, S., and Duszenko, M. (2005). Prostaglandin D_2 induces programmed cell death in *Trypanosoma brucei* bloodstream form. *Cell Death Differ.* **12,** 335–346.

Figarella, K., Uzcategui, N. L., Beck, A., Schoenfeld, C., Kubata, B. K., Lang, F., and Duszenko, M. (2006). Prostaglandin-induced programmed cell death in *Trypanosoma brucei* involves oxidative stress. *Cell Death Differ.* **13,** 1802–1814.

Figueiredo, R. C., Rosa, D. S., and Soares, M. J. (2000). Differentiation of *Trypanosoma cruzi* epimastigotes: Metacyclogenesis and adhesion to substrate are triggered by nutritional stress. *J. Parasitol.* **86,** 1213–1218.

Furuya, T., Kashuba, C., Docampo, R., and Moreno, S. N. (2000). A novel phosphatidylinositol-phospholipase C of *Trypanosoma cruzi* that is lipid modified and activated during trypomastigote to amastigote differentiation. *J. Biol. Chem.* **275,** 6428–6438.

Hamm, B., Schindler, A., Mecke, D., and Duszenko, M. (1990). Differentiation of *Trypanosoma brucei* bloodstream trypomastigotes from long slender to short stumpy-like forms in axenic culture. *Mol. Biochem. Parasitol.* **40,** 13–22.

Hansen, M., Chandra, A., Mitic, L. L., Onken, B., Driscoll, M., and Kenyon, C. (2008). A role for autophagy in the extension of lifespan by dietary restriction in *C. elegans*. *PLoS Genet.* **4,** e24.

Herman, M., Gillies, S., Michels, P. A., and Rigden, D. J. (2006). Autophagy and related processes in trypanosomatids: Insights from genomic and bioinformatic analyses. *Autophagy* **2,** 107–118.

Herman, M., Pérez-Morga, D., Schtickzelle, N., and Michels, P. A. (2008). Turnover of glycosomes during life-cycle differentiation of *Trypanosoma brucei*. *Autophagy* **4,** 294–308.

Hesse, F., Selzer, P. M., Muhlstadt, K., and Duszenko, M. (1995). A novel cultivation technique for long-term maintenance of bloodstream form trypanosomes *in vitro*. *Mol. Biochem. Parasitol.* **70,** 157–166.

Higuchi, K. (1963). Studies on the nutrition and metabolism of animal cells in serum-free media. I. Serum-free monolayer cultures. *J. Infect. Dis.* **112,** 213–220.

Hirsch, J. G., and Fedorko, M. E. (1968). Ultrastructure of human leukocytes after simultaneous fixation with glutaraldehyde and osmium tetroxide and "postfixation" in uranyl acetate. *J. Cell Biol.* **38,** 615–627.

Hirumi, H., Doyle, J. J., and Hirumi, K. (1977). African trypanosomes: Cultivation of animal-infective *Trypanosoma brucei in vitro*. *Science* **196,** 992–994.

Hirumi, H., and Hirumi, K. (1989). Continuous cultivation of *Trypanosoma brucei* blood stream forms in a medium containing a low concentration of serum protein without feeder cell layers. *J. Parasitol.* **75,** 985–989.

Jayaraman, T., and Marks, A. R. (1993). Rapamycin-FKBP12 blocks proliferation, induces differentiation, and inhibits cdc2 kinase activity in a myogenic cell line. *J. Biol. Chem.* **268,** 25385–25388.

Juhasz, G., and Neufeld, T. P. (2006). Autophagy: A forty-year search for a missing membrane source. *PLoS Biol.* **4,** e36.

Jungwirth, H., Bergler, H., and Hogenauer, G. (2001). Diazaborine treatment of Baker's yeast results in stabilization of aberrant mRNAs. *J. Biol. Chem.* **276,** 36419–36424.

Kim, J., and Klionsky, D. J. (2000). Autophagy, cytoplasm-to-vacuole targeting pathway, and pexophagy in yeast and mammalian cells. *Annu. Rev. Biochem.* **69,** 303–342.

Kirkegaard, K., Taylor, M. P., and Jackson, W. T. (2004). Cellular autophagy: Surrender, avoidance and subversion by microorganisms. *Nat. Rev. Microbiol.* **2,** 301–314.

Klionsky, D. J. (2005). The molecular machinery of autophagy: Unanswered questions. *J. Cell Sci.* **118,** 7–18.

Klionsky, D. J. (2006). What can we learn from trypanosomes? *Autophagy* **2,** 63–64.

Landfear, S. M. (2001). Molecular genetics of nucleoside transporters in Leishmania and African trypanosomes. *Biochem. Pharmacol.* **62,** 149–155.

Levine, B., and Klionsky, D. J. (2004). Development by self-digestion: Molecular mechanisms and biological functions of autophagy. *Dev. Cell* **6,** 463–477.

Levine, B., and Yuan, J. (2005). Autophagy in cell death: An innocent convict? *J. Clin. Invest.* **115,** 2679–2688.

Maiuri, M. C., Zalckvar, E., Kimchi, A., and Kroemer, G. (2007). Self-eating and self-killing: Crosstalk between autophagy and apoptosis. *Nat. Rev. Mol. Cell Biol.* **8,** 741–752.

Martelli, A. M., Tazzari, P. L., Evangelisti, C., Chiarini, F., Blalock, W. L., Billi, A. M., Manzoli, L., McCubrey, J. A., and Cocco, L. (2007). Targeting the phosphatidylinositol 3-kinase/Akt/mammalian target of rapamycin module for acute myelogenous leukemia therapy: From bench to bedside. *Curr. Med. Chem.* **14,** 2009–2023.

Marzella, L., Ahlberg, J., and Glaumann, H. (1981). Autophagy, heterophagy, microautophagy and crinophagy as the means for intracellular degradation. *Virchows Arch. B. Cell. Pathol. Incl. Mol. Pathol.* **36,** 219–234.

Mijaljica, D., Prescott, M., and Devenish, R. J. (2007). Different fates of mitochondria: Alternative ways for degradation? *Autophagy* **3,** 4–9.

Moreno, S. N., Vercesi, A. E., Pignataro, O. P., and Docampo, R. (1992). Calcium homeostasis in *Trypanosoma cruzi* amastigotes: presence of inositol phosphates and lack of an inositol 1,4,5-trisphosphate-sensitive calcium pool. *Mol. Biochem. Parasitol.* **52,** 251–261.

Morgan, G. W., Hall, B. S., Denny, P. W., Carrington, M., and Field, M. C. (2002a). The kinetoplastida endocytic apparatus. Part I: A dynamic system for nutrition and evasion of host defences. *Trends Parasitol.* **18,** 491–496.

Morgan, G. W., Hall, B. S., Denny, P. W., Field, M. C., and Carrington, M. (2002b). The endocytic apparatus of the kinetoplastida. Part II: Machinery and components of the system. *Trends Parasitol.* **18,** 540–546.

Morice, W. G., Brunn, G. J., Wiederrecht, G., Siekierka, J. J., and Abraham, R. T. (1993). Rapamycin-induced inhibition of p34cdc2 kinase activation is associated with G1/S-phase growth arrest in T lymphocytes. *J. Biol. Chem.* **268,** 3734–3738.

Munafo, D. B., and Colombo, M. I. (2001). A novel assay to study autophagy: Regulation of autophagosome vacuole size by amino acid deprivation. *J. Cell Sci.* **114,** 3619–3629.

Nakatogawa, H., Ichimura, Y., and Ohsumi, Y. (2007). Atg8, a ubiquitin-like protein required for autophagosome formation, mediates membrane tethering and hemifusion. *Cell* **130,** 165–178.

Ngô, H., Tschudi, C., Gull, K., and Ullu, E. (1998). Double-stranded RNA induces mRNA degradation in *Trypanosoma brucei. Proc. Natl. Acad. Sci. USA* **95,** 14687–14692.

Overath, P., Czichos, J., and Haas, C. (1986). The effect of citrate/cis-aconitate on oxidative metabolism during transformation of *Trypanosoma brucei. Eur. J. Biochem.* **160,** 175–182.

Parsons, M., Furuya, T., Pal, S., and Kessler, P. (2001). Biogenesis and function of peroxisomes and glycosomes. *Mol. Biochem. Parasitol.* **115,** 19–28.

Priest, J. W., and Hajduk, S. L. (1994). Developmental regulation of mitochondrial biogenesis in *Trypanosoma brucei. J. Bioenerg. Biomembr.* **26,** 179–191.

Raught, B., Gingras, A. C., and Sonenberg, N. (2001). The target of rapamycin (TOR) proteins. *Proc. Natl. Acad. Sci. USA* **98,** 7037–7044.

Redmond, S., Vadivelu, J., and Field, M. C. (2003). RNAit: An automated web-based tool for the selection of RNAi targets in *Trypanosoma brucei. Mol. Biochem. Parasitol.* **128,** 115–118.

Reuner, B., Vassella, E., Yutzy, B., and Boshart, M. (1997). Cell density triggers slender to stumpy differentiation of *Trypanosoma brucei* bloodstream forms in culture. *Mol. Biochem. Parasitol.* **90,** 269–280.

Richards, F. F. (1984). The surface of the African trypanosomes. *J. Protozool.* **31,** 60–64.

Rigden, D. J., Herman, M., Gillies, S., and Michels, P. A. (2005). Implications of a genomic search for autophagy-related genes in trypanosomatids. *Biochem. Soc. Trans.* **33,** 972–974.

Rodrigues, J. C., Attias, M., Rodriguez, C., Urbina, J. A., and Souza, W. (2002). Ultrastructural and biochemical alterations induced by 22,26-azasterol, a delta(24(25))-sterol methyltransferase inhibitor, on promastigote and amastigote forms of *Leishmania amazonensis. Antimicrob. Agents Chemother.* **46,** 487–499.

Rogers, M. E., Chance, M. L., and Bates, P. A. (2002). The role of promastigote secretory gel in the origin and transmission of the infective stage of *Leishmania mexicana* by the sandfly *Lutzomyia longipalpis. Parasitol.* **124,** 495–507.

Santa-Rita, R. M., Lira, M., Barbosa, H. S., Urbina, S. A., and de Castro, S. L. (2005). Anti-proliferative synergy of lysophospholipid analogues and ketoconazole against *Trypanosoma cruzi* (Kinetoplastida: Trypanosomatidae): cellular and ultrastructural analysis. *J. Antimicrob. Chemother.* **55,** 780–784.

Sacks, D. L., and da Silva, R. P. (1987). The generation of infective stage *Leishmania major* promastigotes is associated with the cell-surface expression and release of a developmentally regulated glycolipid. *J. Immunol.* **139,** 3099–3106.

Scherz-Shouval, R., Shvets, E., Fass, E., Shorer, H., Gil, L., and Elazar, Z. (2007). Reactive oxygen species are essential for autophagy and specifically regulate the activity of Atg4. *EMBO J.* **26,** 1749–1760.

Shapiro, T. A. (1993). Kinetoplast DNA maxicircles: Networks within networks. *Proc. Natl. Acad. Sci. USA* **90,** 7809–7813.

Shi, H., Djikeng, A., Mark, T., Wirtz, E., Tschudi, C., and Ullu, E. (2000). Genetic interference in *Trypanosoma brucei* by heritable and inducible double-stranded RNA. *RNA* **6,** 1069–1076.

Smith, H. C., Gott, J. M., and Hanson, M. R. (1997). A guide to RNA editing. *RNA* **3,** 1105–1123.

Suzuki, K., and Ohsumi, Y. (2007). Molecular machinery of autophagosome formation in yeast, *Saccharomyces cerevisiae. FEBS Lett.* **581,** 2156–2161.

Tait, A. (1980). Evidence for diploidy and mating in trypanosomes. *Nature* **287,** 536–538.

Tait, A. (1983). Sexual processes in the kinetoplastida. *Parasitol.* **86,** 29–57.

Tomlinson, S., Vandekerckhove, F., Frevert, U., and Nussenzweig, V. (1995). The induction of *Trypanosoma cruzi* trypomastigote to amastigote transformation by low pH. *Parasitol.* **110,** 547–554.

Uzcategui, N. L., Szallies, A., Pavlovic-Djuranovic, S., Palmada, M., Figarella, K., Boehmer, C., Lang, F., Beitz, E., and Duszenko, M. (2004). Cloning, heterologous expression, and characterization of three aquaglyceroporins from *Trypanosoma brucei. J. Biol. Chem.* **279,** 42669–42676.

Uzcátegui, N. L., Carmona-Gutiérrez, D., Denninger, V., Schoenfeld, C., Lang, F., Figarella, K., and Duszenko, M. (2007a). Antiproliferative effect of dihydroxyacetone on *Trypanosoma brucei* bloodstream forms: Cell cycle progression, subcellular alterations, and cell death. *Antimicrob. Agents Chemother.* **51,** 3960–3968.

Uzcátegui, N. L., Denninger, V., Merkel, P., Schoenfeld, C., Figarella, K., and Duszenko, M. (2007b). Dihydroxyacetone induced autophagy in African trypanosomes. *Autophagy* **3,** 626–629.

Vertommen, D., Van Roy, J., Szikora, J. P., Rider, M. H., Michels, P. A., and Opperdoes, F. R. (2008). Differential expression of glycosomal and mitochondrial proteins in the two major life-cycle stages of *Trypanosoma brucei. Mol. Biochem. Parasitol.* **158,** 189–201.

Vickerman, K. (1965). Polymorphism and mitochondrial activity in sleeping sickness trypanosomes. *Nature* **208,** 762–766.

Vickerman, K. (1969). The fine structure of *Trypanosoma congolense* in its bloodstream phase. *J. Protozool.* **16,** 54–69.

Williams, R. A., Tetley, L., Mottram, J. C., and Coombs, G. H. (2006). Cysteine peptidases CPA and CPB are vital for autophagy and differentiation in *Leishmania mexicana. Mol. Microbiol.* **61,** 655–674.

Woodward, R., and Gull, K. (1990). Timing of nuclear and kinetoplast DNA replication and early morphological events in the cell cycle of *Trypanosoma brucei. J. Cell Sci.* **95,** 49–57.

Xie, Z., and Klionsky, D. J. (2007). Autophagosome formation: Core machinery and adaptations. *Nat. Cell Biol.* **9,** 1102–1109.

Yorimitsu, T., and Klionsky, D. J. (2005). Autophagy: Molecular machinery for self-eating. *Cell Death Differ.* **12,** 1542–1552.

Yorimitsu, T., Nair, U., Yang, Z., and Klionsky, D. J. (2006). Endoplasmic reticulum stress triggers autophagy. *J. Biol. Chem.* **281,** 30299–30304.

CHAPTER TWENTY-SIX

Methods to Investigate Autophagy During Starvation and Regeneration in Hydra

Wanda Buzgariu,* Simona Chera,* *and* Brigitte Galliot*,†

Contents

Abstract

In hydra, the regulation of the balance between cell death and cell survival is essential to maintain homeostasis across the animal and promote animal survival during starvation. Moreover, this balance also appears to play a key role during regeneration of the apical head region. The recent finding that autophagy is a crucial component of this balance strengthens the value of the *Hydra* model system to analyze the implications of autophagy in starvation, stress response and regeneration. We describe here how we adapted to *Hydra* some established tools to monitor steady-state autophagy. The ATG8/LC3 marker used in biochemical and immunohistochemical analyses showed a significant increase in autophagosome formation in digestive cells after 11 days of starvation. Moreover, the maceration procedure that keeps intact the morphology of the various cell types allows the quantification of the autophagosomes and autolysosomes in any cell type, thanks to the detection of the MitoFluor or LysoTracker dyes combined with

* Department of Zoology and Animal Biology, Faculty of Sciences, University of Geneva, Geneva, Switzerland
† Corresponding author

Methods in Enzymology, Volume 451
ISSN 0076-6879, DOI: 10.1016/S0076-6879(08)03226-6

the anti-LC3, anti-LBPA, and/or anti-RSK (ribosomal S6 kinase) immunostaining. The classical activator (rapamycin) and inhibitors (wortmannin, bafilomycin A_1) of autophagy also appear to be valuable tools to modulate autophagy in hydra, as daily-fed and starved hydra display slightly different responses. Finally, we show that the genetic circuitry underlying autophagy can be qualitatively and quantitatively tested through RNA interference in hydra repeatedly exposed to double-stranded RNAs.

1. The Value of the Hydra Model System for Investigating Autophagy

Hydra is a freshwater polyp that belongs to Cnidaria (Fig. 26.1A), a phylum positioned as a sister group to the bilaterians, having separated early during animal evolution. Despite their apparent simplicity, cnidarian animals possess a sophisticated neuromuscular system and display an amazing potential for regeneration (Galliot and Schmid, 2002; Holstein *et al.*, 2003;

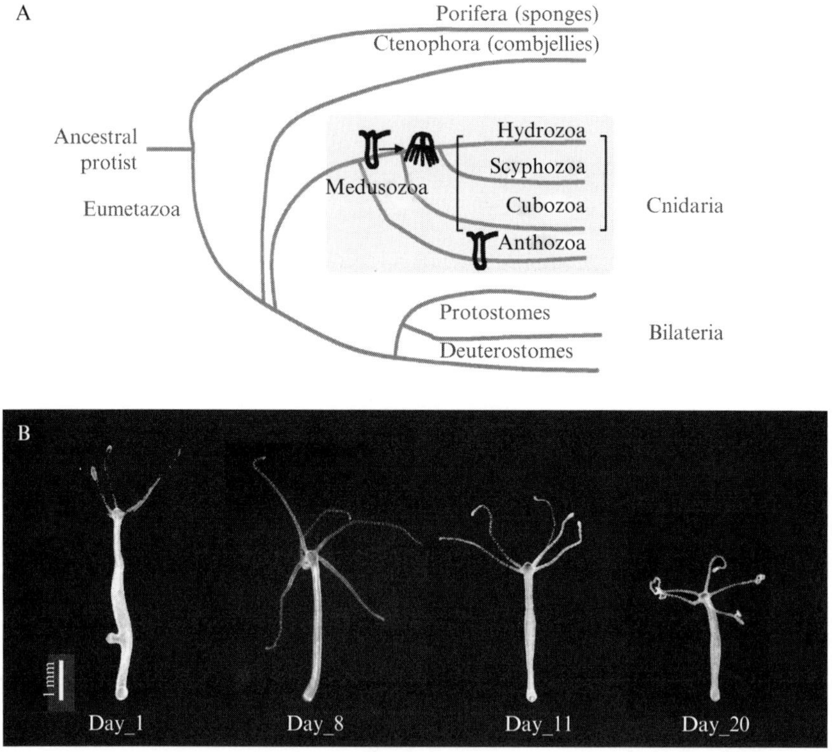

Figure 26.1 (A) Position of hydra (Cnidaria, hydrozoan) in the metazoan tree. (B) Dramatic reduction in hydra size after 20 days of starvation. The animals were kept relaxed at the indicated time points and pictured using a stereomicroscope. Bar: 1 mm.

Bosch, 2007). Since the genomic era, cnidarian species actually currently emerge as new model systems to investigate fundamental biological questions in cell biology, developmental biology, and evolution (Kamm *et al.*, 2006; Momose and Houliston, 2007; Seipel and Schmid, 2006; Technau *et al.*, 2005). Hydra polyps are characterized by a radial symmetry with an oral-aboral polarity along their unique body axis; they display a tube shape with a mouth opening at the top, surrounded by a ring of tentacles, whereas the basal disk allows the animal to attach to substrates (Fig. 26.1B). The body wall comprises two layers, ectodermal and endodermal, which all together contain about a dozen cell types (Lentz, 1966), which derive from three distinct stem cell populations, ectodermal myoepithelial, endodermal myoepithelial, and interstitial cells (Bode, 1996; Dubel *et al.*, 1987; Galliot *et al.*, 2006; Steele, 2002). The myoepithelial cells form continuous epithelia and differentiate into head- and foot-specific myoepithelial cells, whereas the interstitial cells provide different nerve cells, mechanoreceptor cells (stinging cells, named *nematocytes* or *cnidocytes*), gland cells, and gametes when the animals follow the sexual cycle. The spatial distribution of these highly differentiated cells occurs as a consequence of the continuous division of stem cells in the body column and the active migration or displacement of the committed/precursor cells toward the extremities.

Hydra have the amazing ability to regenerate any missing part (foot and head) in several days after amputation, a property identified in the 18th century by Abraham Trembley (1744). In fact, whatever the level and the number of amputation along the body axis, each piece of tissue, except tentacles, has the potential to reform a new hydra. This potential that is dissected at the cellular and molecular levels since decades ago was shown to rely on signaling pathways that are evolutionarily conserved (Galliot and Schmid, 2002; Galliot *et al.*, 2006; Holstein *et al.*, 2003; Miljkovic-Licina *et al.*, 2007; Steele, 2002), although less conserved peptides probably also play a role (Bosch, 2007; Fujisawa, 2003). Moreover hydra can reproduce asexually through budding (Otto and Campbell, 1977) and, after dissociation, are able to reaggregate and rebuild the initial shape (Gierer *et al.*, 1972). This unique developmental plasticity among multicellular organisms has made hydra one of the classical model organisms to investigate the molecular and cellular basis of regeneration. In homeostatic conditions the hydra polyp can survive an extended period without any feeding, up to several weeks. During this time, the starving animals shrink from the adult size to a half size (Fig. 26.1B), while the cell cycle length of epithelial cells lengthen (Bosch and David, 1984). However, an imbalance between the tissue growth rate and cell cycle length was observed: the overproduced cells actually die and are phagocytosed by the neighboring cells, hence providing a nutrient source for the surviving animals (Bosch and David, 1984). This is in contrast with the heavy feeding condition when hydra cells from all three stem-cell compartments continuously divide and the growing rate of the tissues tightly follows the cell cycling rate. Although a systematic study concerning macroautophagy in hydra and autophagy-mediated cell death has not been made yet, this evidence suggests that survival of starving animals relies on the latter. In any case the respective roles of autophagic cell death and apoptosis remain to be established in the future.

In fact, a first analysis of autophagy in hydra has been recently reported in animals where *Kazal1*, an evolutionarily-conserved gene encoding a serine protease inhibitor Kazal-type (SPINK), was knocked down on RNA interference (Chera *et al.*, 2006). This study shows the formation of autophagosomes in both gland cells and digestive cells (endodermal myoepithelial cells) in intact polyps where *Kazal1* transcripts are no longer detected. This autophagy process immediately affects the homeostatic condition as evidenced by the decreased budding rate. In *Kazal1*(RNAi) hydra, the autophagic cells progressively shrink, their number drastically decreases, and the animals die. During regeneration, the expression of *Kazal1* is dramatically enhanced in regenerating tips. *Kazal1* up-regulation immediately after amputation appears to be essential to prevent excessive autophagy, as hydra knocked-down for *Kazal1* display massive autophagy in the regenerating tips as early as 1 h after amputation and, following repeated exposures to *Kazal1* double stranded RNAs (dsRNAs), survive the amputation only a few hours.

This study indicates that the autophagy process in hydra is strictly controlled in both homeostatic and regenerative contexts, as a long-standing increase in autophagy prevents asexual reproduction (budding) and slowly leads to the death of intact animals, whereas the absence of regulation of the amputation-induced autophagy in the regenerative context results in rapid animal death. Moreover, we propose that the cytoprotective role played by *Kazal1* immediately after amputation actually allows the activation of the regenerative process (Galliot, 2006). Therefore, systematic cellular and molecular studies that would dissect the regulative pathways and the function of autophagy in these two contexts are required. One question would be to learn how similar are the autophagy processes in the starvation and regenerative contexts. We anticipate that such studies will no doubt identify some evolutionarily conserved mechanisms that regulate autophagy in most metazoan species when they starve. In addition, such studies could also provide the starting point for comparing the regulation of stress-induced autophagy in regenerating and non-regenerating species. Given the growing importance of autophagy in pathological and homeostatic conditions, there is a need to define valuable comparative tools to measure and study autophagy in different model systems. Here we present and discuss the respective values of some morphological, immunochemical and biochemical methods, which allow for the monitoring of autophagy in hydra. We show that specific autophagy markers as well as pharmacological agents that are widely used in other model systems indeed provide reliable tools in hydra. Moreover the RNA interference (RNAi) method developed in our laboratory is well suited for dissecting in quantitative terms the genetic circuitry underlying the regulation and function of autophagy in hydra.

2. Experimental Paradigms to Follow Autophagy in Hydra

2.1. Biochemical detection of autophagy in hydra

2.1.1. LC3-II as a marker to monitor starvation-induced autophagy in hydra

Although the number of methods used to monitor autophagy is limited, one should consider that some methods monitor the steady-state level of autophagy, whereas others measure the autophagic flux (Klionsky *et al.*, 2008). In this chapter we consider exclusively the former ones. One of the most widely used biochemical methods to characterize autophagy in steady-state conditions is the detection of the LC3/Atg8 protein. LC3 is synthesized as a precursor, proLC3, which is converted to LC3-I by proteolysis and then modified by lipidation to the phosphatidylethanolamine-conjugated form, LC3-II. Once lipidated, LC3-II is anchored to the phagophore and autophagosome membranes. It is well accepted that the amount of LC3-II detected in Western blot analysis correlates with the number of autophagosomes (Kabeya *et al.*, 2000).

In starving and daily-fed hydra, endogenous LC3 is detected as two bands in Western blot analysis: LC3-I at a molecular mass around 17 kD and LC3-II at approximately 14 kD (Fig. 26.2A). The level of LC3-I is very weak compared with the LC3-II isoform, which is not surprising, as the selected antibody more easily detects LC3-II than LC3-I (Settembre *et al.*, 2008). Moreover LC3-I is more labile, as it is more sensitive to the extraction condition than LC3-II; finally, hydra tissues might contain a limited amount of LC3-I protein. Therefore, the amount of LC3-II is normalized against α-tubulin and the samples are compared based on the relative index of LC3-II amount. In starved animals, a progressive increase in LC3-II level can be observed. After 13 days of starvation, the LC3-II value exceeds a 2-fold increase when compared to the LC3-II value measured in daily-fed hydra (Fig. 26.2B). At later time points, a plateau value is recorded. As LC3-II relative index is considered as an indicator of autophagosome formation, this indicates that the autophagic process reaches its maximal rate of turnover within 2 weeks of starvation.

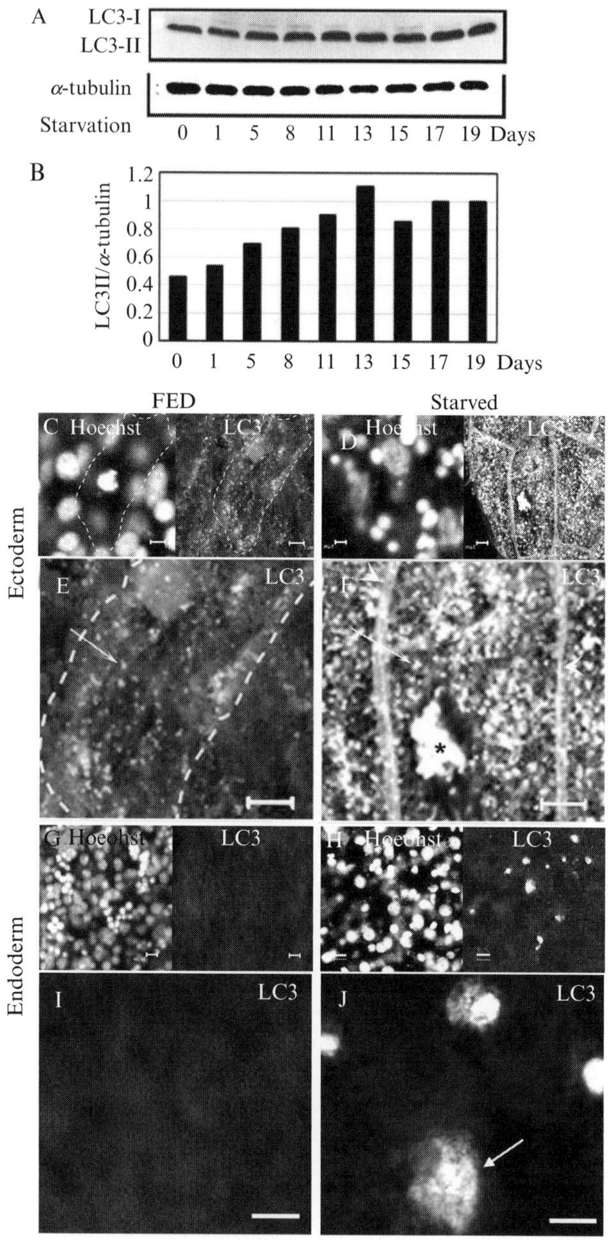

Figure 26.2 Increase in LC3-II expression levels during starvation. (A, B) Western blot analysis showing the progressive increase in LC3-II level during the first 2 weeks of starvation (the LC3-I band is faintly seen above the LC3-II band). During the third week, a plateau level was reached assuming that the maximal rate of LC3-II turnover was attained. (A) The same blot was detected first with anti-LC3 antibody and then with anti-α-tubulin as described in protocol 1. (B) For each time point, the LC3-II and

2.1.2. Hydra culture and starvation conditions

1. Culture about 500 hydra (*H. vulgaris*, Basel strain) in 2-liter square dishes (20 cm long, Roth) covered by 1 liter of hydra medium (HM: 1 m*M* NaCl, 1 m*M* $CaCl_2$, 0.1 m*M* KCl, 0.1 m*M* $MgSO_4$, 1 m*M* Tris, pH 7.6) at 18 °C.
2. Feed animals every day between 9 a.m. and 10 a.m. with freshly hatched nauplii (larvae) of *Artemia salina*.
3. 6–8 h after feeding, transfer the animals in a 2-liter beaker and wash them in 3 successive 800-ml glass beakers containing each 200 ml of HM. Use large glass pipettes to transfer the animals from one beaker to the other, before plating them in a clean HM prefilled culture dish. Under these conditions the cultures have a doubling time of 3.5 days (David and Campbell, 1972).
4. For starvation experiments, select 100 large budless hydra per condition the day following the last feeding to provide a homogeneous animal population and plate 50 hydra per 10-cm diameter plastic dish filled with 50 ml of HM (1 ml HM /hydra).
5. Let them starve for the indicated periods of time (2, 5, 8, 11, 13, 15, 17, and 19 days). During this period, wash the animals every other day and carefully remove any debris or dead animals. Discard the culture in case of yeast or bacterial infection.
6. Stop animal culture on the same day for all conditions to prepare the extracts.

2.1.3. Protocol 1: Whole cell protein extract and LC3-II detection in Western blot

1. Wash 100 hydra per condition in 3 successive 800-ml glass beakers containing each 200 ml of HM as previously. Transfer hydra to 2-ml microcentrifuge tubes and briefly wash them in 2 ml of precooled lysis buffer (LB) containing 50 m*M* HEPES, 150 m*M* NaCl, 2.5 m*M* $MgCl_2$, 0.5 m*M* DTT, 10% glycerol, 1% Triton X100, 25 m*M* NaF, 20 m*M* β-glycerophosphate, 1 m*M* Na_3VO_4, 0.1 mg/ml PMSF, and the complete protease inhibitor cocktail (Roche Applied Science).

α-tubulin bands are quantified with the ImageJ software and the LC3-II amount normalized against α-tubulin is represented (arbitrary units). (C–J) LC3 immunostaining of the ectodermal (C–F) and endodermal (G–J) layers of daily-fed (C, E, G, I) and starved (D, F, H, J) polyps as described in protocol 2. (C–F) Note the weak LC3 staining in the ectodermal epithelial cells of daily-fed polyps (C, E, dashed line) contrasting with the strong LC3 signal detected in the same cells of starved animals (D, F, arrow), particularly in cell membranes (arrowhead) and abundant small cytoplasmic vacuoles (arrow). The asterisk marks an artifact. (G–J) LC3 staining is completely absent in the endodermal layer of daily-fed hydra (G, I) but in starved polyps (H, J) reveals the presence of large LC3-specific vacuoles (J, arrow). Scale C, D, G, H: 10 μm; E, F, I, J: 5 μm.

2. Remove the last wash and resuspend the animals in 200 μl of LB, homogenize on ice through a 2.5-gauge needle 20–30 times, and centrifuge at 14,000*g* for 10 min. Collect the supernatant fractions, aliquot them (25 μl), freeze in liquid nitrogen and store at −80 °C.
3. Measure the protein content on duplicates (1 μl each) with the Bradford method by reading the optical densities of the samples at 595 nm. Extrapolate the protein concentrations from a bovine serum albumin (BSA) calibration curve read at the same time.
4. Mix 17 μg of each extract with 2x loading buffer (62.5 m*M* Tris, pH 6.8, 4% SDS, 20% glycerol, 0.1% bromophenol blue, 200 m*M* β-mercaptoethanol), boil for 3 min, load onto a 15% denaturing SDS-PAGE gel, and migrate 1 h at 200 V.
5. Transfer the proteins onto a Hybond PVDF membrane (GE-Amersham) in a wet-transfer system unit (Invitrogen) for 30 min at 25 V.
6. After transfer, block the membrane for 3 h at room temperature (RT) in PBS, 0.2% Tween 20 (TBS) containing 5% nonfat milk.
7. Incubate the membrane overnight (O/N) at 4 °C with a rabbit polyclonal anti-LC3 antibody 1:1000 (Novus Biological, NB 100-2220).
8. Wash the membrane 4 times in TBS, 5% nonfat milk at RT.
9. Incubate for 1 h at RT with horseradish peroxidase (HRP)-conjugated antirabbit 1:10,000 (Promega W401B).
10. Wash 4 times in TBS, 5% nonfat milk, and 2 times in TBS at RT.
11. Detect the peroxidase activity using the enhanced chemiluminescent (ECL) blotting reagent (GE-Amersham) on hyperfilm ECL (Amersham Bioscience).
12. For α-tubulin detection as a loading control, strip the membrane in 62 m*M* Tris, 0.2% Tween 20, 200 m*M* β-mercaptoethanol for 30 min in 3 steps at RT and reprobe with the α-tubulin 1:5000 (Sigma, T5168) antibody as earlier. Detect with an HRP-coupled antimouse antibody 1:3000 (Biorad, 170-6516).

Scan the films and perform a densitometric analysis with ImageJ software (http://rsb.info.nih.gov/ij/). Quantify the area of LC3 and α-tubulin bands and normalize against α-tubulin. The results can be expressed either as the relative index of LC3-II amount normalized against α-tubulin (Fig. 26.2) or as percentage from the fed control value (Fig. 26.4).

2.2. Analysis of autophagy in hydra tissues

2.2.1. Autophagosome identification on whole-mount hydra

The LC3 immunostaining performed on whole-mount polyps shows clear differences between regularly fed (3 days starvation) and starved polyps (24 days starvation) observed in both ectodermal and endodermal layers (Fig. 26.2C–2J). In the ectodermal layer of regularly fed polyps the LC3 staining is almost absent, with weakly LC3-positive small granules present in the cytoplasm of the epithelial cells (Fig. 26.2C, zoom in 2E, arrow). In contrast, in starved animals, the ectodermal LC3 immunostaining is very intense (Fig. 26.2D, zoom in 2F), strongly labeling membranes (Fig. 26.2F, arrowhead) and cytoplasmic vacuoles (Fig. 26.2F, arrow) of the myoepithelial cells. Some of these cells also display conglomerates of LC3 granules observed in the vicinity of the nucleus (Fig. 26.2F, asterisk). However, the most striking differences in LC3 staining are observed in the endodermal layer of starved and daily-fed polyps. In the endodermal cells of regularly fed polyps we do not record any LC3 staining (Fig. 26.2G right, 2I), whereas in starved polyps we note large conglomerates that correspond to vacuoles (Fig. 26.2H, right, 2J arrow). Furthermore, the endoderm of starving polyps appears thinner, rippled, and disrupted.

2.2.2. Protocol 2: LC3 immunostaining on whole-mount hydra

1. Relax 1–30 polyps for 90 s in 1 ml of 2% urethane/water and fix them O/N at 4 °C in 500 μl of 4% formaldehyde /50% ethanol.
2. Wash out the fixative with 4 × 10 minutes washes in PBS and subsequently block the endogenous peroxidase activity by incubating the animals in 500 μl of 3% H_2O_2 for 30 min at RT.
3. Next, block the samples with 500 μl of 2% BSA/PBS and incubate with the rabbit polyclonal anti-LC3 antibody (Novus Biological, NB910-40752, 1:200) O/N at 4 °C.
4. Wash out the first antibody with 4 × 10 minutes washes in PBS and incubate the hydra for 4 h at RT with the antirabbit HRP 1:100 from the Tyramide Amplification Kit (Molecular Probes, T-30954).
5. Wash the samples 4 × 10 minutes in PBS and incubate them with tyramide as described in the Tyramide Amplification Kit protocol (Molecular Probes, T-30954) for 16–18 min at RT.
6. Immediately wash the hydra 2 × 10 minutes in PBS, counterstain with 500 μl of Hoechst 33342 1 μg/ml in H_2O (Molecular Probes, H3570), and block for 1 h in H_2O_2 3% at RT.
7. Wash out the blocking reagent twice in PBS, once in water, and mount the samples on untreated glass slides in 50 μl of Mowiol 4-88 (Sigma 81381).
8. Picture and scan at the confocal microscope (SP2 AOBS Leica) at the proper wavelength depending on the type of fluorochrome linked to tyramide.

2.2.3. Protocol 3: LysoTracker Red (LTR) staining in fixed whole mounts

LysoTracker Red is a vital fluorescent acidotropic dye that accumulates in acidic vacuoles including functional lysosomes and autolysosomes. However, LTR is not specific to the autophagic process, as it also labels endosomes and phagosomes by virtue of their low internal pH. Therefore, it is necessary to combine this approach with a specific autophagic marker to trace with high fidelity the autophagic process (Bampton *et al.*, 2005). As a first approach to detect autophagic vacuoles in hydra cells, we tested LTR staining in three distinct conditions: on fixed whole mount animals (protocol 3), on macerated tissues (protocol 5) combined with immunostaining (Fig. 26.3), and on live hydra (protocol 7) in the presence or absence of rapamycin (Fig. 26.5). Concerning the first approach, the large LTR+ vacuoles remain stained after fixation with aldehydes and LTR can thus be used on animals stained live and immediately fixed. However, we do not find the results highly informative (not shown) when compared either to anti-LC3 immunostaining (Fig. 26.2) or to LTR live staining with live recording (Fig. 26.5).

1. Incubate in the dark 10 polyps for 15 min in 2.5 μM LysoTracker Red DND-99 (Molecular Probes, L 7528) diluted in 500 μl of HM at RT.
2. Wash out the dye several times with HM; relax the animals in 2% urethane for 90 s and fix them in 4% PFA diluted in HM for 15 min at 4 °C.
3. Wash out the fixative in 3 × 5 minutes PBS washes at RT.
4. Stain the nuclei for 10 min in 1 μg/ml Hoechst 33342 diluted in PBS; wash several times in PBS, once in water, and mount the animals with Mowiol 4-88 on untreated slides.
5. Record the LTR fluorescence with a Leica SP2 AOBS confocal microscope under a 40x oil objective lens. Excitation at 543 nm is provided by an argon/krypton laser.

2.3. Analysis of autophagy in hydra cells

2.3.1. The maceration procedure

Another approach used to monitor autophagy in steady state conditions in hydra is based on the immunocytochemical detection of markers for autophagosome formation on macerated tissues. In fact, the maceration technique developed previously (David, 1973) allows for the identification and quantification of each different cell type as upon maceration with acetic acid and glycerol, and subsequently fixation with paraformaldehyde (PFA), the hydra tissues are completely dissociated into individual cells or small clusters, which retain their *in vivo* morphology. Because only small amounts of tissues are required, specific regions along the axis of the animals can be isolated and macerated as for example the different regions of the regenerating halves (Fig. 26.3A; Chera *et al.*, unpublished). Moreover, this maceration procedure can be combined with dye staining or immunostaining. That way the specific hallmarks of autophagy can be identified and followed in the different cell types undergoing autophagy.

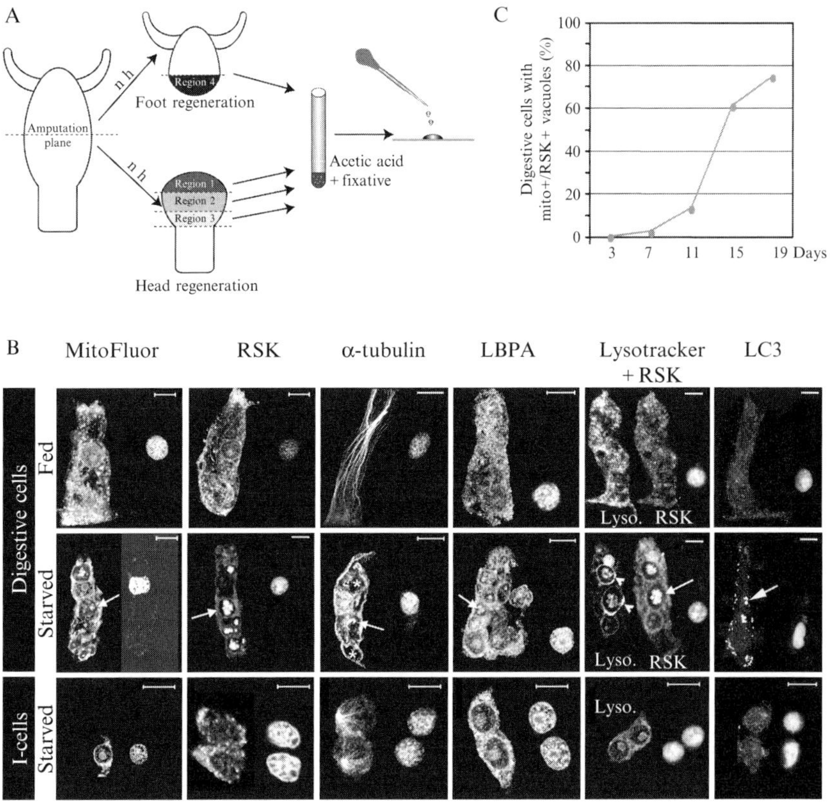

Figure 26.3 Maceration procedure as a tool to monitor autophagy in hydra cells. (A) Scheme depicting the maceration procedure in hydra. Intact or regenerating polyps are incubated with acetic acid and glycerol followed by fixation with PFA as described in protocol 4. For monitoring autophagy during regeneration, hydra are bisected and left to regenerate for the desired amount of time (n hours); the entire regenerative half or the regions of interest are then collected and macerated. (B) Detection of autophagic vacuoles in cells from daily-fed or 14 days starved animals with MitoFluor, anti-RSK, anti α-tubulin, anti-LBPA, LysoTracker (lyso.) combined to anti-RSK (RSK), and anti-LC3 as described in protocol 5. For each cell, the Hoechst staining is shown on the right. Autophagic vacuoles are missing in the digestive cells of fed animals (upper row) as well as in the interstitial cells (i-cells) of starved animals (lower row) but are numerous in the digestive cells of starved hydra (arrows). Note the disturbed α-tubulin network of these cells (third column asterisk) and the strong LysoTracker staining at the margins of the RSK-positive vacuoles (arrowheads). Furthermore, note the large LC3 precipitates in the cytoplasm of starved cells (column 6, arrow). Scale bars: 10 μm. (C) Graph showing the dramatic increase during starvation in the proportion of digestive cells containing large vacuoles whose content was labeled with both MitoFluor and anti-RSK antibody (mito+/RSK+). Approximately 600 digestive cells were counted for each condition. Scale bars: 200 μm (A); 8 μm (B).

2.3.2. Gelatin-coated slides protocol

Prepare the gelatin solution (in water):

1. Prepare a 0.5% gelatin and 0.1% chrome-alum solution in water and mix on a magnetic agitator using a magnetic bar.
2. Microwave for 30 s and mix gently to avoid bubbles, repeat the procedure until the solution is clear (3–5 min).
3. Filter the solution through filter paper. This gelatin solution can be reused up to 1 month.

Preparing the slides:

1. Insert the slides in Universal slide racks and place the racks inside clean black boxes (Roth, T214.1).
2. Wash the slides one time in ethanol 100% for 1 min.
3. Dry the slides and boxes for 30 min using ventilation.
4. Poor the 40 °C gelatin solution in the boxes.
5. Insert the slide racks and incubate them for 2 min with the gelatin solution.
6. Extract the racks from boxes and dry them using ventilation.
7. Store the slides in closed boxes at RT for several weeks.

2.3.3. Protocol 4: Maceration technique (after David, 1973)

1. Wash hydra 3x in HM as described in protocol 1, transfer 10 animals into tubes and incubate at RT in 100 μl 7% glycerol, 7% acetic acid, H_2O. From time to time, vortex the samples mildly until the polyps dissociate and no visible tissue fragments are observed in the tube (30–40 min).
2. Fix the cell suspension by adding an equal volume (100 μl) of 8% PFA. Mix and let them stand in the fixative for 30 min at RT.
3. Add 20 μl of 10% Tween 80 to the cell suspension and spread 50 μl over an area of about 2 cm^2 (2×1) on freshly prepared gelatin-coated slides. Let the cells air-dry on the bench at least 40 h at RT.
4. Once dried, either proceed immediately for immunostaining or store the slides at -20 °C for long periods.

2.3.4. Dye staining and immunostaining of macerated tissues

Vacuoles that contain degraded organelles are characterized as autophagosomes or autolysosomes thanks to the detection of autophagosomal and lysosomal markers, whose presence indicates the induction or maintenance of an autophagic process (Klionsky *et al.*, 2008). To identify reliable markers of autophagy in hydra, we detect in macerates prepared from both starved and fed hydra two organelle dyes, LysoTracker Red DND-99, which accumulates in the acidic compartment of the living cells, and MitoFluor Red 589 (Molecular

Probes), which labels mitochondria and membranes in both live and fixed cells. Moreover, we test several antibodies targeted against the Atg8/LC3 protein (Novus Biological NB100-2220) (Settembre *et al.*, 2008), 6C4-lysobisphosphatidic acid (anti-LBPA, a generous gift from Jean Grünberg) (Kobayashi *et al.*, 1998), ribosomal S6 kinase (anti-RSK, BD Transduction Laboratory, R23820) (Chera *et al.*, 2007), and anti-α-tubulin (Sigma T5168).

As criteria to quantify the induction of the autophagic process during starvation, we quantify the number of digestive cells that display large vacuoles containing MitoFluor-positive organelles and RSK-positive conglomerates as depicted in Fig. 26.3B (first and second columns, arrows) at 3, 7, 11, 15, and 19 days of starvation (Fig. 26.3C). Whereas the number of cells containing such type of vacuoles is very low in regularly fed animals (Fig. 26.3B, upper row), starvation induces a dramatic increase in their number, markedly between 11 and 15 days of starvation, when 60% of the cells appear affected (Fig. 26.3C). This result confirms that autophagy massively occurs after 11 days of starvation as noted in the Western blot analysis of LC3-II (Figs. 26.2A, B). Besides digestive cells, autophagy is also detected in myoepithelial cells from the ectodermal layer (not shown) but not in derivatives from the interstitial cell lineage (Fig. 26.3B, lower row and not shown).

In addition to these two markers, the anti-α-tubulin staining reveals the strong disorganization of the microtubule filaments following prolonged starvation (Fig. 26.3B third column, asterisk). Moreover, the late-endosomal marker LBPA shows a pattern similar to that obtained with MitoFluor in starved cells (Fig. 26.3B fourth column, arrows) labeling endosomal structures located in large vacuoles. The LysoTracker staining provides a sharp staining of the periphery of these large vacuoles (Fig. 26.3B, fifth column, arrowheads), implying that those vacuoles correspond to autolysosomes. When the cells are costained with LysoTracker and the RSK antibody, we note that the content of these LysoTracker-positive vacuoles is also intensely RSK-positive (Fig. 26.3B, fifth column, arrow), indicating that RSK is indeed a reliable marker for detecting the advanced stages of autophagy. Furthermore, the anti-LC3 staining detects numerous sizable LC3 precipitates in the cytoplasm of the epithelial cells of starved polyps (Fig. 26.3B, last column, arrow). These are absent or reduced in size in the cells of regularly fed polyps, confirming that anti-LC3 is a rather specific marker of autophagy in hydra cells, to be used either alone or in combination with other markers (not shown).

2.3.5. Protocol 5: Dye labeling and immunostaining on macerated samples

Note that LysoTracker Red DND-99, used on live animals, and MitoFluor 589, used on fixed cells, emit with similar wavelengths and cannot be combined.

1. Transfer 10 polyps per condition in a 24-well dish and incubate in 2.5 μM LysoTracker Red (Molecular Probes, L 7528) diluted in 500 μl of HM for 15 min at RT.
2. Wash the animals 4 times in 1 ml of HM, and macerate as indicated in protocol 4.
3. Once cells are dried on slides, wash them 2 × 10 minutes in 1 ml of PBS, block in 1 ml of PBS, 2% BSA, for 1 h at RT.
4. Incubate O/N at 4 °C with the primary antibodies: anti-RSK 1:1000 (Transduction Laboratories, cat. 610226), anti-LC3 1:200 (Novus Biological, NB 100–2220), anti-α-tubulin 1:2,000 (Sigma, T5168), and anti-6C4-phospholipd LBPA (1:50) (Kobayashi *et al.*, 1998).
5. Wash the slides several times in PBS, incubate for 2 h at RT in the secondary antibody coupled with AlexaFluor dyes (AlexaFluor 555 or AlexaFluor 488) 1:600.
6. Wash the slides in PBS, stain them in Hoechst 33342 diluted in water 1 μg/ml (Molecular Probes) for 5 minutes, wash 2 × 10 minutes in PBS, 1 × 2 minutes in water, and mount with 30 μl of Mowiol.
7. When required, stain with the dye MitoFluor 589, 200 nM diluted either in water or in PBS (Molecular Probes) for 20 min before Hoechst staining.

2.4. Pharmacological modulations of autophagy in hydra

Autophagic degradation is a multistep process regulated by a number of pathways. As a consequence, the autophagy process can be modulated by pharmacological agents that act at various stages. The inhibitors wortmannin, 3-methyladenine, and LY294002 act at the sequestration step (Petiot *et al.*, 2000), whereas most other inhibitors impair the formation and function of autolysosomes by inhibiting lysosomal enzymes or vesicle fusion or by elevating lysosomal pH (Klionsky *et al.*, 2008). The main limitation in the use of these pharmacological agents is their low specificity; in most cases they are not exclusively targeted to autophagy, implying that the nonautophagic effects should be carefully considered. As a first approach we test several of these modulators on intact hydra that are either daily-fed or maintained starved. For each context, we follow the drug-induced morphological modifications (Fig. 26.4A) and detect the level of LC3-II by Western analysis (Fig. 26.4B). In the case of rapamycin, we also detect the formation of acidic vesicles in live hydra stained with LTR (Fig. 26.5).

2.4.1. Rapamycin, a candidate activator of autophagy in daily-fed hydra

The TOR (target of rapamycin) protein negatively controls autophagy in yeast and mammals. It is a serine/threonine kinase that belongs to the PI3K-related family and exerts a central role in nutrient-sensing signal transduction,

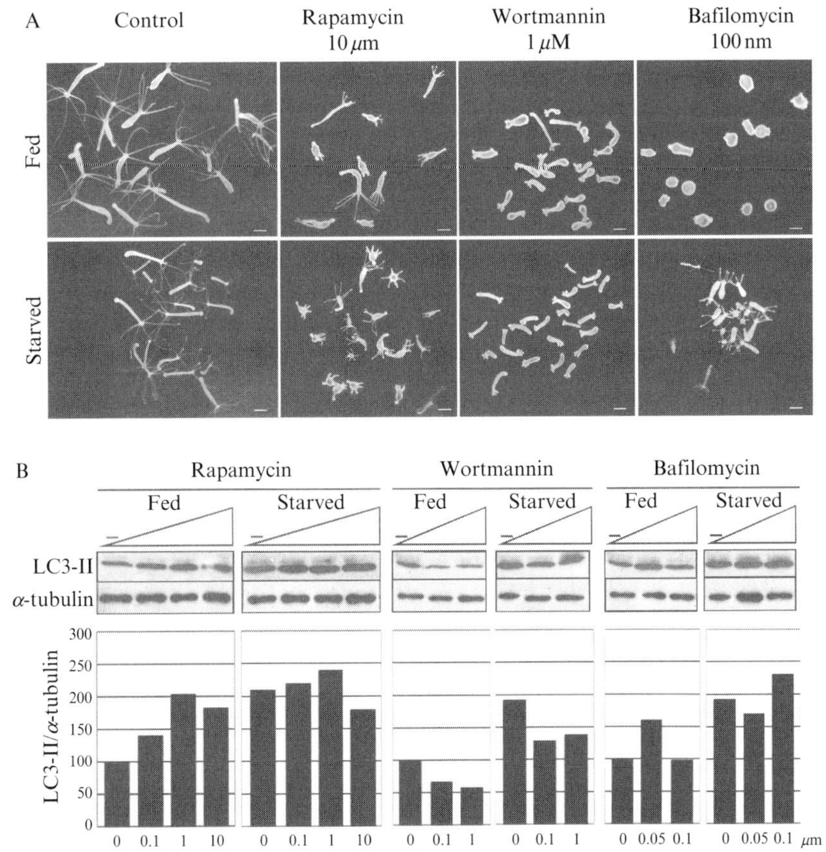

Figure 26.4 Drug-induced modulations of morphology and autophagy in daily-fed and starved hydra. (A) Morphological changes induced in hydra either daily-fed (upper row) or starved for 19 days (lower row) after 12 h exposure to rapamycin (10 μM), wortmannin (1 μM), or bafilomycin A_1 (100 nM), as described in protocol 6. Note the constriction of the lower-body column in rapamycin-treated hydra, the toxic effect of wortmannin and the resistance of starved hydra to the toxic effect of bafilomycin A_1; these are reduced in size but display a normal shape with full tentacles. Scale bar: 1 mm. (B) Western blot analysis of the LC3-II content in daily-fed and 19 days starved hydra exposed to rapamycin (0.1, 1, and 10 μM), wortmannin (0.1 and 1 μM), and bafilomycin (50 and 100 nM) as in (A) (the LC3-I band is faintly seen above the LC3-II band). The relative index of LC3-II amount was normalized against α-tubulin by densitometric analysis with ImageJ software and is expressed as percentage from the fed control value.

regulation of translation, and cell-cycle progression (Blommaart *et al.*, 1995; Kamada *et al.*, 2000; Klionsky *et al.*, 2005; Noda and Ohsumi, 1998). The lipophilic macrolide antibiotic rapamycin induces autophagy by forming a complex with the immunophilin FK506-binding protein 12 (FKBP12), which then binds to and inactivates mTOR, leading to a derepression of

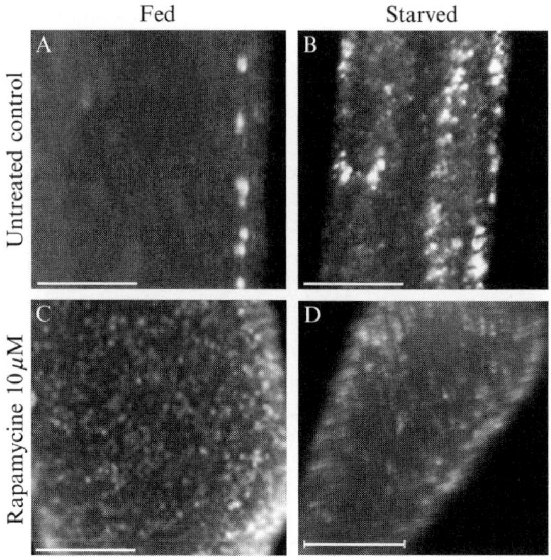

Figure 26.5 LysoTracker detection in live hydra. Live intact hydra either daily-fed (A, C) or starved for 24 days (B, D) were stained with LTR as described in protocol 7. After 12 h rapamycin treatment, both daily-fed (C) and starved (D) hydra showed numerous bright LTR spots spread all over the body. This increase in LTR signal was more evident in daily-fed hydra than in the starved condition. Bars: 200 μm (A, B); 400 μm (C, D).

autophagy. To verify whether rapamycin affects the level of autophagy in hydra, starved and daily-fed animals are exposed for 12 h to different concentrations of rapamycin (0.1 μM, 1 μM, and 10 μM). Surprisingly, we observe in both starved and daily-fed hydra a rather unusual morphological change, as the polyps become strongly constricted in the lower part of the body column (Fig. 26.4A, column 2). This change is even more obvious in starved hydra. In both contexts, hydra are reduced in size but are in good shape, with intact tentacles and reacting to mechanical stimulus by contracting as do untreated hydra. Concerning the LC3-II amount (Fig. 26.4B), fed and starved hydra show a gradual increase up to 1 μM, indicating an increase in the number of autophagosomes, though with a lower increment in starved than in fed polyps. When levels of rapamycin increase to 10 μM, the LC3-II amount decreases in both fed and starved animals. This result was not expected and requires further investigation to determine whether the level of autophagy is indeed reduced in this condition. In fact, the LC3-II content is not a definitive indicator of the total autophagic flux and further experiments where exposure to rapamycin is performed in the presence of lysosomal inhibitors such as pepstatin A, E64d, and leupeptin should provide more information about the LC3 turnover rate. In addition, side effects of high rapamycin concentration on mTOR signaling cannot be excluded, as rapamycin is able to repress protein translation, arrest the cell cycle, and alter cell size. Moreover, on chronic exposure to rapamycin, the rapamycin-insensitive TORC2 complex can be destabilized, affecting downstream signaling through Akt1 (Sarbassov *et al.*, 2006).

2.4.2. Wortmannin, a candidate inhibitor of autophagy in daily-fed and starved hydra

Autophagic flux can be inhibited by targeting the sequestration event, which requires phosphatidylinositol 3-kinase (PI3K) activity. Wortmannin, a furanosteroid metabolite of the fungi *Penicillium funiculosum*, is a specific PI3K inhibitor that inhibits autophagy at the sequestration step (Arcaro and Wymann, 1993; Blommaart *et al.*, 1997). When hydra, either daily-fed or starved, are treated with wortmannin at 1 μM for 12 h, they rapidly show dramatic morphological changes with a reduced body size and loss of their tentacles (Fig. 26.4A, column 3). At lower concentration (0.1 μM), the animals are not affected to the same degree (not shown). In fact, immunoblot analysis shows that the level of the membrane-bound LC3-II is significantly lowered when starved and daily-fed animals are exposed to 0.1 and 1 μM wortmannin (Fig. 26.4B). Interestingly, the starved and daily-fed animals are inhibited in the same manner by wortmannin, at the morphological and biochemical levels, implying a similar sensitivity to this pharmacological agent.

2.4.3. Bafilomycin A_1, a candidate inhibitor of autophagy in daily-fed and starved hydra

Bafilomycin A_1, a macrolide antibiotic isolated from *Streptomyces* sp. is an extremely potent inhibitor of the vacuolar type H^+-ATPases (Bowman *et al.*, 1988). Bafilomycin A_1 plays an important role in maintaining the acidic environment of the endosomes, lysosomes, and other secretory vesicles. At low concentrations (in the nanomolar range), bafilomycin A_1 increases the pH of acidic vesicles by disrupting the proton gradient and may prevent the fusion of autophagosomes with lysosomes. As a consequence, an accumulation of the autophagosomes can be observed (Yamamoto *et al.*, 1998). However, other data indicate that bafilomycin blocks the activity of lysosomal hydrolases but does not affect fusion (Fass *et al.*, 2006).

In contrast to wortmannin, daily-fed and starved hydra display different sensitivities to bafilomycin A_1. In both cases, exposure to 200 nm bafilomycin is rapidly lethal, whereas exposure to 50 nM does not alter the animal size and shape (not shown). However, exposure to 100 nM bafilomycin is clearly more toxic for the daily-fed hydra than for the starved ones, the former becoming dramatically scrubby with short and buttoned tentacles (Fig. 26.4A; compare the upper and lower panels in column 4). At the biochemical level, the level of LC3-II remained unchanged or decreased (other experiment) in daily-fed hydra exposed to 100 n*M* bafilomycin that were actually dying but increased in daily-fed hydra exposed to 50 n*M* bafilomycin (Fig. 26.4B). In starved animals, a significant increase in LC3-II level is noticed only at 100 n*M*. These results suggest that bafilomycin can inhibit autophagy in hydra, though at different concentrations in daily-fed and starved hydra; the former are more sensitive than the latter.

2.4.4. Protocol 6: Hydra treatment with rapamycin, wortmannin, and bafilomycin

1. Dissolve rapamycin, wortmannin, and bafilomycin (LC Laboratories) in DMSO to obtain 10 m*M* (rapamycin, wortmannin) and 1 m*M* (bafilomycin) stock solutions. Aliquot and store at −20 °C. Freshly dilute the stock solutions in HM prior any hydra treatment.
2. Dilute in 10 ml of HM each drug, rapamycin (0.1, 1, 10 μ*M*), wortmannin (0.1, 1, 10 μ*M*), bafilomycin (50, 100, 200 n*M*), and 0.1% DMSO for the control condition. Fill 9 ml of each solution in the wells of 6-well plate.
3. Wash 30 hydra per condition in HM and transfer to 1.5-ml tubes; aspirate HM, and add 1 ml of the diluted drug solution.
4. Dispatch the hydra of each condition in the corresponding wells of the 6-well plate. Keep the dish in the dark for 12 h.
5. After treatment, wash hydra several times in HM, image them live using a stereomicroscope (Leica MZ16FA), and prepare total protein extracts as in protocol 1.

2.4.5. LysoTracker Red staining in live animals

To assess autophagy-induced modulations of the acidic compartment (including all of the previously mentioned vacuoles), starved and daily-fed hydra are loaded with LysoTracker Red and examined with a stereomicroscope after 12 h exposure to 10 μ*M* rapamycin (Fig. 26.5). In daily-fed animals, only a few number of highly fluorescent LTR spots are observed along the body axis, which corresponds to cells with an increased lysosomal-derived activity (Fig. 26.5A). The red intense fluorescence detected in the tentacles is attributed to nematocytes (not shown). In starved animals, the number of LTR-positive vacuoles increases along the body column (Fig. 26.5B). When hydra are exposed to rapamycin, numerous LTR bright spots are observed spread all over the body, in the ectoderm as well as in the endoderm of starved and daily-fed hydra (Fig. 26.5C and D). Therefore, rapamycin exposure significantly increases the number of acidic vacuoles when compared to nontreated animals, but this increase is more pronounced in the daily-fed than in the starved animals. This result is consistant with the level of LC3-II detected in rapamycin-treated hydra after Western analysis.

2.4.6. Protocol 7: LysoTracker Red staining in live animals

1. Treat 10 hydra, either daily-fed or 21 days starved, with rapamycin (10 μM) or 0.1% DMSO alone for 12 h as described in protocol 6.
2. Incubate for 15 min in 1 μM LysoTracker Red diluted in 500 μl of HM.
3. Wash out the dye several times in HM.
4. To image live hydra, keep them relaxed in 0.5% urethane diluted in HM under a fluorescence stereomicroscope (Leica MZ16FA equipped with a DFC300 FX Leica camera). The red fluorescence is measured with a Texas red filter.

2.5. Functional identification of autophagy regulators

2.5.1. RNA interference obtained through hydra dsRNAs feeding is a potent functional tool

The dsRNA-mediated genetic interference approach has been successfully applied to the hydra model system, allowing the possibility to study the signaling pathways and the genetic circuitry regulating homeostasis, stress response, and head regeneration through loss-of-function assays (Galliot *et al.*, 2006, 2007). Among the different RNAi procedures (injection, soaking, electroporation, or feeding), the most suitable for our system is based on the regular feeding of animals with bacteria that produce dsRNAs and are embedded in agarose; these can be administrated to animals over long periods without any toxic effects (Chera *et al.*, 2006). Moreover, we show that gene silencing through RNAi is well adapted for deciphering the epistatic relationships between candidate genes expressed in a given cellular process, thus making it possible to decipher the genetic circuitry underlying its regulation (Miljkovic-Licina *et al.*, 2007).

2.5.2. Knocking down *Kazal1* expression leads to excessive autophagy in hydra

To decipher the regulation and function of the autophagy process in hydra, and especially its potential role during hydra regeneration, we silenced the *Kazal1* gene that encodes a serine protease inhibitor related to the vertebrate SPINK genes. *Kazal1* is expressed along the body column in the gland cells that are located in the endodermal cell layer and secrete the digestive enzymes. *Kazal1* is not expressed in the head and foot regions (Fig. 26.6A, left panel). Recurring feedings with *Kazal1* dsRNA completely abolish *Kazal1* expression in the gland cells of the *Kazal1 (RNAi)* animals (Fig. 26.6A, right panel). Surprisingly, at the cellular level the animals exhibit large vacuoles in their gland cells but also in the neighboring digestive cells, the endodermal epithelial cells. These vacuoles contain cytoplasmic organelles (Chera *et al.*, 2006) and RSK+

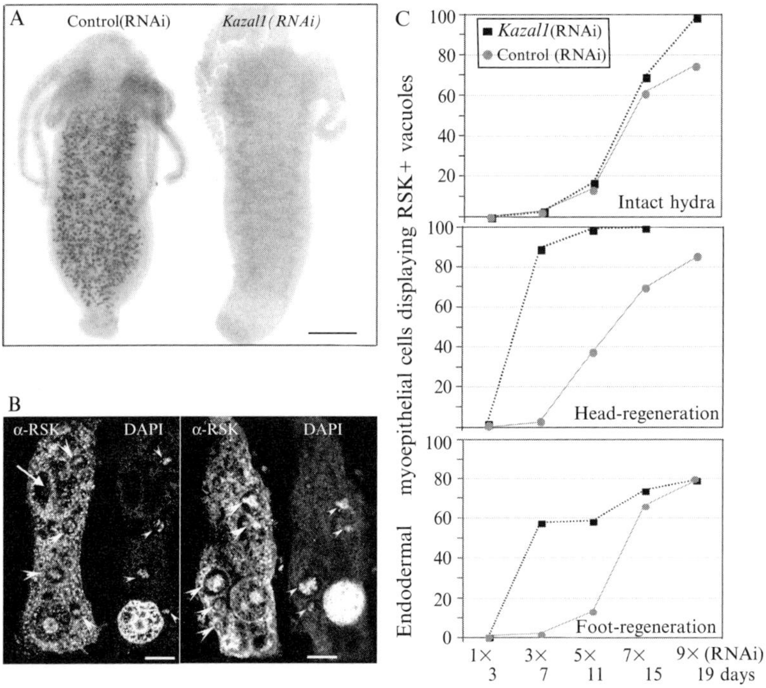

Figure 26.6 Silencing the hydra Kazal-1 serine-protease inhibitor induced autophagy. (A) *Kazal-1* expression detected in gland cells is knocked down in hydra polyps exposed 5 times to *Kazal1* dsRNAs (*Kazal1(RNAi)* 5x, right) but normal in hydra exposed 5 times to control dsRNAs (control(RNAi) 5x, left), as described in protocol 8. (B) Detection of the strongly vacuolated phenotype in the digestive cells from head- and foot-regenerating tips (left and right, respectively) of *Kazal1(RNAi)* 5x hydra. Anti-RSK immunostaining and DAPI staining were performed on tips macerated 1 h after amputation. Note the large intracellular spaces (arrows) that characterize the digestive cells and the presence of numerous RSK+ vacuoles identified as autophagosomes (large arrowheads). Small arrowheads indicate artifactual DAPI staining. (C) Graphs depicting the gradual increase in the proportion of the digestive cells displaying RSK+ vacuoles in intact (upper panel), head-regenerating (middle panel) and foot-regenerating (lower panel) hydra. For each context, 2 hydra populations were compared, the control(RNAi) one, repeatedly exposed to vector dsRNAs (gray curve) that shows the progressive formation of autophagosomes on starvation, and the *Kazal1*(RNAi) one, repeatedly exposed to *Kazal1* dsRNAs (black curve). None of these populations were fed during the experiment but both received every other day the bacterial-agarose mixture and were amputated on alternate days. Note the dramatic increase in the proportion of cells containing autophagosomes after 3 exposures to *Kazal1* dsRNAs in regenerating hydra. For each condition 600 digestive cells were counted. Scale bars: 200 μm (A); 8 μm (B).

conglomerates as shown in Fig. 26.6B, and are thus identified as autophagosomes. The monitoring of digestive cells over a long period of time (19 days) where dsRNAs exposures are repeated (up to 9 feedings) shows a dramatic increase in the number of vacuolated cells in *Kazal1(RNAi)* hydra compared to control polyps maintained under the same starvation conditions (Fig. 26.6C upper panel). In *Kazal1(RNAi)* hydra, we also note additional morphological alterations of the cells that underwent massive autophagy, such as their detachment from the basal membrane, loss of epithelial polarity, disruption of the microtubule cytoplasmic organization, fusion of the autophagic vacuoles forming giant vacuoles, and cell size decrease. All together these changes are interpreted as a progressive process leading to an excessive autophagic degradation and finally to cell death (Chera *et al.*, 2006).

Interestingly, the chronic excessive autophagic event identified in intact *Kazal1(RNAi)* animals is dramatically enhanced 1 h after amputation at the regenerating tip of head- but also foot-regenerating animals exposed at least 3 times to dsRNAs (Fig. 26.6C, middle and lower panels). Moreover, after 7x feedings the *Kazal1*(RNAi) hydra become unable to regenerate, suggesting that the protection of the endodermal cells against excessive autophagy is an absolute requirement for regeneration (Chera *et al.*, 2006; Galliot, 2006). However, the physiological role of autophagy during hydra regeneration was not tested in this study and remains unknown. Preliminary studies suggest that, surprisingly, in regularly fed animals autophagy is not required for either head or foot regeneration (Buzgariu, unpublished). Nevertheless it is possible that the time-point chosen for monitoring the amputation-induced autophagy (1 h after amputation) was not appropriate, as it might be a very rapid and transient event.

2.5.3. Protocol 8: Production of *Kazal1* knocked-down hydra on RNA interference

The RNAi procedure detailed here was initially described in (Chera *et al.*, 2006):

1. The *Kazal-1* cDNA was inserted in the pPD129.36 (L4440) double T7 (http://www.addgene.org/pgvec1?f=c&plasmidid=1654&cmd=viewseq) and transformed into HT115(DE3) bacteria. This strain is tetracycline resistant, it lacks RNAse III and can express IPTG-inducible T7 polymerase (http://wormbase.org/db/gene/strain?name=HT115(DE3); class=Strain). Such constructs should be verified regularly through sequencing as this vector is unstable and frequently recombines.
2. As negative control for the RNAi feeding procedure, use either pPD129.36 (L4440)_(no insert, empty L4440 vector) or pLT61.1 vector (L4440 *unc-22*) (http://www.addgene.org/pgvec1?f=c&identifier=1690&atqx=pLT61.1&cmd=findpl). The pLT61.1 construct produces dsRNAs targeted against the *unc-22* nematode gene. It is considered a standard control by labs that adapted the RNAi feeding strategy from

C. elegans (Timmons *et al.*, 2001) to planarians or hydra (Chera *et al.*, 2006; Newmark *et al.*, 2003).

3. Transform into HT115(DE3) competent cells 2 days before starting the RNAi experiment and plate the transformed bacteria on ampicillin (50 μg/ml) + tetracycline (12.5 μg/ml) agar plates. Do not use glycerol stocks; always retransform.
4. Inoculate 1 colony in 4 ml of starter culture in Terrific Broth (Sambrook and Russell, 2001) containing ampicillin 50 μg/ml, tetracycline 12.5 μg/ml (TBAT) and grow O/N (>16 h) at 37 °C in long glass tubes vigorously shaken (250 rpm).
5. For each construct, start duplicated precultures by adding 200 μl of the starter culture to 2 ml of TBAT in 14-ml tubes (Falcon, 2059). Grow the preculture for 90 min at 37 °C to reach $OD_{595} = 0.4$.
6. Induce with IPTG (400 μM final) and grow for 4 h at 37 °C vigorously shaken. Do not overgrow the culture: dead bacteria do not produce dsRNAs. Do not increase the IPTG concentration as the RNAi phenotypes will become hypomorphic (http://genomebiology.com/2000/2/1/RESEARCH/0002/table/T2).
7. Melt 500 μl/sample 1% low-melting agarose (LMP Qbiogene) in HM and keep it preheated at 37 °C. Alternatively, the agarose solution can be stored in 50-ml tubes at 37 °C for weeks.
8. Transfer the bacterial culture to 2-ml microcentrifuge tubes and spin at 5000 rpm for 10 min at 4 °C in a precooled centrifuge.
9. For each construct, resuspend the pellet of 1 tube in 100 μl HM and transfer into a second tube to resuspend that pellet (that way, the bacteria of both tubes are collected in 100 μl final volume).
10. Add 100 μl of preheated 1% low-melting agarose and keep the bacteria-agarose mixture on ice until it gets solidified (almost immediately).
11. Collect 50–70 healthy hydra starved for 2 days and transfer them into 1.5-ml tubes. Adjust the volume to 250 μl of HM and add 250 μl of HM, 10 mM Tris, pH 7.5.
12. Add glutathione to 100 μM final concentration (2 μl of a 25 mM stock), and incubate for 5 min. Glutathione efficacy can be checked with a stereomicroscope before grinding the agar; if efficient, the hydra polyps have a wide-open mouth.
13. Transfer the bacteria-agarose solidified mix onto the outside of an artemia filter (Hobby Dohse) and grind it with the help of a glass slide. Collect the mixture on the inside of the filter with the edge of a glass slide, and serve it to the animals for 1.5 h.
14. Gently shake the tubes every 10–15 min, as hydra tend to clump on top of one another, decreasing their capability to eat the dsRNA-producing bacteria.
15. After 90 min, extensively wash the hydra with HM until the agarose chunks are removed, usually 10x, first in tubes, then in dishes. Repeat the washing the next day.

16. The day following each feeding, record animal survival, animal shape, animal size, and budding rate (budding rate over 1 week is usually quite informative).
17. Repeat RNAi feeding every other day.
18. Gene knockdown is best proved by RT-PCR or *in situ* hybridization performed after several dsRNAs exposures, as described previously (Miljkovic-Licina *et al.*, 2007).
19. To monitor autophagy in *Kazal1*(RNAi)-regenerating hydra, perform midgastric section the day following each dsRNA exposure, and 1 h after amputation isolate with a scalpel the different regions to be macerated, as depicted in Fig. 26.3A.
20. Detect the autophagy phenotype by macerating either 1 intact hydra or 10 regenerating tips, and stain the cells with dye and anti-RSK antibody.

3. Concluding Remarks

In this work we detail several methods to investigate the autophagy process in daily-fed and starved hydra at three distinct levels: biochemical, tissue, and cellular, as summarized in Table 26.1 Among the different dyes and antibodies tested here, we show that the conventional anti-LC3 antibody used in other model systems is fully appropriate to detect and measure autophagy in hydra, in immunostainings performed either on Western blots (Fig. 26.2, Fig. 26.4) or on whole mounts (Fig. 26.2) or on macerated tissues (Fig. 26.3). The analysis of autophagic vacuoles on macerated tissues actually appears well suited for quantitative analyses, as it is more precise and more reliable than dye staining or immunostainings performed on whole-mount animals, which are nevertheless advantageous for providing information about regional differences in the regulation of autophagy (Fig. 26.2B). The cellular analysis of macerated tissues also appears far more sensitive than Western analysis. For example, a maximal 2-fold increase in LC3-II is detected over the first 13 days of starvation on Western blot, then reaching a plateau value (Fig. 26.2A), whereas the increase in the proportion of digestive cells containing MitoFluor+/RSK+ vacuoles is actually dramatic after 11 days of starvation, starting from 0% at day 3, reaching 10% at day 11, 60% at day 15, and almost 80% at day 19 (Fig. 26.3B). Moreover, the combination of the anti-RSK antibody with the LysoTracker Red dye also provides a very useful tool to identify autolysosomes and distinguish them from autophagosomes. When macerates are stained with LTR and RSK, 88% of the vacuoles display LTR+ outlines and 78% contain RSK+ conglomerates as shown in Fig. 26.3C. We assume that those LTR+ and LTR+/RSK+ vacuoles correspond to autolysosomes, whereas the large vacuoles that are RSK+ only (about 8%) correspond to autophagosomes.

Table 26.1 Tools to monitor autophagy in hydra

Staining	Macerates	Whole-mount	Western	References
Mitochondrial dye MitoFluor Red 589 (Molecular Probes)	1. Label fixed cells for 20 min; sequestration in autophagosomes	2. Label fixed tissues for 10 min; sequestration in autophagosomes	—	1. Chera *et al.* 2006 2. unpublished
Lysosome dye LysoTracker Red (LTR, Molecular Probes)	Label live hydra for 15 min prior to maceration;strongly stains outlines of autolysosomes	Label live hydra for 15 min; Starvation and rapamycin patterns	—	This work
Anti-LC3 antibody (Novus laboratory)	Autophagosomes in myoepithelial cells	Starvation-induced pattern	LC3-I (+/−) LC3-II (++)	This work
6C4 anti-LBPA antibody (Gruenberg's lab)	Late endosomes; sequestered in autophagosomes	nd	nd	Kobayashi *et al.*, 1998; Chera *et al.*, 2006
Anti-RSK antibody (Transduction Laboratoires)	1. autophagosomes 2. autolysosomes	Starvation-specific pattern?	Not tested during starvation	Chera *et al.*, 2006
Anti-α-tubulin antibody (Sigma)	1. Disruption of the tubulin network 2. α-tubulin inside the autophagosomes	Not of interest for autophagy	Not of interest for autophagy	Chera *et al.*, 2006 This work
Anti-RSK antibody + LTR	Autolysosomes	nd	—	This work

A potential problem in the monitoring of autophagy in the digestive cells (endodermal epithelial cells) of regularly fed animals is the presence of digestive vacuoles. These are large vacuoles (1 μm) in intimate contact with the cell membrane, which usually contain fine granules or filaments (Lentz, 1966). As dyes or antibodies easily penetrate into the digestive vacuoles, it is essential that the pattern generated by this nonspecific staining is clearly distinguished. In fluorescence, the digestive vacuoles have no clear outlines and their content is homogeneously fuzzy, as no specific binding of the antibody or dye takes place. For animals exposed to dsRNAs, the identification of digestive vacuoles is easy, as they often contain small chunks of agarose that exhibit mild autofluorescence, making them clearly visible.

A second problem is the presence of apoptotic phagosomes in the cytoplasm of endodermal epithelial cells, as these cells carry out the efferocytosis (phagocytosis of apoptotic cells) of the apoptotic bodies when the surrounding cells die by apoptosis (Chera *et al.*, submitted). The internalized apoptotic bodies have a characteristic spherical shape, containing a highly condensed chromatin surrounded by a thin layer of cytoplasm that is strongly RSK-positive. The RSK staining coats the condensed chromatin of the apoptotic bodies. In contrast the autophagosomes are characterized by a cleft between the outlines of the vacuole and their content. Moreover, RSK appears strongly positive on conglomerates located inside the autophagosome but never stains the autophagosome membrane. Therefore, with these criteria, the distinction between digestive vacuoles, autophagosomes, and apoptotic phagosomes should be quite obvious for the trained microscopist.

The analysis of LC3-II levels after exposure to pharmacological agents shows that rapamycin at moderate concentrations (0.1 μM, 1 μM) induces autophagy more efficiently in daily-fed than in starved hydra, where the level of autophagy is already increased. The decrease in LC3-II levels observed at higher concentration (10 μM) in Western analysis might correspond to a massive autophagy with an extensive degradation of LC3-II bound to autolysosomes, as suggested by the significant increase of the acidic compartment in live hydra exposed to same concentration of rapamycin (Fig. 26.5C, 26.5D). The concomitant increase in the autophagic flux should be tested in the presence of lysosomal protease inhibitors, which partially inhibit LC3-II degradation. In contrast, wortmannin inhibits autophagy similarly in daily-fed and starved hydra when used at 0.1 μM and 1 μM concentrations. Concerning bafilomycin A_1, it is interesting to note that daily-fed hydra appear far less resistant to bafilomycin A_1 toxicity than starved hydra, as if high levels of autophagy are protective in that context. In all cases, autophagy is primarily regulated in the endodermal myoepithelial cells that carry out the digestive function in homeostatic conditions. Nevertheless, we also report here that starvation (Fig. 26.2C–26.2F) and rapamycin treatment (Fig. 26.5) can likely induce autophagy in ectodermal myoepithelial cells.

Given the potential of the hydra model system to highlight the function(s) of autophagy in homeostatic and developmental contexts, there is a clear need for establishing methods that would measure the flux of autophagy. Indeed, as reporter constructs can be expressed in adult polyps through electroporation (Miljkovic *et al.*, 2002), the next step is to more precisely monitor the regulation of autophagy in hydra by using reporter constructs that prove to be highly efficient in other model systems, such as the chimeric GFP-LC3 construct to monitor cleavage of GFP from LC3, or the use of lysosomal protease inhibitors along with LC3-II turnover (Klionsky *et al.*, 2008).

Besides the regulation of autophagy in the context of homeostasis, there is also the need to investigate more precisely the physiological role of autophagy during regeneration. We previously showed that excessive autophagy is deleterious for cell survival after amputation (Chera *et al.*, 2006; Galliot, 2006). More recently we showed that a massive wave of apoptosis is taking place in head-regenerating tips soon after amputation and that this apoptotic event is required to trigger the head-regeneration program (Chera *et al.*, submitted). Therefore, numerous questions are currently pending: Is autophagy induced at any time of the regeneration process in the wild-type context? Is a low level of autophagy helpful or detrimental for the regeneration process? The expression and functional analysis of the genes that regulate the autophagy process should address these questions and provide useful information for understanding the core mechanisms underlying regenerative processes in the animal kingdom.

Abbreviations

BSA:	bovine serum albumin
dsRNAs:	double stranded RNAs
HM:	hydra medium
LTR:	LysoTracker Red
O/N:	overnight
PBS:	phosphate buffered saline
PFA:	paraformaldehyde
PI3K:	phosphatidylinositol 3-kinase
RNAi:	RNA interference
RT:	room temperature
SPINK:	Serine Protease Inhibitor Kazal-type
TBS:	Tween phosphate buffered saline

ACKNOWLEDGMENTS

The authors thank Ovidiu Olteanu for help with the figures, Jean-Claude Martinou for advice, and Jean Gruenberg for the 6C4 anti-LBPA antibody. The laboratory is funded by the Swiss National Foundation, the Geneva State, the Claraz Donation, and the Academic Society of Geneva.

REFERENCES

Arcaro, A., and Wymann, M. P. (1993). Wortmannin is a potent phosphatidylinositol 3-kinase inhibitor: The role of phosphatidylinositol 3,4,5-trisphosphate in neutrophil responses. *Biochem. J.* **296**(Pt. 2), 297–301.

Bampton, E. T., Goemans, C. G., Niranjan, D., Mizushima, N., and Tolkovsky, A. M. (2005). The dynamics of autophagy visualized in live cells: From autophagosome formation to fusion with endo/lysosomes. *Autophagy* **1,** 23–36.

Blommaart, E. F., Krause, U., Schellens, J. P., Vreeling-Sindelarova, H., and Meijer, A. J. (1997). The phosphatidylinositol 3-kinase inhibitors wortmannin and LY294002 inhibit autophagy in isolated rat hepatocytes. *Eur. J. Biochem.* **243,** 240–246.

Blommaart, E. F., Luiken, J. J., Blommaart, P. J., van Woerkom, G. M., and Meijer, A. J. (1995). Phosphorylation of ribosomal protein S6 is inhibitory for autophagy in isolated rat hepatocytes. *J. Biol. Chem.* **270,** 2320–2326.

Bode, H. R. (1996). The interstitial cell lineage of hydra: A stem cell system that arose early in evolution. *J. Cell Sci.* **109,** 1155–1164.

Bosch, T. C. (2007). Why polyps regenerate and we don't: Towards a cellular and molecular framework for Hydra regeneration. *Dev. Biol.* **303,** 421–433.

Bosch, T. C., and David, C. N. (1984). Growth regulation in Hydra: Relationship between epithelial cell cycle length and growth rate. *Dev. Biol.* **104,** 161–171.

Bowman, E. J., Siebers, A., and Altendorf, K. (1988). Bafilomycins: A class of inhibitors of membrane ATPases from microorganisms, animal cells, and plant cells. *Proc. Natl. Acad. Sci. USA* **85,** 7972–7976.

Chera, S., de Rosa, R., Miljkovic-Licina, M., Dobretz, K., Ghila, L., Kaloulis, K., and Galliot, B. (2006). Silencing of the hydra serine protease inhibitor Kazal1 gene mimics the human Spink1 pancreatic phenotype. *J. Cell. Sci.* **119,** 846–857.

Chera, S., Kaloulis, K., and Galliot, B. (2007). The cAMP response element binding protein (CREB) as an integrative HUB selector in metazoans: Clues from the hydra model system. *Biosystems* **87,** 191–203.

David, C. N. (1973). A quantitative method for maceration of hydra tissue. *Wilhelm Roux Archiv.* **171,** 259–268.

David, C. N., and Campbell, R. D. (1972). Cell cycle kinetics and development of Hydra attenuata. I. Epithelial cells. *J. Cell Sci.* **11,** 557–568.

Dubel, S., Hoffmeister, S. A., and Schaller, H. C. (1987). Differentiation pathways of ectodermal epithelial cells in hydra. *Differentiation* **35,** 181–189.

Fass, E., Shvets, E., Degani, I., Hirschberg, K., and Elazar, Z. (2006). Microtubules support production of starvation-induced autophagosomes but not their targeting and fusion with lysosomes. *J. Biol. Chem.* **281,** 36303–36316.

Fujisawa, T. (2003). Hydra regeneration and epitheliopeptides. *Dev. Dyn.* **226,** 182–189.

Galliot, B. (2006). Autophagy and self-preservation: A step ahead from cell plasticity? *Autophagy* **2,** 231–233.

Galliot, B., Miljkovic-Licina, M., de Rosa, R., and Chera, S. (2006). Hydra, a niche for cell and developmental plasticity. *Semin. Cell. Dev. Biol.* **17,** 492–502.

Galliot, B., Miljkovic-Licina, M., Ghila, L., and Chera, S. (2007). RNAi gene silencing affects cell and developmental plasticity in hydra. *C. R. Biol.* **330,** 491–497.

Galliot, B., and Schmid, V. (2002). Cnidarians as a model system for understanding evolution and regeneration. *Int. J. Dev. Biol.* **46,** 39–48.

Gierer, A., Berking, S., Bode, H., David, C. N., Flick, K., Hansmann, G., Schaller, H., and Trenkner, E. (1972). Regeneration of hydra from reaggregated cells. *Nature New Biol.* **239,** 98–101.

Holstein, T. W., Hobmayer, E., and Technau, U. (2003). Cnidarians: An evolutionarily conserved model system for regeneration? *Dev. Dyn.* **226,** 257–267.

Kabeya, Y., Mizushima, N., Ueno, T., Yamamoto, A., Kirisako, T., Noda, T., Kominami, E., Ohsumi, Y., and Yoshimori, T. (2000). LC3, a mammalian homologue of yeast Apg8p, is localized in autophagosome membranes after processing. *EMBO J.* **19,** 5720–5728.

Kamada, Y., Funakoshi, T., Shintani, T., Nagano, K., Ohsumi, M., and Ohsumi, Y. (2000). Tor-mediated induction of autophagy via an Apg1 protein kinase complex. *J. Cell. Biol.* **150,** 1507–1513.

Kamm, K., Schierwater, B., Jakob, W., Dellaporta, S. L., and Miller, D. J. (2006). Axial patterning and diversification in the cnidaria predate the Hox system. *Curr. Biol.* **16,** 920–926.

Klionsky, D. J., Abeliovich, H., Agostinis, P., Agrawal, D. K., Aliev, G., Askew, D. S., Baba, M., Baehrecke, E. H., Bahr, B. A., Ballabio, A., Bamber, B. A., Bassham, D. C., *et al.* (2008). Guidelines for the use and interpretation of assays for monitoring autophagy in higher eukaryotes. *Autophagy* **4,** 151–175.

Klionsky, D. J., Meijer, A. J., and Codogno, P. (2005). Autophagy and p70S6 kinase. *Autophagy* **1,** 59–60.

Kobayashi, T., Stang, E., Fang, K. S., de Moerloose, P., Parton, R. G., and Gruenberg, J. (1998). A lipid associated with the antiphospholipid syndrome regulates endosome structure and function. *Nature* **392,** 193–197.

Lentz, T. L. (1966). The cell biology of hydra North-Holland Publishing, Amsterdam.

Miljkovic, M., Mazet, F., and Galliot, B. (2002). Cnidarian and bilaterian promoters can direct GFP expression in transfected hydra. *Dev. Biol.* **246,** 377–390.

Miljkovic-Licina, M., Chera, S., Ghila, L., and Galliot, B. (2007). Head regeneration in wild-type hydra requires de novo neurogenesis. *Development* **134,** 1191–1201.

Momose, T., and Houliston, E. (2007). Two oppositely localised frizzled RNAs as axis determinants in a cnidarian embryo. *PLoS Biol.* **5,** e70.

Newmark, P. A., Reddien, P. W., Cebria, F., and Sanchez Alvarado, A. (2003). Ingestion of bacterially expressed double-stranded RNA inhibits gene expression in planarians. *Proc. Natl. Acad. Sci. USA* **100**(Suppl. 1), 11861–11865.

Noda, T., and Ohsumi, Y. (1998). Tor, a phosphatidylinositol kinase homologue, controls autophagy in yeast. *J. Biol. Chem.* **273,** 3963–3966.

Otto, J. J., and Campbell, R. D. (1977). Budding in Hydra attenuata: bud stages and fate map. *J. Exp. Zool.* **200,** 417–428.

Petiot, A., Ogier-Denis, E., Blommaart, E. F., Meijer, A. J., and Codogno, P. (2000). Distinct classes of phosphatidylinositol 3′-kinases are involved in signaling pathways that control macroautophagy in HT-29 cells. *J. Biol. Chem.* **275,** 992–998.

Sambrook, J., and Russell, D. W. (2001). Molecular cloning: A laboratory manual. Cold Spring Harbor Laboratory Press, Cold Spring Harbor, New York.

Sarbassov, D. D., Ali, S. M., Sengupta, S., Sheen, J. H., Hsu, P. P., Bagley, A. F., Markhard, A. L., and Sabatinim, D. M. (2006). Prolonged rapamycin treatment inhibits mTORC2 assembly and Akt/PKB. *Mol. Cell.* **22,** 159–168.

Seipel, K., and Schmid, V. (2006). Mesodermal anatomies in cnidarian polyps and medusae. *Int. J. Dev. Biol.* **50,** 589–599.

Settembre, C., Fraldi, A., Jahreiss, L., Spampanato, C., Venturi, C., Medina, D., de Pablo, R., Tacchetti, C., Rubinsztein, D. C., and Ballabio, A. (2008). A block of autophagy in lysosomal storage disorders. *Hum. Mol. Genet.* **17,** 119–129.

Steele, R. E. (2002). Developmental signaling in Hydra: What does it take to build a "simple" animal? *Dev. Biol.* **248,** 199–219.

Technau, U., Rudd, S., Maxwell, P., Gordon, P. M., Saina, M., Grasso, L. C., Hayward, D. C., Sensen, C. W., Saint, R., Holstein, T. W., Ball, E. E., and Miller, D. J. (2005). Maintenance of ancestral complexity and non-metazoan genes in two basal cnidarians. *Trends Genet.* **21,** 633–639.

Timmons, L., Court, D. L., and Fire, A. (2001). Ingestion of bacterially expressed dsRNAs can produce specific and potent genetic interference in Caenorhabditis elegans. *Gene* **263,** 103–112.

Trembley, A. (1744). Mémoires pour servir à l'histoire d'un genre de polypes d'eau douce, à bras en forme de cornes, Leiden.

Yamamoto, A., Tagawa, Y., Yoshimori, T., Moriyama, Y., Masaki, R., and Tashiro, Y. (1998). Bafilomycin A1 prevents maturation of autophagic vacuoles by inhibiting fusion between autophagosomes and lysosomes in rat hepatoma cell line, H-4-II-E cells. *Cell Struct. Funct.* **23,** 33–42.

CHAPTER TWENTY-SEVEN

Autophagy in Freshwater Planarians

Cristina González-Estévez*

Contents

Abstract

Planarians provide a new and emergent *in vivo* model organism to study autophagy. On the whole, maintaining the normal homeostatic balance in planarians requires continuous dynamic adjustment of many processes, including proliferation, apoptosis, differentiation, and autophagy. This makes them very different from other models where autophagy only occurs at very specific times and/or in very specific organs. This chapter aims to offer a general vision of planarians as a model organism, placing more

* Department of Developmental Genetics and Gene Control, Institute of Genetics, Queen's Medical Centre, University of Nottingham, United Kingdom

Methods in Enzymology, Volume 451
ISSN 0076-6879, DOI: 10.1016/S0076-6879(08)03227-8

emphasis on those characteristics related to autophagy and describing how autophagy fits into the processes of body remodeling during regeneration and starvation. We also define exactly what is known about autophagy in these organisms and we discuss the techniques available to study the relevant processes, as well as the techniques that are currently being developed. As such, this chapter will serve as a compilation of the techniques available to investigate autophagy in planarians.

1. Getting Started

When you cut a planarian "in longitudinal pieces less than half the width of the old worm the new tissue increases at the side until it is equal to the old. . .but further that if the worm is not fed the old part decreases as the new part increases, so that the width of both new and old remains about the same as that of the piece when first removed. If the piece is fed, the old part does not decrease in width, but remains about the same as at first, and may subsequently enlarge as the new worm grows bigger." This extract from Morgan (1901) serves to demonstrate the tremendous plasticity exhibited by planarians and the recycling mechanisms that enable the animal to reconstitute parts that are lost, mostly when there is little food around. Such behavior inevitably brings to mind the process of autophagy.

Autophagy is a physiological process of organelle and protein turnover that helps maintain the organism's homeostasis and that can be rapidly up-regulated when the organism is undergoing architectural remodeling or when cells must generate nutrients and energy during starvation (Aubert *et al.*, 1996). Indeed, autophagy is known to be a survival mechanism under conditions of nutrient deprivation in mammalian cells, although the role it plays at the level of the whole organism remains to be determined.

Planarians provide a new and emergent *in vivo* model organism to study autophagy. To understand why we believe planarians may be a great model organism for the study of autophagy, the reader must first understand certain peculiar features of this system. However, since it is not our aim to review the entire planarian bibliography, we will direct the reader toward some of the magnificent reviews that have recently been published (Agata and Umesono, 2008; Cebrià, 2007; Pellettieri and Sánchez Alvarado, 2007; Rossi *et al.*, 2008; Saló, 2006; Sánchez Alvarado, 2006).

2. With Just a Few Strokes of the Brush: The Essence of Planarian as Model System

The freshwater planarian is an organism that is able to regenerate a new individual from a tiny fragment of its own body, no matter whether that fragment belongs to its head, its trunk, or its tail (Fig. 27.1). The minimum fragment size is as small as {1/279}th of a planarian according to the findings of T. H. Morgan (1898), equivalent to approximately 1×10^4 cells (Montgomery and Coward, 1974). The tremendous regenerative capacity of these animals has made them a favored organism for the study of the mechanisms underlying pattern restoration for more than two centuries.

The name *planarian* is generally applied to the free-living members of the phylum Platyhelminthes, clade Lophotrocozoa (Álvarez-Presas *et al.*, 2008; Carranza *et al.*, 1997). They are distributed worldwide and they can be found in lakes, rivers, ponds, soils, and in the sea below stones or living on plants. Among them, the Tricladida freshwater planarians are still the most commonly used turbellarian (a class of Platyhelminthes) in regeneration research due to their ease of culture and handling under laboratory conditions. More complete information about the taxonomy of Platyhelminthes can be found in the extensive bibliography available (Brusca and Brusca,

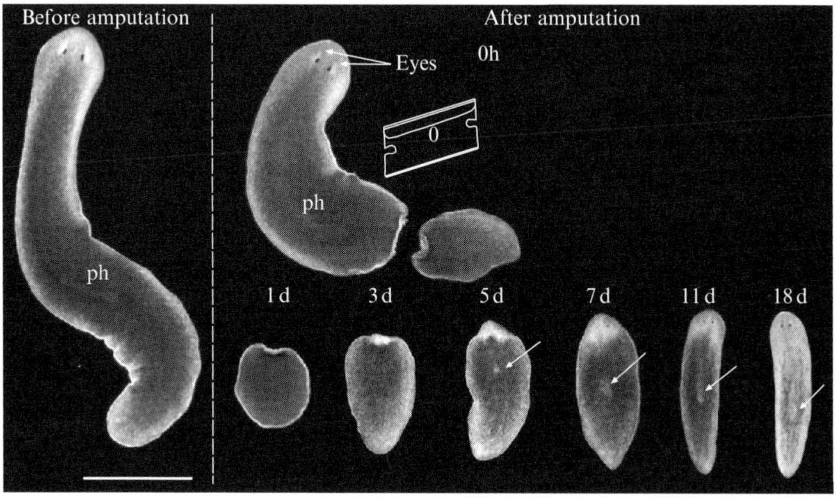

Figure 27.1 The figure shows a *Schmidtea mediterranea* asexual strain before and after performing an amputation (with a razor blade) at the postpharynx level. The process of tail regeneration is followed from 0 hours to 18 days at 20 °C. ph, pharynx; white arrows indicate the pharynx position changing during regeneration due to the remodeling process; scale bar indicates 3 mm.

1990; Cannon, 1986; Hyman, 1951). For simplicity we will use the term *planarians* when referring to triclad freshwater planarians in this work.

All planarian organs are embedded in a mesodermal tissue called *parenchyma*, which consists of several nonproliferating cell types (Romero, 1987) and only one mitotically active cell type, the neoblast, which is believed to be a somatic stem cell (Baguñà and Romero, 1981; Baguñà *et al.*, 1989, reviewed in Saló, 2006; Rossi *et al.*, 2008; Sánchez Alvarado, 2006). Together with their immediate progeny, neoblasts account for approximately 15%–25% of all parenchymal cells (Baguñà, 1976a). The proliferative capacity and multi-potency of neoblasts underlies the extreme tissue plasticity in planarians. Indeed, this enormous plasticity is a singular and distinct characteristic of planarians, setting them apart from more canonical model organisms used in research nowadays that have fixed adult forms in terms of pattern and size. This plasticity can be seen in their capacity to regenerate and their ability to continuously remodel their body pattern (growing and rescaling) depending, for instance, on the availability of food (Fig. 27.2). The term *remodeling* is used here in a very broad sense to refer to the capacity of planarians to rescale their bodies after amputation or under starvation. Remodeling is an important component of tissue homeostasis and regeneration, which occurs constantly in planarians.

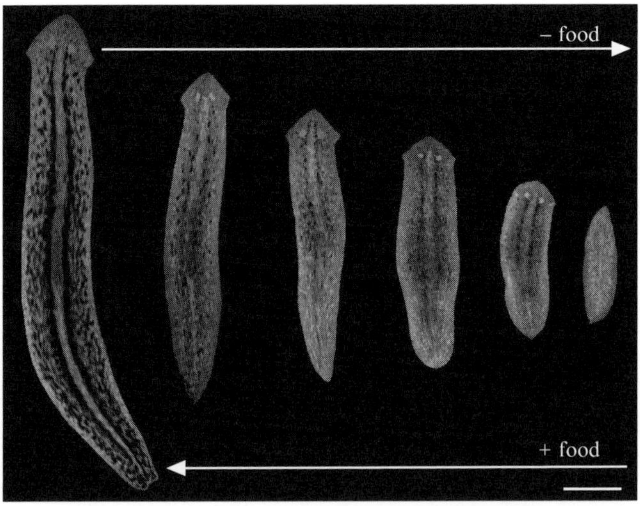

Figure 27.2 The figure shows the process of growing/degrowing in the *Girardia tigrina* species depending on food availability. *G. tigrina* body pigmentation is easy to distinguish; note that the overall body pigmentation becomes reduced during the process of degrowing, thus resembling a rejuvenation process with the smallest sizes indistinguishable from newly hatched juveniles. Scale bar indicates 2 mm.

3. Planarians: A New Model for the Study of Autophagy

Planarians present special characteristics that make them very different from other model organisms used to study autophagy. First of all, using planaria as a model organism to study autophagy means working with the adult organism as a whole, not only with one organ or one system, like the salivary glands of *Drosophila*, for instance. Moreover, it is very likely that autophagy occurs continuously in planarians due to its tremendous plasticity and continual process of change. Planarians offer a very dynamic system that depends on the continued juggling of many processes to maintain their homeostasis, including those of proliferation, apoptosis, differentiation, and probably autophagy. This exquisite equilibrium differs greatly from the situation in other models where autophagy only occurs at specific time points, such as during insect metamorphosis.

During starvation, planarians reduce their size ("degrow") without any impairment in their physiological characteristics or behavior (Fig. 27.2). Interestingly, if we take planarians of about 10 mm and measure their mitotic activity during the process of starvation, mitotic activity not only remains stable but also increases slightly (Baguñà, 1976b). Similarly, if we starve planarians of different sizes and we measure the mitotic index in animals of around 10 mm after different periods of starvation, the mitotic index is the same in all these animals. Thus, while the animal reverses its growth, it retains a basal rate of proliferation, which signifies that the reversal of growth is due to cell loss, not to a decrease in the proliferation rate. In fact, it has been shown that the number of neoblasts increases during starvation, whereas the number of fixed parenchyma cells decreases (Baguñà and Romero, 1981; Romero, 1987). Moreover, the proportion of the remaining cell types is constant, which maintains the animal perfectly in scale and functional, although much smaller. During feeding, this situation reverses and planarians will again grow back to their normal size, reducing both the mitotic index and neoblast density (Baguñà, 1976b) while increasing the number of fixed parenchyma cells (Baguñà and Romero, 1981; Romero, 1987). Indeed, this explains why small planarians or starved planarians regenerate faster than bigger or unstarved ones. The fact that starved planarians resemble juvenile planarians and that they regenerate faster than bigger ones has driven some researchers to consider starved planarians to be younger than well-fed ones, starvation leading to a rejuvenation of the organism (Baguñà, 1976b; Calow, 1978; Child, 1914; Lillie, 1900). It is even more interesting that during starvation, sexual planarians digest (Hyman, 1951) or "resorb" (Berninger, 1911; Schultz, 1904) their bodies starting with the organs that are least useful at that time, the gonads.

This regression has been described at the histological level in *Bdellocephala brunnea* (Hase *et al.*, 2003; Teshirogi and Fujiwara, 1970) and *Dugesia lugubris* (Fedecka-Brunner, 1967a; Grasso 1959). When planarians grow back to their maximum size, they again form these structures in a predetermined order: ovaries, testes, oviducts, yolk glands, and finally copulatory apparatus (Curtis, 1902; Kobayashi and Hoshi, 2002). To maintain the basal rate of proliferation, autophagy of differentiated cells such as fixed parenchyma cells or the gonads may essentially serve as a recycling process in order to supply the correct resources that permit proliferation to stay constant. Growing/degrowing is a continuous process in planarian life and thus there is a continuous rescaling of their bodies depending on food availability and other external parameters. This suggests that a huge amount of autophagy occurs continuously in planarians.

The process of planarian regeneration itself is special since following amputation or injury, the planarian can completely regenerate all the missing structures in about 7 days (Fig. 27.1). At 17 °C, the process of regeneration in *Schmidtea mediterranea* and in *Girardia tigrina* shows a bimodal pattern in which an initial phase of maximal proliferation at 4–8 h of regeneration is followed by a lower level of proliferation on day 1 and a higher second peak on days 2 and 3 of regeneration (Saló and Baguñà, 1984). Interestingly, there is a regression of the gonads as a consequence of amputation in sexual planarians (Fedecka-Brunner, 1967b; Morgan, 1902). After 7 days, the planarian still has to completely finish the rescaling or remodeling of its body to regain the correct proportions. How planarian remodeling is undertaken remains to be fully understood. However, since remodeling proceeds differently if the planarian is fed as opposed to starved, autophagy may play an important role in this process (González-Estévez *et al.*, work in progress).

In light of these amazing characteristics, it is tempting to describe a model whereby autophagy and proliferation in planarians are tightly coupled during stress-induced events. This correlation may be indirect and simply related to the balance between energy supply by autophagy and energy demand created by production of new cells. In such a model, the resources to maintain proliferation will come from food present in the gastrodermal cells and from reserves present in the parenchyma cells at the beginning of starvation. After the depletion of these reserves, nonessential cells (e.g., the cells of the sexual organs) may undergo autophagy. As such, the planarian will decrease in size but will be able to maintain its basal rate of proliferation. If new food is encountered, the starved cells will simply produce new organelles; if no new food is encountered, the cells will reach a point of no return and die.

During regeneration, the formation of the blastema must be a tremendously resource-demanding process due to the proliferation and differentiation of neoblasts that occurs. Moreover, no food can be taken in until a new functional pharynx is formed, and we propose that autophagy plays an essential role in fueling this process. It has already been shown that fixed parenchyma cells establish intercellular gap junctions with neoblasts (Hori, 1991) and that these gap junctions are required for neoblast maintenance (Oviedo *et al.*, 2007). This may explain how resources are supplied to neoblasts from cells that undergo autophagy. As regeneration proceeds, many cells are superfluous or incorrectly situated as the symmetry and proportions are restored, and many cells of different organs undergo cell death to scale the animal to the new proportions. One exciting possibility is that postmitotic differentiating cells (neoblast-like cells) may transdetermine through autophagy.

4. What We Know About Autophagy in Planarians

There has always been certain interest in studying planarian stem cells and the means by which body polarity is established and maintained, leaving aside issues related to body remodeling in which cell death may play an essential role. Attention has often been focused on the importance of cell death processes in planarians (Baguñà and Romero, 1981; Romero, 1987), although technical problems have hindered our progress in this direction. Only a few studies, principally addressing other issues, have suggested the possible importance of cell death during the process of morphallaxis during regeneration and starvation. Accordingly, studies on planarian cell dynamics have confirmed that cell death occurs during planarian regeneration and starvation (Baguñà and Romero, 1981; Romero, 1987). Similarly, studies of the regression of the sexual organs during starvation or regeneration suggest a clear role for cell death during body remodeling (see references in the previous section). In all these and other studies, the existence of cell death is demonstrated without entering into further detail (Coward *et al.*, 1974; de Duve, 1969; Hay and Coward, 1975; Hoff-Jorgensen *et al.*, 1953; Jennings, 1962; Klima, 1961; Osborne and Miller, 1962, 1963).

The view of apoptosis as the sole mechanism underlying programmed cell death has dominated research in this area, leading to a detailed understanding of this form of cell death in many organisms (reviewed in Danial and Korsmeyer, 2004). However, this has implied a clear neglect in the study of other mechanisms of cell death apart from apoptosis (Levine and Klionsky, 2004). Amazingly, the involvement of autophagy in cell death has only unambiguously been recognized by the scientific community in the past 10 years (Klionsky, 2007).

In fact, the only research on cell death processes in planarians has shown that the main mechanisms underlying cell death differs from apoptosis. Such studies were carried out by I. D. Bowen and colleagues in the late 1970s and early 1980s. While focusing on the study of cell death during regeneration and starvation in *Polycelis tenuis* Iijima maintained at 15 °C–17 °C, interestingly they highlighted the extensive autophagy that occurs during regeneration and starvation, and that the cell death observed was not due to apoptosis. All the studies carried out were based on electron microscopy observations of ultrathin sections and on cytochemically enzymatic quantification of acid phosphatase activity.

In adult planarians (Bowen and Ryder, 1974), selective cell autolysis and cell deletion (in actual terms, autophagic cell death) was described that reflects normal cell renewal in the organism, and different degrees of autolysis were observed in the parenchyma of the unstarved planarian. The earliest characteristics involve an increase in cytoplasmic vesiculation and engulfing of the endoplasmic reticulum. These features are followed by an increase in the number of ribosomes, the number of lysosomes, and acid phosphatase activity in the lysosome-like vacuoles, which at later stages shift toward an extracisternal and subsequently a diffuse distribution, leading to selective autolysis even of the nucleus. This process of autolysis is essentially observed in all parenchyma cell types, and although it mostly occurs in gastrodermal phagocytic cells, it is also evident in gland cells (acidophil and cyanophil) and pigment cells. Autolysis is never observed in nerve cells. Moreover, areas of lysis extend into thin prolongations and it is suggested that the lysis of entire cells followed by the possible recycling of cellular material may, like autophagy, be a crucial factor in the ability of these animals to survive starvation.

In planarians starved for a period of about 5 weeks (Bowen *et al.*, 1976), in which planarians decrease their body size by about 32%, a sequence of 4 peaks in acid phosphatase activity are observed. The first is associated with a rapid and immediate reaction of the gastrodermis to food and an increase in phagocytic hydrolase activity. The second peak (6–7 days post-ingestion) is associated with an increase in muscle lysis and autophagy: cells that loose their secretory vesicles and endoplasmic reticulum, an increase in the number of lysosomes is observed, the chromatin becomes relaxed and the cytoplasmic surface diminishes drastically, and that presents many free ribosomes. The third peak (14–15 days post-ingestion) is associated with large-scale lysis of gut cells, whereas the fourth (25–26 days post-ingestion) is due to considerable lysis in the reproductive system. At the structural level, an increase in intracellular vacuolization is detected, mostly in gland and pigment cells, and particularly in fixed parenchyma cells, as well as autophagy, crinophagy, and muscular atrophy. When there are many lysed cells, an increase in the intracellular space and a loss of extracellular matrix, lipids, and glycogen content occurs. No changes in either the nerve cells, the muscles of the pharynx or excretory cells are observed throughout the starvation process. It is suggested that autolysis is completely selective, as all the different cell types in the planarian can still be observed at the end of the experiment. Moreover, it is suggested that this strong autolysis supplies the organism with energy so that it may survive starvation. In conclusion, two different cell populations are observed: one that undergoes autophagy and one that undergoes cell lysis or autophagic cell death.

In regenerating planarians (Bowen *et al.*, 1982), less exhaustive studies were carried out over only 7 days of regeneration in which only 2 peaks of acid phosphatase activity were observed. The first of these was very early and it was higher than the second (0–12 h post-amputation). Phagocytosis of cell debris by intact parenchymal and gastrodermal cells was observed near the cut surface, which was later sealed, and the bulk of autolysis occurred at this time. The second peak was evident at 40–48 h post-amputation, while from 48 h onward cell autolysis was less evident and autophagy was observed. The profiles of acid phosphatase activity were observed in the gastrodermis, parenchymal gland cells, and testis in the wound zone, and in the blastema as regeneration proceeds. These results differed from the peaks of acid phosphatase activity previously described in *Dugesia dorotocephala* at 20 °C (Coward *et al.*, 1974), with a maximum at 12 h and minimal or basal activity at 36 h. The differences between these two studies may at least partially reside in the use of different planarian species kept at different temperatures. As a consequence, further work will be necessary to unravel the importance of autophagy during regeneration.

More than 20 years have lapsed since these studies, during which little work has been carried out on autophagy in planarians. The studies by González-Estévez and colleagues initiated resurgence in research into this topic (González-Estévez *et al.*, 2007a,b; reviewed in Tettamanti *et al.*, 2008). As a result, the function of the planarian gene *Gtdap-1* has been elucidated, this being the ortholog of the human death-associated protein-1 (*DAP-1*: Deiss *et al.*, 2005). Interestingly, *Gtdap-1* is spatiotemporally up-regulated when regeneration occurs during planarian remodeling, that responsible for the correct scaling of the body, and during starvation. However, its expression is restricted to the sex organs in the sexual strain during regeneration and starvation, which clearly indicates that the gene is likely to be involved in remodeling, particularly in the removal of nonessential structures. Moreover, by the fifth day of regeneration when remodeling is at its peak, 44% of all the cells express the gene. These cells are always phenotypically differentiated or differentiating rather than stem cell–like. TEM shows that *Gtdap-1* transcripts are expressed in cells with autophagic morphology. In addition, around 6% of the cells in a day 5 regenerating planarian express cleaved caspase-3 protein but they are never stained by TUNEL. Indeed, the profile of caspase-3 activity throughout regeneration correlates with the profile of *Gtdap-1* expression. Moreover, RNAi silencing of *Gtdap-1* produces not only deficiencies in remodeling but also a decrease in the rate of neoblast proliferation and of caspase-3 activity during regeneration. Finally,

Gtdap-1 gain-of-function experiments induce cell death. Hence, whereas *Gtdap-1* appears to be involved in autophagy, a small percentage of these autophagic cells will undergo cell death at any given time. Whether this cell death is autophagic cell death, and whether it occurs directly or indirectly, remains to be determined. Nevertheless, widespread autophagy appears to be a response both to nutrient deprivation as well as to injury in planarians.

5. Methods Available to Study Autophagy in Planarians

The majority of more recent studies on autophagy have been carried out in cell cultures or in single cell eukaryotic organisms such as yeast, with relatively few performed *in vivo* or in organs. As a consequence, monitoring autophagy at this level is one of the least developed areas in this field and there are no truly adequate methods to detect this process. Autophagy research in planarians is still at a very early phase and hence only a few techniques are available to study autophagy. Part of these methods include how to detect DNA fragmentation by TUNEL and active caspase-3; although they have been always related to apoptosis and in a strict sense they are not detecting autophagy itself, they do detect a different process of cell death that occurs after autophagy is active called autophagic cell death. The methods are outlined subsequently and all the steps carried out in these studies are performed at room temperature, unless otherwise indicated.

5.1. Organisms

The maintenance conditions that we describe are for *S. mediterranea* and for *G. tigrina*. In principle, these can be extended to any freshwater planarian species, although the perfect water composition may change slightly from species to species, and the ideal conditions may need to be found. General maintenance conditions can be found in the literature (Saló and Baguñà, 1984) and are described briefly.

5.1.1. Media

There are several options:

1. Depending on the quality of the tap water, it can be diluted 50% with distilled water or used pure. Prior to dilution, the tap water should be treated with a commercial product for aquarium water (e.g., AquaSafe from TetraAqua following manufacturer's directions).
2. Another possibility is that employed in the 1970s (E. Saló, personal communication) based on the composition of the Foixarda Lake in the

Montjuïc Mountain (Barcelona, Spain), where *S. mediterranea* are usually found:
0.1 m*M* NaCl
1.74 m*M* $CaCl_2$
0.1 m*M* $MgSO_4$
0.1 m*M* KCl
0.9 m*M* $NaHCO_3$ (in powder) pH 7.4 (if necessary, adjust pH to 7.4 with 2 N HCl)

3. Deionized water can be used remineralized to a conductivity of 500 microsievert by the addition of Tropic Marine Sea Salt (available in any pet shop where seawater fish are sold).
4. River water where planarians are found can be used after mechanical filtering to remove large debris.

5.1.2. Maintenance

1. Keep animals at 17 °C–22 °C (regeneration is faster at higher temperatures) in an incubator with 12 h of light. It is important to prevent contamination by fungi, yeast, and bacteria from nearby media for other species (e.g., *Drosophila*).
2. Use glass or plastic containers of about 1500 ml (approximately 250 asexual planarians or 100 sexual) covered with aluminum foil or an opaque lid with small holes that allows water aeration. The aluminium foil or the opaque lid prevents direct light on the planarians. Although planarians need a light cycle, they don not like direct light during long periods of time. Alternatively to an opaque cover, a stone can be introduced into the water to permit planarians to hide below from the light. Feed 5–6 times per month with fresh *Tubifex* (aquarium shop) or with small pieces of liver (chicken or cow, not pork; better if it is biomeat) that can be bought whole, cut it in small pieces, and frozen at −20 °C. On each feeding day, frozen pieces of liver can be cut into slices (1.5 cm long and thin) that are put in the media (approximately 1 piece per 150 planarians). Feed for 1–2 h and remove the leftover food.
3. Pour off the water (planarians stay attach to the surface of the container), clean the container with a brush to remove algal, bacterial growth, and small leftover food without touching the planarians and add new media. Pour off this media and add new one. Repeat the same cleaning procedure the day after. In starvation experiments, change media and clean the container at least once weekly.

5.1.3. Reproduction

Asexual strains reproduce by fission preferentially in the dark.

Sexual strains lay cocoons. Cocoons can be kept together with the adults or carefully removed with a small brush and kept in a separate container with water. The eclosion (hatching) of the juveniles takes 3–4 weeks, after which they can be fed immediately.

5.1.4. Contamination

The first signs of lysis due to fungal or bacterial contamination are the reduction of the head size. If this is observed, the water should immediately be replaced with water containing 10–50 μg/ml of gentamicin sulfate, depending on the severity. Once they have regenerated the head they can again be returned to normal media.

5.1.5. Amputation experiments

Starve animals for 1 week prior to any experiment to prevent the accumulation of food debris in the intestines, which interferes with the resolution of most molecular techniques. Always use animals of similar size.

Amputation is made with a scalpel or a razor blade (Fig. 27.1), and standard procedures to follow the regeneration process involve ante-pharynx and post-pharynx amputations that produce 3 regenerating pieces. Change media the day after amputation.

5.2. Transmission electron microscopy (TEM) observations and TEM *in situ* hybridization

TEM enables cells undergoing autophagy and the process of autolysis to be observed as indicated previously. Moreover, the cell type expressing a gene of interest can be established by TEM *in situ* hybridization and thus whether a gene is expressed in cells with an autophagic morphology (González-Estévez *et al.*, 2007a; Pineda *et al.*, 2002). The ultrastructural descriptions of planarian tissues already available permit the correct identification of each specific cell type (Figs. 27.3 and 27.4) (Auladell, 1990; Bowen *et al.*, 1974; Bowen and Ryder, 1973; Ishii, 1962, 1964, 1965, 1966; Klima, 1961; Pedersen, 1959a,b; Romero, 1987; Török and Röhlich, 1960).

All the reagents used when carrying out such experiments should be EM-grade and the following controls should be included (they control the specificity of the labeling): the sense probe control, a proteinase K control, and a RNase A control.

1. To detect the maximum amount of autophagy, planarians should be used after 3–5 days (d) regeneration or under starvation.
2. 10–15 organisms should be fixed in 50 ml of (falcon tube) 4% paraformaldehyde and 0.1% glutaraldehyde in 0.1 *M* phosphate buffer (pH 7.4) for 1 h at 4 °C.
3. Cryoprotect by using saturated sucrose. Remove the fixative and incubate in graded series of sucrose in Milli-Q water until reaching 2.3 *M*.

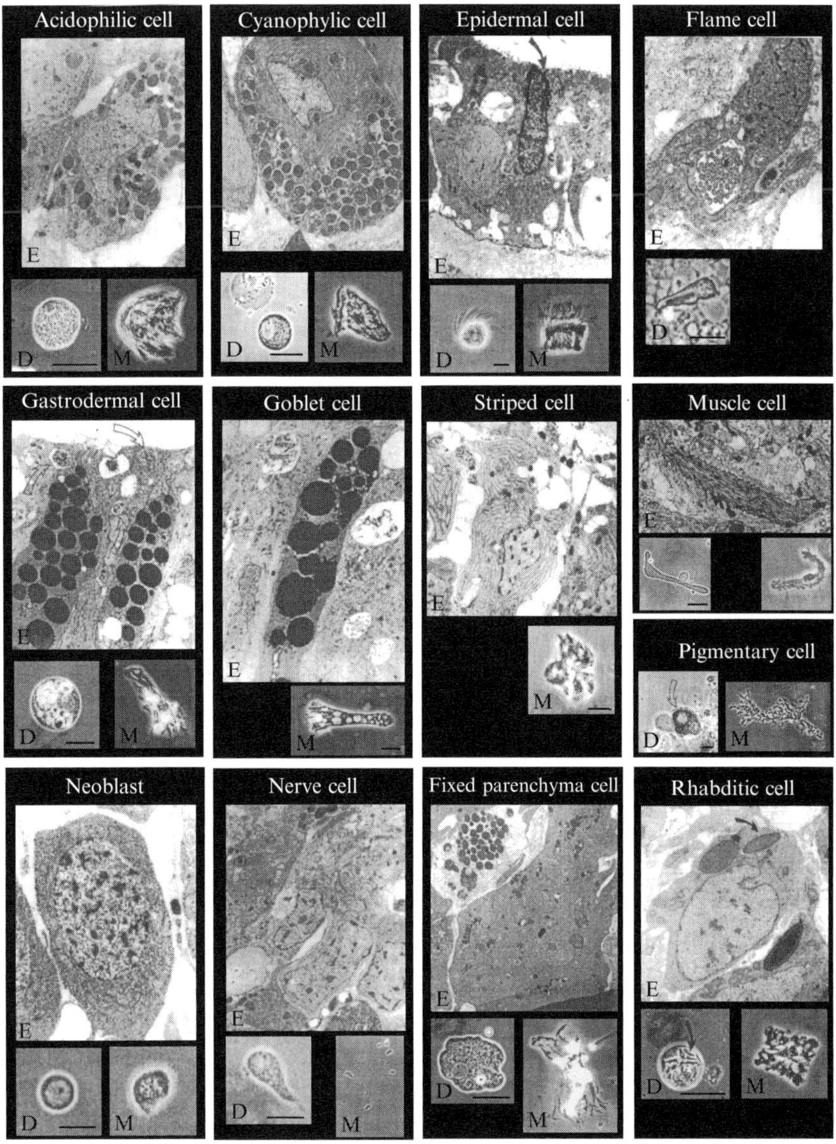

Figure 27.3 The figure shows all the easily recognizable planarian cell types. (E) electron micrograph; (D) cell dissociate; (M) cell macerate; scale bars indicate 50 μm for fixed parenchyma cell, rhabditic cell, muscle cell, pigmentary cell, cyanophilic and acidophilic cell; for other cell types scale bars indicate 100 μm. The figure is adapted from Dr. Rafael Romero's Ph.D. dissertation with his permission (Romero, 1987).

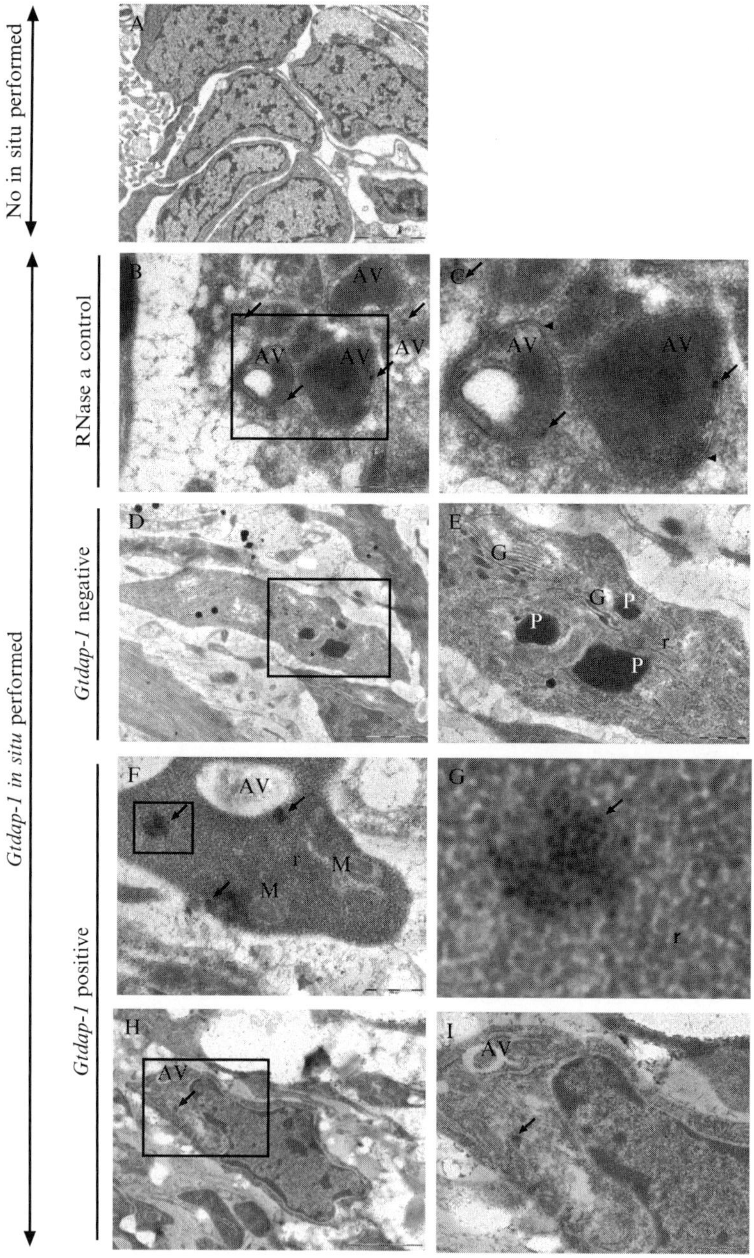

Figure 27.4 Micrographs showing cells at the postblastema level of 5d-regenerating sexual *G. tigrina* species. (A) micrograph showing nerve cells; in this section, no TEM *in situ* hybridization was performed hence the resolution, the contrast, the tissue preservation and morphology of the cells is very good compared to B-I, where TEM *in situ* for *Gtdap-1* was performed; the process of TEM *in situ* hybridization always affects all these

Leave O/N in 2.3 *M* sucrose. Ultrarapid freeze in liquid propane by using an EMCPC Leica fast-freezing workstation.

4. Cryosubstitute the samples with 0.5% uranyl acetate in methanol for 2 d at −90 °C (1 ml per planarian).
5. Embed the specimens in 1 ml of Lowicryl HM20 (polymerization at −50 °C) or Lowicryl K4M (polymerization at −35 °C) resin under UV light (10 w of 220V during 4 d). Lowicryl HM20 is not recommended when performing *in situ* hybridization (enzymatic reactions).
6. Obtain ultrathin sections (50 nm) with a diamond knife on an Ultracut UCT ultramicrotome or similar apparatus, and place sections on Formvar-carbon coated nickel grids. If ultrathin sections are going to be observed go to step 18.
7. To performing *in situ* hybridization on the ultrathin sections: From now on, all the incubations are done by placing the grids facedown on drops that lay on a piece of parafilm. Some control specimens on nickel grids are treated with 2 mg/ml proteinase K in DEPC water (0.1% DEPC) (proteinase K control) for 15 min at 37 °C in a humid chamber to prevent evaporation. The noncontrol sample grids are treated with DEPC Milli-Q water alone.
8. Wash 2 x 5 min, jet wash or extensively rinsed (running a stream of buffer/water from a washing bottle across the grids held with forceps) with DEPC Milli-Q water and air-dry the nickel grids.
9. Treat the RNAse A control nickel grids with 100 μg/ml RNAse A in PBS for 90 min at 37 °C in the humid chamber while the noncontrol sample grids are similarly exposed to DEPC Milli-Q.

parameters and may constitute a problem for a correct interpretation for the inexperienced microscopist. (B) detail of a cell undergoing autophagy; at least 4 autophagic vesicles can be observed (AV); this micrograph represent a RNase A control, hence the gold particles observed in regions where usually there is not mRNA (arrows) are just due to unspecific binding of the anti-digoxigenin-gold antibody; observe that there are only few gold particles per group, which is also a sign of unspecificity. (C) high magnification of the square region in (B) the double membranes of the autophagic vesicles are clearly observed (arrowheads). (D) probably a pigment-forming pigmentary cell which is negative for *Gtdap-1*, since no gold particles are observed. (E) high magnification of the square region in D which shows pigmentary granules (P), ribosomes (r) and a highly functionally active Golgi (G). (F) small cytoplasmic portion rich in free ribosomes (r) with two normal mitochondria (M) and an autophagic vesicle (AV); three groups with multiple gold particles are observed (arrows). (G) high magnification of the square region in (F) it shows that the gold particles (arrow) are more electrodense than free ribosomes (r), thus easily distinguishable. (H) cell in a very initial stage of the autophagic process where a group of gold particles (arrow) and one autophagic vesicle (AV) are observed. (I) high magnification of the square region in (H). Scale bars indicate: A, 2 μm; B, 500 nm; C is a digital magnification from B; D, 2 μm; E, 500 nm; F, 500 nm; G is a digital magnification from F; H, 7 μm; I, 500 nm.

10. Wash 2 × 5 min, jet wash with DEPC Milli-Q water and dry the nickel grids.
11. Hybridization: Prepare the hybridization buffer (50% formamide; 10% dextran sulfate; 4x Saline-Sodium Citrate (SSC) buffer; 400 μg/ml salmon testis DNA and DEPC miliQ water). Preactivate the antisense or sense probes (0.3 ng/μl/kb or at 8 ng/μl for probes of 100 bp) by heating at 70 °C for 10 min and place them on ice. Boil the hybridization buffer for 4 min then place it on ice before adding the probe to the hybridization buffer. Incubate the nickel grids in 50 μl of hybridization buffer containing the probe for 4 h at 37 °C in the humidified chamber (5 × SSC; 50% formamide).
12. Wash 3 × 5 min with 50 μl of PBS per grid.
13. Block with 50 μl of 1% BSA in PBS per grid for 5 min.
14. Detection with 1:40 anti-digoxigenin-gold (10 nm, AURION) in PBS for 30 min.
15. Wash 3 × 5 min in 50 μl PBS.
16. Wash 1 × 1 min and jet wash with Milli-Q water.
17. Air dry O/N.
18. Contrast: Treat the grids with a 10-μl drop of 2% uranyl acetate in Milli-Q water for 30 min, jet wash with Milli-Q water, treat with 10 μl of lead citrate (Reynolds) for 1 min and jet wash with Milli-Q water.
19. Observe on a Jeol 1010 transmission electron microscope or similar (Fig. 27.4).

5.3. Planarian cell dissociation

This protocol permits the dissociation of whole planarians to obtain all the easily recognizable planarian cell types (Fig. 27.3) (Auladell, 1990; Romero, 1987), a protocol adapted from one previously published (González-Estévez *et al.*, 2007a; Salvetti *et al.*, 2000).

1. To detect the maximum amount of autophagy, a 3–5 days (d) regenerating or a starved planarian should be used in the experiment. Place 1–2 planarians in a Petri dish (from 1 planarian 10–15 slides can be obtained). Remove the media and incubate the animals for 1 min in 50–60 ml of 2% L-cysteine hydrochloride-monohydrate (Merck, K23484539) in PBS (adjust to pH 7.0; this step removes the external mucus and it does not kill the animal)
2. Wash planarians in media and transfer them to a new Petri dish with fresh media.
3. Put ice in a new Petri dish, place a microscope slide on the ice, and transfer the planarians to the slide (without media). Cut them into small pieces with a razor blade, removing the mucus from the razor blade when necessary.

4. Transfer the planarian pieces to a 15-ml conical glass tube containing 3 ml of PBS with 4 mg of trypsin and 1 m*M* EDTA with a brush. Pipette the tissue up and down 3–5 times with a drawn out Pasteur pipette to get a homogeneous sample.
5. Shake in a 37 °C shaker for 20 min at 37 °C and then filter through a 100-μm nylon mesh (any manufacturer is fine).
6. Add 3 ml of 8% p-formaldehyde in PBS and incubate at 4 °C for 20 min before adding 6 ml of PBS.
7. Centrifuge the falcon tubes at 50*g* for 3 min at 17 °C.
8. Remove the supernatant fraction and resuspend the pellet in 500 μl of PBS.
9. Put a drop of the cell suspension on a Menzel Superfrost Lysine-coated slide (Menzel-Glaser, Germany) and dry the slide on a 45 °C hot plate.
10. The number of cells in the supernatant fractions can be counted in a counting chamber (Improved Neubauer Hawksley Cristalite).

5.4. TUNEL (terminal deoxynucleotide transferase-mediated dUTP nick-end labeling) on dissociated planarian cells

This technique permits DNA fragmentation to be detected in cells undergoing apoptosis and autophagic cell death (González-Estévez *et al.*, 2007a). All the steps are performed by putting a drop of each solution on the dissociated cells spotted on a slide unless otherwise indicated. All incubations longer than 1 min are performed in a humidified chamber. This protocol is performed with a positive control that consists of cells treated with DNase I.

1. Put a drop of PBS for 1 min on the slide containing dissociated cells.
2. Aspirate off the PBS and incubate for 2 min on ice in a solution containing 0.1% Triton X-100, 0.1% Na-Citrate.
3. Wash slides by filling a staining jar (Glass Coplin Staining Jar; Dynamic Aqua-Supply, Canada) with PBS.
4. Incubate the positive control for 10 min in DNAse I (RNAse free) 10 U/ml diluted in 50 m*M* Tris-HCl, pH 7.4, 10 m*M* MgCl, 1 mg/ml BSA, and then wash with PBS in a staining jar.
5. Incubate for 30 min in 45 μl of TdT buffer (Invitrogen) under a coverslip.
6. Incubate in the TUNEL reaction mix (*In situ* Cell Death Detection Kit, AP; Roche, Basel, Switzerland) at 37 °C for 60 min under a coverslip.
7. Wash twice in the staining jar with PBST (0.1% Triton X-100) for a total of 16 min. If dual TUNEL and immunohistochemistry is to be performed, continue with the immunohistochemistry protocol (section 5.6, from steps 1–6 and then step 9) but using a TRITC-conjugated donkey antirabbit secondary antibody (Jackson ImmunoResearch).
8. Counterstain the nuclei with 1:1000 DAPI in PBS for 10 min.

9. Rinse and wash in PBS for 5 min in the staining jar. Mount slides in Slow-Fade Gold Antifade Reagent (Molecular Probes) or similar, and examine the fluorescence (fluorescein) in a Zeiss Axiophot fluorescence microscope or similar.

5.5. *In situ* hybridization on dissociated planarian cells

This protocol permits the different cell types expressing genes associated with autophagy to be detected, as well as the cytoplasmic characteristics of these cells. The whole-mount *in situ* hybridization protocol for planarians (Umesono *et al.*, 1997) was adapted to dissociated planarian cells (González-Estévez *et al.*, 2007a; Salvetti *et al.*, 2000). Probes should be prepared as for TEM *in situ* hybridization (section 5.2). Unless otherwise indicated, all steps were performed by putting a drop of each solution on the dissociated cells, and all the steps longer than 1 min are performed in a humidified chamber. This protocol is performed with a negative control that consists of hybridization with the sense probe.

1. Incubate for 15 min in PBST (PBS + 0.1% Triton X-100).
2. Prehybridization. Incubate for 60 min at 55 °C in prehybridization solution (50% formamide, 5×SSC, 0.1 mg/ml yeast RNA, 0.1 mg/ml heparin, 0.1% Tween-20, 10 m*M* DTT) under a coverslip.
3. Hybridization. Incubate O/N at 55 °C in hybridization solution (prehybridization solution, 10% dextran sulfate, probe at a concentration of 0.05–0.2 ng/μl previously activated for 10 min at 70 °C and 2 min on ice) under a coverslip.
4. Wash with 50% Formamide/5 × SSC/0.1% Tween-20 for 5, 15, and 30 min at 55 °C in a staining jar. The coverslip will fall from the slide when introduced in the staining jar.
5. Quick wash and wash in Buffer I (MAB: Maleic acid 11.6*g*, NaCl 9.76 g, adjust to pH 7.6 with 2 N NaOH and with 0.1% Triton X-100) for 5 and 20 min in the staining jar.
6. Incubate in Buffer II (1% Blocking Solution for nucleic acid hybridization and detection (Roche) in Buffer I) for 20 min under a coverslip.
7. Incubate with anti-DIG-AP (Roche) diluted 1:1000 in Buffer II for 180 min under a coverslip.
8. Rinse and wash with Buffer I for 5, 30, and 45 min.
9. Detect the probe with ELF 97 Endogenous Phosphatase Detection Kit, Molecular Probes, Invitrogen. Wash 3 x 5 min in the buffer provided and develop using the ELF kit (Molecular Probes) with a 1:100 dilution of the substrate for 10–120 min following the manufacturer's instructions. The ELF 97 substrate is a very strong fluorescent yellow-green precipitate. Alternatively, the probes can be detected by applying 25 μl NBT/BCIP (Roche) in 1 ml of TMN (0.1 *M* Tris-HCl, pH 9.5, 0.1 *M*

NaCl, 0.05 *M* $MgCl_2$, 1% Tween-20, 10% PVA (Polyvinyl alcohol, 98–99% hydrolyzed), which produces a blue precipitate.

To perform double *in situ* hybridization and TUNEL, develop the *in situ* with the ELF kit as described, and continue as follows:

10. Wash for 5 min in PBS in a staining jar.
11. Inactivate alkaline phosphatase by incubating 30 min in 0.1 *M* glycine-HCl, pH 2.2/0.1% Tween-20.
12. Fix for 20 min with 4% p-formaldehyde in Buffer I.
13. Rinse and wash 2 x 5 min in Buffer I in the staining jar.
14. Add 50 μl of Converter AP (anti-fluorescein-AP; included in the ELF kit) and incubate for 30 min at 37 °C under a coverslip.
15. Rinse 3 times with Buffer I in the staining jar.
16. Add 50 μl Fast Red (Roche) and develop the TUNEL reaction for 10–30 min to produce a red precipitate. Alternatively, the TUNEL reaction can be detected by applying 25 μl of NBT/BCIP (Roche) in 1 ml TMN to produce a blue precipitate. Rinse in PBS.
17. Counterstain the nuclei for 10 min with 1:1000 DAPI in PBST in a staining jar.
18. Rinse and wash for 5 min in PBS in the staining jar.
19. Mount slides in Slow-Fade Gold Antifade Reagent (Molecular Probes) or similar, and examine the fluorescence (fluorescein) on a Zeiss Axiophot fluorescence microscope or similar.

5.6. Immunohistochemistry on dissociated planarian cells

This protocol can be used to detect specific autophagy proteins and here we describe how to detect active caspase-3 in planarians (González-Estévez *et al.*, 2007a). Moreover, active caspase-3 detects autophagic cell death and apoptosis. All steps longer than 1 min are performed in a humidified chamber. A negative control is performed by preincubating cleaved caspase-3 antibody (Asp175) with a cleaved caspase-3 (Asp175) blocking peptide (Cell Signaling Technology) prior to the immunohistochemistry according to the manufacturer's instructions. If dual *in situ* hybridization and immunohistochemistry is to be performed, the protocol for *in situ* hybridization on dissociated planarian cells (section 5.5) is followed until step 13, and then continue as follows:

1. Wash in PBST (0.1% Triton X-100)/0.5% BSA for 5, 10, and 15 min in the staining jar.
2. Incubate with the primary antibody, a rabbit antihuman cleaved caspase-3 polyclonal antiserum (Asp175, Cell Signaling Technology) diluted 1:50 in PBST/0.5 % BSA O/N at 4 °C under a coverslip.
3. Wash in PBST/0.5% BSA for 5, 15, and 40 min at 4 °C in the staining jar.

4. Incubate with the FITC-conjugated donkey antirabbit secondary antibody (Jackson ImmunoResearch) diluted 1:200 in PBST/0.5 % BSA for 4 h at 4 °C under a coverslip.
5. Wash in PBS for 5, 10, and 15 min in the staining jar.
6. Add 50 μl of AP Converter (anti-fluorescein-AP), and incubate for 30 min and 20 min at 37 °C under a coverslip.
7. Wash in Buffer I (see section 5.5) for 5 and 10 min.
8. Detect using ELF 97 Endogenous Phosphatase Detection Kit (Molecular Probes, Invitrogen).
9. Follow steps 17, 18, and 19 from the *in situ* hybridization protocol (section 5.5).

5.7. Analysis of caspase-3 activity

This protocol allows caspase-3 activity to be measured in a planarian protein extract (González-Estévez *et al.*, 2007a). As a negative control, pre-incubate 20 μg of protein extract for 15 min at 37 °C with 100 μ*M* Z-DEVD-FMK and Z-VAD-FMK (Calbiochem) independently before performing the caspase-3 activity assay.

1. For each group, 6 planarians are used in the case of the asexual strain or 3 in the case of the sexual one. Wash planarians with 50 ml of media in a Petri dish to remove excess mucus.
2. Put ice in a new Petri dish, place a slide on the ice, and transfer the planarians to the slide (without media). Chop them into small pieces with a razor blade, removing the mucus from the razor blade when necessary.
3. Transfer these pieces into 400 μl of lysis buffer (5 m*M* Tris-HCl at pH 8.0, 20 m*M* EDTA, 0.5% Triton X-100), and homogenize by pipetting and vortexing at 4 °C to get a uniform sample.
4. Clear lysates by centrifuging at 13,000×*g* for 10 min at 4 °C in a 1.7-ml microcentrifuge tube and transfer the supernatant fraction to a new tube at 4 °C. For long-term storage, keep at −20 °C.
5. Determine the protein concentration using the Bradford assay (use 2 μl and 4 μl of sample).
6. Incubate the reaction mixture containing 20 μg of protein extract, the assay buffer (20 m*M* HEPES, pH 7.5, 10% glycerol, 2 m*M* dithiotreitol), and 20 μ*M* caspase-3 substrate Ac-DEVD-AMC (BD Biosciences Pharmingen) for 2 h in the dark at 37 °C.
7. Measure enzyme activity in a luminescence spectrophotometer (Perkin-Elmer LS-50) (1 excitation, 380 nm; 1 emission, 440 nm), whereby one unit of caspase-3 activity is defined as the amount of active enzyme necessary to produce an increase of 1 arbitrary luminescence unit after the 2-h incubation. The results are presented as units of caspase-3 activity per μg of protein.

5.8. Transgenesis

To date, the transgenesis technique has only been possible with the 3xP3-EGFP marker (González-Estévez *et al.*, 2003) and using a cassette where 3xP3 drives the expression of the gene of interest pHer[3xP3-EGFPaf-3xP3-*gene of interest*] in the *G. tigrina* species (González-Estévez *et al.*, 2007a). The 3xP3-EGFP marker is based on an artificial promoter that is responsive to the transcription factor Pax6 from different animal phyla and the enhanced GFP (EGFP) reporter gene (Berghammer *et al.*, 1999). P3-related sequences are found in rhodopsin and other photoreceptor specific genes ranging from flies to humans. The planarians obtained are transgenic mosaics and pure transgenic animals may be obtained by crossing sexual mosaics. The protocol for transformation assays in planarians is as follows:

1. Place some adult planarians and/or the posterior and anterior 3–5 d regenerating planarians in a Petri dish with media (use 150–200 planarians per construct). Put ice in a Petri dish, place a slide or a small piece of card on the ice and transfer the planarians onto it with the dorsal part up (without media). Take great care that planarians do not dry out.
2. Inject the parenchyma of 10-mm-long planarians with 2 μl of plasmid DNA solution that includes 1 μl (1 μg/μl) of the plasmid pHer[3xP3-EGFPaf-3xP3-*gene of interest*] plus 1 μl (1 μg/μl) of the helper plasmid (pKhsp82Hermes). The solution is injected using a nanoject injector (Drummond Scientific, Broomball, PA) 3–7 mm anterior to the pharynx and directing the needle to the anterior part to increase the chances of hitting the ovaries. If the volume is not equally distributed over the body, half of the volume can be injected 2–5 mm posterior to the pharynx). The more efficient way of injection is to press "empty" until the 2 μl are delivered and observing if they are inflated due to the introduction of the plasmids. If the planarians are bigger, increase the amount of DNA injected, whereas if planarians are smaller it can be decreased. *piggyBac* has been shown to have a similar efficiency.
3. Place injected planarians back in the Petri dish with media and allow them to recover for 10–15 min.
4. Electroporate the injected planarians one at a time and for a 10-mm-long *G. tigrina*, a single 15 V 30 ms pulse is appropriate, ideally using CUY701-P5E and CUY701-P5L electrodes (Nepa Gene, Chiba, Japan). For bigger or smaller planarians the ideal parameters are yet to be defined. Two different electrosquare porators have been tested: a CUY21EDIT square wave electroporator (Nepa Gene, Chiba, Japan) which results in 100% planarian transfection (González-Estévez *et al.*, 2003) and TSS20 Ovodyne Electroporator (Intracel, Frederick, MD), which results in a lower transformation efficiency (approximately 32%,

González-Estévez *et al.*, 2007a). Thus, the use of the CUY21EDIT is recommended. Electroporation is performed in five-eighths concentration of Holtfreter's standard saline solution (2.188 g NaCl, 0.031 g KCl, 0.063 g $CaCl_2$, 0.125 g $NaHCO_3$)/1,000 ml; Holtfreter, 1944) and between electroporations, the electrodes should be cleaned with ethanol and/or Milli-Q water to remove the mucus that could interfere in the efficiency of the next electroporation.

5. After each electroporation, place the planarians in a Petri dish with media containing 10 μg/ml gentamicin sulfate and change the media over the following days until the planarians have recovered and the gentamicin sulfate can be removed from the media. After 15 days, cut planarians into 2 or 3 pieces to generate smaller planarians with a faster cell turnover.
6. Check for phenotypes every week for 60 days until the fluorescent photoreceptors are clearly detected on a Leica (MZ FLIII) fluorescence stereomicroscope with GFP3 filters and a Leica camera (DC 300F) or similar.

6. Methods Being Developed to Study Autophagy in Planarians

We are obtaining very promising results regarding autophagy in planarians by screening the *S. mediterranea* genome for all the *Atg* genes, whose homologs are usually required for autophagy in yeast, and which will demonstrate clear evolutionary conservation of genes crucial in autophagy. Moreover, we are also characterizing some of the main elements involved in the TOR pathway to establish some direct relationships with autophagy pathways. Hence, we shall systematically perform whole mount *in situ* hybridization and obtain loss-of-function mutants by RNAi. These two techniques have not been explained in the last section because they are two routine techniques that are reviewed and explained elsewhere. Moreover, we are looking for the best way to perform these experiments since many different approaches are possible.

To understand how loss of gene function affects autophagy, we need to establish specific techniques such as those already available in other model organisms, as well as the techniques described previously:

1. Measuring acid phosphatase activity was performed in other planarian species in the 1970s-1980s (see "What we know about autophagy in planarians" in this chapter), and this should be adapted to *S. mediterranea*.

2. Atg8/LC3 gene amplification should be assessed and perhaps anti-LC3 antibodies produced. Thus, the distribution of this product could be assessed by TEM immunohistochemistry, immunohistochemistry on sections, and whole-mount immunohistochemistry.
3. Although whole-mount immunohistochemistry is another routine planarian technique, we have not yet set up whole mount TUNEL or cleaved caspase-3 staining. We believe these techniques should be established, since the results will be more relevant to the *in vivo* context and the relationships to other processes will be more readily seen. Recently, a review mentioned the possibility of carrying out whole mount TUNEL staining to detect apoptosis in planarians, although the protocol was not described (Pellettieri and Sánchez Alvarado, 2007).
4. Transgenic planarians for GFP-LC3 (Klionsky *et al.*, 2008) would be great to monitor autophagy through regeneration and starvation.
5. The use of acidotropic dyes, although not especially recommended, could complement the use of transgenics, and the use of LysoSensor Blue and LysoTracker Red is under way.

Finally, it is important to point out that we refer to autophagy in this chapter as meaning all different types of autophagy, as nothing is known about the different pathways active in planarians. Work needs to be done to be able to distinguish the different types of autophagy that occur.

7. Concluding Remarks

Throughout this chapter, we have tried to show the potential of planarians as a model organism to study autophagy. The research into autophagy in planarians has been described and the current ability to study autophagy in planarians detailed. Although a body of work is already available, more effort will clearly be necessary over the following years to better define the importance and the mechanics of this process.

Acknowledgments

I would like to specially thank D. A. Felix for critical reading of the manuscript, for comments on my English, and for spending his weekends in discussion. I would like to thank Professor E. Saló and Dr. A. Aboobaker for a critical reading of the manuscript, and Dr. M. Sefton for advice on my English style and critical comments on the manuscript. I thank Dr. R. Romero for permitting me to adapt a figure from his Ph.D. dissertation and Dr. C. López for comments on the TEM *in situ* hybridization protocol. This work was supported by a Beatriu de Pinós fellowship (Generalitat de Catalunya, Catalunya, Spain).

REFERENCES

Agata, K., and Umesono, Y. (2008). Brain regeneration from pluripotent stem cells in planarian. *Philos. Trans. R. Soc. Lond. B Biol. Sci.* **363,** 2071–2078.

Álvarez-Presas, M., Baguñà, J., and Riutort, M. (2008). Molecular phylogeny of land and freshwater planarians (Tricladida, Platyhelminthes): from freshwater to land and back. *Mol. Phylogenet. Evol.* **47,** 555–568.

Aubert, S., Gout, E., Bligny, R., Marty-Mazars, D., Barrieu, F., Alabouvette, J., Marty, F., and Douce, R. (1996). Ultrastructural and biochemical characterization of autophagy in higher plant cells subjected to carbon deprivation: control by the supply of mitochondria with respiratory substrates. *J. Cell Biol.* **133,** 1251–1263.

Auladell, C. (1990). Anàlisi de l'heterogeneïtat estructural i funcional de les cèl.lules indiferenciades (neoblasts) de la planària *Dugesia* (G) *tigrina*. (PhD Thesis). University of Barcelona, Barcelona.

Baguñà, J. (1976a). Mitosis in the intact and regenerating planarian *Dugesia mediterranea* n. sp. II. Mitotic studies during regeneration and a possible mechanism of blastema formation. *J. Exp. Zool.* **195,** 65–80.

Baguñà, J. (1976b). Mitosis in the intact and regenerating planarian *Dugesia mediterranea* n. sp. I. Mitotic studies during growth, feeding and starvation. *J. Exp. Zool.* **195,** 53–64.

Baguñà, J., and Romero, R. (1981). Quantitative analysis of cell types during growth, degrowth and regeneration in the planarians *Dugesia mediterranea* and *Dugesia tigrina*. *Hydrobiologia* **84,** 181–194.

Baguñà, J., Saló, E., and Auladell, C. (1989). Regeneration and pattern formation in planarians III. Evidence that neoblasts are totipotent stem cells and the source of blastema cells. *Development* **107,** 77–86.

Berghammer, A. J., Klingler, M., and Wimmer, E. A. (1999). A universal marker for transgenic insects. *Nature* **402,** 370–371.

Berninger, J. (1911). Über die Einwirkung des Hungers auf Planarien. *Zool. Jahrb.* **30,** 181–216.

Bowen, I., and Ryder, T. (1974). Cell autolysis and deletion in the planarian *Polycelis tenuis* Iijima. *Cell Tissue Res.* **154,** 265–271.

Bowen, I. D., den Hollander, J. E., and Lewis, G. H. (1982). Cell death and acid phosphatase activity in the regenerating planarian *Polycelis tenuis* Iijima. *Differentiation* **21,** 160–167.

Bowen, I. D., and Ryder, T. (1973). The fine structure of the planarian *Polycelis tenuis* Iijima. 1. The pharynx. *Protoplasma* **78,** 223–241.

Bowen, I. D., Ryder, T., and Dark, C. (1976). The effects of starvation on the planarian worm *Polycelis tenuis* Iijima. *Cell Tissue Res.* **169,** 193–209.

Bowen, I. D., Ryder, T., and Thompson, J. A. (1974). The fine structure of the planarian *Polycelis tenuis* Iijima. II. The intestine and gastrodermal phagocytosis. *Protoplasma* **79**.

Brusca, R. C., and Brusca, G. J. (1990). "Invertebrates." Sinauer Associates, Massachusetts USA.

Calow, P. (1978). "Life cycles and evolutionary approach to the physiology of reproduction, development and ageing." London.

Cannon, L. R. G. (1986). "Turbellaria of the world: A guide to families and genera." Queensland Museum, Brisbane.

Carranza, S., Baguñà, J., and Riutort, M. (1997). Are the Platyhelminthes a monophyletic primitive group? An assessment using 18S rDNA sequences. *Mol. Biol. Evol.* **14,** 485–497.

Cebrià, F. (2007). Regenerating the central nervous system: How easy for planarians! *Dev. Genes Evol.* **217,** 733–748.

Child, C. (1914). Starvation, rejuvenescence and acclimation in *Planaria dorotocephala*. *Arch. Entwm. Org.* **38,** 418–446.

Coward, S. J., Bennett, C. E., and Hazlehurst, B. L. (1974). Lysosomes and lysosomal enzyme activity in the regenerating planarian; evidence in support of dedifferentiation. *J. Exp. Zool.* **189,** 133–146.
Curtis, W. (1902). The life history, the normal fission, and the reproductive organs of *Planaria maculata. Proc. Boston Soc. Nat. Hist.* **30,** 515–559.
Danial, N. N., and Korsmeyer, S. J. (2004). Cell death: Critical control points. *Cell* **116,** 205–219.
de Duve, C. (1969). Intracellular localization. *In* "The phosphohydrolases: Their biology, biochemistry and clinical enzymology" (W. H. Fishman, ed.), Vol. 166, pp. 602–603. Ann. N. Y. Acad. Sci., New York.
Deiss, L. P., Feinstein, E., Berissi, H., Cohen, O., and Kimchi, A. (1995). Identification of a novel serine/threonine kinase and a novel 15-kD protein as potential mediators of the gamma interferon-induced cell death. *Genes Dev.* **9,** 15–30.
Fedecka-Bruner, B. (1967a). [Studies on the regeneration of the genital organs of the planaria *Dugesia lugubris*. I. Regeneration of the testes after destruction]. *Bull. Biol. Fr. Belg.* **101,** 255–319.
Fedecka-Bruner, B. (1967b). [Differentiation of the male gonads in the planarian, Dugesia lugubris, during regeneration]. *C. R. Seances Soc. Biol. Fil.* **161,** 21–23.
Fukushima, T., and Matsuda, R. (1991). Experiments with culture media for planarian cells. *Hydrobiologia* **227,** 187–192.
González-Estévez, C., Felix, D. A., Aboobaker, A. A., and Saló, E. (2007a). Gtdap-1 promotes autophagy and is required for planarian remodeling during regeneration and starvation. *Proc. Natl. Acad. Sci. USA* **104,** 13373–13378.
González-Estévez, C., Felix, D. A., Aboobaker, A. A., and Saló, E. (2007b). Gtdap-1 and the role of autophagy during planarian regeneration and starvation. *Autophagy* **3,** 640–642.
González-Estévez, C., Momose, T., Gehring, W. J., and Saló, E. (2003). Transgenic planarian lines obtained by electroporation using transposon-derived vectors and an eye-specific GFP marker. *Proc. Natl. Acad. Sci. USA* **100,** 14046–14051.
Grasso, M. (1959). Fenomerigenerativi e apparato genitale in *Dugesia lugubris. Boll. Zool.* **26,** 523–527.
Hase, S., Kobayashi, K., Koyanagi, R., Hoshi, M., and Matsumoto, M. (2003). Transcriptional pattern of a novel gene, expressed specifically after the point-of-no-return during sexualization, in planaria. *Dev. Genes Evol.* **212,** 585–592.
Hay, E. D., and Coward, S. J. (1975). Fine structure studies on the planarian, *Dugesia*. I. Nature of the "neoblast" and other cell types in noninjured worms. *J. Ultrastruct. Res.* **50,** 1–21.
Hoff-Jorgensen, E., Løvtrup, E., and Løvtrup, S. (1953). Changes in deoxyribonucleic acid and total nitrogen in planarian worms during starvation. *J. Embr. Exp. Morph.* **1,** 161–165.
Hori, I. (1991). Role of fixed parenchyma cells in blastema formation of the planarian *Dugesia japonica. Int. J. Dev. Biol.* **35,** 101–108.
Holtfreter, J. (1944). Neural differentiation of ectoderm through exposure to saline solution. *J. Exp. Zool.* **95,** 307–343.
Hyman, L. (1951). "The Invertebrates: Platyhelminthes and Rhynchocoela- The Acoelomate Bilateria." McGraw-Hill, New York.
Ishii, S. (1962). Electron microscopic observations on the planarian tissues. I. A survey of the pharynx. *Fukushima J. Med. Sci.* **9-10,** 51–73.
Ishii, S. (1964). The ultrastructure of the outer epithelium of the planarian pharynx. *Fukushima J. Med. Sci.* **11,** 109–125.
Ishii, S. (1965). Electron microscopic observations on the planarian tissues. II. The intestine. *Fukushima J. Med. Sci.* **12,** 67–87.

Ishii, S. (1966). The ultrastructure of the insunk epithelium linin the planarian pharyngeal cavity. *J. Ultrastruct. Res.* **14,** 345–355.
Jennings, J. E. (1962). Further studies on feeding and digestion in triclad Turbellaria. *Biol. Bull.* **123,** 571–581.
Klima, J. (1961). Elektronenmikroskopische Studien über die Feinstruktur der Tricladen (Turbellaria). *Protoplasma* **54,** 101–162.
Klionsky, D. J. (2007). Autophagy: From phenomenology to molecular understanding in less than a decade. *Nat. Rev. Mol. Cell Biol.* **8,** 931–937.
Klionsky, D. J., Abeliovich, H., Agostinis, P., Agrawal, D. K., Aliev, G., Askew, D. S., Baba, M., Baehrecke, E. H., Bahr, B. A., Ballabio, A., Bamber, B. A., Bassham, D. C., *et al.* (2008). Guidelines for the use and interpretation of assays for monitoring autophagy in higher eukaryotes. *Autophagy* **4,** 151–175.
Kobayashi, K., and Hoshi, M. (2002). Switching from asexual to sexual reproduction in the planarian Dugesia ryukyuensis: Change of the fissiparous capacity along with the sexualizing process. *Zoolog. Sci.* **19,** 661–666.
Levine, B., and Klionsky, D. J. (2004). Development by self-digestion: Molecular mechanisms and biological functions of autophagy. *Dev. Cell* **6,** 463–477.
Lillie, F. (1900). Some notes on regeneration and regulation in planarians. I. The source of material of new parts and limits of size. *Am. Nat.* **34,** 173–177.
Montgomery, J. R., and Coward, S. J. (1974). On the minimal size of a planarian capable of regeneration. *Trans. Am. Microsc. Soc.* **93,** 386–391.
Morgan, T. H. (1898). Experimental studies of the regeneration of *Planaria maculata. Archiv für Entwickelungsmechanik der organismen* **7,** 364–397.
Morgan, T. H. (1901). "Regeneration." Macmillan, New York.
Morgan, T. H. (1902). Growth and regeneration in *Planaria lugubris. Arch. Entw. Mech. Org.* **13,** 179–212.
Osborne, P. J., and Miller, A. T. (1962). Uptake and intracellular digestion of protein (peroxidase) in planarians. *Biol. Bull.* **123,** 589–596.
Osborne, P. J., and Miller, A. T. (1963). Acid and alkaline phosphatase changes associated with feeding, starvation and regeneration in planarians. *Biol. Bull.* **124,** 285–292.
Oviedo, N. J., and Levin, M. (2007). Smedinx-11 is a planarian stem cell gap junction gene required for regeneration and homeostasis. *Development* **134,** 3121–3131.
Pedersen, K. J. (1959a). Cytological studies on the planarian neoblast. *Z. Zellforsch.* **50,** 799–817.
Pedersen, K. J. (1959b). Some features of the fine structure and histochemistry of planarian subepidermal gland cells. *Z. Zellforsch.* **50,** 121–142.
Pellettieri, J., and Alvarado, A. S. (2007). Cell turnover and adult tissue homeostasis: From humans to planarians. *Annu. Rev. Genet.* **41,** 83–105.
Pineda, D., Rossi, L., Batistoni, R., Salvetti, A., Marsal, M., Gremigni, V., Falleni, A., Gonzalez-Linares, J., Deri, P., and Saló, E. (2002). The genetic network of prototypic planarian eye regeneration is Pax-6 independent. *Development* **129,** 1423–1434.
Romero, R. (1987). Anàlisi cellular quantitativa del creixement i de la reproducció a diferentes espècies de planàries. (Ph.D. dissertation). Universitat de Barcelona, Barcelona.
Rossi, L., Salvetti, A., Batistoni, R., Deri, P., and Gremigni, V. (2008). Planarians, a tale of stem cells. *Cell Mol. Life Sci.* **65,** 16–23.
Sakai, F., Agata, K., Orii, H., and Watanabe, K. (2000). Organization and regeneration ability of spontaneous spernumerary eyes in planarians -eye regeneration field and pathway selection by optic nerves. *Zool. Sci.* **17,** 375–381.
Saló, E. (2006). The power of regeneration and the stem-cell kingdom: Freshwater planarians (Platyhelminthes). *Bioessays* **28,** 546–59.

Saló, E., and Baguñà, J. (1984). Regeneration and pattern formation in planarians. I. The pattern of mitosis in anterior and posterior regeneration in *Dugesia* (G) *tigrina*, and a new proposal for blastema formation. *J. Embryol. Exp. Morphol.* **83,** 63–80.

Salvetti, A., Rossi, L., Deri, P., and Batistoni, R. (2000). An MCM2-related gene is expressed in proliferating cells of intact and regenerating planarians. *Dev. Dyn.* **218,** 603–614.

Sánchez Alvarado, A. (2006). Planarian regeneration: Its end is its beginning. *Cell* **124,** 241–245.

Schultz, E. (1904). Über Reduktionen. I. Über Hungerserscheinungen bei *Planaria lactea. Arch. Entwm. Org.* **18,** 555–577.

Teshirogi, W., and Fujiwara, H. (1970). Some experiments on regression and differentiation of genital organs in a freshwater planarian, *Bdellocephala brunnea. Sci. Rep. Hirosaki Univ.* **17,** 38–49.

Tettamanti, G., Saló, E., González-Estévez, C., Felix, D. A., Grimaldi, A., and de Eguileor, M. (2008). Autophagy in invertebrates: Insights into development, regeneration and body remodeling. *Curr. Pharm. Des.* **14,** 116–125.

Török, L. J., and Röhlich, P. (1960). Electronmikrozkopos vizsgalatok planariak eosinofil miriqysjtjcin. *Biol. Közl.* **8,** 117–124.

Umesono, Y., Watanabe, K., and Agata, K. (1997). A planarian orthopedia homolog is specifically expressed in the branch region of both the mature and regenerating brain. *Dev. Growth Differ.* **39,** 723–727.

CHAPTER TWENTY-EIGHT

Qualitative and Quantitative Characterization of Autophagy in *Caenorhabditis elegans* by Electron Microscopy

Timea Sigmond,* Judit Fehér,† Attila Baksa,† Gabriella Pásti,† Zsolt Pálfia,† Krisztina Takács-Vellai,* János Kovács,† Tibor Vellai,* *and* Attila L. Kovács†

Contents

Abstract

Caenorhabditis elegans has been introduced relatively late into the field of autophagy with no previous results by classical methods. Therefore, it has to be

* Laboratory of Developmental Genetics, Department of Genetics, Eötvös Loránd University, Budapest, Hungary

† Cell Physiology Laboratory, Department of Anatomy, Cell and Developmental Biology, Eötvös Loránd University, Budapest, Hungary

Methods in Enzymology, Volume 451

ISSN 0076-6879, DOI: 10.1016/S0076-6879(08)03228-X

studied in parallel with both traditional electron microscopy and modern molecular approaches. In general, correct identification of autophagic elements by electron microscopy is indispensable to establish a firm basis for our understanding of the process. The principles and the method for identification, applied also for *C. elegans*, are summarized first in this article, to facilitate their utilization both for further studies and the analysis of new cell types and to support researchers new to electron microscopy techniques. Studying autophagy in the worm by electron microscopy has required the development of special handling and sampling techniques in addition to overcoming the general technical difficulties due to the nature of *C. elegans* samples. These are described in detail, together with some initial qualitative and quantitative results obtained by them. The feasibility of the presented method is supported by data which show that in continuously fed worms the autophagic compartment is in the lower range of the $10^{-2}\%$ order of magnitude of the cytoplasmic volume, while immediately after molting or upon starvation in the second larval period, usually more than a 10-fold increase can be measured. In dauer larvae, individual variation of the autophagic compartment is very high. The predauer stage in *daf*-2 mutants does not seem to show significant constitutive autophagic activity. Some autophagy-related gene mutants show characteristic ultrastuctural features, such as autophagosomes with membrane abnormalities (*unc*-51/*Atg*1) or the hypertrophy of multivesicular bodies (*let*-512/*Vps*34, *bec*-1/*Atg*6).

1. Introduction

The discovery and the initial development of the concept of autophagy was mostly based on electron microscopy observations. Although insufficient alone, electron microscopy has continued to be indispensible even in the modern era of autophagy research. The high-resolution pictures promote our understanding of autophagy-related subcellular (ultrastructural) changes and help to validate the specificity of various autophagy markers, both at light and electron microscopy levels. With the exponential expansion of interest in autophagy the need for the introduction of new cell types into the studies is also increasing. In these cases the researchers are faced with problems similar to the pioneers, they have to clearly identify autophagic structures. This has also occurred with us in the case of *C. elegans*. The general features of our approach and the technical details of the applied method are given here with two aims in mind. First, to help researchers without previous practice in electron microscopy to get acquainted with the advantages and limitations of ultrastructural analysis. Second, to describe specific solutions and initial results by the application of electron microscopy in *C. elegans*, a new model organism without any background from the earlier, classical era of autophagy research.

2. The Challenge of Identifying Autophagic Structures by Electron Microscopy: Overview and General Principles

In the mid-1950s, some unusual electron microscopy observations were made in kidney cells which showed that mitochondria may occur in membrane-bordered compartments inside the cytoplasm. The first reference to this type of structure can be traced back to a Ph.D. dissertation in 1954 but the picture was misinterpreted (Rhodin, 1954). In 1957, the electron microscopy image of an unusually placed mitochondrion in a dense body was published first in an international journal (Clark, 1957). The real meaning of kidney cell mitochondria being inside dense bodies was recognized by Alex B. Novikoff (1959), who demontstrated that they contain lysosomal enzymes, and it was suggested that they perform spatially limited lysosomal autolysis of the cell's own material. To describe the sequence of events leading to this focal self-degradation, the inherent limitation of transmission electron microscopy had to be surmounted, which is that four-dimensional conclusions need to be drawn from sections that represent only two dimensions. The cornerstone of reliable results was and still is the correct identification of autophagic elements. This might be one of the most challenging problems for the analysis of the autophagic process (Eskelinen, 2007), especially when new cell types are introduced to the field (Kovács *et al.*, 2004).

The most important criteria for the electron microscopic identification of autophagic elements were developed during the 1960s and are still valid today (Ericsson, 1969a). This identification begins with the recognition of cytoplasmic components completely bordered in the electron microscopic section by ribosome-free membrane. However, one cannot be sure that a membrane that appears closed in a plane is also closed in three dimensions. This problem has different implications in various stages of autophagy. The solution is relatively easy for early autolysosomes. Their content is partially disintegrated but still recognizable (Fig. 28.1C). The fact that the degradation appears to be focal (i.e., confined to a certain membrane-limited area in the plane) is indirect proof of its also being closed in three dimensions (Ashford and Porter, 1962). It is in this stage when simultaneous demonstration of the presence of lysosomal enzymes makes it obvious that we face a lytic compartment in action.

A possible difficulty in correct identification at this stage may come from phagocytosis of cellular fragments. This problem typically occurs in phagocytic types of cells or in areas of tissues with neigboring apoptotic cells. Careful analysis of conditions has to be made to exclude the possibility of heterophagy (Ashford and Porter, 1962; de Duve and Wattiaux, 1966;

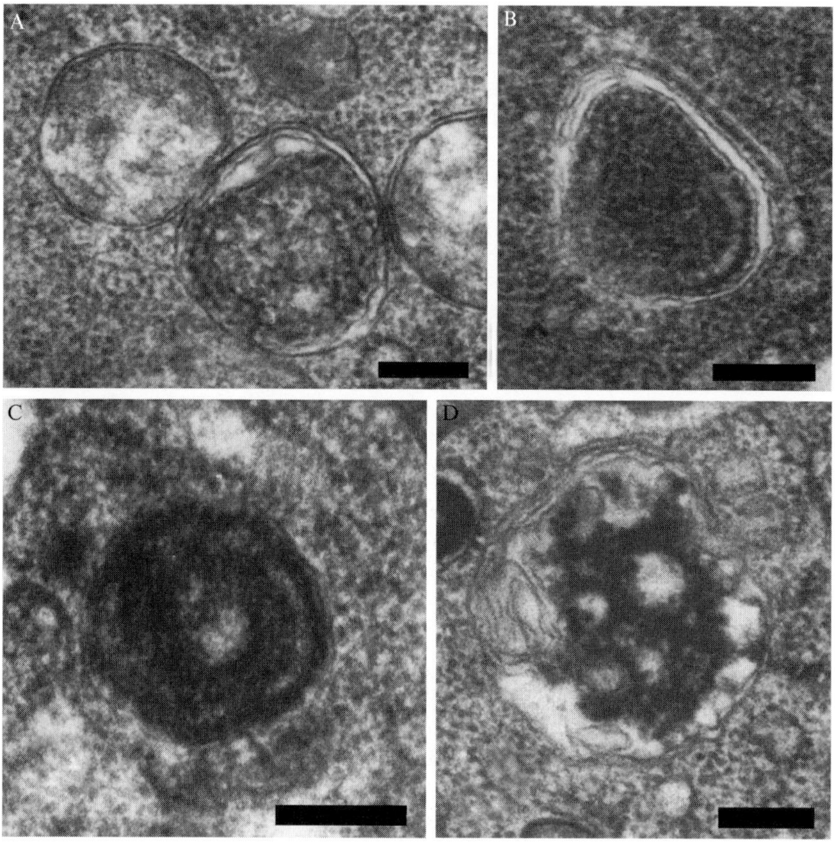

Figure 28.1 Autophagic elements in various stages of progression from gut epithelial cells of *C. elegans*. (A) An autophagosome between two mitochondria. The morphology of the cytoplasm segregated by the smooth double membrane is unchanged. (B) The characteristic empty-looking cleft between the bordering membranes is prominent in this autophagosome while some of the ribosomes look slightly swollen. (C) Only a single membrane can be seen bordering this autolysosome with obvious signs of degradation of the sequestered RER cisterns (general increase in density and strongly swollen ribosomes). (D) A late autolysosome shows barely identifiable dense clumps which can be traced back by series image reconstruction to RER and swollen ribosomes. Scale bars 250 nm.

Ericsson, 1969a). The convergence of hetero- and autophagic routes makes it especially difficult to differentiate the origin of contents at this stage; however, the possibilities are better in the earlier stage (see subsequent sections).

Although autophagic elements are usually in a relatively small quantity in cells, their progression is quick enough for all stages to be present at the same time. That makes it possible to reconstruct series of images which follow the gradual disintegration of contents. Thereby the dimension of time can be added to our knowledge. This approach provides an opportunity to reconstruct the autophagic process from the early autolysosome stage both in the backward and forward direction (Ericsson, 1969a). We can call this method series image reconstruction. Following the morphological changes in the earlier direction means to identify autophagic elements with less or no degradation of the content (Figs. 28.1A and 28.1B). This type of analysis led to the discovery that in the initial stage the autophagic content is bordered by a ribosome-free double membrane, and these autophagic elements are without lysosomal enzymes (Arstila *et al.*, 1972; Arstila and Trump, 1968; Glinsmann and Ericsson, 1966; Pfeifer, 1976). In this prelysosomal stage, which was first hypothetically introduced and called an autophagosome by de Duve (de Duve and Wattiaux, 1966), the morphology of the content is practically unchanged. Autophagosomes are bordered by the thin type of intracellular membranes (Ericsson, 1969b; Klionsky *et al.*, 2008; Kovács *et al.*, 2004), similar in thickness to those of mitochondria and the endoplasmic reticulum. According to our measurements, they appear to be 7–8 nm thick in routine electron microscopy preparations. The two delimiting membranes tend to get separated by an empty cleft when the usual aldehyde fixatives are used (Fig. 28.1B). Although artifactual, this cleft is a practical indicator of the autophagosome (Kovács *et al.*, 2007). However, the lack of morphological deterioration in the encircled area in this early stage cancels the strongest evidence for the bordering membrane being closed in three dimensions. To aggravate the problem, sections of plasma membrane invaginations between neighboring cells (not uncommon; e.g., in *C. elegans*), and the apoptotic bodies immediately after phagocytic uptake, may be misinterpreted as autophagosomes (Kovács *et al.*, 2004). In the case of cell types newly introduced to autophagy research, an initial investment is necessary to see the frequency and distribution of these structures independent of autophagy. This may include serial sectioning and measurement of membrane thickness (Kovács *et al.*, 2004). Interdigitations and phagocytized apoptotic bodies (which are rather big: 1.5–3 μm or even bigger) are usually close to the plasma membrane of neighboring cells. Both interdigitations and phagocytized apoptotic bodies are bordered by a pair of plasma membranes that are of the thick type (approximately 9–10 nm by our measurements) without the empty cleft between them (Ericsson, 1969b; Kovács *et al.*, 2004).

More problems arise with the later stages when the structural deterioration of the content does not allow clear recognition of the originally sequestered material. There are, however, some considerations that may help in these cases too. As mentioned earlier, when intensive autophagic activity occurs, representatives of all stages of progression are generally present simultaneously at least at the initial period of the process. Therefore, by the method of series image reconstruction the structural deterioration can be followed by stepwise changes to otherwise unrecognizable derivatives, making them still identifiable (Figs. 28.1C, 28.1D). Autophagic elements of different ages may coalesce into complex vacuoles with earlier and later forms in them, which helps to identify the content of later stages. It may also help that the speed of morphological degradation varies among organelles (Papadopoulos and Pfeifer, 1986). By applying this series analysis method we can also discover the changes in the heterophagy or secretory routes that lead to heterogenous structures, which may otherwise easily be misinterpreted as late autophagic elements (Kovács *et al.*, 2004). Very late stages of prior autophagic activity may leave vacuoles with homogenous content of various density (totally electron lucent, moderately or strongly dense). In these cases, samples must be taken from earlier phases of the process to arrive at reliable conclusions.

3. *Caenorhabditis* as an Object of Autophagy Studies by Electron Microscopy

Because of its small size and impenetrable cuticle, *C. elegans* is a rather difficult object for electron microscopy in general (Hall, 1995; Kovács *et al.*, 2004). This fact together with the additional problems that arise with studying autophagy in this organism amount to a real challenge. Apart from very special cases, mammalian samples are practically unlimited in size even for a certain cell type of an organ. In *C. elegans*, however, any tissue appears in a rather small quantity. In addition, the whole body, including all cell types, undergoes continuous morphological changes along the antero-posterior body axis, resulting in a marked polarity in the distribution of structural elements. Adding to this the equally continuous change during development shows the real magnitude of the problem. Disregarding the polarity by taking a few samples by cross-sections from certain regions along the antero-posterior axis would lead to unreliable results and conclusions.

The theoretically straightforward solution would be to cut the whole body "from head to toe" to cover the regional differences by serial sectioning. This method has indeed been used in *C. elegans* research (e.g., to describe the details of the nervous system; White *et al.*, 1986). However, it would need impossibly huge investments of time and energy for autophagy studies, when lots of observations have to be made throughout development and under changing experimental conditions. To satisfy the conflicting requirements of investment efficiency and covering all body regions, we applied longitudinal sectioning. The detailed description of our method is presented in the next sections.

3.1. Practical aspects of sample preparation from fixation to sectioning

The cuticle of *C. elegans* is a very efficient barrier for fixatives, which can only get into the cells if the body is cut into pieces (Hall, 1995). Our procedure for fixation is the following:

1. Selected worms are individually carried by a thin platinum wire into a drop of fixative with an approximate volume of 100 μl. The fixative contains 0.2% glutaraldehyde and 2.7% formaldehyde in 0.15 mol/l neutralized cacodylate buffer. Our experience shows that in formaldehyde-containing fixatives the movement of worms slows down and stops within a short time (10–20 s), which allows them to be cut into pieces by, for example, a miniature scalpel or the tip of a hypodermic needle. We found that the use of a piece of razor blade, broken to an appropriate size and shape and then fixed into a holder, is the cheapest and gives the best cuts (Fig. 28.2A).

   The drop of fixative is pipetted out in the lid of a polystyrene Petri dish, which is soft enough to save the edge of the blade for several rounds of cutting. For successful cutting it is important to hold the scalpel or other cutting device like a pen; find a stable support for the hand, so that the cutting action is carried out by the movement of the fingers only.
2. Cut pieces are pipetted into the well of a Terasaki-plate by a 10-μl micropipette tip with the volume of the pipetter set to 1 μl. The pipette tip is cut obliquely (Fig. 28.2B) to help the pipetting maneuvers when transferring the pieces, as well as when removing the fixative and washing buffer later. It is important to keep track of all pieces, as accidental attachment to the wall of the pipette tip may occur.

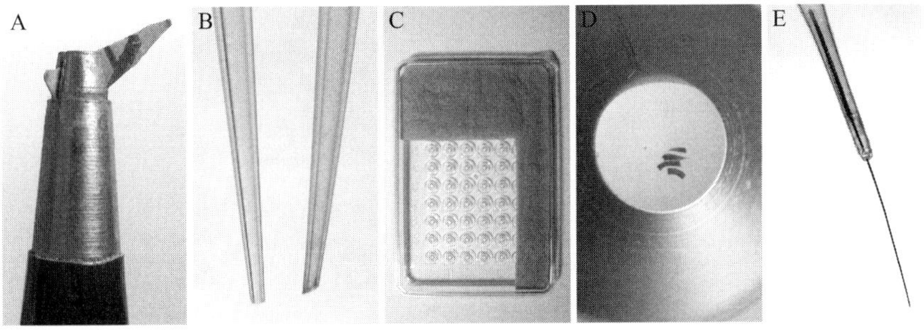

Figure 28.2 Tools for sample handling in electron microscopy of *C. elegans* autophagy. (A) The cutting device from a razor blade in the holder from a repeater pencil. (B) The oblique cutting of a 10-μl pipette tip. (C) The Terasaki plate with moisturizing sponge in it. (D) The four pieces of a cut, fixed and ruthenium-red stained worm sample in washing buffer inside the Terasaki well. (E) The thick hair in the tip of a Pasteur pipette.

Unnoticed removal of pieces previously attached to the wall may cause partial mixing of samples.

3. The successful transfer is followed by completely filling the Terasaki well with the fixative. Inside the plate is placed an appropriately sized sponge, which is filled with 0.5% glutaraldehyde solution in distilled water (Fig. 28.2C). This serves to maintain continuous moisture of the inner space of the plate to prevent excess evaporation, and to kill fungi and bacteria, which otherwise begin to grow in the wet environment.
4. After overnight fixation in a refrigerator, the aldehyde-containing solution is replaced by cacodylate buffer without the fixative for washing. This is carried out by cautiously sucking off most of the liquid from the samples. It is not possible to remove the solution completely and leave the samples in the well at the same time. Therefore, partial replacement should be repeated 3–4 times in each round of washings.

   At the end of the first washing we add 1–2 μl 0.1% ruthenium red also dissolved in washing buffer until the solution becomes deep red. This is done for improving the visibility of samples for later steps of the preparation procedure. The ruthenium red stains the samples in a few hours. At least 1 day has to be devoted to 3–4 rounds of additional washings, while the samples are kept in a refrigerator in the Terasaki plate (Fig. 28.2D). After several days of storage, the ruthenium red changes color from red to gray by reduction, but this does not seem to influence the quality of the electron microscopy images.
5. Because of the very small size of the samples, it is practically impossible to carry out the later steps of the preparation procedure without embedding them into agar gel. We use 1.8%–2% general purpose agar (e.g., from Sigma, cat no. A1296) kept at 56 °C to ensure liquidity. 1.5–2 ml of agar is pipetted into a 5-cm polystyrene Petri dish and spread evenly over the surface. When using covers or bacterial-grade Petri dishes, more volume is needed as their surface is hydrophobic. Tissue-culture-quality bottoms are hydrophilic and can be used with less volume, leading to a thinner agar layer.
6. After spreading, the agar solidifies at room temperature in a short time in the bottom of the Petri dish. This layer is now ready for transferring the fixed samples of the worm onto it from the Terasaki plate. The pieces have to be pipetted out in the smallest possible volume, but still they are scattered in a rather large area in the drop of buffer. They now have to be rounded up, which can be done using a thick strand of hair glued into the tip of a Pasteur-pipette (Fig. 28.2E). Having collected the pieces into a tight group, preferably parallel with each other, the excess liquid must be completely removed by filter paper. The sample is now ready for covering (Fig. 28.3A).
7. A small drop of agar is pipetted on the pieces (Fig. 28.3B), taking care that when falling, it exactly hits the pieces in the center. If too much

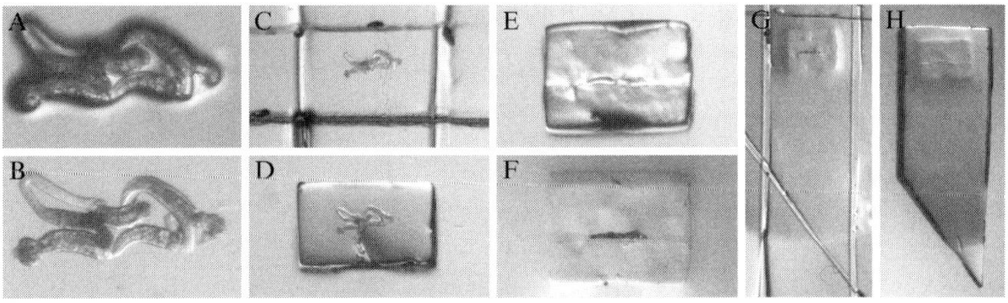

Figure 28.3 Major stages of agar-embedding. (A) The collected worm pieces on the agar surface in the bottom of a polystyrene Petri-dish. (B) The same as (A), but covered with a drop of agar. (C) The sample prism cut out from the agar with razor blade. (D) The separated prism in the original position, and (E) after reorientation by a 90 degree forward turning. (F) The reoriented prism covered with a new drop of agar. (G) Lower magnification picture of the reoriented and newly covered sample to show the second cutting to a shape which helps proper positioning in the embedding mold. (H) The separated sample that enters the dehydration and embedding procedure.

washing buffer was left, or the agar drop lands out of center, the pieces tend to move away from each other, spreading over a larger area and making the later handling of the sample problematic. Pieces of 6–8 worms by clockwise positioning are usually covered this way in a 5-cm Petri dish to increase efficiency.

The samples are now embedded in agar, with the pieces in a plane parallel with the bottom of the Petri dish. It is useful to routinely make light microscopy pictures at this stage, which will greatly aid in orientation during sectioning and in the electron microscope. Pictures might also be very helpful to clarify problems from accidental sample mixing.

8. The next step is to cut the samples out in a small cube or prism of agar (Fig. 28.3C), which we do with a razor blade. The size should be big enough to ensure easy treatment during the subsequent steps of preparation. Further handling depends on the way we want to do the embedding. If we will eventually put the agar cube, with the worm pieces inside, into a flat-bottom embedding capsule, no additional treatment is necessary at this step.

   As it is more efficient, we usually use flat silicone embedding molds. To ensure that the samples will get into the right position for later longitudinal sectioning also in the flat mold, we insert a reorientation step. This is done by turning the first sample cube by 90 degrees and putting it on fresh agar so that the sample plane becomes perpendicular to the new agar surface. In this position the cube is covered again by a drop of agar, and cut out as an approximately 2 mm $\times$ 4 mm flat prism convenient for further handling (Fig. 28.3D–H).
9. The preparation procedure from this point on follows the usual steps of tissue samples for electron microscopy. The agar blocks, with the

embedded worm pieces inside, are put for 1 h into 1 ml of 0.5% osmium tetroxide solution in 0.1 mol/l neutralized cacodylate buffer, in small volume (8 ml) wide-opening glass vials, tightly closed with stoppers. As the osmic acid solution is harmful, we have to work in a ventilated cabinet.

10. The osmic acid treatment is succeded by washing twice with distilled water. Further treatment with 1% uranyl-acetate in water for 0.5 h ensues, followed by dehydration in a graded ethanol series of 70%, 90%, 96% and twice in absolute ethanol for 20 min in each grade. Finally, for complete removal of even the traces of water we use 100% propylene oxide for 30 min.

    From the uranyl-acetate step onward, the vials may remain open, but from the absolute ethanol step they have to be stoppered again. Throughout the whole procedure the liquid is sucked off from the samples which remain in the vials.

11. The impregnation with the diluted embedding resin comes next, which must be done according to the recommended recipe of the chosen resin type. During the dehydration and impregnation procedure the vials are put into a slowly turning rotation drum to facilitate diffusion. The last stage of impregnation is usually carried out with the complete undiluted resin. From the vials, the samples are fished out with a toothpick and put into either the flat bottom capsule or into the depressions of the flat embedding mold, already containing the resin of appropriate quantity. Careful checking of the position of the samples must be done under the dissecting microcope and, if necessary, their position has to be adjusted. In the case of the capsule, a central position should be kept and the plane of the sample pieces must be parallel with the bottom. In the flat embedding mold the sample plane should be close to the upper edge in parallel with it. This is the right position for later longitudinal sectioning (Fig. 28.4A).

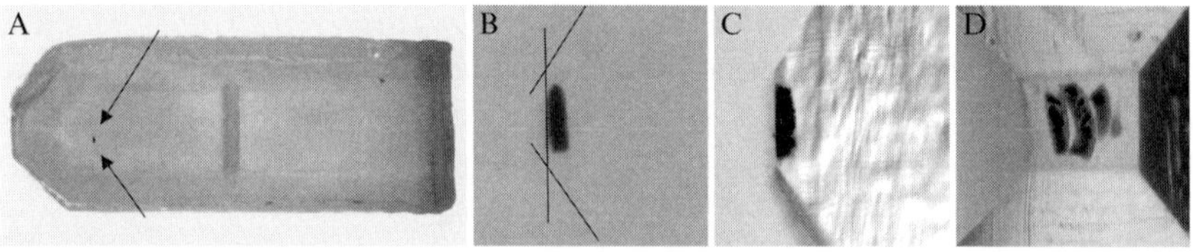

Figure 28.4 The preparation of the resin-embedded sample for sectioning. (A) The sample after flat embedding in the polymerized resin. The arrows point at the small black dot, which is the sample. (B) The magnified picture of the sample inside the resin. The lines show the final plane to be cut by the razor blade. (C) The sample in the tip of the truncated pyramid after cutting is completed under a dissecting microscope. (D) The upper view of the sample in (C).

12. After polymerization according to the prescriptions for the applied resin, the samples are processed for sectioning by the ultramicrotome. A truncated pyramid has to be cut in such a way that the sample gets as close to the surface as possible. In the flat-bottomed capsule this position is automatically ensured. Samples from the flat embedding mold need careful approaching. This we do under the dissecting microscope by hand with the help of razor blades (Fig. 28.4B–D). The impregnation with ruthenium red helps a lot at this stage by making the worm pieces clearly visible. Unfortunately, neither osmic acid no uranyl-acetate treatment gives strong enough staining for the small pieces to be easily recognizable after embedding. Without good visibility, approaching the samples becomes problematic and the danger that they accidentally become cut off increases a lot. The approach by hand must be stopped in a certain distance in any case and completed with the ultramicrotome. If we have made photographs after agar embedding, they can also be utilized well here, to recognize the appearance of the sample pieces in the ultrathin sections already when they are only tiny spots. Alternatively, photographs can also be taken before the ultrathin cutting begins.

3.2. Handling of sections for ultrastructural morphology and sampling for quantitative evaluation

As mentioned earlier, longitudinal sectioning offers the possibility to cover the antero-posteriorly occurring regional differences. 2–5 worm pieces aligned close to each other will usually make it possible to have the whole body in the area of a single section for all sizes from the smallest L1 larva to the largest adult (Ad). This gives a rather efficient approach to get an overall qualitative picture about the occurrence of autophagic elements in various types of cells. The cylindrical pieces of the cut body rotate freely along their longitudinal axis and are fixed in random positions when embedded in agar. By this inherent randomization step, our sampling procedure comes close to satisfying the criteria of unbiased random sampling required for proper morphometry (Howard and Reed, 1998).

It follows from the structure of the body of *C. elegans* that if we ensure taking sections roughly as deep as one-third of the diameter of the body, all major tissue types will likely be represented in them. On the basis of the preceding criteria and conclusions, we apply the following procedure for morphometry: The collection of ultrathin sections starts at random positions when all embedded pieces clearly appear in them. We take a larger area for starting with bigger worms. From a single animal, 70–100 sections are routinely collected on 5 electron microscopy grids. This sampling ensures that the pieces are cut into a depth where all major organs appear in the sections. When using ordinary electron microscopy grids, the mesh of the grid will cover some of the area of the cut samples. In principle the best would be to use coated single-slot grids (Hall, 1995); however, their application is cumbersome and requires special experience. We found that the simplest and still satisfactory solution is to use high-transmission ordinary grids. Before the observations in the electron microscope, the sections on the grids are contrasted in alkaline lead citrate solution according to Reynolds (1963).

For quantitative measurements, two nonneighboring grids are selected from the five (e.g., 1 and 3, 2 and 4, 3 and 5) for making photographs from them. This way the distance between the planes will be long enough for the samples to be considered independent from each other with regard to autophagy. Photographs are taken from the chosen grids to cover all available areas of future measurements. The problem which arises here is that autophagic vacuoles are relatively rare objects in the cells of *C. elegans*. Therefore, the largest possible area has to be evaluated for reliable results, which needs pictures with low magnification. However, autophagic elements are rather small and can only be recognized at relatively high magnification. To solve this controversy we have applied two-level photography. The pictures for the test area are taken at 2700x primary magnification and autophagic elements are identified and photographed individually at 14,000x primary magnification within this test area. Volumetric morphometry is carried out by the point counting method (Howard and Reed, 1998; Williams, 1977) on prints or, recently, on the computer screen.

3.3. Basic quantitative description of autophagy in postembryonic development in three major tissues of wild-type *C. elegans* by electron microscopic morphometry

Autophagic elements and the initial observations on autophagy by electron microscopy in *C. elegans* have already been described (Kovács *et al.*, 2004). Preliminary results from randomly taken worms show a rather rare appearance of autophagic structures even in animals from starving cultures. We then decided to work out the systematic sampling method described above and regularly applied it to cover all larval stages of the postembryonic development of the wild-type (N2) strain. Worms were taken from well-fed cultures either while eating continuously in the intermolt periods or at the end of the so-called lethargus or molting phase, immediately after they had left behind the old cuticle. During the lethargus phase all movements are stopped including pharyngeal pumping; therefore, feeding is also arrested.

Our results in Table 28.1 show that in continuously eating worms the volume fraction of the autophagic compartment (autophagosomes and autolysosomes together) is very small. It is less than 0.03% of the cytoplasmic volume in all measured tissues. However, in worms freshly hatched from

Table 28.1 The cytoplasmic volume fraction of the autophagic compartment during postembryonic development in wild-type *C. elegans* in three major cell types[a]

	Hypodermal cells		Body wall muscle cells		Gut epithelial cells	
Developmental stage	Autophagic compartment (%)	Total (average) test area (μm^2)	Autophagic compartment (%)	Total (average) test area (μm^2)	Autophagic compartment (%)	Total (average) test area (μm^2)
Freshly hatched	**0.180 ± 0.088**	2096 (262 ± 81)	**0.043 ± 0.021**	1452 (182 ± 38)	**0.115 ± 0.041**	3819 (477 ± 80)
L1 eating	**0.011 ± 0.006**	3370 (421 ± 80)	**0.006 ± 0.005**	1267 (158 ± 31)	**0.011 ± 0.007**	4360 (545 ± 139)
L1/L2 lethargus	**0.236 ± 0.071**	2990 (374 ± 62)	**0.239 ± 0.038**	1391 (174 ± 37)	**0.138 ± 0.046**	3500 (438 ± 64)
L2 eating	**0.007 ± 0.007**	2973 (372 ± 95)	**0.000 ± 0.000**	1891 (236 ± 48)	**0.002 ± 0.002**	6494 (812 ± 187)
L2/L3 lethargus	**0.187 ± 0.055**	3118 (390 ± 178)	**0.125 ± 0.043**	2156 (269 ± 48)	**0.117 ± 0.037**	7808 (976 ± 278)
L3 eating	**0.008 ± 0.003**	5980 (748 ± 92)	**0.000 ± 0.000**	2418 (302 ± 27)	**0.004 ± 0.004**	8634 (1079 ± 146)
L3/L4 lethargus	**0.152 ± 0.029**	5504 (668 ± 154)	**0.150 ± 0.049**	2398 (300 ± 66)	**0.151 ± 0.031**	9679 (1210 ± 170)
L4 eating	**0.021 ± 0.009**	6706 (838 ± 112)	**0.000 ± 0.000**	2089 (261 ± 41)	**0.008 ± 0.006**	7699 (962 ± 141)

(*continued*)

Table 28.1 *(continued)*

Developmental stage	Hypodermal cells		Body wall muscle cells		Gut epithelial cells	
	Autophagic compartment (%)	Total (average) test area (μm^2)	Autophagic compartment (%)	Total (average) test area (μm^2)	Autophagic compartment (%)	Total (average) test area (μm^2)
L4/Ad lethargus	**0.250 ± 0.079**	7597 (950 ± 163)	**0.238 ± 0.063**	2507 (313 ± 48)	**0.118 ± 0.018**	9672 (1209 ± 168)
Ad eating	**0.026 ± 0.007**	12214 (1527 ± 232)	**0.000 ± 0.000**	2754 (344 ± 71)	**0.009 ± 0.006**	9837 (1230 ± 217)

[a] Worms were kept in well-fed cultures and randomly taken either when eating in the intermolt or at the end of the molting (lethargus) phases. Eating worms with continuous pharyngeal pumping were picked from the bacterial lawn. Lethargus phase animals were fixed right after they left the old cuticle, and were not allowed to eat any food.

Each of the averages in the table are from measurements on electron microscopy pictures of 8 worms and show the volume of the autophagic compartment (autophagosomes and autolysosomes together) expressed as percentage (± standard error) of the cytoplasmic volume containing them (cytoplasmic volume fraction). The total and average size (± standard deviation) of the test areas are also given to indicate sample behavior and facilitate comparison and reproducibility. The differences between the eating and lethargus+freshly hatched samples are statistically significant ($p < 0.05$) by the Mann–Whitney's U test.

L1, L2, L3, L4 and Ad, are the larval and adult stages of development.

the egg, or from the old cuticle at the end of lethargus phases, the values are one order of magnitude higher, representing statistically significant increases ($p < 0.05$) in all cases.

We also measured the effect of starvation on worms in the L2 larval period when transformation into the dauer stage may take place. As the data in Table 28.2 show, the autophagic compartment significantly enlarges upon strarvation up to about 0.13%. The strongest response is given by the hypodermal cells. The values upon starvation are not higher than those at the end of the lethargus phase and they do not grow with the length of food deprivation in the observed period.

3.4. Measurements on dauer and predauer stages

Several interesting questions arise concerning the possible role of autophagy in both the special enduring (dauer or L3D) larval stage, and in the L2 predauer (PD) phase during which the transformation into L3D takes place. Worms in the L3D stage do not eat and do not age either. L2 larvae undergo a series of structural and functional transformations during PD. We used the above described morphometric approach to look for the role of autophagy in both the dauer and predauer stages.

First, we studied spontaneously developed, randomly taken dauers from N2 cultures. Measurements were made in the 3 selected major cell types (hypodermal, body wall muscle and gut epithelial cells) of 20 animals. In addition to the autophagic compartment we also measured the quantity of glycogen and lipid which are stored to support the long-term survival of dauers. The most striking feature of the data was their extreme variability (Table 28.3). In such cases it is best to present the results individually, as averaging may rather hide than reveal information. Out of 20 we found autophagic elements in 13 worms. Among these the autophagic compartment of any cell type was on the order of magnitude of 0.01% and 0.1% in 8 and 4 animals, respectively. In the latter case the highest value is just above 0.2%. Gut epithelial cells contain autophagic elements most frequently (11 cases of 13). Hypodermal and body wall muscle cells are only represented in 2 and 3 worms, respectively. There was only one individual where either two or three tissues contained autophagic elements. Glycogen and lipid was present in 10^0% or 10^1% order of magnitude, sometimes missing from muscle cells. Although the number of animals was rather small, we performed a pilot correlation analysis. No correlation was shown either between autophagy and lipid, or autophagy and glycogen, or lipid and glycogen content.

Worms with the thermosensitive mutation of insulin-like growth factor receptor *daf-2(e1370)* transform into dauer at 25 °C constitutively even in the presence of food. We took samples from the culture of *daf-2* mutants in a similar random way as we did from the wild type (N2) cultures previously

Table 28.2 The effect of starvation on the cytoplasmic volume fraction of the autophagic compartment in the L2 larval stage of wild-type *C. elegans* in three major cell types[a]

	Hypodermal cells		Body wall muscle cells		Gut epithelial cells	
Developmental stage	Autophagic compartment (%)	Total (average) test area (μm^2)	Autophagic compartment (%)	Total (average) test area (μm^2)	Autophagic compartment (%)	Total (average) test area (μm^2))
L1/L2	**0.147 ± 0.051**	2990 (374 ± 62)	**0.101 ± 0.038**	1391 (174 ± 37)	**0.099 ± 0.036**	3500 (438 ± 64)
L2 eating	**0.007 ± 0.007**	2973 (372 ± 95)	**0.000 ± 0.000**	1891 (236 ± 48)	**0.002 ± 0.002**	6494 (812 ± 187)
L2 10 h starving	**0.134 ± 0.062**	3761 (342 ± 176)	**0.036 ± 0.019**	2213 (201 ± 88)	**0.104 ± 0.035**	6977 (634 ± 278)
L2 25 h starving	**0.127 ± 0.072**	3247 (295 ± 133)	**0.071 ± 0.064**	1925 (175 ± 72)	**0.072 ± 0.052**	7788 (708 ± 307)
L2 60 h starving	**0.026 ± 0.012**	4593 (459 ± 186)	**0.013 ± 0.009**	2151 (215 ± 112)	**0.054 ± 0.031**	7648 (765 ± 332)
L2/L3 lethargus	**0.127 ± 0.055**	3118 (390 ± 178)	**0.172 ± 0.050**	2156 (269 ± 48)	**0.127 ± 0.037**	7808 (976 ± 278)

[a] Eating worms with continuous pharyngeal pumping were randomly taken from the bacterial lawn. Lethargus phase animals were fixed right after they left the old cuticle and were not allowed to eat any food. For later starvation, worms well-fed in L1 were taken as L1/L2 lethargus and put onto empty plates without food.
Each of the averages in the Table are from measurements on electron microscopic pictures of 8 worms. The data show the cytoplasmic volume fraction of the autophagic compartment (autophagosomes and autolysosomes together) expressed as percentage cytoplasm (± standard error) and the total and average size (± standard deviation) of the test areas. The differences between the eating and lethargus+starving samples are statistically significant ($p < 0,05$) by the Mann-Whitney's U test.

Table 28.3 The cytoplasmic volume fraction of the autophagic, lipid and glycogen compartments in three major cell types of wild-type dauer worms[a]

Sample number	Hypodermal cells: Compartment (% cytoplasm) Autophagic	Lipid	Glycogen	TA (μm^2)	Body wall muscle cells: Compartment (% cytoplasm) Autophagic	Lipid	Glycogen	TA (μm^2)	Gut epithelial cells: Compartment (% cytoplasm) Autophagic	Lipid	Glycogen	TA (μm^2)
1	**0**	8.08	1.02	891	**0.21**	1.11	9.61	308	**0**	2.29	4.44	596
2	**0**	21.6	13.4	253	**0**	3.37	8.99	305	**0.20**	5.93	4.23	404
3	**0**	11.3	23.4	483	**0**	1.12	1.04	182	**0.19**	3.20	2.40	428
4	**0.01**	22.6	11.0	997	**0.05**	0	0	284	**0.10**	5.86	5.85	760
5	**0**	12.4	17.4	688	**0**	5.26	15.8	165	**0.11**	3.31	1.98	517
6	**0**	14.0	11.9	490	**0**	0	0	380	**0.09**	2.19	2.10	627
7	**0**	2.13	2.13	327	**0**	1.28	2.56	267	**0.08**	5.05	7.22	948
8	**0**	14.2	5.88	247	**0.08**	13.4	4.17	197	**0**	10.4	1.02	521
9	**0.02**	6.30	24.6	1030	**0**	0	1.19	288	**0.05**	10.7	13.2	414
10	**0**	3.53	10.6	291	**0**	0	8.75	274	**0.06**	1.23	3.70	278
11	**0**	13.5	6.45	1061	**0**	0	0	688	**0.06**	11.3	17.9	519
12	**0**	17.2	8.60	319	**0**	3.97	10.3	432	**0.04**	18.1	1.55	661
13	**0**	7.37	13.8	743	**0**	0	0	373	**0.01**	5.22	2.35	1312
14	**0**	8.58	15.7	1158	**0**	1.07	3.23	638	**0**	15.3	11.3	941
15	**0**	11.5	14.9	398	**0**	9.18	14.3	337	**0**	9.77	10.9	911
16	**0**	16.0	1.60	257	**0**	2.22	12.0	154	**0**	11.3	6.60	394
17	**0**	19.2	17.3	277	**0**	13.6	12.8	238	**0**	5.41	19.7	337
18	**0**	14.7	4.83	292	**0**	4.32	3.42	211	**0**	5.13	6.72	304
19	**0**	7.85	9.33	269	**0**	1.85	2.37	207	**0**	3.41	2.79	285
20	**0**	4.18	3.48	379	**0**	1.08	1.06	274	**0**	4.21	3.78	394

[a] In a standard culture in a 5 cm Petri-dish 20 spontaneously developed dauers were randomly selected for electron microscopic morphometry by the sampling method described in the text. Volume fraction data (% cytoplasm) and the covered cytoplasmic test area (TA) are given for individual worms designated by sample number.

and made the same morphometric analysis on them. The overall conclusion from the data is that autophagy seems to be more intense in these mutant dauers (Table 28.4). We found autophagy in 15 of 20 worms and in 3 cases the magnitude is in the range of 10^0%. 13 values fall into the 10^{-1} order of magnitude, 7 of them being higher than 0.4%. It is again the gut epithelial cells that show the highest activity. Occurrence of autophagic elements in multiple tissues in an individual is also more frequent.

Taking advantage of the constitutive dauer formation in *daf-2* mutants even in the presence of food, we followed their progression during the predauer stage for 28 h at 25 °C, up to the appearance of the typical dauer cuticle. We started with 41 worms from the L1/L2 lethargus stage and took the samples distributed evenly throughout the observation period. For statistical evaluation the data were collected into groups of four time intervals (Table 28.5). In the case of all three tissues the lipid content increases and shows a positive correlation with time. However, the glycogen content either does not change (body wall muscle) or decreases (hypodermis and gut). Averages of the autophagic compartment are usually in the range of 10^{-3}–10^{-2}%, and even the highest average is less than 0.05%. The highest individual value was 0.18%, which remains below those measured in the lethargus phases or in certain dauer individuals. These results suggest that constitutive high level autophagy is not necessary for dauer formation in *C. elegans*.

4. Some Ultrastructural Features of Autophagy-Related Mutations

The autophagy research in *C. elegans* still suffers from the relative scarceness of data and a definite well proven knowledge about the role of the autophagic process itself in the life of this worm. The special methodological difficulties of ultrastructural studies in *C. elegans* autophagy outlined previously can at least partially explain this lagging behind. Attempts have been made to apply the GFP-tagged ortholog of the widely used autophagy marker Atg8/LC3/lgg-1, the GFP::lgg1 fusion protein, in *C. elegans* (Kang *et al.*, 2007; Melendez *et al.*, 2003). There are indications to suggest that processes with no obvious or direct connection to autophagy itself may involve the expression and activity of this marker protein as is summarized in the chapter by Sigmond *et al.* in this volume. Therefore, the limitations and conditions, and the specificity and reliability of the GFP::lgg1 fusion protein as an indicator of functioning autophagic activity have yet to be properly worked out, so that electron microscopy again seems to be indispensable.

Table 28.4 The cytoplasmic volume fraction of the autophagic, lipid and glycogen compartments in three major cell types of *daf-2(e1370)* dauer worms[a]

	Hypodermal cells				Body wall muscle cells				Gut epithelial cells			
	Compartment (% cytoplasm)				Compartment (% cytoplasm)				Compartment (% cytoplasm)			
Sample number	Autophagic	Lipid	Glycogen	TA (μm^2)	Autophagic	Lipid	Glycogen	TA (μm^2)	Autophagic	Lipid	Glycogen	TA (μm^2)
1	**0**	13.8	0	395	**0.03**	3.90	0	408	**2.03**	8.51	0	264
2	**0**	3.74	0	233	**0.42**	1.50	0	183	**1.14**	4.20	0	423
3	**0**	16.7	0	308	**0.03**	1.66	0	442	**1.33**	7.38	0	520
4	**0.06**	18.9	0	1165	**0.10**	4.86	0.70	924	**0.76**	12.6	0	1576
5	**0.08**	5.37	0.27	500	**0.09**	0.33	0	193	**0.62**	15.2	0	671
6	**0**	18.7	0	285	**0**	0	0	184	**0.73**	13.7	0	579
7	**0.04**	26.6	4.1	1424	**0.05**	5.04	0	315	**0.53**	13.4	1.00	1044
8	**0**	4.57	0	203	**0**	0.64	0	171	**0.57**	4.32	0	380
9	**0.06**	29.2	0	398	**0.01**	2.88	0	880	**0.46**	3.62	3.54	1257
10	**0.16**	19.0	2.25	1127	**0.06**	5.95	0	449	**0.24**	11.8	0	2223
11	**0.07**	8.24	2.61	1007	**0**	0.57	0	264	**0.21**	5.54	0	835
12	**0.08**	29.3	7.70	633	**0**	7.09	0.37	959	**0.12**	17.4	1.18	2555
13	**0.07**	15.1	2.41	517	**0**	0.96	1.23	599	**0.07**	13.9	1.66	545
14	**0.01**	14.7	3.54	1233	**0**	7.08	5.18	1089	**0.11**	17.6	0.61	2294
15	**0**	13.5	5.24	397	**0**	2.81	1.84	187	**0.08**	8.96	0	284
16	**0**	15.4	8.47	469	**0**	4.11	15.5	189	**0**	13.7	5.82	377
17	**0**	13.1	2.21	787	**0**	7.91	6.98	445	**0**	26.3	1.17	1099
18	**0**	10.2	0	227	**0**	8.57	0	179	**0**	18.5	5.31	438
19	**0**	12.7	1.42	389	**0**	6.78	2.31	274	**0**	7.49	3.47	496
20	**0**	11.6	0	472	**0**	1.83	0	195	**0**	0.92	0	220

[a] In a standard culture in 5 cm Petri-dish 20 spontaneously developed dauers were randomly selected for electron microscopic morphometry by the sampling method described in the text. Volume fraction data (% cytoplasm) and the covered cytoplasmic test area (TA) are given for individual worms designated by sample number.

Table 28.5 The cytoplasmic volume fraction of the autophagic, lipid and glycogen compartments in three major cell types of *daf-2(e1370)* worms during the predauer period[a]

Time (spent in predauer L2 stage)	Hypodermal cells				Body wall muscle cells				Gut epithelial cells			
	Compartment (% cytoplasm)				Compartment (% cytoplasm)				Compartment (% cytoplasm)			
	Autophagic	Lipid	Glycogen	Average TA μm^2 (worm number)	Autophagic	Lipid	Glycogen	Average TA μm^2 (worm number)	Autophagic	Lipid	Glycogen	Average TA μm^2 (worm number)
1–10 h	**0.019** ± 0.013	4.822 ± 0.941	14.46 ± 2.539	473 ± 339 (13)	**0.010** ± 0.010	1.79 ± 0.491	6.450 ± 1.602	218 ± 133 (13)	**0.002** ± 0.001	7.039 ± 0.757	4.179 ± 0.653	895 ± 357 (13)
11–20 h	**0.012** ± 0.012	6.000 ± 0.792	10.36 ± 2.536	565 ± 321 (10)	**0.000** ± 0.000	6.030 ± 1.091	7.742 ± 0.959	288 ± 121 (10)	**0.002** ± 0.001	12.24 ± 0.860	2.137 ± 0.245	1135 ± 898 (10)
21–24 h	**0.002** ± 0.002	8.852 ± 1.429	11.49 ± 3.369	591 ± 494 (9)	**0.000** ± 0.000	9.468 ± 1.325	8.217 ± 0.193	370 ± 240 (9)	**0.007** ± 0.005	13.41 ± 0.673	2.382 ± 0.775	964 ± 429 (9)
25–28 h	**0.049** ± 0.019	14.56 ± 2.055	4.111 ± 0.558	839 ± 312 (9)	**0.000** ± 0.000	25.32 ± 3.320	8.959 ± 1.376	395 ± 189 (9)	**0.013** ± 0.005	26.79 ± 1.565	1.376 ± 0.435	1334 ± 606 (9)

[a] In a standard culture in a 5 cm Petri-dish we put egg laying mothers at 25 °C. The developing L1 larvae become committed to dauer development at this temperature. After the L1/L2 lethargus phase we took the sample worms at certain time points approximately evenly distributed along the L2 predauer period, up to 28 h when the final dauer cuticle was completed. After morphometric evaluation the data representing the individual worms were grouped into 5 time periods to allow statistical assesment. Volume fraction data (% cytoplasm) are given as average (± standard error) and the test areas (TA) as averge (± standard deviation), in parentheses we indicate the number of worms evaluated.

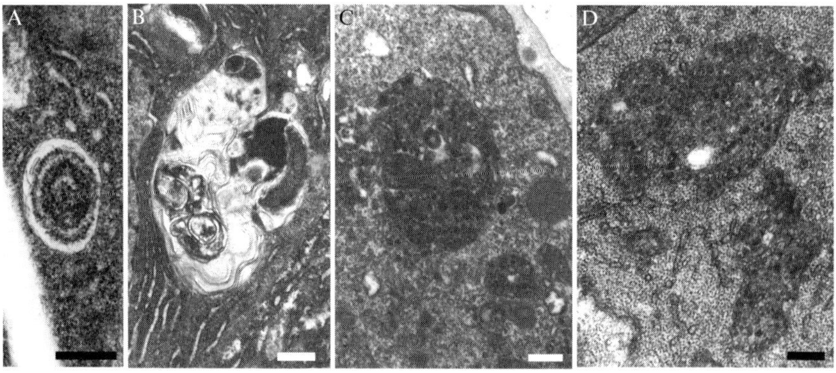

Figure 28.5 Some ultrastructural features of autophagy-related mutants. (A and B) Autophagic elements in an *unc-51(e369)* worm. (A) A normal autophagosome in a gut epithelial cell with the characteristic cleft between the two sequestration membranes. (B) A complex autophagic structure in a hypodermal cell with autophagosomal and autolysosomal content and prominent "myelinated" membrane whorls. (C and D) Huge multivesicular bodies from the hypodermal cells of a *let-512(h510)* (C), or a *bec-1(ok700)* mutant (D). Scale bars 500 nm.

Utilizing our technique in systematic ultrastructural studies of *C. elegans*, we included strains harboring mutant genes important for autophagy in other species. Some of these results follow here to further support the relevance and usefulness of the presented electron microscopy method of sample handling and preparation.

The *unc-51/Atg1* gene codes for a serine/threonine kinase. Worms defective for UNC-51 show uncoordinated movement, stemming from the abnormal development of certain neurons (Brenner, 1974; Ogura and Goshima, 2006; Ogura *et al.*, 1994). In yeast, Atg1 is essential for autophagy (Noda and Ohsumi, 2004) and important for the formation of the phagophore (Cheong *et al.*, 2008), which is the growing sequestration membrane before the completion of the autophagosome (Kovács *et al.*, 2007; Seglen, 1987). Studying the *unc-51(e369)* mutant strain by electron microscopy reveals that contrary to our expectations these worms are not without autophagic elements. Their quantity at the end of the lethargus phases is comparable to that in wild-type worms in hypodermis and gut cells; however, they cannot be found in muscle cells of the body wall (Table 28.6). Autophagic elements in these animals are mostly normal in appearance in gut epithelium (Fig. 28.5A); however, in the hypodermal cells, big complexes with high quantity of "myelinated" membrane whorls may appear, typically in the anterior part of the body (Fig. 28.5B). This structure might indicate disturbances in phagophore formation.

The *let-512/Vps34* and *bec-1/Vps30/Atg6* genes code for the class III phoshphatidylinositol 3-kinase (Roggo *et al.*, 2002; Stack and Emr, 1994)

Table 28.6 The cytoplasmic volume fraction of the autophagic compartment at the end of the lethargus phases in three major cell types of *unc-51(e369)* mutants[a]

	Hypodermal cells		Body wall muscle cells		Gut epithelial cells	
Develop-mental stage	Autophagic compartment (%)	Total (average) test area (μm^2)	Autophagic compartment (%)	Total (average) test area (μm^2)	Autophagic compartment (%)	Total (average) test area (μm^2)
L1/L2 lethargus	**0.596 ± 0.097**	1563 (521 ± 82)	**0**	564 (188 ± 93)	**0.061 ± 0.026**	2232 (744 ± 344)
L2/L3 lethargus	**0.465 ± 0.128**	1719 (573 ± 178)	**0**	480 (160 ± 36)	**0.129 ± 0.051**	2079 (693 ± 96)
L3/L4 lethargus	**0.029 ± 0.011**	3294 (1098 ± 494)	**0**	711 (237 ± 105)	**0.121 ± 0.020**	2001 (667 ± 188)
L4/Ad lethargus	**0.166 ± 0.052**	2247 (749 ± 63)	**0**	495 (165 ± 35)	**0.094 ± 0.032**	2580 (860 ± 244)

[a] Each of the averages in the Table are from measurements on electron microscopy pictures of 3 worms taken at the end of the lethargus phases. Data show the volume of the autophagic compartment (autophagosomes and autolysosomes together) expressed as percentage (± standard error) of the cytoplasmic volume containing them, and total and average size (± standard deviation) of the test areas.

and a coiled-coil protein interacting with the anti-apoptotic Bcl-2 protein (Liang *et al.*, 1998; Melendez *et al.*, 2003), respectively. These proteins interact with each other and are part of complexes, one of which is essential for autophagosome formation, and the other takes part in protein sorting at the *trans*-Golgi network in yeast (Cao and Klionsky, 2007; Stromhaug and Klionsky, 2004; Suzuki and Ohsumi, 2007; Takács-Vellai *et al.*, 2005). Here we demonstrate that a striking hypertophy of the multivesicular bodies can be observed both in worms with the *Vps34/let-512(h510)* or the *Vps30/Atg6/bec-1* mutation (Fig. 28.5C and 28.5D). These observations indicate that in addition to their role in autophagy and protein sorting these proteins may work in yet another membrane mediated process, the functioning of multivesicular bodies.

5. Conclusions and Perspectives

C. elegans is one of the most favored model organisms among multicellular animals for the application of the most advanced molecular genetic techniques. As autophagy research has entered the era of molecular understanding, the worm is an attractive choice for becoming a powerful object also in this field. Electron microscopy has played a major role in autophagy research from the beginning and continues to be indispensable ever since. Although *C. elegans* is an especially demanding object for electron microscopy, the method we have worked out, and the initial results presented with it, show that this technique can provide the essential support for autophagy studies in this model organism too. On the one hand, our technique offers the possibility of extensive and reliable ultrastructural characterization of autophagy-related mutants along the whole body and in all major tissues. On the other hand, the presented method for electron microscopic morphometry provides the opportunity to make quantitative measurements and characterize the level of autophagic activity.

ACKNOWLEDGMENTS

Some nematode strains used in this work were generously provided by Beth Levine (University of Texas Southwestern Medical Center, Dallas), the *Caenorhabditis* Genetics Center funded by the NIH National Center for Research Resources, and the International *C. elegans* Gene Knockout Consortium. We are grateful to Mariann Saródy for excellent technical help. This work was supported by grants from the Ministry of Health (167/2006), the National Office for Research and Technology (NKFP 1A/007/2004), and the Hungarian Scientific Research Foundation (OTKA K68372) to TV, and the Hungarian Scientific Research Foundation (OTKA T047241) to ALK. TV is a grantee of the János Bolyai scholarship.

REFERENCES

Arstila, A. U., Shelburne, J. D., and Trump, B. F. (1972). Studies on cellular autophagocytosis. A histochemical study on sequential alterations of mitochondria in the glucagon-induced autophagic vacuoles of rat liver. *Lab. Invest.* **27,** 317–323.

Arstila, A. U., and Trump, B. F. (1968). Studies on cellular autophagocytosis. The formation of autophagic vacuoles in the liver after glucagon administration. *Am. J. Pathol.* **53,** 687–733.

Ashford, T. P., and Porter, K. R. (1962). Cytoplasmic components in hepatic cell lysosomes. *J. Cell Biol.* **12,** 198–202.

Brenner, S. (1974). The genetics of *Caenorhabditis elegans*. *Genetics* **77,** 71–94.

Cao, Y., and Klionsky, D. J. (2007). Physiological functions of Atg6/Beclin 1: A unique autophagy-related protein. *Cell Res.* **17,** 839–849.

Cheong, H., Nair, U., Geng, J., and Klionsky, D. J. (2008). The Atg1 kinase complex is involved in the regulation of protein recruitment to initiate sequestering vesicle formation for nonspecific autophagy in *Saccharomyces cerevisiae*. *Mol. Biol. Cell* **19,** 668–681.

Clark, S. L., Jr. (1957). Cellular differentiation in the kidneys of newborn mice studied with the electron microscope. *J. Biophys. Biochem. Cytol.* **3,** 349–362.

de Duve, C., and Wattiaux, R. (1966). Functions of lysosomes. *Annu. Rev. Physiol.* **28,** 435–492.

Ericsson, J. L. (1969a). Mechanism of cellular autophagy. *In* "Lysosomes in Biology and Pathology" (J. T. Dingle and H. B. Fell, eds.), Vol. 2, pp. 345–395. North Holland, Amsterdam-London.

Ericsson, J. L. (1969b). Studies on induced cellular autophagy. II. Characterization of the membranes bordering autophagosomes in parenchymal liver cells. *Exp. Cell Res.* **56,** 393–405.

Eskelinen, E.-L. (2007). To be or not to be? Examples of incorrect identification of autophagic compartments in conventional transmission electron microscopy of mammalian cells. *Autophagy* **4.**

Glinsmann, W. H., and Ericsson, J. L. (1966). Observations on the subcellular organization of hepatic parenchymal cells. II. Evolution of reversible alterations induced by hypoxia. *Lab. Invest.* **15,** 762–777.

Hall, D. H. (1995). Electron microscopy and three-dimensional image reconstruction. *Methods Cell. Biol.* **48,** 395–436.

Howard, V., and Reed, M. G. (1998). "Unbiased stereology; three dimensional measurement in microscopy." Bios Scientific Publishers.

Kang, C., You, Y. J., and Avery, L. (2007). Dual roles of autophagy in the survival of *Caenorhabditis elegans* during starvation. *Genes Dev.* **21,** 2161–2171.

Klionsky, D. J., Abeliovich, H., Agostinis, P., Agrawal, D. K., Aliev, G., Askew, D. S., Baba, M., Baehrecke, E. H., Bahr, B. A., Ballabio, A., Bamber, B. A., Bassham, D. C., *et al.* (2008). Guidelines for the use and interpretation of assays for monitoring autophagy in higher eukaryotes. *Autophagy* **4,** 1–25.

Kovács, A. L., Pálfia, Z., Réz, G., Vellai, T., and Kovács, J. (2007). Sequestration revisited: Integrating traditional electron microscopy, de novo assembly and new results. *Autophagy* **3,** 655–662.

Kovács, A. L., Vellai, T., and Müller, F. (2004). Autophagy in *Caenorhabditis elegans*. *In* "Autophagy" (D. J. Klionsky, ed.), pp. 217–223. Eurekah.com, Landes Bioscience, Austin, Georgetown, Texas, USA.

Liang, X. H., Kleeman, L. K., Jiang, H. H., Gordon, G., Goldman, J. E., Berry, G., Herman, B., and Levine, B. (1998). Protection against fatal Sindbis virus encephalitis by Beclin, a novel Bcl-2-interacting protein. *J. Virology* **72,** 8586–8596.

Meléndez, A., Tallóczy, Z., Seaman, M., Eskelinen, E.-L., Hall, D. H., and Levine, B. (2003). Autophagy genes are essential for dauer development and life-span extension in *C. elegans*. *Science* **301,** 1387–1391.
Noda, T., and Ohsumi, Y. (2004). Macroautophagy in yeast. *In* "Autophagy" (D. J. Klionsky, ed.), pp. 70–83. Landes Bioscience, Georgetown, Eurekah.com, Austin, Texas, USA.
Novikoff, A. B. (1959). The proximal tubule cell in experimental hydronephrosis. *J. Biophys. Biochem. Cytol.* **6,** 136–138.
Ogura, K., and Goshima, Y. (2006). The autophagy-related kinase UNC-51 and its binding partner UNC-14 regulate the subcellular localization of the Netrin receptor UNC-5 in *Caenorhabditis elegans*. *Development* **133,** 3441–3450.
Ogura, K., Wicky, C., Magnenat, L., Tobler, H., Mori, I., Muller, F., and Ohshima, Y. (1994). *Caenorhabditis elegans* unc-51 gene required for axonal elongation encodes a novel serine/threonine kinase. *Genes Dev.* **8,** 2389–2400.
Papadopoulos, T., and Pfeifer, U. (1986). Regression of rat liver autophagic vacuoles by locally applied cycloheximide. *Lab. Invest.* **54,** 100–107.
Pfeifer, U. (1976). Lysosomen und Autophagie. *Verh. Dtsch. Ges. Pathol.* **60,** 28–64.
Reynolds, E. S. (1963). The use of lead citrate at high pH as an electron-opaque stain in electron microscopy. *J. Cell Biol.* **17,** 208–212.
Rhodin, J. (1954). Correlation of ultrastructural organization and function in normal and experimentally changed proximal tubule cell of the mouse kidney. Ph.D. dissertation, *Aktiebolaget Godvil, Stockholm,* 1–76.
Roggo, L., Bernard, V., Kovacs, A. L., Rose, A. M., Savoy, F., Zetka, M., Wymann, M. P., and Muller, F. (2002). Membrane transport in *Caenorhabditis elegans*: An essential role for VPS34 at the nuclear membrane. *Embo J.* **21,** 1673–1683.
Seglen, P. O. (1987). Regulation of autophagic protein degradation in isolated liver cells. *In* "Lysosomes: their role in protein breakdown" (H. Glaumann and F. J. Ballard, eds.), pp. 371–414. Academic Press.
Stack, J. H., and Emr, S. D. (1994). Vps34p required for yeast vacuolar protein sorting is a multiple specificity kinase that exhibits both protein kinase and phosphatidylinositol-specific PI 3-kinase activities. *J. Biol. Chem.* **269,** 31552–31562.
Stromhaug, P. E., and Klionsky, D. J. (2004). Cytoplasm to vacuole targeting. *In* "Autophagy" (D. J. Klionsky, ed.), pp. 84–106. Landes Bioscience, Georgetown, Eurekah. com, Austin, Texas, USA.
Suzuki, K., and Ohsumi, Y. (2007). Molecular machinery of autophagosome formation in yeast, *Saccharomyces cerevisiae*. *FEBS Lett.* **581,** 2156–2161.
Takács-Vellai, K., Vellai, T., Puoti, A., Passannante, M., Wicky, C., Streit, A., Kovács, A. L., and Muller, F. (2005). Inactivation of the autophagy gene bec-1 triggers apoptotic cell death in *C. elegans*. *Curr. Biol.* **15,** 1513–1517.
White, J. G., Southgate, E., Thomson, J. N., and Brenner, S. (1986). The structure of the nervous System of the nematode *Caenorhabditis elegans*. *Phil. Trans. R. So. Lond. B, Biol. Sci.* **314,** 1–340.
Williams, M. A. (1977). "Quantitative methods in biology." North-Holland Publishing Company, Amsterdam.

CHAPTER TWENTY-NINE

Monitoring the Role of Autophagy in *C. elegans* Aging

Alicia Meléndez,* David H. Hall,† *and* Malene Hansen‡

Contents

Abstract

Autophagy plays crucial roles in many biological processes, and recent research points to a possibly conserved role for autophagy in the process of organismal aging. Experiments in the nematode C. elegans suggest that autophagy may be required specifically for longevity pathways that are regulated by environmental signals. Known longevity genes can be assigned to four major longevity pathways/processes: insulin/IGF-1 signaling, dietary restriction, protein translation, and mitochondrial respiration. Of these, reduced insulin/IGF-1 signaling and dietary restriction, but not protein translation inhibition, appear to rely on

* Queens College-CUNY, Department of Biology, Flushing, New York, USA
† Center for *C. elegans* Anatomy, Albert Einstein College of Medicine, Bronx, New York, USA
‡ Burnham Institute for Medical Research, Program of Development and Aging, La Jolla, California, USA

Methods in Enzymology, Volume 451
ISSN 0076-6879, DOI: 10.1016/S0076-6879(08)03229-1

autophagy to increase life span. Multiple experimental approaches have been used to study autophagy in the context of aging in *C. elegans*. This chapter describes techniques used to address the link between aging and autophagy in *C. elegans*. Specifically, we summarize how to examine organismal life span in various longevity mutants and how to visually detect autophagy and autolysosomal formation in *C. elegans*.

1. Introduction to Longevity Pathways in *C. elegans*

The nematode *C. elegans* has proved a successful model organism for studying the molecular mechanisms of several biological processes, including aging (Kenyon, 2005). *C. elegans* offer many advantages with regards to studying longevity, including a short life span of approximately 3 weeks. Furthermore, *C. elegans* is experimentally tractable especially in regard to gene inactivation by RNA interference (RNAi). In *C. elegans*, RNAi treatment can be done in multiple ways (Wang and Barr, 2005), but it is particularly easy to do by feeding RNAi in which double-stranded RNA of the gene of interest is expressed in bacteria, the food source of the worm. As the entire genome of *C. elegans* is sequenced, this has allowed for RNAi feeding screens for new longevity genes (Hamilton *et al.*, 2005; Hansen *et al.*, 2005; Chen *et al.*, 2007; Curran and Ruvkun, 2007). Through such approaches, well more than 150 genes have so far been identified to modulate organismal aging in *C. elegans*. The majority of these currently known longevity genes can be assigned to four major longevity pathways/processes: insulin/IGF-1 signaling, dietary restriction, protein translation, and mitochondrial respiration. A brief outline of our current molecular understanding of these pathways/processes with specific focus on links to autophagy is summarized here.

1.1. Insulin/IGF-1-like signaling

The most well-studied longevity pathway in *C. elegans* is the insulin/IGF-1 signaling (IIS) pathway. Decreased activity of the insulin/IGF-1-like receptor, DAF-2, extends the life span more than 2-fold (Kenyon *et al.*, 1993), and similar gene perturbations extend the life span of flies and mice (Bluher *et al.*, 2003; Giannakou and Partridge, 2007; Holzenberger *et al.*, 2003). The canonical IIS pathway in worms involves signaling through a cascade of kinases, including AGE-1/PI3K, PDK-1 and AKT-1/2/SGK-1, via at least three different transcription factors DAF-16, HSF-1, and SKN-1 (Hsu *et al.*, 2003; Lin *et al.*, 1997; Ogg *et al.*, 1997; Tullet *et al.*, 2008). These transcription factors function downstream of *daf-2* and modulate life span by regulating a variety of target genes (Honda and Honda, 1999; Lee *et al.*, 2003a; McElwee *et al.*, 2003; Murphy *et al.*, 2003; Oh *et al.*, 2006). These include genes that are required for stress responses, detoxification of oxidative damage, and resistance to bacterial infection. Insulin/IGF-1 signaling from both the nervous system and the intestine modulates life span (Apfeld and Kenyon, 1998; Libina *et al.*, 2003; Wolkow *et al.*, 2000), and aging is specifically affected by *daf-2* during adulthood (Dillin *et al.*, 2002a). In *C. elegans*, IIS overlaps with several other processes that also have been implicated in aging, including sensory perception, signaling from the reproductive system, JNK signaling, and TGF-beta signaling (Hsin and Kenyon, 1999; Oh *et al.*, 2005; Shaw *et al.*, 2007).

While many studies have focused on elucidating the transcriptional output of IIS in *C. elegans*, recent studies have implicated a role for various cellular processes (e.g., autophagy) in the long life span of *daf-2* mutants. In *C. elegans*, autophagy was first investigated in the context of the developmental process of dauer morphogenesis (Meléndez *et al.*, 2003). *C. elegans* responds to unfavorable environmental conditions (e.g., crowding, starvation, increased temperature) by entering into an alternative third arrested larval stage, called dauer diapause. Autophagy is required for dauer formation in the insulin/IGF-1 receptor *daf-2* mutant, which has elevated levels of autophagy (see subsequent sections). Taken together, these findings suggest that autophagy genes are required for the organismal remodeling that allows *C. elegans* to adapt to environmental stresses, such as starvation. Similar to a role in development, autophagy is also required for the long life span of *daf-2* mutants, and *daf-2* long-lived mutants have increased levels of autophagy (Hars *et al.*, 2007; Meléndez *et al.*, 2003). Autophagy seems to be specifically linked to the aging function of *daf-2*, as inhibition of autophagy genes during adulthood is sufficient to significantly shorten the life span of *daf-2* mutants (Hansen *et al.*, 2008).

What regulates autophagy in *daf-2* mutants? So far no other mutants in the IIS pathway have been investigated in the context of autophagy, yet *daf-2* mutants without the FOXO transcription factor DAF-16 retain high levels of autophagic vesicles (as measured by the LC3/LGG-1 reporter, see subsequent sections) (Hansen *et al.*, 2008). As *daf-16*; *daf-2* double mutants are short lived, this finding suggests that increased levels of autophagy are not sufficient to extend life span in the context of reduced IIS. However, DAF-16/FOXO could be involved in downstream steps of the autophagy process (e.g., the lysosomal processing of the autophagosomal vesicles). Consistent with a role for DAF-16/FOXO in autophagy, Riddle and colleagues show that DAF-16/FOXO regulates DAF-15/Raptor (a TOR-binding protein), which functions in reproductive growth and organismal life span (Jia *et al.*, 2004). Reduction of DAF-15 extends life span via a mechanism that is at least partially dependent on autophagy (Hansen *et al.*, 2008). This finding would be consistent with a possible role for DAF-16/FOXO upstream of autophagy to modulate aging and development. Taken together, more experiments are clearly needed to clarify the role of DAF-16/FOXO in the autophagy process.

Another crucial question would be to address in which tissues, in *daf-2* mutants, autophagy modulates organismal life span. Likewise, it will be interesting to address the role of autophagy in the life span phenotypes mediated by the IIS overlapping processes mentioned previously (e.g., sensory perception and signaling from the reproductive system).

Similar to autophagy genes, genes involved in vesicular sorting were recently identified in a *daf-2* suppressor screen as well as in other independent studies to be required for the long life span of *daf-2* mutants (Hansen *et al.*, 2008; Samuelson *et al.*, 2007). Genes involved in vesicular sorting are important for endocytic trafficking of proteins from the plasma membrane to the lysosomes. It remains to be investigated to what degree these cellular processes of molecular trafficking (vesicular sorting) and/or turnover (autophagy) overlap.

1.2. Dietary restriction

Reduced food intake without malnutrition (here referred to as dietary restriction) has been shown to extend the life span of many organisms ranging from yeast to mice (Guarente and Picard, 2005). Dietary restriction also extends the life span of *C. elegans*, and the process is here often studied in genetic models such as *eat-2* mutants. These mutants lack a pharyngeal-specific nicotinic acetylcholine receptor subunit that is required for pharyngeal pumping (feeding) (Raizen *et al.*, 1995). Similar to worms subjected to direct dietary restriction in liquid media (see below), *eat-2* mutants live approximately 20%–30% longer than well-fed wild-type animals (Lakowski and Hekimi, 1998).

The molecular basis of how dietary restriction extends life span in *C. elegans* is less well understood. An upstream mediator of the nutritional status of the animal might be the nutrient sensor TOR (target of rapamycin) (Hansen *et al.*, 2007; Henderson *et al.*, 2006; Meissner *et al.*, 2004). Similar to dietary restriction, animals with reduced TOR signaling are also long lived (Vellai *et al.*, 2003). As in other organisms, *C. elegans* TOR regulates multiple biological processes, including protein translation and autophagy (Hansen *et al.*, 2007, 2008). Both of these processes have been linked to aging (see subsequent sections). Consistent with a role for TOR in dietary restriction, both dietary-restricted animals and mutants with reduced TOR pathway activity have increased levels of autophagy and require autophagy genes to live long (Hansen *et al.*, 2008; Toth, *et al.*,).

Dietary restriction and TOR inhibition can extend *C. elegans* life span when food intake is reduced during adulthood only (Klass, 1977; Vellai *et al.*, 2003), and inhibition of autophagy specifically during adulthood shortens the long life span of *eat-2* and TOR pathway mutants (Hansen *et al.*, 2008). Interestingly, a gene functioning in vesicular trafficking, the small GTPase *rab-10*, appears to similarly be involved in dietary restriction, and *rab-10* mutants require autophagy gene activity to live long (Hansen *et al.*, 2005, 2008).

In contrast to the IIS pathway, dietary restriction is not dependent on the FOXO transcription factor DAF-16 (Houthoofd *et al.*, 2003; Lakowski and Hekimi, 1998). However, two other transcription factors, the FOXA transcription factor PHA-4 (Panowski *et al.*, 2007) and the oxidative stress-responsive transcription factor SKN-1 (Bishop and Guarente, 2007) were recently found to be required for the long life span of dietary-restricted worms. Whereas SKN-1 also functions in the *daf-2* IIS pathway to modulate longevity (Tullet *et al.*, 2008), PHA-4/FOXA has a specific function in the life span response to dietary restriction (Panowski *et al.*, 2007). Furthermore, PHA-4/FOXA appears to be required for the increased levels of autophagy seen in dietary-restricted *eat-2* mutants (Hansen *et al.*, 2008), suggesting that PHA-4/FOXA is involved in regulating this process in *C. elegans*. Another conserved longevity molecule, the sirtuin SIR-2.1, a member of the Sir-2 family of NAD(+)-dependent protein deacetylases, has been suggested to mediate the longevity response to dietary restriction in several organisms (Guarente and Picard, 2005), including *C. elegans* (Wang and Tissenbaum, 2006). Overexpression of SIR-2.1 in worms extends life span, yet in a *daf-16* dependent fashion (Tissenbaum and Guarente, 2001). Moreover, the life span of *sir-2.1* deletion mutants can be extended by dietary restriction (Hansen *et al.*, 2007; Kaeberlein *et al.*, 2006; Lamming *et al.*, 2005; Lee *et al.*, 2006), and it thus seems more likely that *sir-2.1* affects life span by mechanisms that are not overlapping with dietary restriction in *C. elegans*. In contrast, dietary restriction appears to overlap with another *daf-16* dependent process, namely life span extension by removal of the worm's germline. Worms without germline cells are ~60% longer lived than wild-type animals (Hsin and Kenyon, 1999), and this long life span is not further extended in germline-ablated *eat-2* mutants (Crawford *et al.*, 2007). The basis for this apparent overlap between dietary restriction and signals from the germline is currently unclear, yet it would be interesting to address the role of autophagy in this context. In addition, it would be of wide interest to address the tissue-specificity of the signals that extend life span via dietary restriction, including the autophagic response involved.

1.3. Protein translation

Protein translation is an essential process involved in all aspects of biology. Recently, it was found that reduced levels of protein translation, either by inhibition of the protein translation machinery (i.e., ribosomal proteins, or regulators of translation) or ribosomal S6 kinase and translation initiation factors (eIFs) can extend the life span of *C. elegans* (Hansen *et al.*, 2007;

Henderson *et al.*, 2006; Pan *et al.*, 2007; Syntichaki *et al.*, 2007; Chen *et al.*, 2007; Curran and Ruvkun, 2007). Inhibition of many of these factors also extends the life span of other organisms, including yeast and flies (Kaeberlein *et al.*, 2005; Kapahi *et al.*, 2004).

Animals with reduced protein translation generally have less progeny but are more stress resistant, a phenotype they share with both the long-lived IIS mutants and dietary-restricted animals mentioned previously. In contrast, though, these animals may not live longer due to increased levels of autophagy as S6K mutants have normal levels of autophagy (as measured by the number of LC3/LGG-1-positive foci in the seam cells) and do not require the autophagic gene *bec-1* (*C. elegans* ortholog of yeast *ATG6*/ mammalian *beclin*1) to have an extended life span (Pan *et al.*, 2007; Hansen *et al.*, 2008) in the latter experiment, RT-PCR analysis showed that S6K mutants responded well to the *bec-1* RNAi treatment. This is in contrast to what would be expected if lower levels of mRNA translation simply reduce the levels of damaged proteins as a result of decreased protein turnover. It will be interesting to learn the underlying molecular mechanism behind how reduced protein translation extends life span.

Similarly, *bec-1* RNAi does not significantly shorten the extended life span of *ife-2*/eIF4E mutants (Hansen *et al.*, 2008) or *ifg-1*/eIF4G mutants (Pan *et al.*, 2007). Thus, not all known longevity pathways appear to rely on autophagy to extend life span. Interestingly, the life span extension observed in these animals can be induced in adult animals, indicating a postdevelopmental mechanism (Hansen *et al.*, 2007; Henderson *et al.*, 2006; Pan *et al.*, 2007; Curran and Ruvkun, 2007; Chen *et al.*, 2007). Little is known about the genetic requirements underlying the extended life span of animals with reduced translation. Some RNAi treatments extend life span independently of *daf-16*/FOXO (TOR, S6K, ribosomal proteins), whereas it is presently unclear what the relationship is between eIFs and DAF-16/FOXO, since it has been reported that life span extension of reduced protein translation during adulthood can occur in both a *daf-16*-dependent and *daf-16*-independent fashion (Hansen *et al.*, 2007; Henderson *et al.*, 2006; Pan *et al.*, 2007; Syntichaki *et al.*, 2007). Since this remains controversial, future experiments are needed to fully clarify these issues.

1.4. Mitochondrial respiration

Reduced levels of respiration induced by inactivation of the electron-transport chain (ETC) have been shown to extend life span in worms (Dillin *et al.*, 2002b; Lee *et al.*, 2003b). Several nuclear mutations also extend life span, including the gene *clk-1* (an enzyme involved in ubiquinone synthesis) (Feng, 2001; Lakowski and Hekimi, 1996). Perturbations of this gene also extend the life span of mice (Liu *et al.*, 2005).

Inactivation of mitochondrial genes by RNAi results in small animals with few progeny and the overall structure of the mitochondria in the cells is impaired. Furthermore, the ATP levels of these animals are reduced (Dillin *et al.*, 2002b; Lee *et al.*, 2003b). Knocking down ETC gene function by RNAi is highly dependent on dsRNA concentration. Intermediate dsRNA concentrations give rise to life span extension, whereas higher concentrations cause developmental problems and result in a shortened life span (Rea *et al.*, 2007).

The genetic factors mediating longevity via the mitochondria are far from understood. The transcription factor DAF-16/FOXO is not required for this life span extension (Dillin *et al.*, 2002b; Lee *et al.*, 2003b). It also seems unclear, at present, if the process of autophagy is specifically involved in mitochondrial-mediated life span extension. *C. elegans* goes through four larval stages (L1-L4) before becoming an adult and, interestingly, the mitochondrial signal has to be reduced during development, specifically in the L3/L4 stage, to observe life span extension in the adult animal (Dillin *et al.*, 2002b; Rea *et al.*, 2007). Inhibiting the ETC by RNAi only during adulthood does not change the life span of these animals (Dillin *et al.*, 2002b). Evidence for a longevity role of autophagy was not found in experiments where autophagy genes were inhibited by RNAi in long-lived mitochondrial mutants throughout life (which shortens the life span of not only mitochondrial mutants but also of normal animals) or in adult, reproductive animals (which does not have any effect on the long life span of mitochondrial mutants; Hansen *et al.*, 2008). In contrast, whole-life ETC RNAi cannot fully extend the life span of mutants defective in autophagy (Toth, *et al.*, 2008), suggesting that autophagy is required for reduced mitochondrial respiration to extend life span. However, as autophagy mutants are short lived for reasons that are likely to relate not only to accelerated aging but also to developmental defects, it remains to be experimentally tested if autophagy genes are specifically involved in life span extensions by the mitochondria from the L3/L4 stage on. In this regard, it is interesting to note that *isp-1* (another long-lived mitochondrial mutant, see below) and *clk-1* mutants have slightly elevated levels of autophagy at this stage compared to wild-type animals (Hansen *et al.*, 2008).

1.5. Summary

Several longevity pathways have been identified in *C. elegans*. Of these, reduced insulin/IGF-1 signaling and dietary restriction appear to rely on autophagy to increase life span. A summary of the major longevity pathways in *C. elegans* and their relation to autophagy is outlined in Fig. 29.1.

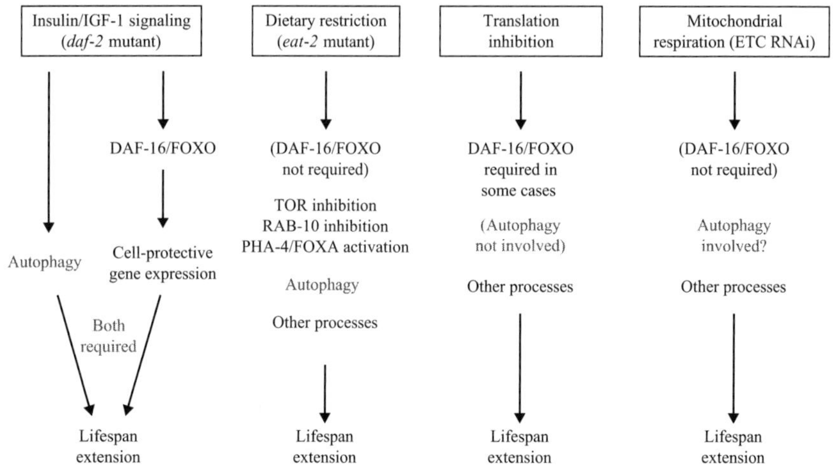

Figure 29.1 Model for the role of autophagy in life span extension in *C. elegans*. This review discusses the role of autophagy in four different processes that affect life span in *C. elegans*, including insulin/IGF-1 signaling, dietary restriction, protein translation, and mitochondrial respiration. See text for details. (Modified after Hansen *et al.*, PLoS Genetics, 2007).

2. Examination of Life Span and Visual Detection of Autophagy and Autolysosome Formation

Multiple experimental approaches have been used to study autophagy in the context of aging in *C. elegans*. We here outline the techniques that have been reported so far to examine life span and detect autophagy in *C. elegans*.

2.1. Life span analysis

2.1.1. Experimental setup and scoring death

Collection of eggs by bleaching

1. The strains to be analyzed should always be well fed on *C. elegans*' regular food source OP50 *E. coli* bacteria (~100 μl bacterial culture [grown overnight at 37 °C] per 6-cm plate) seeded on NGM plates (normal growth media, per liter: 20 g of agar, 2.5 g of bactopeptone, 3 g of NaCl; autoclave. Then add 1 ml of 5 mg/ml cholesterol, 1 ml of 1 *M* $CaCl_2$, 1 ml of 1 *M* $MgSO_4$ and 25 ml of potassium phosphate buffer (0.69 *M* KH_2PO_4, 0.15 *M* K_2HPO_4, pH 6.0) for at least two generations. Depending on the size of the experiment, eggs can either be picked from a growing population by use of a worm pick or a large quantity of eggs can be obtained by bleaching a population of worms.

2. Harvest approximately 1000 fully gravid animals (or more) by washing them off the plate with 5–10 ml M9 buffer (86 m*M* NaCl, 42 m*M* $Na_2HPO_4 \cdot 7H_2O$, 17 m*M* KH_2PO_4, 1 m*M* $MgSO_4$ (add $MgSO_4$ after autoclaving) and spin worms down in 15-ml conical tubes in a clinical centrifuge at ~3000 rpm for approximately 1 min. Wash once with M9 to remove bacteria.
3. Treat animals with bleach solution (sterile filtered mixture of 30 ml of bleach, 15 ml of 5 N KOH, and 55 ml of H_20) until the animals are dissolved (~5–10 min). Quickly recover eggs by centrifugation as noted previously and dissolve in M9. Count the number of eggs recovered (several thousands).

Setup of life span assay

4. Add ~100 eggs to one 6-cm NGM plate for each experimental treatment as well as the control. The experimental plates should be preseeded with ~100-μl bacterial cultures (grown overnight at 37 °C) of either the standard food source OP50 or RNAi bacteria (HT115 expressing double-stranded RNA from inverted T7 promoters) and incubated for 1–2 days at room temperature. The temperature is here an important point to consider, as some strains are temperature-sensitive sterile. Often the life span assays are carried out at 15 °C or 20 °C. If performing feeding RNAi experiments, consider the timing requirements to be investigated (i.e., is this an adult-only RNAi experiment) and induce the bacteria on the plates with 0.5–1 m*M* IPTG before adding eggs to the plates.
5. At the time of adulthood (e.g., 3 days at 20 °C for wild-type animals), transfer animals to new NGM plates to start the actual life span assay. Typically, 10–15 animals are placed on each one of 8–10 6-cm NGM plates to ease the counting of the animals throughout the experiments. For RNAi feeding experiments, the bacteria on the plates are induced prior to the start, with IPTG.

Transfer of reproductive animals early in life span assay

6. As wild-type animals are reproductive for a period of ~3–5 days, the animals have to be transferred away from their progeny to new plates every second day in this time period. Alternatively, addition of the chemical FUDR (2′fluoro-5′deoxyuridine, addition of 100 μl of 18 m*M* per plate should be sufficient to arrest progeny production) to animals at the late L4 stage can be used to inhibit the development of the progeny and thus abrogate the need for transferring. However, as FUDR has been observed to affect different aspects relating to organismal aging, including RNAi-induced longevity phenotypes (see, e.g., "Methods" in Hansen *et al.*, 2008), this should be considered when relying on life span analysis using feeding RNAi.

Scoring of deaths in life span assay

7. The plates are analyzed every second day and the worms are scored to be either alive or dead. An animal is per definition dead when it no longer responds to prodding by a platinum pick. When dead, the animal is removed from the plate. For each time point, the number of alive and dead animals is noted. Furthermore, the number of censored animals should be tabulated. Censoring in *C. elegans* life span analysis includes animals that rupture, bag (i.e., exhibit internal progeny hatching), or crawl off the plates. Wild-type animals typically have a mean life span of ~18 days at 20 °C. The maximal life span at this temperature (the time until the last animal is dead) is usually approximately 30 days. For comparison, the long-lived *daf-2(e1370)* mutant strain has a mean life span of >40 days at 20 °C.

Evaluation of life span data

8. At the end of the experiment, the data are prepared for statistical analysis (see subsequent sections). This includes calculating for each time point how many animals died or were censored. The data are now ready for statistical evaluation (e.g., in the STATA program after the data have been entered into an Excel spreadsheet). The entire experimental protocol is outlined schematically in Fig. 29.2.

2.2.2. Direct dietary restriction

Multiple regimens of food-intake reduction have been described in *C. elegans* that extend life span. Traditionally, reduced food intake has been controlled in *C. elegans* by diluting the amount of bacteria in liquid (Klass, 1983). As this method is the only method that so far has been used when investigating the role of autophagy, only this method will be summarized in detail here.

Preparation of bacteria cultures for feeding worms

1. Start a bacterial culture by inoculating 500 μl of *E. coli* OP50 (O.D.$_{600}$ ~1) into 250 ml of LB medium (10 g of bactotryptone, 5 g of yeast extract, 5 g of NaCl, 10 ml 1 *M* Tris, pH 8.0, per liter).
2. Incubate for 5 h at 37 °C.
3. Resuspend in 25 ml of S-basal medium (0.1 *M* NaCl, 0.05 *M* potassium phosphate, pH 6, 1 ml of 5 mg/ml cholesterol (in ethanol) per liter). This culture is now the stock as well as the fully fed condition.
4. Determine the cell density of the stock by counting DAPI stained cells (add 6 μl of 2 mg/ml stock (in ethanol) to 1 ml of S-basal medium with bacteria) in a Petroff-Hausser counting chamber.
5. Make a dilution series (e.g., 2-fold, 8-fold, 10-fold, 12-fold, 20-fold, 40-fold, 80-fold) in S-basal medium of 100–200 ml. These different

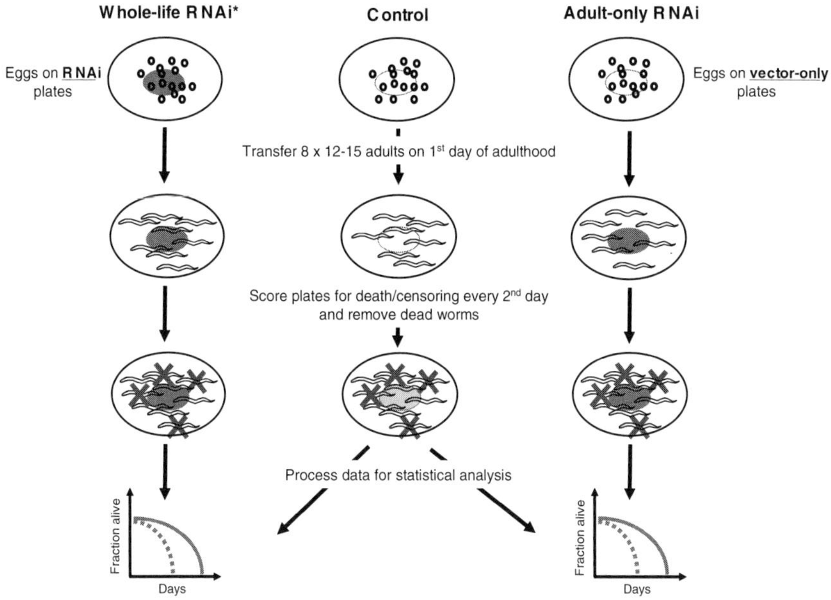

Figure 29.2 *C. elegans* RNAi life span assay. Outlined are crucial steps of a life span assay, performed using feeding RNAi to inactivate gene function. Feeding RNAi allows the researcher to investigate the function of a gene specifically during adulthood (adult-only RNAi) versus throughout life (whole-life RNAi). In both cases, the survival of the population of worms under RNAi treatment is compared to that of animals fed vector-only bacteria (control). (*) The whole-life RNAi versus control comparison is equivalent to life span assays of mutant strains conducted on non-RNAi plates (seeded with the "generic" food source, OP50 bacteria) compared to wild-type N2 animals grown on the same food. Yellow circle indicates normal OP50; green circle indicates OP50 expressing the desired dsRNA. See text for further details.

bacterial solutions will serve as different food conditions in the life span assay (undiluted and 2-fold should be equivalent to fully fed; 8-, 10-, and 12-fold might represent healthy dietary restriction; 40-fold should be bordering on starvation; 80-fold should induce arrest). These solutions can be stored at 4 °C.

Life span assay

1. Add ~60 eggs (to get approximately 30 animals per well, depending on the bleaching procedure) per well of a 24-well plate containing 600 μl of each of the bacteria-supplemented S-basal media prepared above, and culture at 20 °C with shaking. For best results, triplicate samples are recommended.
2. Expect to see a range of phenotypes ranging from dietary-restricted animals (see previous) to animals under direct starvation (larval arrest).

3. Transfer the animals to new media every other day once the eggs have developed into adults, and continue this throughout the experiment (i.e., until all animals are dead).
4. Measure the life span of the adult animals by observing if the animals respond to a gentle touch. Animals subjected to dietary restriction will be longer lived than the fully fed animals.

2.1.3. RNAi during development and during adulthood

Genes often have different functions during development compared to adult animals (e.g., the insulin/IGF-1 receptor *daf-2* is required for dauer formation during development and important for aging during adulthood). Addressing the timing requirement of a gene is very easy in *C. elegans*, when using the RNAi feeding method. Basically, animals can be transferred to bacterial lawns expressing double-stranded RNA at any given time point during their life. As autophagy is important for development (several autophagy mutants are developmentally impaired and RNAi of autophagy genes during development may shorten life span irrespective of the genetic make-up of the animal), inhibition of autophagy beginning in adulthood is important to specifically address a role for autophagy in adult life span. By shifting animals from plates with control bacteria to plates with bacteria expressing double-stranded RNA of genes involved in autophagy (e.g., *bec-1*, at the time after the animals reached adulthood: when they become reproductively active) it was found that the extended life span of long-lived insulin/IGF-1 receptor *daf-2* and dietary-restricted *eat-2* mutants were specifically shortened (Hansen *et al.*, 2008).

2.1.4. Statistics

Life span assays are statistically treated as a survival analysis. Survival analysis is unique in that it includes censored objects when comparing a control population with a population subjected to a treatment (e.g., mutant versus wild-type animals). Censoring in *C. elegans* life span analysis includes animals that rupture, bag (i.e., exhibit internal progeny hatching), or crawl off the plates. Several types of mathematical survival analyses exist, yet the most generally used variant in the *C. elegans* field to estimate *P*-values is the log-rank (sometimes called the Mantel-Cox) test. Theoretically, both the control and the treatment groups should have similar levels of censoring. This is not always true, yet the statistical analysis can be carried out anyway. Several statistical packages are available to perform the analysis, including the STATA package. This package easily analyzes actual life span data plotted in Excel format.

There are several important factors that influence the statistics of the life span experiment, including the degree of censoring and the expected life span change. These factors are important to ensure that enough statistical power is achieved in the experiment. If these factors are suboptimal, the power of the experiment can be improved by increasing the sample size. Typically, on the order of ~80 objects are needed for a life span experiment in which a ~20%–30% change is to be detected with normal censoring rates. If high levels of censoring are expected, the sample size should be increased. In the *C. elegans* longevity field, significant *P* values are generally more stringent than the generally used $P < 0.05$, and are often reported as $P < 0.001$ between the control and the experimental treatment groups.

2.2. Life span genetic models

C. elegans life span can be extended by many different genetic perturbations. Here, we summarize the longevity strains that have been investigated so far in the context of autophagy and list the issues to be considered when using these strains in longevity assays.

2.2.1. Insulin/IGF-1 signaling, *daf-2* mutants

In *C. elegans*, autophagy was first investigated in the context of the developmental process of dauer formation (Meléndez *et al.*, 2003). *C. elegans* responds to unfavorable environmental conditions (e.g., crowding, starvation, increased temperature) by entering into an alternative third larval stage, called dauer diapause. Dauer larvae are arrested and are characterized by radial constriction and elongation of the body and the pharynx, resistance to detergent (i.e., sodium dodecyl sulfate) treatment, increased fat accumulation, and long life span. Dauer development is regulated by different conserved signaling pathways that also regulate fat storage, reproduction and stress responses (Riddle, 1997), including the insulin/IGF-1 and TGF-beta pathways. Autophagy is required for dauer formation in both the insulin/IGF-1 receptor *daf-2* mutant and the TGF-beta/*daf-7* mutant (Meléndez *et al.*, 2003), suggesting that autophagy genes are required for the organismal remodeling that permits *C. elegans* to adapt to environmental stresses, such as starvation.

As noted previously, adult animals with mutations in the insulin/IGF-1 receptor-encoding *daf-2* are also long lived and this life span extension is dependent on autophagy. A frequently used allele in longevity studies addressing the role of autophagy is the *daf-2* allele *e1370*. To measure the life span of adult *daf-2(e1370)* animals several issues must be considered. Because of the alternative dauer state, life span assays with this temperature-sensitive dauer-constitutive mutant have to be conducted at the permissive temperatures (15 °C or 20 °C) (or alternatively, the animals can be shifted to the restrictive temperature (25 °C) after the dauer decision point in the L2 phase (i.e., 1 day at 20 °C followed by a shift to 25 °C). Notably, the development of the mutant is ~1 day delayed compared to wild-type animals and these mutant animals have reduced numbers of progeny. The *daf-2(e1370)* mutant generally shows ~100% life span extension and exhibits a low censoring rates. Accordingly, relatively few animals need to be observed in a single experiment.

2.2.2. Dietary restriction, *eat-2* mutants

Dietary restriction can be modeled in *C. elegans* by the use of certain eating-deficient mutants, called *eat-2* mutants (Lakowski and Hekimi, 1998). As mentioned previously, these mutants lack a pharyngeal-specific nicotinic acetylcholine receptor subunit that is required for pharyngeal pumping (feeding). Similar to worms subjected to direct dietary restriction in liquid media (see earlier), *eat-2* mutants live ~20%–30% longer than well-fed wild-type animals. Likewise, *eat-2* mutants also require the FOXA transcription factor PHA-4 to extend life span, show a characteristic spectrophotometric profile, a slim appearance and a reduced and delayed reproductive period (Crawford *et al.*, 2007; Gerstbrein *et al.*, 2005; Panowski *et al.*, 2007). Importantly, *eat-2* mutants, similar to animals that are directly dietary restricted, also display increased levels of autophagy and require autophagy genes to live long (Hansen *et al.*, 2008, Toth *et al.*, 2008). An often-used *eat-2* allele is *ad1116*, which will be discussed herein.

When performing life span assays on *eat-2(ad1116)* mutants, several issues must be considered. First, *eat-2(ad1116)* mutants are ~9 h developmentally delayed compared to wild-type animals. Second, adult *eat-2 (ad1116)* animals are generally more difficult to find as they are smaller and more transparent. Third, *eat-2*(*ad1116*) life span assays are more cumbersome to perform as transfer of the adult animals away from their progeny is needed later in adult life due to the delayed progeny profile. This might encourage the use of 2′fluoro-5′deoxyuridine (FUDR) to avoid the development of progeny (see previously). However, it has been observed that addition of FUDR in the context of assaying the effect of RNAi of autophagy genes can result in lack of life span extension phenotypes (Hansen *et al.*, 2008). Finally, the censor rate of these animals is often elevated due to animals frequently crawling off the plate. Thus, it is advisable to design longevity experiments with *eat-2* animals that include a large number of animals.

2.2.3. Protein translation, *ife-2* and *rsks-1* mutants

Gene perturbations that reduce the level of protein translation have been shown to extend the life span of multiple model organisms, including *C. elegans*. The mechanism by which this life span extension is mediated is currently not known, yet it does not appear to rely on autophagy. Many of the conditions that have been investigated are induced by the use of RNAi. However, some genetic mutants have also been investigated, including the translation initiation factor *ife-2* (homolog of eIF4E) and the ribosomal protein S6 kinase *rsks-1*. Both of these strains have reduced protein translation (as measured by incorporation of ^{35}S methionine-labeled bacteria) and

~10%–25% longer lived than wild-type animals (Hansen *et al.*, 2007; Pan *et al.*, 2007). The *ife-2(ok306)* strain seems to develop very similarly to wild-type animals, whereas the *rsks-1(sv33)* strain is ~16 h developmentally delayed at 20 °C (Hansen *et al.*, 2007). Note that as these mutants have rather small life span extensions, it is advisable to use larger numbers of animals in the longevity assays of *rsks-1* or *ife-2* mutant animals.

2.2.4. Mitochondrial mutants, *isp-1 and clk-1* mutants

Reduced mitochondrial respiration has been shown to extend the life span of *C. elegans.* Life span is extended by reducing the levels of components of the electron transport chain (ETC), either by RNAi or by genetic mutation in *isp-1*, iron-sulfur protein-1, or in *clk-1*, an enzyme required for ubiquinone synthesis.

When performing life span experiments with *isp-1(qm150)* and the *clk-1 (qm30)* strains, several issues have to be considered. First, both of these strains are significantly developmentally delayed, the *isp-1(qm150)* strain by ~3 days and the *clk-1(qm30)* strain by ~1 day at 20 °C (Feng, 2001; Lakowski and Hekimi, 1996). Second, adult animals can be difficult to identify, as they are small and clear. Third, these animals have a reduced brood size and a delay in their reproductive period, thus late progeny can bias the life span analysis. Last, a large number of these animals are often censored, often due to internal hatching of progeny inside the mother (i.e., "bagging"). Large numbers of worms are therefore generally needed to assay the life span of these mutants, which are generally ~30% longer lived than wild-type animals.

The mitochondria-mediated longevity extension is achieved only if mitochondrial genes (at least the ETC genes) are inhibited during the L3/L4 larval stage (Rea *et al.*, 2007). This is in contrast to the other longevity pathways discussed previously, which can all modulate life span in the adult animal. As the longevity role of autophagy has been addressed by inhibiting autophagy during development (where the life span of all strains analyzed is shortened) or in adult animals only (where no effect was observed, see previously) (Hansen *et al.*, 2008), it is presently not clear if autophagy plays a role in the mitochondria longevity response. This should be specifically addressed by inhibiting mitochondrial genes in L3/L4 animals.

2.3. Pigment assays for aging

2.3.1. Applications

A conserved feature of aging is the accumulation of age pigments or lipofuscin granules (composed of lipids, proteins, and carbohydrates) in lysosomes (Clokey and Jacobson, 1986; Yin, 1996). *In vivo* spectrofluorimetry can be used to quantitate the level of autofluorescence in living animals by use of a spectrofluorimeter (Gerstbrein *et al.*, 2005). This is done by recording broad-range excitation/emission spectra in intact animals over time and comparing it to that of the same number of wild-type animals. A control (TRP signal) measurement of excitation/emission at 290 nm/330 nm corresponds to the fluorescence signal generated by aromatic amino acids such as tryptophan, tyrosine, and phenylalanine (tryptophan being the primary source) and can be used as a measure of overall protein content. The measurement of the peak age pigment fluorescence signal at 340 nm/430 nm, corresponds to age-associated biomarkers such as advanced glycation end (AGE) products and lipofuscin (Gerstbrein *et al.*, 2005; Yin, 1996). Intact animals measurements are normalized by comparing the AGE signal to that of the endogenous TRP scores. Normally, the level of age pigments increases in wild-type animals slowly with age during the reproductive stages and more rapidly during the postreproductive period. Surprisingly, in a population of genetically identical animals, age pigment levels can identify the animals that have aged gracefully from those that have aged poorly (Gerstbrein *et al.*, 2005). Whether high age pigment levels contribute to aging or reflect the consequence of poor aging is not known, but the levels of age pigment accumulation may be used as a highly sensitive reporter of the animal's physiological state of aging (Gerstbrein *et al.*, 2005).

Insulin/IGF-1 signaling and dietary restriction can influence dramatically the levels of age pigment accumulation (Gerstbrein *et al.*, 2005). Mutations that lower insulin signaling and extend life span, lower the accumulation of age pigment. In contrast, the short lived *daf-16*/FOXO mutants display a dramatic increase in age pigment accumulation. Dietary restriction in feedings-impaired *eat-2* mutants moderately decreases the accumulation of age pigments and causes a unique shift in the fluorescent properties of the age pigments, perhaps reflecting changes in the types of molecules that accumulate under dietary restriction (Gerstbrein *et al.*, 2005). However, mutations that increase reactive oxygen species production and shorten life span do not produce elevated levels of age pigment accumulation, suggesting that an increase in mitochondrial stress does not influence age pigment accumulation *in vivo* (Gerstbrein *et al.*, 2005).

As described previously, autophagy has recently been shown to have a direct role in the aging process of *C. elegans* (Hansen *et al.*, 2008; Hars *et al.*, 2007; Jia and Levine, 2007; Meléndez *et al.*, 2003; Toth *et al.*, 2008). However, disrupting genes required for autophagy during adulthood, which significantly shortens life span, does not significantly alter fluorimetric profiles, at least not of dietary-restricted *eat-2* mutants (Hansen *et al.*, 2008). In contrast, Toth and colleagues (2008) measured an earlier accumulation of lipofuscin in autophagy mutants (*unc-51/Atg1* and *bec-1/Atg6*) compared to wild-type animals. As autophagy has distinct roles in development, it is not clear if the pigment accumulation observed in these genetic mutants is the result of developmental or adult functions of autophagy.

Taken together, it remains possible that autophagy is specifically required for certain aspects of life span extension and not for all of the changes associated with aging.

2.3.2. Protocol to measure age-related pigments

1. The spectrofluorimeter (Fluorolog-3, Jobin Yvon Inc.) equipped with a MicroMax 384 plate reader (Jobin Yvon Inc.) has to be spectrally calibrated for excitation and emission following the manufacturer's instructions.
2. A constant number of worms (50–150) should be first cleaned by placing the animals in an unseeded NGM plate and repicked. If feeding RNAi effects will be assayed, eggs from wild-type animals are raised on experimental or control RNAi plates (plates of feeding RNAi bacteria are induced with 0.5–1 m*M* IPTG before adding the eggs or the worms to the plates).
3. The animals are then placed in a 50-μl drop of 10 m*M* NaN_3 in a single well of a 96-well plate (white FluoroNunc plate with a PolySorp Surface, Nunc Brand Products, Nalge Nunc International).
4. Animals are immediately scanned. For best results, scans should be done in triplicate. A control (TRP signal) measurement of excitation/emission at 290 nm/330 nm should be taken. The excitation/emission pair of 290 nm/330 nm corresponds to the fluorescence signal generated by aromatic amino acids such as tryptophan, tyrosine, and phenylalanine (tryptophan being the primary source) and can be used as a measure of overall protein content.
5. Measurement of the peak age pigment fluorescence signal at 340 nm/430 nm can be determined. Peak age pigment fluorescence intensity can be determined by scanning through a range of excitation wavelengths from 280–410 nm, with increments of 2 nm, an integration time of 0.5 s, and an emission wavelength of 430 nm.
6. The age pigment values or the ratio of AGE/TRP values are then scored. The ratio of AGE/TRP has been found to remain constant for any given strain. Data analysis is done using an unpaired, one-tailed *t*-test. Due to the differences between spectrofluorimeter instruments (i.e., detector sensitivity, lamp output, and small variations in the optical path), readings between samples cannot be expected to be identical.

2.4. Autophagosome detection by the GFP-tagged autophagy protein LGG-1/LC3

2.4.1. Applications

The discovery that Atg8 in yeast is a protein that undergoes lipidation and is stably conjugated to phosphatidylethanolamine (PE) in the autophagosomal membrane, has facilitated the detection of autophagosomes *in vivo*. In

C. elegans, the *ATG8* orthologs (MAP1-LC3 in mammals) are *lgg-1*, and *lgg-2*. GFP::LGG-1 is expressed in multiple tissues throughout development, including the nervous system, pharynx, intestine, hypodermis, somatic gonad, and vulva (Meléndez *et al.*, 2003). In the absence of autophagy, GFP::LGG-1 has a diffuse cytoplasmic localization; however, when autophagy is induced, GFP::LGG-1 localization changes into a punctate appearance, presumably reflecting an upconcentration of LGG-1 in pre-autophagosomal/autophagosomal membranes similar to what has been observed for Atg8/LC3 in yeast and mammalian cells (Klionsky *et al.*, 2007).

Autophagosomes are rarely seen in wild-type N2 animals. The increase of GFP::LGG-1 positive puncta in *daf-2(e1370)* demonstrates that *daf-2* mutant animals, raised at either the permissive (15 °C) or the restrictive temperature (25 °C), have increased levels of autophagy compared to wild-type animals. Thus, DAF-2 activity normally inhibits autophagy (Meléndez *et al.*, 2003). These studies using the GFP::LGG-1 transgene were also confirmed by EM studies (Meléndez *et al.*, 2003). By EM, *daf-2* dauer animals accumulate structures (Fig. 29.3) that appear similar to those described in the late stages of autophagy in mammalian cells (Klionsky *et al.*, 2007). In animals that express the human Aβ (1–42) peptide in *C. elegans* muscle cells, GFP::LGG-1 shows an increase in the accumulation of autophagosomes (Florez-McClure *et al.*, 2007). These studies were also confirmed by EM. In both studies, the size, frequency, and distribution of autophagic vesicles, as detected by EM, correspond well with that of the GFP::LGG-1 puncta observed by fluorescence microscopy. Therefore, by monitoring the fluorescently labeled LGG-1 molecules it is possible to effectively and reliably follow autophagy in time and space in live *C. elegans*.

GFP::LGG-1 has now been used to detect the formation of autophagosomes in several different cell types in *C. elegans*, including hypodermal seam cells, intestine, neurons, and muscle (Bamber and Rowland, 2006; Kang *et al.*, 2007; Meléndez *et al.*, 2003; Nicot *et al.*, 2006; Toth *et al.*, 2007). In the presence of RNAi against other autophagy execution genes, LGG-1 can form large aggregates, presumably due to decreased clearance of LGG-1 through the autophagolysosomal pathway (Hansen *et al.*, 2008; Meléndez *et al.*, 2003).

2.4.2. Construct design, transformation, and genetic crosses

2.4.2.1. Construct design The plasmid GFP-LGG-1 was constructed by cloning a 1.7 kb *lgg-1* promoter fragment into pPD117.01 (upstream of GFP) using *Sph*I and *Kpn*I. Downstream of GFP, a 2.2 kb fragment containing the *lgg-1* genomic coding sequences including the 3′UTR was cloned, using *NgoM*IV and *Apa*I sites. Thus, GFP is located in the N-terminus of the LGG-1 protein in this construct; this is critical because a C-terminal GFP is likely to be proteolytically removed from LGG-1,

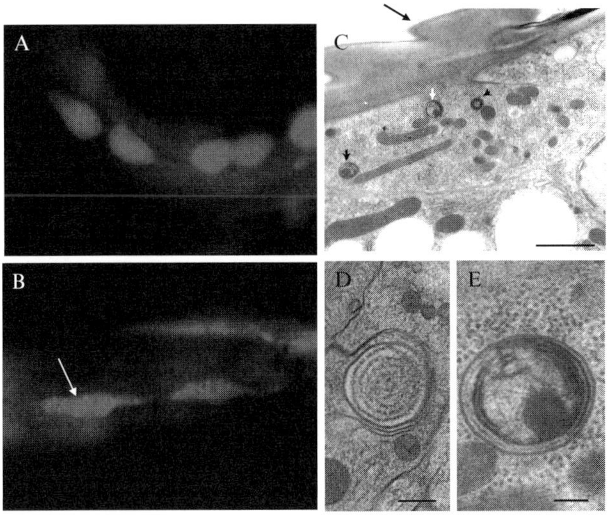

Figure 29.3 Autophagy as detected by the GFP::LGG-1 marker and electron microscopy. An increase in autophagy has been observed in dauers and in long-lived *daf-2 (e1370)* mutant animals. (A) Expression of GFP::LGG-1 in the lateral hypodermal seam cells of wild-type N2 animals. (B) The increase of GFP::LGG-1 punctate areas in *daf-2 (e1370)* dauer mutants indicates an increase in autophagy. Arrowheads show representative GFP-positive punctate areas that label preautophagosomal and autophagosomal structures. (C) A representative electron micrograph of hypodermal seam cells in a *daf-2(e1370)* dauer animal. The black arrowheads denote representative vacuolar structures in the late stages of the autophagolysosomal pathway; the white arrowhead points to the structure shown in higher magnification in *E*, and the arrow denotes the characteristic alae in the cuticle of dauer larvae. (D) A representative early autophagosome and (E) a multilamellar body/autolysosome are shown in a seam cell of a *daf-2* dauer animal. Scale bars, 1 μM in (C), and 0.1 μM in (D) and (E). (Modified from Meléndez *et al.*, 2003.)

similarly to C-terminal processing of Atg8/LC3 by the Atg4 protease ortholog (Levine and Klionsky, 2004).

2.4.2.2. Transformation GFP::LGG-1 was injected at a concentration of 5 μg/ml into wild-type N2 hermaphrodites using pRF4 [*rol-6(su1006)*] at 100 μg/ml as a co-injection marker. Extrachromosomal (*izEx1[lgg-1:: GFP::LGG-1, rol-6(pRF4)*; Meléndez *et al.*, 2003) and integrated (*adIs2122[lgg-1::GFP::LGG-1, rol-6(pRF4)*; Kang *et al.*, 2007) versions of GFP::LGG-1 exist. The integrated version may show less variability in the expression of GFP::LGG-1 among different animals.

2.4.2.3. Genetic crosses To test whether a specific mutation affects the levels of autophagy as measured by the LGG-1 reporter, animals carrying the mutation of interest have to be crossed to animals carrying the GFP:: LGG-1 transgene. Cross-progeny heterozygous for the mutation of interest are identified by selecting hermaphrodites that carry the transgene (i.e., the dominant-negative transgene *rol-6* [which induces a mutant form of a collagen needed for normal forward movement; instead these worms move by rolling] and express GFP) and allowed to self-fertilize, to isolate the homozygous mutant animals that carry the transgene.

2.4.3. Larval staging, tissue dissection, and imaging

2.4.3.1. Larval staging The strains to be imaged should always be well fed for at least two generations. A population of worms can be obtained by allowing adult hermaphrodites to lay eggs for a determined number of hours (2–4 h) or a large quantity of eggs can be obtained by bleaching gravid adults, which will kill the adults but let the eggs survive. Autophagy has been found to increase in dauer animals, and this stage can be determined by the size of the somatic gonad and the presence of dauer-specific lateral ridges (alae). The somatic gonad in dauer animals is similar in size to that of L3 larvae, in which the two arms of the gonad have begun to elongate anteriorly and posteriorly. Other stages have also been scored (i.e. Hansen *et al.* scored adults for GFP::LGG-1-positive puncta in seam cells, Hansen *et al.*, 2008; and Kang *et al.* scored L1 larvae as well as adults for the presence of GFP::LGG-1-positive puncta in pharyngeal muscles, Kang *et al.*, 2007).

2.4.3.2. Tissue dissection GFP::LGG-1-positive punctate regions have been quantified in lateral hypodermal seam cells (Hansen *et al.*, 2008; Toth *et al.*, 2008; Meléndez *et al.*, 2003; Morck and Pilon, 2006). The seam cells are blast cells that form the main hypodermal cell for the body (the skin), which form alae that are visible in L1 larvae, adults, and dauer larvae. These cells undergo several physiological changes during dauer formation, including the secretion of dauer alae. They are easily identified in GFP::LGG-1-expressing animals as a bilateral row of eye-shaped cells. Other tissues in which the number of GFP::LGG-1 puncta have been scored are body wall muscles (Florez-McClure *et al.*, 2007; Rowland *et al.*, 2006) and neurons (Toth *et al.*, 2007). Kang *et al.* also demonstrated that an hyper-activated muscarinic acetylcholine receptor pathway, which normally responds to starvation in the *C. elegans* pharynx, can lead to excessive autophagy in the pharyngeal muscles, as determined by the number of GFP::LGG-1 foci (Kang *et al.*, 2007).

2.4.3.3. Imaging To image the localization of GFP::LGG-1 in live animals, a compound microscope, a deconvolution compound microscope, or a confocal microscope can be used, as long as they are equipped with epifluorescence. Animals are normally imaged at high magnification (630x or 1000x), where the foci can easily be counted. To reduce autofluorescence, Rowland *et al.*, 2006 fixed worms before imaging, and counted only green fluorescent objects larger than 0.5 μm in diameter. GFP::LGG-1 foci were distinguished from autofluorescent gut granules by using different fluorescence filters. The expression of the GFP::LGG-1 quickly bleaches at high magnification, thus it has to be imaged rather quickly. The deconvolution and the laser scanning confocal microscopes can lower the background fluorescence signal and increase the signal-to-noise ratio.

2.5. Electron microscopy to detect autophagosomes and intracellular trafficking defects

2.5.1. Applications

Autophagosomes are characterized ultrastructurally by the presence of double membranes that surround sequestered cytoplasmic material. The electron microscopy (EM) approach to detect autophagy is tedious and difficult to use to assess autophagic activity in different cell types or tissues at the organismal level. Another complication is that autophagosomes are small labile structures and may be difficult to preserve in nematode tissues by standard EM techniques as described earlier (Hall, 1995). In *C. elegans* dauer arrest, a stage with high autophagic activity, autophagosomes and autolysosomes are easily detected by EM (Meléndez *et al.*, 2003). In dauer animals, the presence of a low-permeability cuticle covering the animal makes fixation of *C. elegans* tissue for electron microscopy particularly difficult. To best capture the anatomy of autophagosomes in seam cells of dauer larvae, Meléndez *et al.* (2003) used the technique of high pressure freezing, followed by freeze substitution. This method can fast freeze the live animals in about 15–50 μs, too fast for the organelles to change shape or disappear from the cytoplasm.

The autophagosomes in dauer hypodermal seam cells appear relatively small in *C. elegans* (300–500 nm in diameter) (Meléndez *et al.*, 2003) (compared to yeast autophagosomes, 400–900 nm in diameter, or mammalian autophagosomes, 500–2000 nm) (Mizushima *et al.*, 2004; Takeshige *et al.*, 1992), whereas autophagosomes appear larger in muscle cells (Florez-McClure *et al.*, 2007; Rowland *et al.*, 2006). Whether the size of the autophagic vesicles depends on the cell type where they appear is not known. Florez-McClure *et al.* (2007) were able to use high pressure freezing to detect autophagosomes in muscle cells that overexpress the human Aβ(1-42) peptide. Rowland *et al.* (2006) were also able to detect larger autophagosomes in the cell bodies of body wall muscles after effective denervation in axon guidance mutants, even in animals that had been fixed only by immersion (original samples were from Hedgecock *et al.*, 1990).

A variety of methods for high pressure freezing (HPF) have now been elaborated for *C. elegans* (Muller-Reichert *et al.*, 2007; Weimer, 2006). High-pressure freeze equipment is available from Leica, Baltech, and RMC and several smaller vendors. Leica has produced novel types of specimen holders for their newer HPF devices. Some workers have constructed simple homemade devices for the freeze-substitution step (see McDonald, 1994). There are a variety of methods for loading live worms into the specimen planchette, including the use of yeast paste, mounds of *E. coli*, thick slurries of BSA, or sucking the worms in M9 into short segments of dialysis tubing. As long as the final sample does not contain significant void volumes, one can obtain excellent freezing rates and high pressures to yield superior preservation. In fact, *C. elegans* is an ideal animal for this method, as the whole adult fits simply into the planchette.

2.5.2. High-pressure freezing, staging, tissue dissection, fixation, and sectioning

1. Attention should be given to the temperature and the developmental stage that is required. Animals should be grown at the specific temperature and for the time necessary to get the right developmental stage in the experiment.
2. Scoop the animals off of a culture plate in a coating of *E. coli* using a pick. Place the animals and bacteria into the metal planchette until the interior compartment is full to the brim. Avoid the presence of any air space surrounding the sample within the planchette, to obtain maximum pressure. A thin layer of the fluid 1-hexadecene is often used as a filler to eliminate voids and seems to help release of the frozen material from the planchette in later processing. Place a blank top to seal the compartment and load into the HPF machine and freeze.
3. After the sample is frozen at the temperature of liquid nitrogen, the frozen planchette is transferred to −90 °C in a freeze substitution device for several days to a week.
4. Once inside the freeze substitution device, the sample is exposed to fixative and very slowly warmed. As the planchette warms, individual animals gradually melt into the fixative, allowing water to exit the sample and fixatives to enter but not allowing the structures inside the sample to change their original conformation. This provides superior preservation of membrane-bound organelles (autophagosomes) and plasma membrane. Again, a variety of different fixation solutions have now been tried, including osmium + acetone or methanol, and sometimes including small amounts of aldehydes, tannic acid, or uranyl acetate. Although all regimes use organic solvents, recent methods often include small amounts of water to improve structural quality. Best results seem to derive from the slowest warming regimes, and better staining seems to occur when the sample is held at −90 °C for 4–6 days prior to the beginning of the warming. When the samples reach room temperature (or sometimes −20 °C, or 0 °C), the fixation solution is replaced by pure solvent.

Fixation procedure

a. Keep frozen sample at −90 °C for 3 days in 2.0% OsO_4 , 0.1% uranyl acetate dissolved in 100% acetone as the primary fixative.
b. Warm sample to −74 °C in the same fixative and then to −60 °C over a 14-h period.
c. Warm sample to −30 °C in same fixative over a 14-h period.
d. Warm sample to 0 °C in same fixative over a 1-h period.
e. Transfer sample to 100% acetone three times at room temp, 20 min total.

5. After the worms reach room temperature and are in pure solvent, they are removed from the planchette and infiltrated with plastic resin, placed in an embedding mold, and cured to make hard plastic blocks for thin sectioning.

Infiltration procedure

a. First change is in a dilution of 1 part resin LR Gold resin to 1 part acetone, wait 30 min.
b. Second change is in 2 parts resin to 1 part acetone, wait 30 min.
c. Third change is into pure resin plus accelerator (0.5% benzoin methyl ether) and wait 1 h.
d. Fourth change is into pure resin plus accelerator and hold overnight.
e. Place samples in fresh resin plus accelerator at room temperature and place infiltrated worms in microcentrifuge tubes or gelatin capsules.
f. The gelatin capsules are held approximately 3 inches away from two 15 W ultraviolet lamps (365 nm) within a Pelco Cryobox filled with 20 lbs of dry ice to maintain the samples near −20 °C.

6. The best worms may be cut out of embedment and reembedded in resin at the desired orientation. Thin sections are cut using an ultramicrotome equipped with a diamond knife and picked onto slot grids. The thin sections are usually poststained with 1% buffered (pH 5.2) uranyl acetate and/or diluted Reynolds lead citrate (1:10 to 1:100 dilutions in water, pH 7.2) and stored in Pelco storage boxes prior to examination in the electron microscope.
7. Sections are viewed on the electron microscope, and micrographs are collected on film. Film negatives are then scanned into Adobe Photoshop (Adobe Systems, San Jose, CA) for data analysis.

Several protocols regarding general methods for TEM and immuno EM are available in Wormbook (http://www.wormbook.org/chapters/www_intromethodscellbiology/intromethodscellbiology.html) or at the Center for *C. elegans* Anatomy website (http://neuroscience.aecom.yu.edu/labs/halllab/wormem/methods.html).

3. Conclusion

Autophagy plays crucial roles in many biological processes, and recent research has pointed to a possibly conserved role for autophagy in the process of organismal aging. Experiments in the nematode *C. elegans* suggest that autophagy may be required specifically for longevity pathways that are regulated by environmental signals that reflect the availability of food, such as the insulin/IGF-1 pathway and the responses to dietary restriction. This chapter focused on describing the techniques used to address the link between aging and autophagy in *C. elegans*. Specifically, we here summarized how to examine organismal life span in various genetic backgrounds and visually detect autophagy and autolysosomal formation in *C. elegans*.

ACKNOWLEDGMENTS

The authors wish to thank members of Monica Driscoll's lab for sharing protocols on measurements of age pigments and life span assays in liquid media. Furthermore, we also thank Hannes Bülow for critical comments on the manuscript. Work in the author's laboratories is supported by grants from the NIH (RR 12596 to D.H.H., 3ROI AG024882-0451 to A.M.) and the NSF (0818802 to A.M.). M.H. and A.M. are Ellison Medical Foundation New Scholars on Aging.

REFERENCES

Apfeld, J., and Kenyon, C. (1998). Cell nonautonomy of *C. elegans daf-2* function in the regulation of diapause and life span. *Cell* **95,** 199–210.

Bamber, B. A., and Rowland, A. M. (2006). Shaping cellular form and function by autophagy. *Autophagy* **2,** 247–249.

Bishop, N. A., and Guarente, L. (2007). Two neurons mediate diet-restriction-induced longevity in *C. elegans*. *Nature* **447,** 545–549.

Bluher, M., Kahn, B. B., and Kahn, C. R. (2003). Extended longevity in mice lacking the insulin receptor in adipose tissue. *Science* **299,** 572–574.

Chen *et al. Aging Cell* 2007.

Clokey, G. V., and Jacobson, L. A. (1986). The autofluorescent "lipofuscin granules" in the intestinal cells of *Caenorhabditis elegans* are secondary lysosomes. *Mech. Ageing Dev.* **35,** 79–94.

Crawford, D., Libina, N., and Kenyon, C. (2007). *Caenorhabditis elegans* integrates food and reproductive signals in lifespan determination. *Aging Cell* **6,** 715–721.

Curran and Ruvkun, *PLOS G* 2007.

Dillin, A., Crawford, D. K., and Kenyon, C. (2002a). Timing requirements for insulin/IGF-1 signaling in *C. elegans*. *Science* **298,** 830–834.

Dillin, A., Hsu, A. L., Arantes-Oliveira, N., Lehrer-Graiwer, J., Hsin, H., Fraser, A. G., Kamath, R. S., Ahringer, J., and Kenyon, C. (2002b). Rates of behavior and aging specified by mitochondrial function during development. *Science* **298,** 2398–2401.

Feng, J., Bussiere, F., and Hekimi, S. (2001). Mitochondrial electron transport is a key determinant of life span in *Caenorhabditis elegans*. *Dev. Cell* **1,** 663–644.

Florez-McClure, M. L., Hohsfield, L. A., Fonte, G., Bealor, M. T., and Link, C. D. (2007). Decreased insulin-receptor signaling promotes the autophagic degradation of beta-amyloid peptide in *C. elegans*. *Autophagy* **3**.

Gerstbrein, B., Stamatas, G., Kollias, N., and Driscoll, M. (2005). *In vivo* spectrofluorimetry reveals endogenous biomarkers that report healthspan and dietary restriction in *Caenorhabditis elegans*. *Aging Cell* **4,** 127–137.

Giannakou, M. E., and Partridge, L. (2007). Role of insulin-like signalling in *Drosophila* lifespan. *Trends Biochem. Sci.* **32,** 180–188.

Guarente, L., and Picard, F. (2005). Calorie restriction: The SIR2 connection. *Cell* **120,** 473–482.

Hall, D. H. (1995). Electron microscopy and three-dimensional image reconstruction. *Methods Cell Biol.* **48,** 395–436.

Hamilton, B., Dong, Y., Shindo, M., Liu, W., Odell, I., Ruvkun, G., and Lee, S. S. (2005). A systematic RNAi screen for longevity genes in *C. elegans*. *Genes Dev.* **19,** 1544–1555.

Hansen, M., Chandra, A., Mitic, L. L., Onken, B., Driscoll, M., and Kenyon, C. (2008). A role for autophagy in the extension of lifespan by dietary restriction in *C. elegans*. *PLoS Genet.* **4,** e24.

Hansen, M., Hsu, A. L., Dillin, A., and Kenyon, C. (2005). New genes tied to endocrine, metabolic, and dietary regulation of lifespan from a *Caenorhabditis elegans* genomic RNAi screen. *PLoS Genet.* **1,** 119–128.

Hansen, M., Taubert, S., Crawford, D., Libina, N., Lee, S. J., and Kenyon, C. (2007). Lifespan extension by conditions that inhibit translation in *Caenorhabditis elegans*. *Aging Cell* **6,** 95–110.

Hars, E. S., Qi, H., Ryazanov, A. G., Jin, S., Cai, L., Hu, C., and Liu, L. F. (2007). Autophagy regulates ageing in *C. elegans*. *Autophagy* **3,** 93–95.

Henderson, S. T., Bonafe, M., and Johnson, T. E. (2006). *daf-16* protects the nematode *Caenorhabditis elegans* during food deprivation. *J. Gerontol. A Biol. Sci. Med. Sci.* **61,** 444–460.

Holzenberger, M., Dupont, J., Ducos, B., Leneuve, P., Geloen, A., Even, P. C., Cervera, P., and Le Bouc, Y. (2003). IGF-1 receptor regulates lifespan and resistance to oxidative stress in mice. *Nature* **421,** 182–187.

Honda, Y., and Honda, S. (1999). The *daf-2* gene network for longevity regulates oxidative stress resistance and Mn-superoxide dismutase gene expression in *Caenorhabditis elegans*. *Faseb J.* **13,** 1385–1393.

Houthoofd, K., Braeckman, B. P., Johnson, T. E., and Vanfleteren, J. R. (2003). Life extension via dietary restriction is independent of the Ins/IGF-1 signalling pathway in *Caenorhabditis elegans*. *Exp. Gerontol.* **38,** 947–954.

Hsin, H., and Kenyon, C. (1999). Signals from the reproductive system regulate the lifespan of *C. elegans* [see comments]. *Nature* **399,** 362–366.

Hsu, A. L., Murphy, C. T., and Kenyon, C. (2003). Regulation of aging and age-related disease by DAF-16 and heat-shock factor. *Science* **300,** 1142–1145.

Jia, K., Chen, D., and Riddle, D. L. (2004). The TOR pathway interacts with the insulin signaling pathway to regulate *C. elegans* larval development, metabolism and life span. *Development* **131,** 3897–3906.

Jia, K., and Levine, B. (2007). Autophagy is required for dietary restriction-mediated life span extension in *C. elegans*. *Autophagy* **3**.

Kaeberlein, M., Powers, R. W., III, Steffen, K. K., Westman, E. A., Hu, D., Dang, N., Kerr, E. O., Kirkland, K. T., Fields, S., and Kennedy, B. K. (2005). Regulation of yeast replicative life span by TOR and Sch9 in response to nutrients. *Science* **310,** 1193–1196.

Kaeberlein, T. L., Smith, E. D., Tsuchiya, M., Welton, K. L., Thomas, J. H., Fields, S., Kennedy, B. K., and Kaeberlein, M. (2006). Lifespan extension in *Caenorhabditis elegans* by complete removal of food. *Aging Cell.*

Kang, C., You, Y. J., and Avery, L. (2007). Dual roles of autophagy in the survival of *Caenorhabditis elegans* during starvation. *Genes Dev.* **21,** 2161–2171.

Kapahi, P., Zid, B. M., Harper, T., Koslover, D., Sapin, V., and Benzer, S. (2004). Regulation of lifespan in *Drosophila* by modulation of genes in the TOR signaling pathway. *Curr. Biol.* **14,** 885–890.

Kenyon, C. (2005). The plasticity of aging: Insights from long-lived mutants. *Cell* **120,** 449–460.

Kenyon, C., Chang, J., Gensch, E., Rudner, A., and Tabtiang, R. (1993). A *C. elegans* mutant that lives twice as long as wild type [see comments]. *Nature* **366,** 461–464.

Klass, M. R. (1977). Aging in the nematode *Caenorhabditis elegans*: major biological and environmental factors influencing life span. *Mech. Ageing Dev.* **6,** 413–429.

Klass, M. R. (1983). A method for the isolation of longevity mutants in the nematode *Caenorhabditis elegans* and initial results. *Mech. Ageing De.v* **22,** 279–286.

Klionsky, D. J., Cuervo, A. M., and Seglen, P. O. (2007). Methods for monitoring autophagy from yeast to human. *Autophagy* **3,** 181–206.

Lakowski, B., and Hekimi, S. (1996). Determination of life-span in *Caenorhabditis elegans* by four clock genes. *Science* **272,** 1010–1013.

Lakowski, B., and Hekimi, S. (1998). The genetics of caloric restriction in *Caenorhabditis elegans. Proc. Natl. Acad. Sci. USA* **95,** 13091–13096.

Lamming, D. W., Latorre-Esteves, M., Medvedik, O., Wong, S. N., Tsang, F. A., Wang, C., Lin, S. J., and Sinclair, D. A. (2005). HST2 mediates SIR2-independent life-span extension by calorie restriction. *Science* **309,** 1861–1864.

Lee, G. D., Wilson, M. A., Zhu, M., Wolkow, C. A., de Cabo, R., Ingram, D. K., and Zou, S. (2006). Dietary deprivation extends lifespan in *Caenorhabditis elegans. Aging Cell* **5,** 515–524.

Lee, S. S., Kennedy, S., Tolonen, A. C., and Ruvkun, G. (2003a). DAF-16 target genes that control *C. elegans* life-span and metabolism. *Science* **300,** 644–647.

Lee, S. S., Lee, R. Y., Fraser, A. G., Kamath, R. S., Ahringer, J., and Ruvkun, G. (2003b). A systematic RNAi screen identifies a critical role for mitochondria in *C. elegans* longevity. *Nat. Genet.* **33,** 40–48.

Levine, B., and Klionsky, D. J. (2004). Development by self-digestion: Molecular mechanisms and biological functions of autophagy. *Dev. Cell* **6,** 463–477.

Libina, N., Berman, J. R., and Kenyon, C. (2003). Tissue-specific activities of *C. elegans* DAF-16 in the regulation of lifespan. *Cell* **115,** 489–502.

Lin, K., Dorman, J. B., Rodan, A., and Kenyon, C. (1997). daf-16: An HNF-3/forkhead family member that can function to double the life-span of *Caenorhabditis elegans* [see comments]. *Science* **278,** 1319–1322.

Liu, X., Jiang, N., Hughes, B., Bigras, E., Shoubridge, E., and Hekimi, S. (2005). Evolutionary conservation of the *clk-1*-dependent mechanism of longevity: loss of mclk1 increases cellular fitness and lifespan in mice. *Genes Dev.* **19,** 2424–2434.

McDonald, K. L. (1994). Electron microcopy and EM immunocytochrmistry. *Met. Cell Biol.* **44,** 411–444.

McElwee, J., Bubb, K., and Thomas, J. H. (2003). Transcriptional outputs of the Caenorhabditis elegans forkhead protein DAF-16. *Aging Cell* **2,** 111–121.

Meissner, B., Boll, M., Daniel, H., and Baumeister, R. (2004). Deletion of the intestinal peptide transporter affects insulin and TOR signaling in *Caenorhabditis elegans. J. Biol. Chem.* **279,** 36739–36745.

Meléndez, A., Talloczy, Z., Seaman, M., Eskelinen, E. L., Hall, D. H., and Levine, B. (2003). Autophagy genes are essential for dauer development and life-span extension in *C. elegans*. *Science* **301,** 1387–1391.

Mizushima, N., Yamamoto, A., Matsui, M., Yoshimori, T., and Ohsumi, Y. (2004). *In vivo* analysis of autophagy in response to nutrient starvation using transgenic mice expressing a fluorescent autophagosome marker. *Mol. Biol. Cell* **15,** 1101–1111.

Morck, C., and Pilon, M. (2006). *C. elegans* feeding defective mutants have shorter body lengths and increased autophagy. *BMC Dev. Biol.* **6,** 39.

Muller-Reichert, T., Srayko, M., Hyman, A., O'Toole, E. T., and McDonald, K. (2007). Correlative light and electron microscopy of early *Caenorhabditis elegans* embryos in mitosis. *Methods Cell Biol.* **79,** 101–119.

Murphy, C. T., McCarroll, S. A., Bargmann, C. I., Fraser, A., Kamath, R. S., Ahringer, J., Li, H., and Kenyon, C. (2003). Genes that act downstream of DAF-16 to influence the lifespan of *Caenorhabditis elegans*. *Nature* **424,** 277–283.

Nicot, A. S., Fares, H., Payrastre, B., Chisholm, A. D., Labouesse, M., and Laporte, J. (2006). The phosphoinositide kinase PIKfyve/Fab1p regulates terminal lysosome maturation in *Caenorhabditis elegans*. *Mol. Biol. Cell* **17,** 3062–3074.

Ogg, S., Paradis, S., Gottlieb, S., Patterson, G. I., Lee, L., Tissenbaum, H. A., and Ruvkun, G. (1997). The fork head transcription factor DAF-16 transduces insulin-like metabolic and longevity signals in *C. elegans*. *Nature* **389,** 994–999.

Oh, S. W., Mukhopadhyay, A., Dixit, B. L., Raha, T., Green, M. R., and Tissenbaum, H. A. (2006). Identification of direct DAF-16 targets controlling longevity, metabolism and diapause by chromatin immunoprecipitation. *Nat. Genet.* **38,** 251–257.

Oh, S. W., Mukhopadhyay, A., Svrzikapa, N., Jiang, F., Davis, R. J., and Tissenbaum, H. A. (2005). JNK regulates lifespan in *Caenorhabditis elegans* by modulating nuclear translocation of forkhead transcription factor/DAF-16. *Proc. Natl. Acad. Sci. USA* **102,** 4494–4499.

Pan, K. Z., Palter, J. E., Rogers, A. N., Olsen, A., Chen, D., Lithgow, G. J., and Kapahi, P. (2007). Inhibition of mRNA translation extends lifespan in *Caenorhabditis elegans*. *Aging Cell* **6,** 111–119.

Panowski, S. H., Wolff, S., Aguilaniu, H., Durieux, J., and Dillin, A. (2007). PHA-4/Foxa mediates diet-restriction-induced longevity of *C. elegans*. *Nature* **447,** 550–555.

Raizen, D. M., Lee, R. Y., and Avery, L. (1995). Interacting genes required for pharyngeal excitation by motor neuron MC in *Caenorhabditis elegans*. *Genetics* **141,** 1365–1382.

Rea, S. L., Ventura, N., and Johnson, T. E. (2007). Relationship between mitochondrial electron transport chain dysfunction, development, and life extension in *Caenorhabditis elegans*. *PLoS Biol.* **5,** e259.

Riddle, D. L. (1997). *In* C. elegans II, (D. L. Riddle, T. Blumenthal, B. J. Meyer, and J. R. Priess, eds.), pp. 739–768, Cold Spring Harbor Laboratory, Cold Spring Harbor, NY.

Rowland, A. M., Richmond, J. E., Olsen, J. G., Hall, D. H., and Bamber, B. A. (2006). Presynaptic terminals independently regulate synaptic clustering and autophagy of GABAA receptors in *Caenorhabditis elegans*. *J. Neurosci.* **26,** 1711–1720.

Samuelson, A. V., Carr, C. E., and Ruvkun, G. (2007). Gene activities that mediate increased life span of *C. elegans* insulin-like signaling mutants. *Genes Dev.* **21,** 2976–2994.

Shaw, W. M., Luo, S., Landis, J., Ashraf, J., and Murphy, C. T. (2007). The *C. elegans* TGF-beta Dauer pathway regulates longevity via insulin signaling. *Curr. Biol.* **17,** 1635–1645.

Syntichaki, P., Troulinaki, K., and Tavernarakis, N. (2007). eIF4E function in somatic cells modulates ageing in *Caenorhabditis elegans*. *Nature* **445,** 922–926.

Takeshige, K., Baba, M., Tsuboi, S., Noda, T., and Ohsumi, Y. (1992). Autophagy in yeast demonstrated with proteinase-deficient mutants and conditions for its induction. *J. Cell Biol.* **119,** 301–311.

Tissenbaum, H. A., and Guarente, L. (2001). Increased dosage of a *sir-2* gene extends lifespan in *Caenorhabditis elegans*. *Nature* **410,** 227–230.

Toth, M. L., Sigmond, T., Borsos, E., Barna, J., Erdelyi, P., Takacs-Vellai, K., Orosz, L., Kovacs, A. L., Csikos, G., Sass, M., and Vellai, T. (2008). Longevity pathways converge on autophagy genes to regulate life span in *Caenorhabditis elegans*. *Autophagy* **4,** 330–338.

Toth, M. L., Simon, P., Kovacs, A. L., and Vellai, T. (2007). Influence of autophagy genes on ion-channel-dependent neuronal degeneration in *Caenorhabditis elegans*. *J. Cell Sci.* **120,** 1134–1141.

Tullet, J. M., Hertweck, M., An, J. H., Baker, J., Hwang, J. Y., Liu, S., Oliveira, R. P., Baumeister, R., and Blackwell, T. K. (2008). Direct inhibition of the longevity-promoting factor SKN-1 by insulin-like signaling in *C. elegans*. *Cell* **132,** 1025–1038.

Vellai, T., Takacs-Vellai, K., Zhang, Y., Kovacs, A. L., Orosz, L., and Muller, F. (2003). Genetics: Influence of TOR kinase on lifespan in *C. elegans*. *Nature* **426,** 620.

Wang, J., and Barr, M. M. (2005). RNA interference in *Caenorhabditis elegans*. *Methods Enzymol.* **392,** 36–55.

Wang, Y., and Tissenbaum, H. A. (2006). Overlapping and distinct functions for a Caenorhabditis elegans SIR2 and DAF-16/FOXO. *Mech. Ageing Dev.* **127,** 48–56.

Weimer, R. M. (2006). Preservation of *C. elegans* tissue via high-pressure freezing and freeze-substitution for ultrastructural analysis and immunocytochemistry. *Methods Mol. Biol.* **351,** 203–221.

Wolkow, C. A., Kimura, K. D., Lee, M. S., and Ruvkun, G. (2000). Regulation of *C. elegans* life-span by insulinlike signaling in the nervous system. *Science* **290,** 147–150.

Yin, D. (1996). Biochemical basis of lipofuscin, ceroid, and age pigment-like fluorophores. *Free Radic. Biol. Med.* **21,** 871–888.

CHAPTER THIRTY

Autophagy in *Caenorhabditis elegans*

Tímea Sigmond,* János Barna,* Márton L. Tóth,* Krisztina Takács-Vellai,* Gabriella Pásti,† Attila L. Kovács,† *and* Tibor Vellai*

Contents

Abstract

Autophagy (cellular self-eating) is a highly regulated, lysosome-mediated catabolic process of eukaryotic cells to segregate by a special membrane and subsequently degrade their own constituents during development or starvation. Electron microscopy analysis reveals autophagic elements in various cell types of the nematode *Caenorhabditis elegans*, whose genome contains counterparts of several yeast genes involved in autophagy. Genetic manipulation inactivating autophagy-related genes in *C. elegans* causes defects in development, affects dauer larval morphogenesis, accelerates aging thereby shortening life span, reduces cell size, decreases survival during starvation, promotes apoptotic cell death, and protects neurons from undergoing hyperactive ion channel- or neurotoxin-induced degeneration. These results implicate autophagy in various developmental and cellular functions such as reproductive growth, aging, and cell growth, as well as cell survival and loss. This chapter

* Department of Genetics, Eötvös Loránd University, Budapest, Hungary
† Department of Anatomy, Cell and Developmental Biology, Eötvös Loránd University, Budapest, Hungary

Methods in Enzymology, Volume 451
ISSN 0076-6879, DOI: 10.1016/S0076-6879(08)03230-8

discusses methods of inactivating *C. elegans* autophagy genes by RNA interference, testing the resistance of autophagy-deficient nematodes to starvation-induced stress, handling mutants carrying a deletion in the autophagy pathway, and monitoring autophagic activity by using LysoTracker Red dye or reporters labeled with green fluorescent protein. Such methods may be adaptable to identify additional roles of autophagy in development and cellular function, and may also help to detect the intracellular accumulation of autophagy proteins and monitor autophagosome formation.

1. Introduction

1.1. Autophagosomes in *C. elegans*

Autophagy (used as the synonym of *macroautophagy* in this article) involves the formation of subcellular double-membrane-bound structures called *autophagosomes*, which contain sequestered cytoplasmic materials and deliver them into lysosomes (autolysosomes) for degradation (Klionsky, 2005; Klionsky *et al.*, 2008). Despite intense functional analysis of autophagy-related genes in the genetic model system *Caenorhabditis elegans*, there are only a few papers reporting the electron microscopy of autophagic elements in this species (Kovács *et al.*, 2004, 2007; Meléndez *et al.*, 2003). Autophagosomes with ribosome-free double membrane of the thin type, as well as autolysosomes have been described in various cell types at different developmental stages. Thus, *C. elegans* as a genetically tractable metazoan system offers a good possibility also for studying autophagy. Although electron microscopy is still an indispensable method for visualizing autophagic elements, most *C. elegans* researchers prefer to use green fluorescent protein (GFP)-labeled LGG-1/Atg8 as a reporter (Meléndez *et al.*, 2003) to monitor autophagic activity. However, it remains unproven whether this marker is indeed specific for autophagy or may also label other intracellular compartments unrelated to autophagic degradation.

1.2. Autophagy-related genes in the *C. elegans* genome

To date, no genetic screen has been performed to identify genes involved in autophagy in *C. elegans*. Instead, genetic factors underlying the process in this organism have been uncovered by searching the *C. elegans* genome for orthologs of previously identified yeast autophagy genes (Kovács *et al.*, 2004; Meléndez *et al.*, 2003). Such sequence analysis demonstrates the existence of a core set of nematode genes whose yeast counterparts are involved in the regulation and execution of autophagosome formation (Table 30.1). Interestingly, several yeast autophagy-related genes have no *C. elegans* counterpart, while some are represented by more than one worm

Table 30.1 *C. elegans* orthologs of yeast genes involved in autophagosome formation

Yeast gene	*C. elegans* ORF	*C. elegans gene*	Probability of BLAST hit	Function in yeast
ATG1	Y60A3A.1	*unc-51*	$8e^{-37}$	Ser/Thr kinase, induction of autophagy
ATG3	Y55F3AM.4		e^{-34}	E2-like enzyme, conjugates Atg8 to phosphatidylethanolamine
ATG4	Y87G2A.3		$2e^{-20}$	Cysteine protease, cleaves the C-terminal extension of Atg8
ATG6 (VPS30)	T19E7.3	*bec-1*	$3e^{-9}$	Coiled-coil protein, component of the class III PI 3-kinase (VPS34) complex
ATG7	M7.5	*atg-7*	e^{-110}	E1-like enzyme, activates Atg8 and Atg12
ATG8	C32D5.9	*lgg-1*	$6e^{-30}$	Ubiquitin-like protein, conjugated to the autophagosomal membrane
	ZK593.6	*lgg-2*	$2e^{-15}$	
ATG9	T22H9.2	*atg-9*	$2e^{-47}$	Transmembrane protein, binds to the preautophagosomal membrane
ATG18	F41E6.13	*atg-18*	$8e^{-15}$	Phosphoinositide-binding protein, releases Atg9 from phagophore to cytosol
VPS34	B0025.1	*let-512*	$2.7e^{-68}$	Class III PI 3-kinase, conjugated to Atg6

ORF: open reading frame.

ortholog. These paralogous sequences are predicted to function redundantly during development or in the cellular response to starvation. Thus, genetic inactivation of such factors requires a special experimental design.

1.3. Role of autophagy genes in development, aging, resistance to starvation-induced stress, cell growth, and cell death in *C. elegans*

In *C. elegans*, autophagy genes control several aspects of development and cellular function. For example, nematodes defective for BEC-1 (see Table 30.1) are unable to undergo proper dauer lava morphogenesis (Meléndez *et al.*, 2003). RNA interference-(RNAi)-mediated depletion of other autophagy-related proteins also causes defects in dauer formation. The dauer larval stage is a developmental arrested state specialized to survive adverse conditions. Mutant strains with reduced insulin/IGF-1 (insulin-like growth factor receptor-1) and TGF-β (transforming growth factor-beta) signaling the enter into dauer independently of environmental cues (Riddle, 1997). However, insulin/IGF-1 or TGF-β signaling-defective mutants form abnormal dauers when they are treated with double-stranded RNA specific for an autophagy gene (Meléndez *et al.*, 2003). Thus, the autophagy gene cascade is required for dauer development in this organism.

The insulin/IGF-1 hormonal system also regulates aging in *C. elegans*; loss-of-function mutations in the pathway double the natural life span (Kenyon *et al.*, 1993). Longevity in insulin/IGF-1 signaling mutants relies on the activity of autophagy genes, suggesting that the autophagy pathway acts downstream of, and is inhibited by, insulin/IGF-1 signaling to lengthen life span (Meléndez *et al.*, 2003; Hars *et al.*, 2007; Tóth *et al.*, 2008; Hansen *et al.*, 2008). In addition, autophagy genes are required for life span extension induced by reduced TOR (target of rapamycin) kinase activity, lowered mitochondrial respiration or dietary restriction (Hansen *et al.*, 2008; Jia and Levine, 2007; Tóth *et al.*, 2008; Vellai *et al.*, 2003). These data raise the possibility that several longevity pathways converge on autophagy genes to control the rate at which the tissues age.

Under optimal growth conditions, inhibiting autophagy during development appears to shorten life span (Hars *et al.*, 2007; Meléndez *et al.*, 2003; Tóth *et al.*, 2008). However, the magnitude by which autophagy genes promote survival is more significant in worms that are exposed to prolonged starvation (Kang *et al.*, 2007). Interestingly, only physiological levels of autophagy can promote survival, while both insufficient and excessive levels of autophagy contribute to premature death. This indicates that autophagy acts as a fine-tuned cellular pathway to maintain homeostasis in energy metabolism during periods of nutrient depletion.

In *C. elegans* (mutational) inactivation by a chromosomal mutation of two autophagy genes, *unc-51* and *bec-1*, results in retarded cell growth by interacting with the insulin/IGF-1 and TGF-β signaling pathways (Aladzsity *et al.*, 2007), which influences body length by controlling cell size (McCulloch and Gems, 2003; Morita *et al.*, 1999). Double mutant analysis shows that the small body size of *unc-51* and *bec-1* mutants is epistatic to the giant phenotype of insulin/IGF-1 and TGF-β signaling mutant animals, suggesting that autophagy proteins mediate the effect of the two pathways on cell growth. Moreover, defects in feeding, which are accompanied by the intracellular accumulation of GFP::LGG-1, also result in small body size without affecting cell number (Mörck and Pilon, 2006). Upon these observations one can conclude that, similar to the regulation of organism survival, both excessive and insufficient levels of autophagy interfere with the cell growth process (Vellai *et al.*, 2008). According to the preceding results, autophagy is implicated in cell growth control in nematodes.

Cell survival is also influenced by autophagy genes. Decreased *bec-1* activity triggers apoptotic cell death during embryonic development as well as in the germ-line of adult hermaphrodites (Takács-Vellai *et al.*, 2005). Although autophagy gene products participate in the phagocytotic removal of cell corpses (Qu *et al.*, 2007), massive cell death observed in *bec-1* null mutant embryos affects many more cells than those that normally die during development (Takács-Vellai and Vellai, unpublished results). This argues for a complementary role of autophagy and apoptosis during early development rather than simply accumulating uncleared but otherwise physiologically dying cells in an autophagy-defective background. However, to determine unambiguously whether compromised autophagy is causing an increase in apoptosis and not a lack of apoptotic clearance one should monitor the number of apoptotic cell corpses in *bec-1*; *ced-1* double mutant embryos, in which the phagocytotic removal of apoptotic cell corpses is blocked. Interestingly, autophagy genes also interact with necrotic-like cell loss in *C. elegans*. Neurons undergoing hyperactive ion channel- or neurotoxin-induced (i.e., excitotoxic) degeneration exhibit intense vacuole formation and membrane infolding in the early phase of destruction, and extensive degradation of cytosolic materials in the executing phase of the process (Hall *et al.*, 1997). Reduced activity of autophagy genes suppresses excitotoxic neuronal demise (Samara *et al.*, 2008; Tóth *et al.*, 2007; Vellai *et al.*, 2007), implying their active contribution to degeneration. Depending on the circumstances, autophagy can thus prevent or trigger cell loss by interacting with the apoptotic and necrotic cell death pathways (Takács-Vellai *et al.*, 2006).

In this chapter we describe below protocols for inactivating autophagy genes, testing the effects of such manipulations on survival during long-term starvation, and monitoring the intracellular accumulation of autophagy proteins in *C. elegans*.

2. Inactivation of Autophagy Genes by RNA Interference

To explore the biological roles of autophagy genes in *C. elegans*, one can analyze the phenotypic effects that mutational or RNAi-mediated inactivation of individual autophagy genes has on development and cellular function. At present, however, only a limited number of nematode autophagy genes are available as mutant alleles. Thus, RNAi-mediated gene silencing is often the only effective method to assess their *in vivo* function. However, prior to performing RNAi experiments, one must consider two important seemingly opposing factors. First, genomic coding sequences used as DNA template for fragment amplification must be unique (i.e., they are not allowed to match any paralogous sequences); otherwise RNAi treatment might interfere with additional genes and thereby provide nonspecific results. Second, because certain autophagy genes in the *C. elegans* genome have a paralog(s) with high sequence similarity, down-regulation of such single genes is expected to lead to no or weak reactions as a result of functional redundancy. To overcome this problem, simultaneous silencing of the paralogs could help to uncover their function. Once an RNAi construct expressing a specific double-stranded RNA is generated, its effectiveness can be tested on transgenic strains that accumulate the corresponding GFP-labeled autophagy protein. Abolishment of GFP-glowing upon RNAi treatment indicates that depletion of the target protein has successfully occurred. As an example, we describe below how simultaneous silencing of the two *Atg8* paralogs, *lgg-1* and *lgg-2*, can be achieved (Tóth *et al.*, 2008). The procedure is applicable for knocking down any single (autophagy) gene.

The RNAi experiment for simultaneous depletion of LGG-1 and LGG-2 involves the following main steps: (1) total RNA isolation from worm extract; (2) reverse transcriptase-based RT-PCR amplification of a hybrid cDNA fragment that is specific for both genes; (3) cloning the fragment into (feeding) RNAi vector; (4) feeding nematodes with bacteria expressing the fragment-specific double-stranded RNA.

Isolation of total RNA from mix-staged worms

1. Maintain nematodes in Petri dishes containing standard nematode growth medium (NGM) as described (Brenner, 1974; Sulston and Hodgkin, 1988), on an *Escherichia coli* OP50-based diet. For preparing NGM agar, add the following per liter:

   NaCl: 3 g
   agar: 17 g
   bactopeptone: 2.5 g

cholesterol (5 mg/ml in EtOH): 1 ml
distilled H_2O: 975 ml
Autoclave, then add the following components, with sterile technique:
1 *M* $CaCl_2$: 1 ml
1 *M* $MgSO_4$: 1 ml
1 *M* potassium phosphate buffer, pH 6.0: 25 ml
(To prepare potassium phosphate, per liter, add 136 g KH_2PO_4, 900 ml H_2O, adjust pH to 6 with cc. 50–55 ml of 5 *M* KOH, add H_2O up to 1 liter, autoclave.)

2. Wash mix-staged worms off approximately 30 plates (crowded with animals) with distilled water or M9 buffer (per liter: 3 g KH_2PO_4, 6 g Na_2HPO_4, 5 g NaCl, 1 ml 1 *M* $MgSO_4$, 1 liter H_2O) and collect them in a 15-ml Falcon tube.
3. Concentrate the worm suspension by centrifugation (at 1500 rpm for 1 min) and remove the supernatant fraction by aspiration.
4. Add 4 ml of Trizol (Gibco #15596-018) to the worm pellet (approximately 1 ml in volume) and vortex the tube rigorously. Trizol (monophasic solution of phenol and guanidine isothiocyanate) is used to disclose cells and to separate the protein, DNA and RNA phases. RNA remains in the aqueous phase.
5. Place the tube into liquid nitrogen, and then let the frozen worm/Trizol mixture thaw at room temperature (RT).
6. Split the solution into 1.5-ml microcentrifuge tubes (1250 μl per tube), and let them stand at RT for 5 min.
7. Spin the solution at 14,000 rpm for 10 min at 4 °C.
8. Transfer liquid to new microcentrifuge tubes, and add 200 μl of chloroform to each tube; chloroform reacts strongly with Trizol.
9. Vortex for 15 s (this step is critical!) and incubate at RT for 3 min.
10. Spin solution at 14,000 rpm for 15 min at 4 °C to separate phases.
11. Transfer the upper aqueous phase to new RNase-free tubes, add 500 μl of isopropanol to each tube, mix, and incubate at RT for 10 min to precipitate the RNA.
12. Spin at 14,000 rpm for 10 min at 4 °C to recover RNA. Note: the RNA pellet is white and can be seen along the side of the tube.
13. Carefully remove the aqueous phase from the pellet fraction by aspiration. Wash the pellet with 100 μl of 75% EtOH in diethyl pyrocarbonate (DEPC)-treated water. To prepare DEPC-treated water, add 0.2 ml of DEPC to 100 ml of double distilled H_2O, shake vigorously overnight, and then autoclave the solution.
14. Spin at 7500 rpm for 5 min at 4 °C, remove the supernatant fraction, and air-dry the pellets for 5–10 min.
15. Dissolve the pellets in 20–50 μl of DEPC-treated water. To help dissolve, heat to 50 °C for 5 min. Expect 1–5 mg RNA/gram of

worms. To determine the concentration and purity, dilute 1–5 μl of prep into 1 ml (RNase-free) H_2O, and take determine the A260/A280 ratio. RNA is pure if the ratio is 2.0; if it is < 1.6, the RNA is not completely dissolved or there is protein contamination. Store samples frozen at −80 °C.

RT-PCR amplification of a hybrid cDNA fragment specific for both lgg-1 and lgg-2

Figure 30.1. shows a schematic view of generating a hybrid cDNA fragment consisting of *lgg-1* and *lgg-2* coding sequences (by applying a PCR fusion-based protocol described by Hobert, 2002).

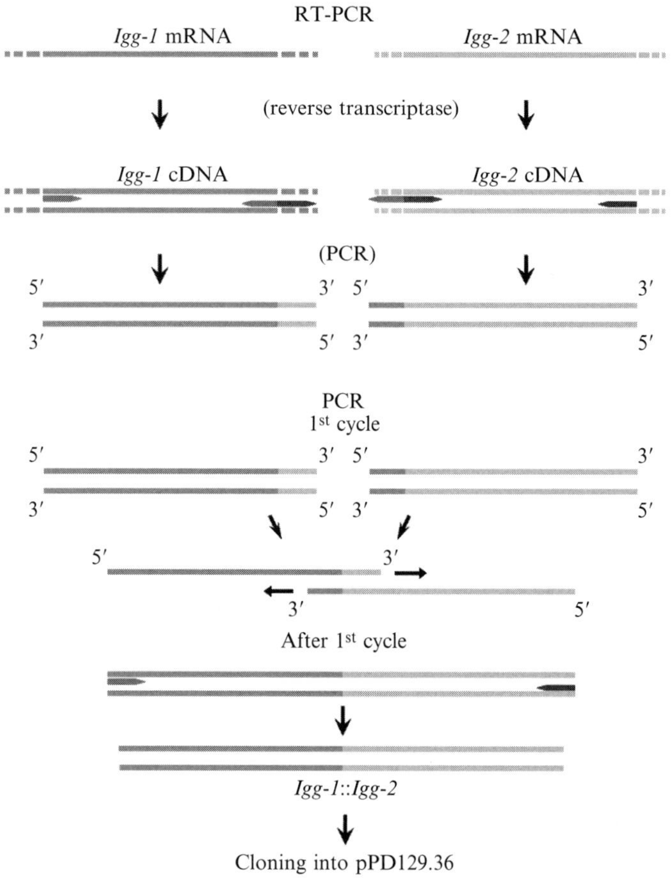

Figure 30.1 Schematic model of a PCR fusion-based approach to create a hybrid cDNA fragment for simultaneous silencing of *lgg-1* and *lgg-2*. The process contains a reverse transcriptase-based RT-PCR amplification of single cDNA fragments, and a subsequent PCR amplification of the chimeric cDNA fragment. Orange and light blue lines indicate DNA double helices, red and dark blue arrows indicate primers. (See Color Insert.)

1. Use the following primers for performing RT-PCR. For *lgg-1*: forward primer 5′-ttt cgt cac tgt agg cga tg-3′, reverse primer 5′-cac gtc tca acc agc att tga agt ggg ctt aca agg agg ag-3′; for *lgg-2*: forward primer 5′-ctc ctc ctt gta agc cca ctt caa atg ctg gtt gag acg tg-3′, reverse primer 5′-aat cgt tcc atc gtt caa gg-3′.
2. Prepare RT-PCR mix and perform RT-PCR reaction by using Titan One Tube RT-PCR System (Roche, #1888382) according to the protocol provided by the manufacturer:

   1 μl of total RNA (1 ug/reaction)
   11 μl of DEPC-treated, RNase-free water
   5 μl of 5x buffer
   1 μl of 20 μM forward primer
   1 μl of 20 μM reverse primer
   2.5 μl of DTT
   1.5 μl of Mg^{2+}
   1 μl of 10 mM dNTP mix
   1 μl of polymerase mix
   Total volume: 25 μl

   Use the following RT-PCR Program:

   a. Cycle at 50 °C for 30 min
   b. Cycle at 94 °C for 2 min
   c. Cycle at 59 °C for 1 min
   d. Cycle at 72 °C for 1 min
   e. Cycle at 93 °C for 35 s

   Note: Repeat cycles from c to e, 35 times.

3. For the PCR reaction, use 1000x-diluted template DNA from RT-PCR and the following primers: 5′-tga gcg tca gaa aga tgt gg-3′ (forward primer); 5′-tct tcc tcg tga tgg tcc tg-3′ (reverse primer).
4. Prepare PCR mix by using Expand High Fidelity PCR System (Roche, #11732641001), and perform a standard PCR reaction:

   1 μl of template DNS (1000x diluted from RT-PCR reactions)
   32 μl of Milli-Q H_2O
   5 μl of 10x buffer
   5 μl of 2 μM forward primer
   5 μl of 2 μM reverse primer
   1 μl of 10 mM dNTP mix
   1 μl of polymerase mix
   Total volume: 50 μl

PCR program:

a. Cycle at 94 °C for 2 min
b. Cycle at 59 °C for 1 min
c. Cycle at 72 °C for 1 min
d. Cycle at 93 °C for 35 sec
Note: Repeat cycles from b to d, 35 times.

Cloning hybrid cDNA fragment into feeding RNAi vector

1. Load PCR products into 0.8 % agarose gel (use freshly washed tanks) and run at 80 mV.
2. Excise bands (migrating at approximately 0.7 kb) from the gel by a razor blade.
3. Isolate DNA from the gel block by using the QIAEX II system (Qiagen, #20021), according to instructions provided by the manufacturer, or an equivalent purification reagent.
4. Ligate DNA fragment using the pGEM-T Easy Vector System (Promega, #A1360; this is a blunt-end cloning system for PCR fragments).
5. Subclone the fragment into the pPD129.36 feeding vector (Timmons and Fire, 1998), using NotI digestion, dephosphorylation of the vector, and ligation.
6. Transform vector into competent *E. coli* HT115(DE3) bacterial cells.

Feeding nematodes with bacteria expressing double-stranded RNA

For RNAi experiments, the so-called feeding method is effective and relatively simple (Kamath *et al.*, 2000; Timmons and Fire, 1998).

1. Prepare NGM plates supplemented with 12.5 μg/ml tetracycline, 50 μg/ml ampicillin, and 1 m*M* isopropylthiogalactoside (IPTG).
2. Grow *E. coli* HT115(DE3) bacteria expressing the empty vector (control) and specific (*lgg-1::lgg-2*) dsRNA in $LB^{TetrAmpIPTG}$ at 37 °C to $OD_{595} = 0.4$, and then spread the bacteria onto the plates (approximately 0.1 ml of bacterial solution per plate).
3. Incubate plates at 37 °C overnight.
4. Transfer 3–5 L4 stage nematodes (P) to plates by hand-picking, and observe the phenotype of the F1 generation (usually after 2–3 days) at 15 °C, 20 °C, and 25 °C. The effects/specificity of RNAi can be tested with quantitative-RT-PCR methods; RNAi treatment should reduce the amount of targeted messenger RNA, as compared with the wild type.

It is likely that simultaneous silencing of *lgg-1* and *lgg-2* results in a strong loss-of-function phenotype, as it eliminates detectable expression of a functional GFP::LGG-1 reporter (Fig. 30.2). Except for interfering with dauer morphogenesis (Meléndez *et al.*, 2003) and necrotic loss of specific

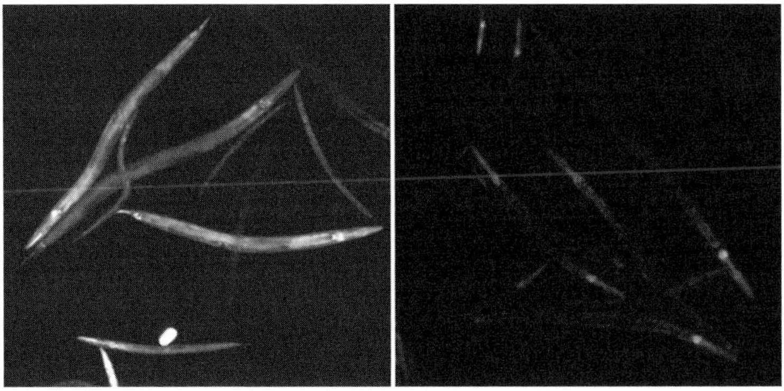

Figure 30.2 *lgg-1::lgg-2* RNAi treatment reduces detectable expression of an integrated GFP::LGG-1 reporter. Left fluorescence image shows GFP::LGG-1 expression in animals treated with control RNAi (expressing the empty plasmid only); right fluorescence image shows GFP::LGG-1 expression in *lgg-1::lgg-2(RNAi)* background. Images were taken with the same exposure time. In *lgg-1::lgg-2(RNAi)* animals, the faint background fluorescence is produced by the GFP vector alone.

neurons (Samara *et al.*, 2008; Tóth *et al.*, 2007), depleting LGG-1 or LGG-2 alone causes no obvious morphological and behavioral changes (Tóth *et al.*, 2008). Only *lgg-1*, but not *lgg-2*, RNAi treatment leads to a weak reduction in life span (Hars *et al.*, 2007; Tóth *et al.*, 2008). Conversely, simultaneous silencing of *lgg-1* and *lgg-2* markedly shortens adult life span (Tóth *et al.*, unpublished results), uncovering a redundant role for the paralogs in life span control. Note that we did not observe lethality after *lgg-1* RNAi, when the dsRNA is delivered by feeding.

Silencing of certain autophagy genes can result in severe defects in *C. elegans* development and behavior. For example, the pleiotropic phenotype of *bec-1(RNAi)* animals includes a highly penetrant development arrest at different embryonic or larval stages, failure to shed the old larval cuticle after molting, and various morphological malformations (Takács-Vellai *et al.*, 2005). *unc-51* RNAi treatment confers small body size and paralyzed (uncoordinated) locomotion. However, it is likely that some of these abnormalities result from effects of these multifunctional proteins that are independent of autophagy.

3. Handling Mutants With a Deletion in the Autophagy Pathway

In *C. elegans*, several autophagy-related genes, including *Atg1/unc-51*, *Atg6/bec-1*, *Atg7/atg-7*, *Atg18/atg-18*, and *vps34/let-512*, have been characterized by loss-of-function mutant alleles (Table 30.1). Except for *unc-51* and *let-512* alleles, these mutations were isolated in reverse genetic screens for deletional derivatives of target genes, and the corresponding mutant strains are available from the collection of the International *C. elegans* Knockout Consortium. Before starting to analyze such a strain, one must backcross it with wild-type males ideally at least 5 times. This step, called isogenization, eliminates background mutations that the strain likely contains due to the mutagenesis process. Interestingly, the morphology and behavior of *atg-7* and *atg-18* mutant strains appear superficially wild type. The only phenotype they display is shortened life span (Tóth *et al.*, 2008). Thus, each backcrossing step of isogenization requires PCR-based genotyping of single worms (i.e., selection for homozygous individuals carrying a deletion in the gene of interest).

C. elegans strains carrying a mutation in the autophagy pathway show reduced survival during prolonged starvation (Fig. 30.3). Therefore, their decreased ability to tolerate food deprivation might be a general characteristic indicating compromised autophagy. We propose that the assay below for measuring resistance to starvation-induced stress can be used to characterize additional mutants suspected to be defective in autophagy. The protocol has been developed to score the survival of nematodes under conditions of prolonged starvation at the first larval (L1) stage, where *C. elegans* arrest development until food is available (Derry *et al.*, 2001).

1. Wash well-fed gravid hermaphrodites off approximately 10–20 NGM plates using M9 buffer, and prepare eggs (embryos) by hypochlorite treatment (Sulston and Hodgkin, 1988). A simple procedure is as follows: wash adults off into M9 buffer, add to the suspension a half volume of alkaline hypochlorite (2 volumes 4 *M* NaOH:3 volumes 10–20% NaOCl, preferably freshly mixed), store for 5 min at room temperature (shake once every minute), spin down briefly (1500 rpm, 1 min), remove the supernatant fraction by pipetting/aspiration, and wash the pellet (eggs) with M9 5 times.

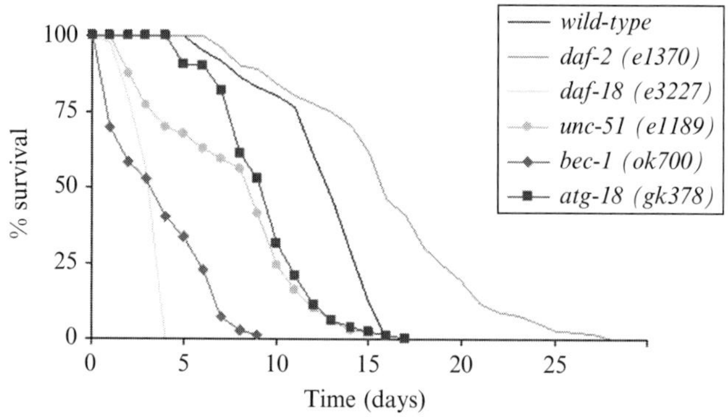

Figure 30.3 Decreased survival of animals with a mutation in the autophagy pathway during prolonged L1 starvation. All autophagy-deficient mutant strains isolated so far exhibit a similar phenotype under identical conditions. (See Color Insert.)

2. Take up eggs in 5 ml of M9 buffer supplemented with cholesterol (10 μg/ml) and ampicillin (50 μm/ml), and let them hatch for 1 day at 20 °C.
3. Transfer aliquots of arrested L1 larvae every 48 h to NGM plates seeded with *E. coli* OP50 bacteria.
4. Determine the fraction of L1 larvae that are able to survive and develop into adulthood, after 2 days of growth at 25 °C.

The effect of prolonged L1 starvation on survival to adulthood of autophagy-deficient animals is shown in Fig. 30.3.

4. Monitoring Autophagy-Related Gene Activities During Development

In *C. elegans*, the most preferably used autophagy marker is a full-length LGG-1 protein, which is fused amino-terminally with GFP and expressed under the control of its own promoter (Meléndez *et al.*, 2003). LGG-1 is orthologous to yeast Atg8 (Table 30.1) and mammalian LC3 (microtubule-associated protein 1, light chain 3), which are small ubiquitin-like proteins required for the formation of autophagosomes. For functionality, Atg8 undergoes a carboxyl-terminal cleavage by the cysteine protease Atg4 to expose a reactive glycine residue, which is activated by the Atg7 (E1-like enzyme) and Atg3 (E2-like enzyme) proteins. The activated Atg8 is then covalently linked to phosphatidylethanolamine (Atg8–PE in yeast and the lipidated LC3-II in mammals) and remains bound to the autophagosome membrane relatively long (Klionsky, 2005). Thus, Atg8 (and LC3) is believed to label preautophagosomal and autophagosomal (and in some cases even autolysosomal) structures fairly specifically (Klionsky *et al.*, 2008; Mizushima, 2004).

GFP::LGG-1 accumulates ubiquitously throughout development in almost all cells (Meléndez *et al.*, 2003; Tóth *et al.*, 2008). Intracellularly, it accumulates in both diffuse and punctate patterns. Although GFP::LGG-1-positive foci are considered as autophagosomal structures, it has not been tested whether this is indeed the case. To this end, we first integrated the reporter into a chromosome to avoid its mosaic expression that results from the instability of the original extrachromosomal array of GFP::LGG-1. Since integration of the extrachromosomal array was achieved by UV irradiation (30,000 mJ/cm^2; $\lambda = 254$ nm), we then eliminated background mutations by backcrossing with wild-type males (5 times) (Kang *et al.*, 2007; Tóth *et al.*, 2008).

A transgenic strain carrying integrated GFP::LGG-1 shows an expression profile (Fig. 30.4A,E,F) that is highly similar to that described previously (Meléndez *et al.*, 2003; Tóth *et al.*, 2007). However, some of the GFP-positive foci in three independent integrant lines are clearly associated with compartments that are distinct from autophagosomes. Note that two independent integrant lines exhibit this pattern. For example, Before Molting, at the time of intense cuticle formation, the lateral seam cells always contain two twin spots roughly as large as the nucleus, and labeled intensively with GFP (Fig. 30.4B,C). Due to their special position and size, they can be identified as hypertrophied Golgi structures by electron microscopy (Fig. 30.4D). The timing of their appearance coincides with the intense cuticular synthesis (Singh and Sulston, 1978). Structures of the same size and in the identical position were described to be Golgi complexes based on Nomarski optics and GFP-tagged Golgi markers (Cho *et al.*, 2005; Singh and Sulston, 1978). This is not really surprising, as an ortholog of Atg8, GATE16 was identified to be a Golgi-associated protein (Sagiv *et al.*, 2000). Many of the GFP::LGG-1-positive smaller structures, however, seem to colocalize with granules that are also stained with LysoTracker Red, especially in intestinal cells (see subsequent sections). Taken together, these results show that in *C. elegans*, like in some other systems (Klionsky *et al.*, 2008; Kuma *et al.*, 2007), the interpretation of GFP::LGG-1-positive spots has to be treated with caution.

In *Drosophila*, LysoTracker Red dye readily stains acidic compartments, including that of the lysosomes (Scott *et al.*, 2004). To stain nematodes with LysoTracker Red, we used the following protocol.

1. Supplement NGM agar with LysoTracker Red (1 m*M* stock solution in DMSO; Molecular Probes, Eugene, OR) at 2 μM.
2. Seed plates with solution of *E. coli* OP50 bacteria, and incubate overnight at room temperature in the dark.
3. Grow nematodes on plates for 2 days at 20 °C in the dark.
4. For microscopy analysis, use appropriate filters.

LysoTracker Red–stained compartments show a strong colocalization with GFP::LGG-1-positive foci, including twin spots (Fig. 30.5). This implies that LysoTracker Red labels acidic compartments that are not exclusively usual lysosomes and not typical autophagic elements either. Therefore, more observations by regular analyses are needed for the exact identification of these LysoTracker Red–positive foci as well. Nevertheless, this dye still might be used as a marker for monitoring autophagic activity in *C. elegans* too, but the conditions have yet to be clarified in a similar way as has been done in *Drosophila*, where the appearance of LysoTracker Red–positive foci was proved to parallel the presence of electron microscopically detectable autophagic elements.

Atg6 (Table 30.1) is conjugated to the phosphatidylinositol 3-kinase Vps34 to form a complex that initiates phagophore nucleation (Klionsky, 2005). Thus, the nematode Atg6 ortholog BEC-1 can potentially be used for monitoring the rate at which autophagosomes are formed. Indeed, BEC-1 shows an accumulation pattern that is reminiscent to that of LGG-1

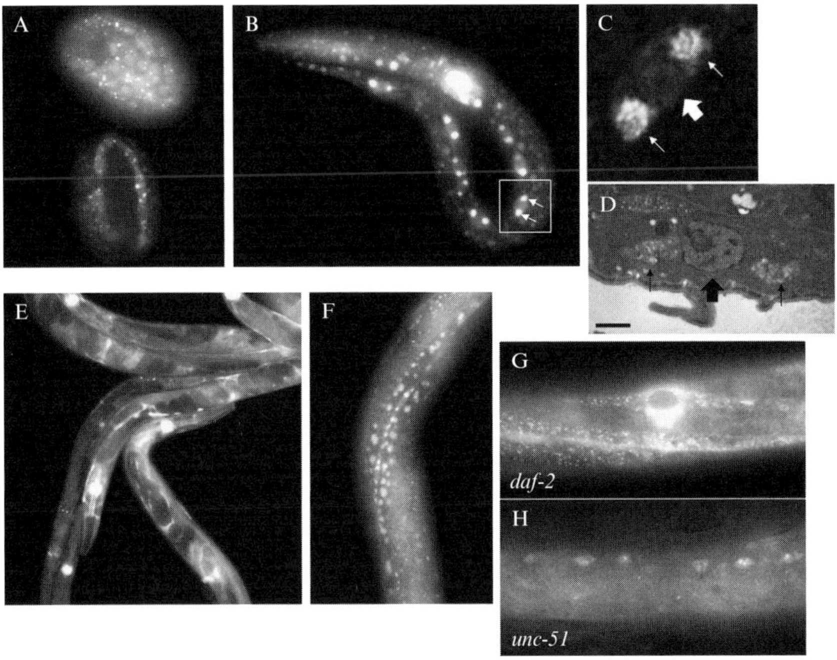

Figure 30.4 Intracellular accumulation of GFP::LGG-1 is not exclusively associated with autophagosomal structures. (A) Fluorescence image showing GFP::LGG-1 accumulation in an early (16-cell)-stage embryo (upper) and in a 3-fold-stage embryo (lower). LGG-1 accumulates in both diffuse and smaller punctate patterns. The latter might label autophagosomal structures. (B) Seam cells along the lateral side of a larva in the L1/L2 lethargus stage display twin-spot-like GFP::LGG-1 accumulation (arrows). (C) Higher magnification confocal image of the twin spot (the section framed in B). Thin arrows indicate the twin spots, the thick arrow shows the nucleus. (D) Electron microscopy image of a lateral seam cell with the twin spot reveals hypertrophied Golgi structures on the two sides of the nucleus (arrows, scale bar 1 μm). Thin arrows indicate Golgi structures, the thick arrow indicates the nucleus. (E) Mostly diffuse GFP::LGG-1 accumulation in adult animals. (F) GFP::LGG-1-positive foci in the hypodermis and intestine of an animal at the L4/adult lethargus stage. (G) GFP::LGG-1 accumulation in a daf-2 mutant hermaphrodite deficient in insulin/IGF-1 signaling. An increase in the number of GFP-positive puncta in the hypodermis is evident, as compared with the wild type. (H) GFP::LGG-1 accumulates mainly in a diffuse pattern in an *unc-51/Atg1* mutant animal. (See Color Insert.)

(Fig. 30.6). It is also localized intracellularly in both a diffuse pattern and in puncta. To describe the construction of an autophagy reporter in *C. elegans*, we summarize, as an example, the generation of a functional (translational fusion) GFP-labeled BEC-1 reporter.

1. Isolate genomic DNA from mixed stage worms as described (Sulston and Hodgkin, 1988). Briefly, collect mixed-stage worms from approximately 10–20 NGM plates into 1 ml of M9 buffer by washing, centrifugation and aspiration, add 0.5 ml of 100 m*M* NaCl, 100 m*M* Tris-Cl (pH 8.5),

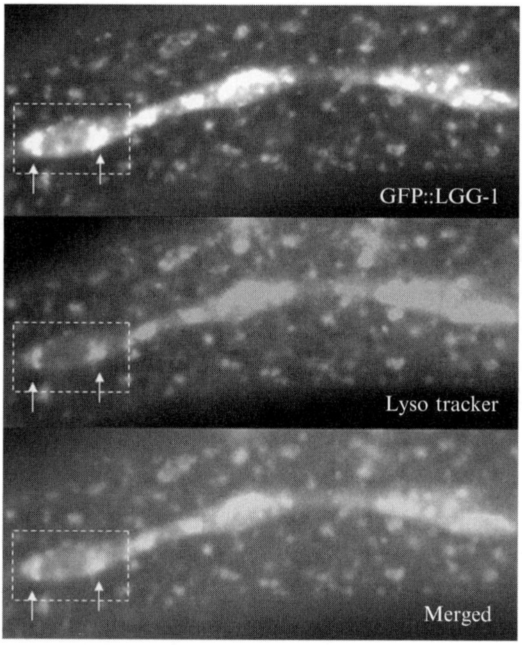

Figure 30.5 LGG-1 accumulates in acidic compartments. GFP::LGG-1 accumulation (green) is colocalized with LysoTracker Red (red)-positive lysosomal compartments. A lateral seam cell is indicated by the box with dotted lines; arrows indicate "twin spots". (See Color Insert.)

50 m*M* EDTA (pH 7.4), 1% SDS, 1% β-mercaptoethanol, and 100 μg/ml proteinase K, and incubate with occasional agitation for 2 h at 65 °C. Extract genomic DNA with 0.5 ml of phenol (mix by hand shaking, separate the phases by centrifugation at 12,000 rpm for 5 min, transfer the supernatant fraction [aqueous phase] into a fresh microcentrifuge tube; DNA remains in the aqueous phase), and 0.5 ml of phenol/chloroform.

2. Using genomic DNA as template, amplify a genomic fragment containing approximately 5 kb of upstream regulatory sequences and the entire coding region of *bec-1*, except from the last three nucleotides (the stop codon), with the following primers: 5′-GCT ACTCCT GCA GGC ATA GCG CGT AAT TAC TAT TGC GTT CTC G-3′ and 5′-CGG GAT CCC GAA TAG GCG ATC TGA GAG CAT CG-3′.
3. Digest the PCR fragment (approximately 9 kb in length) with SbfI and BamHI, and clone it into pPD95.75 (a nematode expression vector, kindly provided by Andrew Fire), in frame with the *gfp* gene.

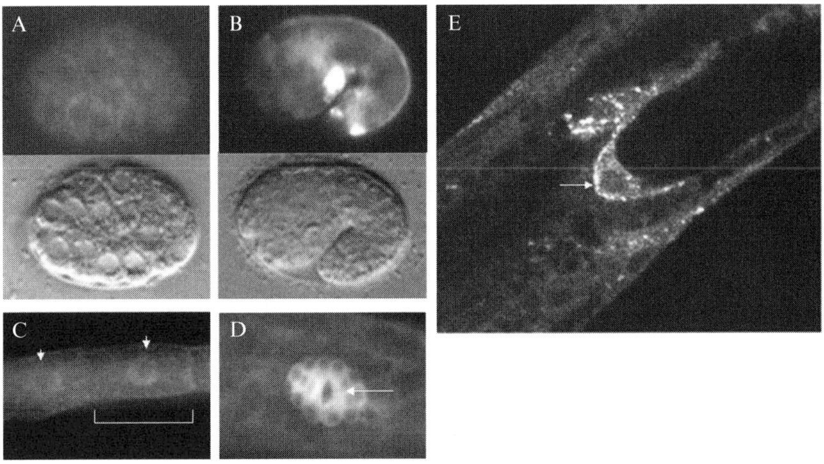

Figure 30.6 BEC-1::GFP accumulation in various stages of development. (A) Mostly diffuse and even expression of BEC-1 in all cells of an early stage embryo. Upper: epifluorescence image; Lower: the corresponding Nomarski image. (B) In a comma-stage embryo, BEC-1::GFP shows stronger accumulation in some regions of the body. Upper: epifluorescence image; Lower: the corresponding Nomarski image. (C) In intestinal cells, BEC-1 shows a characteristic perinuclear accumulation. The bracket indicates an intestinal cell, arrows show the nuclei. (D) Intense BEC-1 accumulation in the vulval cells (arrow) of a hermaphrodite. (E) Confocal image showing BEC-1::GFP-positive puncta. The arrow points to a distal tip cell. (See Color Insert.)

4. Inject the DNA construct at 50 ng/μl with a cotransformation marker (e.g., *rol-6(su1006)* that confers a dominant Roller phenotype) at 30 ng/μg into wild-type hermaphrodites, and establish transgenic lines as described (Mello *et al.*, 1991). Briefly, inject DNA solution into the syncytial distal gonad arm of adult hermaphrodites of the wild-type *C. elegans* strain, N2. Transfer roller (transformant) progeny of injected animals to fresh plates, maximally 5 animals per 6-cm plate. Transfer roller animals at four subsequent generations to establish transformed lines, before analyzing GFP expression.
5. Analyze the expression pattern of BEC-1 with epifluorescence microscopy.

Atg18 (Table 30.1) is implicated in retrieving proteins from the pre-autophagosomal structure (Krick *et al.*, 2006). Using a transcriptional fusion *atg-18::gfp* reporter (provided by the *Caenorhabditis* Genetics Center; the strain BC13209 *sEx13209*[*F41E6.13a::gfp*]), we found that this autophagy gene, similar to *lgg-1* and *bec-1*, is active throughout development in nearly all cells (Fig. 30.7).

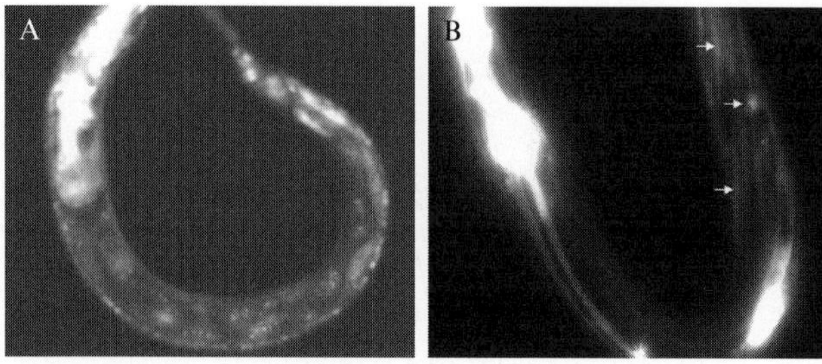

Figure 30.7 *atg-18::gfp* expression. (A) Fluorescence image shows that *atg-18* is expressed in almost every cell of a L3 larva. (B) *atg-18* expression in body wall muscle cells (arrows) of an L4 stage larva. (See Color Insert.)

6. Conclusions and Future Perspectives

C. elegans appears to utilize autophagy to regulate and execute several aspects of development and cellular functions. Therefore, this genetically tractable organism is an ideal metazoan system to uncover additional *in vivo* roles of the autophagy pathway as well as to identify novel autophagy-related genes. For the former, one can deplete known autophagy factors and assess the phenotypic effect caused by RNAi treatment. In addition, PCR-based reverse genetic screens for isolating deletional derivatives of target autophagy genes yet uncharacterized mutationally are also applicable. For the latter, which might provide data important for researchers working on model systems other than *C. elegans*, genomewide RNAi screens can be used to identify genetic backgrounds in which GFP-labeled reporter proteins fail to accumulate in punctate patterns or LysoTracker Red dye does not stain acidic compartments. Exploring novel components of the autophagy pathway in *C. elegans*—some of them might be nematode specific while others are evolutionarily conserved—is an important area for future investigations.

Acknowledgments

Some nematode strains used in this work were provided by Beth Levine (University of Texas Southwestern Medical Center, Dallas), the *Caenorhabditis* Genetics Center funded by the NIH National Center for Research Resources, and the International *C. elegans* Gene Knockout Consortium. We are grateful to Sára Simon and Tünde Pénzes for excellent technical help and to all members of our group for helpful comments on the manuscript. This work was supported by grants from the Ministry of Health (167/2006), the National Office for Research and Technology (NKFP 1A/007/2004), and the Hungarian Scientific Research Foundation (OTKA K68372) to TV, and the Hungarian Scientific Research Foundation (OTKA T047241) to ALK. TV is a grantee of the János Bolyai scholarship.

REFERENCES

Aladzsity, I., Tóth, M. L., Sigmond, T., Szabó, E., Bicsák, B., Regõs, Á., Orosz, L., Kovács, A. L., and Vellai, T. (2007). Autophagy genes *unc-51* and *bec-1* are required for normal cell size in *Caenorhabditis elegans*. *Genetics* **177,** 655–660.

Brenner, S. (1974). The genetics of *Caenorhabditis elegans*. *Genetics* **77,** 71–94.

Cho, J. H., Ko, K. M., Singaravelu, G., and Ahnn, J. (2005). *Caenorhabditis elegans* PMR1, a P-type calcium ATPase, is important for calcium/manganese homeostasis and oxidative stress response. *FEBS Lett.* **579,** 778–782.

Derry, W. B., Putzke, A. P., and Rothman, J. H. (2001). *Caenorhabditis elegans* p53: Role in apoptosis, meiosis, and stress resistance. *Science* **294,** 591–595.

Hansen, M., Chandra, A., Mitic, L. L., Onken, B., Driscoll, M., and Kenyon, C. (2008). A role for autophagy in the extension of lifespan by dietary restriction in *C. elegans*. *PLoS Genet.* **4,** e24.

Hars, E. S., Qi, H., Ryazanov, A. G., Jin, S., Cai, L., Hu, C., and Liu, L. F. (2007). Autophagy regulates aging in *C. elegans*. *Autophagy* **3,** 93–95.

Hobert, O. (2002). PCR fusion-based approach to create reporter gene constructs for expression analysis in transgenic *C. elegans*. *Biotechniques* **32,** 728–730.

Jia, K., and Levine, B. (2007). Autophagy is required for dietary restriction-mediated life span extension in *C. elegans*. *Autophagy* **3,** 597–599.

Kamath, R. S., Martinez-Campos, M., Zipperlen, P., Fraser, A. G., and Ahringer, J. (2000). Effectiveness of specific RNA-mediated interference through ingested double-stranded RNA in *Caenorhabditis elegans*. *Genom Biol.* **2,** 0002.

Kang, C., You, Y., and Avery, L. (2007). Dual roles of autophagy in the survival of *Caenorhabditis elegans* during starvation. *Genes Dev.* **21,** 2161–2171.

Kenyon, C. (1997). Environmental factors and gene activities that influence life span. *In* "*C. elegans II*," (D. L. Riddle, T. Blumenthal, B. J. Meyer, and J. R. Priess, eds.), pp. 791–813. Cold Spring Harbor Laboratory, Cold Spring Harbor, New York.

Kenyon, C., Chang, J., Gensch, E., Rudner, A., and Tabtiang, R. A. (1993). *C. elegans* mutant that lives twice as long as wild type. *Nature* **366,** 461–464.

Klionsky, D. J. (2005). The molecular machinery of autophagy: Unanswered questions. *J. Cell Sci.* **118,** 7–18.

Klionsky, D. J., Abeliovich, H., Agostinis, P., Agrawal, D. K., Aliev, G., Askew, D. S., Baba, M., Baehrecke, E. H., Bahr, B. A., Ballabio, A., Bamber, B. A., Bassham, D. C., *et al.* (2008). Guidelines for the use and interpretation of assays for monitoring autophagy in higher eukaryotes. *Autophagy* **4,** 151–175.

Kovács, A. L., Vellai, T., and Müller, F. (2004). Autophagy in *Caenorhabditis elegans*. *In* "Autophagy." (D. J. Klionsky, ed.), pp. 217–23. Eurekah.com, Landes Bioscience, Austin, Texas, USA.

Kovács, A. L., Pálfia, Z., Réz, G., Vellai, T., and Kovács, J. (2007). Sequestration revisited. Integrating traditional electron microscopy, *de novo* assembly and new results. *Autophagy* **3,** 655–662.

Krick, R., Tolstrup, J., Appelles, A., Henke, S., and Thumm, M (2006). The relevance of the phosphatidylinositolphosphat-binding motif FRRGT of Atg18 and Atg21 for the Cvt pathway and autophagy. *FEBS Lett.* **580,** 4632–4638.

Kuma, A., Matsui, M., and Mizushima, N. (2007). LC3, an autophagosome marker, can be incorporated into protein aggregates independent of autophagy: Caution in the interpretation of LC3 localization. *Autophagy* **3,** 323–328.

McCulloch, D., and Gems, D. (2003). Body size, insulin/IGF-1 signaling and aging in the nematode *Caenorhabditis elegans*. Exp. Gerontol. **38,** 129–136.

Meléndez, A., Tallóczy, Z., Seaman, M., Eskelinen, E.-L., Hall, D. H., and Levine, B. (2003). Autophagy genes are essential for dauer development and life span extension in *C. elegans*. *Science* **301,** 1387–1391.

Mello, C. C., Kramer, J. M., Stinchcomb, D., and Ambros, V. (1991). Efficient gene transfer in *C. elegans*: Extrachromosomal maintenance and integration of transforming sequences. *EMBO J.* **10,** 3959–3970.

Mizushima, N. (2004). Methods for monitoring autophagy. *Int. J. Biochem. Cell Biol.* **36,** 2491–2502.

Morck, C., and Pilon, P. (2006). *C. elegans* feeding defective mutants have shorter body lengths and increased autophagy. *BMC Dev. Biol.* **6,** 39.

Morita, K., Chow, K. L., and Ueno, N. (1999). Regul ation of body length and male tail ray pattern formation of *Caenorhabditis elegans* by a member of TGF-beta family. *Development* **126,** 1337–1347.

Qu, X., Zou, Z., Sun, Q., Luby-Phelps, K., Cheng, P., Hogan, R. N., Gilpin, C., and Levine, B. (2007). Autophagy gene-dependent clearance of apoptotic cells during embryonic development. *Cell* **128,** 931–946.

Riddle, D. L., and Albert, P. S. (1997). Genetic and environmental regulation of dauer larva development. *In* "*C. elegans II*" (D. L. Riddle, T. Blumenthal, B. J. Meyer, and J. R. Priess, eds.), pp. 739–768. Cold SpringHarbor Laboratory, Cold Spring Harbor, New York.

Sagiv, Y., Legesse-Miller, A., Porat, A., and Elazar, Z. (2000). GATE-16, a membrane transport modulator, interacts with NSF and the Golgi v-SNARE GOS-28. *EMBO J.* **19,** 1494–1504.

Samara, C., Syntichaki, P., and Tavernarakis, N. (2008). Autophagy is required for necrotic cell death in *Caenorhabditis elegans*. *Cell Death Diff.* **15,** 105–112.

Scott, R. C., Schuldiner, O., and Neufeld, T. P. (2004). Role and regulation of starvation-induced autophagy in the *Drosophila* fat body. *Dev. Cell* **7,** 167–178.

Singh, R. N., and Sulston, J. E. (1978). Some observations on molting in *Caenorhabditis elegans*. *Nematologica* **24,** 63–69.

Sulston, J., and Hodgkin, J. (1998). Methods. *In* "*The nematode* Caenorhabditis elegans" (W. B. Wood, ed.), pp. 587–606. Cold Spring Harbor, New York.

Takács-Vellai, K., Vellai, T., Puoti, A., Passannante, M., Wicky, C., Streit, A., Kovács, A. L., and Müller, F. (2005). Inactivation of the autophagy gene *bec-1* triggers apoptotic cell death in *C. elegans*. *Curr. Biol.* **15,** 1513–1517.

Takács-Vellai, K., Bayci, A., and Vellai, T. (2006). Autophagy in neuronal cell loss: A road to death. *BioEssays* **28,** 1126–1131.

Timmons, L., and Fire, F. (1998). Specific interference by ingested dsRNA. *Nature* **395,** 854.

Tóth, M. L., Simon, P., Kovács, A. L., and Vellai, T. (2007). Influence of autophagy genes on ion-channel dependent neuronal degeneration in *Caenorhabditis elegans*. *J. Cell Sci.* **120,** 1134–1141.

Tóth, M. L., Sigmond, T., Borsos, É., Barna, J., Erdélyi, P., Takács-Vellai, K., Orosz, L., Kovács, A. L., Csikós, G., Sass, M., and Vellai, T. (2008). Longevity pathways converge on autophagy genes to regulate life span in *Caenorhabditis elegans*. *Autophagy* **4,** 330–338.

Vellai, T., Bicsák, B., Tóth, M. L., Takács-Vellai, K., and Kovács, A. L. (2008). Regulation of cell growth by autophagy. *Autophagy* **4,** 507–509.

Vellai, T., Takács-Vellai, K., Zhang, Y., Kovács, A. L., Orosz, L., and Müller, F. (2003). Influence of TOR kinase on lifespan in *C. elegans*. *Nature* **426,** 620.

Vellai, T., Tóth, M. L., and Kovács, A. L. (2007). Janus-faced autophagy. A dual role of cellular self-eating in neurodegeneration? *Autophagy* **3,** 461–463.

CHAPTER THIRTY-ONE

Chimeric Fluorescent Fusion Proteins to Monitor Autophagy in Plants

Ken Matsuoka*

Contents

* Faculty of Agriculture, Kyushu University, Fukuoka, Japan

Methods in Enzymology, Volume 451
ISSN 0076-6879, DOI: 10.1016/S0076-6879(08)03231-X

Abstract

Autophagy is induced under nutrient-deficient conditions in both growing tobacco BY-2 cultured cells as well as *Arabidopsis* and others intact plants. The fluorescent protein-tagged structural protein for autophagosomes, the Atg8 protein, allows nondestructive detection of autophagy induction in plant cells and tissues by fluorescence microscopy. Using this technique, the general operation of autophagy in growing root cells has been observed. A synthetic cargo protein for autophagy consisting of cytochrome b5 and the red fluorescence protein, DsRed, allows for the quantitative assay of autophagy in tobacco cells. This chapter describes methods for detecting autophagy in these plant cells using fluorescent protein fusions *in situ* with light microscopy, as well as quantification of autophagy.

1. Introduction

Real-time detection of autophagic events is important for the molecular dissection of the process of autophagy. Fluorescent proteins, such as green fluorescent protein (GFP), red fluorescent protein (RFP), and derivatives of these proteins are now widely used to detect intracellular transport events. Yet special caution is necessary to use such fluorescent proteins in differentiated plant cells, including photosynthetic cells in green leaves and cells accumulating secondary metabolites; these cells have compounds that are highly fluorescent and tend to interfere with the detection of fluorescent proteins.

The detection of autophagy induction can be done by monitoring the presence of autophagosomes in cells. The yeast Atg8 protein and its orthologs in higher organisms are widely used to detect autophagosomes, after tagging this protein with GFP and its derivatives (see, e.g., Mizushima and Kuma, 2008). Using this technique in combination with other methods, crucial roles for plant autophagy have been uncovered. These include degradation of oxidized proteins during oxidative stress (Slávikov *et al.*, 2005; Xiong *et al.*, 2007), disposal of protein aggregates (Toyooka *et al.*, 2006), and possibly removal of damaged proteins and organelles during normal growth conditions as a housekeeping function (Slávikov *et al.*, 2005; Yano *et al.*, 2007).

Quantification of autophagic degradation is also important for the elucidation of the autophagic process and its regulation. In yeast cells, activation of an alkaline phosphatase precursor, Pho8Δ60, dependent on the activity of vacuolar proteases was used to quantify autophagy and to isolate mutants in this pathway (Noda *et al*, 1995; see the chapter by Noda and Klionsky in this volume). In plants, we have developed a system to use fluorescent artificial cargo proteins to quantify autophagy in plant cells (Toyooka *et al.*, 2006). In this chapter, I summarize the recent use of fluorescent proteins for the analysis of autophagy in plant cells and provide protocols and notes on these methods.

2. Fluorescent Proteins and Autofluorescence in Plant Cells and Organelles

Developed plant cells tend to have a large quantity of fluorescence compounds, such as chlorophyll and lignin, which tend to interfere with the detection of fluorescent proteins. Chlorophyll is an essential pigment for photosynthesis that produces the green color in leaves. Although this compound is only found in chloroplasts, the volume of this organelle in photosynthetic cells is quite large, and accordingly this affects the analysis of fluorescent proteins in plant photosynthetic cells. In addition, chloroplasts in living cells change intracellular location and angles when cells are illuminated with visible light (Wada M *et al.*, 2003). For example, when photosynthetic cells are illuminated with white light when adjusting the focus of the microscope, the chloroplasts tend to alter their orientation so that they face the plane of the light source to absorb the maximum intensity of light. Such behavior of chloroplast as well as the nature of chlorophyll, which emits red fluorescence with broad excitation wavelengths, tends to prevent the detection of intracellular structures when using a fluorescence microscope. The bottom wavelength of the excitation spectrum is approximately 480 nm (Fig. 31.1). This wavelength is similar to the excitation peak wavelength of GFP and yellow fluorescent protein (YFP). Thus, these proteins can be used without significant loss of detection sensitivity (see Fig. 31.1). Likewise, the red fluorescent protein DsRed has three-excitation peak wavelengths and one of the shortest is approximately 480 nm (Campbell *et al.*, 2002). Thus, this protein can also be used with appropriate filter sets without interfering with the fluorescence of chlorophylls.

The more problematic cells and tissues are lignified tissues and cells accumulating phenolic compounds. Lignin and related compounds have a high intensity of fluorescence with a broad excitation and emission spectrum (Willemse, 1989). Induction of a stress response induces formation of lignin and related compounds in plants. For example, application of the stress-related phytohormone methyl-jasmonate to tobacco BY-2 cells, which usually display very little fluorescence, induces the synthesis of phenolic compounds and the accumulation of a large quantity of fluorescent

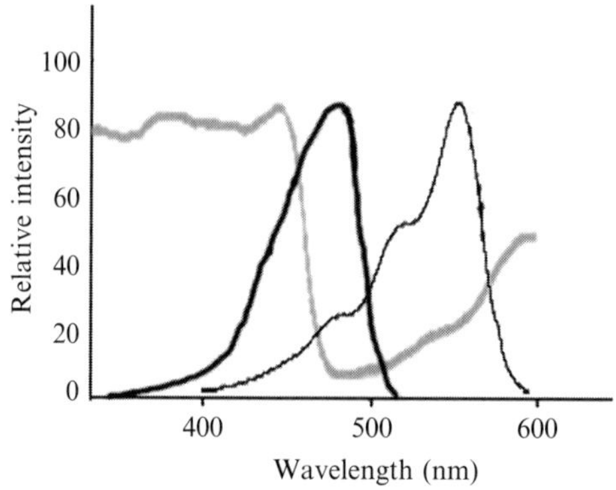

Figure 31.1 Relative intensity of excitation spectra of rice leaf at 700-nm emisson (gray line), EGFP (thick line) and DsRed (thin line).

pigments in vacuoles (Galis *et al.*, 2006; Matsuoka and Galis, 2006). In this case, fluorescent proteins cannot be used to monitor cellular events because of the intense fluorescence of the central vacuoles, which comprise more than 90% (w/v) of the cell volume. Therefore, caution with regard to autofluorescence is needed to monitor autophagy in plant cells, as vacuoles are the lytic organelle for autophagic degradation in plants (Bassham 2007).

3. Visual Detection of Autophagosomes and Autophagic Bodies Using Fluorescent Protein-Tagged Atg8

Plants generally have multiple Atg8 proteins. Arabidopsis have nine Atg8 genes that are expressed differentially in different organs (Slavikova *et al.*, 2005; Yoshimoto *et al.*, 2002), and tobacco BY-2 cells express at least five distinct Atg8 orthologs (Toyooka and Matsuoka, 2006). Cellular functions and biochemical evens of these proteins as well as many of the homologs of other Atg proteins in plants during autophagy progression are similar to those observed in yeast and mammals (Doelling *et al.*, 2002; Fujioka *et al.*, 2008; Ketelaar *et al.*, 2004; Phillips *et al.*, 2008; Su *et al.*, 2006; Thompson *et al.*, 2005; Xiong *et al.*, 2005). Thus, these proteins function as structural proteins for the autophagosome generation and can be used to detect autophagosomes and autophagic transport to the vacuoles in tobacco BY-2 cells, Arabidopsis protoplasts or Arabidopsis root cells after tagging GFP or YFP at their N terminus (Contento *et al.*, 2005; Toyooka *et al.*, 2006; Toyooka and Matsuoka, 2006; Yano *et al.*, 2007; Yoshimoto *et al.*, 2004). Essentially no difference in autophagosome detection was observed using different types of fluorescent protein tags in both Arabidopsis root cells and tobacco BY-2 cells under starvation conditions (Toyooka and Matsuoka, 2006;

Yoshimoto *et al.*, 2004). Thus, chimeric proteins consisting of one of the Atg8 family members and GFP or YFP expressed under the control of a constitutive promoter can be used to monitor autophagy by detecting the fluorescence of such proteins. The following are the protocols for monitoring autophagy induction in transformed tobacco cells that are easy to use for the detection of autophagosomes within a day of nutrient starvation.

3.1. General note for the visual detection of autophagosomes in tobacco BY-2 cells expressing YFP-NtAtg8a under nutrient starvation conditions

Stably transformed tobacco BY-2 cells are subcultured weekly in 95 ml of culture medium as described previously (Toyooka *et al.*, 2006). As cells reach the stationary phase of growth within a week due to the limitation of phosphate and nitrogen sources in the medium, autophagy is already induced in cells at such a growth phase (Toyooka *et al.*, 2006). Thus, cells in logarithmic growth phase (i.e., 3 days after subculture) should be used in all experiments.

3.2. Preparation of culture medium and standard culture conditions

1. Dissolve 30 g of sucrose and the premixed powder Murashige & Skoog salts (Wako Pure Chemicals, Osaka, Japan) in approximately 900 ml of water.
2. Add 10 ml of 20 g/l KH_2PO_4 (pH not adjusted).
3. Add 1 ml of 1000x vitamins (1000x = 100 g/l myo-inositol, 1 g/l thiamin-HCl).
4. Add 20 μl of 2,4-D stock (10 mg/ml 2,4-dichlorophenoxyacetic acid in ethanol).
5. Adjust the pH of the solution to 5.8 with 1 *M* KOH, and bring the volume up to 1000 ml.
6. Pour 95 ml into a 300-ml conical flask, and cover the opening with two layers of aluminum foil.
7. Autoclave at 121 °C for 15 min, cool down to room temperature, and store in the dark.
8. Once a week, transfer 1.5 ml of culture at the stationary phase to 95 ml of fresh medium, and culture at 26.5 °C ±1 °C with rotation of 130 rpm. In some case transformed cells grow slower than nontransformed cells. In this case, a larger volume of culture (up to 5 ml) is transferred to fresh medium.

3.3. Preparation of sugar-starved medium

Use 15 g of mannitol instead of 30g of sucrose. The other conditions are the same as for the standard media described previously.

3.4. Preparation of phosphate or nitrogen-starved medium

1. 1000x B, Mo solution: 6 g/l H_3BO_4, 250 mg/l $Na_2MoO_4 \cdot 2H_2O$.
2. 100x Fe, Mn, Zn solution: 1.7 g/l $MnSO_4 \cdot 7H_2O$, 1.05 g/l $ZnSO_4 \cdot 7H_2O$, 5.125 g/l $FeSO_4 \cdot 7H_2O$, 3.75 g/l Na_2EDTA.
3. 1000x trace elements: 830 mg/l KI, 25 mg/l $CoCl_2 \cdot 6H_2O$, 25 mg/l $CuSO_4 \cdot 5H_2O$.
4. 20x phosphate-free major salts: 33 g/l NH_4NO_3, 38 g/l KNO_3, 8.8 g/l $CaCl_2 \cdot 2H_2O$, 7.4 g/l $MgSO_4 \cdot 7H_2O$.
5. 20x nitrogen-free major salts: 3.4 g/l KH_2PO_4, 8.8 g/l $CaCl_2 \cdot 2H_2O$, 7.4 g/l $MgSO_4 \cdot 7H_2O$.
6. Mix approximately 900 ml of water, 50 ml of appropriate major salts (either phosphate- or nitrogen-free), 1 ml of B, Mo solution, 10 ml of Fe, Mn, Zn solution, 1 ml of trace elements, 30 g of sucrose, 1ml of 1000x vitamins, and 20 μl of 2,4-D stock, and adjust to pH 5.8 with KOH. Sterilize as previously.

3.5. Nutrient starvation and detection of autophagosomes in living cells

Transformed BY-2 cells expressing YFP-Atg8 at the logarithmic growth phase are transferred into nutrient-free MS medium to induce autophagy.

1. Place 10–50 ml of a 3-day-old culture into a 50-ml conical tube.
2. Centrifuge at 100 $\times g$ for 5 min to pellet cells.
3. Remove the supernatant fraction and add the original volume of medium lacking one of the nutrients. Suspend the cells with gentle shaking.
4. Pellet the cells again by centrifugation to wash out the original medium.
5. Remove the supernatant fraction, and then add the same volume of medium lacking one of the nutrients. Suspend the cells.
6. Transfer the cell suspension into a conical flask of appropriate volume. We use 100-ml flasks for 10–30 ml of culture, 200-ml flasks for 30–60 ml of culture, and 300-ml flasks for 60–95 ml of culture.
7. Shake flasks using a rotary shaker at 26.5 °C at a rotation speed 130 rpm.
8. After 24 h, an aliquot of the cell suspension is placed on a slide glass, covered with cover glass and monitored for the presence of autophagosomes using an epifluorescence microscope or laser-scanning confocal microscope with GFP or FITC filter sets. Dotted structures of 100–1000 nm in diameter, which are moving slowly in the cells, are the autophagosomes (Fig. 31.2).

Note: The expression level of YFP-NtAtg8a in tobacco BY-2 cells is not very high. Therefore, when observing the cells with an epifluorescence microscope equipped with the usual mercury lamp, a high-sensitivity CCD camera is necessary to detect autophagosomes. We routinely use an

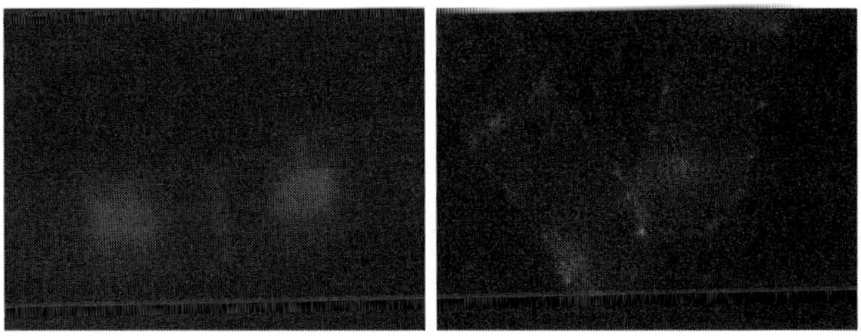

Figure 31.2 Autophagosome formation in transformed tobacco BY-2 cells expressing YFP-NtAtg8a. Logarithmic phase tobacco BY-2 cells expressing the YFP-NtAtg8a construct under the control of the CaMV35S promoter (*left*) and 24 h after induction of autophagy (*right*). Images were collected using an Olympus IX70 inverted fluorescence microscope with NBA filter sets with a Roper CoolSNAP HQ camera. The dotted structures in the left images are the autophagosomes.

Olympus IX70 or IX81 fluorescence microscope equipped with a Roper CoolSNAP HQ camera. Autophagy inhibitors 3-methyladenine or E-64 can be included during the incubation for a negative control. For details, see Takatsuka *et al.* (2004) or the chapter by Moriyasu and Inoue in this volume.

3.6. Colocalization analysis of autophagosomes and other cellular structures

Colocalizaton studies of autophagosomes and other cellular structures are important for the analysis of autophagy targets under various induction conditions. However, as autophagosomes are transient structures during the entire autophagic degradation process, it is difficult to quantify the association of autophagosomes and cellular structures in living cells. Therefore we use acetone-fixed cells retaining YFP fluorescence (or YFP fluorescence and other fluorescent proteins in cells expressing two different fluorescent fusion proteins) to analyze the colocalization.

1. Cool 100% acetone to −20 °C in a freezer.
2. Transfer 2 ml of autophagy-induced BY-2 cell culture into a 15-ml conical tube.
3. Centrifuge at 1000 × *g* for 5 min. Remove culture medium and discard.
4. Suspend cells in 5 ml of PBS (10 m*M* sodium phosphate, 0.138 *M* NaCl, 0.0027 *M* KCl, pH 7.5).
5. Centrifuge as previously and remove as much PBS as possible.
6. Add 5 ml of cold acetone, mix, and keep in a freezer for 10 min. During this incubation most of the cells sediment to the bottom.
7. Remove acetone using a Pasteur pipette.
8. Wash cells by adding 10 ml of PBS, suspend, and collect the cells by centrifugation.

9. Repeat PBS wash twice.
10. Mount cells on slides and analyze the colocalization of the YFP signal and DsRed or other fluorescent protein signals using a fluorescence microscope. Alternatively, fixed cells are further stained with antibodies as described previously to localize organelles (Toyooka *et al.*, 2006).

4. Visual Detection of Autophagic Degradation Using Fluorescent Protein-Tagged Synthetic Cargo

Detection of autophagosomes using fluorescent protein-tagged Atg8 is an easy and reproducible method for the detection of autophagy. Fluorescent autophagosomes tagged with Atg8 fusion proteins are seen as dots. Yet Atg8 and its orthologs tend to also form intracellular dot structures that are not related to autophagy (Kuma *et al.*, 2007). Therefore, an alternative method is needed to confirm that autophagy is actually taking place. One of the systems to detect autophagic degradation is to monitor the relocation of cargo protein from the cytoplasm to the vacuole. As plant vacuoles occupy most of the cell volume, relocation of cytoplasmic protein to the vacuoles can be easily monitored using light microscopy. This strategy was used to detect transport of GFP-Atg8 from the cytoplasm to the vacuole in Arabidopsis root cells (Yoshimoto *et al.*, 2004), as well as to monitor transport of a synthetic autophagic cargo from the cytoplasm to the vacuole in tobacco BY-2 cells (Toyooka *et al.*, 2006). In this section, detailed information for the use of synthetic fluorescent cargo proteins to monitor autophagic degradation is discussed.

4.1. Aggregated fluorescent protein as a reporter protein

Protein aggregates can be easily formed when a protein contains two or more oligomerization domains. Overexpression of such proteins tends to cause aggregates to form in the cell. One such protein is a fusion protein of cytochrome b5 and the original DsRed protein (Toyooka *et al.*, 2006). Cytochrome b5 is an integral endoplasmic reticulum membrane protein, which tend to form into an octamer after solublization (Calabro *et al.*, 1976). DsRed is a tetrameric red fluorescent protein (Baird *et al.*, 2000). Thus, the expression of a fusion construct of these two proteins causes the self-association of a protein with two different affinity sites, namely from both cytochrome b5 and DsRed, and generates protein aggregates that emit strong red fluorescence after excitation with green light in tobacco BY-2 cells (Toyooka *et al.*, 2006). The tetrameric nature of DsRed is essential for the generation of the aggregates, as a fusion protein of cytochrome b5 and a monomeric mutant of DsRed causes the exclusive targeting of this fusion protein to the endoplasmic reticulum (supplemental figure in Toyooka *et al.*, 2006).

4.2. Expression and detection of autophagic degradation in tobacco BY-2 cells

1. Generate a cytochrome b5-DsRed fusion protein construct (Toyooka *et al.*, 2006) under the control of a strong promoter, such as the tandem 35S promoter (Matsuoka and Nakamura, 1991) and introduce into tobacco BY-2 cells.
2. Culture stably transformed cells as described above (*Preparation of culture medium and standard culture conditions*) and observe red fluorescence using a fluorescence microscope with a RFP filter set.
3. At log phase, most of the cells contain punctate protein aggregates with red fluorescence (Fig. 31.3).
4. Induce autophagy as previously using sugar-, phosphate- or nitrogen-free media. Red fluorescence can be seen emitting from most parts of the cells, indicating that some of the RFP fusion protein is targeted to the vacuole and converted to a soluble form (see Fig. 31.3).
5. If required, vacuolar membrane can be stained with a styryl dye, FM 1-43 (Invitrogen, T35356), which gives green fluorescence under blue light excitation (Emans *et al.*, 2002). In this case, an aliquot of stock solution of FM 1-43 in DMSO (20 m*M*) was mixed with the autophagy-induced culture to a final concentration of 20 μM, and further incubated for 2 h at 26.5 °C with shaking.

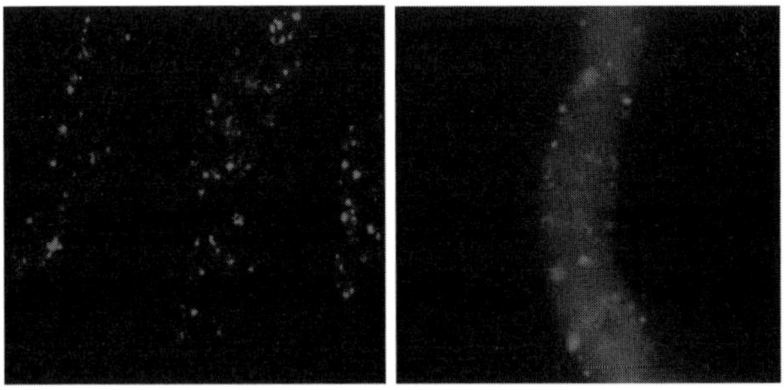

Control Induced

Figure 31.3 Expression and aggregate formation of cytochrome b5-DsRed fusion protein in tobacco BY-2 cells and vacuolar targeting of the cytochrome b5-DsRed fusion protein into vacuoles. Logarithmic phase tobacco BY-2 cells expressing a cytochrome b5-DsRed construct under an enhancer duplicated derivative of the CaMV35S promoter (*left*) and 24 h after the induction of autophagy (*right*). Images were collected using an Olympus IX70 inverted fluorescence microscope with WIG filter sets equipped with an Olympus DP70 digital camera.

6. Likewise, the lumen of the vacuoles can be stained with BCECF to emit green fluorescence (Matsuoka *et al.*, 1997). In this case, an aliquot of stock solution of BCECF-AM in DMSO (Invitrogen, B3051) was mixed with the autophagy-induced culture to a final concentration of 6 m*M*, incubated for 5 min at room temperature and washed the cells in the culture with flesh medium. Fluorescence of BCECF was monitored using fluorescence microscopes with a standard GFP filter setting.

Note: Roots of transformed Arabidopsis plants can be analyzed with the same setting of the microscope.

4.3. Expression and detection of autophagic degradation in leaves of Arabidopsis plants expressing the cytochrome b5-DsRed fusion protein

Green leaves contain a lot of chlorophyll and other pigments for photosynthesis. Thus, choosing the proper wavelengths of both excitation and fluorescence emission is necessary to obtain clear images of RFP fusion proteins.

1. Transform Arabidopsis plants using a dipping method (Zhang *et al.*, 2006) and collect seeds from kanamycin-resistant T1 plants.
2. Germinate kanamycin-resistant seeds on Murashige-Skoog plate as described (Zhang *et al.*, 2006).
3. Observe fluorescence of DsRed in leaves using a Zeiss LSM 510 META or other confocal laser scanning microscope with variable absorption wavelength. Choose 488-nm laser excitation and 580-nm fluorescence emission. Chloroplasts show weak fluorescence at 580 nm under blue light excitation (Terao *et al.*, 1995). Thus, this condition allows the detection of fluorescent aggregates in the presence of weak background fluorescence from chloroplasts in the leaves of transformants, whereas no such structures were seen in control plants without expressing cytochrome b5-DsRed fusion protein (Fig. 31.4).
4. Detach leaves from seedlings and incubate with water for 24 h in the dark to induce autophagy. Under this condition a decrease in the red fluorescent puncta is observed. For control, leaves are incubated with Murashige-Skoog medium containing 1% (w/v) sucrose.

Note: Recording of DsRed fluorescence in roots can also be done with the same condition as for leaves. The emission fingerprinting protocol of the Zeiss LSM 510 META microscope allows more clear separation of chloroplasts and aggregates. However, patterns of chloroplast fluorescence change under growth and starvation conditions. Thus, as a control, the fluorescence patterns of chloroplasts should be determined experimentally using non-transformed Arabidopsis leaves grown or incubated exactly under the same condition as for leaves expressing the cytochrome b5-DsRed fusion protein.

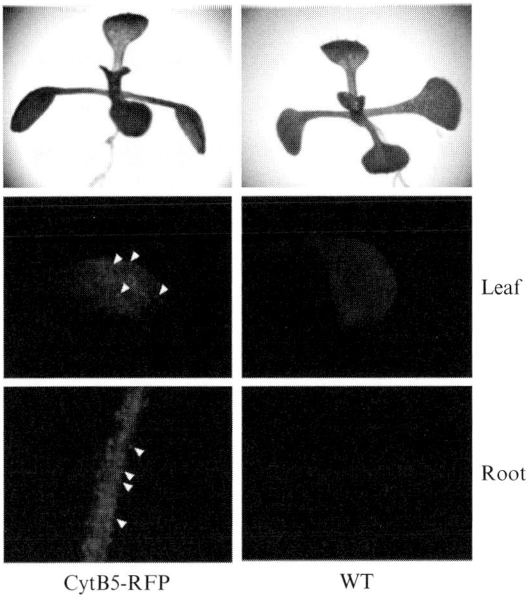

Figure 31.4 Expression and aggregate formation of a cytochrome b5-DsRed fusion protein in Arabidopsis plants. Upper panels show transgenic and control plants. Middle panels show red fluorescence emitted from Arabidopsis leaves after excitation with a 488-nm laser and 580-nm florescence emission using a Zeiss LSM 510 META microscope. Punctate signals of the aggregates of the cytochrome b5-DsRed fusion protein are apparent in the transformants expressing the cytochrome b5-DsRed fusion protein. Lower panels show red fluorescence emitted from Arabidopsis roots. Arrowheads in middle and lower panels indicate some of the aggregates of the cytochrome b5-DsRed fusion protein.

5. Quantification of Fluorescent Fusion Proteins After Separation by Gel Electrophoresis

It is not sufficient to monitor autophagy by fluorescence microscopy; it is also necessary to quantify autophagic activity for the full molecular dissection of autophagy. Here, I describe a method to quantify autophagic degradation using the chimeric fusion proteins described previously.

A basic background on the quantification of autophagy is that, when proteins are transported to a lytic environment, such as the lumen of the vacuole, these proteins are degraded partially or completely by proteases in the organelle. The relatively stable nature of the fluorochromes of DsRed and GFP in the presence of SDS allows us to detect the fusion proteins or derivatives after separation by SDS-polyacrylamide gel electrophoresis (Baird *et al.*, 2000; Shimizu *et al.*, 2005). Thus, vacuolar delivery of cytochrome b5-DsRed and conversion of this protein to a soluble form through proteolytic processing by vacuolar proteases can be detected as the change of relative molecular mass of the RFP in transformed plant cells expressing the fusion construct.

5.1. Protein sample preparation, separation by SDS-PAGE, and detection of the DsRed fusion protein from transformed tobacco BY-2 cells

1. Culture stably transformed tobacco BY-2 cells expressing the fusion protein as previously and induce autophagy.
2. Transfer 1 ml of cells into a 15-ml round-bottomed polypropylene tube (e.g., Falcon #352063).
3. Spin the tube at 1000 × g for 5 min.
4. Collect and measure the volume of the supernatant fraction.
5. Add 1 ml of PBS (pH 7.5 at room temperature) to the cell pellet. Suspend by gentle shaking.
6. Centrifuge as previously, remove supernatant fraction, and discard.
7. Add 1 ml of PBS to cell pellet and suspend cells. Chill on ice for at least 5 min (up to 1 h without any effect on the processing).
8. Sonicate the cell suspension using a probe-type sonicator (UR-20P, TOMY Seiko, Tokyo, Japan) for 30 s, twice with a 1-min interval between sonications.
9. Transfer the disrupted cell suspension into a 1.5-ml microcentrifuge tube and centrifuge at 1000 ×*g* for 5 min at 4 °C.
10. Collect the supernatant fraction. The supernatant fraction may be stored at −20 °C at this point if required.
11. Mix the supernatant fraction (disrupted cells) and an equal volume of 2x SDS-PAGE loading buffer [0.01% (w/v) bromophenol blue, 4% (w/v) SDS, 20% (w/v) glycerol, 100 mM Tris-HCl, pH 8.8].
12. Incubate at room temperature for 10–30 min and apply to a 12.5% (w/v) polyacrylamide minigel of 0.75 mm thickness using the Hoefer SE 260 Mighty Small system or equivalent.
13. Apply 25 mA current per gel until the dye front reaches the bottom of the gel.
14. After the separation of proteins, the gel is separated from the plates and placed into a plastic bag.
15. Put the plastic bag onto the glass stage of a Typhoon 8600 or 9400 image scanner (GE Health Care, London).
16. Scan the gel image using a 532-nm excitation laser at 650V through a 580BP30 emission filter, which allows the detection of DsRed. A typical image of the detection of autophagic degradation of the cytochrome b5-DsRed fusion protein is shown in Fig. 31.5.

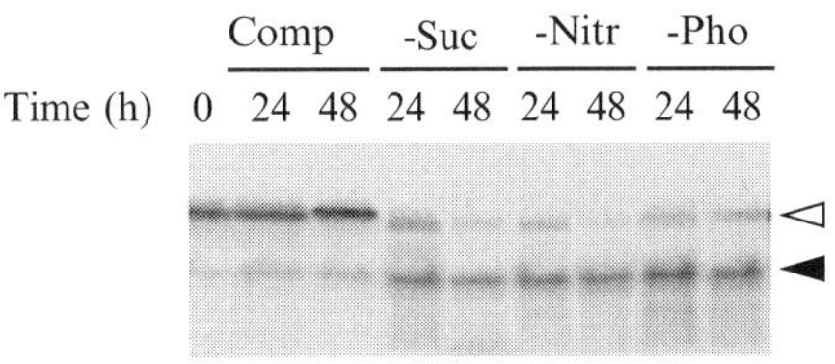

Figure 31.5 Processing of cytochrome b5-DsRed fusion protein under starvation conditions. Rapidly growing transformed tobacco BY-2 cells expressing the cytochrome b5-DsRed fusion protein were incubated for 24 or 48 h in complete culture medium or medium lacking either sucrose, nitrogen containing compounds or phosphate. Proteins prepared from cells were separated by SDS-PAGE, and DsRed-related proteins were detected using a Typhoon 8600 image analyzer. Open arrowhead indicates intact cytochrome b5-DsRed protein. Closed arrowhead indicates processed cytochrome b5-DsRed protein accumulated in the vacuoles.

Note: Prolonged incubation of protein samples with sample buffer or heating samples abolishes the fluorescence. The same quantification procedure can be applied for transformed Arabidopsis leaves and roots. In this case tissues are homogenized with the same buffer using a motor and pestle. After scanning, gels can either be stained with an appropriate dye to detect proteins or can be used for immunoblotting. Fluorescence of GFP or YFP fusion proteins can also be detected using the same scanners with different settings (Shimizu *et al.*, 2005). As GFP or YFP fluorochromes are unstable in acidic conditions, fusions with these proteins and Atg8 tend to lose fluorescence after targeting to vacuoles.

ACKNOWLEDGMENTS

I thank Drs. Kiminori Toyooka in the RIKEN Plant Science Center and Ivan Galis in the Max Planck Institute for Chemical Biology for sharing data presented in this work. This work was supported in part by a grant-in-aid for scientific research from MEXT Japan.

REFERENCES

Baird, G., Zacharias, D., and Tsien, R. (2000). Biochemistry, mutagenesis, and oligomerization of DsRed, a red fluorescent protein from coral. *Proc. Natl. Acad. Sci. USA* **97,** 11984–11989.

Bassham, D. (2007). Plant autophagy: More than a starvation response. *Curr. Opin. Plant Biol.* **10,** 587–593.

Calabro, M., Katz, J., and Holloway, P. (1976). Self-association of cytochrome b5 in aqueous solution. Gel filtration and ultracentirfugational studies. *J. Biol. Chem.* **251,** 2113–2118.

Contento, A., Xiong, Y., and Bassham, D. (2005). Visualization of autophagy in Arabidopsis using the fluorescent dye monodansylcadaverine and a GFP-AtATG8e fusion protein. *Plant J.* **42,** 598–608.

Doelling, J., Walker, J., Friedman, E., Thompson, A., and Vierstra, R. (2002). The APG8/12-activating enzyme APG7 is required for proper nutrient recycling and senescence in Arabidopsis thaliana. *J. Biol. Chem.* **277,** 33105–33114.

Emans, N., Zimmermann, S., and Fischer, R. (2002). Uptake of a fluorescent marker in plant cells is sensitive to brefeldin A and wortmannin. *Plant Cell.* **14,** 71–86.

Fujioka, Y., Noda, N., Fujii, K., Yoshimoto, K., Ohsumi, Y., and Inagaki, F. (2008). *In vitro* reconstitution of plant Atg8 and Atg12 conjugation systems essential for autophagy. *J. Biol. Chem.* **283,** 1921–1928.

Gális, I., Simek, P., Narisawa, T., Sasaki, M., Horiguchi, T., Fukuda, H., and Matsuoka, K. (2006). A novel R2R3 MYB transcription factor NtMYBJS1 is a methyl jasmonate-dependent regulator of phenylpropanoid-conjugate biosynthesis in tobacco. *Plant J.* **46,** 573–592.

Ketelaar, T., Voss, C., Dimmock, S., Thumm, M., and Hussey, P. (2004). Arabidopsis homologues of the autophagy protein Atg8 are a novel family of microtubule binding proteins. *FEBS Lett.* **567,** 302–306.

Kuma, A., Matsui, M., and Mizushima, N. (2007). LC3, an autophagosome marker, can be incorporated into protein aggregates independent of autophagy: caution in the interpretation of LC3 localization. *Autophagy* **3,** 323–328.

Matsuoka, K., and Galis, I. (2006). EST and microarray analysis of tobacco BY-2 cells. *In* "Tobacco BY-2 cells: From cellular dynamics to omics" (T. Nagata, K. Matsuoka, and D. Inze, eds.) Vol. 58, pp. 293–311. Springer Verlag, Berlin.

Matsuoka, K., Higuchi, T., Maeshima, M., and Nakamura, K. (1997). A vacuolar-type H+-ATPase in a nonvacuolar organelle is required for the sorting of soluble vacuolar protein precursors in tobacco cells. *Plant Cell.* **9,** 533–546.

Matsuoka, K., and Nakamura, K. (1991). Propeptide of a precursor to a plant vacuolar protein required for vacuolar targeting. *Proc. Natl. Acad. Sci. USA* **88,** 834–838.

Mizushima, N., and Kuma, A. (2008). Autophagosomes in GFP-LC3 transgenic mice. *Methods Mol. Biol.* **445,** 119–124.

Noda, T., Matsuura, A., Wada, Y., and Ohsumi, Y. (1995). Novel system for monitoring autophagy in the yeast *Saccharomyces cerevisiae. Biochem. Biophys. Res. Commun.* **210,** 126–132.

Phillips, A., Suttangkakul, A., and Vierstra, R. (2008). The ATG12-conjugating enzyme ATG10 Is essential for autophagic vesicle formation in *Arabidopsis thaliana. Genetics* **178,** 1339–1353.

Shimizu, M., Igasaki, T., Yamada, M., Yuasa, K., Hasegawa, J., Kato, T., Tsukagoshi, H., Nakamura, K., Fukuda, H., and Matsuoka, K. (2005). Experimental determination of proline hydroxylation and hydroxyproline arabinogalactosylation motifs in secretory proteins. *Plant J.* **42,** 877–889.

Sláviková, S., Shy, G., Yao, Y., Glozman, R., Levanony, H., Pietrokovski, S., Elazar, Z., and Galili, G. (2005). The autophagy-associated Atg8 gene family operates both under favourable growth conditions and under starvation stresses in Arabidopsis plants. *J. Exp. Bot.* **56,** 2839–2849.

Su, W., Ma, H., Liu, C., Wu, J., and Yang, J. (2006). Identification and characterization of two rice autophagy associated genes, OsAtg8 and OsAtg4. *Mol. Biol. Rep.* **33,** 273–278.

Takatsuka, C., Inoue, Y., Matsuoka, K., and Moriyasu, Y. (2004). 3-methyladenine inhibits autophagy in tobacco culture cells under sucrose starvation conditions. *Plant Cell Physiol.* **45,** 265–274.

Terao, T., Yamashita, A., and Satake, K. (1985). Chlorophyll b-deficient mutants of rice : I. Absorption and fluorescence spectra and chlorophyll a/b ratios. *Plant Cell Physiol.* **26,** 1361–1367.

Thompson, A., Doelling, J., Suttangkakul, A., and Vierstra, R. (2005). Autophagic nutrient recycling in Arabidopsis directed by the ATG8 and ATG12 conjugation pathways. *Plant Physiol.* **138,** 2097–2110.

Toyooka, K., Moriyasu, Y., Goto, Y., Takeuchi, M., Fukuda, H., and Matsuoka, K. (2006). Protein aggregates are transported to vacuoles by a macroautophagic mechanism in nutrient-starved plant cells. *Autophagy* **2,** 96–106.

Toyooka, K., and Matsuoka, K. (2006). Autophagy and non-classical vacuolar targeting in tobacco BY-2 cells. *In* "Tobacco BY-2 cells: From cellular dynamics to omics" (T. Nagata, K. Matsuoka, and D. Inze, eds.) Vol. 58, pp. 167–180. Springer Verlag, Berlin.

Wada, M., Kagawa, T., and Sato, Y. (2003). Chloroplast movement. *Annu. Rev. Plant Biol.* **54,** 455–468.

Willemse, M. T. M. (1989). Cell wall autofluorescence. *In* "Physico-chemical characterisation of plant residues for industrial and feed use" (E. R. Q. A. Chesson, ed.), pp. 50–57. Elsevier Applied Science, London.

Xiong, Y., Contento, A., and Bassham, D. (2005). AtATG18a is required for the formation of autophagosomes during nutrient stress and senescence in *Arabidopsis thaliana. Plant J.* **42,** 535–546.

Xiong, Y., Contento, A., Nguyen, P., and Bassham, D. (2007). Degradation of oxidized proteins by autophagy during oxidative stress in Arabidopsis. *Plant Physiol.* **143,** 291–299.

Yano, K., Suzuki, T., and Moriyasu, Y. (2007). Constitutive autophagy in plant root cells. *Autophagy* **3,** 360–362.

Yoshimoto, K., Hanaoka, H., Sato, S., Kato, T., Tabata, S., Noda, T., and Ohsumi, Y. (2004). Processing of ATG8s, ubiquitin-like proteins, and their deconjugation by ATG4s are essential for plant autophagy. *Plant Cell.* **16,** 2967–2983.

Zhang, X., Henriques, R., Lin, S., Niu, Q., and Chua, N. (2006). Agrobacterium-mediated transformation of *Arabidopsis thaliana* using the floral dip method. *Nat. Protoc.* **1,** 641–646.

CHAPTER THIRTY-TWO

Use of Protease Inhibitors for Detecting Autophagy in Plants

Yuji Moriyasu* *and* Yuko Inoue†

Contents

* Department of Regulatory Biology, Faculty of Science, Saitama University, Saitama, Japan
† Life Sciences Institute, University of Michigan, Ann Arbor, Michigan, USA

Methods in Enzymology, Volume 451
ISSN 0076-6879, DOI: 10.1016/S0076-6879(08)03232-1

Abstract

In cultured tobacco (BY-2) cells, autophagy seems to be induced under nutrient-starvation conditions, whereas in root cells from *Arabidopsis* and barley, it occurs constitutively though is activated under nutrient starvation conditions. In both cases, protease inhibitors such as E-64, E-64c, antipain, and leupeptin block autophagy at the step of degradation of the cytoplasm enclosed in lysosomes/vacuoles, and cause the accumulation of autolysosomes (lysosomes containing parts of the cytoplasm) and/or of many cytoplasmic inclusions in the central vacuoles. Both types of autophagy are inhibited by 3-methyladenine, which is known as a potent inhibitor of autophagy in mammalian cells. Thus, using protease inhibitors and 3-methyladenine provides us with a method useful for analyzing autophagy in plant cells. This chapter describes protocols for detecting autophagic compartments in BY-2 cells and in the root-tip cells of *Arabidopsis* and barley by microscopy.

1. Introduction

Autophagy is a process in which cells degrade their own components. In macroautophagy, a part of the cytoplasm including organelles is first enclosed by a double-membrane-bounded structure, initially called a phagophore and then an autophagosome, which subsequently fuses with a preexisting lysosome. The resulting structure is a lysosome containing a part of the cytoplasm, which is called an *autolysosome*. In yeast cells, autophagosomes fuse with the vacuole and release their inner-membrane-bounded structures into the vacuole. Thus, the resulting structure is the vacuole containing many membrane-bounded parts of the cytoplasm (called *autophagic bodies*). In microautophagy, lysosomes and/or vacuoles directly incorporate bits of the cytoplasm without making autophagosomes. In both types of autophagy, parts of the cytoplasm taken up into lysosomes/vacuoles are eventually degraded by hydrolytic enzyme therein.

Protein is one of the major cytoplasmic components. Thus, the inhibition of vacuolar and/or lysosomal proteases with inhibitors blocks or slows down the degradation of parts of the cytoplasm and causes the accumulation of cytoplasmic materials in these organelles. In mammalian cells, the protease inhibitor leupeptin inhibits lysosomal cysteine proteases, cathepsins B and L, and accumulate parts of the cytoplasm in lysosomes (Kominami *et al.*, 1983). Similarly, the serine protease inhibitor phenylmethanesulfonyl fluoride inhibits proteinase B in the yeast vacuole, which results in the accumulation of autophagic bodies in the vacuole (Takeshige *et al.*, 1992). Yeast mutant cells lacking vacuolar protease activities exhibit the same phenotype. Thus, by treating cells with appropriate protease inhibitors and monitoring the accumulation of undegraded particles of cytoplasmic origin, we can analyze the presence and location of autophagy in the cells.

We have been investigating autophagy in plant cells using cultured tobacco (BY-2) cells and root tips from *Arabidopsis* and barley. When BY-2 cells at the logarithmic growth phase are transferred to a sucrose-free culture medium and further cultured, cellular protein content decreases (Moriyasu and Ohsumi, 1996; Takatsuka *et al.*, 2004). E-64c (Tamai *et al.*, 1986) and other protease inhibitors such as E-64, antipain, and leupeptin added to the culture medium inhibit protein degradation and concomitantly cause the accumulation of autolysosomes (Moriyasu and Ohsumi, 1996). Such accumulation is not significant when BY-2 cells are cultured in a medium containing sucrose, suggesting that autophagy is induced under nutrient-starvation conditions (Inoue and Moriyasu, 2006). In contrast, basal autophagy occurs constitutively, irrespective of the presence or absence of sucrose in the culture media, in root cells from *Arabidopsis* and barley, although it is activated in a sucrose-free medium (Inoue *et al.*, 2006; Moriyasu *et al.*, 2003; Yano *et al.*, 2007). Furthermore, parts of the cytoplasm accumulate in preexisting central vacuoles as well as possibly in newly formed lysosomes in root-tip cells treated with a protease inhibitor (Inoue *et al.*, 2006; Moriyasu *et al.*, 2003). In addition, 3-methyladenine, a potent inhibitor of autophagy in mammalian cells (Gordon and Seglen, 1982; Seglen and Gordon, 1982), inhibits autophagy in BY-2 cells (Takatsuka *et al.*, 2004) and in *Arabidopsis* root-tip cells (Inoue *et al.*, 2006).

We think that using these protease inhibitors and 3-methyladenine provides us with a model useful for analyzing autophagy in plant cells. This chapter describes protocols for detecting autolysosomes and large vacuoles containing many cytoplasmic inclusions in BY-2 cells and in the root-tip cells of *Arabidopsis* and barley by microscopy.

2. Measurement of Protein Degradation and Intracellular Protease in BY-2 Cells

2.1. Culture of BY-2 cells

BY-2 cells are suspension-cultured plant cells, which derive from tobacco (*Nicotiana tabacum*, Bright Yellow 2). They are subcultured in a Murashige and Skoog culture medium (MS medium, which consists of Murashige and Skoog salts mixture, 2 mg/l glycine, 100 mg/l myo-inositol, 0.5 mg/l nicotinic acid, 0.5 mg/l pyridoxine-HCl, and 0.1 mg/l thiamine-HCl; Murashige and Skoog, 1962) containing 30 g/l sucrose and 0.2 mg/l 2,4-dichlorophenoxyacetic acid. There are several ways of preparing stock solutions to make up the MS medium from individual chemicals. Here, we introduce a relatively easy way to make it, which uses commercially available, premixed powder for MS medium.

1. Dissolve 30*g* of sucrose and the premixed powder Murashige & Skoog Medium including Vitamins (M0222.0001, Duchefa Biochemie, Haarlem, The Netherlands) in approximately 800 ml of water.
2. Add 1 ml of 0.2 mg/ml 2,4-dichlorophenoxyacetic acid.
3. Adjust the pH of the solution to 5.8 with 1 *M* KOH, and increase volume to 1000 ml.
4. Pour 80 ml into a 300-ml Erlenmeyer flask, and cover it with aluminum foil.
5. Autoclave at 120 °C for 15 min.
6. Once a week, transfer 1.5–2.0 ml of culture at the stationary phase to 80 ml of fresh medium, and culture at 26 °C ± 1 °C with a rotation of 110 rpm.
7. Use a culture 4 days after transfer (4-day-old cells), which is at the logarithmic phase, for experiments.

2.2. Sucrose starvation

BY-2 cells at the logarithmic growth phase are transferred into a sucrose-free culture medium.

1. Transfer a culture (4-day-old cells, 1–40 ml depending on experiments) to a 15-ml or 50-ml centrifuge tube.
2. Centrifuge at 100×*g* for 5 min to pellet the cells.
3. Remove the supernatant fraction and add the original volume of the culture medium lacking sucrose. Resuspend the cells with gentle shaking.
4. Pellet the cells again by centrifugation (wash step).
5. Remove the supernatant fraction, and add the same volume of the culture medium lacking sucrose. Resuspend the cells.
6. Transfer the cell suspension to a Petri dish (We use a dish of 35 mm in diameter for 1–2 ml culture, and of 60 mm in diameter for 3–4 ml culture). Add one-hundredth of the culture volume of 1 m*M* E-64c (Peptide Institute, Minoh-shi, Osaka, Japan) in methanol. As a solvent control, add the same volume of methanol alone. Thus, the final concentration of methanol in the culture medium is 1%.
7. Culture at 26 °C ±1 °C with a rotation of 110 rpm.

Note: The culture medium lacking sucrose is the MS medium containing 0.2 mg/l 2,4-dichlorophenoxyacetic acid but not sucrose. The autophagy inhibitor 3-methyladenine (M9281, Sigma) is dissolved in water by keeping it at 100 °C for a few minutes to make a 0.1 *M* stock solution. The solution is stored at −20 °C. Dissolve completely by boiling again immediately before use, and add one-twentieth to one-tenth of the culture volume. Thus, the final concentration of 3-methyladenine in the culture medium is 5–10 m*M*. As a solvent control, add the same volume of water to one culture.

2.3. Measurement of cellular total protein in BY-2 cells

1. Filter 1 ml of suspension culture through a glass filter (GF/A, 24 mm in diameter, Whatman) using vacuum filtration. Wash the cells on the filter with 10 ml of water.
2. Transfer the cells with the filter to a Petri dish (60 mm in diameter). Add 3 ml of 0.2 *M* NaOH, which releases the cells from the filter.
3. Transfer the cell suspension in 0.2 *M* NaOH to a plastic tube (16.2 × 103 mm) and homogenize with a homogenizer (Polytron with a generator shaft of 12 mm in diameter; Kinematica, Littau, Switzerland) for 15 s at maximal speed.
4. Centrifuge the homogenate at 2200×*g* for 5 min. The resulting supernatant fraction is used for protein assay according to the protocol in Bensadoun and Weinstein (1976).

2.4. Measurement of protease activity in BY-2 cells using FITC-casein

1. Prepare FITC-labeled casein according to the following protocol, which is slightly modified from the original protocol (Twining 1984).
   a. Dissolve 1*g* of casein (218682, Calbiochem) in 100 ml of 50 m*M* carbonate buffer, pH 9.5, containing 150 m*M* NaCl. Check the pH and readjust if necessary.
   b. Add 40 mg of FITC (fluorescein isothiocyanate isomer I, F7250, Sigma) and stir gently for 1 h at room temperature.
   c. Dialyze (Spectrapor membrane tubing 4, MWCO: 12-14,000) twice against 3 L of water containing 1 g/l activated charcoal at 4 °C (take 1 day for each dialysis).
   d. Dialyze the protein solution against 3 L of 50 m*M* Tris-HCl, pH 7.5, at 4 °C for 1 day.
   e. Dialyze the protein solution against 3 L of 5 m*M* Tris-HCl, pH 7.5, at 4 °C for 1 day.
   f. Adjust the protein concentration to 0.5% (w/v) with 5 m*M* Tris-HCl, pH 7.5.
   g. Freeze in 5-ml aliquots and store at −20 °C. The substrate is stable for several years.

2. Prepare crude enzyme solution from BY-2 cells following the protocol described next.
   a. Collect cells on a glass filter (GF/A, 47 mm in diameter, Whatman) by vacuum filtration of a 3-ml suspension culture in a Petri dish. Wash the Petri dish and the cells on the filter with approximately 30 ml of water.
   b. Scrape the cells off the glass filter with a spatula, and transfer to a mortar on ice.
   c. Homogenize the cells with 0.5 ml of 0.1 *M* acetate-Na buffer, pH 5.0, containing 28 m*M* 2-mercaptoethanol and 0.1*g* of sea sands (107712, Merck).
   d. Transfer all the homogenate to a 1.5 ml-microcentrifuge tube, and centrifuge at 15,000x*g* for 10 min. Transfer the resulting supernatant fraction to a new microcentrifuge tube, and use as a crude enzyme solution.

Note: Cells can be collected by centrifugation instead of by filtration, and they can be homogenized using a Teflon-homogenizer instead of using a mortar and pestle. The following is an alternate protocol we use for preparing a crude enzyme solution from BY-2 cells.

   a. Transfer the entire culture (3 ml) from a Petri dish to a 15-ml centrifuge tube.
   b. Wash the Petri dish with 3 ml of culture medium twice, and add the wash solutions to the centrifuge tube. Thus, the cell suspension in the centrifuge tube is approximately 9 ml.
   c. Centrifuge at 700 rpm for 5 min to pellet the cells.
   d. Remove the supernatant fraction using a vacuum aspirator.
   e. Add 6 ml of culture medium and resuspend the cells.
   f. Centrifuge again at 700 rpm for 5 min.
   g. Remove the supernatant fraction and discard.
   h. Add 1 ml of homogenization solution (0.1 *M* acetate-Na buffer, pH 5.0, 28 m*M* 2-mercaptoethanol), and resuspend the cells.
   i. Transfer the suspension to the vessel of a homogenizer standing on ice.
   j. Wash the centrifuge tube with another 1 ml of homogenization solution, and add the resulting solution to the vessel of the homogenizer.
   k. Homogenize the cells with a motor-driven Teflon pestle.
   l. Transfer the homogenate to microcentrifuge tubes, and centrifuge at 15,000 rpm for 10 min at 4 °C.
   m. Transfer the supernatant fraction to new microcentrifuge tubes on ice, and use as a crude enzyme solution.

3. Measure protease activity in the crude enzyme solution as follows:

   a. Mix 60 μl of crude enzyme solution and 40 μl of 0.5% (w/v) FITC-casein in a microcentrifuge tube on ice.

b. Start to incubate the microcentrifuge tube at 37 °C.
c. Add 100 μl of 10% TCA to stop the reaction at t = 0, 30 and 60 min.
d. Stand the tubes on ice for approximately 30 min.
e. Centrifuge the tubes at 15,000 rpm for 5 min at 4 °C.
f. Transfer 150–170 μl of the supernatant fraction into a new microcentrifuge tube.
g. Centrifuge again at 15,000 rpm for 10 min.
h. Place 2 ml of 0.5 *M* Tris-HCl, pH 7.5, in a test tube (13 × 100 mm).
i. Add 100 μl of the supernatant fraction into the test tube and mix.
j. Measure the fluorescence with excitation wavelength at 490 nm and emission wavelength at 525 nm.

Note: For t = 0, it is easiest to add 100 μl of 10% TCA into the mixture on ice, before starting the incubation at 37 °C. The second centrifugation at step g is very important to completely precipitate small debris of FITC-casein remaining in the supernatant fraction.

3. Detection of the Accumulation of Autolysosomes in BY-2 Cells with Neutral Red and Quinacrine

Autolysosomes that accumulate by E-64c treatment in BY-2 cells can be observed by a light microscope with Nomarski (differential interference contrast) optics (Fig. 32.1). They can also be stained with quinacrine and neutral red (Fig. 32.2). Since these acidotropic reagents are basic by nature and can penetrate cells as their noncharged form, cells should be treated with these reagents in solutions with alkaline pH. Here we introduce our methods of vital staining of autolysosomes with neutral red and quinacrine in BY-2 cells.

3.1. Staining with neutral red

1. Place 100 μl of cell suspension (cells cultured in the sucrose-free culture medium in the presence or absence of E-64c for 1 day) in a microcentrifuge tube, and add 400 μl of 5 m*M* HEPES-Na, pH 7.5, containing 100 m*M* sorbitol.
2. Centrifuge at 500 rpm for 5 min to precipitate the cells.
3. Discard the supernatant fraction and add 400 μl of 5 m*M* HEPES-Na, pH 7.5, containing 100 m*M* sorbitol. Resuspend the cells.
4. Centrifuge again to pellet and wash the cells.
5. Discard the supernatant fraction and add 400 μl of 5 m*M* HEPES-Na, pH 7.5, containing 100 m*M* sorbitol. Resuspend the cells.

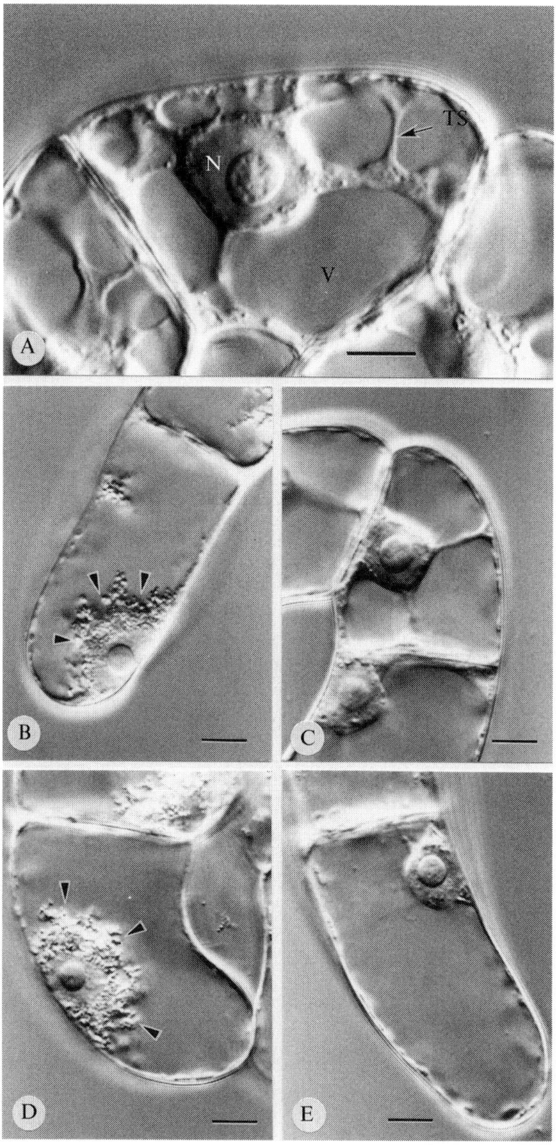

Figure 32.1 Morphological changes in BY-2 cells during sucrose starvation. BY-2 cells were treated with 10 μM E-64c for 0 day (A), 1 day (B), and 2 days (D) of sucrose starvation or with 1% methanol for 1 day (C) and 2 days (E) of sucrose starvation. The cells were observed on a light microscope with Nomarski optics. N, nucleus; TS, transvacuolar strand; V, the central vacuole; arrowheads, the accumulation of autolysosomes. Bar, 10 μm. From Moriyasu and Ohsumi (1996) with permission.

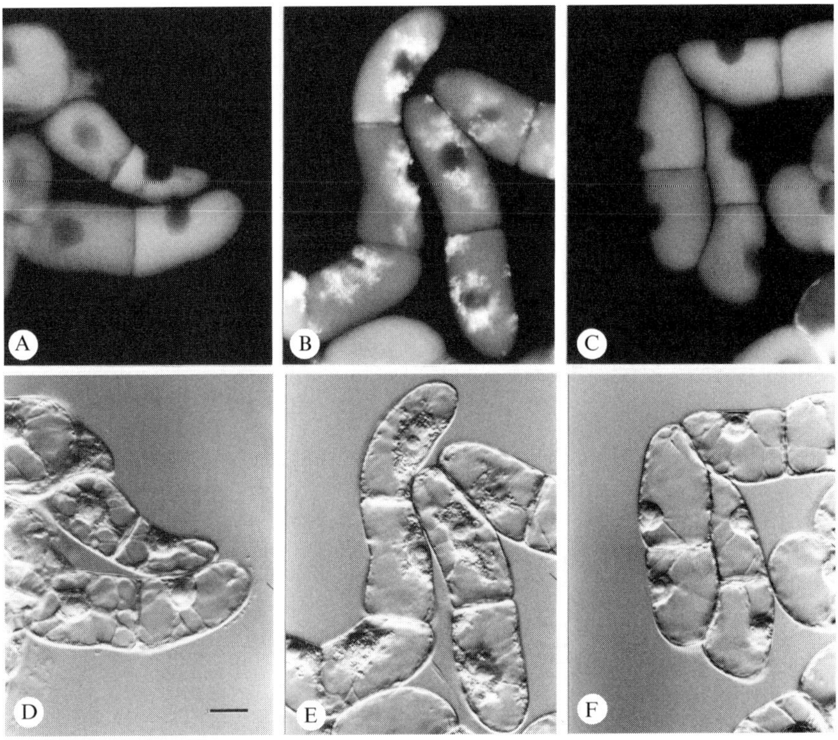

Figure 32.2 Staining of autolysosomes in BY-2 cells with quinacrine. The cells before sucrose starvation (A) and the cells treated with 10 μM E-64c (B) or with 1% methanol (C) for 1 day of sucrose starvation were stained with quinacrine. These cells were observed with a fluorescence microscope. Corresponding Nomarski images (D, E, and F) are also shown below. Bar, 20 μm. From Moriyasu and Ohsumi (1996) with permission.

6. Add 4 μl of 0.1% neutral red (140-00932, Wako Pure Chemical Industries, Osaka, Japan), and mix. Keep the cell suspension for a few minutes at room temperature.
7. Observe the cells by bright-field light microscopy.

Note: MS culture medium contains 30 g/l (i.e., 87 m*M*) sucrose and has its osmotic value near 100 m*M*. To reduce osmotic stress on cells, 100 m*M* sorbitol is used in all of the washes and suspensions.

To stain the cells more quickly and densely, a higher concentration of neutral red can be used as follows:

1. Place 100 μl of cell suspension in a microcentrifuge tube.
2. Pellet the cells by centrifugation at 500 rpm for 5 min.

3. Remove and discard the supernatant fraction.
4. Add 100 μl of 5 m*M* HEPES-Na, pH 7.5, containing 100 m*M* sorbitol, and resuspend the cells.
5. Pellet the cells again by centrifugation, and discard the supernatant fraction.
6. Add 100 μl of 5 m*M* HEPES-Na, pH 7.5, containing 100 m*M* sorbitol, and resuspend the cells.
7. Add 10 μl of 0.1% (w/v) neutral red, and mix.
8. Keep for 2–5 min at room temperature, and observe the cells by bright-field light microscopy.

Note: A long exposure to 5 m*M* HEPES buffer, pH 7.5, seems to be slightly harmful to BY-2 cells. 5 m*M* Mes-Tris, pH 6.5, appears to be better but is less effective for staining the cells.

3.2. Staining with quinacrine

1. Place 100 μl of cell suspension in a microcentrifuge tube.
2. Add 400 μl of 5 m*M* HEPES-Na, pH 7.5, containing 100 m*M* sorbitol.
3. Centrifuge to pellet the cells.
4. Discard the supernatant fraction, and add 400 μl of 5 m*M* HEPES-Na, pH 7.5, containing 100 m*M* sorbitol
5. Centrifuge again.
6. Discard the supernatant fraction, and add 200 μl of 5 m*M* HEPES-Na, pH 7.5, containing 100 m*M* sorbitol.
7. Add 4 μl of 2 m*M* quinacrine (Q3251, Sigma), and mix.
8. Keep the cell for approximately 5 min at room temperature.
9. Add 400 μl of 5 m*M* HEPES-Na, pH 7.5, containing 100 m*M* sorbitol.
10. Centrifuge to pellet the cells.
11. Discard the supernatant fraction, and add 400 μl of 5 m*M* HEPES-Na, pH 7.5, containing 100 m*M* sorbitol.
12. Centrifuge.
13. Discard the supernatant fraction, and add 200 μl of 5 m*M* HEPES-Na, pH 7.5, containing 100 m*M* sorbitol.
14. Observe cells by fluorescence light microscopy.

Note: Quinacrine should be washed out after staining. Also, as noted in the section "Staining with Neutral Red," Mes-Tris, pH 6.5 may be less harmful for the cells than HEPES-Na, pH 7.5, but the staining intensity is reduced.

4. Staining of Autolysosomes in BY-2 Cells by the Use of Endocytosis Markers

In mammalian cells, the autophagic and endocytic pathways converge at the endosomes (Liou *et al.*, 1997). A similar process occurs in plant cells (Herman and Lamb, 1992; Record and Griffing, 1988). Thus, autolysosomes in BY-2 cells are located on the endocytic pathway, which can be traced using a fluorescent marker of endocytosis, FM 4–64 (Fig. 32.3).

4.1. Staining with FM 4-64

1. Transfer 1 ml of 4-day-old BY-2 cells to a sucrose-free culture medium, and culture for 1 day in the presence or absence of 10 μM of E-64c.
2. Transfer 200 μl of each culture to a pre-cooled culture tube (12 × 75 mm, culture tube with cap) on ice.
3. Add 2 μl of 10 mM FM 4-64 (T13320, Molecular Probes) in DMSO. Rotate the culture tubes for 30–60 min, keeping the culture at 0 °C–4 °C.
4. Add 1 ml of precooled sucrose-free MS medium, and centrifuge at 700 rpm for 5 min, keeping the culture at 0 °C–4 °C.
5. Remove as much of the supernatant fraction as possible, keeping the culture at 0 °C.

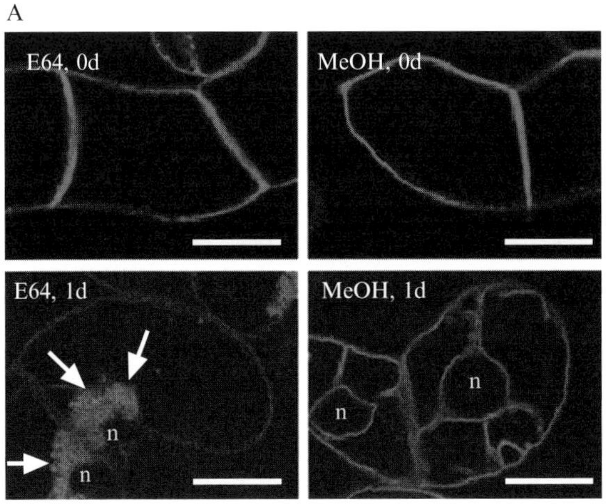

Figure 32.3 (continued)

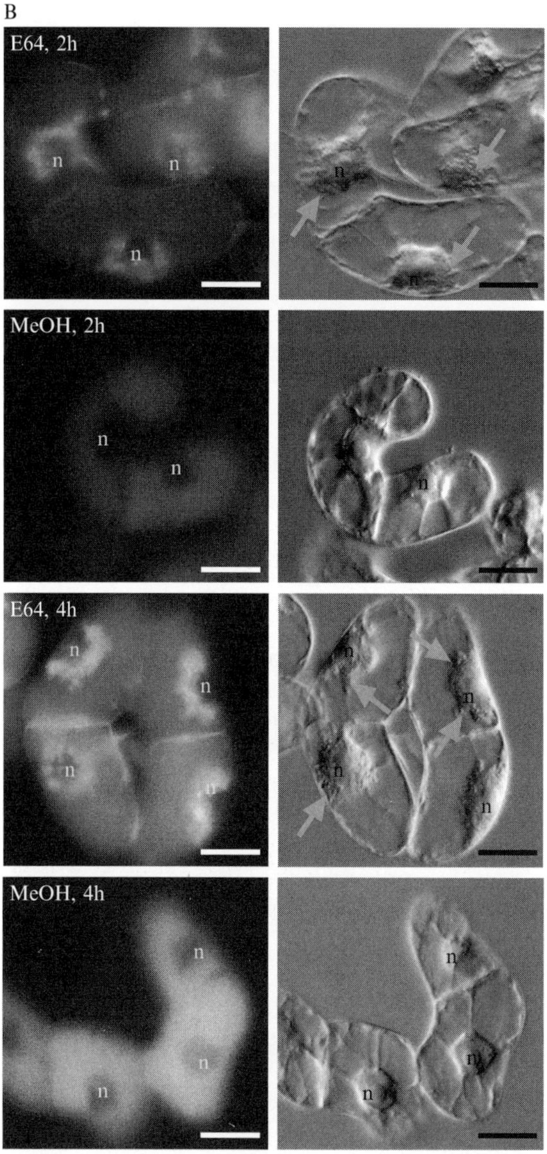

Figure 32.3 Localization of autolysosomes in BY-2 cells on the endocytic pathway by the use of the fluorescent dyes FM 4-64 and Lucifer Yellow CH. A. BY-2 cells were cultured under sucrose starvation conditions for 1 day in the presence of 10 μM E-64c (E64, 0 d and E64, 1 d) or in the presence of 1% (v/v) methanol as a solvent control (MeOH, 0 d and MeOH, 1 d). The plasma membranes of these cells were pulse-labeled with 100 μM FM 4-64. The cells were immediately observed (E64, 0 d and MeOH, 0 d), or cultured for another 1 day and then observed (E64, 1 d and MeOH, 1 d) using a confocal laser microscope to obtain the image of FM 4-64 fluorescence (red). Arrows indicate the

6. Add 200 μl of precooled sucrose-free MS medium, and resuspend the cells.
7. Place 20 μl of culture onto a glass slide for observation by confocal microscopy (t = 0).
8. Warm the rest of the culture to 26 °C.
9. Place 20 μl of culture onto a glass slide for observation after appropriate time intervals. (At t = 30 min, numerous dotted structures [putative endosomes] appear in the peripheral cytoplasm; at t = 24 h, FM 4-64 stains the membrane of the central vacuole and/or autolysosomes).

Note: At 0 °C, endocytosis does not occur (Vida and Emr 1995), and thus only the plasma membrane is stained with FM 4-64.

4.2. Staining with Lucifer Yellow CH

The presence of autolysosomes on the pathway of endocytosis can also be demonstrated using another fluorescent marker, Lucifer Yellow CH (Fig. 32.3). It should be noted that Lucifer Yellow CH is incorporated into the vacuoles in some plant cells not by endocytosis but through anion transporters. In our experience, however, pretreatment of BY-2 cells with 10 μM wortmannin for 2 h prevents the uptake of Lucifer Yellow into the central vacuoles. This confirms that most of the Lucifer Yellow is taken up into the central vacuoles by endocytosis in BY-2 cells under the present experimental conditions.

1. Incubate the same types of cells as used in the FM 4-64 experiment in a sucrose-free culture medium containing 2 mg/ml Lucifer Yellow (L0259, Sigma) in the presence or absence of E-64c for 1 day.
2. Wash the cells through centrifugation and resuspension with the sucrose-free MS medium, and observe by fluorescence microscopy.

Note: As shown in Fig. 32.3, a significant uptake of Lucifer Yellow into autolysosomes and the central vacuole can be observed after 2–4 h of incubation. Alexa Fluor 488 (A10436, Molecular Probes) can be used as an endocytosis marker in the same way as Lucifer Yellow. In this case, prepare a 25 mM stock solution of the dye in water, and add 1 μl to 1 ml of culture.

accumulation of autolysosomes; n denotes the location of the nucleus. Bar represents 20 μm. B. Control and experimental cells, prepared in an identical manner as A, were incubated in a sucrose-free culture medium containing 2 mg/ml Lucifer Yellow CH. At 2 and 4 h, the cells were observed using a conventional epifluorescence microscope fitted with Nomarski optics. For each treatment indicated in the figure, the images of Lucifer Yellow fluorescence are shown on the left; the Nomarski images are on the right. Arrows indicate the accumulation of autolysosomes. n denotes the location of the nucleus. Bar, 20 μm. From Yano *et al.* (2004) with permission.

5. Enzyme Cytochemistry for Acid Phosphatase by Light Microscopy

Acid phosphatase is localized in autolysosomes as well as in the central vacuole. But the enzyme in the central vacuole seems to be free from chemical fixation by aldehyde because the protein concentration of vacuolar sap is low. Thus, in this method, only autolysosomes appear to be stained.

1. Place 1.0 of ml cell suspension (cells cultured in the sucrose-free culture medium in the presence or absence of E-64c for 1 day) in a 2.0-ml microcentrifuge tube with a round bottom, and let the cells sediment spontaneously or by centrifugation.
2. Remove and discard the supernatant fraction.
3. Add 1 ml of fixative (1% (w/v) glutaraldehyde, 1% (w/v) formaldehyde, 86 m*M* NaCl and 0.1 *M* phosphate-Na buffer, pH 7.4) and gently suspend the cells. Store at 4 °C for more than 1 day.
4. The cells sediment spontaneously in the tube. Remove and discard the supernatant fraction, and add 1 ml of 86 m*M* NaCl, 0.1 *M* phosphate-Na buffer, pH 7.4 (wash step). Repeat the wash step more two times.
5. Remove and discard the supernatant fraction, and add 1 ml of 50 m*M* acetate-Na buffer, pH 5.0 (wash step). Repeat this wash step three times in total.
6. Remove and discard the supernatant fraction, and add 1 ml of the staining solution (50 m*M* acetate-Na buffer, pH 5.0, containing 0.1% (w/v) Fast Garnet GBC (F8761, Sigma) and 1 m*M* 1-naphthylphosphate). Keep for 5–10 min at room temperature and observe with a light microscope.

Note: The staining solution should be prepared fresh. Dissolve Fast Garnet GBC in 50 m*M* acetate-Na buffer, pH 5.0, and add one-hundredth of the volume of 0.1 *M* 1-naphthylphosphate in dimethyl sulfoxide to this solution.

6. Neutral Red and LysoTracker Red Staining to Detect Autolysosomes and Cytoplasmic Inclusions in the Central Vacuole in Plant Root-Tip Cells

When the root tips of *Arabidopsis* and barley are incubated with E-64d, an esterified and thus more membrane-permeable form of E-64c, cytoplasmic particles accumulate in large vacuoles, which appear to have preexisted before inhibitor treatment as well as in small vesicles, which appear to be formed *de novo* and to correspond to autolysosomes in BY-2 cells. Since both preexisting vacuoles and newly formed lysosomes are acidic, the acidotropic dyes neutral red and LysoTracker Red are concentrated in the lumen of these organelles and stain cytoplasmic inclusions more densely than the lumen, probably because of adsorption of the dyes.

6.1. Preparation of seedlings from *Arabidopsis* seeds

1. Place the seeds (10 or more) of *Arabidopsis thaliana* (ecotype Columbia or Landsberg) in a 1.5-ml microcentrifuge tube. Add 1 ml of 5% (v/v) commercially available bleach and sterilize the seeds for approximately 5 min.
2. Remove and discard the sterilizing solution after the seeds sediment. Add 1 ml of sterile water and vortex (wash step). Repeat such wash step 4 times.
3. Add 100 μl (or more, depending on the amount of seeds) of sterile 0.1% agar, and disperse the seeds in the agar solution.
4. Transfer the seeds using a Pipetman with a wide-bore yellow tip onto culture medium (0.3% (w/v) Gellan Gum (073-03071, Wako Pure Chemical Industries, Osaka, Japan), one-half diluted Murashige and Skoog salts mixture (392-00591, Wako Pure Chemical Industries, Osaka, Japan; or M5524, Sigma) and 3% (w/v) sucrose) solidified in a Petri dish.
5. Keep the dish at 4 °C for 2–3 days to break dormancy, and then germinate and grow at 22 °C ±2 °C for 6–10 days under continuous light from fluorescent lamps.

Note: 0.3% Gellan Gum can be replaced with 0.8–1.2% agar (016-11875, Wako Pure Chemical Industries, Osaka, Japan).

6.2. Preparation of seedlings from barley seeds

1. Place 10–20 barley seeds in a 50-ml centrifuge tube, and add 40 ml of 16% (v/v) commercially available bleach. Cap the tube and shake for approximately 20 min to sterilize the seeds.
2. Wash the seeds with approximately 40 ml of sterile water 4 times.
3. Pick up the seeds with tweezers and put them on sterile 0.8–1.5% agar prepared in a Petri dish. Grow at 23 °C to 26 °C under light for approximately 3 days. Slant a dish so that the roots grow straight in parallel with the surface of agar.

6.3. Staining of Arabidopsis root tips with neutral red

The roots of *Arabidopsis* are so thin that the accumulation of cytoplasmic inclusions in lysosomes and large vacuoles can be detected by directly observing a whole root tip under a light microscope with Nomarski optics. Cytoplasmic inclusions in lysosomes and vacuoles are, however, more easily detected by staining with neutral red (Fig. 32.4). They can also be observed by LysoTracker Red staining (see Figure 4 in Inoue *et al.*, 2006).

1. Excise root tips (8–12 mm in length) from the seedlings, and incubate them in 2 ml of liquid culture medium consisting of one-half diluted Murashige and Skoog salts mixture and 3% (w/v) sucrose at 22 °C ±2 °C in the dark under agitation at 50–100 rpm.
2. Pick up root tips with tweezers, and put them in 0.01% (w/v) neutral red in 50 m*M* Mes-Tris, pH 6.5, for 2–5 min.

Note: When sucrose is omitted from the culture medium for starvation treatment, 1.6% (w/v) mannitol should be added to keep the osmotic value constant.

6.4. Staining of barley root tips with LysoTracker Red

After treatment with E-64d, root tips from barley are stained with 1 μM LysoTracker Red (Molecular Probes) and fixed with formaldehyde. To observe the accumulation of cytoplasmic inclusion in lysosomes and vacuoles, the root tips are treated with cellulase, and their epidermal cells are released (Fig. 32.5). Almost all cells isolated by this method are immature and do not contain large vacuoles.

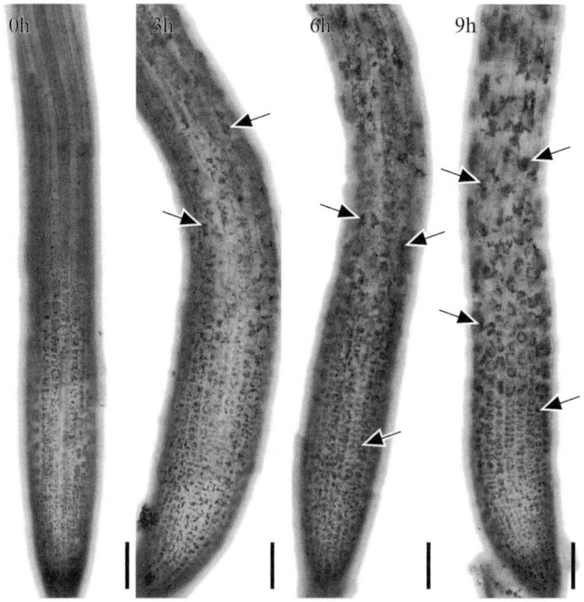

Figure 32.4 Effect of the membrane-permeable cysteine protease inhibitor E-64d on the morphology of *Arabidopsis* root tips. Root tips were excised from the seedlings of *Arabidopsis* (Columbia). Root tips were incubated in culture medium containing 100 μM E-64d for various hours as indicated in each image, and stained with neutral red. Arrows point to vacuolar inclusions. Bar, 50 μm. From Inoue *et al.* (2006) with permission.

1. Cut roots (approximately 1 cm from the tips) from the seedlings.
2. Incubate approximately 12 roots in 2 ml of MS medium containing 3% (w/v) sucrose and 100 μM E-64d with a rotation of 80–100 rpm at room temperature for 1 day.

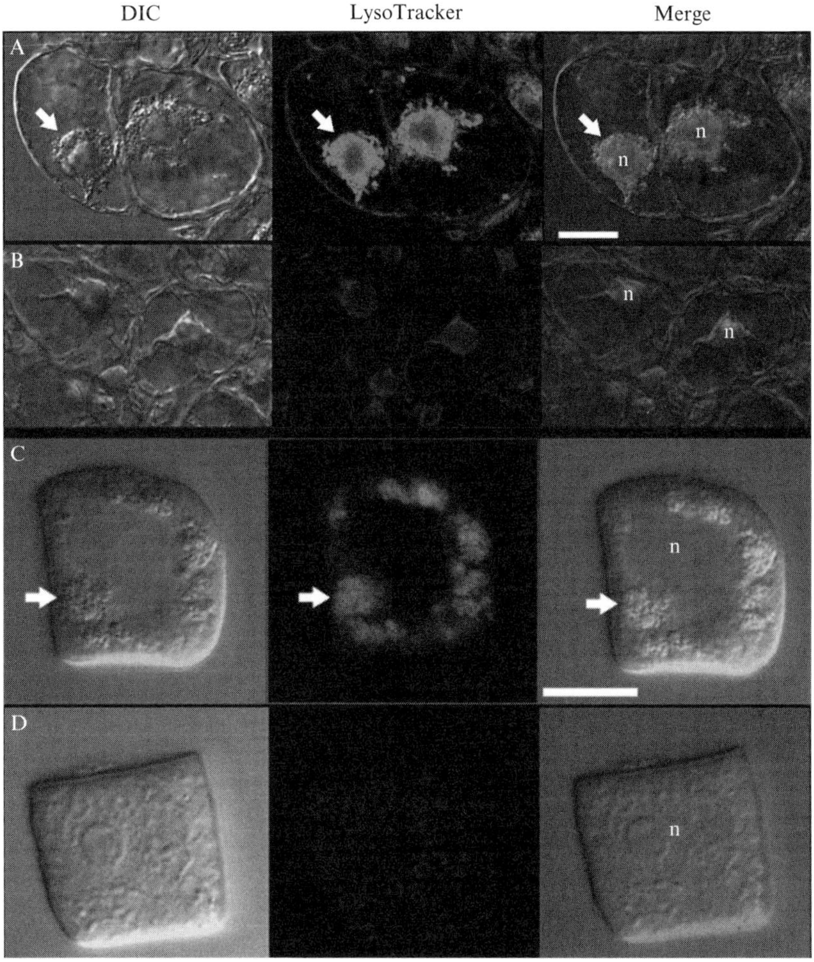

Figure 32.5 Concentration of LysoTracker Red in autolysosomes. (A, B) BY-2 cells were kept under sucrose starvation conditions for 24 h in the presence (A) or absence (B) of E-64c, and then labeled with LysoTracker Red. After fixation the cells were observed by confocal microscopy. In the cells treated with E-64c, autolysosomes accumulated and concentrated LysoTracker Red as indicated by the arrow. (C, D) Cells from barley root tips incubated in MS medium containing sucrose in the presence (C) or absence (D) of E-64d; the arrow indicates organelles with an appearance similar to that of autolysosomes that concentrate LysoTracker Red. n = nucleus; bar = 20 μm (A, B); 10 μm (C, D). From Moriyasu *et al.* (2003) with permission. (See Color Insert.)

3. Transfer the root tips to 10 m*M* HEPES-Na, pH 7.0, containing 1 μM LysoTracker Red (L-7528, Molecular Probes) and incubate for 1 h with rotation.

Note: LysoTracker Red is available as a 1 m*M* solution in DMSO. Prepare 100 μM solution in DMSO from this solution, and add 10 μl to 1 ml of HEPES-Na, pH 7.0.

4. Transfer the root tips to fixative (3.7% (w/v) formaldehyde in P-EGTA buffer (50 m*M* phosphate-Na buffer, pH 7.0, 5 m*M* EGTA, 0.02%(w/v) NaN_3)), and fix the tissues at 4 °C for more than 1 day.

Note: The root tips of *Arabidopsis* can be stained with LysoTracker Red in the same way.

5. Isolate cells from root tips according to the following protocol:
   a. Take out several root tips onto a glass slide, cut tips (1–2 mm) with a razor blade, and then put them in a microcentrifuge tube containing 0.5 ml of P-EGTA buffer. Mix briefly.
   b. Carefully remove P-EGTA buffer, and add 0.5 ml of fresh P-EGTA buffer (wash step). Repeat the wash step more two times, and then keep root tips in fresh P-EGTA buffer for 1 h at room temperature.
   c. Remove P-EGTA buffer, and add 0.5–1.0 ml of 1% cellulase (Cellulysin, Calbiochem) in P-EGTA buffer (i.e., 0.05g in 5 ml P-EGTA buffer).
   d. Keep at room temperature for 20 min with gentle shaking.
   e. Wash the root tips once for 5 min with P-EGTA buffer (wash step).
   f. Remove all the solution and add a small amount (100-200 μl, depending on the amount of root tips) of fresh P-EGTA buffer.
   g. Squash the root tips with a pellet pestle or punch with a Pipetman yellow (100–200 μl) tip in P-EGTA buffer to release root-tip cells (almost all are immature epidermal cells).
   h. Transfer P-EGTA buffer containing root-tip cells into a fresh microcentrifuge tube. Try not to suck up the large debris of root tips.
   i. Pellet the cells by centrifugation at 500 rpm for 5 min, and resuspend the cells in 100–200 μl of P-EGTA buffer.
   j. Place 5–10 μl of suspension onto a glass slide and allow to dry slightly.
   k. At the same time, place 5–10 μl of Mowiol mounting medium onto a cover glass.
   l. Put the cover glass on the glass slide.
   m. Observe by fluorescence microscopy.

Note: Mowiol mounting medium can be prepared from Mowiol 4-88 (475904, Calbiochem) and glycerol according to the protocol in Sambrook and Russell (2001).

7. ImmunoStaining of Lysosomes/Vacuoles in Barley Root-Tip Cells

In barley root-tip cells, it was reported that antibodies against alpha- and gamma-TIP differentially bind to the membranes of various kinds of vacuoles (Paris *et al.*, 1996). The membrane of the lysosomes/vacuoles that are accumulated by treatment with E-64d can be stained with alpha-TIP antibodies (Fig. 32.6).

1. Treat root-tip cells (step 5.i in the previous section) with 0.5% (w/v) Triton X-100 for 5 min at room temperature.

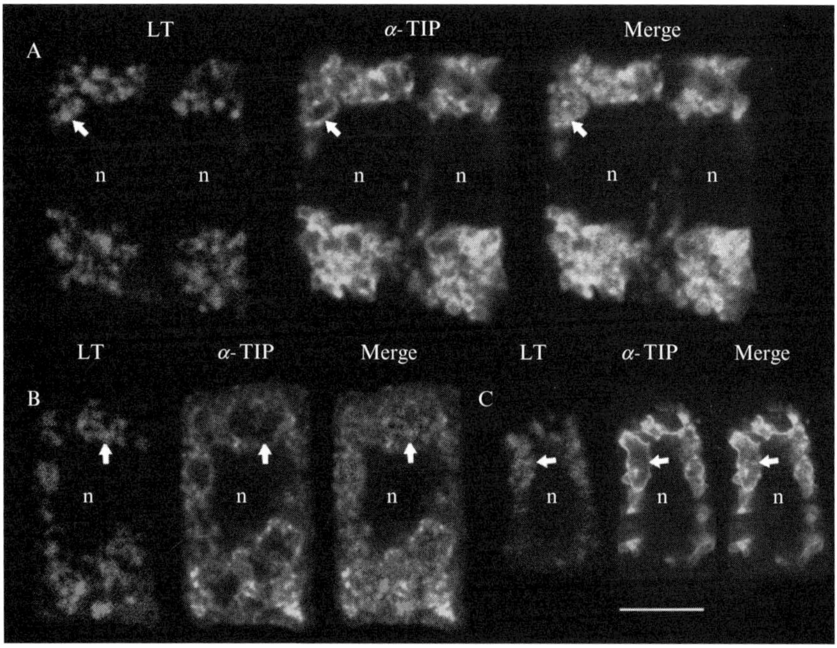

Figure 32.6 Barley root-tip cells labeled with LysoTracker Red and anti-alpha-TIP antibody. Each set (A, B, and C) presents LysoTracker (LT, red) and anti-alpha-TIP antibody (α-TIP, green) images individually, and the two images combined (Merge); cells were incubated with E-64d for 24 h. A. Two cells where predominantly individual autolysosomes appear to be surrounded by alpha-TIP-containing membrane; arrow: a larger structure containing multiple red autolysosomal inclusions. B. A cell where multiple inclusions (arrow) are incorporated into large vacuoles marked by alpha-TIP in their membranes. C. A cell where some vacuoles appear to contain only individual autolysosomal inclusions, while others appear to contain multiple inclusions (arrow). For all: n, position of nucleus; bar = 5 μm. From Moriyasu *et al.* (2003) with permission. (See Color Insert.)

2. Keep them in a blocking buffer consisting of 0.25% (w/v) BSA, 0.25% (w/v) gelatin, 0.05% (w/v) Nonidet P-40 and 0.02% (w/v) NaN_3 in phosphate-buffered saline (PBS) for 30 min at room temperature.
3. Keep the cells in the blocking buffer containing an antibody against alpha-TIP at 4 °C overnight.
4. Wash with the blocking buffer three times each for 5 min and once for 30 min.
5. Keep in PBS containing a secondary antibody at room temperature for 1 h.
6. Wash with PBS overnight.
7. Follow the steps from 5.j to 5.m in the previous section.

8. Electron Microscopy of Autolysosomes and Vacuoles Containing Cytoplasmic Inclusions in Plant Cells

In BY-2 cells, autolysosomes can be detected as membrane vesicles containing electron-dense particles by conventional electron microscopy (Fig. 32.7). Electron-dense particles are scarcely seen in the central vacuole (Fig. 32.7). In contrast, in *Arabidopsis* and barley root cells, small vesicles containing electron-dense particles are seen and the central vacuole also seems to have electron-dense particles (Fig. 32.8). The former seems to correspond to autolysosomes in BY-2 cells, whereas the latter structure

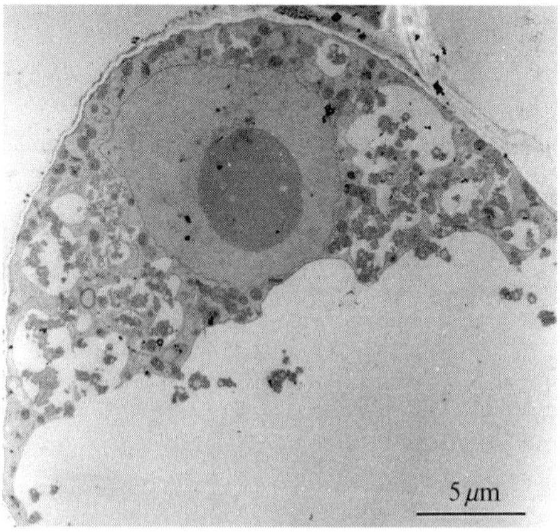

Figure 32.7 Autolysosomes in BY-2 cells. BY-2 cells were cultured in a sucrose-free culture medium containing 10 μM E-64c for 1 day, and observed by electron microscopy.

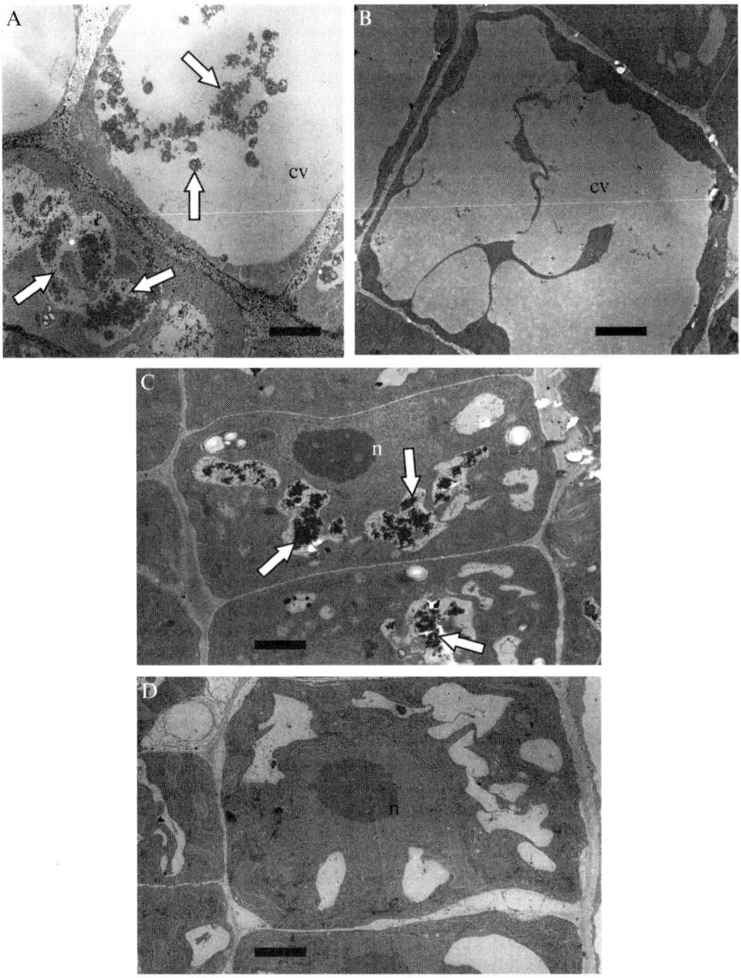

Figure 32.8 Electron micrographs of the cells of *Arabidopsis* root tips cultured in the presence of E-64d. Root tips were excised from the seedlings of *Arabidopsis* (Landsberg) grown under nutrient-sufficient conditions and incubated in culture medium containing sucrose in the presence (A and C) or absence (B and D) of E-64d for 1 day. They were fixed and embedded in plastic blocks. Sections made from these blocks were stained with uranyl acetate and lead nitrate, and observed by electron microscopy. Images of relatively young root-tip cells (C and D) and mature vacuolated cells (A and B) are shown. Arrows point to vacuolar inclusions. CV, the central vacuoles; n, nucleus. Bar, 2 μm. From Inoue *et al.* (2006) with permission.

seems to be formed by the fusion between autolysosomes and the central vacuole. Here we introduce our method for fixing BY-2 cells for electron microscopy. A similar procedure can be used for the fixation of *Arabidopsis* and barley root tips.

8.1. Electron microscopy of autolysosomes in BY-2 cells

1. Take 1 ml of BY-2 cell suspension (cells cultured in the sucrose-free culture medium in the presence or absence of E-64c for 1 day) in a 2.0-ml microcentrifuge tube with a round bottom.
2. Sediment the cells as gently as possible by centrifugation at a low speed.
3. Remove as much of the supernatant fraction as possible. Add a fixative (2% (w/v) glutaraldehyde and 1% (w/v) formaldehyde in 100 m*M* sodium cacodylate-HCl buffer, pH 6.9, and resuspend the cells gently. Keep the tube at room temperature for 1 h.
4. Remove the fixative (centrifugation is not needed because the cells sediment in the tube during the fixation). Add 1 ml of a fresh fixative, suspend the cells gently, and store at 4 °C overnight.
5. Remove as much of the fixative as possible, add 1 ml of 100 m*M* sodium cacodylate-HCl buffer, pH 6.9, and suspend the cells gently (wash step). Perform the wash step more two times each for 5 min.
6. Remove as much of the wash buffer as possible, and add approximately the same volume as the cell pellet of 2% (w/v) osmium tetroxide (thus, the concentration of osmium tetroxide becomes approximately 1%). Suspend the cells gently, and keep the tube at room temperature for 2 h.
7. Remove as much of the osmium solution as possible, add 1 ml of water, and suspend the cells gently (wash step). Repeat the wash step more two times each for 5 min.
8. Remove as much of the water as possible, and add 0.5 ml of 2% (w/v) uranyl acetate for en block staining. Suspend the cells gently, and keep the tube at room temperature for 2 h.
9. Remove as much of the uranyl acetate solution as possible, add 1 ml of water, and suspend the cells gently (wash step). Repeat the wash step more two times each for 5 min.
10. Remove as much of the water as possible. Dehydrate specimens through a water/ethanol series (1 ml of 30%, 50%, 70%, 80% 90%, 95%, and 99.5% ethanol each for 20 min, and 100% ethanol three times each for 10 min), and transfer the whole content to a glass vial with a screw cap (e.g., 24 × 45 mm, 10 ml), which is resistant to propylene oxide.
11. Dehydrate the cells through ethanol/propylene oxide (1:1) for 10 min and 100% propylene oxide twice each for 30 min. Close the cap of the vial during each step, and use a glass Pasteur pipette for changes of the solutions.
12. Remove the propylene oxide, and add 8–10 ml of propylene oxide/Spurr's resin (1:3) mixture (i.e., 25% Spurr's resin in propylene oxide) to the vial. Open the cap and evaporate propylene oxide at room temperature overnight (thus, the concentration of the resin increases up to 100% after evaporation). Rotate the vial gently to efficiently infiltrate the resin into the cells.

13. Cap the vial, and centrifuge at 500–1000 rpm for 5–10 min to sediment the cells. Remove the resin in the supernatant fraction, and add fresh resin. Perform such a change of resin three times in total in a day.
14. Transfer the cells dispersed in the resin to suitable plastic molds, and polymerize the resin at 70 °C for 8 h.
15. Make sections, and stain with uranyl acetate for 5 min and with lead nitrate for 5 min.

Note: 100% ethanol can be prepared by adding molecular sieve (0.3 nm, Merck) to 99.5% ethanol. Sections can be stained with the mixture (2:1) of 4% (w/v) uranyl acetate and 1% (w/v) $KMn0_4$ instead of with uranyl acetate and lead nitrate. Fixation without formaldehyde (for example, using only 2.5% glutaraldehyde in 100 m*M* phosphate-Na buffer, pH 7.4, and 86 m*M* NaCl) works well.

9. Enzyme Cytochemistry for Acid Phosphatase by Electron Microscopy

1. Fix cells (cells cultured in the sucrose-free culture medium in the presence or absence of E-64c for 1 day) overnight with 1% (w/v) glutaraldehyde and 1% (w/v) formaldehyde in 0.1 *M* sodium cacodylate-HCl buffer, pH 7.2, at 4 °C.
2. Wash the cells three time with 8% (w/v) sucrose in 0.1 *M* sodium cacodylate-HCl buffer, pH 7.2, and subsequently three times with 8% (w/v) sucrose in 50 m*M* acetate-Na buffer, pH 5.0.
3. Keep the cells in 45.5 m*M* acetate-Na buffer, pH 5.0, 3 m*M* lead nitrate, 10 m*M* ß-glycerophosphate, and 230 m*M* sucrose for approximately 20 min at room temperature.
4. Wash the cells with 8% (w/v) sucrose in 0.1 *M* sodium cacodylate-HCl buffer, pH 7.2.
5. Postfix the cells with 1% (w/v) osmium tetroxide at room temperature for 2 h.
6. Dehydrate the specimens in a water/ethanol series, followed by propylene oxide, and embed in Spurr's resin as described previously.
7. Prepare thin sections of resin-embedded cells and stain with uranyl acetate and lead nitrate as described previously.

References

Bensadoun, A., and Weinstein, D. (1976). Assay of proteins in the presence of interfering materials. *Anal. Biochem.* **70,** 241–250.

Gordon, P. B., and Seglen, P. O. (1982). 6-Substituted purines: A novel class of inhibitors of endogenous protein degradation in isolated rat hepatocytes. *Arch. Biochem. Biophys.* **217,** 282–294.

Herman, E. M., and Lamb, C. J. (1992). Arabinogalactan-rich glycoproteins are localized on the cell surface and in multivesicular bodies. *Plant Physiol.* **98,** 264–272.

Inoue, Y., and Moriyasu, Y. (2006). Autophagy is not a main contributor to the degradation of phospholipids in tobacco cells cultured under sucrose starvation conditions. *Plant Cell Physiol.* **47,** 471–480.

Kominami, E., Hashida, S., Khairallah, E. A., and Katunuma, N. (1983). Sequestration of cytoplasmic enzymes in an autophagic vacuole-lysosomal system induced by injection of leupeptin. *J. Biol. Chem.* **258,** 6093–6100.

Liou, W., Geuze, H. J., Geelen, M. J. H., and Slot, J. W. (1997). The autophagic and endocytic pathways converge at the nascent autophagic vacuoles. *J. Cell Biol.* **136,** 61–70.

Moriyasu, Y., and Ohsumi, Y. (1996). Autophagy in tobacco suspension-cultured cells in response to sucrose starvation. *Plant Physiol.* **111,** 1233–1241.

Moriyasu, Y., Hattori, M., Jauh, G.-Y., and Rogers, J. C. (2003). Alpha tonoplast intrinsic protein is specifically associated with vacuole membrane involved in an autophagic process. *Plant Cell Physiol.* **44,** 795–802.

Murashige, T., and Skoog, F. (1962). A revised medium for rapid growth and bio assays with tobacco tissue cultures. *Physiol. Plant* **15,** 473–497.

Paris, N., Stanley, C. M., Jones, R. L., and Rogers, J. C. (1996). Plant cells contain two functionally distinct vacuolar compartments. *Cell* **85,** 563–572.

Record, R., and Griffing, L. (1988). Convergence of the endocytic and lysosomal pathways in soybean protoplasts. *Planta* **176,** 425–432.

Sambrook, J., and Russell, D. W. (2001). Molecular cloning: A laboratory manual 3rd ed. Cold Spring Harbor Laboratory Press, Cold Spring Harbor, New York.

Seglen, P. O., and Gordon, P. B. (1982). 3-Methyladenine: Specific inhibitor of autophagic/lysosomal protein degradation in isolated rat hepatocytes. *Proc. Natl. Acad. Sci. USA* **79,** 1889–1892.

Takatsuka, C., Inoue, Y., Matsuoka, K., and Moriyasu, Y. (2004). 3-Methyladenine inhibits autophagy in tobacco culture cells under sucrose starvation conditions. *Plant Cell Physiol.* **45,** 265–274.

Takeshige, K., Baba, M., Tsuboi, S., Noda, T., and Ohsumi, Y. (1992). Autophagy in yeast demonstrated with proteinase-deficient mutants and conditions for its induction. *J. Cell Biol.* **119,** 301–311.

Tamai, M., Matsumoto, K., Omura, S., Koyama, I., Ozawa, Y., and Hanada, K. (1986). *In vitro* and *in vivo* inhibition of cysteine proteinases by EST, a new analog of E-64. *J. Pharmacobio-Dyn.* **9,** 672–677.

Twining, S. S. (1984). Fluorescein isothiocyanate-labeled casein assay for proteolytic enzymes. *Anal. Biochem.* **143,** 30–34.

Vida, T. A., and Emr, S. D. (1995). A new vital stain for visualizing vacuolar membrane dynamics and endocytosis in yeast. *J. Cell Biol.* **128,** 779–792.

Yano, K., Suzuki, T., and Moriyasu, Y. (2007). Constitutive autophagy in plant root cells. *Autophagy* **3,** 360–362.

CHAPTER THIRTY-THREE

Lysosomes and Autophagy in Aquatic Animals

Michael N. Moore,* Angela Kohler,† David Lowe,* *and* Aldo Viarengo‡

Contents

* Plymouth Marine Laboratory, Prospect Place, The Hoe, Plymouth, United Kingdom
† Alfred Wegener Institute for Polar and Marine Research, Department of Ecotoxicology and Ecophysiology, Bremerhaven, Germany
‡ Department of Environmental and Life Science, Universita di Piemonte Orientale (Amadeo Avogaddro), Alessandria, Italy

Methods in Enzymology, Volume 451
ISSN 0076-6879, DOI: 10.1016/S0076-6879(08)03233-3

Abstract

The lysosomal-autophagic system appears to be a common target for many environmental pollutants, as lysosomes accumulate many toxic metals and organic xenobiotics, which perturb normal function and damage the lysosomal membrane. In fact, autophagic reactions frequently involving reduced lysosomal membrane integrity or stability appear to be effective generic indicators of cellular well-being in eukaryotes: in social amoebae (slime mold), mollusks and fish, autophagy/membrane destabilization is correlated with many stress and toxicological responses and pathological reactions. Prognostic use of adverse lysosomal and autophagic reactions to environmental pollutants can be used for predicting cellular dysfunction and health in aquatic animals, such as shellfish and fish, which are extensively used as sensitive bioindicators in monitoring ecosystem health; and also represent a significant food resource for at least 20% of the global human population. Explanatory frameworks for prediction of pollutant impact on health have been derived encompassing a conceptual mechanistic model linking lysosomal damage and autophagic dysfunction with injury to cells and tissues. Methods are described for tracking *in vivo* autophagy of fluorescently labeled cytoplasmic proteins, measuring degradation of radiolabeled intracellular proteins and morphometric measurement of lysosomal/cytoplasmic volume ratio. Additional methods for the determination of lysosomal membrane stability in lower animals are also described, which can be applied to frozen tissue sections, protozoans and isolated cells *in vivo*. Experimental and simulated results have also indicated that nutritional deprivation (analogous in marine mussels to caloric restriction)-induced autophagy has a protective function against toxic effects mediated by reactive oxygen species (ROS). Finally, coupled measurement of lysosomal-autophagic reactions and simulation modelling is proposed as a practical toolbox for predicting toxic environmental risk.

1. Overview: Autophagic and Lysosomal Responses in Cell Physiology and Pathological Reactions Induced by Environmental Stress

Autophagy is the degradation of cellular components in lysosomes, and it is implicated in many disease processes, cell death and adaptive responses (Cuervo, 2004). Regulation of this highly conserved group of cellular processes appears to be very similar in eukaryotic organisms ranging from yeasts to man (Cuervo, 2004; Klionsky and Emr, 2002; Klionsky *et al.*, 2007; Levine, 2005).

Pathological reactions involving the lysosomal system are often linked to augmented autophagic sequestration of cellular components; and numerous studies have shown that many environmental stressors including pollution are capable of inducing cellular autophagy in the cells of social amoebae (slime mold, myxomycetae), aquatic mollusks and fish (Figs. 33.1, 33.2, and 33.3; Dondero *et al.*, 2006; Einsporn and Kohler, 2008a,b, submitted; Kohler *et al.*, 2002; Moore *et al.*, 2004a; 2006a, 2007). Lower animals such as protozoans, earthworms, aquatic mollusks and flatfish are now widely used in monitoring the health of the environment; and fish and shellfish also provide approximately 20% of global animal protein consumed by mankind (Deutsch *et al.*, 2007).

Given the increasing demands for environmentally sustainable food production, it is essential to have effective indicators of the health and growth of the prospective food animals, whether these are farmed (aquaculture), or from the wild. Indicators of animal well-being and potential for growth are provided by cellular biomarkers of lysosomal and autophagic processes in both mollusks and fish (Kohler *et al.*, 2002; Moore, 1985; Moore *et al.*, 2006a, 2007). Other studies have shown that lysosomal biomarkers can also be used in protozoans, coelenterates, annelid worms and crustaceans (Brown *et al.*, 2004; Dondero *et al.*, 2006; Galloway *et al.*, 2006; Moore and Stebbing, 1976; Sforzini *et al.*, 2008; Svendsen and Weeks, 1995; Svendsen *et al.*, 2004).

The social amoeba *Dictyostelium discoideum* has been utilized recently as an experimental model to study the biological effects of adverse environmental conditions. In particular, the cellular changes induced by inorganic as well as organic pollutants on the activity of the lysosomal vacuolar system have been studied (Dondero *et al.*, 2006). The results show that nanomicromolar concentrations of mercury or benzo[a]pyrene are able to destabilize the lysosomal membranes in the amoebae, thus confirming the high sensitivity of the neutral red retention assay (Dondero *et al.*, 2006; Lowe and Pipe, 1994; Lowe *et al.*, 1992; Moore *et al.*, 2004b) utilized in these experiments.

The relationship between the destabilization of lysosomal membranes and the autophagic rate has been established utilizing recombinant amoebae constitutively expressing a destabilized Green Fluorescent Protein (d2EGFP) variant under the control of an endogenous actin promoter (Dondero *et al.*, 2006). The half-life of the recombinant protein, calculated in the presence of cycloheximide, is approximately 4.5 h, indicating a sufficiently rapid turnover. This strain has been utilized to verify the possible effects induced by pollutant concentrations able to destabilize lysosomal membranes. In the amoebae, the results confirm that the destabilization of lysosomal membranes is associated with a parallel increase in the catabolic rate of d2EGFP. This clearly indicates that the activation of the lysosomal/vacuolar system in amoebae maintained in a standard medium suitable to support cell growth and division, represents the initial event of an increased autophagic activity itself representing an important aspect of the physiological response of the protozoan to environmental stressors (Dondero *et al.*, 2006).

Autophagy is often considered to be primarily a survival strategy in multicellular organisms, which may be initiated by stressors (e.g., restricted nutrients, hyperthermia, hypoxia and salinity increase; Cuervo, 2004; Klionsky and Emr,

2000; Moore *et al.*, 2006a,b,c). However, recent evidence indicates that autophagy is much more than just a survival process and is, in fact, intimately involved in cell physiology (Cuervo, 2004; Lockshin and Zakeri, 2004; Moore, 1988, 2004; Moore *et al.*, 1980, 2006a). Augmented autophagy is apparently controlled by switching off the mTOR (mammalian target of rapamycin) kinase: mTOR signalling is involved in many aspects of cell growth-regulation and has been implicated in some cancers (Asnaghi *et al.*, 2004; Levine, 2005; Proud, 2002). mTOR kinase is also coupled with a nutrient sensing pathway; and is switched off by lack of nutrients (see review by Proud, 2002). This kinase is evolutionarily conserved in eukaryotes and has been variously described in yeast, nematodes, mollusks, insects, crustaceans and mammals (Cammalleri *et al.*, 2003; Beaumont *et al.*, 2001; Levine, 2005; Klionsky and Emr, 2000; Klionsky *et al.*, 2007). The mTOR signalling system is classically switched off by nutrient deprivation, with resultant up-regulation of autophagy in mammals, which has also been described in mussels and marine and terrestrial snails (Bayne *et al.*, 1978, 1979; Bergamini *et al.*, 2003; Cuervo, 2004; Moore and Halton, 1973, 1977; Moore *et al.*, 1978a,b, 1979, 1985, 1986; Proud *et al.*, 2002). Hypoxia also switches off the mTOR protein and has previously been demonstrated to induce autophagy in eukaryotes including mussels (Hipkiss, 2006; Moore *et al.*, 1979; Proud *et al.*, 2002; Wouters *et al.*, 2005); as have both increases and decreases in salinity (Moore *et al.*, 1980; Pipe and Moore, 1985).

Physiological responses and pathological reactions to environmental stressors in molluskan hepatopancreatic digestive cells (hepatocyte analogues), phagocytic hemocytes (blood cells) and fish hepatocytes frequently involve destabilizing changes in the lysosomal membrane and the induction of autophagy (Axiak *et al.*, 1988; Cajaraville *et al.*, 1995a,b; Da Ros *et al.*, 2000; Kohler *et al.*, 1992, 2002; Marigómez and Baybay-Villacorta, 2003; Marigómez *et al.*, 2005; Moore, 1976, 1980, 1985, 1990, 2002, 2004; Moore and Clarke, 1982; Moore and Halton, 1973, 1977; Moore *et al.*, 1978a,b, 1979, 1984, 1985, 1986, 1987, 1996, 2006a,b,c, 2007).

Other related lysosomal perturbations can also occur, such as lysosomal swelling, accumulation of lipid (lipidosis) and age pigment or lipofuscin (lipofuscinosis): all of these changes have been described in molluskan hepatopancreatic digestive cells (Figs. 33.1, 33.2 and 33.4; Cajaraville *et al.*, 1995a,b, 2000; Domouhtsidou and Dimitriadis, 2001, Etxeberria *et al.*, 1995; Krishnakumar *et al.*, 1994; Lowe *et al.*, 1981; Marigómez and Baybay-Villacorta, 2003; Marigómez *et al.*, 1996, 2005; Moore, 1988; Moore *et al.*, 1978a,b, 1979, 2006a; Nott and Moore, 1987; Regoli, 1992).

Damaged cellular constituents and redundant products are removed by lysosomal autophagy, but this process is also critically involved in the continuous basal turnover of intracellular components (Cuervo, 2004; Levine, 2005; Hawkins and Day, 1996; Ryazanov and Nefsky, 2002; Tavernarakis and Driscoll, 2002). Autophagy is up-regulated in times of stress or physiological change by breaking down longer-lived proteins and organelles, and recycling the products into protein-synthesis and energy-production pathways: the process allows cells to be temporarily self-sustaining during periods when nutrients are restricted (Bergamini *et al.*, 2003; Cuervo, 2004; Levine, 2005). New evidence also indicates that autophagy may have a protective role in the context of oxidative stress (Bergamini *et al.*, 2003; Cuervo, 2004; Moore, 2008; Moore *et al.*, 2006b,c, 2007). Nutrient deprivation-induced autophagy in mollusks appears to confer some resistance to the toxicity of both ROS-generating polycyclic aromatic hydrocarbons (PAH) and copper (Moore, 2004; Moore *et al.*, 2006a,b,c; Viarengo *et al.*, 1987). Normal tidal fluctuations in salinity, food and oxygen, however, do not induce a stress syndrome (Bayne *et al.*, 1978, 1979; Moore, 1980; Moore *et al.*, 1979, 1982, 1987; Widdows *et al.*, 1981, 1982).

In eukaryotic cells, the first tier of defense against oxidative damage is provided by xenobiotic transporters, biotransformation enzymes and antioxidant protection enzymes, such as superoxide dismutase and catalase (Livingstone, 2001; Moore, 2004). Lysosomal autophagy provides a second line of defense, by removing oxidatively damaged proteins, inappropriately folded glycosylated proteins and impaired organelles, and even portions of the nucleus and DNA (Bergamini *et al.*, 2003; Cuervo, 2004; Ji and Kaplowitz, 2006). Consequently, when the first line of defensive systems is overcome, autophagy protects the cell against the harmful effects of damaged and malfunctioning proteins, which can form aggregates that will accumulate irreversibly in cells (Brunk and Terman, 2002; Grune *et al.*, 2004; Kiffen *et al.*, 2004; Moore, 2004, 2008; Moore *et al.*, 2006a,b,c, 2007). Autophagy may also serve as a third tier of defense: when autophagic capabilities are compromised, the autophagic system can trigger programmed cell death (Lockshin and Zakeri, 2004) to remove irreversibly damaged cells in order to maintain organ integrity.

This chapter describes a number of methods that have been use to study autophagic and related lysosomal processes in cells of lower animals: particularly, social amoebae, aquatic mollusks and flatfish.

2. Visual Detection of Autophagic Responses

2.1. Tracking *in vivo* autophagy using fluorescein-isothiocyanate diacetate (FITC diacetate)

Autophagic (self-eating) processes play an important role in the degradation of intracellular proteins, particularly under conditions of stress or induced cell injury. In fact, there is mounting evidence that lysosomal protein catabolism may be regulated by the flux of proteins into the lysosomal compartment and that proteinases, such as cathepsins B and L, are unlikely to be limiting factors (Yucel-Lindberg *et al.*, 1991).

There is abundant evidence of stress-induced autophagy in animal cells, including those of invertebrates, fish and mammals (Klionsky *et al.*, 2007; Moore *et al.*, 2007). In particular, the exposure of animals to pollutant chemicals (both metals and organic xenobioitics) is known to induce cell injury and ensuing pathological change, which frequently involves autophagy and lysosomal alterations (Moore, 1990; Lowe *et al.*, 1995a,b). These latter include increased fragility of the lysosomal membrane, enlargement of the lysosome and in some cases lipidoses. Enhanced catabolism of cytosolic proteins is also indicated in some experimental studies (Hawkins and Day, 1996; Moore and Viarengo, 1987; Viarengo *et al.*, 1992). However, there is still considerable debate concerning the relative importance of lysosomal as opposed to nonlysosomal pathways of intracellular protein catabolism. A great variety of xenobiotics and nanoparticles are taken up by lysosomes (Einsporn and Kohler, 2008a,b and submitted; Kohler *et al.*, 2008; Moore, 2006; Rashid *et al.*, 1991), and a pertinent question is whether lysosomotropic chemicals enhance traffic of intracellular proteins to the degradative lysosomal compartment (Figs. 33.2 and 33.5)?.

To answer this question, an *in situ* study using FITC-adducts of intracellular proteins enabled the effects of the lysosomotropic xenobiotics chloroquine and chlorpromazine on the transport of proteins into the degradative compartment to be characterized (Fig. 33.6; Moore *et al.*, 1996).

This investigation used the lysosome-rich phagocytic blood cells (hemocytes) from the common marine mussel (*Mytilus edulis*); an animal used globally for assessing impact by environmental pollutants. These phagocytic cells are involved in cellular immune functions including defense against pathogens and removal of moribund cells during the frequent tissue reorganizations that accompany the reproductive cycle in mussels (Bayne *et al.*, 1978; Grundy *et al.*, 1996a,b). The hemocytes can be readily maintained *in vitro* and lend themselves to experimental manipulation. Cells were exposed to membrane permeant FITC-diacetate, which is rapidly metabolized to reactive FITC (Kolb *et al.*, 1998). This latter forms fluorescent adducts with intracellular proteins, thus facilitating visual tracking of the subsequent transport and compartmentalization of these proteins by either epifluorescence or confocal microscopy (Fig. 33.6; Moore *et al.*, 1996). Fluorescent lysosomotropic probes (neutral red or cresyl violet) were used simultaneously to identify acidic compartments, autophagic vesicles and late endosomes, as well as to provide an indicator of cell viability and functional integrity of lysosomes (Fig. 33.6; Rashid *et al.*, 1991).

These results show that *in situ* FITC labeling of intracellular proteins can be used to follow the transport of these proteins into an autophagic degradative compartment (Fig. 33.6). The presence of fluorescent adducts in vacuoles, which did not also take up the lysosomotropic probes indicates

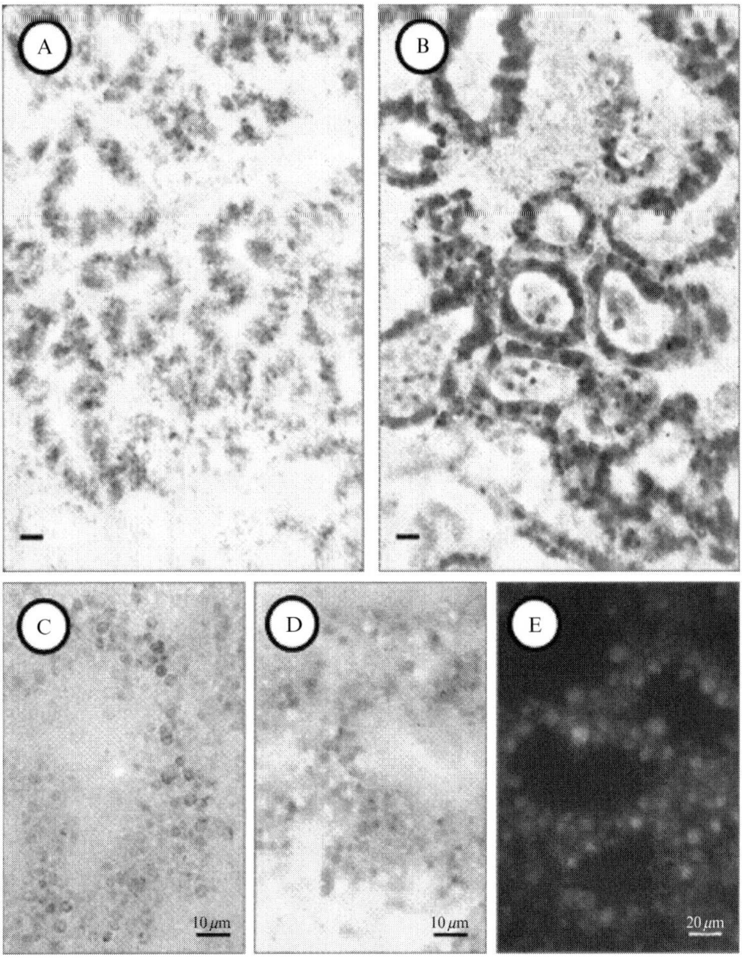

Figure 33.1 Micrographs of frozen sections of formalin-fixed mussel hepatopancreas showing lysosomes reacted for β–N-acetylhexosaminidase in the epithelial digestive cells of fed controls (A) and enlarged autophagolysosomes in mussels deprived of food for 15 days (B) indicative of starvation-induced autophagy (Moore 2004). Unfixed frozen sections of hepatopancreatic epithelial digestive cells from copper treated mussel (40 μg/l for 3 days) showing dithionite reaction for copper in lysosomes (C) Shikata reaction for autophagocytosed metallothionein in lysosomes (D) and UV-fluorescence of 3-methylcholanthrene (3-MC at a daily dose of 150 μg/l for 7 days) in swollen lysosomes (E). Experimental details are available in Moore (2004), Moore *et al.* (2006a) and Viarengo *et al.* (1985). Scale bar in A, B and E = 10 μm and in C and D = 10 μm. (See Color Insert.)

that at least part of the sequestration of intracellular proteins is routed through a prelysosomal autophagic pathway (Moore *et al.*, 1996).

Chloroquine and chlorpromazine accelerate the autophagic traffic of intracellular proteins into the lysosomal degradative compartment.

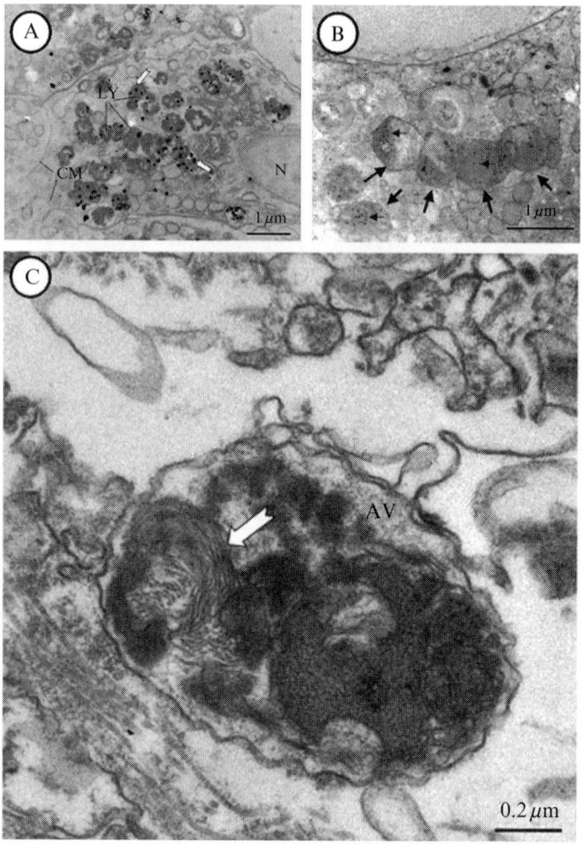

Figure 33.2 (A) Lysosomal accumulation of lead in a mussel gill epithelial cell visualized by electron immunocytochemistry (Einsporn and Kohler, submitted). LY – lysosomes; CM – plasma membrane; N – nucleus. (B) Autophagic vesicles following phenanthrene exposure in mussel hepatopancreatic digestive cell. Small arrows indicate electron immunocytochemical detection of phenanthrene, large arrows indicate autophagic vesicles (Einsporn and Kohler, 2008a,b and unpublished; and see Nott and Moore, 1987). (C) Hepatopancreatic digestive cell autophagic vesicle (AV) following phenanthrene exposure without antibody treatment (arrow indicates autophagocytosed intracellular membranes (Einsporn, Kohler, 2008a,b; Kohler, unpublished; and see Nott and Moore, 1987).

The reasons for this enhanced autophagy are not clear; however, chloroquine is a potent inducer of autophagy in rat liver cells. Yucel-Lindberg *et al.* (1991), have demonstrated that chloroquine induces the formation of autophagic vesicles and stimulates lysosomal proteolysis after 60 min of treatment.

In summary, treatment of the hemocytes with chloroquine (10 μg/ml) and chlorpromazine (10 μg/ml), following fluorescent FITC-labeling of

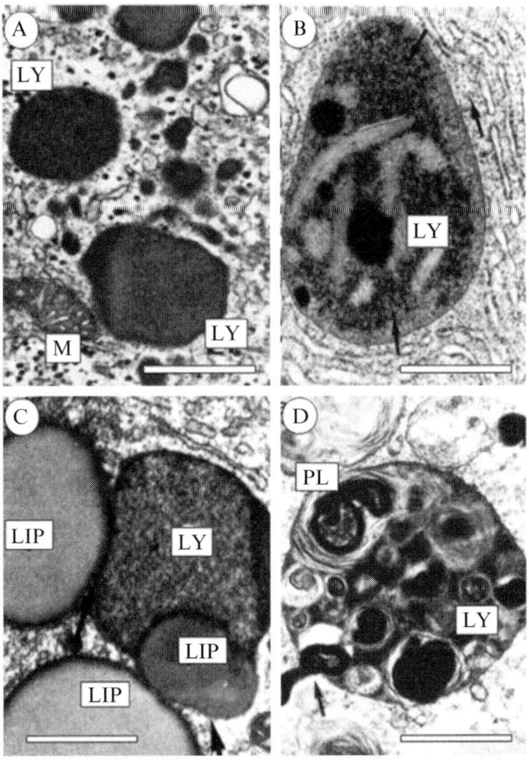

Figure 33.3 Electron microscopy photographs of characteristic morphological changes of lysosomes during the progression of toxicopathic lesions in flounder liver (*Platichthys flesus*). (A) Normal small lysosomes with homogenous structure. (B) Typical lysosomes in a reversibly altered liver with high induction of CYP1A1 activity (EROD), which also contains fibrillar elements, black lipofuscin granules and fields of digested ribosomes (arrows). (C) Lysosomes that have taken up a lipid droplet (arrows), which had accumulated in the cytoplasm. (D) Uptake and accumulation of phospholipid whorls into a lysosome (arrow), which has given rise to eosinophilic granules in the cytoplasm as seen at the light microscopy level (compare to Fig. 33.1D). LY- lysosomes, M- mitochondria, LIP- lipid droplets, PL- phospholipid whorls. With ICES permission from Moore *et al.* (2004b). Scale bar = 1 μm.

cytoplasmic proteins, results in a considerable increase in the rate of transfer of FITC-protein adducts into the autophagic-lysosomal compartment (Fig. 33.6; Moore *et al.*, 1996). After 60–80 min there was substantial vacuolar fluorescence for FITC and about half of these vacuoles (1-4 μm diameter) were also flourescent for the lysosomotropic probes. After 2 h, all of the FITC was present only in vacuoles, which were also predominantly lysosomes. Simultaneous treatment of the hemocytes with 5 and 10 m*M* 3-methyladenine (an inhibitor of autophagy, Yucel-Lindberg *et al.*, 1991)

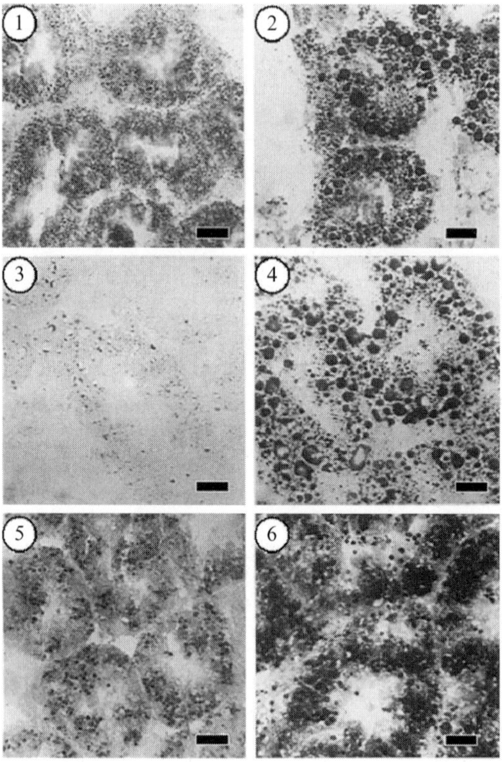

Figure 33.4 Cryostat sections (10 μm) of unfixed hexane quenched (−70 °C) digestive gland of the marine mussel sampled from clean and contaminated sites. (1) Normal appearance of the digestive tubules showing lysosomes reacted for N-acetyl-β-hexosaminidase in the digestive cells (clean site). (2) Abnormally enlarged lysosomes reacted as in 1 (contaminated site). (3) Lipid droplets localized in digestive cells using oil red-O (clean site). (4) Unsaturated neutral lipid accumulation in pathologically enlarged lysosomes, together with a general increase in lipid droplets, reacted as in 3 (contaminated site). (5) Lipofuscin in secondary and tertiary lysosomes localized using the Schmorl reaction (clean site). (6) Enhanced lipofuscin content (contaminated site). With ICES permission from Moore *et al.* (2004b). Scale bar = 20 μm.

inhibited the transfer of labeled proteins to lysosomes by a factor of 60%–80% (Moore *et al.*, 1996).

This relatively nonintrusive fluorescent labeling of intracellular proteins can be used for tracking labeled proteins and their transport to the autophagosomes and lysosomal compartment (Moore *et al.*, 1996). Given that the flux of proteins into the lysosomes is believed to be important in regulating protein catabolism, the fluorescent microscopy technique described could provide a useful experimental tool for studying intracellular protein traffic and catabolism (Glaumann *et al.*, 1986; Yucel-Lindberg *et al.*, 1991). These findings also show that chlorpromazine is an effective enhancer of

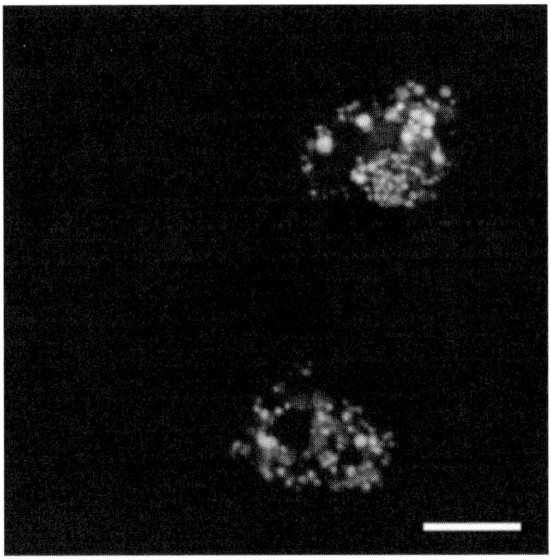

Figure 33.5 Mussel blood cell showing strong fluorescence (orange) for lysosomal accumulation of acridine orange (exposed to 1 μg/l for 15 min). Weaker fluorescence (green) is present in probable pre-lysosomal compartments with possibly some mitochondrial fluorescence. Nuclear staining of DNA is generally not apparent using this concentration of acridine orange. Blue light (FITC) excitation; With ICES permission from Moore *et al.* (2004b). Scale Bar = 10 μm. (See Color Insert.)

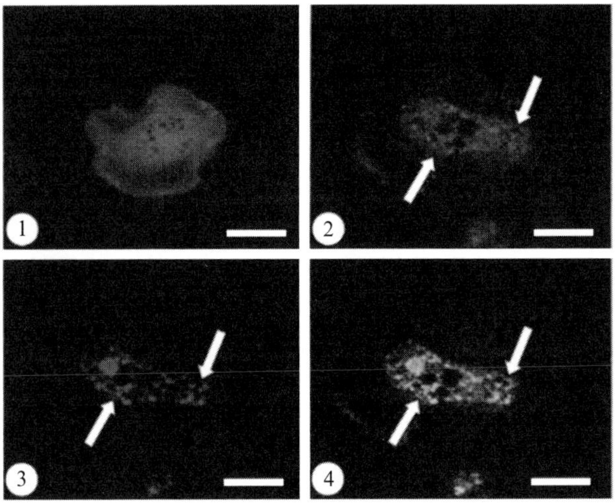

Figure 33.6 Confocal images of mussel blood cells treated with FITC diacetate showing even fluorescent labeling of cellular proteins after 30 min (1) and fluorescent vacuolar distribution of autophagocytsed FITC labeled protein after a further 3 h (2). The distribution of fluorescent lysosomes in the same cell as in (2) labeled with neutral red (rhodamine excitation and merged image of (2) and (3) showing that lysosomes and fluorescent vacuolar fluorescence are predominantly at the same sites (4) (arrows). See Moore *et al.*, (1996) for experimental details. Scale bar = 10 μm. (See Color Insert.)

autophagy, as is chloroquine in rat hepatocytes (Fig. 33.7). This is probably due to similarity in the structure-activity characteristics of these two lysosomotropic xenobiotics: both are amphiphilic cations and become trapped as protonated forms within the lysosomes (Lüllmann-Rauch, 1979; Rashid *et al.*, 1991). Chloroquine and chlorpromazine both have structural similarities to polycyclic aromatic hydrocarbons (PAHs), which are ubiquitous environmental pollutants (Lüllmann-Rauch, 1979). They induce similar effects to PAHs on the lysosomes of molluskan cells, including autophagy (Moore, 1990; Moore *et al.*, 1996, 2006a, 2007); thus, it appears reasonable to extrapolate that pollutant-induced autophagy may involve a similar process to that described previously.

2.1.1. Method

Mussel hemocytes are extracted and allowed to attach to glass microscope slides as described below. Exposure of hemocytes attached to glass microscope slides to fluorescein isothiocyanate (FITC) diacetate (Molecular Probes, 30 μg/ml in physiologically buffered saline; see subsequently) for 30 min at 15 °C results in even fluorescent labeling of the cytoplasm (Fig. 33.6; Kolb *et al.*, 1998). This pattern of fluorescence remains largely unchanged for up to 2 h when there are only a few fluorescent vacuoles

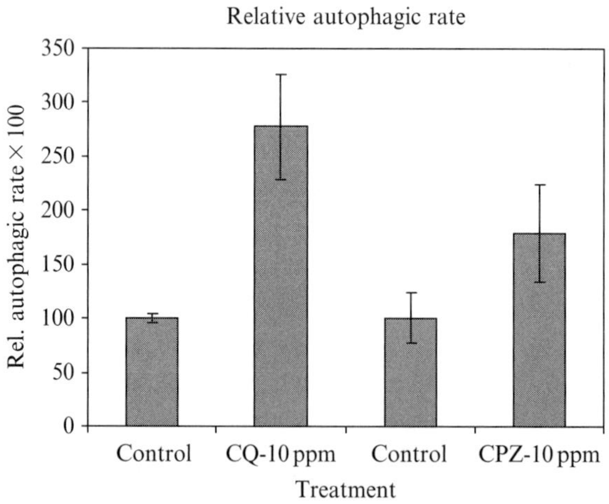

Figure 33.7 Effects of 10 ug/ml (ppm) chloroquine (CQ) and chlorpromazine (CPZ) on lysosomal autophagy of *in situ* FITC-labeled intracellular proteins (mean ± 95% CL, CQ replicates n = 6, CPZ replicates n = 4). Relative rate of autophagy in the drug treatments is the time of endpoint for control/time of endpoint for drug treatment × 100, with paired exposures of blood cell samples for controls and drug treatments. The autophagic rate in the control treatments is is determined as the time of endpoint for control/time of endpoint for control × 100, hence a relative autophagic rate of 100. Adapted from Moore *et al.* (1996).

(1–4 μm diameter) present. Most of these latter also take up the fluorescent lysosomotropic probes neutral red and cresyl violet (both 10 μg/ml, visualized using rhodamine excitation), indicating that those compartments not fluorescing with these dyes are probably a prelysosomal autophagic compartment (Lowe *et al.*, 1995a; Rashid *et al.*, 1991). Between 2 and 3 h, the vacuolar FITC fluorescence increases substantially and most of the FITC-positive vacuoles also sequester the lysosomotropic probes (Fig. 33.6). The time required for evident vacuolar fluorescence in >50% of the cells observed in 10 different fields of microscopic view (approx. 200–400 cells) is taken as the end point for quantification of autophagic transfer of the FITC-labeled cytoplasmic proteins.

Hemolymph (blood) extraction

1. The mussel valves should be carefully pried apart along the ventral surface, using a solid scalpel, which should remain in position in order to keep the valves apart (see Technical Note 1). Allow any water retained within the shell cavity to drain out before attempting to withdraw any hemolymph.
2. Withdraw 0.1 ml of hemolymph from the posterior adductor muscle by using a 1-ml (or 2-ml) hypodermic syringe fitted with a 25-gauge needle and containing 0.1 ml of physiological saline.
3. Having now obtained a sample of hemolymph, remove the needle from the syringe and transfer the contents into a 1.5- to 2.0-ml siliconized microcentrifuge tube.
4. Gently invert the tubes to mix the contents and pipette 50 μl of hemolymph and physiological saline mixture onto each slide, using a clean pipette tip for each sample.
5. Place slides in a light-proof humidity chamber and incubate at 15 °C for 15–20 min to allow attachment to occur (see Technical Note 2).
6. Drain off excess suspension and carefully wipe around the area containing adhered cells to remove any remaining excess fluid.

Technical notes

1. The blade width of a solid scalpel should be sufficient to hold the valves apart to insert a hypodermic needle.
2. It is most important that the cell preparations on microscope slides are maintained at 15 °C throughout the period of cell attachment and FITC diacetate incubation in a light-proof humidity chamber. The slides are placed on racks allowing sufficient space (approximately 3 cm) for air to circulate.
3. When applying the FITC diacetate solution, do not drop the solution onto the cells; touch the surface of the slide with the pipette tip and slowly eject the dye onto the cells.

Preparation of mussel physiological saline

20 m*M* HEPES
436 m*M* NaCl
53 m*M* $MgSO_4$
10 m*M* KCl
10 m*M* $CaCl_2$

The physiological saline should be stored in a refrigerator at 4 °C, raised to 15 °C prior to use and the pH checked and adjusted to 7.36 with 1 *M* NaOH.

3. Autophagic Protein Degradation

3.1. Protein turnover method using *in vivo* ^{14}C-radiolabeled protein

An improved technique for the evaluation of protein catabolism in mussel tissues has been developed by Viarengo *et al.* (1992). The procedure, which requires the calculation of the protein turnover rate from a decay curve (Waterlow *et al.*, 1978), was devised to minimize errors due to isotope reutilization. In fact, the method of the decay curve, which involves the treatment of the animals with a pulse of radiolabeled amino acids, is reliable only when there is no significant reutilization of labeled amino acids resulting from protein degradation.

The procedure takes into account the fact that mussels are filter-feeding mollusks which are able to accumulate high concentrations of various soluble compounds in their cells, such as amino acids, sugars, metals, and so on (Moore and Viarengo, 1987; Viarengo *et al.*, 1992). This capacity has been utilized to rapidly obtain high levels of ^{14}C-leucine in the cells of the animals exposed to the radioactive amino acid, and subsequently (i.e., when significant labeling of protein occurred) to induce a rapid and drastic reduction of the specific activity of the precursor pool by adding an excess of nonradioactive leucine to the medium.

In this way, an accurate estimation of the mean half-life of the cytosolic proteins in the digestive gland of the mussels can be achieved. This method has been employed to evaluate the effects of sublethal concentrations of phenanthrene, an organic xenobiotic compound known to adversely affect the stability of the lysosomal membrane in molluskan digestive cells (Moore, 1979; Nott and Moore, 1987). Lysosomes are known to play an important role in the catabolism of macromolecules (Glaumann and Ballard, 1987; Grinde, 1985; Hawkins and Day, 1996), and the half-life of the cytosolic proteins from both the control and phenanthrene-treated mussels was directly correlated with the stability of the lysosomal membranes (Fig. 33.8; Moore and Viarengo, 1987).

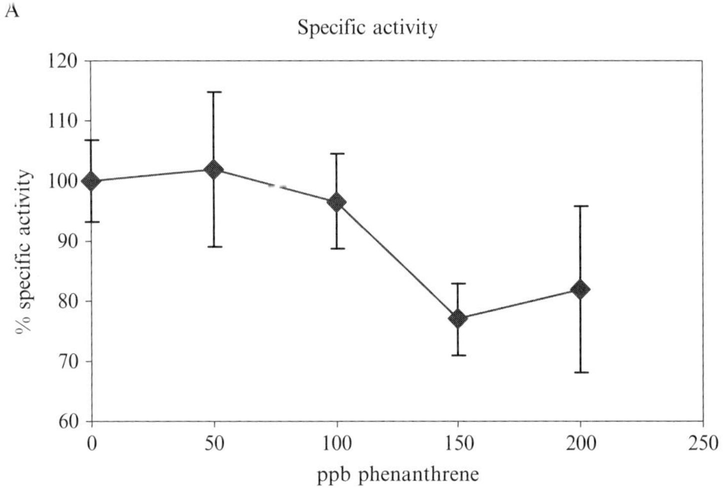

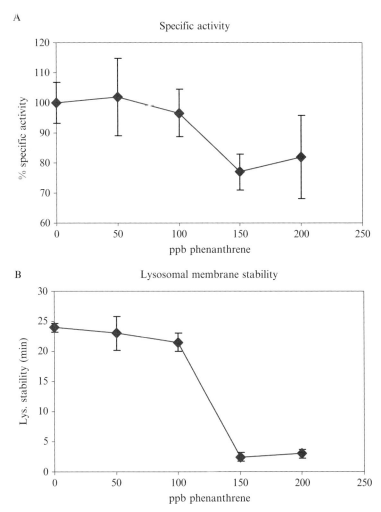

Figure 33.8 (A) The effect of 24-h treatment with phenanthrene (at varying concentration) in seawater on specific activity of ^{14}C-labeled cytosolic proteins (as percent of control value) in the hepatopancreas (each point is the mean ± SE of at least 5 replicate experiments). (B) The effect of phenanthrene on lysosomal membrane stability in hepatopancreatic digestive cells (each point is the mean ± SE of at least 5 replicate experiments). Seawater was spiked with a single aliquot of phenanthrene dissolved in acetone at the start of the experiment to give the ascending treatment concentrations. Values at initial concentrations of phenanthrene of 150 and 200 ppb exposure for 24 h are significant at $P < 0.05$. Lysosomal membrane stability is strongly correlated with the autophagic degradation of proteins ($R^2 = 0.988$, $P < 0.001$). See experimental details in Moore and Viarengo (1987).

3.1.1. Method

3.1.1.1. Experimental animals Mussels (4–6 cm shell length) are maintained at 15 °C, in aerated, artificial seawater (Viarengo *et al.*, 1992), at 35 psu (practical salinity units) and one liter of seawater per animal for 4 days. The animals are fed with a continuous supply of a mixed algal food (30 mg dry weight algae/animal/day) as previously described (Viarengo *et al.*, 1987). During the experimental treatments the water is changed daily.

Evaluation of the rate of protein catabolism

1. Groups of 6–8 mussels are exposed to ^{14}C-leucine (specific activity = 300 mCi/mmole) for 24 h (1 pCi/animal).
2. Subsequently, a cold chase is used whereby the animals are maintained for 22 h in seawater containing a high concentration of unlabeled leucine (100 mg/l/animal). At the end of the 22-h cold-chase period, a set of mussels is collected and utilized as reference (time 0) for the protein radioactivity decay experiments.
3. The mollusks are then exposed to experimental treatments to evaluate the effect of the treatment on the loss of radioactivity from the cytosolic proteins of the digestive gland. Mussels are exposed to unlabeled leucine (50 mg/1/animal) during the treatment.

Evaluation of the specific activity of the cytosolic proteins

The autophagic protein degradation is determined by following the decrease in acid-precipitable radiolabeled protein over time. This analysis ensures that the cold chase condition selected is suitable to minimize the reutilization in newly synthesized proteins of the ^{14}C-leucine utilized in the experiments.

1. The mussel digestive glands from 8–9 mussels are homogenized in 10 volumes of 0.5 *M* sucrose containing 0.15 *M* NaCl, buffered with 20 m*M* Tris-HCl, pH 8.6, with 139 m*M* trypsin inhibitor, 0.5 m*M* PMSF (phenylmethylsulphonyl fluoride), and 3 μg/ml leupeptin.
2. The homogenate is centrifuged at 30,000*g* for 20 min; the supernatant is then centrifuged at 100,000*g* for 90 min at 4 °C to obtain the cytosolic fraction.
3. Aliquots of 25 μl of the cytosolic fractions are used to analyze the protein content (see subsequent sections). Caution must be exercised, as the samples are radioactive.
4. Aliquots of 100 μl of the cytosolic fractions are spotted onto Whatman 3 *MM* filter paper squares (2.3-cm edge length). These are utilized to evaluate the incorporation of the labeled precursor into proteins, essentially as described by Yu and Feigelson (1971). The squares are extracted 4 times in 5% TCA (15 ml/square) for 20 min at 0 °C to remove the acid-soluble radioactivity. The TCA is discarded as radioactive waste.

5. Filters are successively extracted for 20 min in 5% TCA at 90 °C, and then further washed twice with cold 5% TCA, twice with ethanol-ether 1:1 (v/v) and then with diethyl ether before being dried.
6. Radioactivity is measured in a liquid scintillation spectrometer. Each sample was counted in 5 ml of Instagel for enough time to assure a counting error no greater than 5%. Radioactivity values, due to the incorporation of the labeled amino acid into the protein fraction, are expressed per mg of cytosolic proteins.

Evaluation of the specific activity of the free leucine cellular pool

1. The digestive gland is homogenized in 10 volumes of methanol and then the preparation is centrifuged at 15,000*g* at 4 °C for 15 min.
2. The supernatant is filtered on a Millipore 0.022-mm filter (Millex).
3. The concentration of soluble leucine in the methanol extract is evaluated by HPLC separation of the *o*-phthaldialdehyde (OPT) derivatives on a reverse-phase column (Nucleosil RP-18.5 pm), which uses 0.1 *M* Na-phosphate buffer, pH 7.4, as a mobile phase (essentially as described by Lindroth and Mopper, 1979).
4. The presence of the OPT derivatives in the eluates is detected using a spectrofluorimeter (Exc = 370 nm; Em = 418 nm). The leucine radioactivity is determined by direct evaluation of the radioactivity present in 100 pl of methanol extract (corresponding to the extracted leucine).

3.1.1.2. Protein assay Protein is evaluated according to Harthree's method (1972) using albumin fraction V from bovine serum (VWR, Cat. No. 1.12018.0025) as a standard.

4. Morphometric Methods

4.1. Lysosomal/cytoplasmic volume ratio

It has been previously demonstrated by Langton (1975) that in mussel hepatopancreas (digestive gland) the cells may be in any one of four different phases (i.e., resting, digestive, excretory, and reconstituting).

Lowe (1988) reported that in mussels exposed to contaminants for a long period of time, the digestive gland tubules contain the majority of cells in the reconstituting phase. In this case the tubules show a very thin epithelium (Cajaraville *et al.*, 1990, 1992). These results may be easily interpreted by taking into account that, as reported earlier, the exposure of mollusks to environmental stressors causes a destabilization of lysosomal membranes. Such activation of the lysosomal vacuolar system is usually associated with an increase in the lysosomal volume due to the formation of active secondary autophagic lysosomes. Image analysis of tissue sections prepared for evaluation of lysosomal membrane stability, as described in the next section, enables measurement of changes in the lysosomal/cytoplasmic volume ratio. An increase in this value clearly indicates that the digestive gland cells are becoming catabolic, which may alter the functionality of the organism (Figs. 33.4 and 33.9; Lowe *et al.*, 1981, Moore and Clarke, 1982; Moore *et al.*, 2006a, 2007).

A drastic decrease in lysosomal membrane stability (as usually happens in pollutant-exposed mollusks) is, therefore, associated with a significant increase in autophagy and this, with time, can cause a loss of cell cytoplasm; and consequently a reduction in their physiological status (Lowe,1988; Lowe *et al.*, 1981; Moore, 1988; Moore and Clarke, 1982; Moore *et al.*, 2007).

The evaluation of the lysosome/cytoplasm volume ratio and the thickness of the digestive gland epithelium may therefore represent two simple ways to follow the changes in autophagy rate and its effects on the digestive gland cells of mollusks.

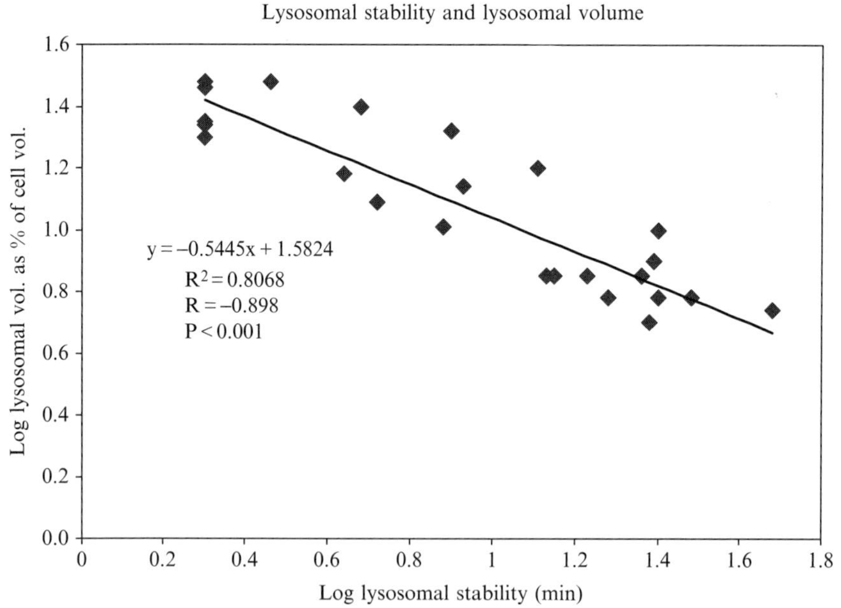

Figure 33.9 Inverse linear relationship between lysosomal stability (determined cytochemically) and lysosomal volume in mussel hepatopancreatic digestive cells using log-transformed data. Each point on the graph is the mean for a sample of at least 10 animals. Original data used from Clarke and Moore (1983), Lowe and Moore (1981), Moore and Clarke (1982), Moore, previously unpublished data and Widdows *et al.* (1982).

4.1.1. Method

Mussel digestive glands are extracted from the animals and processed to obtain 10-μm cryostatic sections as described subsequently (see section "Preparation of tissue sections for enzyme cytochemistry"). Sections are then processed for the evaluation of lysosomal membrane stability using, as marker, the lysosomal enzyme N-acetyl β hexosaminidase using the cytochemical method described herein (Moore, 1976; Moore *et al.*, 2004b).

The lysosome/cytoplasm volume ratio is then determined in the stained sections associated with the peak of enzyme activity as used for the evaluation of the lysosomal membrane stability (see section on lysosomal stability herein). Finally, the lysosome/cytoplasm volume ratio is calculated using image analysis (Scion Image freeware).

5. Autophagy-Related Lysosomal Membrane Stability Methods

5.1. Cytochemical methods for lysosomal membrane stability

Cytochemistry and histochemistry have been used as the main tools in the study of environmentally induced alterations in lysosomes of lower animals for several reasons (Moore, 1990; Moore and Simpson, 1992). Frequently, the tissue samples are very small and cytochemistry lends itself well to dealing with this problem. In addition there is a requirement to be able to relate functional changes in the tissues and cells to alterations in their structure; once again cytochemistry is highly appropriate in this context. The cytochemical tests used in these investigations involve procedures for lysosomal hydrolases, lipofuscin and lipid (Moore, 1976, 1988; Fig. 33.4). Another advantage of the cytochemical approach is that changes can be detected in particular target cells and lesion types, thus potentially increasing the sensitivity by many orders of magnitude as compared with the more disruptive analytical procedures involving homogenization and cell fractionation.

When fish and mollusks, such as mussels or snails, are exposed to contaminant chemicals, the lysosomes in the hepatocytes and digestive gland epithelial cells show fairly rapid and characteristic pathological alterations as described above (Figs. 33.2, 33.3 and 33.4 Broeg *et al.*, 1999a,b, 2002, 2005; Kohler, 1989a,b, 1990, 1991; Kohler *et al.*, 1992, 2002; Lowe, 1988; Moore, 1988, 1990). These include swelling of the digestive cell and hepatocyte lysosomes (Figs. 33.3 and 33.4), which is correlated with increased fragility of the lysosomal membrane (Fig. 33.9; Moore *et al.*, 2006a), excessive buildup of unsaturated neutral lipid (lipidosis) in the lysosomal compartment (Figs. 33.3 and 33.4) and accumulation of lipofuscin (lipofuscinosis)

(Figs. 33.3 and 33.4). These changes are accompanied by atrophy of the digestive epithelium, apparently involving augmented autophagic processes, although there is also evidence of increased cell deletion (analogous to apoptosis in mammals) and the relationship between the two processes, if any, is unclear (Lowe, 1988; Pipe and Moore, 1985). For instance, do the autophagic-type changes predispose the cells to deletion by programmed cell death (PCD)? Programmed cell death is divided into apoptosis (PCD Type I) and autophagy-prominent cell death (PCD Type II): autophagic cell death appears to be a phylogenetically ancient phenomenon and occurs in both physiological and disease states (Bursch, 2001; Zhao *et al.*, 2001). Lysosomal changes are involved in both types of cell death and they should not be considered as mutually exclusive processes (Bursch, 2001).

Linked biochemical and cytochemical investigations have demonstrated that increased fragility of the lysosomes, induced by phenanthrene, corresponds directly with increased catabolism of cytosolic proteins (Fig. 33.8; Moore and Viarengo, 1987; Viarengo *et al.*, 1992).

Experimental studies have clearly demonstrated that the lysosomal alterations described earlier can be induced by single toxicants such as copper (although lipidosis is less prominent with toxic metals) and polycyclic aromatic hydrocarbons (Moore, *et al.*, 2006a, 2007; Viarengo *et al.*, 1985).

In fish liver, lysosomal changes comprise membrane fragility, enlargement, and accumulation of lipids (unsaturated neutral lipids, phospholipids) and lipofuscin. These changes are closely linked to toxico-pathological alterations of the fish liver and have clear prognostic value for cell death, and are also correlated with concentrations of lipophilic compounds and some heavy metals such as cadmium (Kohler *et al.*, 2002). Interestingly, lysosomal membrane stability breakdown coincides with induction of cytochrome P-450 (CYP1A1) (Kohler and Pluta, 1995). It is likely that reactive free radicals (reactive oxygen species and reactive xenobiotic derivatives) produced during biotransformation contribute to the damaging effects on the lysosomal membrane and build up of lipofuscin (Kirchin *et al.*, 1992; Winston *et al.*, 1991, 1996). Lipofuscin is an end product of oxidative attack on lipids and proteins and is also an indicator of autophagy (Brunk and Terman, 2002; Moore, 2008).

At first glance these findings are perhaps surprising given that many thousands of individual toxic chemicals are often present in a contaminated situation (Fig. 33.4). Lysosomal destabilization is essentially very generalized and can also be induced by nonchemical stressors such as hypoxia, hyperthermia, osmotic shock, dietary depletion, and various combinations of these (Moore, 1985). Consequently, it would appear that many adverse conditions are capable of inducing autophagic-type changes; and that this nonspecificity of the lysosomal reactions is only of value as a general indicator of deterioration in the health of the animal. However, differences in the lysosomal response can be used to identify the causative agency. Specifically, patterns of lysosomal change can be used to distinguish between the effects induced by lipophilic organic xenobiotics, metals and nonchemical stressors. These include lysosomal swelling and lipid accumulation induced by lipophilic xenobiotics but not metals; and accumulation of metallothionein in lysosomes induced by particular metals (Fig. 33.1D; Kohler *et al.*, 2002; Moore, 1988, 1990; Viarengo *et al.*, 1985). Considered as a package, the use of cytochemical tests as subcellular pathological probes can provide relatively specific information on autophagic processes (Moore, 1990).

The types of cytochemical tests described herein have been used in a range of environmental situations. The more widely used tests have been those for lysosomal membrane fragility; this has been applied to both molluskan and fish species and is based either on the demonstration of latency of lysosomal hydrolases, or the retention of the amphiphilic cationic dyes such as neutral red and acridine orange (Fishelson *et al.*, 1999; Lowe *et al.*, 1992, 1995a; Moore, 1990; Moore and Simpson, 1992). Exposure to a variety of contaminant effluents such as sewage sludge, pulp-mill waste, oil spillages and mixed wastes from industry have all been found to increase the fragility of molluskan digestive cell lysosomes as well as fish hepatocyte lysosomes (Broeg *et al.*, 1999a,b, 2002, 2005; Cajaraville *et al.*, 2000; Kohler, 1989; Kohler, 1991a,b, 2004; Kohler *et al.*, 1992; Lowe *et al.*, 1992, 1995a; Moore, 1985, 1988; von Landwüst *et al.*, 1996; Wahl *et al.*, 1995; Wedderburn *et al.*, 2000). In general, the reduction in lysosomal stability is accompanied by enlargement or swelling. Fatty change is also a frequent reaction to xenobiotics in the digestive cells and fish hepatocytes, leading to apparent autophagic uptake of the unsaturated neutral lipid into the often already enlarged lysosomes (Figs. 33.1E, 33.3, and 33.4; Kohler *et al.*, 2002; Moore, 1988).

Clearly then, the cytochemical probes are providing data, which are entirely consistent with data obtained using biochemical, histopathological and physiological approaches (Moore *et al.*, 1980, 2006a, 2007; Viarengo *et al.*, 1992). Such good agreement provides strong support for the validity of the cytochemical data, particularly as a logical conceptual framework can be devised linking pathological changes at the molecular and subcellular levels of organization to impairment of the physiological performance of the whole animal (Bayne *et al.*, 1988). The inference here is that lysosomal stability/fragility is a prognostic indicator or biomarker for putative pathologies and as such is an integrated pathophysiological indicator of health status (Fig. 33.10; Moore, 1990, 2002; Moore *et al.*, 2004a,b, 2006a).

The advantages of using cytochemical approaches in pollutant effects assessment are several. First, cytochemistry is capable of providing information that can shed light on the molecular and subcellular mechanisms of pathological alteration induced by the contaminants. Second, cytochemistry can be applied to very small tissue samples and to sections obtained from a single tissue sample, and these can be readily varied to meet the particular

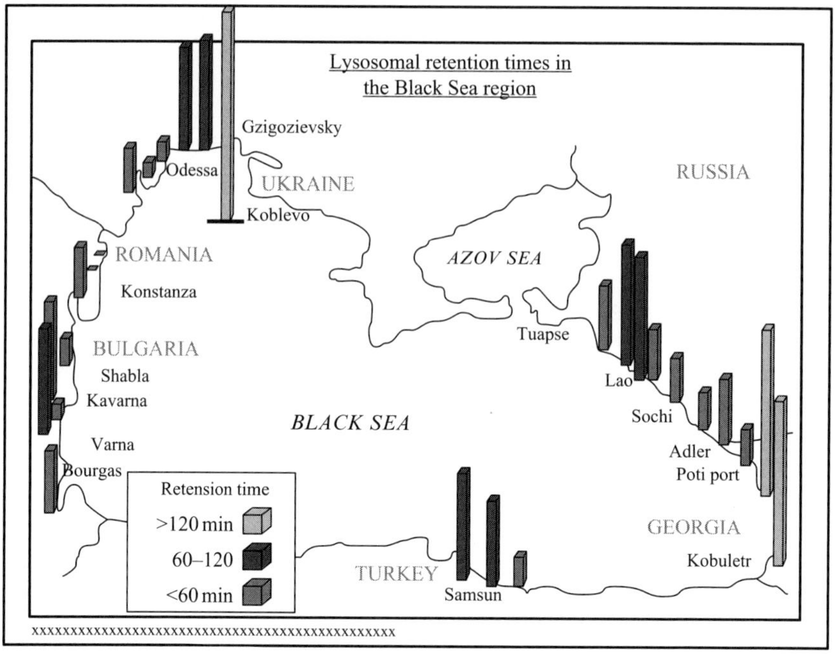

Figure 33.10 Summary results of the UNESCO-IOC Biological Effects Mussel Watch Programme in the Black Sea. Lysosomal stability was determined in mussels sampled from the coasts of the six littoral countries surrounding the Black Sea using the neutral red retention method. Lysosomal retention values of less than 60 min indicate severely impaired health, and it is apparent that much of the coastline is subject to environmental degradation on the basis of the lysosomal stability data. With ICES permission from Moore *et al.* (2004b). This study was a component of the GEF-Black Sea Environment Programme. (See Color Insert.)

requirements of the situation. Finally, most of the cytochemical tests used in the studies described above can be readily quantified by microdensitometry, image analysis or categorical assessment using a ranked series of photomicrographs (Chieco *et al.*, 2001; Moore, 1988).

5.1.1. Method

Preparation of tissue sections for enzyme cytochemistry

1. For cytochemical examination, small pieces (5 mm × 5 mm × 5 mm) of freshly excised molluskan digestive gland tissues or fish liver (i.e., the midportion of the organ) from 10 animals are placed on metal cryostat chucks (e.g., up to 5 pieces of tissue in a straight row across the center).
2. Each chuck is then placed for 1 min in a small bath of *n*-hexane (aromatic hydrocarbon-free; boiling range 67 °C–70 °C) that has been

precooled to −70 °C (using a surrounding bath of liquid nitrogen or, alternatively, a mixture of crushed solid CO_2, and acetone). For fish liver and digestive gland from marine mussels and snails, the tissue can also be deep-frozen directly in liquid nitrogen.

3. The metal chuck plus the quenched (supercooled) solidified tissues are then sealed with aluminum foil and by double wrapping in parafilm and stored at −30 °C or, preferably, at −70 °C until required for sectioning. Tissues may be stored for 6–12 months at −70 °C. By following this procedure there is no evident formation of large ice crystals and hence no structural damage to the subcellular components (Moore, 1976).
4. Cryostat sections (10 μm) are cut in a cryostat (preferably with motorized cutting), with the cabinet temperature below −25 °C and with the haft of the knife cooled with crushed solid carbon dioxide (dry ice).
5. The sections are transferred to warm slides (i.e., 20 °C, or room temperature), which effectively flash-dries them (Moore, 1976), and the slides can be stored in the cryostat for at least 4 h before use. Cryostat sections that are required for concurrent structural or nonenzymic cytochemistry (e.g., lipid and lipofuscin) can be fixed in Baker's calcium formol or 10% neutral formalin (+2.5% NaCl w:v).

5.1.2. Demonstration of latent activity of lysosomal hydrolases for assessment of lysosomal stability

Latent lysosomal activity of the lysosomal enzymes N-acetyl-β-hexosaminidase and β-glucuronidase can be demonstrated in the digestive cells of bivalve mollusks using naphthol AS-BI substrates and post-coupling with diazonium salts to prevent inhibition by the coupler. For fish liver we recommend the use of either N-acetyl-β-hexosaminidase, β-glucuronidase or acid phosphatase.

Technique for N-Acetyl-β-hexosaminidase

1. Serial cryostat sections (in duplicate on the same slide) prepared as described previously are pretreated in a histological staining jar by immersion in 0.1 *M* citrate buffer (pH 4.5) containing 2.5% NaCl (w:v) at 37 °C to labilize (controlled permeablization) the lysosomes (Moore, 1976). Sections on slides are incubated in a sequentially timed pretreatment series starting at 30 min for mollusks down to 2 min (i.e., 30, 25, 20, 15, 10, 5, and 2 min) and 50 min for fish down to 2 min (i.e., 50, 40, 30, 25, 20, 15, 10, 5, and 2 min). Two (mollusks) or three (fish) minutes is used as the minimal pretreatment time since sections that have undergone zero pretreatment may sometimes show stronger staining than short-term pretreated sections (Moore, 1976). This staining activity is believed to be largely due to nonmembrane bound acid hydrolase (e.g., from damaged lysosomes at the cut surfaces of the section) that can be lost by diffusion from the section when no polypeptide stabilizer is present.

The enzyme activity is frequently localized in large secondary lysosomes or digestive vacuoles, which may be damaged in sectioning. Due to this complicating factor, the zero preincubation is usually omitted and the 2-min pretreatment is taken as representing the free lysosomal enzyme activity.

2. Following this pretreatment sequence, the slides are transferred to the substrate incubation medium; this contains 20 mg of naphthol AS-BI N-acetyl-β-glucosaminide (Sigma) dissolved in 2.5 ml of 2-methoxyethanol, which is made up to 50 ml with 0.1 *M* citrate buffer, pH 4.5, containing 2.5% NaCl (w:v) and 3.5*g* of low viscosity polypeptide (Sigma, POLYPEP P5115) to act as a section stabilizer (Moore, 1976). Incubation time is 20 min at 37 °C in a staining jar, preferably in a shaking water bath.
3. The slides are subsequently rinsed in 3.0% NaCl at 37 ° C for 2 minutes before being transferred to 0.1 *M* phosphate buffer, pH 7.4, containing a diazonium coupler (1 mg/ml) at room temperature for 10 min. Suitable diazonium salts are fast violet B (Sigma), fast red violet LB (Difco), fast garnet GBC (Sigma), fast blue BB (Sigma) and fast blue RR (Sigma). Our experience has been that fast violet B is the most suitable.
4. The slides are then rinsed rapidly in running tap water, fixed for 10 minutes in Baker's calcium formol (10% formalin + 1.0% $CaCl_2$) containing 2.5% NaCl (w:v) at 4 °C, rinsed in distilled water, and glass coverslips are mounted using aqueous mounting medium (e.g., Shandon Immuno-mount, Cat. No. 9990402).

5.1.3. Technique for β-glucuronidase

The method for the demonstration of latent activity of lysosomal β-glucuronidase (Moore, 1976) is essentially similar to the method described previously, but with the following exceptions: the pretreatment to labilize the lysosomal membranes is carried out using 0.1 *M* acetate buffer, pH 4.5, containing 2.5% NaCl (w:v), and the substrate incubation uses 14 mg of naphthol AS-BI β-D-glucuronide (Sigma) as substrate dissolved in 0.6 ml of 50 m*M* $NaHCO_3$, which is made up to 50 ml with 0.1 *M* acetate buffer, pH 4.5, containing 2.5% NaCl (w:v) and 3.5*g* of polypeptide (Sigma POLYPEP P5115) at 37 °C for 20 minutes.

Rinsing and coupling solutions for β-glucuronidase are the same as those used for N-acetyl-β-hexosaminidase.

5.1.4. Determination of lysosomal labilization period (i.e., permeablization time for latent hydrolase)

The labilization period is the time of pre-treatment required to labilize the lysosomal membranes fully, resulting in maximal staining intensity for the enzyme being assayed (Figs. 33.11 and 33.12).

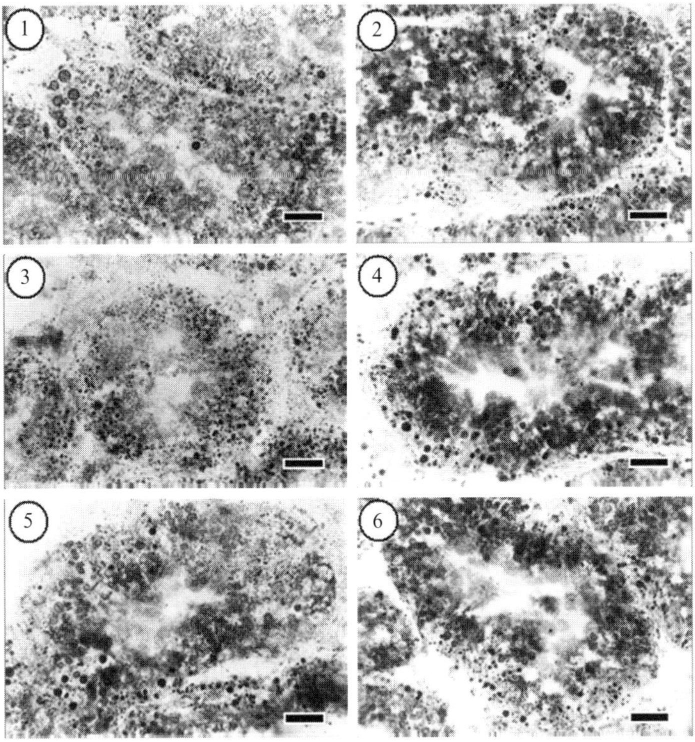

Figure 33.11 Serial cryostat sections of the digestive gland stained to show N-acetyl-β-hexosaminidase reactivity in the lysosomal/vacuolar system of digestive cells in a digestive tubule of a mussel. (1–6) Sections pretreated at pH 4.5 and 37 °C for 2–25 minutes (2, 5, 10, 15, 20, 25 min). Section (5) pretreated for 20 min shows maximal lysosomal staining intensity: this time of pretreatment represents the labilization period. Section (6) pretreated for 25 min shows a decrease in staining intensity indicating a probable loss of enzyme by diffusion from fully labilized lysosomes. With ICES permission from Moore *et al.* (2004b). Scale Bar = 20 μm.

The staining intensity can be assessed visually using microscopy examination or else measured using a scanning integrating microdensitometer or image analyzer to obtain an activity plot as shown in Fig. 33.12 (Moore, 1976). If the animal is stressed, then the peak of activity will be moved toward the y-axis and the decreased labilization period can be determined from the x-axis (Fig. 33.12).

Our experience has shown that a microdensitometer is not necessary for accurate determination, and that the labilization period can be effectively measured by microscopy assessment of the maximum staining intensity in the pre-treatment series (Fig. 33.11). For this procedure, each tissue section should be divided into four roughly equal areas for assessment. This can be done by means of drawing a cross on the cover slide overlaying each section

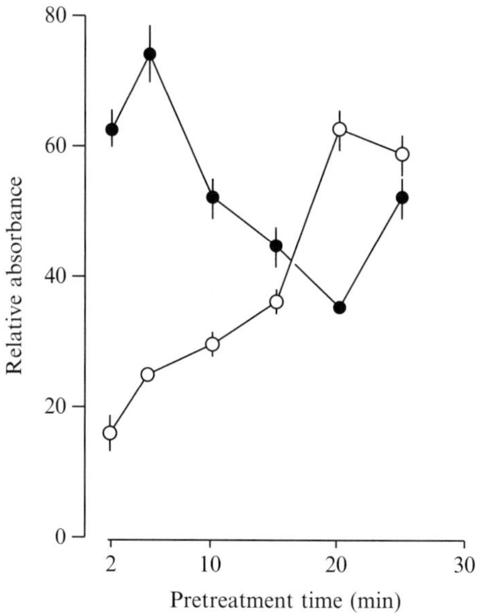

Figure 33.12 Microdensitometric determination of N-acetyl-β-hexosaminidase activity in sequentially pretreated (labilized) tissue sections of mussel digestive gland. Healthy cells: open circles; stressed (hyperthermia) cells: filled circles (Moore, 1976). With ICES permission from Moore *et al.* (2004b). Means $\pm$ SE, n = 10.

with a very fine marker pen, thus giving 4 quadrants. The position and orientation of the cross on all sections should be the same.

All assessments should be carried out on *duplicate sections for each digestive gland or liver* at each pretreatment time. A mean or median value is obtained for each set of duplicate sections from the average of the assessments in each of the four subdivisions (i.e., quadrant 1 from all sections in the sequence, then 2, 3, and 4). These data can be statistically tested (i.e., test data compared with references or baseline data) using the nonparametric Mann-Whitney *U*-test or Kruskal-Wallis test (Siegel, 1956). The Tukey *t*-test or analysis of variance can also be used with log-transformed data.

5.1.5. Problems in assessment of labilization period

Determination of labilization period is usually quite straightforward, but a complicating situation occasionally arises in which the pre-treatment series shows two peaks of staining intensity (Moore *et al.*, 1978a,b), possibly due to differential latent properties of the subpopulations of lysosomes (Fig. 33.11). In this situation the first peak of activity is used to determine the labilization period as in our experience it has been the most responsive (Fig. 33.12). In fish liver, two peaks are also frequently observed; and the first peak of activity is strongly correlated with the degree of liver lesions (Kohler *et al.*, 2002; Moore *et al.*, 2004b).

5.1.6. Additional methods for autophagy-related lipidosis and lipofuscinosis

Neutral lipid and lipofuscin can be localized following the protocols for the Oil Red O and Schmorl methods, respectively (Bayliss High, 1984; Moore, 1988); and quantified microdensitometrically as described by Moore (1988).

6. *In vivo* Neutral Red Retention Method for Lysosomal Stability (Cellular Dye Retention)

Lysosomes have a remarkable capability for accumulating a diverse range of toxic metals and organic chemicals (Figs. 33.1, 33.2, and 33.5; Einsporn and Kohler, 2008a,b and submitted; Moore, 1985, 1990; Rashid *et al.*, 1991; Viarengo *et al.*, 1985). However, this concentration of toxic contaminants results in lysosomal damage and cell injury; and possible leakage of contaminants into the cytosol. The fact that lysosomes accumulate this very wide range of xenobiotics, dyes and drugs has been used to advantage in the development of an *in vivo* cytochemical method for determining lysosomal membrane damage (Dondero *et al.*, 2006; Fishelson *et al.*, 1999; Lowe *et al.*, 1992, 1995a,b).

Babich and Borenfreund (1987) published a method in which alterations in the capacity of cells to take up the dye neutral red (NR) was used as an indicator of cell damage. The rationale here was that healthy cells could take up and retain larger quantities of the dye than damaged cells. The method involved exposing cells with the test medium and then incubating them in a neutral red solution. Following incubation, the dye remaining in the cells was extracted and measured spectrophotometrically. Lowe *et al.* (1992) reasoned that if the dye could be measured with a spectrophotometer it could also be visualized by using a microscope. Indeed, if the lysosomally accumulated dye could be visualized, then the progress of dye uptake into the cells and, in the case of damaged cells, leakage back into the cytosol could be determined and quantified using the lysosomal retention time as a sensitive measure of effect (Fig. 33.10). The methods developed for fish hepatocytes, mussel digestive gland cells and oyster digestive gland cells use microscopy to assess neutral red retention; and involve sacrificing the animals followed by enzymatic digestion of tissues (Lowe *et al.*, 1992; Lowe and Pipe, 1994, Lowe *et al.*, 1995a,b; Ringwood *et al.*, 1998a,b, 1999). NR retention has also been used successfully in the social amoeba *Dictyostelium discoideum* by Dondero *et al.* (2006).

These approaches are valuable for concurrent studies with other cellular function studies, and particularly when small animals or protozoans are used. In contrast, blood cells, which are generally easy to obtain without harming the host, offer a sensitive but robust lysosome-rich cell type that can be studied using *in vitro* methods; and this provides the opportunity for further contaminant effect studies (Lowe *et al.*, 1995a; Grundy *et al.*, 1996a,b). In their role as components of the innate immune system, blood cell lysosomes can release acid hydrolases that are able to degrade circulating pathogens (Grundy *et al.*, 1996a,b). However, unscheduled or inappropriate release of acid hydrolases may have disastrous consequences for the functional integrity of the cell.

6.1. Method

The neutral red retention technique for blood cell lysosomes is nondestructive; hence, if the animals under test are not unduly stressed during collection they can be returned to their habitat following careful extraction of a blood sample. Mussel byssal threads should be cut from the substrate, since pulling the animals from the rocks can result in damage to internal tissues. Extremes of temperature during transport are also to be avoided and the animals must be maintained in a moist environment during transport to the laboratory.

Stock solutions of physiological saline and neutral red should be prepared in advance and stored in a refrigerator. The neutral red stock solution will solidify in the refrigerator and should also be raised to room temperature for dilution to the working strength.

6.1.1. Hemolymph (blood) extraction (as described in the *in vivo* autophagy section)

Technical notes

1. It is most important that the slide preparations are kept cool throughout the period of cell attachment and neutral red incubation (approx. 15 °C). This can be achieved by having a thin layer of water ice in the light-proof humidity chamber. The slides must not be in direct contact with the ice and should be placed on racks allowing sufficient space (approximately 3 cm) for the chilled air to circulate.
2. When applying the neutral red working solution do not drop the solution onto the cells, touch the surface of the slide with the pipette tip and slowly eject the dye onto the cells.
3. Neutral red is a photosensitizer, therefore all slides should receive the same exposure to light under the microscope; and the light intensity should be kept as low as possible.

6.2. Neutral red stock solution

Prepare a 100 m*M* stock solution of neutral red by dissolving 28.8 mg of dye powder in 1 ml of DMSO and store in the refrigerator prior to use. However, the solution will solidify in the refrigerator and should be raised to room temperature for dilution to the working strength stock. The stock solution will last for about 2–3 weeks when stored in this way.

The working solution of neutral red dye is prepared by diluting 10 μl of stock neutral red in 5 ml of mussel physiological saline (see *Visual detection of autophagic responses* above) The working solution will last about 4 h before the dye begins to precipitate out. Neutral red dye powder is commercially available in a range of purities and strengths. If possible, the highest strength/purity dye (Sigma/Aldrich) should be used. However, what is most important is that only dye batches of similar quality and concentration are used when making comparisons between sites/treatment in an experiment or monitoring exercise. Different grades of dye will have a different effect on the lysosomes depending on their purity and strength.

Neutral red incubation

1. Pipette 40 μl of NR working solution onto the haemocytes, wait 15 minutes to allow neutral red to penetrate cells (see Technical Notes 2 and 3).
2. Gently apply a cover glass. Systematically examine slides under a light microscope after 15 min and then again after a further 15 min. Subsequent examination should be made at intervals of 30 min up to 120 min. The final examination should be made after 180 min of incubation. If possible the whole slide should be scanned and replaced in the chamber as quickly as possible: ideally 1 min viewing time per slide maximum (see Note 4).
3. Cells should be examined for both structural abnormalities and NR probe retention time. Conditions should be recorded in a table at each time increment. The retention time of the NR probe by the lysosomes is recorded by estimating the proportion of cells displaying leakage from the lysosomes into the cytosol and/or exhibiting abnormalities in lysosomal size and color. Cell shape may also change as a consequence of contaminant impact.

6.2.1. Determination of neutral red retention endpoint

The end point is attained when 50% or more of the cells, based on either a visual or digital photographic determination (see below), exhibit lysosomal leakage or show abnormalities such as enlargement (Fig. 33.13). A more objective approach, that would be appropriate for certain types of studies, which is used in some laboratories, is to photograph fields of view, using a digital camera, and then make detailed counts of cells exhibiting dye loss at a later point in time. However, this does remove the capability for real time results.

To minimize the length of time the cells are exposed to light under the microscope it is possible only to make a visual estimate of the condition of the lysosomes; as this approach is potentially open to bias, it is recommended that whenever possible samples are read 'blind'. A typical blood

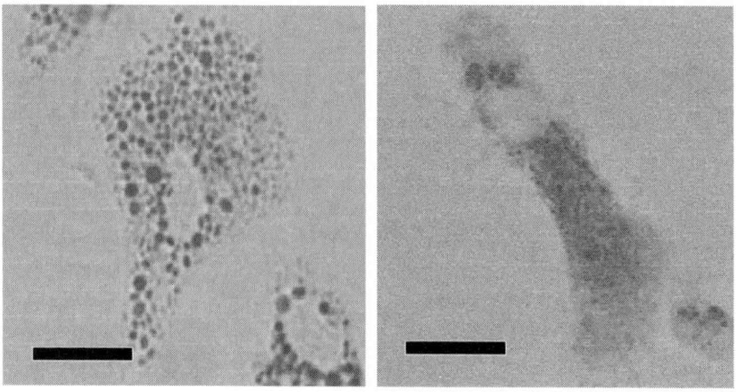

Figure 33.13 Mussel blood cells (hemocytes) showing uptake of neutral red in lysosomes. Healthy cells showing the retention of neutral red within the lysosomal compartment (left-hand micrograph). Fluoranthene treated cells showing loss of neutral red into the cytosol (Lowe *et al.*, 1995b; right-hand micrograph). With ICES permission from Moore *et al.* (2004b). Scale Bar = 5 μm.

sample of 50 μl generates approximately 20 fields of view of attached cells, by quickly rastor scanning the preparation under the microscope it should be possible, with practice, to obtain a visual estimate of the condition of the lysosomes for the entire sample in one minute or less. The number of blood cells in individual mussels is highly variable and a field of view, using a x25 objective lens, may contain anything between 20 and 50 cells, therefore the analysis assesses the lysosomal membrane status of between 400 and 1000 cells.

The test is terminated after 180 min, since for most healthy animals the neutral red itself becomes a toxic xenobiotic stress factor, regardless of the previous contaminant history of the mussels under study. Appropriate statistical tests should then be applied to the data (e.g., Tukey *t*-test on log-transformed data, Mann-Whitney *U*-test or Kruskal-Wallis test).

7. Concluding Remarks: Application of Lysosomal-Autophagic Reactions to Evaluation of the Health of the Environment

7.1. Lysosomal-autophagic reactions as ecotoxicological biomarkers of the harmful impact of pollutants

Cellular changes related to autophagy, such as reduced lysosomal membrane stability, lipofuscin (age or stress pigment) accumulation and other lysosomally related assays, are good indicators of cell injury and animal health status (Broeg *et al.*, 1999a,b, 2002, 2005; Moore *et al.*, 2007). In a situation where exposure to environmental stressors is likely to be sustained, lysosomal biomarkers can be used to predict that further pathological changes will occur. Lysosomal and autophagic functional perturbations also appear to have potential as measures of damage to ecological health (Moore, 2008; Moore *et al.*, 2004a, 2006a). The potential prognostic use of lysosomal reactions to environmental pollutants has been explored in relation to predicting animal health in marine mussels and flatfish (dab and flounder), based on diagnostic biomarker data. Integration of multiple biomarker data has been achieved using multivariate statistics and then mapped onto health status space by using lysosomal membrane stability as a measure of cellular well-being (Moore *et al.*, 2006a). This is viewed as a crucial step toward the derivation of explanatory frameworks for prediction of pollutant impact on animal health; and has facilitated the development of a conceptual mechanistic model linking lysosomal damage and autophagic dysfunction with injury to cells, tissues, and the whole animal (Fig. 33.14; Moore, 2008; Moore *et al.*, 2000a,b, 2007). This model has also complemented the creation and use of cell-based bioenergetic computational models of molluskan hepatopancreatic cells that simulate lysosomal and cellular reactions to pollutants (Allen and McVeigh, 2004; McVeigh *et al.*, 2004, 2006).

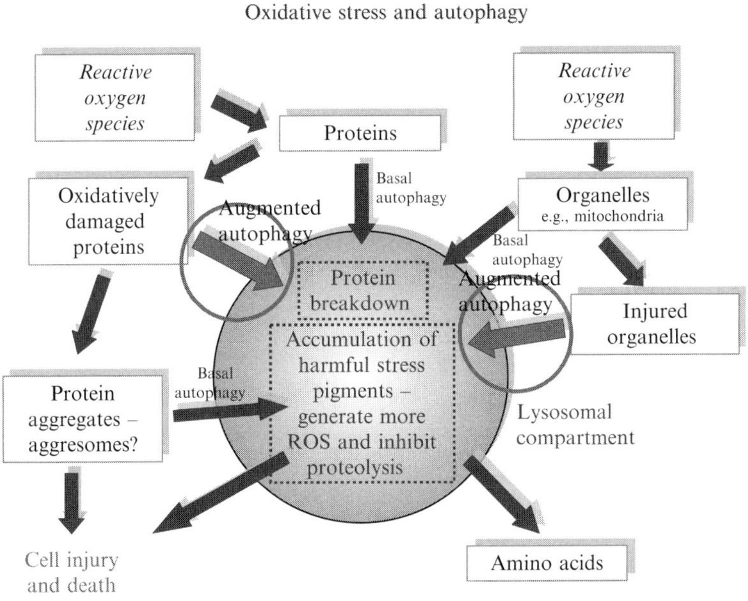

Figure 33.14 Conceptual mechanistic model for effects of autophagy and ROS on proteins and cellular organelles. The model shows the normal autophagic turnover of proteins and organelles (arrows) with superimposed augmented autophagy (large arrows, circled), which is postulated as having a protective role against oxidative stress. If augmented autophagy of damaged organelles and proteins is impaired as occurs in cells with lysosomal damage and autophagic degradative dysfunction, then harmful products can accumulate contributing to cell injury (see Moore *et al.*, 2006a,b, 2007). Modified from Moore *et al.* (2007).

The use of coupled empirical measurements of biomarker reactions and modelling is proposed as a practical approach to the development of an operational toolbox for predicting the health of the environment (Allen and McVeigh, 2004; Moore *et al.*, 2006a). Current assessment methods are largely indicative of exposure to chemical contamination but do not necessarily indicate harmful effects on the health of sentinel animals. In contrast, we have clearly demonstrated a mechanistic link between responses of lysosomal and autophagic biomarkers, oxidative stress and early onset pathology. The lysosomal stability biomarker has also been shown to be significantly correlated with an indicator of ecological status (Moore *et al.*, 2006c).

7.2. Simulation modelling of cellular autophagy and an expert system for data analysis and interpretation

The lysosomal system occupies a central and crucial role in cellular food degradation (intracellular digestion), toxic responses and internal turnover (autophagy) in cells (Cuervo, 2004; Klionsky and Emr, 2000; Moore *et al.*, 2006a, 2007). Understanding the dynamic response of this system requires factors affecting performance (conceived as a function of the throughput, degradative efficiency and membrane stability) to be defined and quantified. A previous version of the carbon/nitrogen flux model has been augmented by separately identifying lysosomal target material (e.g., autophagocytosed or endocytosed proteins, carbohydrates and lipids) and internal material (e.g., digestive enzymes and lipid membrane components) (Allan and McVeigh, 2004; McVeigh *et al.*, 2004, 2006; Moore and Allen, 2002). Additionally, the whole cell energetic costs for maintaining lysosomal pH and productions of these internal components have been incorporated; as has the potentially harmful effect of generation of lipofuscin on the transitory and semipermanent lysosomal constituents. Inclusion of the three classes of nutrient organic compounds at the whole cell level allows for greater range in the simulated response, including deamination of amino acids to provide molecules as a source of energy, as well as controlling nitrogen and carbon concentrations in the cytosol. Coupled with a more functional framework of pollutant-driven reactive oxygen and nitrogen species (RNOS) production and antioxidant defense, the separate and combined effects of three stressors (nutritional quality, nutrient quantity and a contaminant polycyclic aromatic hydrocarbon-phenanthrene) on the digestive cell have been simulated and compared with real data (McVeigh *et al.*, 2004, 2006).

In tandem with this modelling effort, a new decision support system (expert system), able to integrate biomarker data, has been developed by Dagnino *et al.* (2007) that integrates a suite of biomarkers to interpret complex biological data by transforming it into a relatively simple, easy to understand and objective evaluation of the changes in the organism physiology induced by pollutants. This expert system is based on a classification scale that considers the various characteristics of the biological responses to environmental stressors and can be used to categorize harmful environmental impact. A feature of this system is that it incorporates an indicator or biomarker of cellular or tissue health such as lysosomal stability (Dagnino *et al.*, 2007).

8. Conclusions

Autophagic responses and decreased lysosomal membrane stability are effective predictive indicators for cell injury and pathology in the aquatic animals tested previously; and supporting evidence indicates that these parameters are probably generic in the animal kingdom (Moore *et al.*, 2006a,b,c, 2007). The complicated problem of evaluating and predicting health of environmental sentinel animals, such as social amoebae, blue and green mussels, snails (periwinkles) and the flatfish flounder and dab, has been addressed through a coupled biomarker testing and modelling approach in the wider context of forecasting risk (Allen and Moore, 2004; Kohler *et al.*, 2002; Moore and Noble, 2004; Moore *et al.*, 2006a). This approach facilitates the validation, and furthers the essential new development of robust diagnostic and prognostic tools that can be used along with other chemical, biological and ecological tools as indices of environmental sustainability.

REFERENCES

Allen, J. I., and McVeigh, A. (2004). Towards computational models of cells for environmental toxicology. *J. Mol. Histol.* **35,** 697–706.

Allen, J. I., and Moore, M. N. (2004). Environmental prognostics: Is the current use of biomarkers appropriate for environmental risk evaluation? *Mar. Environ. Res.* **58,** 227–232.

Asnaghi, L., Bruno, P., Priulla, M., and Nicolin, A. (2004). mTOR: A protein kinase switching between life and death. *Pharmacol. Res.* **50,** 545–549.

Axiak, V., George, J. J., and Moore, M. N. (1988). Petroleum hydrocarbons in the marine bivalve *Venus verrucosa:* Accumulation and cellular responses. *Mar. Biol.* **97,** 225–230.

Babich, H., and Borenfreund, E. (1987). Fathead minnow FHM cells for use in *in vitro* cytotoxicity assays of aquatic pollutants. *Ecotox. Environ. Safety* **14,** 78–87.

Bayliss High, O. (1984). Lipid histochemistry (Royal Microscopical Society, Microscopy Handbooks, No. 6) Oxford: Oxford University Press, Oxford.

Bayne, B. L., Addison, R. F., Capuzzo, J. M., Clarke, K. R., Gray, J. S., Moore, M. N., and Warwick, R. M. (1988). An overview of the GEEP workshop. *Mar. Ecol. Prog. Ser.* **46,** 235–243.

Bayne, B. L., Holland, D. L., Moore, M. N., Lowe, D. M., and Widdows, J. (1978). Further studies on the effects of stress in the adult on the eggs of *Mytilus edulis. J. mar. biol. Ass. UK* **58,** 825–841.

Bayne, B. L., Moore, M. N., Widdows, J., Livingstone, D. R., and Salkeld, P. N. (1979). Measurement of the responses of individuals to environmental stress and pollution: Studies with bivalve molluscs. *Phil. Trans. R. Soc. (Lond)* **286B,** 563–581.

Bergamini, E., Cavallini, G., Donati, A., and Gori, Z. (2003). The ant-ageing effects of caloric restriction may involve stimulation of macroautophagy and lysosomal degradation, and can be intensified pharmacologically. *Biomed. Pharmacother.* **57,** 203–208.

Beaumont, V., Zhong, N., Fletcher, R., Froemke, R. C., and Zucker, R. S. (2001). Phosphorylation and local presynaptic protein synthesis in calcium- and calcineurin-dependent induction of crayfish long-term facilitation. *Neuron* **32,** 489–501.

Broeg, K., Kohler, A., and von Westernhagen, H. (2002). Disorder and recovery of environmental health monitoring by means of lysosomal stability in liver of European flounder (*Platichthys flexus* L.). *Mar. Environ. Res.* **54,** 569–573.

Broeg, K., von Westernhagen, H., Zander, S., Körting, W., and Kohler, A. (2005). The "bioeffect assessment index" - a concept for the quantification of effects of marine pollution by an integrated biomarker approach. *Mar. Poll Bull* **50,** 495–503.

Broeg, K., Zander, S., Diamant, A., Körting, W., Krüner, G., Paperna, I., and von Westernhagen, H. (1999a). The use of fish metabolic, pathological and parasitological indices in pollution monitoring I. *North Sea. Helgol. Mar. Res.* **53,** 171–194.

Broeg, K., Zander, S., Diamant, A., Körting, W., Krüner, G., Paperna, I., and von Westernhagen, H. (1999b). The use of fish metabolic, pathological and parasitological indices in pollution monitoring I. *Red Sea and Mediterranean Sea. Helgol. Mar. Res.* **53,** 195–208.

Brown, R. J., Galloway, T. S., Lowe, D., Browne, M. A., Dissanayake, A., Jones, M. B., and Depledge, M. H. (2004). Differential sensitivity of three marine invertebrates to copper assessed using multiple biomarkers. *Aquatic Toxicol.* **66,** 267–278.

Brunk, U. T., and Terman, A. (2002). Lipofuscin: Mechanisms of age-related accumulation and influence on cell function. *Free Rad. Biol. Med.* **33,** 611–619.

Bursch, W. (2001). The autophagosomal-lysosomal compartment in programmed cell death. *Cell Death Differentiat.* **8,** 569–581.

Cajaraville, M. P., Abascal, I., Etxeberria, M., and Marigómez, I. (1995a). Lysosomes as cellular markers of environmental pollution: Time- and dose-dependant responses of the digestive lysosomal system of mussels after petroleum hydrocarbon exposure. *Environ. Toxicol. Water Qual.* **10,** 1–8.

Cajaraville, M. P., Bebianno, M. J., Blasco, J., Porte, C., Sarasquete, C., and Viarengo, A. (2000). The use of biomarkers to assess thew impact of pollution in coastal environments of the Iberian peninsula: A practical approach. *Sci. Tot. Environ.* **247,** 295–311.

Cajaraville, M. P., Marigomez, J. A., and Angulo, E. (1990). Short-term toxic effects of 1-naphthol on the digestive gland–gonad complex of the marine prosobranch *Littorina littorea* (L): A light microscopic study. *Arch. Environ. Contam. Toxicol.* **19,** 17–24.

Cajaraville, M. P., Marigomez, J. A., Diez, G., and Angulo, E. (1992). Comparative effects of the water accommodated fraction of three oils on mussels. 2. Quantitative alterations in the structure of the digestive tubules. *Comp. Biochem. Physiol.* **102C,** 113–123.

Cajaraville, M. P., Robledo, Y., Etxeberria, M., and Marigómez, I. (1995b). Cellular biomarkers as useful tools in the biological monitoring of environmental pollution: Molluskan digestive lysosomes. *In* "Cell Biology in Environmental Toxicology" (M. P. Cajaraville, ed.), pp. 29–55. University of the Basque Country Press Service, Bilbo.

Cammalleri, M., Lütjens, R., Berton, F., King, A. R., Simpson, C., Francesconi, W., and Sanna, P. P. (2003). Time-restricted role for dendritic activation of the mTOR-p70S6K pathway in the induction of late-phase long-term potentiation in the CA1. *Proc. Natl. Acad. Sci. USA* **100,** 14368–14373.

Chieco, P., Jonker, A., and Van Noorden, CJ. F. (2001). Image cytometry. *In* "Royal Microscopical Society, Microscopy Handbooks." (A. Kohler and M. Rainsford, eds.), p. 116. ßios, Oxford46.

Clarke, K. R., and Moore, M. N. (1983). Unbiased linear property estimation for spheres, from sections exhibiting over projection and truncation. *J. Microsc.* **131,** 311–322.

Cuervo, A. M. (2004). Autophagy: In sickness and in health. *TRENDS Cell Biol.* **14,** 70–77.

Da Ros, L., Meneghetti, F., and Nasci, C. (2000). Field application of lysosomal destabilisation indices in the mussel Mytilus galloprovincialis: Biomonitoring and transplantation in the Lagoon of Venice (North-East Italy). *Mar. Environ. Res.* **54,** 817–822.

Dagnino, A., Allen, J. I., Moore, M., Broeg, K., Canesi, L., and Viarengo, A. (2007). Integration of biomarker data into an organism health index: Development of an expert system and its validation with field and laboratory data in mussels. *Biomarkers* **12,** 155–172.

Deutsch, L., Gräslund, S., Folke, C., Troell, M., Huitric, M., Kautsky, N., and Lebel, L. (2007). Feeding aquaculture growth through globalization: Exploitation of marine ecosystems for fishmeal. *Global Environ. Change* **17,** 238–249.

Domouhtsidou, G. P., and Dimitriadis, V. K. (2001). Lysosomal and lipid alterations in the digestive gland of mussels, *Mytilus galloprovincialis* (L.) as biomarkers of environmental stress. *Environ. Pollut.* **115,** 123–137.

Dondero, F., Jonsson, H., Rebelo, M., Pesce, M., Berti, E., Pons, G., and Viarengo, A. (2006). Cellular responses to environmental contaminants in amoebic cells of the slime mould *Dictyostelium discoideum. Comp. Biochem. Physiol.* **143C,** 150–157.

Einsporn S, Kohler A (2008a) Immuno-Localisations (GSSP) of subcellular accumulation sites of phenanthrene, Aroclor 1254 and lead (Pb) in Relation to cytopathologies in the gills and digestive gland of the mussel *Mytilus edulis. Mar. Environ. Res.* **66,** 185–186.

Einsporn, S., and Kohler, A. (2008b). Electronmicroscopic localization of lipophilic chemicals by an antibody-based detection system using the blue mussel Mytilus edulis as a model system. *Environ. Toxicol. Chem.* **27,** 554–560.

Etxeberria, M., Cajaraville, M. P., and Marigómez, I. (1995). Changes in digestive cell lysosomal structure as biomarkers of environmental stress in the Urdaibai Estuary (Biscay Coast, Iberian Peninsula). *Mar. Pollut. Bull* **30,** 599–603.

Fishelson, L., Bresler, V., Manelis, R., Zuk-Rimon, Z., Dotan, A., Hornung, H., and Yawetz, A. (1999). Toxicological aspects associated with the ecology of *Donax trunculus* (Bivalvia, Molluska) in a polluted environment. *Sci. Tot. Environ.* **226,** 121–131.

Galloway, T. S., Brown, R. J., Browne, M. A., Dissanayake, A., Lowe, D., Depledge, M. H., and Jones, M. B. (2006). The ECOMAN project: A novel approach to defining sustainable ecosystem function. *Mar. Poll. Bull.* **53,** 186–194.

Glaumann, H., and Ballard, F. J. (eds.), (1987). Lysosomes: Their Role in Protein Breakdown. Academic Press, London, New York.

Glaumann, H., Ahlberg, J., Berkenstam, A., and Henell, F. (1986). Rapid isolation of rat liver secondary lysosomes – autophagic vacuoles – following chloroquine administration. *Exp. Cell Res.* **163,** 151–158.

Grinde, B. (1985). Autophagy and lysosomal proteolysis in the liver. *Experientia* **41,** 1089–1230.

Grundy, M. M., Ratcliffe, N. A., and Moore, M. N. (1996a). Immune inhibition in marine mussels by polycyclic aromatic hdyrocarbons. *Mar. Environ. Res.* **42,** 187–190.

Grundy, M. M., Moore, M. N., Howell, S. M., and Ratcliffe, N. A. (1996b). Phagocytic reduction and effects on lysosomal membranes of polycyclic aromatic hydrocarbons, in haemocytes of *Mytilus edulis. Aquatic. Toxicol.* **34,** 273–290.

Grune, T., Jung, T., Merker, K., and Davies, KJ. A. (2004). Decreased proteolysis caused by protein aggregates, inclusion bodies, plaques, lipofuscin, ceroid, and "aggresomes" during oxidative stress, ageing and disease. *Internat. J. Biochem. Cell Biol.* **36,** 2519–2530.

Harthree, E. F. (1972). Determination of protein. A modification of the Lowry method that gives a linear photometric response. *Analyt. Biochem.* **48,** 422–427.

Hawkins, A. J. S., and Day, A. J. (1996). The metabolic basis of genetic differences in growth efficiency among marine animals. *J. Exp. Mar. Biol. Ecol.* **203,** 93–115.

Hipkiss, A. (2006). Accumulation of altered proteins and ageing: Causes and effects. *Exp. Gerontol.* **41,** 464–473.

Ji, C., and Kaplowitz, N. (2006). ER stress: Can the liver cope? *J. Hepatol.* **45,** 321–333.

Kiffen, R., Christian, C., Knecht, E., and Cuervo, A. M. (2004). Activation of chaperone-mediated autophagy during oxidative stress. *Mol. Biol. Cell* **15,** 4829–4840.

Kirchin, M. A., Moore, M. N., Dean, R. T., and Winston, G. W. (1992). The role of oxyradicals in intracellular proteolysis and toxicity in mussels. *Mar. Environ. Res.* **34,** 315–320.

Klionsky, D. J., and Emr, S. D. (2000). Autophagy as a regulated pathway of cellular degradation. *Science* **290,** 1717–1721.

Klionsky, D. J., Cuervo, A. M., and Seglen, P. O. (2007). Methods for monitoring autophagy from yeasts to human. *Autophagy* **3,** 181–206.

Kohler, A. (1989a). Cellular effects of environmental contamination in fish from the river Elbe and the North Sea. *Mar. Environ. Res.* **28,** 417–424.

Kohler, A. (1989b). Experimental studies on the regeneration of contaminant-induced liver lesions in flounder from the Elbe estuary: Steps towards the identification of cause-effect relationships. *Aquatic. Toxicol.* **14,** 203–232.

Kohler, A. (1990). Identification of contaminant-induced cellular and subcellular lesions in the liver of flounder (*Platichthys flesus* L.) caught at differently polluted estuaries. *Aquatic. Toxicol.* **16,** 271–294.

Kohler, A. (1991). Lysosomal perturbations in fish liver as indicators for toxic effects of environmental pollution. *Comp. Biochem. Phys.* **100C,** 123–127.

Kohler, A. (2004). The gender-specific risk to liver toxicity and cancer of flounder (*Plathichthys flesus*, L.) at the German Wadden Sea coast. *Aquatic. Toxicol.* **70,** 257–276.

Kohler, A., and Pluta, H. J. (1995). Lysosomal injury and MFO activity in the liver of flounder (*Platichthys flesus* L.) in relation to histopathology of hepatic degeneration and carcinogenesis. *Mar. Environ. Res.* **39,** 255–260.

Kohler, A., Deisemann, H., and Lauritzen, B. (1992). Histological and cytochemical indices of toxic injury in the liver of dab *Limanda limanda. Mar. Ecol. Prog. Ser.* **91,** 141–153.

Kohler, A., Marx, U., Broeg, K, Bahns, S., and Bressling, J. (2008). Effects of nanoparticles in *Mytilus edulis* gills and hepatopancreas – a new threat to marine life? *Mar. Environ. Res.* **66,** 12–14.

Kohler, A., Wahl, E., and Söffker, K. (2002). Functional and morphological changes of lysosomes as prognostic biomarkers of toxic liver injury in a marine flatfish (*Platichthys flesus* L.). *Environ. Toxicol. Chem.* **21,** 2434–2444.

Kolb, B., Gibb, R., Gorny, G., and Whishaw, I. Q. (1998). Possible regeneration of rat medial frontal cortex following neonatal frontal lesions. *Behav. Brain Res.* **91,** 127–141.

Krishnakumar, P. K., Casillas, E., and Varanasi, U. (1994). Effect of environmental contaminants on the health of *Mytilus edulis* from Puget Sound, Washington, USA. I. Cytochemical measures of lysosomal responses in the digestive cells using automatic image analysis. *Mar. Ecol. Prog. Ser.* **106,** 249–261.

Langton, R. W. (1975). Synchrony in the digestive diverticula of *Mytilus edulis* L. *J. mar. boil. Assoc. UK* **55,** 221–229.

Levine, B. (2005). Eating oneself and uninvited guests: Autophagy-related pathways in cellular defense. *Cell* **120,** 159–162.

Lindroth, P., and Mopper, K. (1979). High-performance liquid chromatography determination of subpicomole amounts of amino acids by precolumn fluorescence derivatization with o-phthaldialdehyde. *Analyt. Chem.* **51,** 1667–1674.

Livingstone, D. R. (2001). Contaminant-stimulated reactive oxygen species production and oxidative damage in aquatic organisms. *Mar. Pollut. Bull.* **42,** 656–666.

Lockshin, R. A., and Zakeri, Z. (2004). Apoptosis, autophagy, and more. *Internat. J. Biochem. Cell Biol.* **36,** 2405–2419.

Lowe, D. M. (1988). Alterations in the cellular structure of *Mytilus edulis* resulting from exposure to environmental contaminats under field and experimental conditions. *Mar. Ecol. Prog. Ser.* **46,** 91–100.

Lowe, D. M., and Pipe, R. K. (1994). Contaminant induced lysosomal membrane damage in marine mussel digestive cells: An *in vitro* study. *Aquatic. Toxicol.* **30,** 357–365.

Lowe, D. M., Fossato, V. U., and Depledge, M. H. (1995a). Contaminant induced lysosomal membrane damage in blood cells of mussels *Mytilus galloprovincialis* from the Venice Lagoon: An *in vitro* study. *Mar. Ecol. Prog. Ser.* **129,** 189–196.

Lowe, D. M., Moore, M. N., and Clarke, K. R. (1981). Effects of oil on digestive cells in mussels: Quantitative alterations in cellular and lysosomal structure. *Aquatic. Toxicol.* **1,** 213–226.

Lowe, D. M., Moore, M. N., and Evans, B. M. (1992). Contaminant impact on interactions of molecular probes with lysosomes in living hepatocytes from dab *Limanda limanda. Mar. Ecol. Prog. Ser.* **91,** 135–140.

Lowe, D. M., Soverchia, C., and Moore, M. N. (1995b). Lysosomal membrane responses in the blood and digestive cells of mussels experimentally exposed to fluoranthene. *Aquatic. Toxicol.* **33,** 105–112.

Lüllmann-Rauch, R. (1979). Drug-induced lysosomal storage disorders. *In* "Lysosomes in Applied Biology and Therapeutics." (J. T. Dingle, P. J. Jacques, and I. H. Shaw, eds.), vol. 6, pp. 49–130. Elsevier, Amsterdam.

McVeigh, A., Allen, J. I., Moore, M. N., Dyke, P., and Noble, D. (2004). A carbon and nitrogen flux model of mussel digestive gland epithelial cells and their simulated response to pollutants. *Mar. Environ. Res.* **58,** 821–827.

McVeigh, A., Moore, M. N., Allen, J. I., and Dyke, P. (2006). Lysosomal responses to nutritional and contaminant stress in mussel hepatopancreatic digestive cells: A modelling study. *Mar. Environ. Res. 62 Suppl* **1,** S433–S438.

Marigómez, I., and Baybay-Villacorta, L. (2003). Pollutant-specific and general lysosomal responses in digestive cells of mussels exposed to model organic chemicals. *Aquatic. Toxicol.* **64,** 235–257.

Marigómez, I., Orbea, A., Olabarrieta, I., Etxeberria, M., and Cajaraville, M. P. (1996). Structural changes in the digestive lysosomal system of sentinel mussels as biomarkers of environmental stress in Mussel-Watch programmes. *Comp. Biochem. Physiol.* **113C,** 291–297.

Marigómez, I., Lekube, X., Cajaraville, M. P., Domouhtsidou, G., and Dimitriadis, V. (2005). Comparison of cytochemical procedures to estimate lysosomal biomarkers in mussel digestive cells. *Aquatic. Toxicol.* **75,** 86–95.

Moore, M. N. (1976). Cytochemical demonstration of latency of lysosomal hydrolases in digestive cells of the common mussel, *Mytilus edulis* and changes induced by thermal stress. *Cell Tissue Res.* **175,** 279–287.

Moore, M. N. (1979). Cellular responses to polycyclic aromatic hydrocarbons and phenobarbital in *Mytilus edulis. Mar. Environ. Res.* **2,** 255–263.

Moore, M. N. (1980). Cytochemical determination of cellular responses to environmental stressors in marine organisms. *Rapp P-v Reun Cons. Int. Explor. Mer.* **170,** 7–15.

Moore, M. N. (1985). Cellular responses to pollutants. *Mar. Pollut. Bull.* **16,** 134–139.

Moore, M. N. (1988). Cytochemical responses of the lysosomal system and NADPH-ferrihemoprotein reductase in mollusks to environmental and experimental exposure to xenobiotics. *Mar. Ecol. Prog. Ser.* **46,** 81–89.

Moore, M. N. (1990). Lysosomal cytochemistry in marine environmental monitoring. *Histochem. J.* **22,** 187–191.

Moore, M. N. (2002). Biocomplexity: The post-genome challenge in ecotoxicology. *Aquatic. Toxicol.* **59,** 1–15.
Moore, M. N. (2004). Diet restriction induced autophagy: A protective system against oxidative- and pollutant-stress and cell injury. *Mar. Environ. Res.* **58,** 603–607.
Moore, M. N. (2006). Do nanoparticles present ecotoxicological risks for the health of the aquatic environment? *Environ. Internat.* **32,** 967–976.
Moore, M. N. (2008). Autophagy as a second level protective process in conferring resistance to environmentally-induced oxidative stress. *Autophagy* **4,** 254–256.
Moore, M. N., and Allen, J. I. (2002). A computational model of the digestive gland epithelial cell of the marine mussel and its simulated responses to aromatic hydrocarbons. *Mar. Environ. Res.* **54,** 579–584.
Moore, M. N., and Clarke, K. R. (1982). Use of microstereology and cytochemical staining to determine the effects of crude oil-derived aromatic hydrocarbons on lysosomal structure and function in a marine bivalve mollusk, *Mytilus edulis. Histochem. J.* **14,** 713–718.
Moore, M. N., and Halton, D. W. (1973). Histochemical changes in the digestive gland of *Lymnaea truncatula* infected with *Fasciola hepatica. Z. Parasitkde* **43,** 1–16.
Moore, M. N., and Halton, D. W. (1977). Cytochemical localization of lysosomal hydrolases in the digestive cells of littorinids and changes induced by larval trematode infection. *Z. Parasitkde* **53,** 115–122.
Moore, M. N., and Noble, D. (2004). Computational modelling of cell and tissue processes and function. *J. Mol. Histol.* **35,** 655–658.
Moore, M. N., and Viarengo, A. (1987). Lysosomal membrane fragility and catabolism of cytosolic proteins: Evidence for a direct relationship. *Experientia* **43,** 320–323.
Moore, M. N., and Simpson, M. G. (1992). Molecular and cellular pathology in environmental impact assessment. *Aquatic. Toxicol.* **22,** 313–322.
Moore, M. N., and Stebbing, A. R. D. (1976). The quantitative cytochemical effects of three metal ions on a lysosomal hydrolase of a hydroid. *J. mar. biol. Assoc. UK* **56,** 995–1005.
Moore, M. N., Allen, J. I., and McVeigh, A. (2006a). Environmental prognostics: An integrated model supporting lysosomal stress responses as predictive biomarkers of animal health status. *Mar. Environ. Res.* **61,** 278–304.
Moore, M. N., Allen, J. I., McVeigh, A., and Shaw, J. (2006b). Lysosomal and autophagic reactions as diagnostic and predictive indicators of environmental pollutant toxicity in aquatic animals. *Autophagy* **2,** 217–220.
Moore, M. N., Allen, J. I., and Somerfield, P. J. (2006c). Autophagy: Role in surviving environmental stress. *Mar. Environ. Res.* **62**(Suppl 1), S420–S425.
Moore, M. N., Depledge, M. H., Readman, J. W., and Leonard, P. (2004a). An integrated biomarker-based strategy for ecotoxicological evaluation of risk in environmental management. *Mutat. Res.* **552,** 247–268.
Moore, M. N., Koehn, R. K., and Bayne, B. L. (1980). Leucine aminopeptidase (aminopeptidase-1), N-acetyl-β-hexosaminidase and lysosomes in the mussel *Mytilus edulis* L., in response to salinity changes. *J. Exp. Zool.* **214,** 239–249.
Moore, M. N., Livingstone, D. R., Widdows, J., Lowe, D. M., and Pipe, R. K. (1987). Molecular, cellular and physiological effects of oil-derived hydrocarbons on molluscs and their use in impact assessment. *Phil. Trans. R. Soc. (Lond)* **316B,** 603–623.
Moore, M. N., Lowe, D. M., and Fieth, P. E. M. (1978a). Responses of lysosomes in the digestive cells of the common mussel, *Mytilus edulis* to sex steroids and cortisol. *Cell Tiss. Res.* **188,** 1–9.
Moore, M. N., Lowe, D. M., and Fieth, P. E. M. (1978b). Lysosomal responses to experimentally injected anthracene in the digestive cells of *Mytilus edulis. Mar. Biol.* **48,** 297–302.
Moore, M. N., Lowe, D. M., and Kohler, A. (2004b). Measuring lysosomal membrane stability. ICES Techniques in Marine Environmental Sciences, No. 36. ICES, Copenhagen.

Moore, M. N., Lowe, D. M., and Moore, S. L. (1979). Induction of lysosomal destabilisation in marine bivalve molluscs exposed to air. *Mar. Biol. Letters* **1,** 47–57.

Moore, M. N., Mayernick, J. A., and Giam, C. S. (1985). Lysosomal responses to a polynuclear aromatic hydrocarbon in a marine snail: effects of exposure to phenanthrene and recovery. *Mar. Environ. Res.* **17,** 230–233.

Moore, M. N., Pipe, R. K., and Farrar, S. V. (1982). Lysosomal and microsomal responses to environmental factors in *Littorina littorea* from Sullom Voe. *Mar. Poll. Bull.* **13,** 340–345.

Moore, M. N., Pipe, R. K., Farrar, S. V., Thomson, S., and Donkin, P. (1986). Lysosomal and microsomal responses to oil-derived hydrocarbons in *Littorina littorea. In* "Oceanic Processes in Marine Pollution - Biological Processes and Waste in the Ocean." (J. M. Capuzzo and D. R. Kester, eds.), Vol 1, pp. 89–96. Krieger Publishing, Melbourne, FL.

Moore, M. N., Soverchia, C., and Thomas, M. (1996). Enhanced lysosomal autophagy of intracellular proteins by xenobiotics in living molluskan blood cells. *Acta. Histchem. Cytochem.* **29**(Suppl), 947–948.

Moore, M. N., Viarengo, A., Donkin, P., and Hawkins, AJ. S. (2007). Autophagic and lysosomal reactions to stress in the hepatopancreas of blue mussels. *Aquatic. Toxicol.* **84,** 80–91.

Moore, M. N., Widdows, J., Cleary, J. J., Pipe, R. K., Salkeld, P. N., Donkin, P., Farrar, S. V., Evans, S. V., and Thompson, P. E. (1984). Responses of the mussel *Mytilus edulis* to copper and phenanthrene: Interactive effects. *Mar. Environ. Res.* **14,** 167–183.

Nott, J. A., and Moore, M. N. (1987). Effects of polycylic aromatic hydrocarbons on molluskan lysosomes and endoplasmic reticulum. *Histochem. J.* **19,** 357–368.

Pipe, R. K., and Moore, M. N. (1985). Ultrastructural changes in the lysosomal-vacuolar system in digestive cells of *Mytilus edulis* as a response to increased salinity. *Mar. Biol.* **87,** 157–163.

Proud, C. G. (2002). Regulation of mammalian translation factors by nutrients. *FEBS J.* **269,** 5338–5349.

Rashid, F., Horobin, R. W., and Williams, M. A. (1991). Predicting the behaviour and selectivity of fluorescent probes for lysosomes and related structures by means of structure-activity models. *Histochem. J.* **23,** 450–459.

Regoli, F. (1992). Lysosomal responses as a sensitive index in biomonitoring heavy metal pollution. *Mar. Ecol. Prog. Ser.* **84,** 63–69.

Ringwood, A. H., Conners, D. E., and Di Novo, A. (1998a). The effects of copper exposures on cellular responses in oysters. *Mar. Environ. Res.* **46,** 591–595.

Ringwood, A. H., Conners, D. E., and Hoguet, J. (1998b). Effects of natural and anthropogenic stressors on lysosomal destabilization in oysters *Crassostrea virginica. Mar. Ecol. Prog. Ser.* **166,** 163–171.

Ringwood, A. H., Conners, D. E., and Keppler, C. J. (1999). Cellular responses of oysters, *Crassostrea virginica*, to metal-contaminated sediments. *Mar. Environ. Res.* **48,** 427–437.

Ryazanov, A. G., and Nefsky, B. S. (2002). Protein turnover plays a key role in ageing. *Mech. Ageing Develop.* **123,** 207–213.

Sforzini, S., Dagnino, A., Torrielli, S., Dondero, F., Fenoglio, S., Negri, A., Boatti, L., and Viarengo, A. (2008). Use of highly sensitive sublethal stress responses in the social amoeba *Dictyostelium discoideum* for an assessment of freshwater quality. *Sci. Tot. Environ.* **395,** 101–108.

Siegel, S. (1956). Non-parametric Statistics for the Behavioural Sciences. McGraw-Hill, New York.

Svendsen, C., and Weeks, J. M. (1995). The use of a lysosome assay for the rapid assessment of cellular stress from copper to the freshwater snail *Viviparus contectus* (Millet). *Mar. Poll. Bull.* **31,** 139–42.

Svendsen, C., Spurgeon, D. J., Hankard, P. K., and Weeks, J. M. (2004). A review of lysosomal membrane stability measured by neutral red retention: Is it a workable earthworm biomarker. *Ecotox. Environ. Safety* **57,** 20–29.

Tavernarakis, N., and Driscoll, M. (2002). Caloric restriction and lifespan: A role for protein turnover? *Mech. Ageing Develop.* **123,** 215–229.

Viarengo, A., Moore, M., Mancinelli, G., Mazzucotelli, G., Pipe, R. K., and Farrar, S. V. (1987). Metallothioneins and lysosomes in metal toxicity and accumulation in marine mussels: The effect of cadmium in the presence and absence of phenanthrene. *Mar. Biol.* **94,** 251–257.

Viarengo, A., Moore, M. N., Pertica, M., Mancinelli, G., Zanicchi, G., and Pipe, R. K. (1985). Detoxification of copper in the cells of the digestive gland of mussel: The role of lysosomes and thioneins. *Sci. Tot. Environ.* **44,** 135–145.

Viarengo, A., Moore, M. N., Pertica, M., Mancinelli, G., and Accomando, R. (1992). A simple procedure for evaluating the protein degradation rate in mussel *(Mytilus galloprovincialis* Lam.) tissues and its application in a study of phenanthrene effects on protein catabolism. *Comp. Biochem. Physiol.* **103B,** 27–32.

von Landwüst, C., Anders, K., Holst, S., Möller, H., Momme, M., von Neuhoff, N., Lee, K., Scharenberg, W., Weis, N., Cameron, P., Kohler, A., von Westernhagen, H., *et al.* (1996). Fischkrankheiten in der Nordsee. Umweltbundesamt Texte 57/96, ISSN 0722–186X, Umweltbundesamt, Berlin. p. 557.

Wahl, E., Cameron, P., Kohler, A., von Westernhagen, H., Anders, K., Möller, H., Büther, H., Dethlefsen, V., Söffker, K., Hansen, P. D., Pluta, H. J., and Harms, U. (1995). Fischkrankheiten im Wattenmeer. Umweltbundesamt Texte 51/95, ISSN 0722–186X, Umweltbundesamt, Berlin, Germany, p. 216.

Waterlow, J. C., Garlick, P. J., and Milward, D. J. (1978). Protein Turnover in Mammalian Tissues and in the Whole Body. Amsterdam, North-Holland.

Wedderburn, J., McFadzen, I., Sanger, R. C., Beesley, A., Heath, C., Hornsby, M., and Lowe, D. (2000). The field application of cellular and physiological biomarkers, in the mussel *Mytilus edulis*, in conjunction with early life stage bioassays and adult histopathology. *Mar. Pollut. Bull.* **40,** 257–267.

Widdows, J., Bakke, T., Bayne, B. L., Donkin, P., Livingstone, D. R., Lowe, D. M., Moore, M. N., Evans, S. V., and Moore, S. L. (1982). Responses of *Mytilus edulis* L. on exposure to the water accommodated fraction of North Sea oil. *Mar. Biol.* **67,** 5–31.

Widdows, J., Bayne, B. L., Donkin, P., Livingstone, D. R., Lowe, D. M., Moore, M. N., and Salkeld, P. N. (1981). Measurements of the responses of mussels to environmental stress and pollution in Sullom Voe: A baseline study. *Proc. R. Soc. Edin.* **80B,** 323–338.

Winston, G. W., Moore, M. N., Straatsburg, I., and Kirchin, M. (1991). Lysosomal stability in *Mytilus edulis* L.: Potential as a biomarker of oxidative stress related to environmental contamination. *Arch. Environ. Contam. Toxicol.* **21,** 401–408.

Winston, G. W., Moore, M. N., Kirchin, M. A., and Soverchia, C. (1996). Production of reactive oxygen species (ROS) by hemocytes from the marine mussel, *Mytilus edulis. Comp. Biochem. Physiol.* **113C,** 221–229.

Wouters, B. G., van den Beucken, T., Magagnin, M. G., Koritzinsky, M., Fels, D., and Koumenis, C. (2005). Control of the response through regulation of mRNA translation. *Seminars Cell Devel. Biol.* **16,** 487–501.

Yu, F. L., and Feigelson, P. (1970). Paper disc estimation on radioactive RNA: Studies on the presence and elimination of metabolically generated artifacts from labeled purine and pyrimidine precursors. *Analyt. Biochem.* **39,** 319–321.

Yucel-Lindberg, T., Jansson, H., and Glaumann, H. (1991). Proteolysis in isolated autophagic vacuoles from the rat pancreas. Effects of chloroquine administration. *Virchows Arch. B. Cell Pathol. Incl. Mol. Pathol.* **61,** 141–5.

Zhao, M., Brunk, U. T., and Eaton, J. W. (2001). Delayed oxidant-induced cell death involves activation of phospholipase A2. *FEBS Letters* **509,** 399–404.

CHAPTER THIRTY-FOUR

Autophagy in Ticks

Rika Umemiya-Shirafuji,* Tomohide Matsuo,† *and* Kozo Fujisaki*

Contents

Abstract

The generation time of ticks is estimated at several years and most ticks spend more than 95% of their life off the host. They seem to have a unique strategy to endure the off-host state for a long period. We focused on autophagy that is induced by starvation and is essential for extension of the life span in model organisms. Autophagy may occur in ticks that can survive extended periods of starvation. Although little research has been done on autophagy in ticks, recently, we showed the existence of an *ATG* gene homolog, *HlATG12,* in the 3-host tick *Haemaphysalis longicornis*. We have also examined the expression

* Laboratory of Emerging Infectious Diseases, Department of Frontier Veterinary Medicine, Kagoshima University, Kagoshima, Japan
† Department of Infectious Diseases, Kyorin University School of Medicine, Tokyo, Japan

Methods in Enzymology, Volume 451
ISSN 0076-6879, DOI: 10.1016/S0076-6879(08)03234-5

patterns of *HlATG12,* from nymphal to adult stages of this tick and revealed the localization of the HlAtg*12,* protein within midgut epithelial cells of unfed adult ticks. However, autophagy in ticks is a new field, so methods for monitoring this phenomenon in ticks are still to be established. This chapter discusses protocols for the detection of *HlATG12,* gene/HlAtg*12,* protein and the observation of the midgut epithelial cells using an electron microscope during the nonfeeding period of *H. longicornis* ticks. These methods can be adapted and modified for the study autophagy in other hard ticks.

1. Introduction

Ticks are obligate hematophagous (blood feeding) arthropods found in almost every region of the world (Sonenshine, 1991). They not only infest every class of terrestrial vertebrates, including mammals, birds, various reptiles and amphibians, but also are able to transmit various diseases to these animals. All ticks have 4 stages: the egg and three active stages (larva, nymph and adult). Each developmental stage except for the egg of most ticks feeds on a host, and then drops off for development or reproduction in the natural environment. *Haemaphysalis longicornis* is the most dominant tick in Japan, belongs to the hard tick family, and is categorized as a 3-host tick, which feeds on three different hosts in each stage. Engorged larvae drop off and molt to nymphs, which then find a second host animal on which to engorge and drop off again to molt into an adult. After new adults emerging from engorged nymphs experience severe starvation, they attach to a third host animal (Fig. 34.1). Three-host ticks spend more than 95% of their life in the wild off the host, so their life cycle is characterized by a long period without feeding, starvation (months or years), and an on-host parasitic period lasting only a few weeks (Anderson, 2002). In particular, the period of host-seeking appears to be the longest in the adult stage of their life cycle.

Ticks are referred to as gorging-fasting organisms (Needham and Teel, 1991), the only nourishment source for them is vertebrate blood. Unless blood feeding is fulfilled, ticks cannot develop and leave offspring for the next generation. Once they take a blood meal, a variety of genes are up-regulated and begin functioning. As a result, many investigators have studied fed ticks to develop novel vaccines against ticks or tick-borne pathogens. However, their astonishing toughness during the nonfeeding stage is extremely interesting. Details of the mechanisms by which ticks can survive such long periods are unknown, but it is suspected that ticks obtain nutrients by a unique mechanism that allows them to endure long-term starvation. Therefore, we have focused on autophagy, which is induced by starvation, and hypothesize that ticks can adapt to long-term starvation by the mechanism of autophagy (*autophagy* meaning macroautophagy in this chapter).

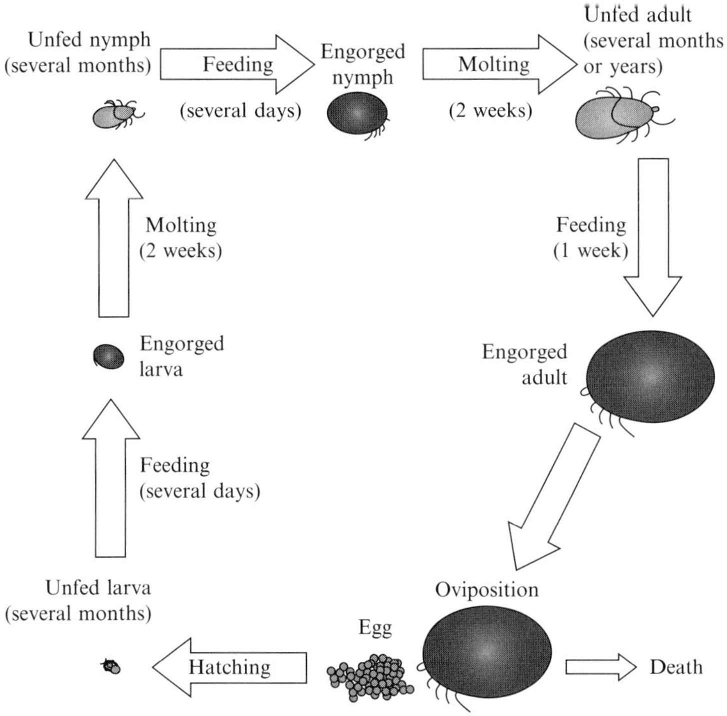

Figure 34.1 Life cycle of a 3-host tick, *H. longicornis*. The life cycle consists of 4 stages, the egg and three active stages: larva, nymph and adult. Larvae and nymphs feed once for several days to engorgement and then molt whereas adults feed for approximately 1 week. The engorged larvae drop off, molt to a nymph stage and find a second host on which to engorge and drop off again to molt to an adult. The adult female ticks attach to and engorge on a third host, and then finally die after oviposition, which occurs off the host. Note that most of their life cycle is characterized by a long nonfeeding period, starvation.

Autophagy in ticks has not been reported, but genes encoding ubiquitin have been reported in the transcriptome analysis of tick salivary glands (Alarcon-Chaidez *et al.*, 2007; Francischetti *et al.*, 2005). The presence of autophagic vacuoles in the midgut, the digestive organ of ticks, has been detected by electron microscopy (Tarnowski and Coons, 1989; Walker and Fletcher, 1987), but only the fine structure of the vacuoles was determined. The midgut cells of unfed ticks have three types of autophagic vacuoles; type 1 with spheroid inclusions, type 2 with lamellate inclusions, and type 3 with excretory inclusions (Raikhel, 1983). These structures seem to appear only during the unfed stages. However, questions about their function(s) and how they are related to autophagy remain unanswered.

We first isolated homologs of autophagy-related (*ATG*) genes and then cloned a homologue of an *ATG* gene (*ATG12*), designated as *HlATG12*, from the hard tick *H. longicornis* (Ummeiya *et al.*, 2007b), a vector of *Babesia* and *Theileria* parasites, and rickettsia. In this chapter, we describe methods for the detection of an autophagy-related gene/protein (*HlATG12*/HlAtg12) and the observation of autophagosome- or autolysosome-like structures in the midgut epithelial cells of the 3-host tick, *H. longicornis*.

2. Rearing of the 3-Host Tick *Haemaphysalis Longicornis*

Adult parthenogenetic *H. longicornis* ticks (Okayama strain) were maintained by feeding on the ears of Japanese white rabbits (3.0 kg, female; Japan Laboratory Animals, Tokyo, Japan) protected with cotton bags. Engorged ticks that dropped off the animals were collected and maintained to allow them to molt or lay eggs. Tick-rearing was performed in an incubator at 25 °C, 100% relative humidity and continuous darkness (Fujisaki, 1978).

3. Autophagy-Related Genes of *H. longicornis*

3.1. Cloning and identification of *H. longicornis ATG genes*

Although tick research has also entered the genomic era, genome sequencing of only a few tick species is in progress because of unexpectedly large tick genomes (Jongejan *et al.*, 2007) similar in size or several times larger than the human genome. For tick sequence analysis, the EST gene indices are the best sources for identifying potential homologs at present because sequencing of the genome is still far from completion. Therefore, the construction of cDNA libraries and the generation of EST databases for *H. longicornis* were obtained from Hitachi High-Tech Manufacturing & Service (Ibaraki, Japan) using proprietary protocols. We isolated 4 *ATG* gene homologs for *H. longicornis* from these databases (Umemiya *et al.*, 2007b; Umemiya-Shirafuji *et al.*, unpublished results). The protocol to construct a cDNA library by the vector-capping method (Kato *et al.*, 2005) is simply outlined subsequently.

1. Extract total RNAs (5 μg) from each organ of partially fed adult females using a TRI reagent (Sigma, MO, USA) as follows:
   a. Homogenize tissue samples with a mortar and pestle (AS ONE, Osaka, Japan) using liquid nitrogen and resuspend in TRI reagent (1 ml per 50–100 mg of tissue).

b. Centrifuge the extracts for 10 min, 4 °C at top speed. Transfer aqueous phase to a new tube.
c. Add chloroform (0.2 ml per 1 ml of TRI reagent) to the suspension. Shake for 15 s and allow to stand for 15 min at room temperature. Centrifuge the resulting mixture at top speed for 15 min at 4 °C.
d. Transfer the aqueous phase to a new tube and add isopropanol (0.5 ml per 1 ml of TRI reagent) and mix. Allow the sample to stand for 10 min at room temperature. Centrifuge at top speed for 10 min at 4 °C.
e. Remove the supernatant fraction and wash the RNA pellet by adding 75% ethanol (1 ml per 1 ml of TRI reagent). Vortex the sample and then centrifuge for 5 min, 4 °C at top speed.
f. Remove the supernatant fraction and dry the RNA pellet for 5 min by air-drying.
g. Add an appropriate volume of nuclease-free water to the pellet.

2. Synthesize cDNAs and ligate into the pGCAP1 plasmid vector (Hitachi High-Tech Manufacturing & Service).
3. Transform into *Escherichia coli* (DH12S strain).
4. Select bacterial colonies and inoculate into 384 well plates containing approximately 0.1 ml of LB medium.
5. Isolate and sequence plasmid DNAs.
6. Perform annotation by searching public protein databases using the BLAST analysis (National Center for Biotechnology Information; NCBI) to identify a provisional function. (See Kato *et al.*, 2005, for experimental details, especially for the procedure from steps 2–6.

Finally, the EST databases of *H. longicornis* were constructed. The Microsoft Office Excel software was adopted for using the databases, which are available at the Laboratory of Emerging Infectious Diseases, Kagoshima University. The databases have not been accessible but will be available online in the future. To isolate pure plasmids from *E. coli*, a miniprep kit such as a QIAprep Spin Miniprep Kit (Qiagen, Hilden, Germany) is useful. The following tools are helpful for characterization of cDNAs: GENETYX version 7 software (Genetyx, Tokyo, Japan), BLAST analysis (http://www.ncbi.nlm.nih.gov/blast/Blast.cgi; NCBI), ClustalW2 (http://www.ebi.ac.uk/Tools/clustalw2/index.html; European Bioinformatics Institute (EMBL-EBI)), ExPASy Proteomics Server (http://kr.expasy.org/; Swiss Institute of Bioinformatics (SIB)), and SignalP 3.0 (http://www.cbs.dtu.dk/services/SignalP/; Center for Biological Sequence Analysis (CBS)).

In addition, some novel genes have been previously isolated from *H. longicornis* by using the databases (Alim *et al.*, 2007, 2008; Boldbaatar *et al.*, 2006; Gong *et al.*, 2007; Harnnoi *et al.*, 2007; Liao *et al.*, 2007a,b; Motobu *et al.*, 2007; Tanaka *et al.*, 2007; Umemiya *et al.*, 2007a,b; Zhou *et al.*, 2006, 2007). The identification of numerous genes from the databases will help in the elucidation of the mystery by which ticks survive for long periods without feeding.

3.2. Expression and purification of recombinant HlAtg protein

The protocol for expression of a recombinant HlAtg protein, specifically HlAtg12 (Umemiya *et al.*, 2007b), for generation of antibodies is described subsequently.

1. DNA for subcloning into the vector plasmid pGEX-4T-3 (GE Healthcare, Chalfont St. Giles, UK) is prepared by conventional methods, utilizing the polymerase chain reaction (PCR) and digestion with restriction enzymes. After subcloning, recombinant plasmids should be confirmed by sequence analysis.
2. The plasmid is then transformed into *E. coli* (BL21 strain) to express the recombinant protein from pGEX-4T-3. The transformants are selected by the antibiotic ampicillin.
3. Synthesis of recombinant glutathione S-transferase (GST)-fused HlAtg12 (GST-HlAtg12) is induced with 0.05 m*M* isopropyl-β-D-thiogalactopyranoside for 4 h at 37 °C.
4. After centrifugation at 8000*g* for 20 min at 4 °C, 20–30 ml of TBST (50 m*M* Tris-HCl, pH 7.5, 150 m*M* NaCl, 1% Triton X-100) containing 2 m*M* dithiothreitol and a proteinase inhibitor (Complete Mini; Roche Diagnostics, Mannheim, Germany) per 1 liter of culture is added to the bacterial pellet.
5. The suspension is sonicated on ice using a sonicator (10 s × 3), and centrifuged at 10,000*g* for 10–30 min at 4 °C.
6. The supernatant fraction is transferred to a new tube and filtered through sterile 5- and 0.22-micron filters (Toyo Roshi, Tokyo, Japan).
7. Soluble GST-HlAtg12 is obtained using batch purification as follows:
   a. 0.5–1 ml of the 50% slurry of Glutathione Sepharose® 4B (GE Healthcare) equilibrated with TBST per 1 liter of culture is added to the cleared lysate and mixed gently using a rotator at 4 °C, overnight.
   b. The mixture is centrifuged at 500*g* for 5 min at 4 °C.
   c. After removal of the supernatant fraction, the sedimented Glutathione Sepharose® 4B is washed with 10 bed volumes of TBST 3 times.
   d. The supernatant fraction is discard and then 16 m*M* reduced glutathione in 50 m*M* Tris-HCl, pH 8.0 (1 ml per bed volume of Glutathione Sepharose 4B) is added to the sedimented matrix.
   e. The mixture is incubated at room temperature for 10 min, then centrifuged at 500*g* for 5 min.
   f. The supernatant fraction containing GST-HlAtg12 is transferred to a new tube.

3.3. Generation of anti-HlAtg12 antibody for immunoblotting and immunofluorescence staining

Purified GST-HlAtg12 is dialyzed against Tris-buffered saline adjusted to pH 7.4. The protein concentration is determined by a bicinchoninic acid assay (Pierce Biotechnology, IL, USA). One hundred micrograms of the GST-HlAtg12 in Freund's complete adjuvant (Sigma-Aldrich, MO, USA) are used to immunize mice (ddY, 6 weeks old, female; Japan Laboratory Animals). The same antigen in Freund's incomplete adjuvant (Sigma-Aldrich) is repeatedly injected intraperitoneally into the mice on days 14 and 28. Sera are collected from these mice 10 days after the last immunization. The antibody titer can be confirmed by an enzyme-linked immunosorbent assay or Western blot analysis.

4. Expression Profiles of *HlATG12* from Nymphal to Adult Stages

4.1. Semiquantitative reverse transcription-polymerase chain reaction (RT-PCR)

4.1.1. RNA extraction and cDNA synthesis

Figure 34.2 shows the outline of sampling.

1. Nymphal ticks infesting on rabbits are detached and used on days 1, 2, and 3 after attachment. In addition, engorged nymphs are collected and then a part of them are maintained in a moist chamber at 25 °C. Some maintained ticks are collected at 10 days (premolting), and the others are

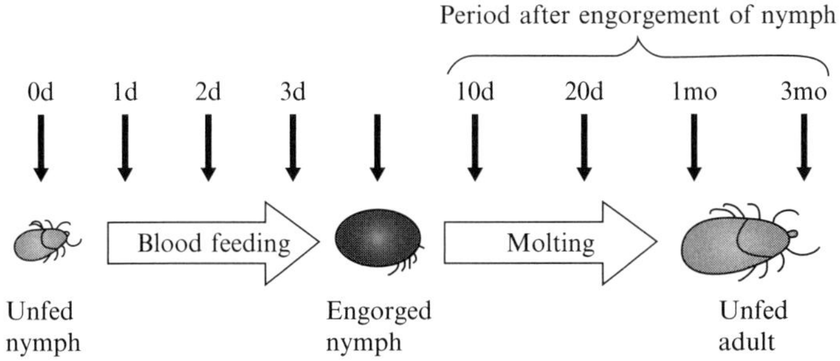

Figure 34.2 Outline for sampling of nymphal and adult stages. Black arrows indicate the timing of sample collection of ticks. Ticks were collected unfed, 1-, 2-, and 3-days-fed, and 10 days (premolting) after engorgement and the others were allowed to molt under the same conditions. Molted new emerging adult ticks were collected at 20 days (postmolting), 1 and 3 months after engorgement in the nymphal stage. d, day(s); mo, month(s).

allowed to molt in a moist chamber at 25 °C. Molted ticks are collected 20 days, 1 month, and 3 months after engorgement.

2. Ten ticks collected in each stage are homogenized with a mortar and pestle (AS ONE, Osaka, Japan) using liquid nitrogen and are resuspended in TRI reagent (SIGMA) as described earlier.
3. Total RNA is extracted from the suspensions, and in order to remove genomic DNA, the sample is treated with DNase (TURBO DNA-*free* Kit; Applied Biosystems, CA, USA). Briefly, DNase (2 units for up to 10 μg of RNA) are added to the RNA solution and mixed gently, then incubated at 37 °C for 20–30 min.
4. Single-strand cDNA is generated from the treated total RNA (10 ng–5 μg) by reverse transcription using Transcriptor First Strand cDNA Synthesis Kit (Roche Diagnostics, Mannheim, Germany). This kit contains oligo(dT)$_{18}$ primer, RNase inhibitor, deoxynucleotide mix, and trascriptor reverse transcriptase which is able to synthesize long cDNA products (up to 14 kDa). The reaction proceeds as follows: 65 °C for 10 min, 50 °C for 60 min, and 85 °C for 5 min.

4.1.2. Polymerase chain reaction (PCR) for amplification of *HlATG12*

PCR is carried out with the appropriate dilutions of synthesized cDNAs and *HlATG12*-specific primers (sense primer, 5′-ATGTCCGATGAAACT GAAGGCTGTGCGACTGCG-3′; antisense primer, 5′-TTAGCCC CATGCGTGACTTTTTGCATAGTGCAGAG-3′). PCR conditions are 94 °C for 30 s, 60 °C for 30 s, and 72 °C for 1 min for 30 cycles. Control amplification is carried out using the specific primers (sense primer, 5′-CCAA CAGGGAGAAGATGACG-3′; antisense primer, 5′-ACAGGTCCT TACGGATGTCC-3′) designed from *H. longicornis actin* (Da Silva Vaz Jr. *et al.*, 2005). The PCR products are electrophoresed on a 1.5% agarose gel and stained with ethidium bromide, and then the gel image is digitized for densitometry analysis by Luminous Imager Software (version 2.0; Aisin Cosmos R&D, Aichi, Japan). The results are expressed as a ratio of the density of the *HlATG12* products to the density of the *actin* products from the same template.

5. Detection of HlAtg Proteins in Midgut Cells

5.1. Immunoblotting (Miyoshi *et al.*, 2004; Umemiya *et al.*, 2007b)

5.1.1. Extraction of soluble proteins from whole ticks

1. Homogenize 5 unfed adult ticks with a mortar and pestle (AS ONE) in liquid nitrogen and resuspend in 0.1 ml of phosphate-buffered saline

(PBS, pH7.4) containing a proteinase inhibitor (Complete Mini; Roche Diagnostics, Mannheim, Germany).

2. Transfer the lysate to a centrifuge tube and sonicate for 30 s using a water-bath sonicator.
3. Centrifuge the extracts at 26,000*g* for 30 min using a high-speed microcentrifuge (Hitachi Koki, Tokyo, Japan). Store supernatants at −30 °C until used for immunoblotting.

5.1.2. Cell fractionation of the midgut by differential centrifugation

The midgut of ticks demonstrates multifunctional activity. Digestion of host blood begins during the slow feeding period (a few days after attachment) and digestion is accomplished slowly within epithelial cells of the midgut (Sonenshine, 1991). However, the protein content in ticks gradually decreases until the tick encounters the next suitable host. In addition to the main digestive function, the midgut serves as the major deposit of nutritional reserves represented by intracellular inclusions of host blood hemoglobin, lipids and carbohydrates. In female ticks, the yolk proteins are synthesized in the midgut as well as the fat body. Unlike insects, many corresponding metabolic processes probably occur in the midgut cells of ticks (Raikhel, 1983). Moreover, the midgut is the first organ that pathogens invade and develop within. There are various genes that show enhanced expression in response to pathogens in the midgut, for example the tick receptor for OspA (*TROSPA*) within the midgut of *Ixodes scapularis* (Pal *et al.*, 2004). Consequently, the midgut is a very important organ for examining autophagy in ticks. The protocol to examine the distribution of HlAtg12 protein within the midgut cells is explained subsequently.

1. Dissect 10 unfed adult ticks in 3–5 ml of PBS containing a proteinase inhibitor (Complete Mini; Roche Diagnostics).
2. Homogenize the midguts using a glass homogenizer (AS ONE) in 0.1 ml of PBS containing a proteinase inhibitor with 0.25 *M* sucrose on ice.
3. Centrifuge at 1000*g* for 7 min at 4 °C to generate a pellet (P1; the fraction-containing the nucleus and unhomogenized cells).
4. Centrifuge the resulting supernatant fraction at 2000*g* for 30 min at 4 °C to generate a pellet (P2; the fraction containing the mitochondria and lysosomes).
5. Transfer the supernatant fraction to a special tube for the ultracentrifuge and centrifuge at 105,000*g* for 60 min at 4 °C to generate a pellet (P105; the fraction containing membranes and ribosomes) using a Micro Ultracentrifuge (Hitachi Koki, Tokyo, Japan).
6. Transfer the supernatant fraction (S105; the fraction containing cytosol) to a new tube. Store samples at −30 °C until used for immunoblotting.

5.1.3. SDS-PAGE and Immunoblotting for detection of HlAtg12

The tick proteins extracted by the preceding methods are separated by SDS-PAGE under nonreducing and/or reducing (in the presence of 2-mercaptoethanol) conditions on a 5%–20% gradient acrylamide gel (e•PAGEL®; ATTO, Tokyo, Japan) and analyzed by Western blot.

1. Separated proteins are transferred from gels onto a PVDF membrane by semidry blotting.
2. The membrane is incubated with 3% (w/v) skim milk in PBS for 1 h at room temperature.
3. After washing with PBS containing 0.1% (v/v) Tween 20 (PBST), the membrane is incubated with a suitable dilution of mouse polyclonal anti-GST-HlAtg12 antibody (Umemiya *et al.*, 2007b) for 1 h at room temperature.
4. The membrane is thoroughly washed with >4 ml/cm^2 of PBST 3 times and the binding of antibody is detected with horseradish peroxidase-conjugated polyclonal goat antimouse immunoglobulin (Dako, Glostrup, Denmark) at a dilution of 1:10000 and an ECL Advance Western Blotting Detection Kit (GE Healthcare). Preliminary measurements must be taken to determine the proper dilution because the ECL advance kit is highly sensitive.
5. The images are analyzed using VersaDoc Model 500 (Bio-Rad Laboratories, Tokyo, Japan).

5.2. Immunofluorescence staining for frozen sections of the midgut

1. Dissect and excise the midgut from unfed and 4-day-fed adult female ticks.
2. Fix with fixative (4% [w/v] paraformaldehyde including 0.1% glutaraldehyde in PBS) in each 10-ml sample bottle at 4 °C, overnight by shaking gently.
3. Wash the midguts with PBS (5 min × 3).
4. Soak in 5%, 10%, 15%, and 20% sucrose in PBS at 4 °C.

Note: The samples should be soaked in each concentration of sucrose solution until they sink as much as possible (more than 4 h).

5. Make a container matching the size of the midgut with aluminum foil, and pour chilled Tissue-Tek O.C.T. Compound (Sakura Finetek Japan, Tokyo, Japan) into the container up to about 80%. Submerge the fixed midgut in the compound. Embed each midgut in Tissue-Tek O.C.T. Compound with liquid nitrogen. Keep frozen blocks at −80 °C.

Note: The samples should be submerged in Tissue-Tek O.C.T. Compound for at least 30 min before freezing.

6. Cut frozen sections (approximately 14 μm thick) on a cryostat (Leica CM 3050; Leica Microsystems, Wetzlar, Germany) and then place them on aminosilanes-coated glass slides (Matsunami Glass, Osaka, Japan).
7. Dry for 30 min or more by air-drying.
8. Wash with PBS (5 min × 3) in a jar. The amount of PBS depends on the size of the jar used.
9. Block with 3%–5% (w/v) skim milk in PBS at room temperature for 1 h on the glass slide or in a jar.
10. Immunostain
    a. Incubate with a suitable dilution of primary antibody (mouse anti-GST-HlAtg12 antibody) overnight at room temperature for 1 h or at 4 °C. Because the dilution factor depends on the antibody titer, preliminary testing is needed.
    b. Wash with PBS (5 min × 3) in a jar.
    c. Incubate with Alexa Fluor 594 conjugated goat anti-mouse IgG (1:1000; Molecular Probes, OR, USA) at room temperature for 1–2 h.
    d. Wash with PBS (5 min × 3).
11. Observe with a confocal laser-scanning microscope (TCS NT, Leica Microsystems) or an immunofluorescence microscope.

The anti-GST-HlAtg12 antibody reacts with the cytoplasm of the midgut epithelial cells, which seem to be digestive cells, during immunofluorescence staining. Moreover, positive reactions appear as some aggregations of red-colored dots in digestive cells of unfed adults. In contrast, the dots are not observed in the digestive cells of 4-day-fed adults even though the anti-GST-HlAtg12 antibody slightly reacts with the cytoplasm. Positive dots observed in unfed samples are very small (less than 1 μm in diameter), so we recommend the observation of the midgut cells by electron microscopy as described in the following methods.

5.3. Immuno-electron microscopy

5.3.1. Preparation of samples from unfed ticks

1. Dissect unfed female ticks as described previously.
2. Fix small pieces of midguts with 4% paraformaldehyde containing 0.1% glutaraldehyde in PBS (pH 7.4) in each 10-ml sample bottle at 4 °C, overnight, by shaking gently.

Note: It is difficult for the fixative to infiltrate into the tick midgut. To deal with this problem, if possible, air should be removed from the midgut with a vacuum pump or an evaporator. Such treatment for more than 30 min will ensure the fixation buffer reaches inside the lumen. This method is also effective for the infiltration of resin during embedding.

3. Remove the fixative and wash with PBS in the bottle at 4 °C by shaking gently (≥30 min × 3).
4. Remove the PBS and dehydrate with an ethanol series shown below in the bottle.
   a. 30% ethanol for ≥10 min at 4 °C
   b. 50% ethanol for ≥10 min at 4 °C
   c. 70% ethanol for ≥10 min at 4 °C
   d. 80% ethanol for ≥10 min at 4 °C
   e. 90% ethanol for ≥10 min at −20 °C
   f. 99% ethanol for ≥10 min at −20 °C
   g. 100% ethanol for ≥20 min × 3 at −20 °C
5. Remove the 100% ethanol and embed in LR gold resin (Polysciences, PA, USA) as indicated subsequently.
   a. The mixture of 100% ethanol and LR gold resin (1:2) at −20 °C, overnight
   b. The mixture of 100% ethanol and LR gold resin (1:1) at −20 °C, overnight
   c. The mixture of 100% ethanol and LR gold resin (2:1) at −20 °C, overnight
   d. Transfer the midgut sample to a gelatin capsule (Nisshin EM, Tokyo, Japan) where 1 drop of pure LR gold resin was placed with a 1-ml syringe.
   e. Fill the capsule with LR gold resin and incubate at −20 °C for 2–3 days (until polymerization of the resin is completed) under UV light.
6. Cut ultrathin sections (approximately 70-nm thick).
7. Place the sections on nickel grids (Nisshin EM).

5.3.2. *Immunostain*

1. Block with 5% (w/v) skim milk in PBS.
2. Incubate the sections overnight by floating the grids on a 100-μl drop of a suitable dilution of primary antibody (anti-GST-HlAtg12 antibody) at 4 °C.
3. Wash three times with 100 μl of PBS.
4. Incubate the grids in a 100-μl drop of 10-nm gold-conjugated goat antimouse IgG + IgM (1:30; GE Healthcare) at room temperature for 2 h.
5. Wash three times with 100 μl of PBS.
6. Fix with a 100-μl drop of 3% (w/v) glutaraldehyde in PBS for 15 min.
7. Wash three times with 100 μl of distilled water.
8. Stain with a 100-μl drop of 5% uranyl acetate in 50% ethanol for 5 min.
9. Observe the sections with a JEM-1010 electron microscope (JEOL, Tokyo, Japan).

By using the preceding methods, gold particles indicate the positive sites showing cytoplasmic distribution, and the anti-GST-HlAtg12 antibody reacts with the exterior of the granulelike structures. These structures are small and comparatively electron dense. The size of most granulelike structures is approximately 500 nm or less (Umemiya *et al.*, 2007b).

6. Ultrastructural Observation of Autophagosome- and Autolysosome-like Structures in Midgut Cells of Unfed Ticks

It is essential to observe the autophagic structures with an electron microscope to be able to describe autophagy in ticks.

1. Dissect unfed female ticks as described previously.
2. Fix small pieces of midguts overnight with 3% glutaraldehyde in 0.1 *M* cacodylate buffer (pH 7.4) in each 10-ml sample bottle at 4 °C by shaking gently.

Note: If possible, air should be removed from the midgut with a vacuum or an evaporator as described earlier.

3. Remove the fixative and wash with 0.1 *M* cacodylate buffer in the bottle by shaking gently (≥30 min × 3).
4. Postfix with approximately 3 ml of 1% OsO_4 in the same buffer for 2 h at 4 °C.
5. Dehydrate with an ethanol series as here.
   a. 30% ethanol for ≥10 min at 4 °C
   b. 50% ethanol for ≥10 min at 4 °C
   c. 70% ethanol for ≥10 min at 4 °C
   d. 80% ethanol for ≥10 min at 4 °C
   e. 90% ethanol for ≥10 min at room temperature
   f. 99% ethanol for ≥10 min at room temperature
   g. 100% ethanol for ≥20 min × 3 at room temperature
   h. 100% propylene oxide
6. Remove the 100% propylene oxide and embed in Epon resin (Nisshin EM) as follows:
   a. The mixture of 100% propylene oxide and Epon resin (1:2) for half a day at room temperature
   b. The mixture of 100% propylene oxide and Epon resin resin (1:1) for half a day at room temperature
   c. The mixture of 100% propylene oxide and Epon resin resin (2:1) for half a day at room temperature
   d. Transfer the midgut sample to the BEAM capsule (Nisshin EM) where 1 drop of pure Epon resin was placed with a 1-ml syringe.

e. Fill the capsule with Epon resin and incubate at 60 °C for 2 days (until polymerization of the resin is completed).
7. Cut ultrathin sections (approximately 80 nm thick) and place on copper grids (Nisshin EM).
8. Double stain with 100 μl of 5% uranyl acetate in 50% ethanol and lead citrate (Reynold's method) for 5 min.
9. Observe the sections with an electron microscope (JEM-1010; JEOL).

The preceding techniques reveal some types of autophagic organelles in digestive cells of unfed females (Figs. 34.3 and 34.4). Most of these structures are approximately 1–2 μm or smaller, surrounded by a single membrane (Fig. 34.3A, arrows), and include almost normal cytoplasm with small dense granules. This appearance seems like an amphisome rather than an autophagosome. Amphisomes are also autophagic organelles surrounded by a single membrane that are formed by the fusion of autophagosomes and early or late endosomes (Berg *et al.*, 1998; Klionsky *et al.*, 2007). Comparison with previous observations of tick midgut cells indicates the structure shown in Fig. 34.3B is a type 1 autophagic compartment (arrowheads). The type 1 inclusions (0.5-2.0 μm in diameter) have a fine granular material of low density and a dense rim under the surrounding membrane (Raikhel, 1983). On the other hand, the structures containing a layerlike form (Fig. 34.3B, arrows) appears to be consistent with myelinosiderome,

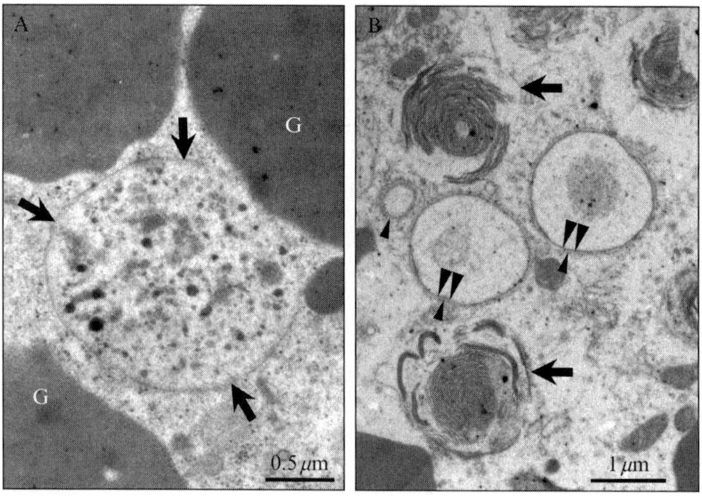

Figure 34.3 Autophagic vacuoles in the midgut epithelial cells of unfed female adult ticks. (A) The structure is surrounded by a single membrane (arrows). The content is almost normal cytoplasm. (B) Two kinds of autophagic vacuoles, one a lucent vacuole (arrowheads), and the other a structure that contains layers of various densities (arrows). Note the double membrane surrounding the lucent structures. Single arrowheads and double arrowheads indicate the outer and inner membranes, respectively. G, granules.

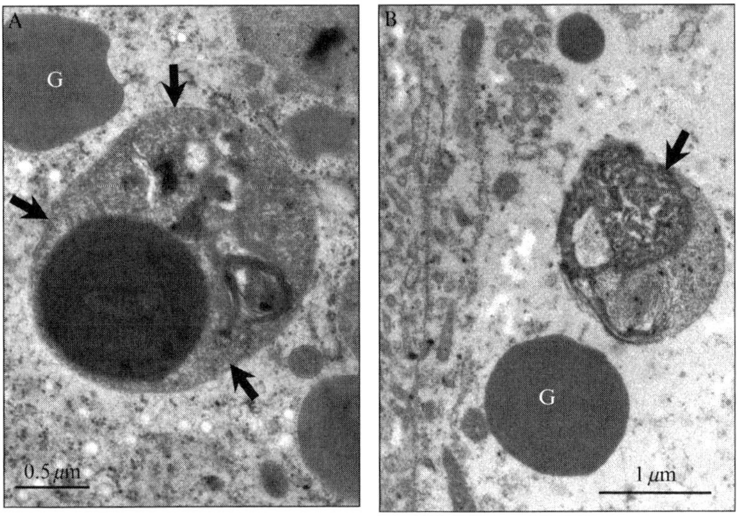

Figure 34.4 Autolysosome-like structures containing recognizable remnants of cytoplasmic elements in the midgut epithelial cells of unfed female adult ticks. (A) The structure is surrounded by a single membrane (arrows) and contains relatively large electron-dense granules. (B) Arrow indicates a single membrane of the autolysosome-like structure. The cytoplasmic contents appear to be denatured. G, granules.

which are residual bodies containing myelin figures and hemosiderin from hemoglobin degradation (Williams *et al.*, 1985). Additionally, the autolysosome-like structures were found in the cytoplasm of the midgut, which is surrounded by a single membrane and contains relatively large electron-dense granules (Fig. 34.4). Although these unique structures are often found in midgut cells of unfed ticks, their function(s) are still unknown. Comparative observations on unfed midgut cells by both electron and immunoelectron microscopy will lead to better understanding as to whether there is a relationship between the HlAtg proteins and autophagic vesicles.

7. Conclusion

The protocols described in this chapter can be used for investigation of autophagy in ticks. Observations using electron microscopy are the most suitable method for detection of autophagic organelles in the midgut of ticks. No standard methods are established for ticks because the study of autophagy in ticks has just begun. There are many useful and convenient methods that can be used to monitor macroautophagy in yeast, but relatively few in other model systems, and there is much confusion regarding acceptable methods to measure macroautophagy in higher eukaryotes

(Klionsky *et al.*, 2008). In addition, autophagy is a dynamic, multistep process that can be both positively and negatively modulated at several steps. If midgut cells can be cultured *in vitro*, the autophagic processes may be monitored in more detail. Phagophores (isolation membranes) are easily recognized as ultrastructurally compressed (electron dense), curving cisternae during the process of enclosing of the cytoplasm in mammalian cells (Klionsky *et al.*, 2007). Unfortunately, the phagophore-like structure has not yet been identified in tick midgut cells.

Autophagy is rapidly induced by starvation in the larval fat body (Scott *et al.*, 2004), which is known to be the nutrient storage organ. While lysosomes in *Drosophila* fat body cells are small and few in number under fed conditions, the lysosomes increase and enlarge, and autolysosomes rapidly form during starvation. Lysosomal staining is useful in *Drosophila* as a proxy method for monitoring autophagy but this method is not always possible in ticks because ticks have an intracellular digestive system for degradation of host blood. The digestion via lysosomes takes place actively in the cytoplasm of midgut cells during and after a blood meal because the midgut is the principle digestive organ (Mendiola *et al.*, 1996). Therefore, it is not always possible to use changes in lysosomes in order to monitor autophagy in ticks.

We have identified homologs of *ATG3*, *ATG4* and *ATG8* genes as well as *ATG12* from ESTs constructed from a cDNA library of *H. longicornis* (Umemiya *et al.*, 2007b; Umemiya-Shirafuji *et al.*, unpublished results). Particularly, further investigation of the *ATG8* homolog, which is a marker for autophagosomes, will provide additional insight into the roles of these molecules in the blood-feeding physiology of ticks (Umemiya *et al.*, 2008), but further trials are required to establish the appropriate methods to monitor autophagy in ticks.

ACKNOWLEDGMENTS

This study was supported by the Bio-oriented Technology Research Advancement Institution (BRAIN), a Grant-in-Aid for Scientific Research (A) from the Japan Society for the Promotion of Science, and a grant from the 21st Century COE program (A-1), the Ministry of Education, Sports, Science, and Technology of Japan. The first author was supported by a Grant-in-Aid for JSPS Fellows from the Japan Society for the Promotion of Science (JSPS). We also thank Dr. DeMar Taylor (University of Tsukuba) for special advice with regard to manuscript improvement.

REFERENCES

Alarcon-Chaidez, F. J., Sun, J., and Wikel, S. K. (2007). Transcriptome analysis of the salivary glands of *Dermacentor andersoni* Stiles (Acari: Ixodidae). *Insect Biochem. Mol. Biol.* **37,** 48–71.

Alim, M. A., Tsuji, N., Miyoshi, T., Islam, M. K., Huang, X., Motobu, M., and Fujisaki, K. (2007). Characterization of asparaginyl endopeptidase, legumain induced by blood feeding in the ixodid tick *Haemaphysalis longicornis*. *Insect Biochem. Mol. Biol.* **37,** 911–922.

Alim, M. A., Tsuji, N., Miyoshi, T., Islam, M. K., Huang, X., Hatta, T., and Fujisaki, K. (2008). HlLgm2, a member of asparaginyl endopeptidases/legumains in the midgut of the ixodid tick *Haemaphysalis longicornis*, is involved in blood-meal digestion. *J. Insect Physiol.* **54,** 573–585.

Anderson, J. F. (2002). The natural history of ticks. *Med. Clin. North Am.* **86,** 205–218.

Berg, T. O., Fengsrud, M., Strømhaug, P. E., Berg, T., and Seglen, P. O. (1998). Isolation and characterization of rat liver amphisomes: Evidence for fusion of autophagosomes with both early and late endosomes. *J. Biol. Chem.* **273,** 21883–21892.

Boldbaatar, D., Sikasunge, C. S., Battsetseg, B., Xuan, X., and Fujisaki, K. (2006). Molecular cloning and functional characterization of an aspartic protease from the hard tick *Haemaphysalis longicornis*. *Insect Biochem. Mol. Biol.* **36,** 25–36.

Da Silva Vaz, I., Imamura, S., Nakajima, C., De Cardoso, F. C., Ferreira, C. A., Renard, G., Masuda, A., Ohashi, K., and Onuma, M. (2005). Molecular cloning and sequence analysis of cDNAs encoding for *Boophilus microplus, Haemaphysalis longicornis* and *Rhipicephalus appendiculatus* actins. *Vet. Parasitol.* **127,** 147–155.

Francischetti, I. M., My-Pham, V., Mans, B. J., Andersen, J. F., Mather, T. N., Lane, R. S., and Ribeiro, J. M. (2005). The transcriptome of the salivary glands of the female western black-legged tick *Ixodes pacificus* (Acari: Ixodidae). *Insect Biochem. Mol. Biol.* **35,** 1142–1161.

Fujisaki, K. (1978). Development of acquired resistance and precipilating antibody in rabbits experimentally infested with females of *Haemaphysalis longicornis* (Ixodoidea: Ixodidae). *Natl. Inst. Anim. Health Q (Tokyo)* **18,** 27–38.

Gong, H., Zhou, J., Liao, M., Hatta, T., Harnnoi, T., Umemiya, R., Inoue, N., Xuan, X., and Fujisaki, K. (2007). Characterization of a carboxypeptidase inhibitor from the tick *Haemaphysalis longicornis*. *J. Insect Physiol.* **53,** 1079–1087.

Harnnoi, T., Sakaguchi, T., Nishikawa, Y., Xuan, X., and Fujisaki, K. (2007). Molecular characterization and comparative study of 6 salivary gland metalloproteases from the hard tick, *Haemaphysalis longicornis*. *Comp. Biochem. Physiol. B, Biochem. Mol. Biol.* **147,** 93–101.

Jongejan, F., Nene, V., de la Fuente, J., Pain, A., and Willadsen, P. (2007). Advances in the genomics of ticks and tick-borne pathogens. *Trends Parasitol.* **23,** 391–396.

Kato, S., Ohtoko, K., Ohtake, H., and Kimura, T. (2005). Vector-capping: a simple method for preparing a high-quality full-length cDNA library. *DNA Res.* **12,** 53–62.

Klionsky, D. J., Cuervo, A. M., and Seglen, P. O. (2007). Methods for monitoring autophagy from yeast to human. *Autophagy* **3,** 181–206.

Klionsky, D. J., Abeliovich, H., Agostinis, P., Agrawal, D. K., Aliev, G., Askew, D. S., Baba, M., Baehrecke, E. H., Bahr, B. A., Ballabio, A., Bamber, B. A., Bassham, D. C., *et al.* (2008). Guidelines for the use and interpretation of assays for monitoring autophagy in higher eukaryotes. *Autophagy* **4,** 151–175.

Liao, M., Zhou, J., Hatta, T., Umemiya, R., Miyoshi, T., Tsuji, N., Xuan, X., and Fujisaki, K. (2007a). Molecular characterization of *Rhipicephalus* (*Boophilus*) *microplus Bm86* homologue from *Haemaphysalis longicornis* ticks. *Vet. Parasitol.* **146,** 148–157.

Liao, M., Hatta, T., Umemiya, R., Huang, P., Jia, H., Gong, H., Zhou, J., Nishikawa, Y., Xuan, X., and Fujisaki, K. (2007b). Identification of three protein disulfide isomerase members from *Haemaphysalis longicornis* tick. *Insect Biochem. Mol. Biol.* **37,** 641–654.

Mendiola, J., Alonso, M., Marquetti, M. C., and Finlay, C. (1996). *Boophilus microplus*: Multiple proteolytic activities in the midgut. *Exp. Parasitol.* **82,** 27–33.

Miyoshi, T., Tsuji, N., Islam, M. K., Kamio, T., and Fujisaki, K. (2004). Cloning and molecular characterization of a cubilin-related serine proteinase from the hard tick *Haemaphysalis longicornis*. *Insect Biochem. Mol. Biol.* **34,** 799–808.

Motobu, M., Tsuji, N., Miyoshi, T., Huang, X., Islam, M. K., Alim, M. A., and Fujisaki, K. (2007). Molecular characterization of a blood-induced serine carboxypeptidase from the ixodid tick *Haemaphysalis longicornis*. *FEBS Lett.* **274,** 3299–3312.

Needham, G. R., and Teel, P. D. (1991). Off-host physiological ecology of ixodid ticks. *Annu. Rev. Entomol.* **36,** 659–681.

Pal, U., Li, X., Wang, T., Montgomery, R. R., Ramamoorthi, N., Desilva, A. M., Bao, F., Yang, X., Pypaert, M., Pradhan, D., Kantor, F. S., Telford, S., Anderson, J. F., and Fikrig, E. (2004). TROSPA, an *Ixodes scapularis* receptor for *Borrelia burgdorferi*. *Cell* **119,** 457–468.

Raikhel, A. S. (1983). The intestine. *In*: Balashov, Y. S. (Ed.), An atlas of ixodid tick ultrastructure. Special publication of the Entomological Society of America, English translation (in Russian, 1979).

Scott, R. C., Schuldiner, O., and Neufeld, T. P. (2004). Role and regulation of starvation-induced autophagy in the *Drosophila* fat body. *Dev. Cell* **7,** 167–178.

Sonenshine, D. E. (1991). Biology of ticks. Vol. 1. Oxford University Press, New York.

Tanaka, M., Liao, M., Zhou, J., Nishikawa, Y., Xuan, X., and Fujisaki, K. (2007). Molecular cloning of two caspase-like genes from the hard tick *Haemaphysalis longicornis*. *J. Vet. Med. Sci.* **69,** 85–90.

Tarnowski, B. I., and Coons, L. B. (1989). Ultrastructure of the midgut and blood meal digestion in the adult tick *Dermacentor variabilis*. *Exp. Appl. Acarol.* **6,** 263–289.

Umemiya, R., Hatta, T., Liao, M., Tanaka, M., Zhou, J., Inoue, N., and Fujisaki, K. (2007a). *Haemaphysalis longicornis*: molecular characterization of a homologue of the macrophage migration inhibitory factor from the partially fed ticks. *Exp. Parasitol.* **115,** 135–142.

Umemiya, R., Matsuo, T., Hatta, T., Sakakibara, S., Boldbaatar, D., and Fujisaki, K. (2007b). Cloning and characterization of an autophagy-related gene, *ATG12*, from the three-host tick *Haemaphysalis longicornis*. *Insect Biochem. Mol. Biol.* **37,** 975–984.

Umemiya, R., Matsuo, T., Hatta, T., Sakakibara, S., Boldbaatar, D., and Fujisaki, K. (2008). Autophagy-related genes from a tick, *Haemaphysalis longicornis*. *Autophagy* **4,** 79–81.

Walker, A. R., and Fletcher, J. D. (1987). Histology of digestion in nymphs of *Rhipicephalus appendiculatus* fed on rabbits and cattle naïve and resistant to the ticks. *Int. J. Parasitol.* **17,** 1393–1411.

Williams, J. P., Barker, D. M., Sauer, J. R., Hair, J. A., Ownby, C., and Koch, H. (1985). Ultrastructural changes in the midgut epithelium of unfed lone star ticks with increasing age. *Ann. Entomol. Soc. Am.* **78,** 62–69.

Zhou, J., Liao, M., Hatta, T., Tanaka, M., Xuan, X., and Fujisaki, K. (2006). Identification of a follistatin-related protein from the tick *Haemaphysalis longicornis* and its effect on tick oviposition. *Gene* **372,** 191–198.

Zhou, J., Liao, M., Ueda, M., Gong, H., Xuan, X., and Fujisaki, K. (2007). Sequence characterization and expression patterns of two defensin-like antimicrobial peptides from the tick *Haemaphysalis longicornis*. *Peptides* **28,** 1304–1310.

CHAPTER THIRTY-FIVE

Quantitative Analysis of Autophagic Activity in *Drosophila* Neural Tissues by Measuring the Turnover Rates of Pathway Substrates

Robert C. Cumming,* Anne Simonsen,† *and* Kim D. Finley‡

Contents

Abstract

The process of macroautophagy occurs in most eukaryotic cells and serves as the main recycling mechanism for the elimination of excess cytoplasmic components. The pathway is upregulated under a wide range of stress-related conditions and basal levels of autophagy are critical for the clearance of age-associated cellular damage, which can accumulate in long-lived, nondividing cells such as neurons. Traditionally, activation of autophagy has been measured by the microscopic observation of newly formed autophagosomes or by monitoring the further modification of the LC3-I protein to the LC3-II isoform by Western blot analysis. However, using these methods to quantitatively determine autophagic activity that occurs in complex tissues over an entire life span has been a technical challenge and difficult to consistently reproduce. We have shown that Western analysis of protein

* Department of Biology, University of Western Ontario, London, Ontario, Canada
† Center for Cancer Biomedicine, University of Oslo and Department of Biochemistry, Norwegian Radium Hospital, Oslo, Norway
‡ BioScience Center, San Diego State University, San Diego, California, and Cellular Neurobiology Laboratory, Salk Institute for Biological Studies, La Jolla, California, USA

Methods in Enzymology, Volume 451
ISSN 0076-6879, DOI: 10.1016/S0076-6879(08)03235-7

substrates normally cleared by the pathway can be used to make quantitative estimates of autophagy occurring in tissues such as the adult *Drosophila* nervous system. By examining the profile of insoluble ubiquitinated proteins (aggregated proteins) we have found that an age-dependent decline in pathway flux or genetic defects in critical autophagic genes can result in the concomitant buildup of substrates that are normally targeted by autophagy to the lysosome. Further, we have found that increasing *Atg8a* expression (a key rate-limiting component of the pathway) during the time in which autophagy is normally suppressed prevents the age-dependent accumulation of insoluble ubiquitinated proteins in neurons. This technique, as well as the detection of proteins damaged by reactive carbonyl groups, can also be used to measure autophagic activity in both normal and genetically altered flies during the aging process or following their acute exposure to oxidants.

1. Introduction

Autophagy ("self-eating") involves the lysosomal import and degradation of cytosolic material and consists of three main pathways that include macroautophagy, microautophagy, and chaperone-mediated autophagy (Klionsky *et al.*, 2008). This chapter will focus on the quantitative assessment of macroautophagy (henceforth referred to as *autophagy*) and the role this key biological process plays in maintaining the *Drosophila* nervous system. Autophagy is a highly conserved trafficking pathway that sequesters cytoplasmic material into double-membrane vesicles called autophagosomes. The newly formed vesicles quickly fuse with lysosomes where their internal cargo is degraded (Reggiori and Klionsky, 2005). Autophagy occurs in response to starvation and environmental stress and has been genetically well characterized in yeast (Kim *et al.*, 2002; Klionsky and Emr, 2000; Ohsumi, 2001). Recent studies in higher eukaryotes have shown that autophagy is also involved with several other physiological processes that include: hormone-triggered cell death during development; clearance of cytotoxic aggregated proteins; and the paradoxical prodeath and prosurvival functions in cancer cells (Cuervo, 2006; D'Andrea *et al.*, 2004; Huang and Klionsky, 2007; Juhasz *et al.*, 2007; Levine, 2007; Simonsen *et al.*, 2008). Genetic studies in mice and *Drosophila* show that suppressing autophagy in the nervous system results in progressive trafficking and morphological defects along with the accumulation of protein aggregates containing ubiquitin (Finley *et al.*, 2003; Hara *et al.*, 2006; Komatsu *et al.*, 2006; Nezis *et al.*, 2008; Simonsen *et al.*, 2008). These findings indicate that the continuous turnover of damaged macromolecules and organelles is a critical function of the pathway, and its essential role in the long-term survival of nerve cells is being characterized in greater detail (Cuervo, 2006; Finley *et al.*, 2003; Hara *et al.*, 2006; Komatsu *et al.*, 2006; Simonsen *et al.*, 2008).

Electron microscopy and confocal imaging studies have suggested that functional decline in autophagy occurs in liver and neuronal cells as normal animals grow older (Cataldo *et al.*, 1996; Cuervo *et al.*, 2005; Donati, 2006; Donati *et al.*, 2001; Simonsen *et al.*, 2008). However, determining the level of autophagy using transmission electron microscopy (TEM), confocal fluorescence imaging studies or Western blot analysis of LC3/Atg8 processing (cytosolic LC3-I becomes membrane-bound LC3-II upon induction of autophagy) has been at best difficult to perform in complex neural tissues (Klionsky *et al.*, 2008). Furthermore, these techniques can fail to distinguish between the robust activation of the pathway and defects in autophagosome-lysosomal trafficking and fusion. Both types of conditions can result in a marked buildup in the number of autophagosomes, alterations in the normal ratios of vesicle subpopulations as well as changes to the LC3-I to LC3-II ratios. Therefore, development of a novel detection method was required that would measure pathway activity that is occurring in complex tissues experiencing a diverse complement of physiological-, genetic-, stress- or age-related conditions (Finley *et al.*, 2003; Simonsen *et al.*, 2008).

To quantitatively measure the level of autophagy in *Drosophila* neural tissues we have developed a two-part protein extraction and Western analysis technique (Finley *et al.*, 2003; Simonsen *et al.*, 2008). The initial step involves the sequential fractionation of proteins obtained from whole tissues. Soluble proteins are first extracted using a mild nonionic detergent (Triton-X100). The remaining proteins, contained within the pellet then undergo an additional more stringent extraction using a harsher ionic detergent (SDS). The second step in the procedure involves Western blot analysis of the SDS-extracted fraction to detect those proteins that carry the molecular marker ubiquitin. Ubiquitin is a protein "tag" that is covalently linked to proteins, which under normal conditions are targeted for elimination by both the proteasome and endosomal/lysosomal systems. Ubiquitinated proteins found in the SDS soluble fraction represent aggregate-like substrates, which are sequestered and cleared by autophagy and not by other clearance pathways that involve ubiquitination. We have termed this group of proteins as the insoluble ubiquitinated protein (IUP) fraction and have used the following technique to detect autophagic substrates and assess any age-related or genetic-based changes that occur to the pathway within the mature *Drosophila* nervous system (Finley *et al.*, 2003; Lindmo *et al.*, 2008; Simonsen *et al.*, 2008). We have found that the rate of autophagy in complex neuronal tissues shows a striking correlation with the accumulation of IUP in adult *Drosophila* brains (Finley *et al.*, 2003; Simonsen *et al.*, 2008). Both the age-related reduction in autophagy (Fig. 35.1) and loss-of-function mutations to key pathway members, like the *blue cheese* (*bchs*, see original publication) and *Atg8a* genes (Fig. 35.2), promotes the premature accumulation of ubiquitinated proteins and the formation of neural

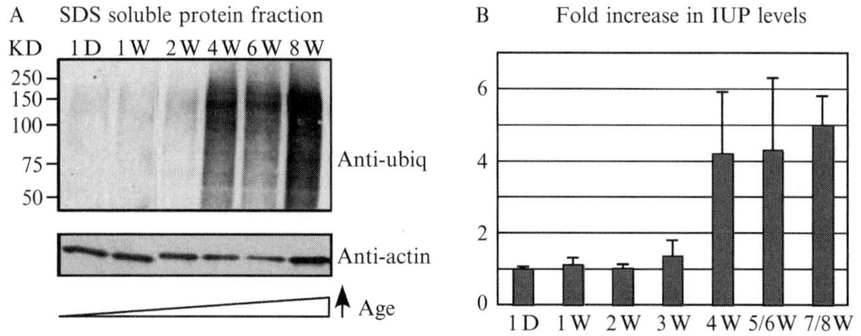

Figure 35.1 Age-related changes to insoluble ubiquitinated protein profiles. (**A**) As autophagic activity decreases in *Drosophila* in aging brains, there is a reciprocal buildup of ubiquitinated proteins found in the SDS soluble protein fraction. Substrates that have biochemical properties of aggregates, partition into this fraction and are mainly cleared by autophagy. Soluble ubiquitinated proteins are first extracted and removed in Triton-X100 buffer. A subsequent SDS extraction concentrates insoluble proteins and Western analysis detects those proteins that carry the ubiquitin tag and are targeted for elimination. (**B**) Densitometric scanning and quantification of IUP and actin levels establishes a relative quantitative value for protein clearance. Please see Simonsen *et al.*, 2008, for original figures.

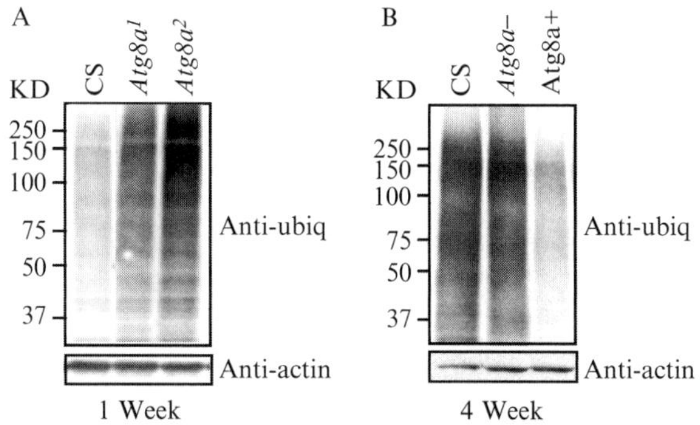

Figure 35.2 IUP profiles from flies with suppressed or enhanced neuronal autophagy. (**A**) When compared to 1-week old Canton-S controls (CS), age-matched *Atg8a* mutants (*atg8a^{1}* or *atg8a^{2}*) show a pronounced accumulation of ubiquitinated proteins in SDS soluble extracts. (**B**) At 4 weeks, control (CS) and *Atg8a*- mutant flies (weak allelic combination, *atg8a^{1}/atg8a^{1}*) show age-appropriate IUP levels. IUP accumulation is suppressed in flies that have enhanced neuronal expression of *Atg8a* (Atg8a+, *EP-UAS-Atg8a/APPL-Gal4*). When quantified and corrected for load (actin) there is a 12-fold decrease in total IUP levels in Atg8a+ flies. Please see Simonsen *et al.*, 2008, for original figures.

aggregates (Finley *et al.*, 2003; Hara *et al.*, 2006; Komatsu *et al.*, 2006; Simonsen *et al.*, 2008). Conversely, we found that when autophagy is enhanced *in vivo*, the normal age-dependent accumulation of ubiquitinated proteins and neural aggregates is substantially suppressed (Fig. 35.2) (Simonsen *et al.*, 2008). The overall biological consequence of promoting basal autophagy in the aging Drosophila nervous system is to increase resistance to oxidant exposure and to extend the average adult life span by over 50% (Simonsen *et al.*, 2008).

As a second marker of autophagic activity we also examined the clearance of carbonylated proteins from *Drosophila* neural tissues. Reactive carbonyl groups are produced as a result of exposing organic molecules to oxidants like free radicals or reactive oxygen species (ROS). These carbonyl groups can then go on to react with and damage a wide range of macromolecules, including proteins (Simonsen *et al.*, 2008). Our findings support the hypothesis that an increase in damage due to reactive modification, which are typically associated with aging or oxidant exposure, may lead to protein misfolding and the formation of insoluble protein aggregates. In most cases, defective macromolecules are then primarily cleared by autophagic/lysosomal function within the cell (Simonsen *et al.*, 2008).

Western blot analysis of insoluble ubiquitinated proteins (IUP), or the direct analysis of protein carbonylation, allows the detection and quantitative assessment of substrates that are preferentially cleared by the pathway (Fig. 35.3) (Das *et al.*, 2001; Levine *et al.*, 1990; Simonsen *et al.*, 2008). By comparing IUP or carbonylated protein profiles found in control and genetically modified or aged animals, the relative levels of autophagic activity can be readily investigated in complex tissues when other methods of examining the pathway are generally considered impractical (Simonsen *et al.*, 2008).

2. Detection of Insoluble Ubiquitinated Protein (IUP) Substrates

2.1. Materials

1. World wide, *Drosophila* stocks and genetic lines are readily available from several public centers and in the United States can be obtained primarily from the Bloomington Stock Center at Indiana University (Bloomington, IN, USA). Information regarding fly genetics, transgenics techniques and stocks can be accessed and individual fly lines obtained using the *Drosophila* database at Flybase (flybase.bio.indiana.edu).

Note: Care should be taken to establish the genetic backgrounds and the baseline husbandry techniques required for the different *Drosophila* strains.

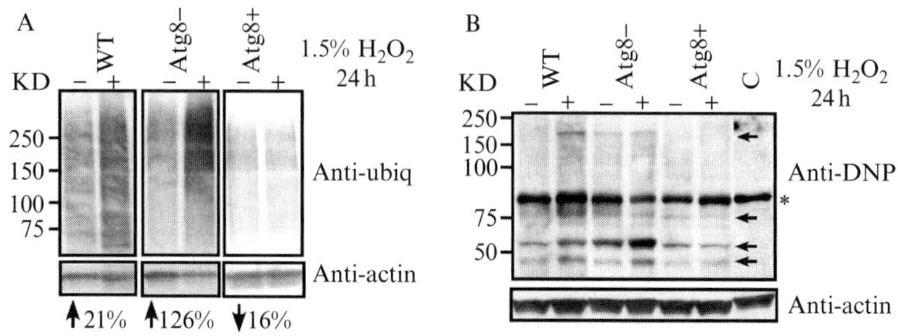

Figure 35.3 Western blot analysis of IUP and carbonylated protein substrates in flies exposed to following hydrogen peroxide. (**A**) IUP Westerns from flies exposed for 24 h to control media (−) or media containing 1.5% H_2O_2 (+). Wild-type control flies show a 20% increase in IUP levels (WT, Canton-S), while *Atg8a* mutant flies (Atg8a−, *Atg8a¹/Atg8a²*) demonstrate a significant accumulation of neuronal IUP (126%) following 24 h of H_2O_2 exposure. In contrast, flies with enhanced *Atg8a* expression have suppressed IUP levels (Atg8+, *EP-Atg8a/APPL-Gal4*). (**B**) Control (WT) flies exposed to the same oxidant conditions show similar alterations to carbonylated protein profiles (arrows). An increase in carbonylation, following H_2O_2 exposure, is observed for a number of protein bands (arrows) and is particularly elevated in *Atg8a* mutant animals (Atg8a−, *Atg8a¹/Atg8a²*). The accumulation of several carbonylated protein bands are suppressed when *Atg8a* is upregulated (Atg8a+, *EP-Atg8a/APPL-Gal4*) (Simonsen *et al.*, 2008). Lane C contains a control protein extract that is not derivatized with DNPH (* band represents protein(s) having nonspecific cross reactivity with the anti-DNP antibody). See Simonsen *et al.*, 2008, for original illustrations.

For most studies we typically outbreed the different genotypes being examined with established stock strains that have been developed and maintained within our laboratory (Canton-S and w^{1118}) (Simonsen *et al.*, 2007). For aging experiments, wild-type control flies (Canton-S) and adult flies from different genetic backgrounds are collected soon after hatching. Male and virgin female flies are aged separately with no more than 25 individuals per vial. To maintain healthy living conditions essential for extended aging studies, individual cohorts of flies should be placed in vials containing fresh food at least once or twice a week. For most types of analysis, which examine the turnover rates of substrate (i.e., IUP), the average life span profiles of individual genotypes should be characterized in detail for a particular temperature (i.e., 25 °C) and culturing condition (Finley *et al.*, 2003; Simonsen *et al.*, 2008; Simonsen *et al.*, 2007). This basic information is then used to estimate the general timing associated with aggregate accumulation and the formation of neurological defects for a particular genotype or treatment condition.

2. Standard medium for genetic crosses and aging experiments consists of yeast, corn meal, molasses, yeast extract, and agar mixture.
3. Liquid nitrogen or dry ice.

4. Precast SDS-PAGE electrophoresis gels, 18-well, 4–20% gradient gels (BioRad, Hercules, CA). Depending on the size range of proteins being examined, other types of SDS-PAGE electrophoresis gels can also be used.
5. Immobilon-P protein transfer membrane (Millipore Corp., Bedford, MA).
6. Protease Inhibitor Cocktail Tablets (Roche Applied Science, Mannheim, Germany).
7. Antibodies: anti-ubiquitin antibody (monoclonal, Cell Signaling), anti-Actin5c monoclonal antibody (JLA20, Developmental Studies Hybridoma Bank NICHD, The University of Iowa, IA), goat-antimouse IgG (H+L)-HRP antibody (secondary antibody, BioRad, Hercules, CA).

2.2. Stock solutions and reagents

1. Lowry Protein Assay (SDS detergent compatible, BioRad, Hercules, CA).
2. 4x Loading buffer: 6 gm SDS, 16 ml of 1 *M* Tris, pH 6.8, 20 gm glycerol, 10 ml β-mercaptoethanol solution (βME) to a final volume of 50 ml.
3. Electorophoresis running buffer: 190 m*M* glycine (14.26 gm), 25 m*M* Tris base (3.03 gm), 5.0 ml of 20% SDS, to a final volume of 1 liter in water (the pH is not adjusted).
4. Electrophoresis transfer buffer: 190 m*M* glycine (14.26 gm), 25 m*M* Tris base (3.03 gm), 100 ml methanol, in water to a final volume of 1 liter in water (the pH is not adjusted).
5. 1x TTBS: 10 m*M* Tris base. pH 7.4, 150 m*M* NaCl (8.76 gm), 1 ml Tween-20, to a final volume of 1 liter in water.
6. ReBlot Western Blot Recycling Kit (Chemicon, Temecula, CA)

Fresh solutions

1. Soluble protein extraction buffer: 1% Triton-X100 (vol./vol.), 1x phosphate buffered saline (PBS) and protease inhibitors.
2. Insoluble protein extraction buffer: 2% SDS, 50 m*M* Tris, pH 7.4 and protease inhibitors.
3. 1x TTBS containing 5% BSA or 5% dry milk (wt./vol.).

Collection and preparation of *Drosophila* neural tissues

1. Typically between 20 and 30 adult *Drosophila* from select genotypes or from different ages are collected under CO_2 anesthesia, placed into 1.5-ml microcentrifuge tubes and flash frozen in liquid nitrogen or on dry ice (Finley *et al.*, 2003; Simonsen *et al.*, 2008).

2. To separate heads from the rest of the body the individual tubes are vortexed several times at high speed before being returned to liquid nitrogen or dry ice.
3. The severed heads are quickly separated from thoraxes and abdomens by passing the heads through a standard tea-sieve. Cohorts of heads are collected direct into fresh 1.5-ml microcentrifuge tubes. These tubes are then returned to liquid nitrogen or dry ice until all samples are ready for further processing (Finley *et al.*, 2003; Simonsen *et al.*, 2008).

3. Sequential Detergent Fraction of *Drosophila* Proteins

1. Freshly prepared 1% Triton-X100 buffer (75 μl) is added to each tube and the heads (20–30 per tube) are gently homogenized using a small pestle, specifically designed for microcentrifuge tubes (Finley *et al.*, 2003; Simonsen *et al.*, 2008).
2. Before removal, the pestle is rinsed with an additional 75 μl of Triton-X100 buffer and the samples placed on ice (final volume 150 μl).
3. The tissue homogenate is centrifuged (14,000 rpm) for 10 min at 4 °C. The individual supernatant fractions are collected into fresh tubes and saved as the Triton-X100 soluble fraction (Finley *et al.*, 2003; Simonsen *et al.*, 2008).
4. The remaining protein pellets are washed once in an additional 100 μl of 1.0% Triton-X100 buffer and centrifuged (14,000 rpm) for 10 min at 4 °C. The wash buffer is then carefully removed from the pellet and discarded.
5. 100 μl of 2% SDS buffer is added to the remaining protein pellets, which are sonicated briefly and centrifuged (14,000 rpm) for 10 min at 4 °C.
6. The resulting supernatant fractions are collect and saved as the SDS soluble fractions (Finley *et al.*, 2003; Simonsen *et al.*, 2008).
7. The protein concentration of each sample is determined using a detergent compatible Lowry protein assay (BioRad, Hercules, CA) (Finley *et al.*, 2003; Simonsen *et al.*, 2008).
8. Protein loading buffer (4x) is added to each sample before they are boiled for 10 min at 100 °C.
9. The samples can be analyzed immediately or stored frozen at −80 °C.
10. For most applications the Triton-X100 and SDS soluble samples are examined separately on Western blots as distinct data sets.
11. Between 15 and 30 μg of total protein for each sample is loaded per lane and the proteins resolved on 4%–20% SDS/polyacrylamide gradient gels, using 33 mAmps at constant current.

12. Gels are electroblotted onto membranes for 2 h using 200 mAmps constant current at 4 °C (Immobilon-P, Millipore Corp., Bedford, MA).
13. Western membranes are blocked with gentle shaking in 1x TTBS containing 5% milk for one hour. They are rinsed twice (5–10 min each) in 1x TTBS and then probed with anti-ubiquitin antibodies overnight at 4 °C (1:1000 dilution in 1x TTBS, 5% BSA).
14. The following day, the blots are washed 3 times (5–10 min each) in 1x TTBS and hybridized in antimouse HRP secondary antibody for one hour at room temperature (anti-mouse 1:10,000 dilution in 1x TTBS, 5% milk).
15. The blots are then rinsed 3 times (5–10 min each) in 1x TTBS and developed using standard ECL reagents and autoradiographs or digital imaging systems.
16. Autoradiographs are digitally scanned using a GS-800 Calibrated Desitometer and Quantity One imaging analysis software (BioRad).
17. The blots are stripped using the Chemicon ReBlot striping kit, reblocked and probed with an anti-actin antibody (1:200 dilution, JLA20) (Simonsen *et al.*, 2008).
18. The relative amounts of IUP proteins from individual samples are quantified and corrected using actin as a loading control (Simonsen *et al.*, 2008).
19. Antibodies directed against the *Drosophila* histone-2B or the alpha-tubulin protein have also been used successfully as loading controls.
20. By dissecting heads individually, scaling back on the extraction buffer volumes and omitting the Triton-X100 buffer wash step, the IUP profiles from as few as 3 adult *Drosophila* heads can be examined (Nezis *et al.*, 2008).

4. Detection of Carbonlyated Protein Substrates

4.1. Materials

1. H_2O_2 *Drosophila* culturing medium: 1% sugar, 1% dry yeast extract, 1.2% agar (w/v) supplemented with or without a final concentration of 1.5% H_2O_2 (v/v). Culturing medium is heated in a microwave until the agar is melted and there is an even suspension of material. The medium is equilibrated to 37 °C before the H_2O_2 is added to a final concentration of 1.5%.
2. Precast 12% SDS-PAGE electrophoresis gels, 18-well (BioRad, Hercules, CA).
3. 2,4-Dinitrophenylhydrazine (DNPH, Sigma, St. Louis, MO).
4. Anti-dinitrophenyl (DPN) monoclonal antibody (Sigma, St. Louis, MO).

4.2. Solutions and reagents

1. 12% SDS solution.
2. 2 *M* Tris, pH 7.4, 30% glycerol.

4.2.1. Fresh solutions

1. Homogenization buffer: 200 m*M* sodium phosphate, pH 6.5, 1 m*M* EDTA, 1% SDS and protease inhibitors.
2. 20 m*M* 2,4-dintrophenylhydrazine (DNPH) dissolved in 2 *M* HCl.
3. Blocking solution: 3% BSA, 1% milk in TTBS.

Exposing *Drosophila* to oxidants

1. To assess the *neuronal* response of autophagy to oxidants, flies from specific genotypes are collected within 4 hours of eclosion (adult hatching) and maintained on standard fly food for 4 to 7 days (25 flies per vial).
2. Groups of flies representing different genotypes (25 per vial) are exposed to medium for 24 s that contains 1% sugar, 1% dry yeast extract, 1.2 % agar without H_2O_2 or supplemented with a final concentration of 1.5% H_2O_2 (vol/vol) (Simonsen *et al.*, 2008).

DNPH derivatization and analysis of cabonylated proteins

1. The next day following the 24-h exposure, the different cohorts of flies are individually collected using CO_2, flash frozen, vortexed and ≈25 heads for each condition and genotype are collected into 1.5-ml microcentrifuge tubes. Heads are quickly separated from the rest of the fly bodies as described previously.
2. Heads are weighed and homogenization buffer added to each tube (4x volume of buffer added per weight of each group of heads).
3. The heads are sonicated briefly and then centrifuged at 14,000 rpm for 10 min at 4 °C (Finley *et al.*, 2003; Simonsen *et al.*, 2008).
4. The supernatant fractions are transferred to fresh 1.5-ml tubes and 35 μl of each sample is added to a fresh tube containing 35 μl of a premixed solution consisting of 12% SDS and 70 μl of 20 m*M* 2,4-dinitrophenylhydrazine (DNPH, dissolved in 2 *M* HCl) (Das *et al.*, 2001; Levine *et al.*, 1990; Simonsen *et al.*, 2008).
5. To detect nonspecific interactions of the primary antibody, a negative control is prepared that consists of 35 μl of a reserved sample, treated with 2 *M* HCl in the absence of DNPH. This control sample is placed into and resolved in a separate lane on the SDS-PAGE gel (Das *et al.*, 2001; Levine *et al.*, 1990; Simonsen *et al.*, 2008).
6. Samples are incubated for 1 h at room temperature followed by neutralization with 52.5 μl of 2 *M* Tris in 30% glycerol (Das *et al.*, 2001; Simonsen *et al.*, 2008).

7. The Lowry Assay (DC protein assay, BioRad) is used to determine protein concentrations and afterwards bromphenol blue dye is added to each sample (0.01% w/v final concentration).
8. Protein samples (20 μg) are resolved in a 12% SDS-PAGE gel (BioRad) and electroblotted onto Immobilon-P membrane.
9. Western blots are blocked in 3% BSA and 1% milk in TTBS and hybridized overnight with anti-DNP antibody at 4 °C (a 1:200 dilution in 5% BSA, TTBS).
10. The Western blots are washed 3 times in TTBS (10 min each) and incubated with secondary antibody (1:10,000 dilution, goat anti-mouse HRP conjugate) in 3% BSA, 1% milk in TTBS for 1 h.
11. The blots are washed an additional 3 times (5–10 min each) in TTBS and developed using standard ECL reagents, autoradiographs and imaging techniques.
12. Westerns blots are stripped (ReBlot stripping solution, Chemicon) and probed for actin as a loading control (Das *et al.*, 2001; Levine *et al.*, 1990; Simonsen *et al.*, 2008).
13. Individual DNP positive bands can be quantified as described above and corrected for load, using quantified actin values for each lane. Direct comparisons can be made between genotypes containing altered autophagy levels, different ages or flies exposed to a range of oxidant conditions. Alterations in DNP-positive protein profiles indicate a change in pathways needed to clear cellular components that have been damaged due in part to reactive oxygen species.

5. Conclusions

Using the type of studies described above, we have demonstrated that Western blot analysis of IUP and carbonylated proteins can consistently detect changes in substrate profiles that are linked to the age-related decline of autophagic activity in *Drosophila* neural tissues. Wild-type Canton-S flies show a significant increase in neural IUP levels starting between 3.5 and 4 weeks of age (Fig. 35.1). This dramatic downward shift in aggregate-like substrate (IUP) clearance is concomitantly associated with an age-dependent suppression of autophagy. This was previously shown using quantitative RT-PCR and Western analysis to characterize the expression levels of several autophagy genes across the entire adult *Drosophila* life span (Simonsen *et al.*, 2008). We have found, using IUP Western blot analysis and imaging studies, that functional loss of the *Drosophila bchs* gene results in progressive neural defects, which suggests a role for *bchs* in autophagic clearance of aggregate-like substrates in the nervous system (Finley *et al.*, 2003). The accelerated buildup of age-dependent neuronal damage in both *bchs* and *Atg8a* mutant flies (Fig. 35.2) not only confirms that these genes are essential for full autophagic function but also demonstrates that basal levels of the pathway are critical to prevent the accelerated accumulation of age-related damage in the mature nervous system (Finley *et al.*, 2003; Simonsen *et al.*, 2008).

Recently we have begun to expand the use of these techniques to examine autophagic activity in a broader genetic and tissue-specific context. A recent study of the PI 3-kinase regulator kinase, *vps15*, shows that functional loss of the gene results in early autophagic trafficking defects in larval fat body cells (Lindmo *et al.*, 2008). Larvae that are homozygous for the Δ*vps15* mutation die early during development (L2 stage), which makes the analysis of autophagy in the mutant adults impossible. However, using confocal and TEM imaging along with Western analysis techniques, these larvae show clear defects in the ability of autophagy to eliminate ubiquitin-positive proteins or inclusions in fat body cells (Lindmo *et al.*, 2008). The implication from this and other studies is that an increasing number of genetic/molecular interactions are required for full autophagic function to occur under a diverse set of physiological and cellular conditions. While additional studies on autophagy are beyond the scope of this chapter, we are building upon these detection methods. Our goal is to develop other quantitative techniques that also measure the level of pathway activity found in *Drosophila* tissues under diverse conditions as well as in tissue culture and mammalian and human model systems.

REFERENCES

Cataldo, A. M., Hamilton, D. J., Barnett, J. L., Paskevich, P. A., and Nixon, R. A. (1996). Properties of the endosomal-lysosomal system in the human central nervous system: Disturbances mark most neurons in populations at risk to degenerate in Alzheimer's disease. *J. Neurosci.* **16,** 186–199.

Cuervo, A. M. (2006). Autophagy in neurons: It is not all about food. *Trends Mol. Med.* **12,** 461–464.

Cuervo, A. M., Bergamini, E., Brunk, U. T., Droge, W., Ffrench, M., and Terman, A. (2005). Autophagy and aging: The importance of maintaining "clean" cells. *Autophagy* **1,** 131–140.

D'Andrea, M. R., Cole, G. M., and Ard, M. D. (2004). The microglial phagocytic role with specific plaque types in the Alzheimer disease brain. *Neurobiol. Aging* **25,** 675–683.

Das, N., Levine, R. L., Orr, W. C., and Sohal, R. S. (2001). Selectivity of protein oxidative damage during aging in *Drosophila melanogaster. Biochem. J.* **360,** 209–216.

Donati, A. (2006). The involvement of macroautophagy in aging and anti-aging interventions. *Mol. Aspects Med.* **27,** 455–470.

Donati, A., Cavallini, G., Paradiso, C., Vittorini, S., Pollera, M., Gori, Z., and Bergamini, E. (2001). Age-related changes in the autophagic proteolysis of rat isolated liver cells: Effects of antiaging dietary restrictions. *J. Gerontol. A Biol. Sci. Med. Sci.* **56,** B375–B383.

Finley, K. D., Edeen, P. T., Cumming, R. C., Mardahl-Dumesnil, M. D., Taylor, B. J., Rodriguez, M. H., Hwang, C. E., Benedetti, M., and McKeown, M. (2003). Blue cheese mutations define a novel, conserved gene involved in progressive neural degeneration. *J. Neurosci.* **23,** 1254–1264.

Hara, T., Nakamura, K., Matsui, M., Yamamoto, A., Nakahara, Y., Suzuki-Migishima, R., Yokoyama, M., Mishima, K., Saito, I., Okano, H., and Mizushima, N. (2006). Suppression of basal autophagy in neural cells causes neurodegenerative disease in mice. *Nature* **441,** 885–889.

Huang, J., and Klionsky, D. J. (2007). Autophagy and human disease. *Cell Cycle* **6,** 1837–1849.

Juhasz, G., Puskas, L. G., Komonyi, O., Erdi, B., Maroy, P., Neufeld, T. P., and Sass, M. (2007). Gene expression profiling identifies FKBP39 as an inhibitor of autophagy in larval Drosophila fat body. *Cell Death Differ.* **14,** 1181–1190.

Kim, J., Huang, W.-P., Stromhaug, P. E., and Klionsky, D. J. (2002). Convergence of multiple autophagy and cytoplasm to vacuole targeting components to a perivacuolar membrane compartment prior to de novo vesicle formation. *J. Biol. Chem.* **277,** 763–773.

Klionsky, D. J., Abeliovich, H., Agostinis, P., Agrawal, D. K., Aliev, G., Askew, D. S., Baba, M., Baehrecke, E. H., Bahr, B. A., Ballabio, A., Bamber, B. A., and Bassham, D. C. (2008). Guidelines for the use and interpretation of assays for monitoring autophagy in higher eukaryotes. *Autophagy* **4,** 151–175.

Klionsky, D. J., and Emr, S. D. (2000). Autophagy as a regulated pathway of cellular degradation. *Science* **290,** 1717–1721.

Komatsu, M., Waguri, S., Chiba, T., Murata, S., Iwata, J., Tanida, I., Ueno, T., Koike, M., Uchiyama, Y., Kominami, E., and Tanaka, K. (2006). Loss of autophagy in the central nervous system causes neurodegeneration in mice. *Nature* **441,** 880–884.

Levine, B. (2007). Cell biology: Autophagy and cancer. *Nature* **446,** 745–747.

Levine, R. L., Garland, D., Oliver, C. N., Amici, A., Climent, I., Lenz, A. G., Ahn, B. W., Shaltiel, S., and Stadtman, E. R. (1990). Determination of carbonyl content in oxidatively modified proteins. *Methods Enzymol.* **186,** 4644–78.

Lindmo, K., Brech, A., Finley, K. D., Gaumer, S., Contamine, D., Rusten, T. E., and Stenmark, H. (2008). The PI 3-kinase regulator Vps15 is required for autophagic clearance of protein aggregates. *Autophagy* **4,** 500–506.

Nezis, I. P., Simonsen, A., Sagona, A. P., Finley, K., Gaumer, S., Contamine, D., Rusten, T. E., Stenmark, H., and Brech, A. (2008). Ref(2)P, the *Drosophila melanogaster* homologue of mammalian p62, is required for the formation of protein aggregates in adult brain. *J. Cell Biol.* **180,** 1065–1071.

Ohsumi, Y. (2001). Molecular dissection of autophagy: two ubiquitin-like systems. *Nat. Rev. Mol. Cell Biol.* **2,** 211–216.

Reggiori, F., and Klionsky, D. J. (2005). Autophagosomes: biogenesis from scratch? *Curr. Opin. Cell Biol.* **17,** 415–422.

Simonsen, A., Cumming, R. C., Brech, A., Isakson, P., Schubert, D. R., and Finley, K. D. (2008). Promoting basal levels of autophagy in the nervous system enhances longevity and oxidant resistance in adult Drosophila. *Autophagy* **4,** 176–184.

Simonsen, A., Cumming, R. C., and Finley, K. D. (2007). Linking lysosomal trafficking defects with changes in aging and stress response in Drosophila. *Autophagy* **3,** 499–501.

CHAPTER THIRTY-SIX

Genetic Manipulation and Monitoring of Autophagy in *Drosophila*

Thomas P. Neufeld*

Contents

Abstract

Drosophila melanogaster provides a model system useful for many aspects of the study of autophagy *in vivo*. These include testing and validation of genes potentially involved in autophagy, discovery of novel genes through genetic screening for mutations that affect autophagy, and analysis of potential roles of autophagy in specific developmental or physiological processes. In recent years, a number of techniques and transgenic and mutant fly strains have been developed to facilitate autophagy analysis in this system. Here, protocols are described for activating or inhibiting autophagy in *Drosophila*, and for examining the progression of autophagy *in vivo* through imaging-based assays. The goal of this chapter is to provide a resource both for autophagy investigators with limited familiarity with fly genetics, as well as for experienced *Drosophila* biologists who wish to test for connections between autophagy and a given gene, pathway or process.

* Department of Genetics, Cell Biology and Development, University of Minnesota, Minneapolis, Minnesota, USA

Methods in Enzymology, Volume 451
ISSN 0076-6879, DOI: 10.1016/S0076-6879(08)03236-9

1. Introduction

The *Drosophila* system has attracted researchers from a wide range of fields, due on the one hand to its powerful genetic capabilities, simple and economic culture, and short generation time, and on the other to its relatively high similarity to human physiology that extends from organ systems to signaling pathways to individual molecules. This combination of the simple with the complex has provided fertile ground for making key discoveries in many areas of cell and developmental biology, as well as insights into mechanisms of disease. In the field of autophagy, recent contributions from this system include establishment of the roles of metazoan orthologs of many of the autophagy-related (Atg) genes identified in yeast, delineation of the signaling networks that regulate autophagy, and analysis of the roles of autophagy in development, aging, cell growth and death, and neurodegeneration (Table 36.1). In addition, these studies have led to the development of *Drosophila* lines carrying transgenes or mutations that either inhibit or activate autophagy, or that allow it to be monitored in living or fixed tissues.

Autophagy has been directly visualized in a number of tissues and organ systems in *Drosophila*, including the larval fat body, gut, imaginal discs and salivary glands, and in neurons, wing epithelium, and ovarian nurse cells of adult flies (Butterworth and Forrest, 1984; Juhász *et al.*, 2003; Kimura *et al.*, 2004; Lee and Baehrecke, 2001; Lee *et al.*, 2002, 2007; Pandey *et al.*, 2007; Rusten *et al.*, 2004; Rusten *et al.*, 2007; Scott *et al.*, 2007; Scott *et al.*, 2004; Velentzas *et al.*, 2007). Of these tissues, several attributes of the larval fat body make it an especially attractive experimental system: as the primary nutrient storage organ of the larva, the fat body is specialized to produce a robust autophagic burst in response to nutrient starvation; it consists of a monolayer of large, polyploid cells which are ideal for imaging-based techniques (Fig. 36.1); due to its low baseline level of lysosomal activity under nutrient-replete conditions, expansion and acidification of the autolysosomal compartment in response to autophagy induction can be specifically visualized using lysotropic dyes such as LysoTracker Red. In addition, fat body tissues can readily be isolated in quantities sufficient for small-scale biochemical analysis.

In the following sections, simple methods are described for inducing autophagy through nutrient starvation or rapamycin treatment, and for monitoring its progression in the larval fat body using LysoTracker dyes or fluorescent marker proteins such as GFP-Atg8. The final section of this chapter describes genetic manipulations that allow transgenic or mutant strains of Drosophila to be used in conjunction with these protocols.

Table 36.1 Evidence for participation of *Drosophila Atg* orthologs in autophagy

Gene	Requirement in starvation-induced autophagy	Requirement in developmental cell death	Promotion of longevity/ neuroprotection	Stimulation of autophagy by overexpression	Localization to autophagic vesicles
Atg1/CG10967	(Lee et al., 2007; Scott *et al.*, 2007; Scott *et al.*, 2004)	(Berry and Baehrecke, 2007)	(Berger *et al.*, 2006)	(Scott *et al.*, 2007)	
Atg2/CG1241	(Scott *et al.*, 2004)	(Berry and Baehrecke, 2007)			
Atg3/CG6877	(Juhász *et al.*, 2003; Scott *et al.*, 2007)	(Berry and Baehrecke, 2007)			
Atg4/CG4428					
Atg5/CG1643	(Scott *et al.*, 2004)				(Juhász *et al.*, 2008; Rusten *et al.*, 2004)
Atg6/CG5429	(Lee *et al.*, 2007; Scott *et al.*, 2004)	(Berry and Baehrecke, 2007)			
Atg7/CG5489	(Juhász *et al.*, 2007; Scott *et al.*, 2004)	(Berry and Baehrecke, 2007; Juhász *et al.*, 2007)	(Juhász *et al.*, 2007)		
Atg8a/ CG32672	(Scott *et al.*, 2007)	(Berry and Baehrecke, 2007)	(Simonsen *et al.*, 2008)	(Simonsen *et al.*, 2008)	(Rusten *et al.*, 2007; Scott *et al.*, 2004)

(continued)

Table 36.1 (*continued*)

Gene	Requirement in starvation-induced autophagy	Requirement in developmental cell death	Promotion of longevity/ neuroprotection	Stimulation of autophagy by overexpression	Localization to autophagic vesicles
Atg8b/ CG12334					(Scott *et al.*, 2004)
Atg9/CG3615					
Atg10/ CG12821					
Atg12/ CG10861	(Scott *et al.*, 2004)	(Berry and Baehrecke, 2007)			
Atg13/CG7331					
Atg16/ CG31033					
Atg18a/ CG7986	(Scott *et al.*, 2004)	(Berry and Baehrecke, 2007)			
Atg18b/ CG11975					
Atg18c/ CG8678					

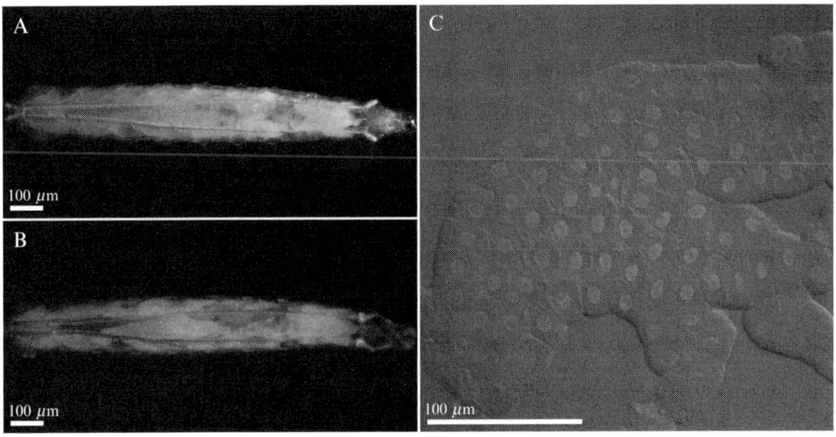

Figure 36.1 Distribution and morphology of the *Drosophila* larval fat body. (A) In intact larvae, the fat body is evident as an opaque, white, bilaterally symmetrical tissue running the length of the animal, just beneath the larval cuticle. (B) The extent of the fat body is evident by expression of GFP using the fat body-specific driver fb-GAL4 (Gronke *et al.*, 2003). (C) A dissected fat body lobe, stained with DAPI and visualized with direct interference contrast optics. Note the large polyploid cells arranged as a monolayer. (See Color Insert.)

2. Methods

2.1. Autophagy induction

Materials

Drosophila larvae of appropriate stage, typically mid-third instar (approx. 96 h after egg laying at 25 °C).
Standard *Drosophila* media
20% sucrose in water
Glass or plastic fly culture vials

Protocol

To induce a starvation response, larvae are removed from food and transferred to a vial containing a few milliliters of sucrose solution (the density of which allows larvae to float at the surface) or to a vial containing moistened filter paper. Either treatment elicits a strong autophagic response in the larval fat body that peaks within 3–4 h, and continues at a lower level for an extended period (24 h or more)

Notes

1. As an alternative to starvation, larvae can be treated with rapamycin, a specific inhibitor of the TOR signaling pathway (1 μM in fly food, 24 h treatment)

2. As autophagy rates in the larval fat body are exquisitely sensitive to nutrient conditions, it is important to pay close attention to crowding conditions during larval culture prior to the experiment. To ensure optimal nutrient conditions, 10–20 larvae should be transferred to a vial containing fresh standard *Drosophila* media 24 h prior to dissection (generally 72 h after embryo collection at 25 °C). This is the case whether autophagy will ultimately be assayed under either fed or starved conditions.
3. During the final 24 h of the larval phase (96–120 h of development at 25 °C), increases in 20-hydroxyecdysone concentrations gradually lead to a rise in the basal rate of autophagy ("developmental autophagy"), even under well-fed conditions.

2.2. LysoTracker staining

Materials

Larvae (starved or otherwise treated; fed controls)
PBS (137 m*M* NaCl, 2.7 m*M* KCl, 1.5 m*M* KH2 PO_4, 6.5 m*M* $NaHPO_4$)
Staining solution, made fresh daily: 100 n*M* LysoTracker Red DND-99 (Invitrogen), 1 μg/ml DAPI or Hoechst 33342 (Sigma) in PBS

Protocol

1. 4–6 larvae are placed in a dissection dish containing PBS. Using fine forceps under a dissecting stereomicroscope, larvae are bisected and inverted.
2. Carcasses are transferred to a small volume (100–200 μl) of staining solution in a dissection dish and incubated 2–3 min at room temperature.
3. During this incubation, the fat body (one major lobe per animal; Fig. 36.1C) is excised with fine forceps, and the remaining carcass is discarded.
4. Fat body lobes are then transferred individually to a small drop of PBS on a glass slide, covered, and immediately photographed on a standard compound fluorescence microscope.

Notes

1. LysoTracker Red has the strongest signal and lowest background in unfixed tissue; however, the preceding protocol can be modified to include a fixation step (1 min in 4% paraformaldehyde/PBS following the incubation in staining solution). Although this step significantly reduces signal intensity, it provides considerably longer preservation of cell morphology than is the case for the nonfixed samples, which show significant degradation within 10 min.

2. During the wandering stage of larval development (final $\approx$ 12 h), larval serum proteins are taken up from the hemolymph into fat body cells by endocytosis (Levenbook, 1985; Sass and Kovacs, 1977). The accumulating endocytic vesicles and protein granules can also incorporate LysoTracker dyes, resulting in a decrease in the usefulness of this reagent as a specific marker of autophagy in this late larval stage.
3. The fluorescent excitation and emission spectra of LysoTracker Red are well separated from those of GFP, and this difference can be exploited for double labeling purposes (see subsequent sections). LysoTracker dyes with other fluorescent properties are available, and may also be useful.

2.3. Assaying autophagy through GFP-Atg8 localization

The fluorescent protein GFP-Atg8 and its mammalian counterpart GFP-LC3 represent particularly useful markers of autophagy *in vivo*, labeling all parts of the autophagic pathway from the earliest stages of autophagosome formation to its fusion with the endolysosomal compartment (Kim *et al.*, 2001; Mizushima *et al.*, 2004; Suzuki *et al.*, 2001). Under nutrient replete conditions, GFP-Atg8 generally has a diffuse appearance by light microscopy, which changes to a highly punctate cytoplasmic localization in response to autophagy induction. It should be noted that mutations or conditions that block autophagosome-lysosome fusion can also result in a punctate appearance of this marker, as can high levels of expression associated with cell transfection (Klionsky *et al.*, 2008). In cultured cells, these effects can be distinguished from increased autophagosome formation and autophagic flux through the use of lysosome fusion inhibitors such as bafilomycin. This type of protocol has not been developed in whole animal models such as *Drosophila*. Instead, the combination of GFP-Atg8 and LysoTracker analysis can be used to determine flux through the autophagic pathway.

Drosophila possesses two Atg8 orthologs, both of which localize to autophagic structures (Scott *et al.*, 2004). Loss-of-function mutations have been characterized for Atg8a but not *Atg8b*, and these disrupt starvation-induced autophagy. GFP-Atg8a and GFP-Atg8b transgenes under both hsp70 and UAS control have been described (Juhasz *et al.*, 2008; Scott *et al.*, 2004; Rusten *et al.*, 2007). In addition, *Drosophila* transgenic lines expressing GFP fused to human LC3 or to *Drosophila* Atg5 have also been used to monitor autophagy. These lines are available from individual laboratories or the Bloomington Drosophila Stock Center at Indiana University.

Materials

Transgenic larvae of appropriate genotype (see below), starved or well fed
PBS

Paraformaldeyhde (4% in PBS)
DAPI (1 μg/ml in PBS)
Fluoroguard (BioRad)

Protocol

1. Using fine forceps under a dissecting stereomicroscope, 10–12 larvae are bisected and inverted in PBS.
2. Larval carcasses are transferred to a microfuge tube containing 1 ml of paraformaldehyde/PBS solution, and fixed with gentle rocking 2 h at room temperature or overnight at 4 °C.
3. Carcasses are washed three times for 20 min in PBS, incubated 1 h in DAPI/PBS, then washed again in PBS and transferred to a dissection dish in PBS.
4. Using fine forceps under a dissecting stereomicroscope, fat body lobes (one per animal) are excised and individually transferred to a slide in a few microliters of Fluoroguard or other suitable mounting agent.
5. Samples are covered with a coverslip, sealed with nail polish, and stored at −20 °C prior to imaging.

2.4. Genetic manipulation and monitoring of autophagy using mosaic analysis

The genetic technique of mosaic analysis is an essential element of modern fly genetics. Through this process, animals comprised of a mixture of cells bearing distinct genotypes are generated through flippase (FLP)/FLP-recognition target (FRT)-mediated recombination, typically generating cells either heterozygous versus homozygous for a given mutation (Xu and Rubin, 1993) or wild type versus transgene-expressing (Brand and Perrimon, 1993). Cells of different genotypes are distinguished on the basis of their differential expression of a marker protein such as GFP. This technique confers several important advantages: it allows the phenotypes of early lethal mutations to be examined at later stages of development, it provides a means to assess cell autonomy, and it provides an ideal juxtaposition of experimental and control cells within the same tissue, organ, and animal. This last feature is particularly helpful for studies involving autophagy, which can show variability from one animal to the next due to minor differences in developmental state or nutrient microenvironment.

Two features of the fat body have posed a challenge to standard mosaic analysis. First, as a postmitotic, endoreplicative tissue, the fat body is resistant to FLP/FRT-mediated induction of mitotic recombination, which is commonly used to generate loss of function mosaic clones. Second, many standard cell markers driven by ubiquitously expressed promoters *(ubiquitin, actin5c, hsp70)* show weak and nonuniform expression in the fat body, which makes them less reliable as genotypic markers. These challenges have been largely surmounted by inducing clones very early in embryonic development (Manfruelli *et al.*, 1999), prior to initiation of endoreplicative cell cycles, and by using the GAL4/UAS system to specifically express uniformly high levels of fluorescent marker proteins using fat body-specific drivers (Fig. 36.1B) such as Cg-GAL4, fb-GAL4, or r4-GAL4 (Asha *et al.*, 2003; Gronke *et al.*, 2003; Lee and Park, 2004).

2.4.1. LysoTracker Red staining of mosaic fat body containing homozygous mutant clones

In this procedure, a mutation of interest is carried on a chromosome containing a centromere-proximal FRT site at the base of the appropriate chromosome arm. This balanced stock is crossed to a line containing the following four P element-derived transgenes: heatshock-inducible FLP, fat body-specific GAL4, the homologous FRT insertion, and a UAS-GFP transgene located more distally on this chromosome arm. An example for the *Vps34* locus on chromosome 2R is as follows:

Cross 1

FRT42D Vps34$^{\Delta m22}$ ♂ x ♀ hsp70-flp; Cg-GAL4 FRT42D UAS-GFP/SM6-TM6B Tb

Relevant progeny

hsp70-flp/+ ; FRT42D Vps34$^{\Delta m22}$/Cg-GAL4 FRT42D UAS-GFP

FLP-induced recombination between the homologous FRT loci will give rise to genetically distinct daughter cells in the following mitosis, in which homozygous mutant cells can be identified by their lack of GFP expression (Fig. 36.2A). This is a stochastic event, and generally occurs in only a few percent of cells in response to heat shock-induced flippase expression. These cells undergo at most 2–3 rounds of cell division prior to mitotic exit, resulting in small GFP- clones of a few cells. In the cross 1, the Tb marker on the compound balancer SM6-TM6B is used to identify appropriate progeny in the larval stage. Any combination of GAL4 and UAS-GFP transgenes can be used, as long as they express well in the fat body and at least one of the two components is located on the appropriate chromosome arm, 2R in the preceding example.

Mitotic clones in the fat body can be efficiently induced only during the first 6–8 h of embryonic development. Therefore, embryos from this cross are obtained over a 6-h collection period, followed immediately by a 1-h heat shock at 37 °C.

Using this approach, suppression of autophagy can be visualized by LysoTracker Red staining in clones homozyogous for mutant alleles of several of the *Drosophila* Atg orthologs listed in Table 36.1, and in mutations in signaling molecules that promote autophagy such as Vps34 (Fig. 36.2A). Conversely, clonal disruption of suppressors of autophagy, such as

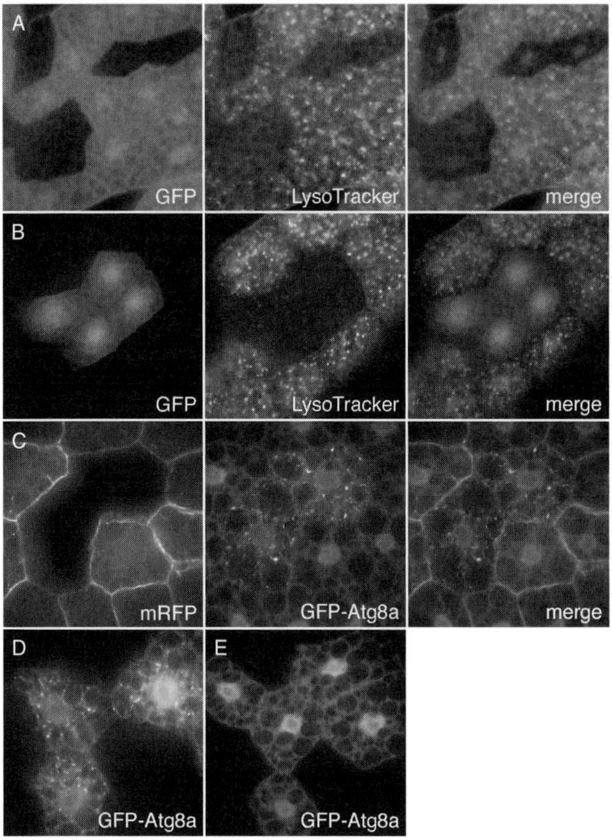

Figure 36.2 Clonal inactivation of autophagy in mosaic larval fat body. (A) Homozygous loss of *Vps34* disrupts starvation-induced accumulation of autophagosomes labeled with LysoTracker Red. Clones of *Vps34* null mutant cells are marked by lack of GFP expression. (B) Clonal expression of kinase-defective Vps34 disrupts starvation-induced accumulation of autophagosomes labeled with LysoTracker Red. A four-cell clone of cells expressing $Vps34^{KD}$ is marked by coexpression of GFP. (C) Homozygous mutation of *Vps25* leads to cell-autonomous accumulation of GFP-Atg8a-labelled autophagosomes in well-fed animals, due to a defect in autophagosome-lysosome fusion. A two-cell *Vps25* mutant clone is marked by lack of mRFP expression; all cells express GFP-Atg8a. D, E. Clonal expression of GFP-Atg8a under starvation conditions labels autophagic vesicles in control animals (D) but not in animals coexpressing $Vps34^{KD}$ (E). Images are modified from Juhász *et al.* (2008). (See Color Insert.)

components of the TOR signaling pathway, results in constitutive activation of autophagy. This protocol provides a sensitive means of detecting even subtle effects of a given mutation on autophagy, as small differences between mutant (GFP-negative) and adjacent control (GFP-positive) cells can be readily observed.

2.4.2. LysoTracker Red staining of fat body containing transgene-expressing clones

In this protocol, clonal activation of the GAL4-UAS system is used to induce GFP-marked cells expressing one or more transgenes (Fig. 26.2B). This system can be used for overexpression studies or to express mutant versions of genes (e.g., constitutively active or dominant negative). In addition, extensive libraries of UAS lines that mediate RNA-based silencing of most *Drosophila* genes have recently been established (Vienna *Drosophila* Research Center, Austia; National Institute of Genetics, Japan), allowing a readily accessible means of assaying gene function. Four transgenes are utilized in this scheme: heatshock-inducible flippase; Act>★CD2>GAL4, in which the constitutive Actin 5C promoter is separated from coding sequences for the GAL4 transcription factor by an FRT [>]-flanked transcription termination sequence [★] and *CD2* marker gene (which is not used in this protocol); UAS-GFP; and a transgene containing the UAS-regulated genetic element of interest. *Vps34* is again used as an example:

Cross 2

hsp70-flp; UAS-Vps34^{KD} ♀ x ♂ Act>★CD2>GAL4 UAS-GFP

Relevant progeny

hsp70-flp/+ ; UAS-Vps34^{KD}/Act>★CD2>GAL4 UAS-GFP

FLP-mediated excision (flip out) of the >★CD2> cassette results in stable expression of GAL4 in a low percentage of cells, which give rise to small GFP-marked clones that express one or more UAS-regulated transgenes. In contrast to most other larval tissues, clonal activation of the GAL4 cassette occurs spontaneously in the fat body in the absence of heat shock, presumably due to leaky expression of the hsp70-flp transgene (Britton *et al.*, 2002). LysoTracker staining is performed as described previously, following dissection of fat body from either well fed or starved animals, as desired.

In the example shown above, the hsp70-FLP transgene has been previously recombined with the UAS-driven gene of interest. Alternatively, hsp70-FLP, Act>★CD2>GAL4 and UAS-GFP can be combined into one stock that can be used to cross directly to a given UAS line. While more convenient, the presence of a hsp70-FLP transgene and a FLP-excisable cassette in the germline of such stocks frequently leads to heritable loss of the cassette and constitutive expression of GAL4.

This versatile system can be used to suppress autophagy by overexpression of a negative autophagy regulator (e.g., the Class I phosphatidylinositol 3-kinase p110 or the small GTPase Rheb), by expression of dominant negative alleles of positive autophagy regulators (e.g., kinase-defective Vps34) or by expression of elements that induce RNA-mediated interference against core components of the autophagy machinery. Autophagy can be activated by this system through GAL4/UAS-mediated overexpression of positive autophagy regulators (e.g., PTEN or Tsc1 and Tsc2), or in some cases by overexpression of Atg genes (Atg1, Atg8a). As an alternative to the mosaic approach described here, other GAL4 lines can be used to express these genes in specific tissues, allowing assessment of the role of autophagy in a given tissue or developmental process.

2.4.3. GFP-Atg8 analysis in mosaic tissues containing homozygous mutant clones

To use GFP-Atg8 in a genetically mosaic tissue, red-fluorescent markers such as dsRed, mRFP, or mCherry can be used as genotypic markers, while GFP-Atg8a expression is maintained at uniform levels in both genotypes (Fig. 36.2C). The relevant transgenes can be arranged on a single chromosome, as in this example using a mutation in the *Vps25* locus on chromosome 2R:

Cross 3

FRT42D Vps25^{A3}/SM6-TM6B Tb ♂ x ♀ hsp70-flp; Cg-GAL4 UAS-GFP-Atg8a FRT42D UAS-mRFP

Relevant progeny

hsp70-flp/+ ; FRT42D Vps25^{A3}/Cg-GAL4 UAS-GFP-Atg8a FRT42D UAS-mRFP

Alternatively, the GAL4 and UAS inserts can be distributed across two or more chromosomes:

Cross 4

Atg1$^{\Delta 3D}$ FRT80B/TM6B Tb ♂ x ♀ hsp70-flp; Cg-GAL4 UAS-GFP-Atg8a; UAS-mRFP FRT80B

Relevant progeny

hsp70-flp/+ ; Cg-GAL4 UAS-GFP-Atg8a/+ ; Atg1$^{\Delta 3D}$ FRT80B/ UAS-mRFP FRT80B

In either case, appropriate arrangement of only the genotypic marker (UAS-mRFP) on the FRT-distal chromosome arm ensures that GFP-Atg8a expression is unaffected by the flp/FRT-mediated chromosomal recombination events. As described above, induction of mitotic clones is achieved through a 1-h heat shock of a 0–6 h embryo collection. Larvae are transferred to fresh food at low density for 24 h, and then either starved 4 h or dissected immediately. Dissected carcasses are fixed overnight in 4% paraformaldehyde in PBS at 4 °C, counterstained with DAPI (1 μg/ml), and fat body is mounted in a glycerol-based mounting agent such as Fluoroguard (BioRad).

This procedure can be modified for use in most other tissues including imaginal discs. For such experiments, appropriate GAL4 drivers that express either ubiquitously or in the tissue of interest are used, and animals are subjected to heat shock early in larval development to induce FLP expression.

2.4.4. GFP-Atg8 analysis in mosaic tissues containing transgene-expressing clones

The same UAS-GFP-Atg8a lines can be readily adapted to monitor autophagy in clones of transgene-expressing cells:

Cross 5

hsp70-flp; UAS-Vps34^{KD} ♀ x ♂ Act>*CD2>GAL4 UAS-GFP-Atg8a

Relevant progeny

hsp70-flp/+ ; UAS-Vps34^{KD}/Act>*CD2>GAL4 UAS-GFP-Atg8a

In this case, GFP-Atg8a is used both to identify the GAL4-positive cells and as a marker of autophagic vesicles. Because the neighboring wild-type cells lack the autophagy marker, the effects of transgene expression must be determined relative to GFP-Atg8a-expressing clones in separate control animals that lack the experimental transgene (Fig. 36.2D, 36.E). Although this negates one of the major advantages of the mosaic approach (side-by-side comparison of experimental and control cells in the same animal), this approach may often be a better choice than organwide or ubiquitous expression, which is more likely to lead to confounding indirect effects.

3. Conclusions

One of the most useful applications of the tools and techniques described here is as a test of the potential role of candidate genes in autophagy. Genes and proteins identified in other assays in *Drosophila* or in orthologous systems can be efficiently characterized through these methods. Although the labor-intensive dissection steps limits their applicability as a primary assay in high throughput screens, this may become possible through further development of autophagy markers whose levels or fluorescence intensity are highly sensitive to autophagic activity.

REFERENCES

Asha, H., Nagy, I., Kovacs, G., Stetson, D., Ando, I., and Dearolf, C. R. (2003). Analysis of Ras-induced overproliferation in *Drosophila* hemocytes. *Genetics* **163,** 203–15.

Brand, A. H., and Perrimon, N. (1993). Targeted gene expression as a means of altering cell fates and generating dominant phenotypes. *Development* **118,** 401–15.

Britton, J. S., Lockwood, W. K., Li, L., Cohen, S. M., and Edgar, B. A. (2002). Drosophila's insulin/PI3-kinase pathway coordinates cellular metabolism with nutritional conditions. *Dev. Cell* **2,** 239–49.

Butterworth, F. M., and Forrest, E. C. (1984). Ultrastructure of the preparative phase of cell death in the larval fat body of *Drosophila melanogaster. Tissue Cell* **16,** 237–50.

Gronke, S., Beller, M., Fellert, S., Ramakrishnan, H., Jackle, H., and Kuhnlein, R. P. (2003). Control of fat storage by a *Drosophila* PAT domain protein. *Curr. Biol.* **13,** 603–6.

Juhász, G., Csikós, G., Sinka, R., Erdélyi, M., and Sass, M. (2003). The *Drosophila* homolog of Aut1 is essential for autophagy and development. *FEBS Lett.* **543,** 154–8.

Kim, J., Huang, W. P., and Klionsky, D. J. (2001). Membrane recruitment of Aut7p in the autophagy and cytoplasm to vacuole targeting pathways requires Aut1p, Aut2p, and the autophagy conjugation complex. *J. Cell Biol.* **152,** 51–64.

Kimura, K., Kodama, A., Hayasaka, Y., and Ohta, T. (2004). Activation of the cAMP/PKA signaling pathway is required for post-ecdysial cell death in wing epidermal cells of *Drosophila melanogaster. Development* **131,** 1597–606.

Klionsky, D. J., Abeliovich, H., Agostinis, P., Agrawal, D. K., Aliev, G., Askew, D. S., Baba, M., Baehrecke, E. H., Bahr, B. A., Ballabio, A., Bamber, B. A., Bassham, D. C., *et al.* (2008). Guidelines for the use and interpretation of assays for monitoring autophagy in higher eukaryotes. *Autophagy* **4,** 151–75.

Lee, C. Y., and Baehrecke, E. H. (2001). Steroid regulation of autophagic programmed cell death during development. *Development* **128,** 1443–55.

Lee, C. Y., Cooksey, B. A., and Baehrecke, E. H. (2002). Steroid regulation of midgut cell death during *Drosophila* development. *Dev. Biol.* **250,** 101–11.

Lee, G., and Park, J. H. (2004). Hemolymph sugar homeostasis and starvation-induced hyperactivity affected by genetic manipulations of the adipokinetic hormone-encoding gene in *Drosophila melanogaster. Genetics* **167,** 311–23.

Lee, S. B., Kim, S., Lee, J., Park, J., Lee, G., Kim, Y., Kim, J. M., and Chung, J. (2007). ATG1, an autophagy regulator, inhibits cell growth by negatively regulating S6 kinase. *EMBO Rep.* **8,** 360–5.

Levenbook, L. (1985). Insect storage proteins Oxford: Pergamon Press, Oxford.

Manfruelli, P., Reichhart, J. M., Steward, R., Hoffmann, J. A., and Lemaitre, B. (1999). A mosaic analysis in *Drosophila* fat body cells of the control of antimicrobial peptide genes by the Rel proteins Dorsal and DIF. *EMBO J.* **18,** 3380–91.

Mizushima, N., Yamamoto, A., Matsui, M., Yoshimori, T., and Ohsumi, Y. (2004). In vivo analysis of autophagy in response to nutrient starvation using transgenic mice expressing a fluorescent autophagosome marker. *Mol. Biol. Cell* **15,** 1101–11.

Pandey, U. B., Batlevi, Y., Baehrecke, E. H., and Taylor, J. P. (2007). HDAC6 at the intersection of autophagy, the ubiquitin-proteasome system and neurodegeneration. *Autophagy* **3,** 643–645.

Rusten, T. E., Lindmo, K., Juhasz, G., Sass, M., Seglen, P. O., Brech, A., and Stenmark, H. (2004). Programmed autophagy in the *Drosophila* fat body is induced by ecdysone through regulation of the PI3K pathway. *Dev. Cell* **7,** 179–92.

Rusten, T. E., Vaccari, T., Lindmo, K., Rodahl, L. M., Nezis, I. P., Sem-Jacobsen, C., Wendler, F., Vincent, J. P., Brech, A., Bilder, D., and Stenmark, H. (2007). ESCRTs and Fab1 Regulate Distinct Steps of Autophagy. *Curr. Biol.* **17,** 1817–25.

Sass, M., and Kovacs, J. (1977). The effect of ecdysone on the fat body cells of the penultimate larvae of Mamestra brassicae. *Cell Tissue Res.* **180,** 403–9.

Scott, R. C., Juhasz, G., and Neufeld, T. P. (2007). Direct induction of autophagy by Atg1 inhibits cell growth and induces apoptotic cell death. *Curr. Biol.* **17,** 1–11.

Scott, R. C., Schuldiner, O., and Neufeld, T. P. (2004). Role and regulation of starvation-induced autophagy in the *Drosophila* fat body. *Dev. Cell* **7,** 167–78.
Suzuki, K., Kirisako, T., Kamada, Y., Mizushima, N., Noda, T., and Ohsumi, Y. (2001). The pre-autophagosomal structure organized by concerted functions of APG genes is essential for autophagosome formation. *EMBO J.* **20,** 5971–81.
Velentzas, A. D., Nezis, I. P., Stravopodis, D. J., Papassideri, I. S., and Margaritis, L. H. (2007). Mechanisms of programmed cell death during oogenesis in *Drosophila virilis*. *Cell Tissue Res.* **327,** 399–414.
Xu, T., and Rubin, G. M. (1993). Analysis of genetic mosaics in developing and adult *Drosophila* tissues. *Development* **117,** 1223–37.

CHAPTER THIRTY SEVEN

Monitoring Autophagy in Insect Eggs

Ioannis P. Nezis*,† *and* Issidora Papassideri*

Contents

Abstract

Oogenesis is a fundamental physiological process in insects. Successful oogenesis is critical for evolutionary success by transferring genetic information to the next generation. This is achieved by the normal maturation of the egg chamber (egg), which is accomplished through cell death of the cells that accompany the oocyte. Recent studies demonstrate that autophagy contributes to this cell death process. Hence, comprehension of the mechanisms that implicates autophagy during cell death in insect eggs is very important. Herein, we describe some experimental approaches that can be used to monitor autophagy in insect eggs.

* Faculty of Biology, Department of Cell Biology and Biophysics, University of Athens, Athens, Greece
† Centre for Cancer Biomedicine, University of Oslo and Institute for Cancer Research, Department of Biochemistry, The Norwegian Radium Hospital, Montebello, N-0310, Oslo, Norway

Methods in Enzymology, Volume 451
ISSN 0076-6879, DOI: 10.1016/S0076-6879(08)03237-0

1. Introduction

Studying autophagy in yeast and in cultured mammalian cells has driven our understanding of the mechanisms that regulate this important cellular function (Klionsky, 2007; Mizushima, 2007). The application of this information to living organisms is a challenge that will illuminate our knowledge of how these mechanisms are implemented during tissue morphogenesis. Genetic studies in the model invertebrate organism *Drosophila melanogaster* have been very critical for understanding significant insights of autophagy (Neufeld and Baehrecke, 2008; Sass, 2008; Tettamanti *et al.*, 2008). An important physiological process for such studies is oogenesis. Oogenesis is the central preparatory process for embryogenesis. It consists of a series of processes by which the oocyte is supplied (1) with the nutrients needed for embryonic development, (2) with the informational content necessary so that the interactions of the zygote nucleus with the egg cytoplasm can produce normal embryogenesis, and (3) with protective coats, which enable the egg to survive in the environment (Mahowald and Kambysellis, 1980; Margaritis, 1985; Margaritis and Mazzini, 1998; Trougakos and Margaritis, 2002).

2. Overview of Oogenesis in Insects

In insects, the ovary consists of two lobes that increase in cell number during the larval stages and differentiate at the pupal stage resulting in the appearance of cylindrical pillars of cells that will develop into ovarioles (Figure 37.1A) (Postlethwait and Giorgi, 1985; Trougakos and Margaritis, 2002). Each ovariole contains a linear array of developmentally ordered egg chambers (Figures 37.1B,C). Two types of ovarioles can be distinguished in insects: panoistic and meroistic (Buning, 1994; Trougakos and Margaritis, 2002). In the panoistic ovariole (found mainly in roaches and crickets), each egg chamber consists of an oocyte surrounded by a layer of follicle cells. Each ovariole matures a single oocyte at a time, and the necessary materials for development are produced through selective gene amplification in the nucleus of the oocyte (Cave, 1982). In contrast, in the meroistic ovariole the egg chamber is composed of three cell types: the oocyte, nurse cells, and follicle cells. In this type of ovariole, it is the nurse cells that supply the oocyte with the nutrients necessary for development. Meroistic ovaries themselves are of two subtypes. In the first subtype (found in the bugs and some beetles) called *telotrophic meroistic ovary*, the cluster of nurse cells remains at the tip of each ovariole, called a *tropharium*, which is connected to the oocyte through the trophic cords. In the second subtype, the

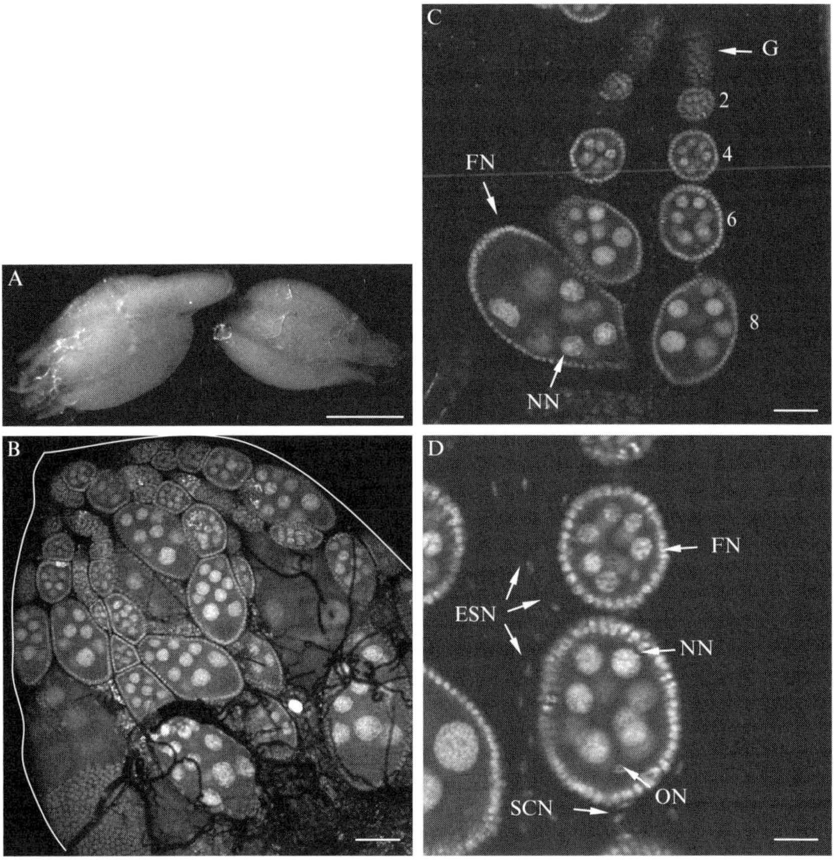

Figure 37.1 Morphology of the ovary in *Drosophila melanogaster*. (A) Light micrograph of a pair of ovaries as it is shown under the stereoscope. (B–D) Confocal micrographs of ovarian tissue of *D. melanogaster* after propidium iodide staining to visualize the nuclei. (B) Middle section of one ovary (outlined). Egg chambers of different developmental stages are evident. (C) Isolated ovarioles containing egg chambers of different developmental stages. The numbers next to the egg chambers of the ovariole at the right side show the developmental stage. G: germarium, FN: follicle cell nuclei, NN: nurse cell nucleus. (D) Higher magnification of figure 1B where the follicle cell nuclei, the nurse cell nuclei, the oocyte nucleus (ON), the nuclei of the epithelial sheath cells (ESN) and the nuclei of stalk cells (SCN) are evident. Scale bars: (A) 250 μm, (B) 100 μm, (C) 50 μm, (D) 25 μm.

polytrophic meroistic ovary (found in flies, bees and moths), the nurse cells are directly connected to the oocyte by intercellular bridges, called *ring canals*, through which nurse cells transfer their cytoplasm to the oocyte during a process known as cytoplasmic dumping (Buning, 1994; Mahajan-Miklos and Cooley 1994; Trougakos and Margaritis, 2002).

In *Drosophila melanogaster* and other higher Diptera, the germline cystoblasts are located at the germarium at the tip of the ovary. They undergo 4 mitotic divisions to generate 16 cystocytes. Only one cell will become the oocyte, and the remaining 15 become the nurse cells. Both cell types are encapsulated by an epithelial layer of somatically derived follicle cells, forming the egg chamber (egg) (Figs. 37.1C,D), which is the structural and functional unit of ovaries in insects. Nurse cells provide the developing oocyte with RNA, proteins and organelles that are necessary for proper development. On the other hand, the follicle cells differentiate into distinct subpopulations, move to specific positions of the egg chamber, modify their surfaces, participate in the oocyte polarity formation, and during the late stages of oogenesis secrete the complex eggshell that enhances survival of the embryo (Margaritis, 1985; Margaritis and Mazzini, 1998; Trougakos and Margaritis, 2002). Fourteen stages of oogenesis have been described according to morphological criteria, including egg chamber size, the proportion of the egg chamber occupied by the oocyte, the position of the follicle cells and the appearance of the eggshell coverings (King, 1970; Margaritis, 1985, 1986; Spradling, 1993). Stage 1 represents the 16-cell syncytium immediately after encapsulation by the follicle cells, while stage 14 refers to the mature egg chamber where the nurse cells have degenerated and the eggshell is completed (Figure 37.2).

It has been previously demonstrated that programmed cell death during Diptera oogenesis occurs in the germarium, and during mid-oogenesis and late oogenesis (Drummond-Barbosa and Spradling, 2001; Nezis *et al.*, 2000, 2003, 2006a; for reviews, see Baum *et al.*, 2005; McCall, 2004). Cell death during mid-oogenesis, known as follicular atresia, has been sporadically observed during physiological Diptera mid-oogenesis (Giorgi and Deri, 1976; Nezis *et al.*, 2000, 2003, 2006a; Uchida *et al.*, 2004), but also as a response to nutritional deprivation, ecdysone signaling inhibition, treatment with chemotherapeutic drugs and ectopic death of follicle cells in *Drosophila* (Buszczak *et al.*, 1999; Chao and Nagoshi, 1999; DeLorenzo *et al.*, 1999; Soller *et al.*, 1999; Nezis *et al.*, 2000; Terashima and Bownes, 2004). The atretic egg chambers were found to contain degenerated nurse cells, mainly characterized by condensed chromatin, fragmented DNA and disorganized actin cytoskeleton (Buszczak *et al.*, 1999; Chao and Nagoshi, 1999; DeLorenzo *et al.*, 1999; Soller *et al.*, 1999; Nezis *et al.*, 2000, 2003, 2006a). Importantly, programmed cell death of nurse and follicle cells is also required for the normal maturation of the developing follicles during the late stages of *D. melanogaster*, *Dacus oleae*, *Ceratitis capitata*, and *Bombyx mori* oogenesis (Cavaliere *et al.*, 1998; Foley and Cooley, 1998; Nezis *et al.*, 2000, 2001, 2002, 2003, 2005, 2006b,c,d; Mpakou *et al.*, 2006, 2008; for reviews, see Baum *et al.*, 2005; McCall, 2004). Autophagy has been shown to participate in this cell death process, functioning cooperatively with apoptosis for the most efficacious elimination of the degenerated nurse cells

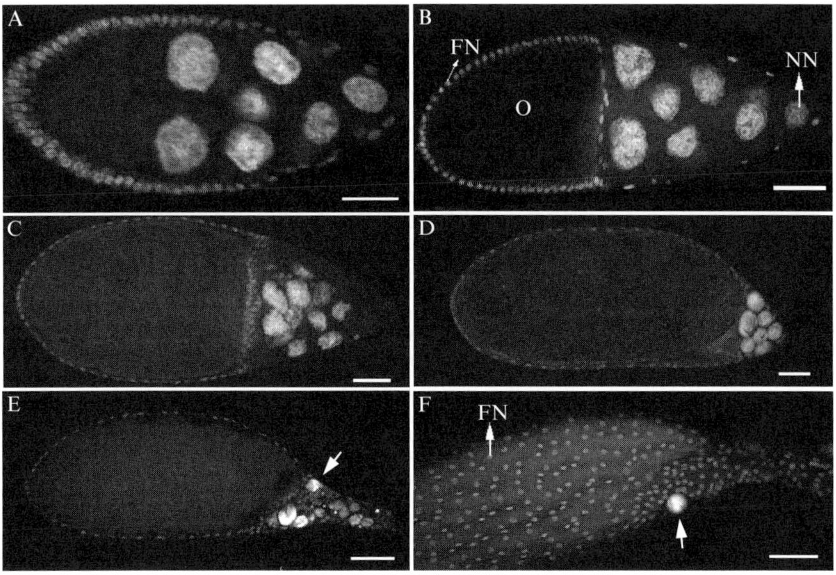

Figure 37.2 Confocal micrographs of isolated egg chambers of *D. melanogaster* after propidium iodide staining to visualize the nuclei. The late stages of oogenesis 9–14 are shown. (A) Stage 9, (B) Stage 10, (C) Stage 11, (D) Stage 12, (E) Stage 13, (F) Early stage 14. Note that during these stages the volume of the oocyte (O) increases when the volume of the nurse cells decreases (NN: nurse cell nucleus, FN: follicle cell nuclei). The nurse cells after stage 12 start to degenerate and at stage 13 the nurse cell nuclei are highly condensed (arrow). During stage 14, only nurse cell nuclear remnants are evident (arrow). Scale bars: 50 μm.

(Mpakou *et al.*, 2006, 2008; Velentzas *et al.*, 2007a,b,c). Autophagy is also required for the degeneration of the follicular epithelium in a caspase-independent manner (Nezis *et al.*, 2006a,b). In this chapter, we will describe protocols for monitoring autophagy in insect eggs.

3. Methods to Study Autophagy in Insect Eggs

3.1. Sample preparation

Flies are conditioned at 25 °C under a 12-h light-dark period. They should be kept together in uncrowded conditions (use equal numbers of 2- to 5-day-old male and female flies). To obtain ovarioles containing egg chambers at all developmental stages, flies should be transferred to fresh food supplemented with wet yeast paste every day, in which they are maintained for 3–4 days before dissection.

3.2. Ovary dissection

1. Anesthetize the flies under CO_2 or diethyl ether.
2. Place the females in deep-well glass slides and dissect them in 100 μl of Ringer's solution (130 m*M* NaCl, 4.7 m*M* KCl, 1.9 m*M* CaCl2, 10 m*M*, HEPES, pH 7.0).
3. Hold the fly between thorax and abdomen with forceps and pull out the terminal part of the abdomen with another pair of forceps. Gently squeeze the abdomen with the forceps to release the ovaries and other organs from the abdomen (Figure 37.1A).
4. Tease ovaries away from debris and separate ovarioles from one another with very fine forceps or tungsten needles.
5. It is important to remove the muscle sheath from the ovarioles. To achieve this, gently squeeze the egg chambers until they exit the epithelial sheath.
6. Transfer the isolated egg chambers to the appropriate solutions using glass pipettes or plastic pipettes that have been previously rinsed in PBT (PBS containing 0.1% Triton-X 100). This helps the isolated egg chambers not to stick on the pipette.

3.3. Conventional transmission electron microscopy

Conventional transmission electron microscopy is the most reliable and one of the most widely used methodologies to monitor autophagy. Electron microscopy is also a time-consuming method and requires substantial expertise for the interpretation of the electron micrographs, since sometimes it is difficult to distinguish autophagic from heterophagic structures or other membrane structures that resemble autophagic compartments (Eskelinen 2005, 2008; Klionsky *et al.*, 2008; Mizushima, 2004). Autophagy is the process by which a portion of the cytosol and organelles are sequestered in a random manner by phagophores (also called *isolation membranes*). The phagophore engulfs portions of the cytoplasm and forms a double-membrane-layered (or sometimes multimembrane-layered) organelle called the autophagosome. Usually the content of the autophagosome has the same morphology and electron density as the surrounding cytoplasm (Fig. 37.3). The autophagosome can fuse with a lysosome and generate the autolysosome that has a single limiting membrane, where the cellular components of the vacuole are degraded. The content of autolysosomes has increased electron density compared to the content of autophagosomes (Fig. 37.3A) (Eskelinen, 2005, 2008; Klionsky *et al.*, 2008). Electron microscopy has been successfully used to monitor autophagy in insect eggs (Mpakou *et al.*, 2006, 2008; Nezis *et al.*, 2006a,b; Velentzas *et al.*, 2007a,b,c). In the following text, a protocol for conventional transmission electron microscopy for insect egg chambers is described.

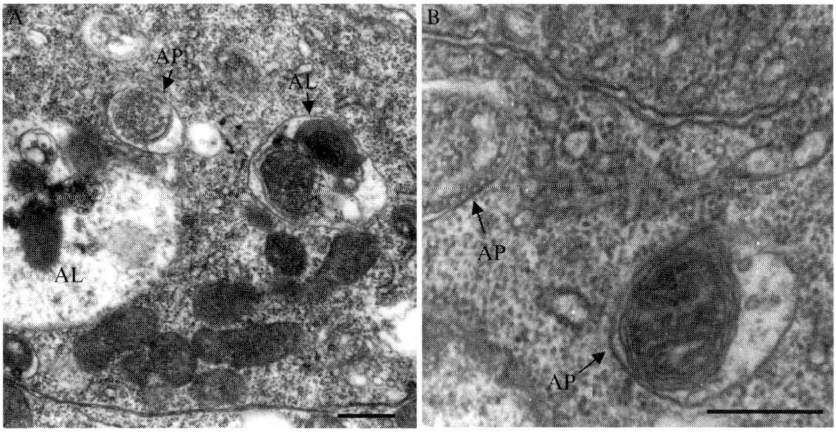

Figure 37.3 Ultrastructure of autophagic compartments in the follicular epithelium of a stage 14 egg chamber in *D. melanogaster*. Autophagosomes (AP) and late autophagic compartments, autolysosomes (AL) are evident. Scale bars: (A–B) 500 nm.

1. Fix the isolated egg chambers (see section 3.2) in 2% glutaraldehyde in PBS [(phosphate-buffered saline): 137 m*M* NaCl, 2.7 m*M* KCl, 10 m*M*, Na_2HPO_4, 2 m*M* KH_2PO_4, pH 7.4] for 1.5 h at room temperature. It is suitable to fix the tissue and proceed in the following steps using 4-ml glass sample bottles with plastic snap cap. The volume of each solution used in this and the following steps could be 1–2 ml.
2. Wash 3 times with PBS (5 min each) at room temperature, by carefully removing the previous solution (using a plastic pipette) and adding the washing solution.
3. Wash 3 times with distilled water (5 min each) at room temperature.
4. Postfix in 2% osmium tetroxide, 1,5% potassium ferricyanide in dH_20 for 1 h at 4 °C.
5. Wash 3 times with distilled water (5 min each) at room temperature.
6. Stain *en bloc* in 4% uranyl acetate (diluted in distilled water) for 30 min in the dark, at room temperature.
7. Wash 3 times with distilled water (5 min each) at room temperature.
8. Dehydrate through a graded series of ethanol concentrations (30%, 50%, 70%, 85%, 95%, 100%, 100% absolute dehydrated) for 10 min each at 4 °C.
9. Infiltrate in propylene oxide two times (15 min each) (It is important to have the glass vials tightly sealed because propylene oxide evaporates easily).
10. Infiltrate in a mixture of Resin A: propylene oxide (volumes 1:3) for 1 h in rotation at room temperature using sealed vials [Resin A: 25*g* of Epon 812, 20*g* of Araldite, 60*g* of DDSA (dodecenyl succinic anhydride) rigorously mixed in 100-ml plastic disposable beakers (Electron Microscopy Sciences, Cat# 60950) using an electrical mixer].

11. Infiltrate in a mixture of Resin A:propylene oxide (volumes 1:1) for 1 h in rotation at room temperature using sealed vials.
12. Infiltrate in a mixture of Resin A:propylene oxide (volumes 3:1) for 1 h in rotation at room temperature using sealed vials.
13. Infiltrate in pure Resin A overnight in rotation at room temperature.
14. Infiltrate in pure Resin B for 4 h (2 times for 2 h) in rotation at room temperature. [Resin B: 10*g* of Resin A supplemented with 8 drops of the epoxy accelerator DMP-30 (2,4,6-Tris dimethylaminomethyl phenol)].
15. Fill gelatin capsules or flat embedding molds with Resin B and place the egg chambers in the proper position (according to which part of the egg chamber you want to section).
16. Incubate the samples for 48 h at 60 °C.
17. Section the samples using an ultramicrotome. Mount the ultrathin sections (60–80 nm) on uncoated copper grids (Gilder, Cat#G200-Cu).
18. Stain with 7% uranyl acetate for 10 min and 0.4% lead citrate for 1 min.
19. Observe under an electron microscope.

3.4. Immunoelectron microscopy

Immunoelectron microscopy can be used as an additional method to conventional transmission electron microscopy when it is uncertain whether autophagic compartments are present in the tissue. The availability of specific antibodies of autophagic markers is required to achieve this. Unfortunately, there are no specific antibodies available for autophagic markers in *Drosophila* and in insects in general, except Ref(2)P the *Drosophila melanogaster* homolog of mammalian p62 (Nezis *et al.*, 2008). Alternatively, several GFP-tagged autophagic proteins can be traced, using anti-GFP antibodies (Lindmo *et al.*, 2006, 2008; Nezis *et al.*, unpublished data; Rusten *et al.*, 2004, 2007; Scott *et al.*, 2004, 2007). In the following text, a protocol for immunoelectron microscopy for insect ovaries is described.

Tissue processing for ultracryosectioning

1. Fix the samples with 4% formaldehyde in PBS, or with 4% formaldehyde, 0.1 % glutaraldehyde in PBS for 1 h maximum at room temperature (or at 4 °C depending the antigenicity; use the same containers and volumes as in section 3.3).
2. Wash 3 times with PBS (5 min each).
3. Briefly (approximately 3–4 min) stain the fixed tissue with 0.5% toluidine blue to make it more visible.
4. Incubate the tissue in 2.1 *M* sucrose overnight at 4 °C in rotation.
5. Mount the tissue on aluminum specimen holders using a drop of 2.1 *M* sucrose solution (have a cold plate under the samples while working). Try to orient them so that they stand the right way (the part you want to examine should be at the top).

6. Freeze the mounted samples in liquid nitrogen by submersion.
7. Section the frozen samples using ultramicrocryotome.
8. Retrieve the ultrathin cryosections from the cryochamber of the microtome using a droplet of a 1:1 mixture of 2.1 *M* sucrose/2% methyl cellulose created on a stainless-steel loop attached to a wooden stick. Move the droplet toward the face of the sections until the contact causes the sections to adhere to the droplet.
9. Place the thawed ultrathin cryosections (50–60 nm) on carbon- and Formvar-coated copper grids (Gilder, cat. no. G100-Cu).
10. Store sections in a Petri dish at 4 °C until the immunogold labeling.

Immunogold labeling

1. Lift the grids with a pair of forceps and transfer them onto 50-μl droplets of PBS containing 0.2 *M* glycine laid out on a sheet of parafilm.
2. Wash 3 times, each time by floating the grids for 5 min on fresh 50-μl droplets of PBS/0.2 *M* glycine.
3. Incubate the grids for 20 min (twice for 10 min each) with blocking solution [PBS containing 0.2 *M* glycine, 0.8% BSA (bovine serum albumin) and 1% FCS (fetal calf serum)] at room temperature.
4. Incubate the grids for 1 h (or more depending on the antibody) with the primary antibody diluted in blocking solution. (The appropriate antibody dilution for immunogold labeling is usually 10 times less dilution compared to the one used for immunofluorescence, or 100 times less dilution compared to the one used for Western blot analysis. This of course has to be tested first).
5. Wash 3 times, each time by floating for 5 min on droplets of blocking solution.
6. When the primary antibody does not interact with protein A (e.g., if it is a mouse monoclonal), incubate grids with a secondary (bridging) antibody (e.g., rabbit anti-mouse) that binds protein A, diluted in blocking solution, for 30 min.
7. Wash grids 3 times, each time for 5 min with blocking solution.
8. Incubate grids for 30 min with 10 nm protein A-gold particle suspensions in blocking buffer (usually diluted one-sixtieth to one-eightieth).
9. Wash grids 5 times, each time for 5 min with PBS.
10. Wash grids 5 times, each time for 5 min with distilled water.
11. Stain the grids in 0.45% uranyl acetate in a 1.7% aqueous solution of methyl cellulose in the dark on ice for 10 min.
12. Retrieve the grids with a stainless steel loop of diameter slightly larger than the grid, attached to a pipette tip.
13. Drain away the excess of methyl cellulose/uranyl acetate solution by touching the loop at an angle of 45 °C to a filter paper.
14. Air-dry the grids at room temperature by placing the pipette tips upside-down in a pipette-tip rack. A film of methyl cellulose/uranyl acetate is

left on the grid after drying. This is important to give optimal contrast and to preserve the integrity of membrane structures.
15. Observe under an electron microscope.

3.5. Acidotropic dyes

3.5.1. Monodansylcadaverine (MDC) and LysoTracker Red staining

The fluorescent compound monodansylcadaverine (MDC) has been proposed as a marker for autophagic vacuoles (Biederbick *et al.*, 1995). Subsequent studies have suggested that this and other acidotropic dyes such as LysoTracker Red, are not specific markers for early autophagosomes, but rather label later stages in the degradation process (Klionsky *et al.*, 2008; Mizushima, 2004). However, MDC has been used as a marker for autophagy in *Drosophila* and silk moth egg chambers combined with additional assays, such as electron microscopy (Mpakou *et al.*, 2006, 2008, Nezis *et al.*, 2006a,b; Velentzas *et al.*, 2007a,b,c) (Fig. 37.4) and in *Drosophila* salivary glands combined with GFP-Atg8/LC3 fluorescence (Akdemir *et al.*, 2006). In the following text protocols for MDC and LysoTracker Red staining for insect eggs are described.

Monodansylcadaverine (MDC) staining

1. Rinse the isolated egg chambers (see section 3.2) briefly with 200 μl of PBS. The procedure is performed using deep-well glass slides.
2. Transfer them to 200 μl of PBS containing 0.05 m*M* MDC (Fluka, Sigma-Aldrich, Germany) for 10 min at 25 °C in the dark.
3. Wash 4 times with PBS (5 min each) in the dark.
4. Immediately observe under a Nikon Eclipse TE 2000-S fluorescence microscope using the UV-2E/C (DAPI) filter.

LysoTracker Red staining

1. Rinse the isolated egg chambers briefly with 200 μl of PBS. The procedure is performed using deep-well glass slides.
2. Transfer them into 200 μl of LysoTracker Red solution (diluted 1:200 in PBS) for 2 min at room temperature in the dark.
3. Wash for 5 min with PBS in the dark.
4. Mount with 20 μl of glycerol and cover with a coverslip.
5. Immediately observe under a fluorescence microscope.

3.6. Fluorescence confocal microscopy to detect special proteins involved in autophagy

As described previously, several GFP-tagged autophagic proteins are available and can be used for monitoring autophagy (Lindmo *et al.*, 2006, 2008; Nezis *et al.*, unpublished data; Rusten *et al.*, 2004, 2007; Scott *et al.*, 2004, 2007).

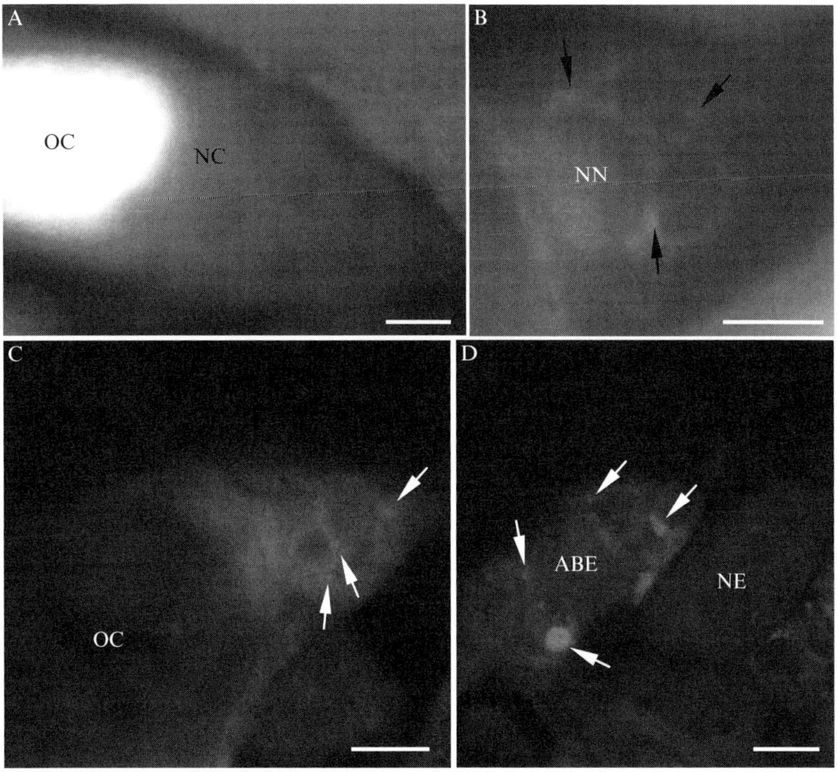

Figure 37.4 Nurse cells undergoing programmed cell death during middle (D) and late (A–C) oogenesis in *Drosophila* exhibit MDC (A–B) and LysoTracker (C–D) staining. During stage 10(A) no positive signal for MDC is detectable (NC nurse cells, OC oocyte), while during stage 12 (B) bright spots are detectable around nurse cell nuclei (NN) (arrows). Similar staining is observed using LysoTracker during stage 12 (C) (arrows) and during mid-oogenesis in abnormal egg chambers (ABE) (arrows), (NE-normal egg chamber). The fluorescence in (A) is due to oocyte autofluorescence. Scale bars: 50 μm.

The tissues from the transgenic flies expressing GFP-tagged autophagic markers can be directly analysed for a GFP signal after a short fixation (long fixation tend to destroy the fluorescence). Usually the GFP signal is enhanced and becomes easier to study when an anti-GFP antibody is used. In the following text, a protocol for immunofluorescence microscopy for insect egg chambers is described.

1. Fix the isolated egg chambers (see section 3.2) in 200 μl of PBS containing 4% formaldehyde for 45 min at room temperature. The procedure is performed using deep-well glass slides.
2. Wash 3 times with 200 μl of PBS (5 min each).

3. Incubate in 200 μl of PBS containing 0.1% Triton X-100 for 20 min.
4. Wash 3 times with 200 μl of PBS (5 min each).
5. Incubate in 200 μl of PBS containing 20 m*M* glycine and 1.5% BSA (blocking solution) for 45 min.
6. Incubate in the primary antibody diluted in blocking solution for 2–3 h at room temperature or overnight at 4 °C.
7. Wash 3 times with blocking solution (10 min each) at room temperature.
8. Incubate in the secondary antibody conjugated with either FITC or rhodamine diluted in blocking solution for 2 h at room temperature in the dark.
9. Wash 3 times with 200 μl of blocking solution (10 min each) at room temperature.
10. Mount with 90% glycerol containing 1.4-diazabicyclo (2.2.2) octane to avoid fading.
11. Put a coverslip over the mounted sample and seal the edges with nail polish.
12. Observe under a fluorescence microscope (conventional or confocal laser-scanning microscope).

4. Concluding Remarks

Reviewing all the methods described, it should be obvious that no single method is perfect or sufficient, and it is better to combine multiple approaches for a rigorous evaluation of autophagic mechanisms in insect eggs. In this chapter, we described some basic protocols to study autophagy in insect eggs. We believe that new methods for monitoring autophagy will continue to be developed soon in this field using the knowledge of other better-known models. Current research from our group is in progress to understand the role of autophagy during cell death of the germline cells and the follicular epithelium.

ACKNOWLEDGMENTS

We are very grateful to Thanassis Velentzas for help with Fig. 37.4 and to our colleagues in Papassideri and Stenmark lab. This work was supported by grants from the Special Account for Research Grants of the University of Athens to Professor I. S. Papassideri.

REFERENCES

Akdemir, F., Farkas, R., Chen, P., Juhasz, G., Medvedová, L., Sass, M., Wang, L., Wang, X., Chittaranjan, S., Gorski, S. M., Rodriguez, A., and Abrams, J. M. (2006). Autophagy occurs upstream or parallel to the apoptosome during histolytic cell death. *Development* **133,** 1457–1465.

Baum, J. S., St. George, J. P., and McCall, K. (2005). Programmed cell death in the germline. *Semin. Cell Dev. Biol.* **16,** 245–259.

Biederbick, A., Kern, H. F., and Elsasser, H. P. (1995). Monodansylcadaverine (MDC) is a specific *in vivo* marker for autophagic vacuoles. *Eur. J. Cell Biol.* **66,** 3–14.

Büning, J. (1994). The insect ovary: Ultrastructure, previtellogenic growth and evolution New York: Chapman and Hall, New York.

Buszczak, M., Freeman, M. R., Carlson, J. R., Bender, M., Cooley, L., and Segraves, W. A. (1999). Ecdysone response genes govern egg chamber development during mid-oogenesis in *Drosophila. Development* **126,** 4581–4589.

Cavaliere, V., Taddei, C., and Gargiulo, G. (1998). Apoptosis of nurse cells at the late stages of oogenesis of *Drosophila melanogaster. Dev. Genes Evol.* **208,** 106–112.

Cave, M. D. (1982). Morphological manifestations of ribosomal DNA amplification during insect oogenesis. *In* "Insect ultrastructure," (R. C. King and H. Akai, eds.), Vol. 1, pp. 86–117. Plenum Press, New York.

Chao, S., and Nagoshi, R. N. (1999). Induction of apoptosis in the germline and follicle layer of *Drosophila* egg chambers. *Mech. Dev.* **88,** 159–172.

De Lorenzo, C., Strand, D., and Mechler, B. M. (1999). Requirement of *Drosophila* I(2)gl function for survival of the germline cells and organization of the follicle cells in a columnar epithelium during oogenesis. *Int. J. Dev. Biol.* **43,** 207–217.

Drummond-Barbosa, D., and Spradling, A. C. (2001). Stem cells and their progeny respond to nutritional changes during *Drosophila* oogenesis. *Dev. Biol.* **231,** 265–278.

Eskelinen, E.-L. (2005). Maturation of autophagic vacuoles in mammalian cells. *Autophagy* **1,** 1–10.

Eskelinen, E.-L. (2008). To be or not to be? Examples of incorrect identification of autophagic compartments in conventional transmission electron microscopy of mammalian cells. *Autophagy* **4,** 257–260.

Foley, K., and Cooley, L. (1998). Apoptosis in late stage *Drosophila* nurse cells does not require genes within the H99 deficiency. *Development* **125,** 1075–1082.

Giorgi, F., and Deri, P. (1976). Cell death in ovarian chambers of *Drosophila melanogaster. J. Embryol. Exp. Morphol.* **35,** 521–533.

King, R. C. (1970). Origin and development of the egg chamber within the adult ovarioles. *In* "Ovarian development in *Drosophila melanogaster,*" pp. 38–54. Academic Press, New York.

Klionsky, D. J. (2007). Autophagy: from phenomenology to molecular understanding in less than a decade. *Nat. Rev. Mol. Cell Biol.* **8,** 931–937.

Klionsky, D. J., Abeliovich, H., Agostinis, P., Agrawal, D. K., Aliev, G., Askew, D. S., Baba, M., Baehrecke, E. H., Bahr, B. A., Ballabio, A., Bamber, B. A., and Bassham, D. C. (2008). Guidelines for the use and interpretation of assays for monitoring autophagy in higher eukaryotes. *Autophagy* **4,** 151–175.

Lindmo, K., Simonsen, A., Brech, A., Finley, K., Rusten, T. E., and Stenmark, H. (2006). A dual function for deep orange in programmed autophagy in the *Drosophila melanogaster* fat body. *Exp. Cell Res.* **312,** 2018–2027.

Lindmo, K., Brech, A., Finley, K. D., Gaumer, S., Contamine, D., Rusten, T. E., and Stenmark, H. (2008). The PI 3-kinase regulator Vps15 is required for autophagic clearance of protein aggregates. *Autophagy* **4,** in press.

Mahajan-Miklos, S., and Cooley, L. (1994). Intercellular cytoplasm transport during *Drosophila* oogenesis. *Dev. Biol.* **165,** 336–351.

Mahowald, A. P., and Kambysellis, M. P. (1980). *In* "Oogenesis: The genetics and biology of Drosophila," (M. Ashburner and T. R. F. Wright, eds.), Vol. 2, pp. 141–224. Academic Press, New York.

Margaritis, L. H. (1985). Structure and physiology of the eggshell. *In* "Comprehensive insect biochemistry, physiology and pharmacology," (L. I. Gilbert and G. A. Kerkut, eds.), Vol. 1, pp. 151–230. Pergamon Press, Oxford.

Margaritis, L. H. (1986). The eggshell of *Drosophila melanogaster*. New staging characteristics and fine structural analysis of choriogenesis. *Can. J. Zool.* **64,** 2152–2175.

McCall, K. (2004). Eggs over easy: Cell death in the *Drosophila* ovary. *Dev. Biol.* **274,** 3–14.

Mizushima, N. (2004). Methods for monitoring autophagy. *Int. J. Biochem. Cell Biol.* **36,** 2491–2502.

Mizushima, N. (2007). Autophagy: Process and function. *Genes Dev.* **21,** 2861–2873.

Mpakou, V. E., Nezis, I. P., Stravopodis, D. J., Margaritis, L. H., and Papassideri, I. S. (2006). Programmed cell death of the ovarian nurse cells during oogenesis of the silkmoth *Bombyx mori. Dev. Growth Differ.* **48,** 419–428.

Mpakou, V. E., Nezis, I. P., Stravopodis, D. J., Margaritis, L. H., and Papassideri, I. S. (2008). Different modes of programmed cell death during oogenesis of the silkmoth *Bombyx mori. Autophagy* **4,** 97–100.

Neufeld, T. P., and Baehrecke, E. H. (2008). Eating on the fly: Function and regulation of autophagy during cell growth, survival and death in *Drosophila. Autophagy,* in press.

Nezis, I. P., Stravopodis, D. J., Papassideri, I., Robert-Nicoud, M., and Margaritis, L. H. (2000). Stage-specific apoptotic patterns during *Drosophila* oogenesis. *Eur. J. Cell Biol.* **79,** 610–620.

Nezis, I. P., Stravopodis, D. J., Papassideri, I., and Margaritis, L. H. (2001). Actin cytoskeleton reorganization of the apoptotic nurse cells during the late developmental stages of oogenesis in *Dacus oleae. Cell Motil. Cytoskeleton* **48,** 224–233.

Nezis, I. P., Stravopodis, D. J., Papassideri, I., Robert-Nicoud, M., and Margaritis, L. H. (2002). The dynamics of apoptosis in the ovarian follicle cells during the late stages of *Drosophila* oogenesis. *Cell Tissue Res.* **307,** 401–409.

Nezis, I. P., Modes, V., Mpakou, V., Stravopodis, D. J., Papassideri, I. S., Mammali, I., and Margaritis, L. H. (2003). Modes of programmed cell death during *Ceratitis capitata* oogenesis. *Tissue Cell* **35,** 113–119.

Nezis, I. P., Stravopodis, D. J., Papassideri, I. S., Stergiopoulos, C., and Margaritis, L. H. (2005). Morphological irregularities and features of resistance to apoptosis in the *dcp-1/pita* double mutated egg chambers during *Drosophila* oogenesis. *Cell Motil. Cytoskeleton* **60,** 14–23.

Nezis, I. P., Stravopodis, D. J., Margaritis, L. H., and Papassideri, I. S. (2006a). Follicular atresia during *Dacus oleae* oogenesis. *J. Insect Physiol.* **52,** 282–290.

Nezis, I. P., Stravopodis, D. J., Margaritis, L. H., and Papassideri, I. S. (2006b). Programmed cell death of follicular epithelium during the late developmental stages of oogenesis in the fruit flies *Bactrocera oleae* and *Ceratitis capitata* (Diptera, Tephritidae) is mediated by autophagy. *Dev. Growth Differ.* **48,** 189–198.

Nezis, I. P., Stravopodis, D. J., Margaritis, L, H., and Papassideri, I. S. (2006c). Chromatin condensation of ovarian nurse and follicle cells is regulated independently from DNA fragmentation during *Drosophila* late oogenesis. *Differentiation* **74,** 293–304.

Nezis, I. P., Stravopodis, D. J., Margaritis, L. H., and Papassideri, I. S. (2006d). Autophagy is required for the degeneration of the ovarian follicular epithelium in higher Diptera. *Autophagy* **2,** 297–298.

Nezis, I. P., Simonsen, A., Sagona, A. P., Finley, K., Gaumer, S., Contamine, D., Rusten, T. E., Stenmark, H., and Brech, A. (2008). Ref(2)P, the *Drosophila melanogaster* homologue of mammalian p62, is required for the formation of protein aggregates in adult brain. *J. Cell Biol.* **180,** 1065–1071.

Postlethwait, J. H., and Giorgi, F. (1985). Vitellogenesis: Insects. *In* "Developmental biology: A comprehensive synthesis," (L. W. Browder, ed.), Plenum, New York.

Rusten, T. E., Lindmo, K., Juhász, G., Sass, M., Seglen, P. O., Brech, A., and Stenmark, H. (2004). Programmed autophagy in the *Drosophila* fat body is induced by ecdysone through regulation of the PI3K pathway. *Dev. Cell.* **7,** 179–192.

Rusten, T. E., Vaccari, T., Lindmo, K., Rodahl, L. M. W., Nezis, I.,P., Sem-Jacobsen, C., Wendler, F., Vincent, J. P., Brech, A., Bilder, D., and Stenmark, H. (2007). ESCRTs and Fab1 regulate distinct steps of autophagy. *Curr. Biol.* **17,** 1817–1825.

Sass, M. (2008). Autophagy research on insects. *Autophagy* **4,** 265–267.

Scott, R. C., Schuldiner, O., and Neufeld, T. P. (2004). Role and regulation of starvation-induced autophagy in the *Drosophila* fat body. *Dev. Cell.* **7,** 167–178.

Scott, R. C., Juhasz, G., and Neufeld, T. P. (2007). Direct induction of autophagy by Atg1 inhibits cell growth and induces apoptotic cell death. *Curr. Biol.* **17,** 1–11.

Soller, M., Bownes, M., and Kubli, E. (1999). Control of oocyte maturation in sexually mature *Drosophila* females. *Dev. Biol.* **208,** 337–351.

Spradling, A. C. (1993). Developmental genetics of oogenesis. *In* "The development of *Drosophila melanogaster*," (M. Bate and A. Martinez-Arias, eds.), Vol. I, pp. 1–70. Cold Spring Harbor Press, Cold Spring Harbor, New York.

Terashima, J., and Bownes, M. (2004). Translating available food into the number of eggs laid by *Drosophila melanogaster*. *Genetics* **167,** 1711–1719.

Tettamanti, G., Saló, E., González-Estévez, C., Felix, D. A., Grimaldi, A., and de Eguileor, M. (2008). Autophagy in invertebrates: Insights into development, regeneration and body remodeling. *Curr. Pharm. Des.* **14,** 116–125.

Trougakos, I. P., and Margaritis, L. H. (2002). Novel morphological and physiological aspects of insect eggs. *In* "Chemoecology of insect eggs and egg deposition," (M. Hilker and T. Meiners, eds.), pp. 3–36. Blackwell Wissenschaftsverlag, Berlin.

Velentzas, A. D., Nezis, I. P., Stravopodis, D. J., Papassideri, I. S., and Margaritis, L. H. (2007a). Stage-specific regulation of programmed cell death during oogenesis of the medfly *Ceratitis capitata* (Diptera, Tephritidae). *Int. J. Dev. Biol.* **51,** 57–66.

Velentzas, A. D., Nezis, I. P., Stravopodis, D. J., Papassideri, I. S., and Margaritis, L. H. (2007b). Mechanisms of programmed cell death during oogenesis in *Drosophila virilis*. *Cell Tissue Res.* **327,** 399–414.

Velentzas, A. D., Nezis, I. P., Stravopodis, D. J., Papassideri, I. S., and Margaritis, L. H. (2007c). Apoptosis and autophagy function cooperatively for the efficacious execution of programmed nurse cell death during *Drosophila virilis* oogenesis. *Autophagy* **3,** 130–132.

Uchida, K., Nishizuka, M., Ohmori, D., Ueno, T., Eshita, Y., and Fukunaga, A. (2004). Follicular epithelial cell apoptosis of atretic follicles within developing ovaries of the mosquito *Culex pipiens pallens*. *J. Insect Physiol.* **50,** 903–912.

CHAPTER THIRTY-EIGHT

In Vitro Methods to Monitor Autophagy in Lepidoptera

Gianluca Tettamanti* *and* Davide Malagoli†

Contents

Abstract

Autophagy is attracting growing interest, especially in relation to increasing evidence of the importance of autophagic processes in animal development, as well as in human cancer progression. In holometabolous insects (i.e., that undergo four distinct life cycle stages, including embryo, larva, pupa and imago), such as flies, butterflies, bees and beetles, autophagy has been found to play a fundamental role in metamorphosis, and given the high degree of conservation of the genes and the basic mechanisms of autophagy, attention

* Department of Structural and Functional Biology, University of Insubria, Varese, Italy
† Department of Animal Biology, University of Modena and Reggio Emilia, Modena, Italy

Methods in Enzymology, Volume 451
ISSN 0076-6879, DOI: 10.1016/S0076-6879(08)03238-2

to these relatively simple models has increased significantly. Together with *Drosophila*, Lepidoptera larvae are among the most common invertebrate models in studies concerning the protective action of starvation-induced autophagy or the possible role of autophagy as a programmed cell death process.

In this chapter, we provide experimental methods developed for, or applicable to, the study of the autophagic process in the IPLB-LdFB cell line derived from the fat body of the caterpillar of the gypsy moth, *Lymantria dispar*.

1. Background

1.1. Features of the autophagic process in traditional Lepidoptera models

Insects are excellent material for investigating autophagic processes, and in the past decades, several species belonging to Lepidoptera have been chosen as animal models for analyses of this self-eating process (Tettamanti *et al.*, 2008b). There are two main reasons for this choice. First, Lepidoptera larva is a convenient and useful tool in biochemical, endocrinological and electrophysiological studies. Second, the holometabolous insects provide an extensive source of larval organs that are strongly remodeled through autophagic processes at metamorphosis (Levine and Klionsky, 2004; Müller *et al.*, 2004). Thus, these organisms are ideal not only to study the protective role of starvation-induced autophagy in selected larval tissues (e.g., the fat body; Scott *et al.*, 2004) but also to gain insights into autophagy as a programmed cell death process (Levine and Klionsky, 2004).

Pioneering work describing the intervention of autophagy during Lepidoptera metamorphosis was undertaken by Locke and Collins. The two authors examined the modifications of fat body cells during the prepupal instar and reported mitochondria autolysis after their isolation "by paired membranes in bodies that fused to become large autophagic vacuoles" (Locke and Collins, 1968). Since that time, a deeper knowledge of these processes has been gained, and the ongoing silkworm and Lepidoptera genome projects (Mita *et al.*, 2004; Papanicolaou *et al.*, 2008) will hopefully provide new bioinformatic and molecular instruments to close the gap relative to *Drosophila melanogaster*, the best-characterized insect model.

Experiments on autophagy in Lepidoptera metamorphosis have mainly been performed using *in vivo* models, and attention has been focused on the fat body, midgut, prothoracic and labial glands, and nervous system. In all these tissues, autophagic processes are triggered by a 20-hydroxyecdysone (20E) surge, an event strictly related to pupation commitment (Tettamanti *et al.*, 2008b). This 20E induction has also been demonstrated *in vitro*: administration of the ecdysteroid to prothoracic gland cell cultures is able to activate cell mortality that displays features of autophagic programmed cell death (Dai and Gilbert, 1999).

It is interesting that despite experimental evidence suggesting the contribution of autophagy during the remodeling of tissues and organs at metamorphosis (Müller *et al.*, 2004; Tettamanti *et al.*, 2008b), some indications of an involvement of apoptotic death have also been reported in some Lepidoptera models (Dai and Gilbert, 1999; Tettamanti *et al.*, 2007a). These data are not unusual, as in other insects, too, the disappearance of larval salivary glands and midgut during metamorphosis is characterized by features typical of both apoptotic and autophagic cell death, demonstrating that the two mechanisms are not mutually exclusive, and that common pathways underlie morphologically distinct forms of programmed cell death (PCD; Lee and Baehrecke, 2001; Tettamanti *et al.*, 2007a,b; Velentzas *et al.*, 2007; Zacarin, 2007).

A detailed analysis of the cell death events observable in these invertebrate models is therefore necessary to clarify any possible functional interactions between the pathways controlling the two PCD mechanisms. In this context, lepidopteran *in vitro* systems could provide powerful experimental models to identify apoptosis- and autophagy-specific genes and to dissect the regulatory pathways shared by these two PCD forms.

1.2. The IPLB-LdFB cell line

Very few data regarding autophagic cell death in Lepidoptera are derived from *in vitro* experiments, so that the IPLB-LdFB cell line represents a new and very promising model (Malagoli, 2008). The IPLB-LdFB cell line was obtained from the larval fat body of the gypsy moth, *Lymantria dispar*, about 20 years ago (Lynn *et al.*, 1988). It is easy to grow and has proved to be a fruitful model for the characterization of cell death features in Lepidoptera (Malagoli, 2008). Together with the obvious advantages usually related to an established *in vitro* system (i.e., large amounts of material, low maintenance cost and no individual variability), one of the main characteristics of IPLB-LdFB cells is the possibility to address them selectively toward apoptosis (Malagoli *et al.*, 2002) or autophagy (Tettamanti *et al.*, 2006) using oxidant stressors or the F_1 F_o-ATPase inhibitor oligomycin A, respectively. This feature also makes IPLB-LdFB cells an interesting model in view of the earlier-mentioned overlap between the two forms of PCD in other insects, as well as the still unclear relationship between apoptotic and autophagic cell death (Berry and Baehrecke, 2007; Lee and Baehrecke, 2001; Velentzas *et al.*, 2007). The possibility to induce the occurrence of these two processes in the same *in vitro* model offers an opportunity to discriminate relationships and differences between apoptosis and autophagic cell death in Lepidoptera more clearly.

IPLB-LdFB cells have been utilized for several experimental approaches from ultrastructural analysis (Tettamanti *et al.*, 2006, 2008b) to proteomics (Malagoli *et al.*, unpublished data). The only current limitation to their application is the lack of data concerning gene and protein sequences in *L. dispar* that makes an approach based on a molecular biology platform rather difficult. However, since genome sequencing of the silkworm, *Bombyx mori* has been completed (Mita *et al.*, 2004), and new lepidopteran databases have been published (Papanicolaou *et al.*, 2008), some basic information for molecular biology–based studies in the IPLB-LdFB is now also available.

This chapter describes the methods utilized to analyze cell death in the IPLB-LdFB cell line, with special attention to those that have revealed their usefulness in highlighting autophagic cell death.

2. Methods

2.1. Transmission electron microscopy

Transmission electron microscopy (TEM) is an essential tool to complement biochemical and molecular biology techniques. Although widely used in the past, nowadays it is often disregarded and underestimated for several reasons. For example, sample preparation for TEM observation is time consuming, and the protocol must be carefully set up to avoid artifacts and in accordance with the specific characteristics of the sample under investigation. Moreover, skillful evaluation of results is necessary to avoid dangerous misinterpretations (Eskelinen, 2008).

Recovery of IPLB-LdFB cells after a 2 h treatment with oligomycin A reveals massive autophagy in the majority of cells. After 96 h of treatment, more than 80% of cells show autophagic features. These cells are easily distinguishable from those that undergo apoptotic cell death immediately after oligomycin administration (Tettamanti *et al.*, 2006). Thus, TEM allows a quick evaluation of the percentages of autophagic and apoptotic cells within the population treated with oligomycin A, while a quantitative and qualitative analysis of the changes in the autophagic structures is also possible (Klionsky *et al.*, 2008) (Figs. 38.1A and 38.1A′).

Sample preparation for TEM

1. Fix cells (10^5) in 1 ml of 0.1 *M* cacodylate buffer, pH 7.2, containing 1% glutaraldehyde for 10 min in a sterile 1.7-ml plastic tube.
2. Centrifuge cells at 200×*g* for 10 min.
3. Discard supernatant and fix cells for further 20 min in 1 ml of 0.1 *M* cacodylate buffer containing 2% glutaraldehyde.
4. Centrifuge cells at 200×*g* for 10 min.

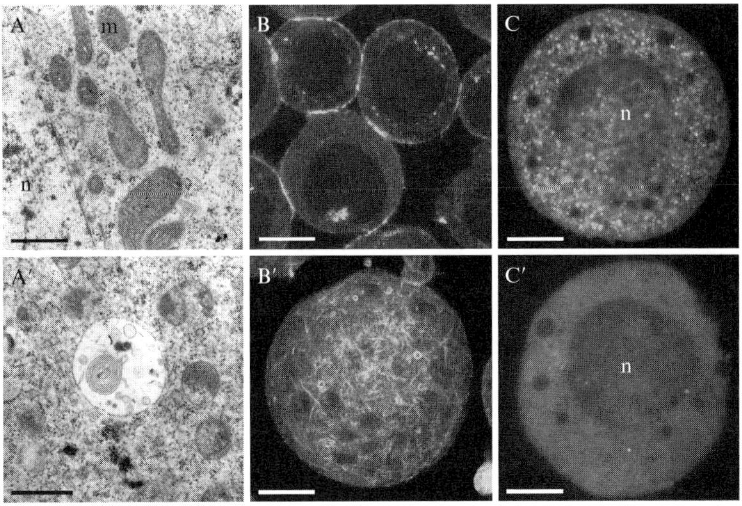

Figure 38.1 Different morphological features observed in control IPLB-LdFB cells (A, B, C) and 22 h (A′) or 46 h (B′, C′) after a 2 h treatment with 10 μM oligomycin A. TEM observations display autophagic structures within the cytoplasm of treated cells (A′), whereas they are absent in control cells (A). TRITC-labeled phalloidin staining and confocal laser scanning microscopy show a ring of F-actin underneath the inner membrane in control cells (B), while a cytoskeletal rearrangement occurs in treated cells (B′). Rhodamine 123 mitochondrial staining and confocal laser scanning microscopy evidence numerous mitochondria in control cells (C) whereas in treated cells (C′) a diminished number of mitochondria, together with a progressive diffusion of fluorescence ascribable to mitochondrial membrane depolarization, is observed. n: nucleus; m: mitochondrion. Scale bars: (A, A′) 500 nm, (B) 6 μm, (B′, C, C′) 3 μm. (See Color Insert.)

5. Wash by resuspending cells in 1 ml of 0.1 *M* cacodylate buffer (3 washes, 1 min each). After each wash, centrifuge cells at 200×*g* for 10 min.
6. Discard supernatant and post-fix with 0.5 ml of 0.1 *M* cacodylate buffer containing 1% osmium tetroxide for 20 min.
7. Centrifuge cells at 200×*g* for 10 min.
8. Discard supernatant and wash with 1 ml of 0.1 *M* cacodylate buffer (3 washes, 1 min each).
9. Centrifuge cells at 200×*g* for 10 min.
10. Discard supernatant and wash cells in 1 ml of ethanol 20% for 10 min.
11. Centrifuge cells at 200×*g* for 10 min.
12. Discard supernatant and wash cells in 1 ml of ethanol 50% for 10 min.
13. Centrifuge cells at 600×*g* for 10 min.
14. Discard supernatant and wash cells in 1 ml of ethanol 70% for 10 min.

    Note: Cells can be stored in ethanol 70% at 4 °C until further processing.

15. Centrifuge cells at 800×*g* for 10 min.
16. Discard supernatant and wash cells in 1 ml of ethanol 90% for 10 min.

17. Centrifuge cells at 1000×*g* for 10 min.
18. Discard supernatant and wash cells in 1 ml of ethanol 100% for 10 min.
19. Centrifuge cells at 1000×*g* for 10 min.
20. Discard supernatant and incubate cells in 1 ml of a 50:50 propylene oxide/Epon-Araldite 812 mixture for 30 min.
21. Centrifuge cells at 4400×*g* for 10 min.
22. Discard supernatant and leave the tube open under a fume hood to allow evaporation of residual propylene oxide (3–4 min should be sufficient).
23. Resuspend cells in 1 ml of Epon-Araldite 812 mixture and stand for 10 min.
24. Centrifuge cells at 5700×*g* for 10 min and allow the resin to polymerize overnight at 70 °C.
25. Cut thin (70 nm) sections with an ultramicrotome.
26. Stain sections using uranyl acetate (UA) and lead citrate (LC) (4 min LC, 8 min UA, 4 min LC).

Note: Although the use of Pasteur glass pipettes with capillary-shaped endings allows complete removal of solutions from the cell pellet at the bottom of the plastic tube, this procedure is not recommended, since small glass residues hamper the subsequent sample sectioning with a diamond knife. Thus, it is better to use disposable plastic tips, always applying gentle aspiration to avoid excessive cell loss. In addition, a progressive increase of centrifugation speed during the procedure ensures good pelleting without damaging the cells.

2.2. Immunofluorescence for cytoskeletal elements

Reorganization of the cytoskeleton certainly represents an important phenomenon both at the onset and during late events of apoptosis (Boldogh and Pon, 2006), but in the last years different observations have suggested that the cytoskeleton is also involved in the process of autophagy (Blommaart *et al.*, 1997). Furthermore, the analysis of the fate of the cytoskeleton during autophagic and apoptotic PCD in cell culture systems, revealed profound differences in the two death mechanisms, with depolymerization and degradation of cytoskeletal components during apoptosis and their redistribution in autophagic cell death (Bursch *et al.*, 2000).

A role of cytoskeleton in autophagic PCD has been evaluated also in insects. In *Drosophila*, a microarray approach that analyzed steroid-triggered autophagic cell death in the fruit fly identified several genes encoding small GTPases able to regulate the assembly and organization of the actin cytoskeleton (Lee *et al.*, 2003). In addition, Jochova and coworkers demonstrated that in *Drosophila* a rearrangement of tubulin and actin cytoskeleton precedes the extensive autophagic activities in dying salivary glands (Jochova *et al.*, 1997), and a subsequent work by Martin and Baehrecke (2004) in the same experimental model system has shown that cytoskeletal reorganization is necessary to drive the autophagic PCD in this organ, although the assessment of the exact role of these modifications needs further investigation.

IPLB-LdFB cells usually display a not particularly organized actin cytoskeleton comprising a ring of F-actin underneath the inner membrane leaflet and around the nuclear membrane (Fig. 38.1B). These two rings are usually connected by a thin web of filaments crossing the cytoplasm (Malagoli *et al.*, 2006). However, both epifluorescence and fluorescence confocal microscopy have revealed that when IPLB-LdFB cells are committed toward autophagic cell death, they present several grouped actin filaments in the cytoplasm forming a basket that surrounds the nucleus and the autophagosomes (Tettamanti *et al.*, 2008a; Fig. 38.1B′).

1. Harvest the cells and centrifuge at 120×*g* for 8 min to form a solid pellet.
2. Eliminate the supernatant and re-suspend the cells in fresh medium to obtain a final concentration of approximately 10^5 cells/ml.
3. Load approximately 80 μl on the cytocentrifuge (Cytospin II, Shandon Instrument, UK) and rotate at 200 rpm for 1 min.

Note: The cytocentrifuge applies a monolayer of cells onto a precise position on the slide, while maintaining cell integrity. If too many cells are loaded, their morphology will be altered. If more than 100 μl of cell suspension are loaded, the spot could be moist and not well defined. Conversely, if less than 70 μl are loaded, the resulting spot may be unbalanced, with more cells on one side. It is advisable to check the quality of some trial spots at the microscope and, if necessary, adjust the volume loaded before continuing with the procedure.

4. Let the spot dry at room temperature (RT) for 3 min.
5. Fix for 10 min in 4% paraformaldehyde in phosphate-buffered saline (PBS) (137 m*M* NaCl, 2.7 m*M* KCl, 10 m*M* Na_2HPO_4, 10 m*M* NaH_2PO_4, pH 7.3 ± 0.1).
6. Incubate for 30 min in a solution of PBS containing 30 n*M* FITC-labeled phalloidin and 0.3% Triton X-100 at 37 °C.
7. Wash for 5 min in PBS at RT.
8. Counterstain nuclei for 5 min at RT with 100 ng/ml 4′,6-diamidino-2-phenylindole, dihydrochloride (DAPI).
9. Wash for 5 min in PBS at RT.
10. Mount the slides in MOWIOL (Calbiochem, Darmstadt, Germany).
11. Analyze the staining with Ex/Em filters at wavelength 495/519 nm (fluorescein) and 345/455 nm (DAPI).

Note: It is important to spin the mounting medium after thawing, in order to remove completely any microbubbles that would interfere with subsequent observation of the slides.

The preceding protocol is identical if phalloidin marked with other fluorochromes (e.g., TRITC) is used.

2.3. Evaluation of mitochondrial loss, mitochondrial membrane potential and ROS production

Mitochondria have been demonstrated to be pivotal regulators of both apoptosis and autophagy (Bras *et al.*, 2005; Lemasters *et al.*, 1998). They are continuously generated and degraded within the cell, while their integrity is fundamental for ATP production. As these organelles are the major source of reactive oxygen species (ROS) production, they are particularly prone to ROS damage. Excessive accumulation of these highly reactive molecules endangers the functionality of mitochondria through a genotoxic effect on mtDNA and possibly an opening of nonspecific, high-conductance permeability transition pores in the mitochondrial membrane (Kim *et al.*, 2007; Scherz-Shouval and Elazar, 2007).

Although the production of ROS has been previously shown to trigger mitochondrial-mediated apoptosis (Bras *et al.*, 2005), several studies have demonstrated that ROS can also be involved in the induction of autophagy (Chen *et al.*, 2007; Scherz-Shouval and Elazar, 2007; Scherz-Shouval *et al.*, 2007). In particular, the different autophagic pathways that can be activated within the cell seem to be dependent on the degree of oxidative stress (Scherz-Shouval and Elazar, 2007). The final outcome ranges from the selective degradation of damaged and deenergized mitochondria (with consequent cell survival) to the activation of the death pathway.

Treatment of IPLB-LdFB cells with oligomycin A provokes a clear damage to mitochondria. This is ascertainable not only through TEM observations, but also using specific probes for intact mitochondria. We previously evaluated the uptake of different mitochondria-specific stainings to assess mitochondrial membrane potential (Tettamanti *et al.*, 2008a). While control cells display a spotted fluorescence (Fig. 38.1C), cells undergoing active autophagy following treatment with oligomycin A show an enhanced and diffused fluorescence due to the membrane damage affecting these organelles (Fig. 38.1C′).

The heavy deterioration of mitochondria causes a massive leakage of ROS toward the cytoplasm, as demonstrated using a ROS-specific probe (Tettamanti *et al.*, 2008a): starting from 22 h after addition of oligomycin A, the intensity of fluorescence progressively increases throughout the whole cytoplasm.

Here we describe the protocol for three different molecules to stain mitochondria (Rhodamine 123, MitoTracker Green and 3,3′-dihexyloxacarbocyanine iodide (DiOC6(3)) stainings) and a protocol used to evaluate

ROS production (2′,7′-dichlorodihydrofluorescein diacetate (H_2DCFDA) staining).

Rhodamine 123 is one of the conventional fluorescent stainings for mitochondria that specifically reacts with biological membranes (Johnson *et al.*, 1980). The fluorescent, mitochondria-specific dye, MitoTracker Green (MTG), selectively accumulates in the matrix, where it becomes fluorescent in the lipid environment of mitochondria (Keij *et al.*, 2000). DiOC6(3) incorporates into mitochondria; the dye is taken up by healthy mitochondria, and the fluorescence is precisely localized in the mitochondria with an undisturbed membrane potential (Koning *et al.*, 1993).

Whereas Rhodamine 123 and MTG specifically stain mitochondria, DiOC6(3) seems to be selective for mitochondria only if used at low concentrations. Indeed at higher concentrations, the dye stains other internal membranes, such as the endoplasmic reticulum (Koning *et al.*, 1993). Therefore, a careful preliminary evaluation of the working concentration is necessary to avoid nonspecific staining. To reduce potential artifacts due to overloading when using MTG, the concentration of the dye should be kept as low as possible, whereas it is possible to use a longer incubation to obtain a good final signal. It should be noted that Rhodamine 123 can be easily washed out of cells if mitochondria undergo a loss in membrane potential.

H_2DCFDA is a fluorogenic probe commonly used to detect cellular production of ROS. It is a compound that readily crosses cell membranes and after oxidation becomes the highly fluorescent 2′,7′-dichlorofluorescein (LeBel *et al.*, 1992).

MTG staining is performed on cells on glass coverslips, while Rhodamine 123, DiOC6(3) and H_2DCFDA stainings are performed on suspension cells.

Note: In each assay, incubation with the fluorescent dye and the subsequent steps must be carried out in a dark room. The MTG, DiOC6 (3) and H_2DCFDA staining protocols are modified versions of those suggested by the manufacturer.

Rhodamine 123 staining

1. Centrifuge cells (10^5 cells/ml) at 100×*g* for 5 min.
2. Discard supernatant and re-suspend cells once in 1 ml of fresh Ex-Cell 400 serum-free medium (JRH Biosciences, Andover, UK).
3. Mix 100 μl of cells with 900 μl of 1% sucrose in PBS and 0.5 μl of Rhodamine 123 (1 mg/ml in distilled water) (Sigma, St. Louis, MO, USA).
4. Incubate for 10–15 min under gentle rotation at 24 °C–28 °C.
5. Centrifuge at 100×*g* for 1–2 min.
6. Discard supernatant and resuspend the pellet in 200 μl of PBS.
7. Repeat steps 5 and 6.

8. Cytocentrifuge onto slides using a Cytospin II (200 rpm for 2 min).
9. Mount cells with Citifluor (Citifluor Ltd, London, UK).
10. Analyze the staining with Ex/Em filters at wavelength 550/590.

MTG staining

1. Prepare a 24-well plate by inserting a previously sterilized, 12 mm glass coverslip into each well.
2. Transfer cells (10^5 cells/ml) in the wells in 1 ml of Ex-Cell 400 medium. Stand overnight to allow cell deposition onto glass coverslips.
3. Discard the culture medium and wash twice with PBS.
4. Wash out PBS and incubate cells with MTG (Molecular Probes, Eugene, OR, USA) (0.2 μm in PBS) for 45 min at 26 °C.
5. Discard the solution and wash with PBS (3 washes, 5 min each).
6. Counterstain nuclei with DAPI (100 ng/ml in PBS) for 5 min.
7. Discard DAPI solution and wash with PBS (3 washes, 5 min each).
8. Mount the coverslip on a glass slide with Citifluor.
9. Analyze the staining with a scanning confocal microscope (Ex/Em wavelength 490/516).

DiOC6(3) staining

1. Wash cells once in fresh medium (Ex-Cell 400).
2. Discard medium and stain cells with DiOC6(3) (Molecular Probes) (20 n*M* in Ex-Cell 400) for 20 min at 26 °C.
3. Wash with PBS (3 washes, 5 min each).
4. Cytocentrifuge onto slides using a Cytospin II at 200 rpm for 1 min.
5. Mount cells with Citifluor.
6. Analyze the staining with a scanning confocal microscope (Ex/Em wavelength 492-495/517-527).

H_2DCFDA staining

1. Wash cells and resuspend in PBS.
2. Discard PBS and incubate cells with H_2DCFDA (Molecular Probes) (10 μM in Ex-Cell 400) for 30 min at 26 °C.
3. Discard H_2DCFDA solution and wash with PBS (3 washes, 5 min each).
4. Cytocentrifuge onto slides using a Cytospin II (200 rpm for 1 min).
5. Mount cells with Citifluor.
6. Analyze the staining with a scanning confocal microscope (Ex/Em wavelength 492-495/517-527).

2.4. Lysosomal staining

Considering the sequence of events that accompany the formation of the autolysosome (or autophagolysosome), that is, phagophore, autophagosome, and autolysosome (Klionsky *et al.*, 2008), detection of lysosomes *per se* cannot be considered as a method to study autophagy. However, since the autolysosome is an acidic compartment that derives from the fusion of autophagosomes with lysosomes (Klionsky *et al.*, 2008), lysosome detection can be used as a late marker of autophagy. In insects, the demise of the salivary gland in the blowfly, *Calliphora vomitoria*, has been associated with increased autophagic and acid phosphatase activity (Bowen *et al.*, 1996). In the honeybee, *Apis mellifera*, positivity for acid phosphatase has been seen to extend also to lysosomes, vesicular structures and the Golgi complex (Silva-Zacarin *et al.*, 2007). It is important to remember that the indications obtained by experiments with lysosomal markers should always be accompanied by other approaches also able to demonstrate early stages of autophagy.

Lysosomal staining of IPLB-LdFB cells has been previously performed to highlight the effects of a marine biotoxin by means of neutral red, acridine orange and acid phosphatase activity stainings (Malagoli *et al.*, 2006). Given the relative nonspecificity of neutral red and acridine orange for lysosomes, we suggest accompanying these stainings with a more specific marker of lysosomal content, such as acid phosphatase activity. However, the suggested protocols for IPLB-LdFB cells give almost overlapping results, indicating that the neutral red staining proposed here specifically marks the acid phosphatase-positive organelles in the IPLB-LdFB cells.

Detection of acid phosphatase activity

1. Harvest the cells and centrifuge at $120\times g$ for 8 min to form a solid pellet.
2. Eliminate the supernatant and resuspend the cells in fresh medium to obtain a final concentration of approximately 10^5 cells/ml.
3. Load 80 μl on the cytocentrifuge Cytospin II and spin at 200 rpm for 1 min.
4. Let the spot dry at RT for 3 min.
5. Incubate for 30 min in a 0.1 *N* sodium acetate-acetic acid buffer, pH 5.0–5.2 (25 °C), containing 0.01% naphthol phosphate AS-BI (Sigma), 2% N-N-dimethylformamide, 0.06% Fast Red Violet LB (Sigma) and 0.5 m*M* $MnCl_2$ at 37 °C.

Note: The components of the incubation solution must be prepared just before their use and mixed in the order given, although the $MnCl_2$ can be stored as a 10% stock solution for a long time. The incubation solution will be cloudy, but should not be filtered. Naphthol phosphate AS-BI must be dissolved in N-N-dimethylformamide, because the powder is not soluble in aqueous solutions. Previously, we reported an incubation time of 4 h (Malagoli *et al.*, 2006). However, further experiments realized with a new Fast Red Violet LB (Sigma) gave a perfect signal after just 30–45 min. To avoid overexposure that would result in a brownish (rather than a pink-violet) signal, we recommend monitoring the reaction at the microscope, especially during the first experiments. Negative controls can be performed by omitting the specific naphthol phosphate AS-BI substrate.

6. Mount the slides in an aqueous mounting medium (e.g., MOWIOL or 50% glycerol; Silva-Zacarin *et al.*, 2007). Because dehydration could result in a loss of signal, avoid organic mounting media.

Neutral red staining

1. Harvest the cells and centrifuge at 120×*g* for 8 min to form a solid pellet.
2. Eliminate the supernatant and re-suspend the cells in fresh medium to obtain a final concentration of approximately 10^5 cells/ml.
3. Put 990 μl of cell suspension in a well of a 24 multi-well plate.
4. Incubate for 1 h in a 50 μg/ml solution of neutral red (Sigma) at 26 °C.

Note: Stock neutral red solution (5 mg/ml) can be prepared either in medium or in sterile PBS, but it should be used within 1 week of preparation.

5. Load 100 μl on the cytocentrifuge Cytospin II and spin at 200 rpm for 1 min.
6. Let the spot dry at RT for 3 min.
7. Do not use MOWIOL for mounting, since the positivity will fade quickly. To obtain optimal images, unmounted slides should be microphotographed immediately after staining. However, at the reported concentration, the positivity is persistent and high-quality images can also be obtained after 24 h.

2.5. Evaluation of DNA integrity

Autophagy has been described as a housekeeping process that prevents DNA damage and chromosomal instability (Mathew and White, 2007). In terms of programmed cell death, apoptosis is associated with characteristic chromatin condensation and DNA fragmentation, whereas autophagy is linked to autolysosomes in which the cellular content is segregated and degraded. Even though DNA damage has not been reported to be typically associated with autophagic cell death, DNA damage has been observed by electrophoresis in IPLB-LdFB cells that were dying as a consequence of uninterrupted autophagy following a short incubation with oligomycin A (Tettamanti *et al.*, 2006). DNA damage has also been evaluated in IPLB-LdFB cells by morphological approaches, such as the TdT-mediated-dUTP Nick-End Labeling (TUNEL) assay (Malagoli *et al.*, 2006) and the fluorescent Hoescht 33342 (Ho-33342) staining (Malagoli *et al.*, 2005). These protocols are routinely used to reveal apoptotic rather than autophagic

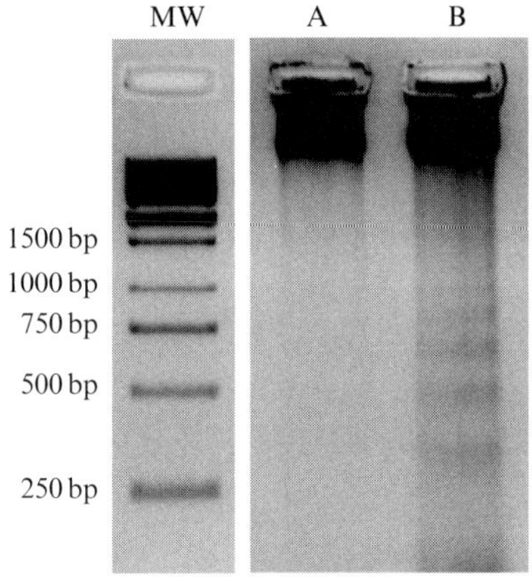

Figure 38.2 Evaluation of DNA integrity and visualization of an apoptotic ladder (B) in IPLB-LdFB cells after treatment with the marine biotoxin yessotoxin (YTX). Genomic DNA was extracted and run on a 3% agarose gel as described in detail in Section 2.5. (A) control; (B) YTX-treated cells.

features. However, given the increasing evidence that the two phenomena can coexist at least in insects (Velentzas *et al.*, 2007; Zacarin, 2007), the evaluation of DNA integrity may represent a useful tool to provide a better characterization of the type of cell death under study. A protocol describing the extraction of nuclear DNA for evaluation (horizontal gel electrophoresis) is proposed here, as this is the only technique capable of demonstrating the apoptotic DNA ladder (Fig. 38.2).

TUNEL Assay

Note: This protocol is a modified version of the one proposed for the application of the DeadEnd Fluorimetric TUNEL system (Promega, Madison, WI, USA). Small adjustments in times and volumes may be necessary when kits from other sources are utilized.

1. Harvest the cells and centrifuge at $120 \times g$ for 8 min to form a solid pellet.
2. Discard the supernatant and wash the cells in PBS solution.
3. Fix for 5 min in a solution of 4% paraformaldehyde in PBS at 4 °C.

   Note: A volume of 500 μl of fixative solution is sufficient for up to 5×10^5 cells.

4. Wash the fixed cells in PBS solution for 5 min and adjust PBS volumes to obtain a final concentration of approximately 10^5 cells/ml.
5. Load 80 μl on the cytocentrifuge Cytospin II and spin at 200 rpm for 1 min.
6. Permeabilize for 5 min in a solution of PBS containing 0.2% Triton X-100 at RT.

7. Wash for 5 min in PBS at RT.
8. Carefully remove the PBS around the cell spots with a soft paper towel, paying close attention not to touch the spot itself.
9. Cover each spot with 100 μl of equilibration buffer (200 m*M* potassium cacodylate, pH 6.6, 25 m*M* Tris-HCl, pH 6.6, 0.2 m*M* DTT, 2.5 m*M* cobalt chloride) and let the cell spot equilibrate for 5 min.

Note: Equilibration can be performed for up to 10 min without influencing the result.

10. Cover each spot with 50 μl of rTDT incubation buffer (45 μl of equilibration buffer + 5 μl of nucleotide mix [50 μ*M* fluorescein-12-dUTP, 100 μ*M* dATP, 10 m*M* Tris-HCl, pH 7.6, 1 m*M* EDTA] + 1 μl of recombinant Terminal Deoxynucleotidyl Transferase).

Note: At this stage, it is critical not to let the spot dry up. Therefore, accurate distribution of the rTDT incubation buffer over the cell spot is recommended.

11. Place a coverslip over each wet spot, then incubate in a humid chamber for 60 min at 37 °C.
12. Remove and discard the coverslips, then put the slides into a Coplin jar containing a 2X SSC solution for 15 min at RT.

Note: Usually SSC solution is maintained as a 20x SSC stock solution prepared as follows: 87.7 g NaCl, 44.1 g sodium citrate, dissolve in 400 ml of bidistilled water, adjust pH to 7.2 with 10 N NaOH and bring to a final volume of 500 ml. The desired dilution of SSC can be obtained just before use with bidistilled water. To ensure a complete dissolution of salts, the stock solution should be warmed to RT before any dilution.

13. Wash slides in a Coplin jar containing PBS for 5 min at RT. Repeat this washing 3 times to remove completely any fluorescein-12-dUTP that has not been incorporated.
14. Counterstain nuclei with 1 mg/ml propidium iodide (PI) or 100 ng/ml DAPI solution.

Note: Nuclear counterstaining may hamper the count of positive nuclei. In our experience, this step is not fundamental to obtain good results and a correct count of positive cells. However, we recommend counterstaining at least some slides during preliminary experiments to ensure that the operator is immediately aware of the proper localization of the signal carried by fluorescein.

15. Wash the spots with bidistilled water for 5 min at RT. Repeat this washing three times to remove completely the excess fluorochrome.
16. Dry the spot with a soft paper towel and mount the slides with an ordinary antifading or an aqueous mounting medium.
17. Observation should be performed at the following Ex/Em wavelengths: 495/519 nm (fluorescein), 536/617 nm (PI), 345/455 nm (DAPI).

DNA extraction and evaluation

The following protocol is a modified version of the classic phenol/chlororform purification method proposed by Blin and Stafford (Blin and Stafford, 1976). Given their high lipid and protein content, the extraction of highly pure DNA from IPLB-LdFB cells needs special attention.

1. Harvest 5×10^5–10^6 cells and centrifuge at $120\times g$ for 8 min to form a solid pellet in a 1.5-ml polypropylene tube.
2. Add 200 μl of HB solution (0.1 *M* Tris-HCl, pH 8, 10 m*M* EDTA, pH 8, 0.35 *M* NaCl, 7 *M* Urea) to the cell pellet and pestle thoroughly into the 1.5-ml tube with a plastic DNase-, RNase-, and PCR inhibitors-free pestle (Sigma).
3. Add a further 380 μl of HB solution plus 20 μl of 10% SDS, pH 7-7.5, then mix by tapping for 3 min.
4. Add 10 μl of proteinase K (stock solution 10 mg/ml) and 4 μl of RNase A (stock solution 100 mg/ml), then mix by inversion for 3 min.
5. Incubate at 57 °C overnight.
6. Add 600 μl of a chloroform/isoamyl alcohol solution (24:1) to the cell lysate, then mix by inversion for 4 min at RT.
7. Centrifuge at $13000\times g$ for 5 min at RT and recover the aqueous phase.

Note: Steps 6 and 7 are important to obtain a good DNA/protein ratio and, therefore, for subsequent enzymatic reactions. If only DNA integrity is to be evaluated, these two steps can be avoided.

8. Mix the aqueous phase with 600 μl of a phenol/chloroform (1:1) solution, then mix by inversion for 5 min at RT.

Note: The phenol must be pH 8.0, and the phenol/chloroform (1:1) solution must be prepared in advance (e.g. 30 min prior), to obtain a homogeneous solution before use.

9. Centrifuge at $13000\times g$ for 5 min at RT and recover the aqueous phase.
10. Repeat steps 8 and 9 three times.
11. Mix the aqueous phase with 400 μl of a chloroform/isoamyl alcohol solution (24:1), then mix by inversion for 2 min at RT.
12. Centrifuge at $13000\times g$ for 5 min at RT and recover the aqueous phase.
13. Add 1 ml (minimum) of ice-cold absolute ethanol and 30 μl of 3 *M* sodium acetate, pH 7.
14. Let the DNA precipitate for at least 3 h.

Note: Even if precipitation of large amounts of DNA is almost immediate, a minimal precipitation time of 3 h is recommended to collect DNA fragments as well. In our experience, overnight precipitation neither reduces nor improves the quality of the result.

15. Centrifuge at 13000×*g* for 5 min at RT and discard the ethanol.
16. Wash the DNA pellet three times with ice-cold 70% ethanol.
17. After removal of the 70% ethanol, let the DNA pellet dry and then dissolve it in 50–100 μl of nuclease-free water.

Note: Pay attention to overdrying, as this greatly reduces DNA solubility and could create artifactual DNA damage.

18. Load 2–3 μg of DNA on a 3% agarose gel containing 0.5 μg/ml ethidium bromide and run for at least 1 h at low voltage (i.e., 60–70 V) to obtain a good band resolution.

Note: Additional treatment with RNase A (for concentrations refer to point 4) may be necessary, especially if large amounts of cells have been collected.

2.6. Evaluation of membrane integrity

Autophagy is first of all a housekeeping process (Cuervo, 2004); therefore, cell membrane damage should not occur during the process. Even if autophagy can result in cell death if it proceeds uninterrupted (Klionsky *et al.*, 2008), membrane damage is a point of no return that definitely leads cells to decay and can be considered as a marker of unavoidable death (Majno and Joris, 1995). Therefore, when studying autophagic cell death as a typology of PCD (Baehrecke, 2003; Berry and Baehrecke, 2007), protocols providing evidence of effective death may be useful. The Hoechst 33342-propidium iodide (Ho 33342-PI) staining (Dartsch *et al.*, 2002; Eguchi *et al.*, 1997) is a simple and quick method to ascertain both nucleus as well as membrane integrity. Ho 33342 can cross intact membranes and specifically binds to nuclear material. Although having the same target as Ho 33342, PI is unable to cross an intact membrane barrier. This means that if a cell is positive to Ho 33342 but negative to PI, it has an intact membrane (Dartsch *et al.*, 2002; Eguchi *et al.*, 1997). Conversely, cells positive to both Ho 33342 and PI can be considered as dead cells. The advantage of the application of the double labeling with Ho 33342 and PI is that it allows not only the identification of samples with a damaged cell membrane (as other vital dyes; e.g., trypan blue) but also gives information about nucleus integrity.

1. Put 1 ml of cell suspension (10^5 cells/ml) in a 24-well plate.
2. Incubate with 10 μ*M* Ho 33342 in the dark for 10 min at 26 °C.
3. Centrifuge the cell suspension at 120×*g* for 5 min at RT and discard the supernatant.
4. Wash the cell pellet with 1 ml of fresh medium (Ex-Cell 400), centrifuge at 120×*g* for 5 min at RT, and discard the supernatant.
5. Resuspend the cells in 1 ml of fresh medium and load 80 μl on the cytocentrifuge Cytospin II, then spin at 200 rpm for 1 min.

6. Put the slides in a Coplin jar containing a solution of 10 μM PI for 5 min at 26 °C.
7. Wash the cell spot with PBS for 5 min at RT.
8. Dry the spot with a soft paper towel and mount the slides with an ordinary antifading or an aqueous mounting medium.
9. Observation should be performed using the following Ex/Em wavelengths: 343/483 nm (Ho 33342), 536/617 nm (PI).

2.7. Evaluation of ATP content

The induction of autophagic processes is usually associated with starvation, and in insect larvae the fat body utilizes autophagy to maintain the proper nutrient concentration in the hemolymph (Grewal and Saucedo, 2004). In IPLB-LdFB cells, ATP deprivation following oligomycin A treatment has been correlated with autophagic cell death (Tettamanti *et al.*, 2006). Furthermore, given the link observed between Ca^{2+}-mobilizing factors (including ATP) and the induction of autophagy (Hoyer-Hansen *et al.*, 2007), it may be important to monitor ATP levels in studies of autophagy.

1. Before proceeding with cell lysis, it is advisable to prepare the ATP standard solutions by serial dilution of an ATP stock solution. Concentrations of both stock and reference solutions may need to be changed depending on instrument sensitivity and the kit adopted. As an approximation, consider a stock solution of 10^{-3} M ATP and serial dilutions between 10^{-5} and 10^{-10} M. Keep the reference solutions in the dark and on ice until use.
2. Prepare/thaw luciferase solution and keep it in the dark at 4 °C (or on ice).
3. Centrifuge 2×10^5 cells/ml at 120×g for 5 min at RT and discard the supernatant.
4. Suspend the cell pellet in 10 μl of PBS.
5. Add 90 μl of boiling TE (100 mM Tris, 4 mM EDTA, pH 7.75) to the cell suspension and pipette two or three times to obtain a homogeneous mixture.
6. Boil the mixture for 2 min to obtain complete cell lysis.
7. Centrifuge cell lysates at 120×g for 3 min at RT, collect the supernatant and keep it on ice in the dark for subsequent ATP determination.
8. Add 50 μl of luciferase enzyme to 50 μl of each ATP standard solution and read the emitted light with a luminometer. The light signal should be integrated for 10 s after a delay of 1 s following the addition of the luciferase enzyme. Use the collected values to track the reference curve that will be necessary to determine the ATP concentration of the lysates.
9. Add 50 μl of luciferase enzyme to 50 μl of cell lysate and read the emitted light with a luminometer. The light signal should be integrated

for 10 s after a delay of 1 s following the addition of the luciferase enzyme.

Note: The protocol can be modified depending on the available instruments. Some luminometers discourage the injection of the enzyme solution, since the removal of the solution from the injection circuit can be troublesome. In these cases, instead of the automatic injection of enzyme solution into the samples, the operator will manually pipette the proper volume, paying attention to prepare and immediately read sample after sample. Volumes might also require some adjustments depending on the luminometer (e.g., tube or plate reader). However it is sufficient to maintain the above-mentioned proportions (i.e., 1:9 for cell pellet and TE, 1:1 for cell lysate and enzyme solution) to obtain good results. The preparation of serial dilutions (e.g., 1:2 and 1:10) of the cell lysate is also recommended, as, in our experience, the best results are obtained by diluting the samples rather than by changing the delay and integration values of the luminometer.

2.8. Evaluation of caspase-3 activity

As stated previously, apoptotic and autophagic cell death present some common features, even if the extent of this overlap has not yet been determined (Lee and Baehrecke, 2001; Velentzas *et al.*, 2007). Historically, caspases have been linked to apoptosis, and processed caspase-3 and caspase-dependent DNA fragmentation have been observed during autophagic cell death of *Drosophila* salivary glands (Lee and Baehrecke, 2001; Martin and Baehrecke, 2004). However, the recent combined inhibition of caspases and autophagy in *Drosophila* salivary glands suggests they may function as independent degradation pathways (Berry and Baehrecke, 2007). In IPLB-LdFB cells, caspase-3-like activity has been demonstrated to correlate with apoptotic cell death (Malagoli *et al.*, 2005). The absence of information on *L. dispar* caspases means that kits designed for mammals must be used. However, these also work properly in our lepidopteran model.

Note: This protocol is a modified version of the one proposed for the application of the "Caspase 3 Assay kit, Colorimetric" (Sigma). Small adjustments of times and volumes may be necessary when kits from other sources are used.

1. Centrifuge 10^7 cells at 100×*g* for 5 min at RT and discard the supernatant.
2. Wash the cell pellet twice with 1 ml of sterile-filtered PBS.
3. Suspend the cell pellet in 100 μl of lysis buffer (50 m*M* HEPES, pH 7.4, 5 m*M* CHAPS, 5 m*M* DTT).
4. Incubate the mixture on ice for 20 min in the dark.

Note: During this time, thaw all the other components required for the analysis.

5. Centrifuge the cell lysates at 16000×*g* for 20 min at 4 °C.
6. Recover the supernatant and place it in a sterile tube.
7. Add 10 μl of cell lysate to 980 μl of assay buffer (20 m*M* HEPES, pH 7.4, 2 m*M* EDTA, 0.1% CHAPS, 5 m*M* DTT) and 10 μl of Ac-Asp-Glu-Val-Asp-p-Nitroaniline (Ac-DEVD-pNA).
8. Incubate the reaction mixture for up to 16 h at 26 °C.
9. Put the reaction mixture in a quartz cuvette and read the optical density (OD) at 405 nm in a spectrophotometer. If the reaction is performed in a multiwell plate, OD values can be read directly with a plate reader.

Note: If the enzyme activity of two samples is to be compared, normalization of the OD values with respect to protein content is recommended. After subtraction of the background signal and normalization to protein content, we have registered an OD value of less than 0.1 in control cells and an OD value of approximately 0.2 in treated cells (Malagoli *et al.*, 2005) by using a Helios *β* spectrophotometer (Spectronic Unicam, Cambridge, UK).

2.9. Protein extraction and preparation for one-dimensional electrophoresis (1DE) and two-dimensional electrophoresis (2DE)

Together with ultrastructural and functional assays, Western blot for proteins specifically linked to autophagy (e.g., LC3-II) has also been recently recommended as a method to verify the proceedings of autophagic processes (Klionsky *et al.*, 2008). Even if the Western blot for the protein homolog to LC3-II has not yet been demonstrated to work in *Drosophila* (Klionsky *et al.*, 2008), it has been successfully applied in the IPLB-LdFB cell line for other well-conserved molecules (e.g., cytochrome c; Malagoli *et al.*, 2002). High-throughput proteomic studies of autophagic cell death in *Drosophila* have provided information that was not detected using whole-genome DNA, microarray and serial analysis of gene expression (Martin *et al.*, 2007). In Lepidoptera, proteomic analyses have been mainly related to development (Li *et al.*, 2006) and PCD (Jia *et al.*, 2007), and to our knowledge, there are no studies such as the autophagy studies reported for *Drosophila*. By means of 2DE, we have seen that an imaginal disk growth factor-like protein and several other factors are correlated with autophagic cell death in the IPLB-LdFB cell line (Malagoli *et al.*, unpublished data). However, these results have been obtained not from the cells themselves, but rather from the conditioned medium. Since the protocols of both 1DE and 2DE used in our studies are absolutely standard for electrophoresis under denaturing conditions, we will focus on the preparation of the sample, while readers are invited to refer to the literature for details about gel preparation, electrophoretic run, gel staining, Western blot, protein sequencing and data mining (Bjellqvist *et al.*, 1993; Blake *et al.*, 1984; Boraldi *et al.*, 2007; Bradford, 1976; Gharahdaghi *et al.*, 1999; Hochstrasser *et al.*, 1988; Laemmli, 1970; Shevchenko *et al.*, 1996; Towbin *et al.*, 1979).

Sample preparation for 1DE

Note: The following protocol has been developed to study either cytosolic and soluble components (soluble fraction) or membrane-linked molecules (microsomal fraction). For some specific studies (e.g., isolation of mitochondrial protein) adjustments may be needed. In our experience, standard protocols developed from mammalian cells need small adjustments for use with insect cells, especially given the high concentration of lipids in the IPLB-LdFB cell line.

1. Harvest 5×10^5–10^6 cells and centrifuge at $120 \times g$ for 8 min to form a solid pellet.
2. Discard the medium and wash the pellet twice in PBS.
3. Add 100 μl of ice-cold protease inhibitor (e.g., Sigma cocktail protease inhibitor) and pestle thoroughly while keeping the sample on ice. A viscous solution should form.
4. Add a further 100 μl of ice-cold protease inhibitor and pestle on ice for 2–3 min.
5. Vortex the sample, put on ice for 2–3 min, vortex the sample again, place at 45 °C–50 °C for 2–3 min, vortex the sample and put on ice for 2–3 min.
6. Centrifuge at $16000 \times g$ for 30 min at 4 °C.
7. Recover the supernatant (soluble fraction) and determine the protein content (e.g., following Bradford, 1976). If the microsomal fraction is required, proceed with step 9.
8. Add 1 vol of a standard 2x Laemmli buffer (Laemmli, 1970) to the protein lysate, boil for 4 min, and spin down immediately before loading the sample on an electrophoresis gel.

Note: Denatured samples can be stored at −80 °C indefinitely, while samples kept at −20 °C should be used within 5–6 months. After thawing the samples, a second boiling is not necessary, however, this does not affect the sample.

9. Add 100 μl of 2x Laemmli buffer (Laemmli, 1970) to the pellet and boil for 4 min. Centrifuge the mixture at $16000 \times g$ for 30 min at 4 °C. The supernatant contains the largest part of the protein of the microsomal fraction.

Note: In our experience, even if Laemmli buffer is added to the pellet without the bromophenol blue tracking dye, protein quantification of samples containing Laemmli buffer using Bradford's method (Bradford,

1976) is not accurate. We usually perform a more accurate quantification directly on the gel stained with Coomassie blue 0.1%, while we have never used the bicinchoninic acid (BCA) assay (Stoscheck, 1990).

Sample preparation for 2DE

Note: The following protocol has been developed to study protein content from Ex-Cell 400 medium conditioned by IPLB-LdFB cells and not to analyze the protein content of the cells themselves.

1. Incubate 10^6 cells in 5 ml of Ex-Cell 400 for 24 h at 26 °C.
2. Centrifuge the cell suspension at $120\times g$ for 8 min at RT.
3. Collect the supernatant and filter it with a 0.2-μm syringe filter.
4. Add 2% protein inhibitor cocktail (e.g., Sigma cocktail protease inhibitor) to the filtered medium.
5. Put 5 ml of conditioned medium into dialysis tubes (porosity 3.5 kDa; Spectrum Laboratories, Rancho Dominguez, CA, USA) and dialyze in doubledistilled water for 24 h at 4 °C by maintaining in constant agitation.

Note: We recommend dialyzation in a volume of water at least 100 times greater than the volume of medium. We usually put two 5-ml dialysis tubes in 2 liters of bidistilled water, changing the water at least 4 times during a 24 h dialysis. If more than one tube is put into the same container, carefully avoid the overlapping of the tubes, as dialysis is performed in constant agitation.

6. After dialysis, retain a small aliquot (e.g., 100 μl) for protein quantification, lyophilize the dialyzed medium and store it at −20 °C until use in 2DE.

Note: Due to the relatively low amount of proteins released in the conditioned medium, up to 80 ml of medium could be necessary for a preparative 2DE. Since the dialysis of such volumes of medium may be problematic, it is possible to dialyze and lyophilize 10 ml of medium each time and store lyophilized parcels at −20 °C. Lyophilized medium can be stored at −20 °C for up to 3 weeks without influencing the results. However, it should be remembered that lyophilization is only intended for protein conservation, and is not a mandatory step in preparing the samples to use in 2DE.

ACKNOWLEDGMENTS

The authors are grateful to Prof. Magda de Eguileor (University of Insubria, Varese, Italy) and Prof. Enzo Ottaviani (University of Modena and Reggio Emilia, Modena, Italy) for critically reviewing this manuscript and for their helpful comments.

REFERENCES

Baehrecke, E. H. (2003). Autophagic programmed cell death in *Drosophila*. *Cell Death Differ.* **10,** 940–945.

Berry, D. L., and Baehrecke, E. H. (2007). Growth arrest and autophagy are required for salivary gland cell degradation in *Drosophila*. *Cell* **131,** 1137–1148.

Bjellqvist, B., Pasquali, C., Ravier, F., Sanchez, J. C., and Hochstrasser, D. (1993). A nonlinear wide-range immobilized pH gradient for two-dimensional electrophoresis and its definition in a relevant pH scale. *Electrophoresis* **14,** 1357–1365.

Blake, M. S., Johnston, K. H., Russell-Jones, G. J., and Gotschlich, E. C. (1984). A rapid, sensitive method for detection of alkaline phosphatase-conjugated anti-antibody on Western blots. *Anal. Biochem.* **136,** 175–179.

Blin, N., and Stafford, D. W. (1976). A general method for isolation of high molecular weight DNA from eukaryotes. *Nucleic Acids Res.* **3,** 2303–2308.

Blommaart, E. F., Luiken, J. J., and Meijer, A. J. (1997). Autophagic proteolysis: Control and specificity. *Histochem. J.* **29,** 365–385.

Boldogh, I. R., and Pon, L. A. (2006). Interactions of mitochondria with the actin cytoskeleton. *Biochim. Biophys. Acta* **1763,** 450–462.

Boraldi, F., Annovi, G., Carraro, F., Naldini, A., Tiozzo, R., Sommer, P., and Quaglino, D. (2007). Hypoxia influences the cellular cross-talk of human dermal fibroblasts. A proteomic approach. *Biochim. Biophys. Acta* **1774,** 1402–1413.

Bowen, I. D., Mullarkey, K., and Morgan, S. M. (1996). Programmed cell death during metamorphosis in the blow-fly *Calliphora vomitoria*. *Microsc. Res. Tech.* **34,** 202–217.

Bradford, M. M. (1976). A rapid and sensitive method for the quantification of microgram quantities of protein utilizing the principle of protein-dye binding. *Anal. Biochem.* **72,** 248–254.

Bras, M., Queenan, B., and Susin, S. A. (2005). Programmed cell death via mitochondria: Different modes of dying. *Biochemistry (Mosc.)* **70,** 231–239.

Bursch, W., Hochegger, K., Torok, L., Marian, B., Ellinger, A., and Hermann, R. S. (2000). Autophagic and apoptotic types of programmed cell death exhibit different fates of cytoskeletal filaments. *J. Cell. Sci.* **113,** 1189–1198.

Chen, Y., McMillan-Ward, E., Kong, J., Israels, S. J., and Gibson, S. B. (2007). Mitochondrial electron-transport-chain inhibitors of complexes I and II induce autophagic cell death mediated by reactive oxygen species. *J. Cell. Sci.* **120,** 4155–4166.

Cuervo, A. M. (2004). Autophagy: Many paths to the same end. *Mol. Cell Biochem.* **263,** 55–72.

Dai, J. D., and Gilbert, L. I. (1999). An *in vitro* analysis of ecdysteroid-elicited cell death in the prothoracic gland of *Manduca sexta*. *Cell Tissue Res.* **297,** 319–327.

Dartsch, D. C., Schaefer, A., Boldt, S., Kolch, W., and Marquardt, H. (2002). Comparison of anthracycline-induced death of human leukemia cells: programmed cell death versus necrosis. *Apoptosis* **7,** 537–548.

Eguchi, Y., Shimizu, S., and Tsujimoto, Y. (1997). Intracellular ATP levels determine cell death fate by apoptosis or necrosis. *Cancer Res.* **57,** 1835–1840.

Eskelinen, E. L. (2008). To be or not to be? *Autophagy* **4,** 1–4.

Gharahdaghi, F., Weinberg, C. R., Meagher, D. A., Imai, B. S., and Mische, S. M. (1999). Mass spectrometric identification of proteins from silver-stained polyacrylamide gel: A method for the removal of silver ions to enhance sensitivity. *Electrophoresis* **20,** 601–605.

Grewal, S. S., and Saucedo, L. J. (2004). Chewing the fat: Regulating autophagy in Drosophila. *Dev. Cell* **7,** 148–150.

Hochstrasser, D. F., Patchornik, A., and Merril, C. R. (1988). Development of polyacrylamide gels that improve the separation of proteins and their detection by silver staining. *Anal. Biochem.* **173,** 412–423.

Hoyer-Hansen, M., Bastholm, L., Szyniarowski, P., Campanella, M., Szabadkai, G., Farkas, T., Bianchi, K., Fehrenbacher, N., Elling, F., Rizzuto, R., Mathiasen, I. S., and Jaattela, M. (2007). Control of macroautophagy by calcium, calmodulin-dependent kinase kinase-beta, and Bcl-2. *Mol. Cell* **25,** 193–205.

Jia, S. H., Li, M. W., Zhou, B., Liu, W. B., Zhang, Y., Miao, X. X., Zeng, R., and Huang, Y. P. (2007). Proteomic analysis of silk gland programmed cell death during metamorphosis of the silkworm *Bombyx mori*. *J. Proteome Res.* **6,** 3003–3010.

Jochova, J., Zakeri, Z., and Lockshin, R. A. (1997). Rearrangement of the tubulin and actin cytoskeleton during programmed cell death in *Drosophila* salivary glands. *Cell Death Differ.* **4,** 140–149.

Johnson, L. V., Walsh, M. L., and Chen, L. B. (1980). Localization of mitochondria in living cells with rhodamine 123. *Proc. Natl. Acad. Sci. USA* **77,** 990–994.

Keij, J. F., Bell-Prince, C., and Steinkamp, J. A. (2000). Staining of mitochondrial membranes with 10-nonyl acridine orange, MitoFluor Green, and MitoTracker Green is affected by mitochondrial membrane potential altering drugs. *Cytometry* **39,** 203–210.

Kim, I., Rodriguez-Enriquez, S., and Lemasters, J. J. (2007). Selective degradation of mitochondria by mitophagy. *Arch. Biochem. Biophys.* **462,** 245–253.

Klionsky, D. J., Abeliovich, H., Agostinis, P., Agrawal, D. K., Aliev, G., Askew, D. S., Baba, M., Baehrecke, E. H., Bahr, B. A., Ballabio, A., Bamber, B. A., Bassham, D. C., *et al.* (2008). Guidelines for the use and interpretation of assays for monitoring autophagy in higher eukaryotes. *Autophagy* **4,** 151–175.

Koning, A. J., Lum, P. Y., Williams, J. M., and Wright, R. (1993). DiOC6 staining reveals organelle structure and dynamics in living yeast cells. *Cell Motil. Cytoskeleton* **25,** 111–128.

Laemmli, U. K. (1970). Cleavage of structural proteins during the assembly of the head of bacteriophage T4. *Nature* **227,** 680–685.

LeBel, C. P., Ischiropoulos, H., and Bondy, S. C. (1992). Evaluation of the probe 2′,7′-dichlorofluorescin as an indicator of reactive oxygen species formation and oxidative stress. *Chem. Res. Toxicol.* **5,** 227–231.

Lee, C. Y., and Baehrecke, E. H. (2001). Steroid regulation of autophagic programmed cell death during development. *Development* **128,** 1443–1455.

Lee, C. Y., Clough, E. A., Yellon, P., Teslovich, T. M., Stephan, D. A., and Baehrecke, E. H. (2003). Genome-wide analyses of steroid- and radiation-triggered programmed cell death in *Drosophila*. *Curr. Biol.* **13,** 350–357.

Lemasters, J. J., Nieminen, A. L., Qian, T., Trost, L. C., Elmore, S. P., Nishimura, Y., Crowe, R. A., Cascio, W. E., Bradham, C. A., Brenner, D. A., and Herman, B. (1998). The mitochondrial permeability transition in cell death: A common mechanism in necrosis, apoptosis and autophagy. *Biochim. Biophys. Acta* **1366,** 177–196.

Levine, B., and Klionsky, D. J. (2004). Development by self-digestion: Molecular mechanisms and biological functions of autophagy. *Dev. Cell* **6,** 463–477.

Li, X. H., Wu, X. F., Yue, W. F., Liu, J. M., Li, G. L., and Miao, Y. G. (2006). Proteomic analysis of the silkworm (*Bombyx mori* L.) hemolymph during developmental stage. *J. Proteome Res.* **5,** 2809–2814.

Locke, M., and Collins, J. V. (1968). Protein uptake into multivesicular bodies and storage granules in the fat body of an insect. *J. Cell Biol.* **36,** 453–483.

Lynn, D. E., Dougherty, E. M., McClintock, J. T., and Loeb, M. (1988). Development of cell lines from various tissues of Lepidoptera. *In* "Invertebrate and fish tissue culture" (Y. Kuroda, E. Kurstak, and K. Maramorosch, eds.), pp. 239–242. Japan Scientific Societies Press, Tokyo.

Majno, G., and Joris, I. (1995). Apoptosis, oncosis, and necrosis. An overview of cell death. *Am. J. Pathol.* **146,** 3–15.

Malagoli, D. (2008). Cell death in the IPLB-LdFB insect cell line: Facts and implications. *Curr. Pharm. Des.* **14,** 126–130.

Malagoli, D., Conte, A., and Ottaviani, E. (2002). Protein kinases mediate nitric oxide-induced apoptosis in the insect cell line IPLB-LdFB. *Cell Mol. Life Sci.* **59,** 894–901.

Malagoli, D., Iacconi, I., Marchesini, E., and Ottaviani, E. (2005). Cell-death mechanisms in the IPLB-LdFB insect cell line: A nuclear located Bcl-2-like molecule as a possible controller of 2-deoxy-D-ribose-mediated DNA fragmentation. *Cell Tissue Res.* **320,** 337–343.

Malagoli, D., Marchesini, E., and Ottaviani, E. (2006). Lysosomes as the target of yessotoxin in invertebrate and vertebrate cell lines. *Toxicol. Lett.* **167,** 75–83.

Martin, D. N., and Baehrecke, E. H. (2004). Caspases function in autophagic programmed cell death in *Drosophila*. *Development* **131,** 275–284.

Martin, D. N., Balgley, B., Dutta, S., Chen, J., Rudnick, P., Cranford, J., Kantartzis, S., DeVoe, D. L., Lee, C., and Baehrecke, E. H. (2007). Proteomic analysis of steroid-triggered autophagic programmed cell death during *Drosophila* development. *Cell Death Differ.* **14,** 916–923.

Mathew, R., and White, E. (2007). Why sick cells produce tumors: The protective role of autophagy. *Autophagy* **3,** 502–505.

Mita, K., Kasahara, M., Sasaki, S., Nagayasu, Y., Yamada, T., Kanamori, H., Namiki, N., Kitagawa, M., Yamashita, H., Yasukochi, Y., Kadono-Okuda, K., Yamamoto, K., *et al.* (2004). The genome sequence of silkworm, *Bombyx mori*. *DNA Res.* **11,** 27–35.

Müller, F., Adori, C., and Sass, M. (2004). Autophagic and apoptotic features during programmed cell death in the fat body of the tobacco hornworm (*Manduca sexta*). *Eur. J. Cell Biol.* **83,** 67–78.

Papanicolaou, A., Gebauer-Jung, S., Blaxter, M. L., Owen McMillan, W., and Jiggins, C. D. (2008). ButterflyBase: A platform for lepidopteran genomics. *Nucleic Acids Res.* **36,** D582–D587.

Scherz-Shouval, R., and Elazar, Z. (2007). ROS, mitochondria and the regulation of autophagy. *Trends Cell Biol.* **17,** 422–427.

Scherz-Shouval, R., Shvets, E., Fass, E., Shorer, H., Gil, L., and Elazar, Z. (2007). Reactive oxygen species are essential for autophagy and specifically regulate the activity of Atg4. *EMBO J.* **26,** 1749–1760.

Scott, R. C., Schuldiner, O., and Neufeld, T. P. (2004). Role and regulation of starvation-induced autophagy in the *Drosophila* fat body. *Dev. Cell* **7,** 167–178.

Shevchenko, A., Wilm, M., Vorm, O., and Mann, M. (1996). Mass spectrometric sequencing of proteins silver-stained polyacrylamide gels. *Anal. Chem.* **68,** 850–858.

Silva-Zacarin, E. C., Tomaino, G. A., Brocheto-Braga, M. R., Taboga, S. R., and De Moraes, R. L. (2007). Programmed cell death in the larval salivary glands of *Apis mellifera* (Hymenoptera, Apidae). *J. Biosci.* **32,** 309–328.

Stoscheck, C. M. (1990). Quantitation of protein. *Methods Enzymol.* **182,** 50–69.

Tettamanti, G., Grimaldi, A., Casartelli, M., Ambrosetti, E., Ponti, B., Congiu, T., Ferrarese, R., Rivas-Pena, M. L., Pennacchio, F., and Eguileor, M. (2007a). Programmed cell death and stem cell differentiation are responsible for midgut replacement in *Heliothis virescens* during prepupal instar. *Cell Tissue Res.* **330,** 345–359.

Tettamanti, G., Grimaldi, A., Pennacchio, F., and de Eguileor, M. (2007b). Lepidopteran larval midgut during prepupal instar: digestion or self-digestion? *Autophagy* **3,** 630–631.

Tettamanti, G., Malagoli, D., Marchesini, E., Congiu, T., de Eguileor, M., and Ottaviani, E. (2006). Oligomycin A induces autophagy in the IPLB-LdFB insect cell line. *Cell Tissue Res.* **326,** 179–186.

Tettamanti, G., Malagoli, D., Ottaviani, E., and de Eguileor, M. (2008a). Oligomycin A and the IPLB-LdFB insect cell line: Actin and mitochondrial responses. *Cell Biol. Int.* **32,** 287–292.

Tettamanti, G., Salo, E., Gonzalez-Estevez, C., Felix, D. A., Grimaldi, A., and de Eguileor, M. (2008b). Autophagy in invertebrates: insights into development, regeneration and body remodeling. *Curr. Pharm. Des.* **14,** 116–125.

Towbin, H., Staehelin, T., and Gordon, J. (1979). Electrophoretic transfer of proteins from polyacrylamide gels to nitrocellulose sheets: Procedure and some applications. *Proc. Natl. Acad. Sci. USA* **76,** 4350–4354.

Velentzas, A. D., Nezis, I. P., Stravopodis, D. J., Papassideri, I. S., and Margaritis, L. H. (2007). Mechanisms of programmed cell death during oogenesis in *Drosophila virilis*. *Cell Tissue Res.* **327,** 399–414.

Zacarin, E. C. (2007). Autophagy and apoptosis coordinate physiological cell death in larval salivary glands of *Apis mellifera* (Hymenoptera: Apidae). *Autophagy* **3,** 516–518.

Author Index

B

D

E

H

I

J

K

M

P

Q

R

S

T

U

V

W

X

Y

Z

Subject Index

F

L

M

W

Y

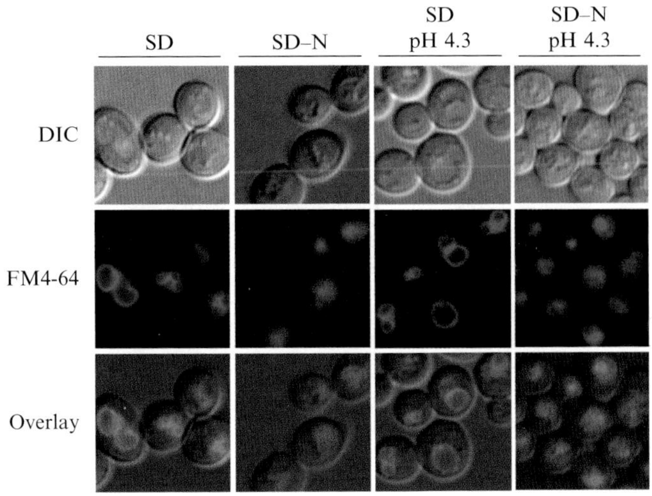

Dikla Journo *et al.*, Figure 7.2 Medium acidification enhances the observation of autophagy by FM 4-64 staining. Wild-type *S. cerevisiae* cells were treated as described in Fig. 7.1 except that resuspension was done in unacidified media (pH ~ 6; left panels) versus media buffered at pH 4.3 using 10 m*M* citrate (right panels).

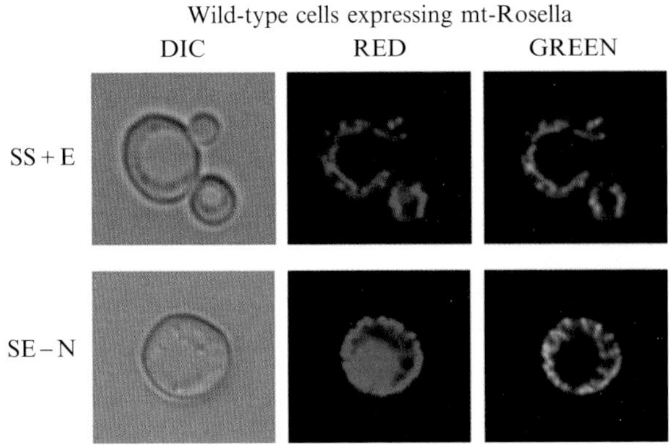

Nadine Camougrand *et al.*, Figure 8.1 The Rosella biosensor targeted to mitochondria (mt-Rosella) is delivered to the vacuole under conditions of nitrogen starvation. DIC and fluorescence images are shown for wild-type cells under growing conditions (SS + E) and after 6 h under nitrogen starvation (SE-N).

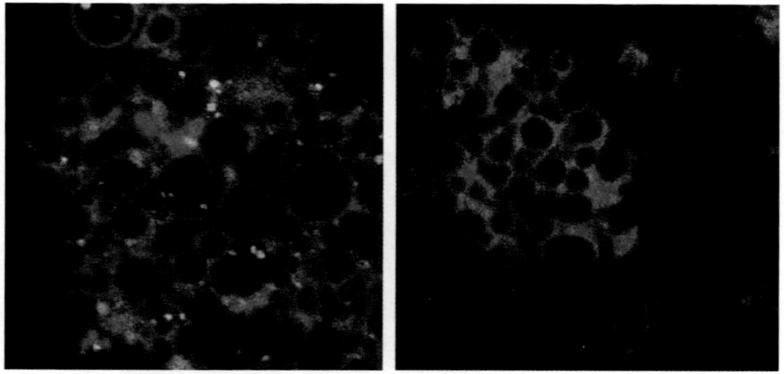

Andreas Mayer, Figure 11.2 Example of uptake of FITC-dextran (green) into vacuoles (red) *in vitro.* Uptake happens in the presence of ATP (left) but not in its absence (right).

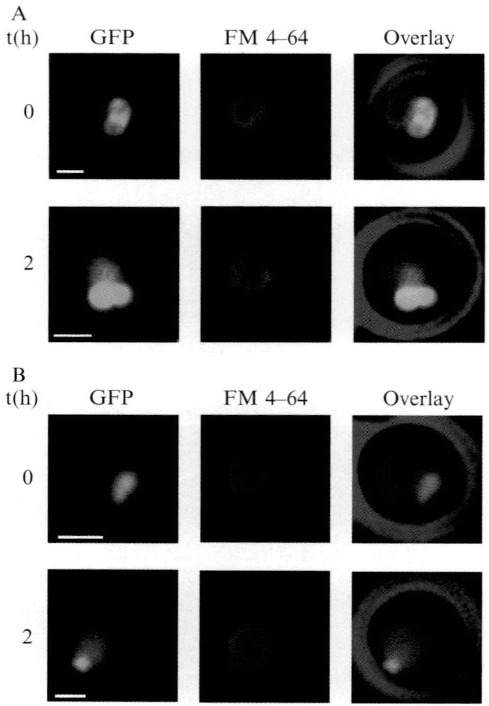

Tim van Zutphen *et al.*, Figure 14.4 Fluorescence microscopy analysis of glucose-induced macropexophagy. (A) In wild-type *H. polymorpha* cells producing GFP.SKL peroxisomes are visualized by GFP fluorescence, the vacuoles by FM 4-64. At T = 0 the vacuoles lack GFP fluorescence, which is evident after 2 h of incubation in the presence of glucose, confirming that macropexophagy has occurred. (B) In *pex13* cells peroxisome remnants are marked by N50.Pex3.GFP. Also these structures are subject to degradation under macropexophagy conditions, as is demonstrated by the presence of GFP fluorescence in the vacuole after 2 h of incubation. The bar represents 1 μm.

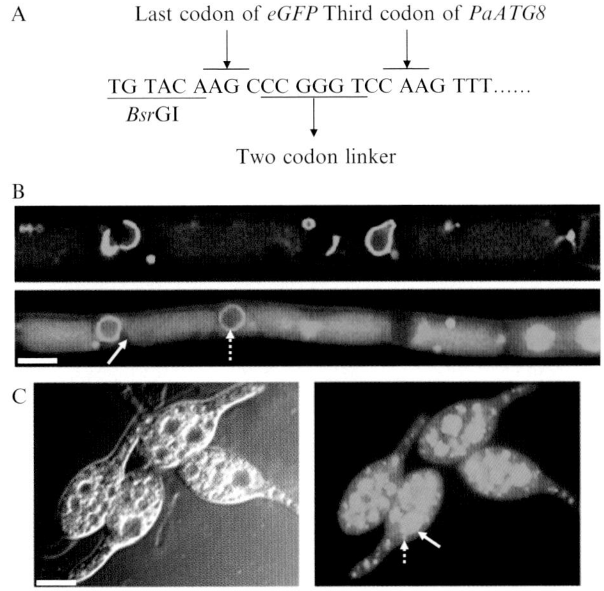

Bèrangère Pinan-Lucarré and Corinne Clave, Figure 18.1 The GFP-PaATG8 autophagy marker. (A) Fusion region between *egfp* and *PaATG8* ORFs in the *egfp-PaATG8* fusion gene. (B) Autophagosomes and vacuoles labeled by GFP-PaATG8 during cell death by incompatibility: the SI *egfp-PaATG8* strain was observed after transfer to 26 °C for 4 h. Adapted from Pinan-Lucarré *et al.* (2007). (C) Autophagosomes and vacuoles labelled by GFP-PaATG8 upon ascospore formation: ascospores were obtained in the progeny of a homozygous cross of a compatible *egfp-PaATG8* strain and observed before melanization of the cell wall, as fluorescence was not observable in mature black ascospores. The solid arrows point to a vacuole; the dotted arrows indicate an autophagosome. Scale bar: 3 μm.

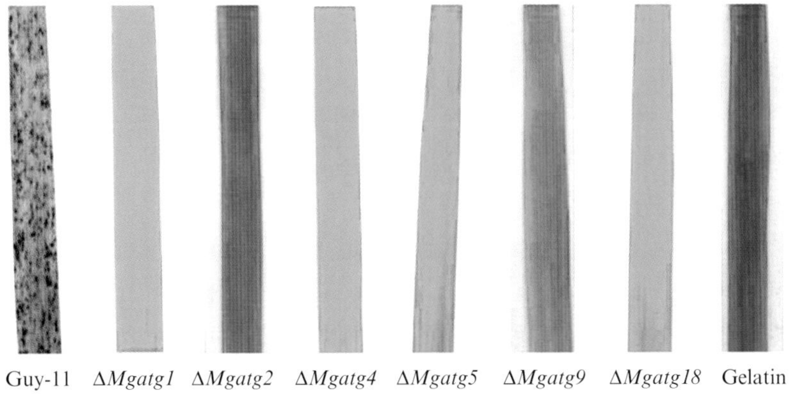

Xiao-Hong Liu *et al.*, Figure 19.2 atg mutants lose the ability to penetrate the host plants. Leaves from CO-39 were spray-inoculated individually with conidia from the wild-type Guy-11 strain, or the Δ*MgATG1*, Δ*MgATG2*, Δ*MgATG4*, Δ*MgATG5*, Δ*MgATG9*, Δ*MgATG18* mutants. Disease symptoms were allowed to develop for 7 days.

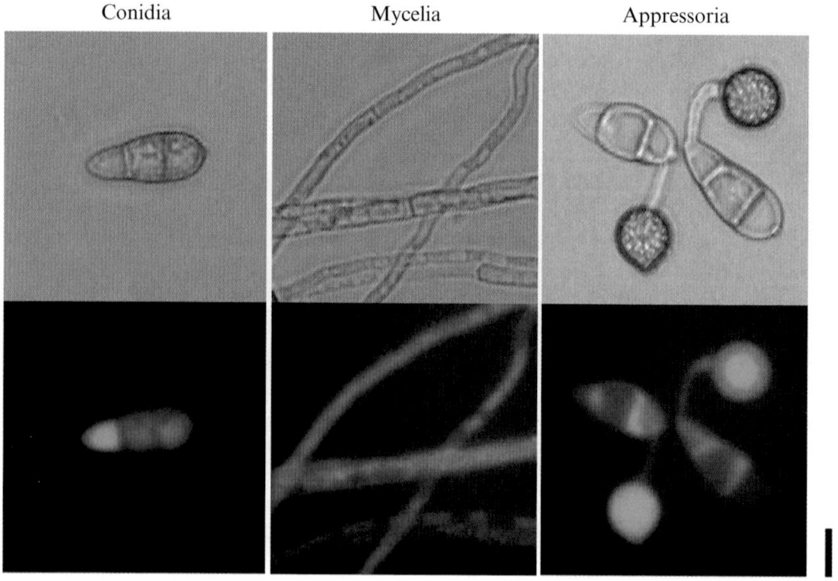

Xiao-Hong Liu *et al.*, Figure 19.4 Localization of MgATG1 in *M. oryzae*. The MgATG1-GFP protein appeared in the cytoplasm of *M. oryzae* strain NGA7. Bright-field (upper) and fluorescence (bottom) are shown. Bar = 10 μm.

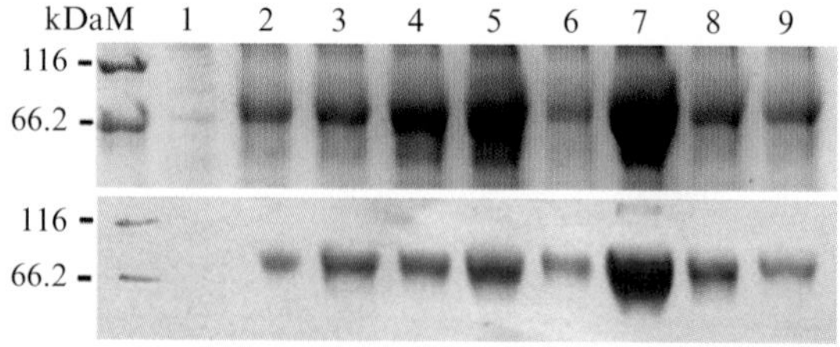

Xiao-Hong Liu *et al.*, Figure 19.6 Expression of MgATG4 in Pichia pastoris. The supernatant of the transformants was prepared as described above (see section 8D). Proteins in the supernatant were separated by SDS-PAGE, and assayed by immunoblotting with an anti-his antibody. Lane 1: The negative control, transformant transformed with the empty vector pPICZαA; Lane 2–9: The supernatants of eight transformants transformed with pPICZαA-ATG4 expressing 70KD MgATG4 protein glycosylied.

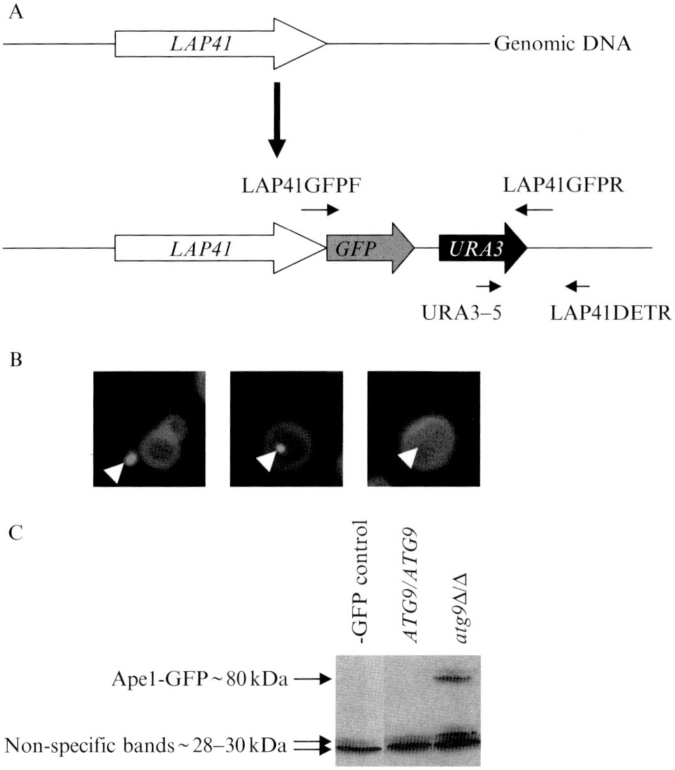

Glen E. Palmer, Figure 21.3 Localization and processing of Ape1-GFP. *LAP41* encodes *C. albicans* Ape1. (A) Construction of *LAP41-GFP* fusion. Megaprimers LAP41GFPF and LAP41GFPR are used to amplify a *GFP-URA3* cassette with flanking sequences to direct the in-frame integration at the 3′ end of the *LAP41* ORF. *C. albicans* transformants are selected using the *URA3* marker, and correct integration confirmed by PCR detection with the primer set URA3-5 and LAP41DETR. (B) Ape1-GFP localization in wild-type *C. albicans*. Cells harboring *LAP41-GFP* were pulse-labeled with FM 4-64 to label the vacuolar membrane. Cells were observed using an epifluorescence microscope with FITC-C and TRIT-C filter sets and 100x objective. GFP (green) and FM 4-64 (red) images were merged. Three distinct localizations are observed, often within the same cell: (1) intense GFP spot outside of the vacuole (*left*); (2) intense spot in the vacuole lumen (*center*); and (3) diffuse staining of the vacuole lumen (*right*). (C) Western blot analysis of Ape1-GFP. Protein extracts from $ATG9^+$ and *atg9*Δ *C. albicans* strains harboring the *LAP41-GFP* fusion and grown in YEPD medium were probed with anti-GFP. A strain without the fusion was used as a negative control (−GFP control).

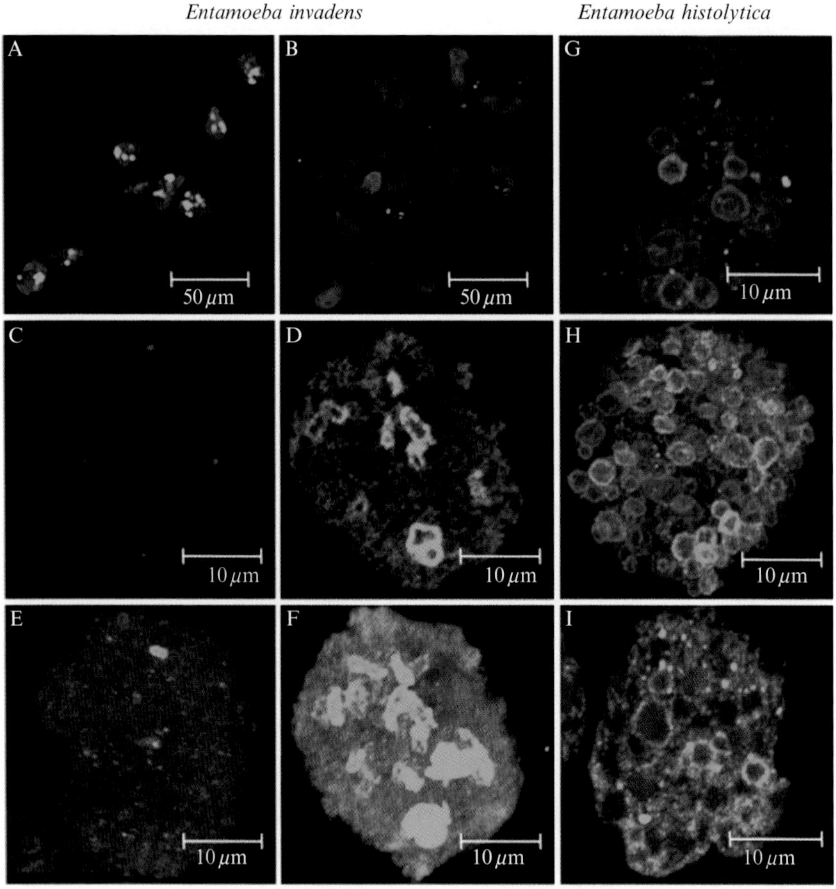

Karina Picazarri *et al.*, Figure 24.2 Immunofluorescence images of autophagy in *E. invadens* and *E. histolytica* on a confocal microscope. (A, B) Autophagosomes in the proliferating *E. invadens* trophozoites at 1 (logarithmic phase, A) and 2 weeks after the initiation of the culture (stationary phase, B). (C–F) Autophagosomes in the encysting *E. invadens* trophozoites at 0 (C, D) and 24 h postencystation induction (D, F). Single slices (C, E) and maximum projections of 20 slices taken at 1-μm intervals on the z-axis (E, F) are shown. (G, H) Autophagosomes in the proliferating *E. histolytica* trophozoites at days 1 (logarithmic phase, G) and 5 (stationary phase, H) (maximum projection). (I) Colocalization of Atg8 (green) and the lysosome marker, LysoTracker Red (red) in an *E. histolytica* trophozoite (day 3).

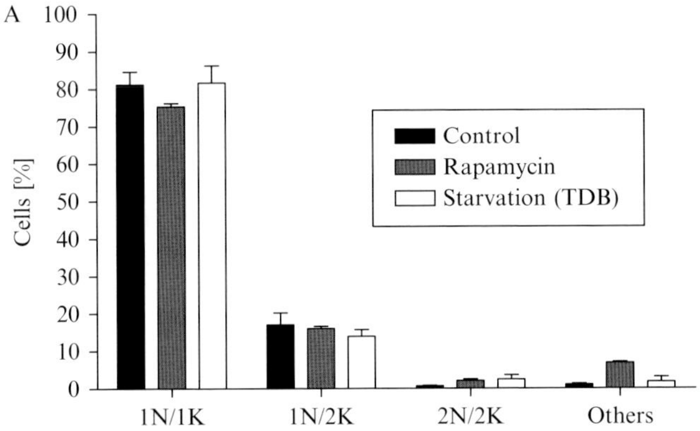

B

Control

Sub	10.91
G1	28.20
S	12.90
G2	41.83
Poly	10.08

Rapamycin

Sub	11.50
G1	22.97
S	11.51
G2	37.43
Poly	20.25

Starvation

Sub	5.82
G1	33.05
S	16.64
G2	38.83
Poly	11.37

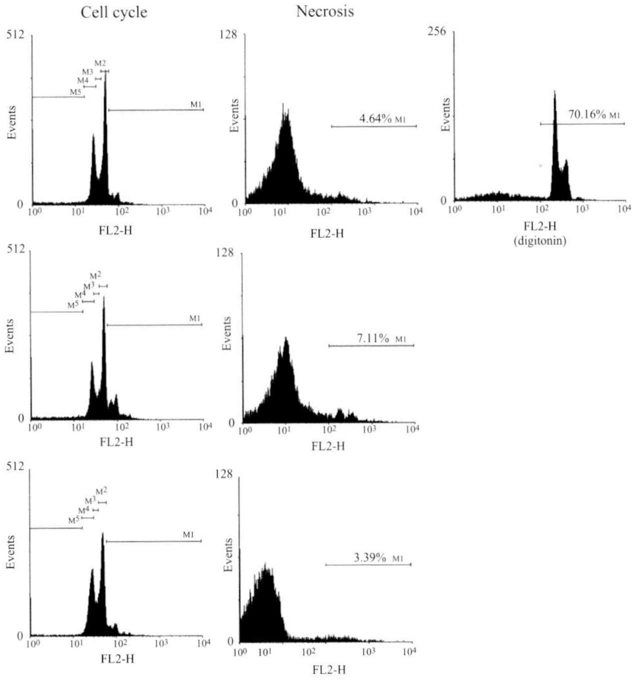

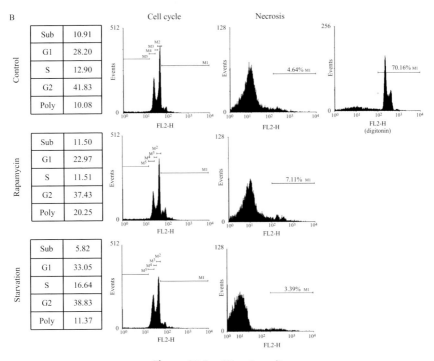

Figure 25.2 (Continued)

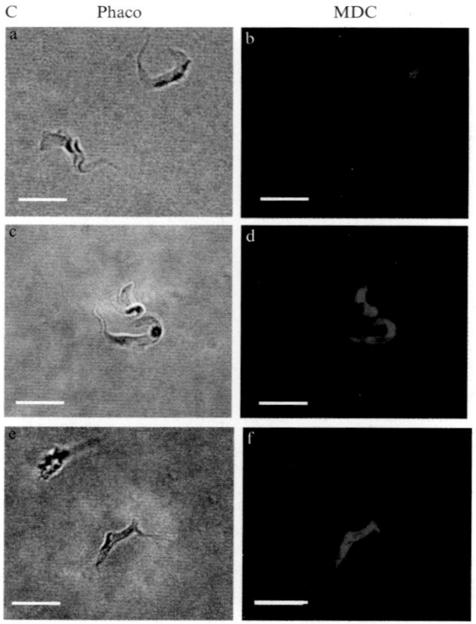

Viola Denninger *et al.*, Figure 25.2 Light microscopy and FACS analysis. (A) Indicated is the proportion of cells in different cell cycle phases out of three independent experiments after bisbenzimide staining of their nucleus and kinetoplastid. N = nucleus, K = kinetoplast. 1N/1K corresponds to G1-phase, 1N/2K and 2N/2K correspond to cells at different stages of cell division. (B) FACS analyses. Measurement of the cells' DNA-content with propidium iodide revealed no cell cycle arrest or DNA-degradation. Rapamycin induced a slight increase of polynucleated cells, which could be verified by electron microscopy. Under all conditions no necrosis could be observed by propidium iodide staining of the whole cells. Digitonin treatment served as a positive control. (C) Monodansylcadaverine (MDC) staining of autophagic vacuoles. (A, B) control cells (C, D) rapamycin 6.5 μM (E, F) starved cells. Bar = 10 μm. Phaco, phase contrast.

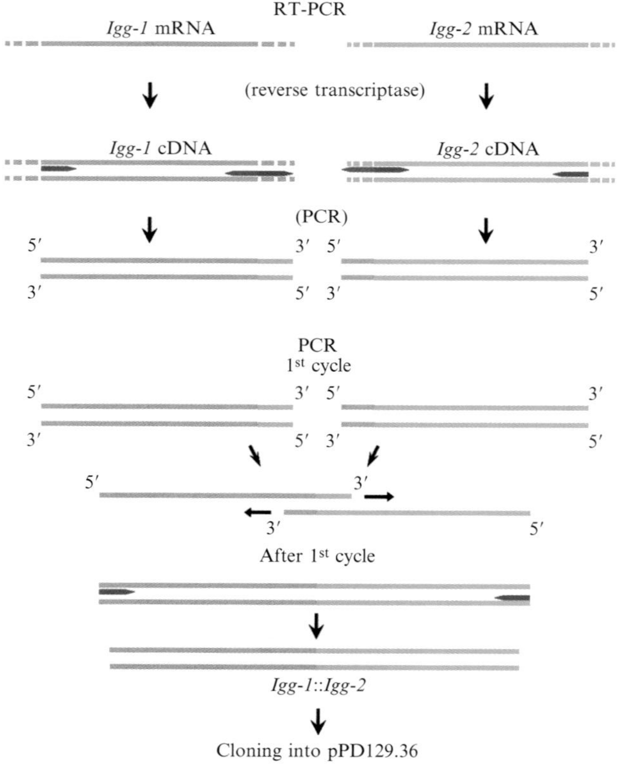

Tímea Sigmond *et al.*, Figure 30.1 Schematic model of a PCR fusion-based approach to create a hybrid cDNA fragment for simultaneous silencing of *lgg-1* and *lgg-2*. The process contains a reverse transcriptase-based RT-PCR amplification of single cDNA fragments, and a subsequent PCR amplification of the chimeric cDNA fragment. Orange and light blue lines indicate DNA double helices, red and dark blue arrows indicate primers.

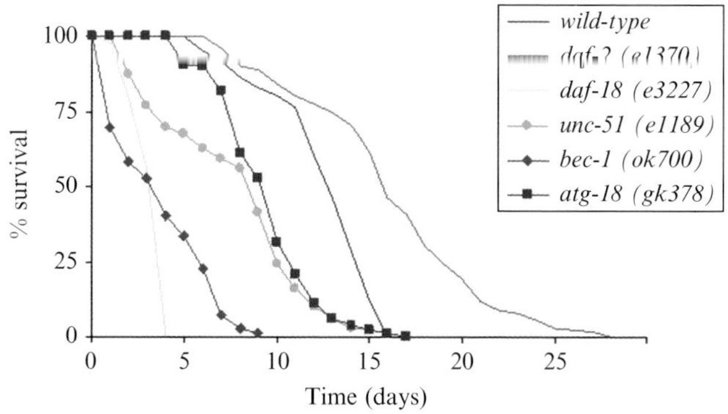

Tímea Sigmond *et al.*, Figure 30.3 Decreased survival of animals with a mutation in the autophagy pathway during prolonged L1 starvation. All autophagy-deficient mutant strains isolated so far exhibit a similar phenotype under identical conditions.

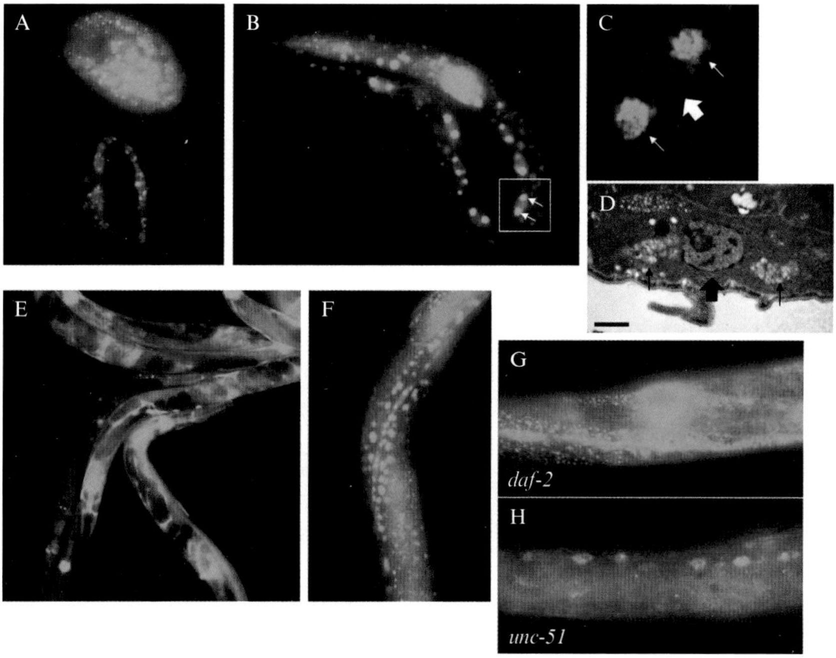

Tímea Sigmond *et al.*, Figure 30.4 Intracellular accumulation of GFP::LGG-1 is not exclusively associated with autophagosomal structures. (A) Fluorescence image showing GFP::LGG-1 accumulation in an early (16-cell)-stage embryo (upper) and in a 3-fold-stage embryo (lower). LGG-1 accumulates in both diffuse and smaller punctate patterns. The latter might label autophagosomal structures. (B) Seam cells along the lateral side of a larva in the L1/L2 lethargus stage display twin-spot-like GFP::LGG-1 accumulation (arrows). (C) Higher magnification confocal image of the twin spot (the section framed in B). Thin arrows indicate the twin spots, the thick arrow shows the nucleus. (D) Electron microscopy image of a lateral seam cell with the twin spot reveals hypertrophied Golgi structures on the two sides of the nucleus (arrows, scale bar 1 μm). Thin arrows indicate Golgi structures, the thick arrow indicates the nucleus. (E) Mostly diffuse GFP::LGG-1 accumulation in adult animals. (F) GFP::LGG-1-positive foci in the hypodermis and intestine of an animal at the L4/adult lethargus stage. (G) GFP::LGG-1 accumulation in a daf-2 mutant hermaphrodite deficient in insulin/IGF-1 signaling. An increase in the number of GFP-positive puncta in the hypodermis is evident, as compared with the wild type. (H) GFP::LGG-1 accumulates mainly in a diffuse pattern in an *unc-51/Atg1* mutant animal.

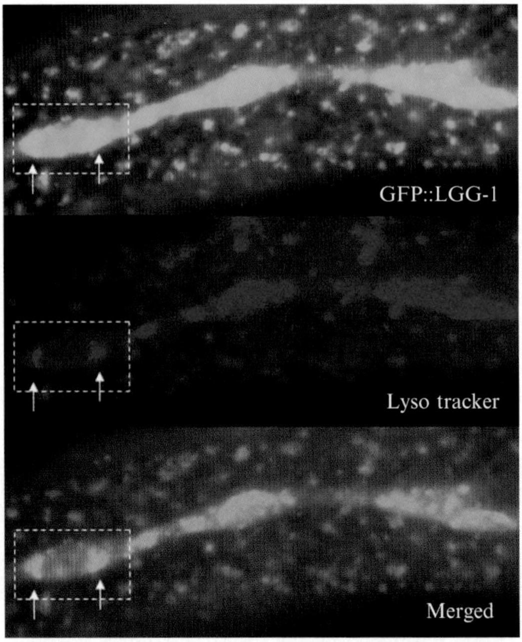

Tímea Sigmond *et al.*, Figure 30.5 LGG-1 accumulates in acidic compartments. GFP::LGG-1 accumulation (green) is colocalized with LysoTracker Red (red)-positive lysosomal compartments. A lateral seam cell is indicated by the box with dotted lines; arrows indicate "twin spots".

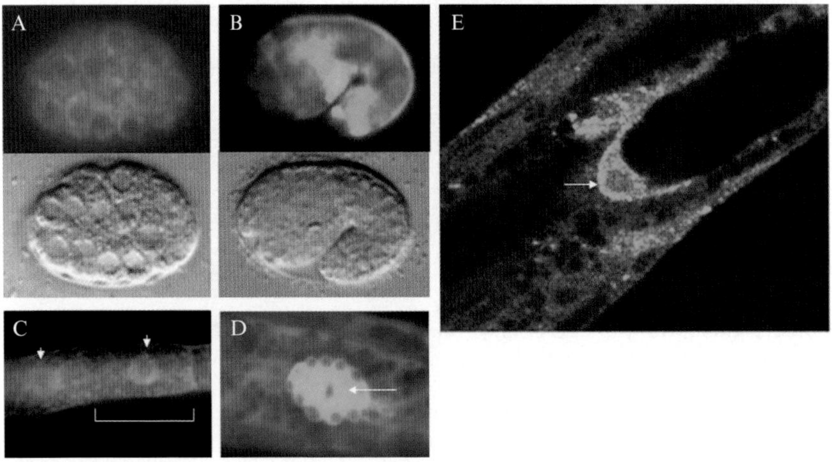

Tímea Sigmond *et al.*, Figure 30.6 BEC-1::GFP accumulation in various stages of development. (A) Mostly diffuse and even expression of BEC-1 in all cells of an early-stage embryo. Upper: epifluorescence image; Lower: the corresponding Nomarski image. (B) In a comma-stage embryo, BEC-1::GFP shows stronger accumulation in some regions of the body. Upper: epifluorescence image; Lower: the corresponding Nomarski image. (C) In intestinal cells, BEC-1 shows a characteristic perinuclear accumulation. The bracket indicates an intestinal cell, arrows show the nuclei. (D) Intense BEC-1 accumulation in the vulval cells (arrow) of a hermaphrodite. (E) Confocal image showing BEC-1::GFP-positive puncta. The arrow points to a distal tip cell.

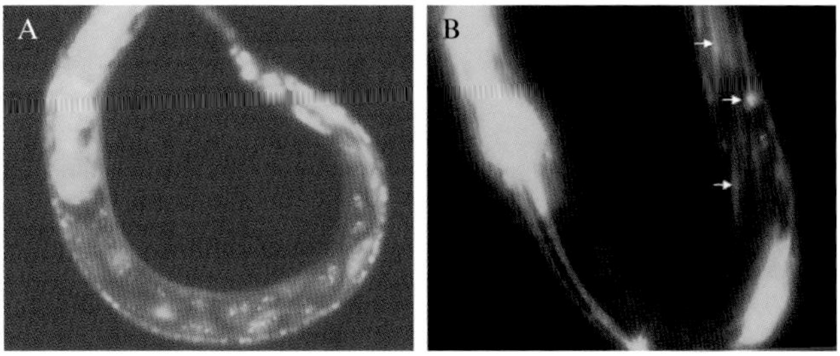

Tímea Sigmond *et al.*, Figure 30.7 *atg-18::gfp* expression. (A) Fluorescence image shows that *atg-18* is expressed in almost every cell of a L3 larva. (B) *atg-18* expression in body wall muscle cells (arrows) of an L4 stage larva.

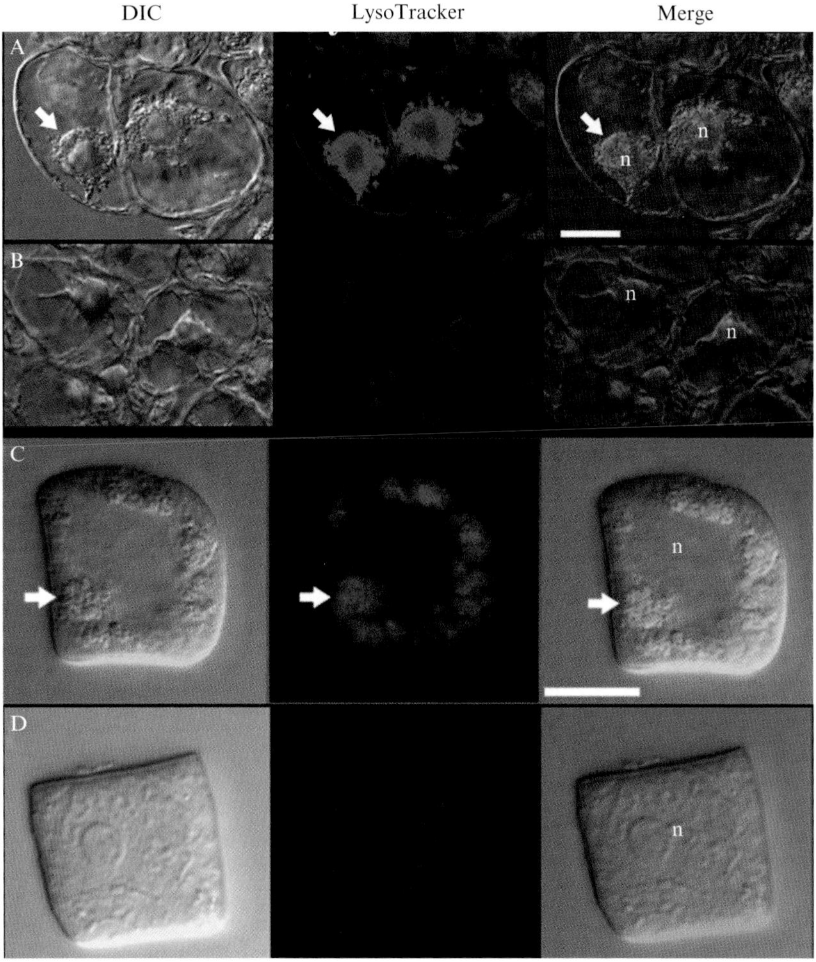

Yuji Moriyasu and Yuko Inoue, Figure 32.5 Concentration of LysoTracker Red in autolysosomes. (A, B) BY-2 cells were kept under sucrose starvation conditions for 24 h in the presence (A) or absence (B) of E-64c, and then labeled with LysoTracker Red. After fixation the cells were observed by confocal microscopy. In the cells treated with E-64c, autolysosomes accumulated and concentrated LysoTracker Red as indicated by the arrow. (C, D) Cells from barley root tips incubated in MS medium containing sucrose in the presence (C) or absence (D) of E-64d; the arrow indicates organelles with an appearance similar to that of autolysosomes that concentrate LysoTracker Red. n = nucleus; bar = 20 μm (A, B); 10 μm (C, D). From Moriyasu *et al.* (2003) with permission.

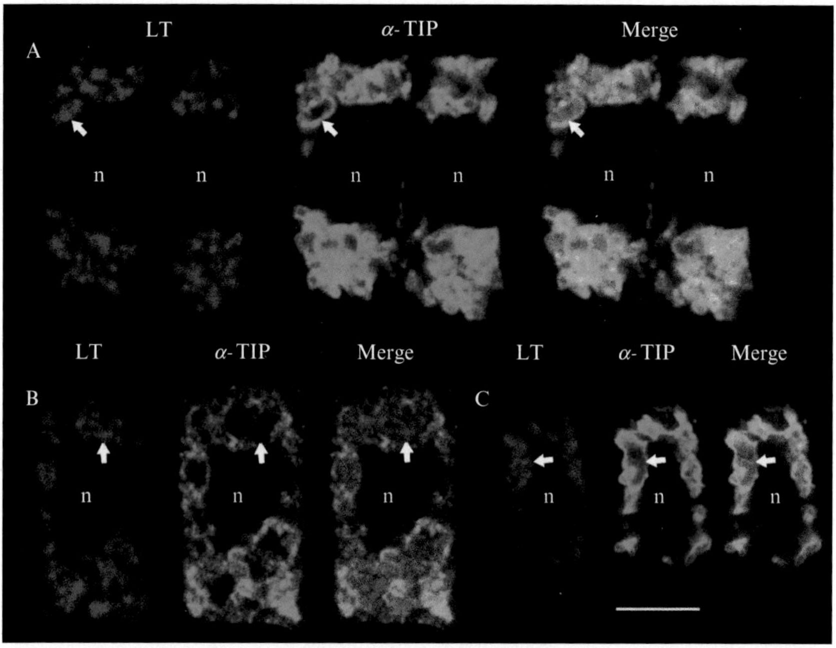

Yuji Moriyasu and Yuko Inoue, Figure 32.6 Barley root-tip cells labeled with LysoTracker Red and anti-alpha-TIP antibody. Each set (A, B, and C) presents LysoTracker (LT, red) and anti-alpha-TIP antibody (α-TIP, green) images individually, and the two images combined (Merge); cells were incubated with E-64d for 24 h. A. Two cells where predominantly individual autolysosomes appear to be surrounded by alpha-TIP-containing membrane; arrow: a larger structure containing multiple red autolysosomal inclusions. B. A cell where multiple inclusions (arrow) are incorporated into large vacuoles marked by alpha-TIP in their membranes. C. A cell where some vacuoles appear to contain only individual autolysosomal inclusions, while others appear to contain multiple inclusions (arrow). For all: n, position of nucleus; bar = 5 μm. From Moriyasu *et al.* (2003) with permission.

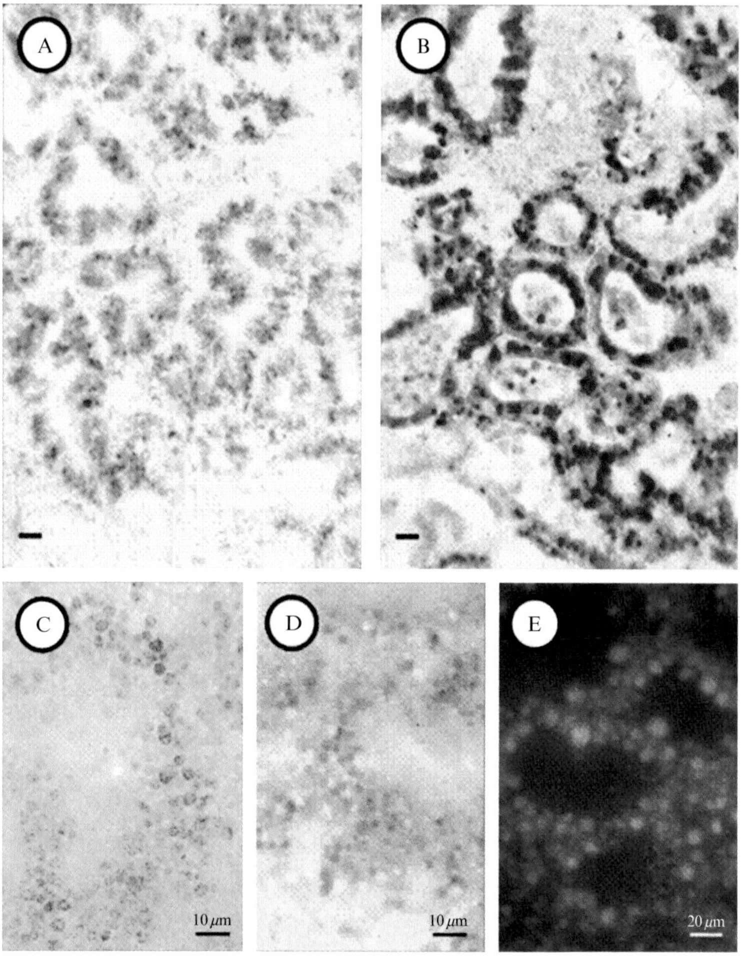

Michael N. Moore *et al.*, Figure 33.1 Micrographs of frozen sections of formalin-fixed mussel hepatopancreas showing lysosomes reacted for β–N-acetylhexosaminidase in the epithelial digestive cells of fed controls (A) and enlarged autolysosomes in mussels deprived of food for 15 days (B) indicative of starvation-induced autophagy (Moore 2004). Unfixed frozen sections of hepatopancreatic epithelial digestive cells from copper-treated mussel (40 μg/l for 3 days) showing dithionite reaction for copper in lysosomes (C), Shikata reaction for autophagocytosed metallothionein in lysosomes (D), and UV-fluorescence of 3-methylcholanthrene (3-MC at a daily dose of 150 μg/l for 7 days) in swollen lysosomes (E). Experimental details are available in Moore (2004), Moore *et al.* (2006a) and Viarengo *et al.* (1985). Scale bar in A, B and E = 20 μm and in C and D = 10 μm.

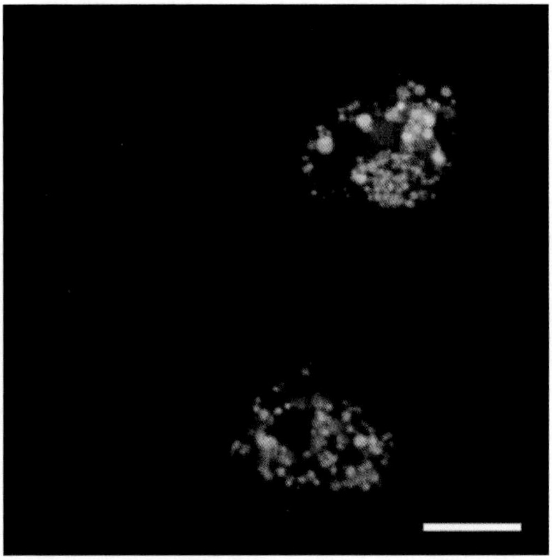

Michael N. Moore *et al.*, Figure 33.5 Mussel blood cell showing strong fluorescence (orange) for lysosomal accumulation of acridine orange (exposed to 1 μg/l for 15 min). Weaker fluorescence (green) is present in probable pre-lysosomal compartments with possibly some mitochondrial fluorescence. Nuclear staining of DNA is generally not apparent using this concentration of acridine orange. Blue light (FITC) excitation; With ICES permission from Moore *et al.* (2004b). Scale Bar = 10 μm.

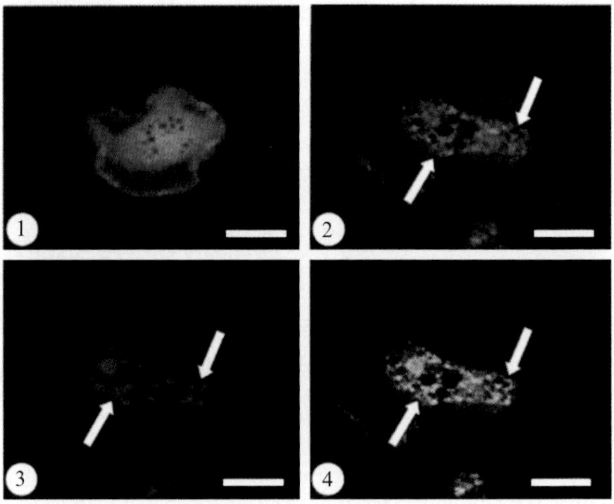

Michael N. Moore *et al.*, Figure 33.6 Confocal images of mussel blood cells treated with FITC diacetate showing even fluorescent labeling of cellular proteins after 30 min (1) and fluorescent vacuolar distribution of autophagocytosed FITC-labeled protein after a further 3 h (2). The distribution of fluorescent lysosomes in the same cell as in (2) labeled with neutral red (rhodamine excitation, 3) and merged image of (2) and (3) showing that lysosomes and fluorescent vacuolar fluorescence are predominantly at the same sites (4) (arrows). See Moore *et al.*, (1996) for experimental details. Scale bar = 10 μm.

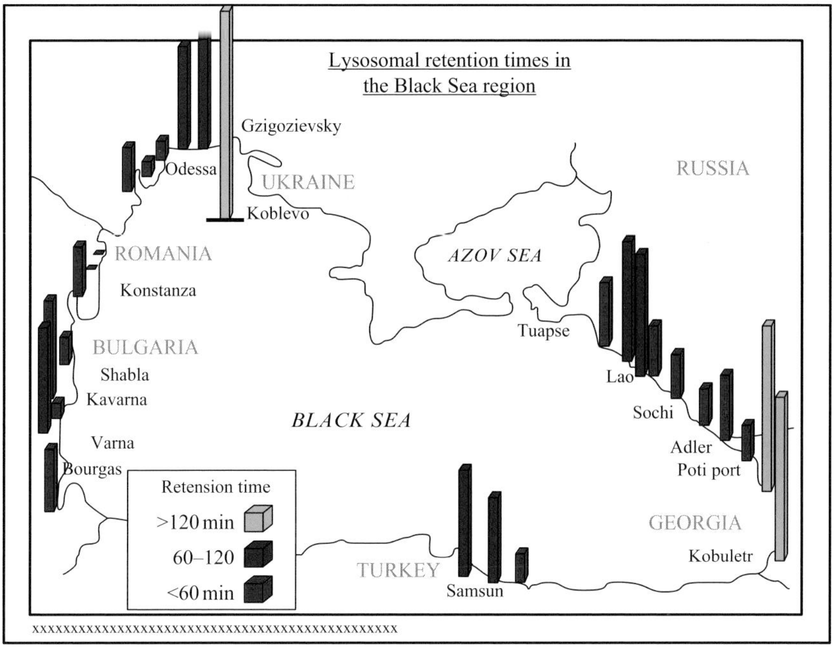

Michael N. Moore *et al.*, Figure 33.10 Summary results of the UNESCO-IOC Biological Effects Mussel Watch Programme in the Black Sea. Lysosomal stability was determined in mussels sampled from the coasts of the six littoral countries surrounding the Black Sea using the neutral red retention method. Lysosomal retention values of less than 60 min indicate severely impaired health, and it is apparent that much of the coastline is subject to environmental degradation on the basis of the lysosomal stability data. With ICES permission from Moore *et al.* (2004b). This study was a component of the GEF-Black Sea Environment Programme.

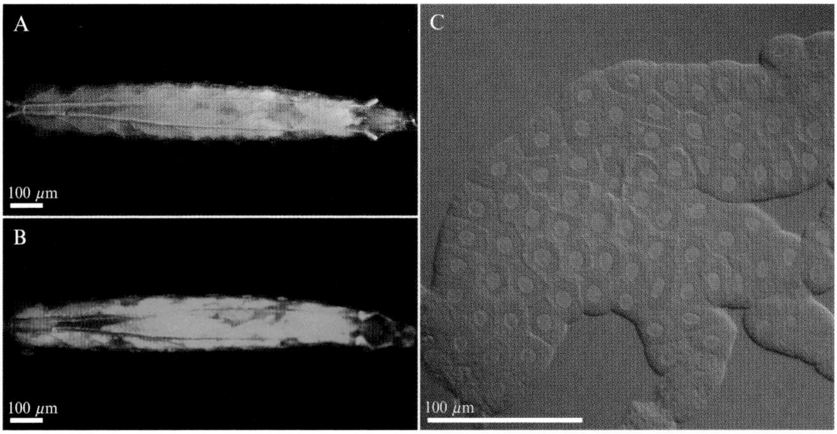

Thomas P. Neufeld, Figure 36.1 Distribution and morphology of the *Drosophila* larval fat body. (A) In intact larvae, the fat body is evident as an opaque, white, bilaterally symmetrical tissue running the length of the animal, just beneath the larval cuticle. (B) The extent of the fat body is evident by expression of GFP using the fat body-specific driver fb-GAL4 (Gronke *et al.*, 2003). (C) A dissected fat body lobe, stained with DAPI and visualized with direct interference contrast optics. Note the large polyploid cells arranged as a monolayer.

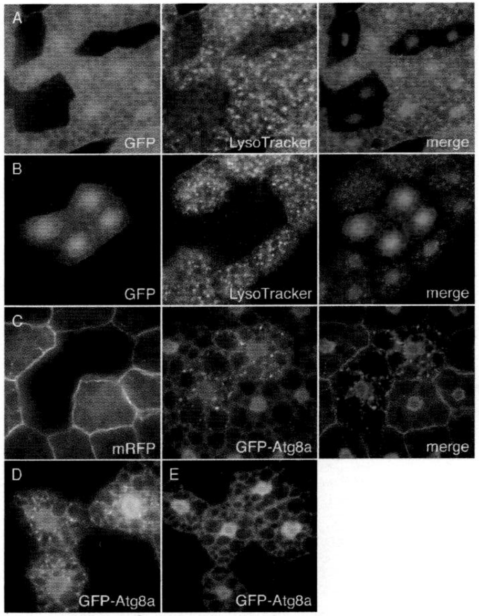

Thomas P. Neufeld, Figure 36.2 Clonal inactivation of autophagy in mosaic larval fat body. (A) Homozygous loss of *Vps34* disrupts starvation-induced accumulation of autophagosomes labeled with LysoTracker Red. Clones of *Vps34* null mutant cells are marked by lack of GFP expression. (B) Clonal expression of kinase-defective Vps34 disrupts starvation-induced accumulation of autophagosomes labeled with LysoTracker Red. A four-cell clone of cells expressing $Vps34^{KD}$ is marked by coexpression of GFP. (C) Homozygous mutation of *Vps25* leads to cell-autonomous accumulation of GFP-Atg8a-labelled autophagosomes in well-fed animals, due to a defect in autophagosome-lysosome fusion. A two-cell *Vps25* mutant clone is marked by lack of mRFP expression; all cells express GFP-Atg8a. D, E. Clonal expression of GFP-Atg8a under starvation conditions labels autophagic vesicles in control animals (D) but not in animals coexpressing $Vps34^{KD}$ (E). Images are modified from Juhász *et al.* (2008).

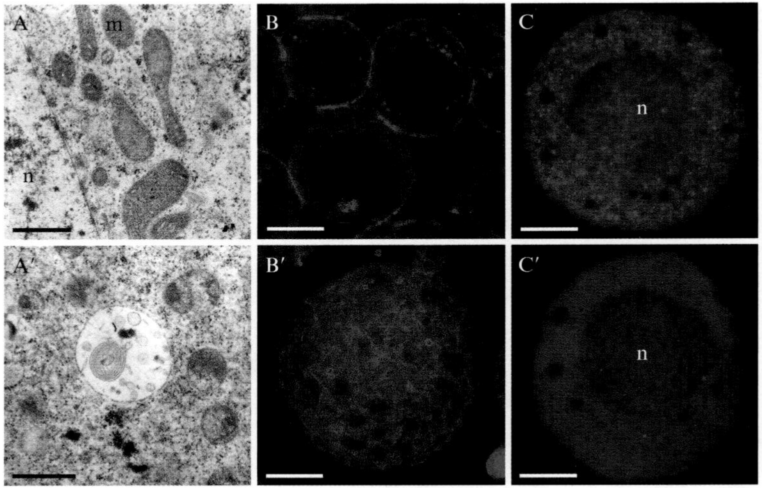

Gianluca Tettamanti and Davide Malagoli, Figure 38.1 Different morphological features observed in control IPLB-LdFB cells (A, B, C) and 22 h (A′) or 46 h (B′, C′) after a 2-h treatment with 10 μM oligomycin A. TEM observations display autophagic structures within the cytoplasm of treated cells (A′), whereas they are absent in control cells (A). TRITC-labeled phalloidin staining and confocal laser-scanning microscopy show a ring of F-actin underneath the inner membrane in control cells (B), while a cytoskeletal rearrangement occurs in treated cells (B′). Rhodamine123 mitochondrial staining and confocal laser scanning microscopy evidence numerous mitochondria in control cells (C) whereas in treated cells (C′) a diminished number of mitochondria, together with a progressive diffusion of fluorescence ascribable to mitochondrial membrane depolarization, is observed. n: nucleus; m: mitochondrion. Scale bars: (A, A′) 500 nm, (B) 6 μm, (B′, C, C′) 3 μm.